TRIGONOMETRY

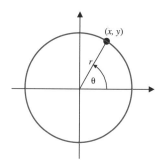

$$\sin\theta = \frac{y}{r}$$

$$\cos\theta = \frac{x}{r}$$

$$\tan\theta = \frac{y}{x}$$

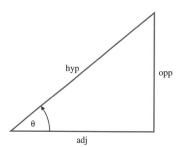

$$\sin\theta = \frac{\text{opp}}{\text{hyp}}$$

$$\cos\theta = \frac{\text{adj}}{\text{hyp}}$$

$$\tan\theta = \frac{\text{opp}}{\text{adj}}$$

RECIPROCALS

$$\cot\theta = \frac{1}{\tan\theta} \qquad \sec\theta = \frac{1}{\cos\theta} \qquad \csc\theta = \frac{1}{\sin\theta}$$

DEFINITIONS

$$\cot\theta = \frac{\cos\theta}{\sin\theta} \qquad \sec\theta = \frac{1}{\cos\theta} \qquad \csc\theta = \frac{1}{\sin\theta}$$

PYTHAGOREAN

$$\sin^2\theta + \cos^2\theta = 1 \qquad \tan^2\theta + 1 = \sec^2\theta \qquad 1 + \cot^2\theta = \csc^2\theta$$

COFUNCTION

$$\sin\left(\frac{\pi}{2} - \theta\right) = \cos\theta \qquad \cos\left(\frac{\pi}{2} - \theta\right) = \sin\theta \qquad \tan\left(\frac{\pi}{2} - \theta\right) = \cot\theta$$

EVEN/ODD

$$\sin(-\theta) = -\sin\theta \qquad \cos(-\theta) = \cos\theta \qquad \tan(-\theta) = -\tan\theta$$

DOUBLE-ANGLE

$$\sin 2\theta = 2\sin\theta\cos\theta \qquad \cos 2\theta = \cos^2\theta - \sin^2\theta \qquad \cos 2\theta = 1 - 2\sin^2\theta$$

HALF-ANGLE

$$\sin^2\theta = \frac{1 - \cos 2\theta}{2} \qquad\qquad \cos^2\theta = \frac{1 + \cos 2\theta}{2}$$

ADDITION

$$\sin(a + b) = \sin a \cos b + \cos a \sin b \qquad \cos(a + b) = \cos a \cos b - \sin a \sin b$$

SUBTRACTION

$$\sin(a - b) = \sin a \cos b - \cos a \sin b \qquad \cos(a - b) = \cos a \cos b + \sin a \sin b$$

SUM

$$\sin u + \sin v = 2 \sin\frac{u + v}{2} \cos\frac{u - v}{2} \qquad \cos u + \cos v = 2 \cos\frac{u + v}{2} \cos\frac{u - v}{2}$$

PRODUCT

$$\sin u \sin v = \tfrac{1}{2}[\cos(u - v) - \cos(u + v)]$$
$$\cos u \cos v = \tfrac{1}{2}[\cos(u - v) + \cos(u + v)]$$
$$\sin u \cos v = \tfrac{1}{2}[\sin(u + v) + \sin(u - v)]$$
$$\cos u \sin v = \tfrac{1}{2}[\sin(u + v) - \sin(u - v)]$$

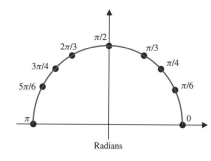

Radians

$\sin(0) = 0$	$\cos(0) = 1$
$\sin\left(\frac{\pi}{6}\right) = \frac{1}{2}$	$\cos\left(\frac{\pi}{6}\right) = \frac{\sqrt{3}}{2}$
$\sin\left(\frac{\pi}{4}\right) = \frac{\sqrt{2}}{2}$	$\cos\left(\frac{\pi}{4}\right) = \frac{\sqrt{2}}{2}$
$\sin\left(\frac{\pi}{3}\right) = \frac{\sqrt{3}}{2}$	$\cos\left(\frac{\pi}{3}\right) = \frac{1}{2}$
$\sin\left(\frac{\pi}{2}\right) = 1$	$\cos\left(\frac{\pi}{2}\right) = 0$
$\sin\left(\frac{2\pi}{3}\right) = \frac{\sqrt{3}}{2}$	$\cos\left(\frac{2\pi}{3}\right) = -\frac{1}{2}$
$\sin\left(\frac{3\pi}{4}\right) = \frac{\sqrt{2}}{2}$	$\cos\left(\frac{3\pi}{4}\right) = -\frac{\sqrt{2}}{2}$
$\sin\left(\frac{5\pi}{6}\right) = \frac{1}{2}$	$\cos\left(\frac{5\pi}{6}\right) = -\frac{\sqrt{3}}{2}$
$\sin(\pi) = 0$	$\cos(\pi) = -1$
$\sin(2\pi) = 0$	$\cos(2\pi) = 1$

CALCULUS

Multivariable

Second Edition

Robert T. Smith

Millersville University of Pennsylvania

Roland B. Minton

Roanoke College

Mc
Graw
Hill

Boston   Burr Ridge, IL   Dubuque, IA   Madison, WI   New York   San Francisco   St. Louis
Bangkok   Bogotá   Caracas   Kuala Lumpur   Lisbon   London   Madrid   Mexico City
Milan   Montreal   New Delhi   Santiago   Seoul   Singapore   Sydney   Taipei   Toronto

McGraw-Hill Higher Education ✖

*A Division of The **McGraw-Hill** Companies*

CALCULUS: MULTIVARIABLE, SECOND EDITION

Published by McGraw-Hill, a business unit of The McGraw-Hill Companies, Inc., 1221 Avenue of the Americas, New York, NY 10020. Copyright © 2002, 2000 by The McGraw-Hill Companies, Inc. All rights reserved. No part of this publication may be reproduced or distributed in any form or by any means, or stored in a database or retrieval system, without the prior written consent of The McGraw-Hill Companies, Inc., including, but not limited to, in any network or other electronic storage or transmission, or broadcast for distance learning.

Some ancillaries, including electronic and print components, may not be available to customers outside the United States.

This book is printed on acid-free paper.

1 2 3 4 5 6 7 8 9 0 VNH/VNH 0 9 8 7 6 5 4 3 2 1

ISBN 0–07–059251–9
ISBN 0–07–112270–2 (ISE)

Publisher: *William K. Barter*
Senior sponsoring editor: *Maggie Rogers*
Developmental editor: *Michelle Munn*
Editorial assistant: *Allyndreth Cassidy*
Marketing manager: *Mary K. Kittell*
Project manager: *Jill R. Peter*
Production supervisor: *Laura Fuller*
Designer: *K. Wayne Harms*
Cover/interior designer: *John Rokusek*
Photo research coordinator: *John C. Leland*
Photo research: *Feldman and Associates, Inc.*
Senior supplement producer: *Stacy A. Patch*
Media technology lead producer: *Steve Metz*
Compositor: *Interactive Composition Corporation*
Typeface: *10/12 Times Roman*
Printer: *Von Hoffmann Press, Inc.*

The credits section for this book begins on page 421 and is considered an extension of the copyright page.

Library of Congress has cataloged the main title as follows:

Smith, Robert T. (Robert Thomas), 1955–
    Calculus / Robert T. Smith, Roland B. Minton. — 2nd ed.
        p.   cm.
    Includes bibliographical references and index.
    ISBN 0–07–239848–5
    1. Calculus.   I. Minton, Roland B., 1956–.   II. Title.

QA303. S6542   2002
515—dc21   2001030156
                                                            CIP

INTERNATIONAL EDITION ISBN 0–07–112270–2
Copyright © 2002. Exclusive rights by The McGraw-Hill Companies, Inc., for manufacture and export. This book cannot be re-exported from the country to which it is sold by McGraw-Hill. The International Edition is not available in North America.

www.mhhe.com

CALCULUS

ABOUT THE AUTHORS

Robert T. Smith is Professor of Mathematics and Chair of the Department of Mathematics at Millersville University of Pennsylvania, where he has taught since 1987. Prior to that, he was on the faculty at Virginia Tech. He earned his Ph.D. in mathematics from the University of Delaware in 1982.

Dr. Smith's mathematical interests are in the application of mathematics to problems in engineering and the physical sciences. He has published a number of research articles on the applications of partial differential equations as well as on computational problems in x-ray tomography. He is a member of the American Mathematical Society, the Mathematical Association of America and the Society for Industrial and Applied Mathematics.

Professor Smith lives in Lancaster, Pennsylvania, with his wife Pam, his daughter Katie and his son Michael. When time permits, he enjoys playing volleyball, tennis and softball. In his spare time, he also coaches youth league soccer. His present extracurricular goal is to learn the game of golf well enough to not come in last in his annual mathematicians/statisticians tournament.

Roland Minton is Professor of Mathematics at Roanoke College, where he has taught since 1986. Prior to that, he was on the faculty at Virginia Tech. He earned his Ph.D. from Clemson University in 1982. He is the recipient of the 1998 Roanoke College Exemplary Teaching Award.

Dr. Minton has supervised numerous student research projects in such topics as sports science, complexity theory and fractals. He has published several articles on the use of technology and sports examples in mathematics, in addition to a technical monograph on control theory. He has received grants for teacher training from the State Council for Higher Education in Virginia. He is a member of the Mathematical Association of America, American Mathematical Society and other mathematical societies.

Professor Minton lives in Salem, Virginia, with his wife Jan, daughter Kelly and son Greg. He enjoys playing golf when time permits and watching any sport on TV (even if time doesn't permit). Jan also teaches mathematics at Roanoke College and is very active in mathematics education. A busy family schedule includes band performances and la crosse and soccer matches. Favorite entertainers include the Marx Brothers, guitarist Danny Gatton and mystery novelist Kinky Friedman.

In addition to the Premiere Edition of *Calculus*, Professors Smith and Minton have together published three earlier books with McGraw-Hill: *Discovering Calculus with the HP-28 and the HP-48*, *Discovering Calculus with the TI-81 and the TI-85* and *Discovering Calculus with the Casio fx-7700 and the Casio fx-8700*. The Premiere edition of *Calculus* has also been translated into Spanish.

To Pam, Katie and Michael
To Jan, Kelly and Greg
And our parents
Thanks for your love and inspiration.

TABLE OF CONTENTS

Preface xi

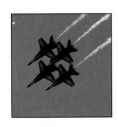

PREFACE

The wide-ranging debate brought about by the calculus reform movement has had a significant impact on the calculus textbook market. In response to many of the questions and concerns surrounding this debate, we have written a modern calculus textbook, intended for students majoring in mathematics, physics, chemistry, engineering and related fields.

This text is intended for the average student, that is, one who does not already know the subject, whose background is somewhat weak in spots and who requires significant motivation to study the subject. Our intention is that students should be able to read our book, rather than merely use it as an encyclopedia filled with the facts of calculus. The book has been written in a conversational style that reviewers have compared to listening to a good lecture. Our sense of what works well with students has been honed by teaching mathematics for more than 20 years at a variety of colleges and universities, both public and private, ranging from a small liberal arts college to large engineering schools.

In an effort to ensure that this textbook successfully addresses our concerns about the effective teaching of calculus, as well as others' concerns, we have continually asked instructors around North America for their opinions on the calculus curriculum, the strengths and weaknesses of current textbooks and the strengths and weaknesses of our own text. In preparing this second edition, as with the Premiere Edition, we enjoyed the benefit of countless insightful comments from a talented panel of reviewers that was carefully selected to help us with this project. Their detailed reviews of our materials and their opinions about the teaching of calculus were invaluable to us during our development of the Premiere Edition and the preparation of this edition. We are deeply indebted to them for their time and effort.

■ OUR PHILOSOPHY

We agree with many of the ideas that have come out of the calculus reform movement. In particular, we believe in the **Rule of Four:** that concepts should be presented **graphically, numerically, algebraically** and **verbally,** whenever these are appropriate. In fact, we would add **physically** to this list, since the modeling of physical problems is an important skill that students need to develop. We also believe that, while the calculus curriculum has been in need of reform, we should not throw out those things that already work well. Our book thus represents an updated approach to the traditional topics of calculus. We follow an essentially traditional order of presentation, while integrating technology and thought-provoking exercises throughout.

One of the thrusts of the calculus reform movement has been to place greater emphasis on problem solving and to present students with more realistic applications, as well as open-ended problems. We have incorporated meaningful writing exercises and extended, open-ended problems into **every** problem set. You will also find a much wider range of applications than in most traditional texts. We make frequent use of applications from within students' experience to both motivate the development of new topics and to illustrate concepts we have already presented. In particular, we have included numerous examples from the physics of sports to give students a familiar context in which to think of various concepts.

We believe that a conceptual development of the calculus must motivate the text. Although we have integrated technology throughout, we have not allowed the technology to drive the book. We have also not given in to the temptation to show off what technology can do, except where this has a direct bearing on learning the calculus. Our goal is to use the available technology to help students reach a conceptual understanding of the calculus as it is used today.

Perhaps the most difficult task when preparing a new calculus text is the actual *writing* of it. We have endeavored to write this text in a manner that combines an appropriate level of informality with an honest discussion regarding the difficulties that students commonly face in their study of calculus. In addition to the concepts and applications of calculus, we have also included many frank discussions about what is practical and impractical, and what is difficult and not so difficult to students of calculus. We have attempted to provide clarity of presentation in the creation of every example, application and exercise.

The book that we have written represents substantive change. By virtue of integrating technology throughout, utilizing a lively presentation style and incorporating a wider variety of problems, we believe that we meet many of the objectives of the calculus reform movement. At the same time, our relatively traditional outline retains the central strengths of mainstream calculus, enabling instructors to teach from a familiar body of material while integrating technology and modern applications.

■ DEVELOPMENT OF CONCEPT

We have endeavored to carefully reconsider the best way in which to present each traditional calculus concept. Our primary objective is to keep students focused on the central concepts and motivated in their learning.

Again serving this end purpose, we have augmented a simple algebraic presentation of selected ideas with numerical methods. For instance, when we introduce the notion of area, we emphasize the computation of area as a limit of a Riemann sum, but use regular partitions exclusively. We do not introduce the notion of the norm of a partition until Chapter 13, when we develop multiple integrals. By that point, students should already be comfortable with the concept of the definite integral as a limit of a sum and this refinement should only enhance their understanding. We are careful to point out that (without the Fundamental Theorem of Calculus) the limit of Riemann sums can be computed directly only for a very small number of functions. In addition, we allow students to explore the same ideas numerically. We are not restricted to polynomials of low degree and students can observe numerical values of Riemann sums approaching a limit. With this approach, students get to see the same problem from several different viewpoints, thus improving the likelihood that they will grasp the underlying concept. Additionally, students are given a useful tool (numerical integration) that they can bring to bear on a wide variety of problems.

■ CALCULUS AND TECHNOLOGY

We do not view technology as a gimmick to be artificially appended to the same old calculus curriculum. Nor do we believe that a calculus course should be a course in how to use technology. We believe that technology can and should be introduced as a natural part of a coherent development of calculus. In our presentation, technology has been seamlessly integrated throughout, but only where appropriate. Users are expected to have access to a graphing calculator or a computer algebra system and to use it routinely. We employ a generic use of technology, using those features that are shared by virtually all graphing calculators and computer algebra systems.

A common concern regarding the substantial use of technology is that our students will become mindless button-pushers. We guard against this by making the technology secondary to understanding and by pointing out errors that can be made by an overreliance on technology. We use technology so that students can focus on the difficult and sometimes subtle connections among the different concepts of the calculus. A student who has mastered these connections will be a much more effective user of the calculus than will a student who is proficient at algebraic methods alone. By engaging students on several different levels, using different approaches, we hope to improve their understanding and empower them to try new problems on their own.

We have chosen not to separate out "technology" exercises in the exercise sets found at the end of each section and chapter. This decision was made quite carefully, so that students would learn when to use the technology. We feel that placing an icon next to technology-based exercises allows students to avoid making this decision on their own. Throughout the text, we provide advice and guidance on the proper use of technology and provide tools to help students determine when the use of technology is most appropriate. These sections are called out by a technology icon. We do, however, make recommendations for technology use in the homework sets in the *Instructor's Resource Manual,* to help instructors plan their homework assignments accordingly.

We assume that students have access to calculator- or computer-generated graphs, allowing us to routinely use graphs as the first step in solving a problem or as a check on the reasonableness of an answer. Being able to visualize a problem is an invaluable aid to students, and we try to take full advantage of this. One benefit of readily available graphics is the ability to solve more realistic application problems. Functions associated with realistic problems are often not mathematically simple, but we can approximate zeros or extrema graphically and numerically. Further, concepts such as the convergence of Taylor series are more meaningful when graphs are used to illustrate this convergence. This same graphical approach benefits our presentation of Fourier series, which is an important tool for understanding much of our digitally enhanced world.

■ CONTENTS WITH COMMENTARY

The vast majority of the topics found in our book are part of the standard calculus curriculum that has defined the mainstream for the last 30 years or so. We believe that this curriculum still has validity in terms of both mathematical precision and student learning. Nevertheless, we have made a small number of significant changes in the table of contents. In the following, you will find brief explanations of each chapter and its focus.

Chapter 10: Vectors and the Geometry of Space

Chapter 10 introduces students to a third dimension of graphing and calculations. Computer graphics are a valuable aid in this chapter, and are used extensively. A discussion of Magnus force relates vectors to a variety of sports applications, while providing students with practice at thinking in three-dimensional space.

Chapter 11: Vector-Valued Functions

Chapter 11 develops the calculus of vector-valued functions. As the graphs become more complicated, our use of computer graphics increases. To keep students thinking and not simply pushing buttons, several of the examples and exercises involve matching functions and graphs, with students using the properties of functions to identify the graphs. Section 11.5 includes the important derivation of Kepler's laws of planetary motion.

Chapter 12: Functions of Several Variables and Partial Differentiation

Chapter 12 presents the calculus of functions of two or more variables. Given the increasing difficulty of visualizing the mathematics in this chapter, the Rule of Four is particularly useful. We use a variety of graphics options (e.g., wireframe and parametric plots) in this chapter so that students can see the traces, and not lose the details that shaded graphs tend to obscure. Where appropriate, three-dimensional graphs are augmented with contour plots and density plots. Numerically, a steepest ascent (descent) algorithm is presented, requiring some computer assistance, but reinforcing several important concepts of the calculus of functions of several variables.

Chapter 13: Multiple Integrals

Chapter 13 introduces double and triple integrals. Considerable emphasis is placed on helping students develop insight into the proper coordinate system and order of integration to use to simplify a given multiple integral. Applications involving the design of rockets and baseball bats are used to enliven the discussion of moments and centers of mass.

Chapter 14: Vector Calculus

Chapter 14 introduces the vector calculus that is essential to an understanding of fluid mechanics and applications in electricity and magnetism. Numerous graphs of vector fields are included, as well as a thorough discussion of various interpretations of these graphs.

■ CHANGES FOR THE SECOND EDITION

Every section of the text and every pedagogical feature was carefully scrutinized during the development of this Second Edition. In preparing our revisions, we had the benefit of comments from a very large panel of reviewers, including both users and nonusers of the Premiere Edition of the text. Based on our analysis and the reviewers' comments, we have made countless changes for this edition of *Calculus*. These include:

Graphs and Tables: All graphs within this edition have been newly rendered to take advantage of the use of color in interpreting figures. This program has been painstakingly poured over to ensure accuracy and consistency throughout.

Examples: Many new examples have been included in this edition, particularly in places where reviewers suggested that additional explanation would be beneficial.

Exercises: Each exercise set was carefully reexamined to ensure that exercises of all levels were included. In particular, we significantly increased the number of exercises of moderate difficulty.

Presentation: All explanations, proofs and derivations have been carefully reconsidered, with countless revisions made to improve clarity.

Historical Notes: Biographical information about prominent mathematicians and their contributions to the development of calculus have been added to this edition. Several dozen mathematicians are profiled in marginal boxes spread throughout the text.

Technology: We have enhanced our discussions of technology in a number of places, including additional tables and functions specified by data.

Design and Color: The text has been completely redesigned to allow for the pedagogical use of color to highlight important results and remarks, as well as for visual emphasis in the graphs and tables.

■ ACKNOWLEDGMENTS

First and foremost, we want to express our appreciation to our sponsoring editor, Maggie Rogers for the encouragement and guidance to keep us going through this very long and challenging project. We also wish to thank our original editor, Jack Shira, for having the confidence in us to start the project. We must also thank our developmental editors, Michelle Munn and Kris Swanson for their hard work and constant support. We could never have completed this project without Maggie's, Michelle's and Kris' tireless efforts and professionalism. We could truly not think of better people with whom to work. We also wish to thank our publishers, Bill Barter, J. P. Lenney and Denise Schanck, for their strong support over the last six years.

A project of this magnitude requires the collaboration of an incredible number of talented people. Our developmental editor team for the Premiere Edition, Glenn and Meg Turner of Burrston House, ran the project through their demanding schedule of reviews and meetings. Our understanding of the project and our ability to create a quality book improved tremendously through their efforts. Our production team at McGraw-Hill, particularly our project manager, Jill Peter, kept the project on schedule and helped us to produce a well-designed text. The problem solvers at Laurel Technical Services, led by Carrie Mallery, and Elka Block and Frank Purcell of Twin Prime Editorial made numerous suggestions to improve the exercise sets. The TEX experts at Interactive Composition Corporation produced the current book design from our original TEX files. George Morris and the staff of Scientific Illustrators created each figure from scratch for the Second Edition. Their diligence in following our design requirements and their skill are greatly appreciated.

A project of this scope could not have been completed without the many insightful comments from our reviewer panels, both past and present. During the development of the manuscript, we continually asked our reviewers for advice on the placement and treatment of everything from major concepts to figures. Thanks to their often unerring judgment, we have made numerous improvements to the text.

With many thanks to our previous edition reviewer panel:

David Anderson, *University of Tennessee*
Wilma Anderson, *University of Nebraska–Omaha*
Robert Beezer, *University of Puget Sound*
Neil Berger, *University of Illinois*
Mike Bonnano, *Suffolk Community College*
George Bradley, *Duquesne University*
Moody Chu, *North Carolina State University*
Raymond Clapsadle, *University of Memphis*
Joe Diestel, *Kent State University*
Dan Drucker, *University of Puget Sound*
Eugene Enneking, *Portland State University*

Ronald Grimmer, *Southern Illinois University*
Mel Hausner, *New York University*
Johnny Henderson, *Auburn University*
Robert Horvath, *El Camino College*
Gail Kaufmann, *Tufts University*
Hadi Kharaghani, *University of Lethbridge (Alberta)*
Masato Kimura, *College of William and Mary*
Robert Knott, *University of Evansville*
Jon Lee, *University of Kentucky*
John Maginnis, *Kent State University*

Chris McCord, *University of Cincinnati*

Remigijus Mikulevicius, *University of Southern California*

Mike Montano, *Riverside Community College*

Christina Pereyra, *University of New Mexico*

Linda Powers, *Virginia Tech*

Joe Rody, *Arizona State University*

Rod Smart, *University of Wisconsin–Madison*

Jerry Stonewater, *Miami University of Ohio*

Juan Tolosa, *Richard Stockton College*

Paul Weichsel, *University of Illinois*

Marvin Zeman, *Southern Illinois University*

And special appreciation to those who aided in the second edition of *Calculus:*

Alisher S. Abdullayev, *American River College*

Edward Aboufadel, *Grand Valley State University*

Shair Ahmad, *University of Texas at San Antonio*

Tom Akers, *University of Missouri–Rolla*

Tuncay Aktosun, *North Dakota State University*

Gerardo Aladro, *Florida International University*

Ariyadasa Aluthge, *Marshall University*

Michael R. Anderson, *West Virginia State College*

Tamas Antal, *Ohio State University*

Seth Armstrong, *Arkansas State University*

Leon Arriola, *Western New Mexico University*

Nuh Aydin, *Ohio State University*

Prem N. Bajaj, *Wichita State University*

Robert Bakula, *Ohio State University*

Rachel Belinsky, *Morris Brown College*

Chris Black, *Seattle University*

Karen Bolinger, *Clarion University of Pennsylvania*

Robert Brabenec, *Wheaton College*

Dave Bregenzer, *Utah State University*

C. Allen Brown, *Wabash Valley College*

Linda K. Buchanan, *Howard College*

James Caggiano, *Arkansas State University*

Jorge Alberto Calvo, *North Dakota State University*

James T. Campbell, *University of Memphis*

Jianguo Cao, *University of Notre Dame*

Florin Catrina, *Utah State University*

Deanna M. Caveny, *College of Charleston*

Maurice J. Chabot, *University of Southern Maine*

Wai Yuen Chan, *University of Science and Arts of Oklahoma*

Mei-Chu Chang, *University of California–Riverside*

Benito Chen, *University of Wyoming*

Karin Chess, *Owensboro Community College*

Dominic P. Clemence, *North Carolina Agricultural and Technical State University*

Barbara Cortzen, *DePaul University*

Julane B. Crabtree, *Johnson County Community College*

Ellen Cunningham, *Saint Mary-of-the-Woods College*

Daniel J. Curtin, *Northern Kentucky University*

Sujay Datta, *Northern Michigan University*

Gregory Davis, *University of Wisconsin–Green Bay*

Shusen Ding, *Seattle University*

Michael Dorff, *University of Missouri–Rolla*

Michael M. Dougherty, *Penn State Berks*

Judith Downey, *University of Nebraska at Omaha*

Tevian Dray, *Oregon State University*

Bennett Eisenberg, *Lehigh University*

Alan Elcrat, *Wichita State University*

Sherif T. El-Helaly, *Catholic University of America*

David L. Fama, *Germanna Community College*

Judith Hanks Fethe, *Pellissippi State Technical Community College*

Earl D. Fife, *Calvin College*

Jose D. Flores, *University of South Dakota*

Teresa Floyd, *Mississippi College*

William P. Francis, *Michigan Technological University*

Michael Frantz, *University of LaVerne*

Chris Gardiner, *Eastern Michigan University*

Charles H. Giffen, *University of Virginia*

Kalpana Godbole, *Michigan Technological University*

Michael Green, *Metropolitan State University*

Harvey Greenwald, *California Polytechnic State University*

Laxmi N. Gupta, *Rochester Institute of Technology*

Joel K. Haack, *University of Northern Iowa*

H. Allen Hamilton, *Delaware State University*

John Hansen, *Iowa Central Community College*

John Harding, *New Mexico State University*

John Haverhals, *Bradley University*

Sue Henderson, *Georgia Perimeter College*

Guy T. Hogan, *Norfolk State University*

Jack Howard, *Clovis Community College*

Cornelia Wang Hsu, *Morgan State University*

Shirley Huffman, *Southwest Missouri State University*

Hristo V. Kojouharov, *Arizona State University*

Emanuel Kondopirakis, *Cooper Union*

Kathryn Kozak, *Coconino County Community College*

Kevin Kreider, *University of Akron*

Tor A. Kwembe, *Chicago State University*

Joseph Lakey, *New Mexico State University*

Melvin D. Lax, *California State University–Long Beach*

James W. Lea, *Middle Tennessee State University*

William L. Lepowsky, *Laney College*

Fengshan Liu, *Delaware State University*

Yung-Chen Lu, *Ohio State University*

Stephen A. MacDonald, *University of Southern Maine*

Michael Maller, *Queens College*

Nicholas A. Martin, *Shepherd College*

Paul A. Martin, *University of Wisconsin Colleges*

Alex Martin McAllister, *Centre College*

Daniel McCallum, *University of Arkansas at Little Rock*

Philip McCartney, *Northern Kentucky University*

Michael J. McConnell, *Clarion University of Pennsylvania*

David McKay, *California State University, Long Beach and Orange Coast College, Costa Mesa*

Aaron Melman, *University of San Francisco*

Gordon Melrose, *Old Dominion University*

Richard Mercer, *Wright State University*

Scott Metcalf, *Eastern Kentucky University*

Allan D. Mills, *Tennessee Technological University*

Jeff Mock, *Diablo Valley College*

Laura Moore-Mueller, *Green River Community College*

Shahrooz Moosavizadeh, *Norfolk State University*

Kandasamy Muthevel, *University of Wisconsin–Oshkosh*

Kouhestani Nader, *Prairie View A & M University*

Sergey Nikitin, *Arizona State University*

Terry A. Nyman, *University of Wisconsin–Fox Valley*

Altay Özgener, *Elizabethtown Community College*

Bent E. Petersen, *Oregon State University*

Cyril Petras, *Lord Fairfax Community College*

Donna Pierce, *Washington State University*

Jim Polito, *North Harris College*

Yiu Tong Poon, *Iowa State University*

Evelyn Pupplo-Cody, *Marshall University*

Anthony Quas, *University of Memphis*

Doraiswamy Ramachandran, *California State University–Sacramento*

William C. Ramaley, *Fort Lewis College*

W. Ramasinghage, *Ohio State University*

M. Rama Mohana Rao, *University of Texas at San Antonio*

Nandita Rath, *Arkansas Tech University*

S. Barbara Reynolds, *Cardinal Stritch University*

Errol Rowe, *North Carolina Agricultural and Technical State University*

Harry M. Schey, *Rochester Institute of Technology*

Charles Seebeck, *Michigan State University*

George L. Selitto, *Iona College*

Shagi-Di Shih, *University of Wyoming*

Mehrdad Simkani, *University of Michigan–Flint*

Eugenia A. Skirta, *University of Toledo*

Scott Smith, *Columbia College*

Alex Smith, *University of Wisconsin–Eau Claire*

Frederick Solomon, *Warren Wilson College*

V. K. Srinivasan, *University of Texas at El Paso*

Mary Jane Sterling, *Bradley University*

Adam Stinchcombe, *Adirondack Community College*

Jeff Stuart, *University of Southern Mississippi*

D'Loye Swift, *Nunez Community College*

Randall J. Swift, *Western Kentucky University*

Lawrence Sze, *California Polytechnic State University*

Wanda Szpunar-Lojasiewicz, *Rochester Institute of Technology*

Fereja Tahir, *Eastern Kentucky University*

J. W. Thomas, *Colorado State University*

Juan Tolosa, *Richard Stockton College of New Jersey*

Michael M. Tom, *Louisiana State University*

William K. Tomhave, *Concordia College*

Stefania Tracogna, *Arizona State University*

Jay Treiman, *Western Michigan University*

Patricia Treloar, *University of Mississippi*

Thomas C. Upson, *Rochester Institute of Technology*

Richard G. Vinson, *University of South Alabama*

David Voss, *Western Illinois University*

Mu-Tao Wang, *Stanford University*

Richard A. Weida, *Lycoming College*

Michael Weiner, *Penn State Altoona*

Alan Wilson, *Kaskaskia College*

Michael Wilson, *University of Vermont*

Jim Wolper, *Idaho State University*

Jiahong Wu, *University of Texas at Austin*

DaGang Yang, *Tulane University*

Xiao-Dong Zhang, *Florida Atlantic University*

Jianqiang Zhao, *University of Pennsylvania*

In addition, a number of colleagues graciously gave of their time and energy to help create or improve portions of the manuscript. We would especially like to thank Bill Ergle, Ben Huddle, Jack Steehler, Deana Carideo, Jan Minton, Chris Lee, Jeff Spielman and Richard Grant of Roanoke College for sharing their expertise in calculus and related applications; Tom Burns of General Electric for help with an industrial application; Dorothee Blum of Millersville University, who helped to field test an early version of the manuscript; Chuck Denlinger and Ron Umble of Millersville University for many conversations about the calculus; Bruce Ikenaga of Millersville University for generously sharing his expertise in TeX and Corel Draw, and Pam Vercellone-Smith, who lent us her expertise with many of the biological applications. We also wish to thank Dorothee Blum, Bob Buchanan, Chuck Denlinger, Bruce Ikenaga, Kit Kittappa, Tim McDevitt and Zhoude Shao of Millersville University for sharing numerous helpful suggestions for improvements from the Premiere Edition. In addition, we would like to thank all of our students through the years who have (sometimes unknowingly) field-tested innumerable ideas, some of which worked and the rest of which will not be found in this book. Finally, we gratefully acknowledge the support of the Millersville University Faculty Grants Committee and the Millersville University Faculty Professional Development Committee for the award of three released-time grants and the Millersville University administration for a sabbatical leave grant. These grants provided critical blocks of time for developing the Premiere Edition of this text and the authors are most grateful.

Ultimately, this book is for our families. We want to thank them for their love and inspiration throughout the years that we have worked on this project. Their understanding, in both the technical and the personal sense, was essential to us. Writing a book of this magnitude could simply not have happened without our families' strong support. Our families provide us with the reason why we do all of the things we do. So, it is fitting that we especially thank our wives, Pam Vercellone-Smith and Jan Minton, our children, Katie and Michael Smith and Kelly and Greg Minton, and our parents, Anne Smith and Paul and Mary Frances Minton.

Robert T. Smith   Roland B. Minton
Lancaster, Pennsylvania   Salem, Virginia

VISUAL GUIDE: EXPANDING UPON THE TEXT

TOOLS FOR LEARNING

Real-World Emphasis

Real-world examples are an important aid to the understanding of calculus. We introduce each chapter with a brief application related to the mathematical concepts being developed in order to put each chapter into a larger problem-solving context. Subsequently, both examples within the text and exercises are used to further demonstrate the importance of calculus within the world.

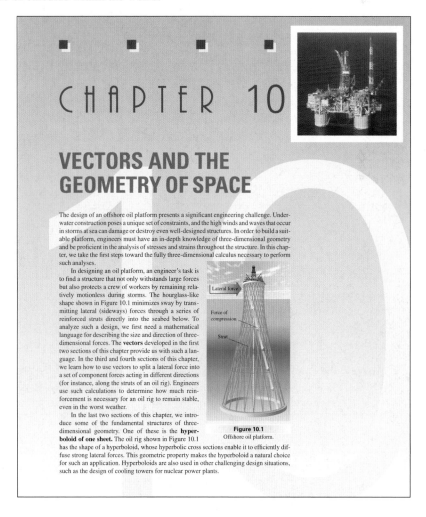

CHAPTER 10

VECTORS AND THE GEOMETRY OF SPACE

The design of an offshore oil platform presents a significant engineering challenge. Underwater construction poses a unique set of constraints, and the high winds and waves that occur in storms at sea can damage or destroy even well-designed structures. In order to build a suitable platform, engineers must have an in-depth knowledge of three-dimensional geometry and be proficient in the analysis of stresses and strains throughout the structure. In this chapter, we take the first steps toward the fully three-dimensional calculus necessary to perform such analyses.

In designing an oil platform, an engineer's task is to find a structure that not only withstands large forces but also protects a crew of workers by remaining relatively motionless during storms. The hourglass-like shape shown in Figure 10.1 minimizes sway by transmitting lateral (sideways) forces through a series of reinforced struts directly into the seabed below. To analyze such a design, we first need a mathematical language for describing the size and direction of three-dimensional forces. The **vectors** developed in the first two sections of this chapter provide us with such a language. In the third and fourth sections of this chapter, we learn how to use vectors to split a lateral force into a set of component forces acting in different directions (for instance, along the struts of an oil rig). Engineers use such calculations to determine how much reinforcement is necessary for an oil rig to remain stable, even in the worst weather.

In the last two sections of this chapter, we introduce some of the fundamental structures of three-dimensional geometry. One of these is the **hyperboloid of one sheet.** The oil rig shown in Figure 10.1 has the shape of a hyperboloid, whose hyperbolic cross sections enable it to efficiently diffuse strong lateral forces. This geometric property makes the hyperboloid a natural choice for such an application. Hyperboloids are also used in other challenging design situations, such as the design of cooling towers for nuclear power plants.

Figure 10.1
Offshore oil platform.

Lateral force

Force of compression

Strut

Definitions, Theorems and Proofs

All formal definitions and theorems are clearly boxed within the text for easy visual reference. Selected proofs are provided for reference. Proofs of some results are found in Appendix A.

Definition 2.3

The **derivative** $\mathbf{r}'(t)$ of the vector-valued function $\mathbf{r}(t)$ is defined by

$$\mathbf{r}'(t) = \lim_{\Delta t \to 0} \frac{\mathbf{r}(t + \Delta t) - \mathbf{r}(t)}{\Delta t}, \qquad (2.2)$$

for any values of t for which the limit exists. When the limit exists for $t = a$, we say that $\mathbf{r}$ is **differentiable** at $t = a$.

Fortunately, you will not need to learn any new differentiation rules, as the derivative of a vector-valued function is found directly from the derivatives of the individual components, as we see in the following result.

Theorem 2.2

Let $\mathbf{r}(t) = \langle f(t), g(t), h(t) \rangle$ and suppose that the components f, g and h are all differentiable for some value of t. Then $\mathbf{r}$ is also differentiable at that value of t and its derivative is given by

$$\mathbf{r}'(t) = \langle f'(t), g'(t), h'(t) \rangle. \qquad (2.3)$$

Proof

From the definition of derivative of a vector-valued function (2.2), we have

$$\mathbf{r}'(t) = \lim_{\Delta t \to 0} \frac{\mathbf{r}(t + \Delta t) - \mathbf{r}(t)}{\Delta t}$$

$$= \lim_{\Delta t \to 0} \frac{1}{\Delta t}[\langle f(t + \Delta t), g(t + \Delta t), h(t + \Delta t) \rangle - \langle f(t), g(t), h(t) \rangle]$$

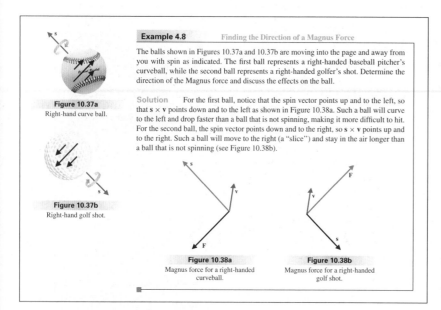

Figure 10.37a
Right-hand curve ball.

Figure 10.37b
Right-hand golf shot.

Example 4.8   Finding the Direction of a Magnus Force

The balls shown in Figures 10.37a and 10.37b are moving into the page and away from you with spin as indicated. The first ball represents a right-handed baseball pitcher's curveball, while the second ball represents a right-handed golfer's shot. Determine the direction of the Magnus force and discuss the effects on the ball.

Solution   For the first ball, notice that the spin vector points up and to the left, so that $\mathbf{s} \times \mathbf{v}$ points down and to the left as shown in Figure 10.38a. Such a ball will curve to the left and drop faster than a ball that is not spinning, making it more difficult to hit. For the second ball, the spin vector points down and to the right, so $\mathbf{s} \times \mathbf{v}$ points up and to the right. Such a ball will move to the right (a "slice") and stay in the air longer than a ball that is not spinning (see Figure 10.38b).

Figure 10.38a
Magnus force for a right-handed curveball.

Figure 10.38b
Magnus force for a right-handed golf shot.

Examples

Each chapter contains a large number of worked examples, ranging from the simple and concrete to more complex and abstract. A thorough understanding of the initial problem presented and the step-by-step solution will greatly enhance your problem-solving capabilities and your further study of the subject.

Use of Graphs and Tables

Being able to visualize a problem is an invaluable aid in understanding the concept presented. To this purpose, we have integrated more than 1500 computer-generated graphs throughout the text. You should use them routinely to aid in solving most problems, even if only as a check on the reasonableness of an answer. Each graph and table has been created very carefully to ensure that the ideas presented are clear and accurate. In many places, we have included multiple graphical perspectives, such as with the contour and density plots found in Chapter 12.

COMMENTARY AND GUIDANCE

In order to help you best interpret the material presented, we have included many side elements and notes for your reference.

3.1   LINEAR APPROXIMATIONS AND L'HÔPITAL'S RULE

For what purpose do you use a scientific calculator? If you think about it, you'll discover that there are two distinctly different jobs that calculators do for you. First, they perform arithmetic operations (addition, subtraction, multiplication and division) much faster than any of us could hope to do them. It's not that you don't know how to multiply 1024 by 1673, but rather that it is time-consuming to carry out this (albeit well-understood) calculation with pencil and paper. For such problems, calculators are a tremendous convenience, which none of us would like to live without. Perhaps more significantly, we also use our calculators to compute values of transcendental functions such as sine, cosine, tangent, exponentials and logarithms. In the case of these function evaluations, the calculator is much more than a mere convenience.

 If asked to calculate sin(1.2345678) without a calculator, you would probably draw a blank. Don't worry, there's nothing wrong with your background. (Also, don't worry that anyone will ever ask you to do this without a calculator.) The problem is that the sine function is not *algebraic*. That is, there is no formula for sin x involving only the arithmetic operations. So, how does your calculator "know" that sin(1.2345678) $\approx$ 0.9440056953? In short, it doesn't *know* this at all. Rather, the calculator has a built-in program that generates **approximate** values of the sine and other transcendental functions.

In this section, we take a small step into the (very large) world of approximation by developing a simple approximation method. Although somewhat crude, it points the way

Technology Guidance

We hope that by using this text you will become proficient at knowing when and how technology is appropriate within calculus. Rather than providing key-punching instructions, we provide advice and guidance on the proper use of technology, empowering you to explore new problems on your own. We also provide comments on possible errors and pitfalls that can be caused due to an over reliance on technology. This guidance is frequently given within an example or explanation of a specific technique. However, an icon is used when this guidance appears within the general discussion.

Remarks Text

Remarks boxes, found both as marginal and text references, provide a summary or overview of techniques. These boxes also foreshadow future ideas related to the topic being discussed.

Remark 1.2

If the level curves in a contour plot are plotted for equally spaced values of z, observe that a tightly packed region of the contour plot will correspond to a region of rapid change in the function. Alternatively, blank space in the contour plot corresponds to a region of slow change in the function.

A **density plot** is closely related to a contour plot, in that they are both two-dimensional representations of a surface in three dimensions. For a density plot, the $(x\text{-}y)$ graphing window is divided into rectangles (like pixels, although usually much larger). Each rectangle is shaded according to the size of the function value of a representative point in the rectangle, ranging from white (maximum function value) to black (minimum function value). In a density plot, notice that level curves can be seen as curves formed by a specific shade of gray.

Example 1.9   Matching Functions and Density Plots

Match the density plots in Figures 12.11a–11c with the functions $f_1(x, y) = \dfrac{1}{y^2 - x^2}$, $f_2(x, y) = \dfrac{2x}{y - x^2}$ and $f_3(x, y) = \cos(x^2 + y^2)$.

Notice that one consequence of the result $\mathbf{F}(t) = -m\omega^2\mathbf{r}(t)$ from example 3.3 is that the magnitude of the force increases as the rotation rate ω increases. You have experienced this if you have been on a roller coaster with tight turns or loops. The faster you are going, the stronger the force that your seat exerts on you.

Just as we did in the one-dimensional case, we can use Newton's second law of motion to do more than simply identify the force acting on an object with a given position function. It's much more important to be able to determine the position of an object given only a knowledge of the forces acting on it. For instance, one of the most significant problems faced by the military is how to aim a projectile (e.g., a missile) so that it will end up hitting its intended target. This problem is harder than it sounds, even when the target is standing still. When the target is an aircraft moving faster than the speed of sound, the problem presents significant challenges. In any case, the calculus can be used to arrive at a meaningful solution. We present the simplest possible case (where neither the target nor the source of the projectile are moving) in the following example.

Example 3.4   Analyzing the Motion of a Projectile

A projectile is launched with an initial speed of 140 feet per second from ground level at an angle of $\frac{\pi}{4}$ to the horizontal. Assuming that the only force acting on the object is gravity (i.e., there is no air resistance, etc.), find the maximum altitude, the horizontal range and the speed at impact of the projectile.

Lightbulb Icon

A lightbulb icon indicates the relaying of an important idea in clear, nontechnical language. This icon is also used to highlight a special problem-solving tip within the text.

Johannes Kepler (1571–1630)
German astronomer and
mathematician whose discoveries
revolutionized Western science.
Kepler's remarkable mathematical
ability and energy produced
connections among many areas of
research. A study of observations of
the moon led to work in optics that
included the first fundamentally
correct description of the operation
of the human eye. Kepler's model
of the solar system used an
ingenious nesting of the five
Platonic solids to describe the orbits
of the six known planets. (Kepler
invented the term "satellite" to
describe the moon, which his

Historical Notes

These notes provide a historical context for
the development of calculus. Biographical
information about prominent mathematicians and their contributions to the development of calculus are given to put the
subject matter in perspective.

Notes Text

Notes boxes, found within the margin,
serve to step back from a concept to gain
perspective by adding details, making
connections to previous material, or noting
when an example is indicative of a general
pattern.

In this case, $\|P\|$ is the longest diagonal of any elementary polar region in the inner parti-
tion. More generally, we have the following result, which holds regardless of whether or
not $f(r, \theta) \geq 0$ on R.

NOTES

Theorem 3.1 says that to write a
double integral in polar coordinates,
we write $x = r \cos \theta$, $y = r \sin \theta$, find
the limits of integration for r and θ
and replace dA by $r\, dr\, d\theta$. Be certain
not to omit the factor of r in
$dA = r\, dr\, d\theta$; this is a very common
error.

Theorem 3.1 (Fubini's Theorem)
Suppose that $f(r, \theta)$ is continuous on the region $R = \{(r, \theta) | \alpha \leq \theta \leq \beta$ and $g_1(\theta) \leq r \leq g_2(\theta)\}$. Then,

$$\iint_R f(r, \theta)\, dA = \int_\alpha^\beta \int_{g_1(\theta)}^{g_2(\theta)} f(r, \theta)\, r\, dr\, d\theta. \qquad (3.3)$$

CONCEPTUAL UNDERSTANDING THROUGH PRACTICE

We have written this text with a strong problem-solving emphasis, including the introduction of many topics from graphical, numerical and algebraic points of view. In many
instances, the emphasis on graphical and numerical methods for solving problems frees us
to consider more realistic and complex problems than are usually presented in calculus.
Furthering this emphasis, we have included a variety of exercise types to strengthen your
problem-solving skills.

EXERCISES 10.2

1. Visualize the circle $x^2 + y^2 = 1$. With three-dimensional axes oriented as in Figure 10.15a, describe how to sketch this circle in the plane $z = 0$. Then, describe how to sketch the parabola $y = x^2$ in the plane $z = 0$. In general, explain how to translate a two-dimensional curve into a three-dimensional sketch.

2. It is difficult, if not impossible, for most people to visu-alize what points in four dimensions would look like. Nevertheless, it is easy to generalize the distance formula to four dimensions. Describe what the distance formula looks like in general dimension n, for $n \geq 4$.

3. It is very important to be able to quickly and accurately visualize three-dimensional relationships. In three di-mensions, describe how many lines are perpendicular to the unit vector $\mathbf{i}$. Describe all lines that are perpendicular to $\mathbf{i}$ and that pass through the origin. In three dimensions, describe how many planes are perpendicular to the unit vector $\mathbf{i}$. Describe all planes that are perpendicular to $\mathbf{i}$ and that contain the origin.

4. In three dimensions, describe all planes that contain a given vector $\mathbf{a}$. Describe all planes that contain two given vectors $\mathbf{a}$ and $\mathbf{b}$ (where $\mathbf{a}$ and $\mathbf{b}$ are not parallel.) Describe all planes that contain a given vector $\mathbf{a}$ and pass through the ori-gin. Describe all planes that contain two given (nonparallel) vectors $\mathbf{a}$ and $\mathbf{b}$ and pass through the origin.

In exercises 7 and 8, sketch the third axis to make xyz a right-handed system.

7.

8.

In exercises 9–14, find the distance between the given points.

9. $(2, 1, 2)$, $(5, 5, 2)$ 10. $(1, 2, 0)$, $(7, 10, 0)$

11. $(-1, 0, 2)$, $(1, 2, 3)$ 12. $(3, 1, 0)$, $(1, 3, -4)$

13. $(0, 0, 0)$, $(2, 3, 5)$ 14. $(2, 1, -4)$, $(0, 0, 0)$

In exercises 15–20, compute $\mathbf{a} + \mathbf{b}$, $\mathbf{a} - 3\mathbf{b}$ and $\|4\mathbf{a} + 2\mathbf{b}\|$.

15. $\mathbf{a} = \langle 2, 1, -2 \rangle$, $\mathbf{b} = \langle 1, 3, 0 \rangle$

16. $\mathbf{a} = \langle 2, -1, 2 \rangle$, $\mathbf{b} = \langle 1, 3, 0 \rangle$

17. $\mathbf{a} = \langle -1, 0, 2 \rangle$, $\mathbf{b} = \langle 4, 3, 2 \rangle$

42. $x^2 + (y - 1)^2 + (z - 4)^2 = 2$

43. $x^2 + y^2 - 2y + z^2 + 4z = 4$

44. $x^2 + 4x + y^2 - 6y + z^2 = 3$

45. $x^2 - 2x + y^2 + z^2 - 4z = 0$

46. $x^2 + x + y^2 - y + z^2 = \frac{7}{2}$

equidistant from $A = (0, 1, 0)$ and $C = (5, 2, 3)$.

66. In this exercise, you will try to identify the three-dimensional surface defined by the equation $a(x - 1) + b(y - 2) + c(z - 3) = 0$ for nonzero constants a, b and c. First, show that $(1, 2, 3)$ is one point on the surface. Then, show that any point which is equidistant from the points $(1 + a, 2 + b, 3 + c)$ and $(1 - a, 2 - b, 3 - c)$ is on the surface. Use this geometric fact to identify the surface.

End-of-Section Exercises

Each exercise set has been carefully
constructed to reinforce both the concepts and mechanics of calculus, while
encouraging individual exploration. Our
goal is to create original and imaginative
exercises that provide an appropriate
review of the topics covered in each
section while reinforcing the basic skills
needed to master the concept.

Writing Exercises

Each exercise set begins with a variety of
writing exercises. These exercises can be
used as springboards for discussion and
are intended to give you an opportunity
to carefully consider important mathematical concepts and ideas and express
these in your own words. Learning to
verbalize mathematical structures is a
key skill in mastering concepts.

60. In this exercise, we look at the ability of fireflies to synchronize their flashes. (To see a remarkable demonstration of this ability, see David Attenborough's video series *Trials of Life*.) Let the function $f(t)$ represent an individual firefly's rhythm, so that the firefly flashes whenever $f(t)$ equals an integer. Let $e(t)$ represent the rhythm of a neighboring firefly, where again $e(t) = n$, for some integer n, whenever the neighbor flashes. One model of the interaction between fireflies is $f'(t) = \omega + A\sin[e(t) - f(t)]$ for constants ω and A. If the fireflies are synchronized ($e(t) = f(t)$), then $f'(t) = \omega$, so the fireflies flash every $1/\omega$ time units. Assume that the difference between $e(t)$ and $f(t)$ is less than π. Show that if $f(t) < e(t)$, then $f'(t) > \omega$. Explain why this means that the individual firefly is speeding up its flash to match its neighbor. Similarly, discuss what happens if $f(t) > e(t)$.

61. The HIV virus attacks specialized T cells that trigger the human immune system response to a foreign substance. If $T(t)$ is the population of uninfected T cells at time t (days) and $V(t)$ is the population of infectious HIV in the bloodstream, a model that has been used to study AIDS is given by the following **differential equation** that describes the rate at which the population of T cells changes.

$$T'(t) = 10\left[1 + \frac{1}{1 + V(t)}\right] - 0.02T(t) + 0.01\frac{T(t)V(t)}{100 + V(t)} - $$

$$0.000024T(t)V(t).$$

Exploratory Exercises

Each exercise set concludes with exploratory exercises that are project-like assignments designed for you to obtain a sense of the ongoing, exciting nature of mathematics-related research. These exercises offer excellent opportunities for group work or discussion, depending on the structure of your individual calculus course.

Chapter Review Exercises

Chapter Review Exercise sets are provided as an overview of the chapter and will test your understanding prior to continuing with the text.

CHAPTER REVIEW EXERCISES

In exercises 1 and 2, find the linear approximation to $f(x)$ at x_0.

1. $f(x) = e^{3x}$, $x_0 = 0$
2. $f(x) = \sqrt{x^2 + 3}$, $x_0 = 1$

In exercises 3 and 4, use a linear approximation to estimate the quantity.

3. $\sqrt[3]{7.96}$
4. $\sin 3$

In exercises 5 and 6, use Newton's method to find an approximate root.

5. $x^3 + 5x - 1 = 0$
6. $x^3 = e^{-x}$

7. Explain why Newton's method fails on $x^3 - 3x + 2 = 0$ with $x_0 = 1$.

8. Show that the approximation $\frac{1}{(1-x)} \approx 1 + x$ is valid for "small" x.

In exercises 9–18, do the following by hand. (a) Find all critical numbers, (b) identify all intervals of increase and decrease, (c) determine whether each critical number represents a local maximum, local minimum or neither, (d) determine all intervals of concavity and (e) find all inflection points.

9. $f(x) = x^3 + 3x^2 - 9x$
10. $f(x) = x^4 - 4x + 1$
11. $f(x) = x^4 - 4x^3 + 2$
12. $f(x) = x^3 - 3x^2 - 24x$
13. $f(x) = xe^{-4x}$
14. $f(x) = x^2 \ln x$
15. $f(x) = x\sqrt{x^2 - 4}$
16. $f(x) = (x^2 - 1)^{2/3}$
17. $f(x) = \dfrac{x}{x^2 + 4}$
18. $f(x) = \dfrac{x}{\sqrt{x^2 + 2}}$

In exercises 19–22, find the absolute extrema of the function on the interval.

19. $f(x) = x^3 + 3x^2 - 9x$ on $[0, 4]$
20. $f(x) = x^3 + 3x^2 - 9x$ on $[-4, 0]$
21. $f(x) = x^{4/5}$ on $[-2, 3]$
22. $f(x) = x^2 e^{-x}$ on $[-1, 4]$

In exercises 23–26, find the x-coordinates of all local extrema.

23. $f(x) = x^3 + 4x^2 + 2x$
24. $f(x) = x^4 - 3x^2 + 2x$
25. $f(x) = x^5 - 2x^2 + x$
26. $f(x) = x^5 + 4x^2 - 4x$

27. Sketch a graph of a function with $f(-1) = 2$, $f(1) = -2$, $f'(x) < 0$ for $-2 < x < 2$, $f'(x) > 0$ for $x < -2$ and $x > 2$.

28. Sketch a graph of a function with $f'(x) > 0$ for $x \neq 0$, $f'(0)$ undefined, $f''(x) > 0$ for $x < 0$ and $f''(x) < 0$ for $x > 0$.

In exercises 29–38, sketch a graph of the function and completely discuss the graph.

29. $f(x) = x^4 + 4x^3$
30. $f(x) = x^4 + 4x^2$
31. $f(x) = x^4 + 4x$
32. $f(x) = x^4 - 4x^2$
33. $f(x) = \dfrac{x}{x^2 + 1}$
34. $f(x) = \dfrac{x}{x^2 - 1}$
35. $f(x) = \dfrac{x^2}{x^2 + 1}$
36. $f(x) = \dfrac{x^2}{x^2 - 1}$
37. $f(x) = \dfrac{x^3}{x^2 - 1}$
38. $f(x) = \dfrac{4}{x^2 - 1}$

39. Find the point on the graph of $y =$

40. Show that the line through the two pendicular to the tangent line to y

41. A city is building a highway from 4 miles east and 6 miles south of p of point A is swamp land, where t way is $6 million per mile. On dr per mile. Find the point on the bou land to which the highway should be built to minimize the total cost.

42. Repeat exercise 41 with a cost of $16 million per mile on swamp land. Explain why the optimal point in this exercise is west of the optimal point found in exercise 41.

43. A soda can in the shape of a cylinder is to hold 16 ounces. Find the dimensions of the can that minimizes the surface area of the can.

44. Suppose that $C(x) = 0.02x^2 + 4x + 1200$ is the cost of manufacturing x items. Show that $C'(x) > 0$ and explain in business terms why this has to be true. Show that $C''(x) > 0$ and explain why this indicates that the manufacturing process is not very efficient.

45. The charge in an electrical circuit at time t is given by $Q(t) = e^{-3t}\sin 2t$ coulombs. Find the current.

46. If the concentration $x(t)$ of a chemical in a reaction changes according to the equation $x'(t) = 0.3x(t)[4 - x(t)]$, find the

where a and b are positive constants. If $n(0) = a/b$, what is $n'(0)$? Based on this calculation, would $n(t)$ increase, decrease or neither? If $n(0) > a/b$, is $n'(0)$ positive or negative? Based on this calculation, would $n(t)$ increase, decrease or neither? If $n(0) < a/b$, is $n'(0)$ positive or negative? Based on this calculation, would $n(t)$ increase, decrease or neither? Putting this information together, conjecture the limit of $n(t)$ as $t \to \infty$. Repeat this analysis under the assumption that $a < 0$. [Hint: Because of its definition, $n(t)$ is positive, so ignore any negative values of $n(t)$.]

52. One way of numerically approximating a derivative is by computing the slope of a secant line. For example, $f'(a) \approx \dfrac{f(b) - f(a)}{b - a}$, if b is close enough to a. In this exercise, we will develop an analogous approximation to the second derivative. Graphically, we can think of the secant line as an approximation of the tangent line. Similarly, we can match the second derivative behavior (concavity) with a parabola. Instead of finding the secant line through two points on the curve, we find the parabola through three points on the curve. The second derivative of this approximating parabola will serve as an approximation of the second derivative of the curve. The first step

BEYOND THE TEXT. ▪   ▪   ▪   ▪   ▪

Access to Calculus: An Interactive Text *is included with each copy of this book. Based on the text itself, this program takes key examples and figures from the text and puts them into an interactive format for further practice. Examples and figures that are expanded on are indicated by an icon* ↟ *within the text. We guarantee that this program will become one of the best study partners that you can find.*

Within Calculus: An Interactive Text *you will find:*

Online Text

This study partner provides your entire *Calculus* text online for easy reference from any computer. In addition to the textual elements themselves, all figures, examples, theorems, and definitions have been compiled into individual libraries for quick access. A complete online glossary of terms has also been provided. These items can be accessed individually or through links within the text, where appropriate.

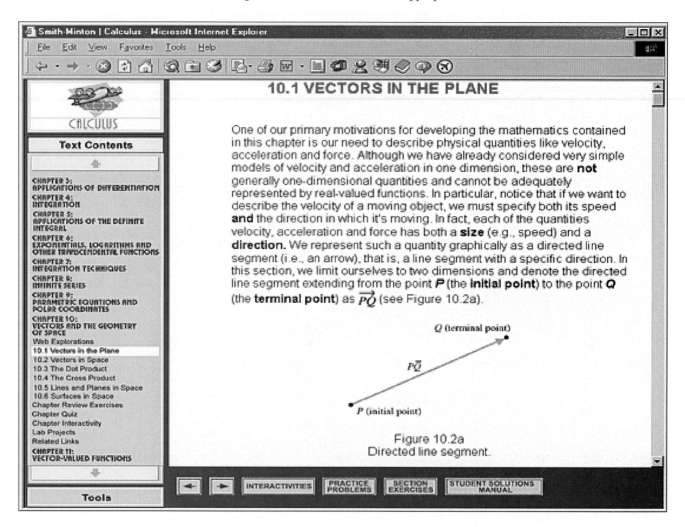

10.1 VECTORS IN THE PLANE

One of our primary motivations for developing the mathematics contained in this chapter is our need to describe physical quantities like velocity, acceleration and force. Although we have already considered very simple models of velocity and acceleration in one dimension, these are **not** generally one-dimensional quantities and cannot be adequately represented by real-valued functions. In particular, notice that if we want to describe the velocity of a moving object, we must specify both its speed **and** the direction in which it's moving. In fact, each of the quantities velocity, acceleration and force has both a **size** (e.g., speed) and a **direction.** We represent such a quantity graphically as a directed line segment (i.e., an arrow), that is, a line segment with a specific direction. In this section, we limit ourselves to two dimensions and denote the directed line segment extending from the point P (the **initial point**) to the point Q (the **terminal point**) as $\overrightarrow{PQ}$ (see Figure 10.2a).

Figure 10.2a
Directed line segment.

Interactivities

Key figures within each section have been turned into interactive Java applets for your further exploration. By interacting with the figure, you will be able to clearly see the usefulness and limitations of the concept being presented. Additionally, a Chapter Interactivity has been provided to help conceptualize the material developed throughout the chapter. These Chapter Interactivities often use real-world visualization to enhance the meaning behind the concepts being presented. There are over 200 Java applets provided within this program for your use.

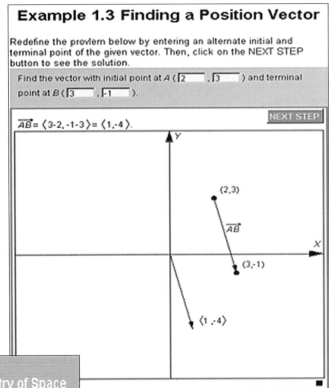

Example 1.3 Finding a Position Vector

Redefine the problem below by entering an alternate initial and terminal point of the given vector. Then, click on the NEXT STEP button to see the solution.

Find the vector with initial point at A (2 , 3) and terminal point at B (3 , -1).

NEXT STEP

$\overrightarrow{AB}= \langle 3\text{-}2, \text{-}1\text{-}3 \rangle = \langle 1,\text{-}4 \rangle.$

Calculus

Chapter 10   Vectors and the Geometry of Space

Section 10.1   Vectors in the Plane

Problem 1   For vectors a = (8, 1) and b = (3, -2), compute 2a + 3b.

    ○ (20 ,-4)   ○ (14 ,-4)   ○ (2 ,-4)

    ○ (25,-4)   ○ (4 ,-4)   ○ none of these   ✓

GIVE ME A HINT	SHOW ME THE SOLUTION	PRINT THIS PROBLEM	VIEW NEXT PROBLEM

Practice Problems

Several key examples per section have been used to create algorithmically-generated practice problems, developed to further your problem-solving skills and to clarify your understanding of concepts. If you are having difficulties understanding an example, these practice problems will walk you through the problem, providing hints and if necessary, will provide you with the fully worked solution and textual reference to help explain the concept. You can then request a similar problem to try again until you feel comfortable with your understanding. A Chapter Quiz consisting of a random sampling of problems from each section can be used to test your understanding prior to a quiz or test. Rather than walking you through the solution of the problems, this quiz will record your answers and provide detailed feedback at the end of the quiz. There are over 400 practice problems available for your use.

Additional Features

Student Solutions Manual

Fully worked solutions for select odd-numbered exercises within the text can be accessed directly through the End-of-Section and End-of-Chapter exercises within the online text. A print version of this manual is available for purchase for those who choose that option.

Lab Projects

A technology-based lab project is provided for your further exploration. These real-world projects will enable you to practice your problem-solving skills by applying technology as appropriate within a controlled environment.

Web Explorations and Web Links

The World Wide Web has become an invaluable research tool, expanding real-world emphasis within calculus much further than could ever be done within a text. We have provided fun exploratory exercises accessing interesting real-world data as well as monthly updated chapter-specific links for further research of topics covered within the chapter.

Tools

A fully functional online graphing calculator is provided for quick access while working with the program. In addition, access to NetTutor™—your online tutorial service—and the text-specific Online Learning Center is provided.

ANCILLARIES

■ INSTRUCTOR TOOLS

PageOut®

Create a custom course website with **PageOut**, free to instructors using a McGraw-Hill textbook.

To learn more, contact your McGraw-Hill publisher's representative or visit www.mhhe.com/solutions.

PageOut with Syllabus Builder

Customize your syllabus on the Web by creating your own course website. No need for HTML coding, graphic design, or a thick how-to book. Simply fill out a series of on-line forms and click on one of McGraw-Hill's professional designs. In minutes, your course is online with a website that is tailored to your individual needs.

You will also be provided complete access to the text-specific Online Learning Center and *Calculus: An Interactive Text* that can be linked to your syllabi.

Online Learning Center (www.mhhe.com/smithminton)

The *Calculus* Online Learning Center provides tools to help in the instruction and understanding of calculus. Instructors can access sample syllabi, a complete Image Bank of figures from the text to customize your presentation and teaching suggestions from the authors. Online solutions to the accompanying technology lab series can be accessed directly through the website.

Instructor's Resource Manual

Worked-out solutions to all exercises in the text (except the writing exercises), in addition to sample chapter tests, midterms, and final exams with corresponding answer keys are provided. A technology exercise index is also included.

McGraw-Hill's ESATest Pro Computerized Testing Software

The algorithmically generated ESATest Pro testing software enables instructors to create printed or online (Web or network) tests for student assessment. Instructors can print tests and answer keys as well as edit the included test questions or add their own. Online tests are automatically graded and student results are stored in the program's course management area and gradebook.

Print Test Bank

This print test bank contains over 1300 problems to use for testing and practice, including true/false, multiple-choice, short answer and free response questions.

Course Integration Guide

The course integration guide provides instructors with detailed suggestions for the most effective way to take advantage of the ancillary materials available with the Smith/Minton *Calculus* text.

■ STUDENT RESOURCES

Calculus: An Interactive Text

This browser-based electronic supplement provides access to the entire *Calculus* text in an interactive format. Features include over 200 text-specific JAVA applets and more than 400 algorithmically generated practice problems that demonstrate key concepts and examples from the text. The electronic Student Solutions Manual is integrated for the complete comprehension of exercises.

This program is available via the Internet (accessed through the text specific Online Learning Center) or on CD-ROM.

Online Learning Center (www.mhhe.com/smithminton)

The *Calculus* Online Learning Center provides help in the instruction and understanding of calculus. Students can access automatically graded practice quizzes, additional explanation of difficult topics and keystroke guides for selected CAS systems and TI Calculators. Live tutorial assistance via NetTutor™ is also available.

Insights into Calculus Using Technology Series

Insights into Calculus Using Maple
Insights into Calculus Using DERIVE
Insights into Calculus Using Mathematica
Insights into Calculus Using TI Calculators: 83 Plus, 86, 89, 92, 92 Plus

Help your students become effective users of technology for calculus problem solving. These text-specific exploratory student workbooks present activities and instructions for the most popular graphing technologies.

NetTutor™

This Web-based "homework hotline" provides immediate text-specific tutorial assistance via an Internet whiteboard during regularly scheduled NetTutor™ hours, or provides 24-hour response to all posted questions. Available free via the *Calculus* website.

Student's Solutions Manual

Worked-out solutions for select odd-numbered problems found in the text. A print version is also available. An electronic version is included in *Calculus: An Interactive Text.*

CHAPTER 10

VECTORS AND THE GEOMETRY OF SPACE

The design of an offshore oil platform presents a significant engineering challenge. Underwater construction poses a unique set of constraints, and the high winds and waves that occur in storms at sea can damage or destroy even well-designed structures. In order to build a suitable platform, engineers must have an in-depth knowledge of three-dimensional geometry and be proficient in the analysis of stresses and strains throughout the structure. In this chapter, we take the first steps toward the fully three-dimensional calculus necessary to perform such analyses.

In designing an oil platform, an engineer's task is to find a structure that not only withstands large forces but also protects a crew of workers by remaining relatively motionless during storms. The hourglass-like shape shown in Figure 10.1 minimizes sway by transmitting lateral (sideways) forces through a series of reinforced struts directly into the seabed below. To analyze such a design, we first need a mathematical language for describing the size and direction of three-dimensional forces. The **vectors** developed in the first two sections of this chapter provide us with such a language. In the third and fourth sections of this chapter, we learn how to use vectors to split a lateral force into a set of component forces acting in different directions (for instance, along the struts of an oil rig). Engineers use such calculations to determine how much reinforcement is necessary for an oil rig to remain stable, even in the worst weather.

In the last two sections of this chapter, we introduce some of the fundamental structures of three-dimensional geometry. One of these is the **hyperboloid of one sheet.** The oil rig shown in Figure 10.1

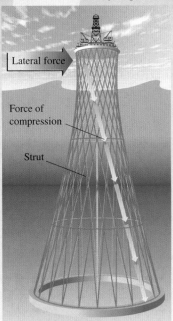

Figure 10.1
Offshore oil platform.

has the shape of a hyperboloid, whose hyperbolic cross sections enable it to efficiently diffuse strong lateral forces. This geometric property makes the hyperboloid a natural choice for such an application. Hyperboloids are also used in other challenging design situations, such as the design of cooling towers for nuclear power plants.

This chapter represents a crossroads from the primarily two-dimensional world of first-year calculus to the three-dimensional world of many important scientific and engineering problems. The rest of the calculus we develop in this book builds directly on the basic ideas developed here.

10.1   VECTORS IN THE PLANE

One of our primary motivations for developing the mathematics contained in this chapter is our need to describe physical quantities like velocity, acceleration and force. Although we have already considered very simple models of velocity and acceleration in one dimension, these are **not** generally one-dimensional quantities and cannot be adequately represented by real-valued functions. In particular, notice that if we want to describe the velocity of a moving object, we must specify both its speed **and** the direction in which it's moving. In fact, each of the quantities velocity, acceleration and force has both a **size** (e.g., speed) and a **direction.** We represent such a quantity graphically as a directed line segment (i.e., an arrow), that is, a line segment with a specific direction. In this section, we limit ourselves to two dimensions and denote the directed line segment extending from the point P (the **initial point**) to the point Q (the **terminal point**) as $\overrightarrow{PQ}$ (see Figure 10.2a).

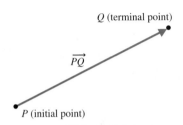

Figure 10.2a

Directed line segment.

We refer to the length of a directed line segment $\overrightarrow{PQ}$ as its **magnitude,** denoted $\|\overrightarrow{PQ}\|$. Mathematically, we consider all directed line segments with the same magnitude and direction to be equivalent, regardless of the location of their initial point. We use the term **vector** to describe any quantity that has both a magnitude and a direction. Vectors are usually represented by directed line segments. We should emphasize that the location of the initial point is not relevant; only the magnitude and direction matter. In other words, if $\overrightarrow{PQ}$ is the directed line segment from the initial point P to the terminal point Q, then the corresponding vector **v** represents $\overrightarrow{PQ}$ as well as every other directed line segment having the same magnitude and direction as $\overrightarrow{PQ}$. In Figure 10.2b, we indicate three vectors that are all considered to be equivalent, even though their initial points are different. In this case, we write

$$\mathbf{a} = \mathbf{b} = \mathbf{c}.$$

When considering vectors, it is often helpful to think of them as representing some specific physical quantity. For instance, when you see the vector $\overrightarrow{PQ}$, you might imagine moving an object from the initial point P to the terminal point Q. In this case, the magnitude of the vector would represent the distance the object is moved and the direction of the vector would point from the starting position to the final position.

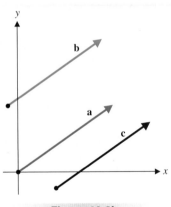

Figure 10.2b

Equivalent vectors.

NOTATION

In this text, we usually denote vectors by boldface characters such as **a**, **b** and **c**, as seen in Figure 10.2b. In the case where the initial and terminal points are specified as P and Q, we denote the vector as $\overrightarrow{PQ}$. Since you will not be able to write in boldface, you should use the arrow notation (e.g., $\vec{a}$). When discussing vectors, we refer to real numbers as **scalars.** It is **very important** that you begin now to carefully distinguish between vector and scalar quantities. This will save you immense frustration both now and as you progress through the remainder of this text.

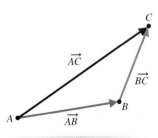

Figure 10.3a

Resultant vector.

Look carefully at the three vectors shown in Figure 10.3a. If you think of the vector $\overrightarrow{AB}$ as representing the displacement of a particle from the point A to the point B, notice

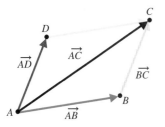

Figure 10.3b

Sum of two vectors.

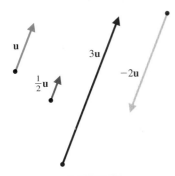

Figure 10.4

Scalar multiplication.

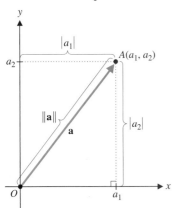

Figure 10.5

Position vector $\mathbf{a} = \langle a_1, a_2 \rangle$.

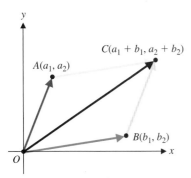

Figure 10.6

Adding position vectors.

that the end result of displacing the particle from A to B (corresponding to the vector $\overrightarrow{AB}$), followed by displacing the particle from B to C (corresponding to the vector $\overrightarrow{BC}$) is the same as displacing the particle directly from A to C, which corresponds to the vector $\overrightarrow{AC}$ (called the **resultant vector**). We call $\overrightarrow{AC}$ the **sum** of $\overrightarrow{AB}$ and $\overrightarrow{BC}$ and write

$$\overrightarrow{AC} = \overrightarrow{AB} + \overrightarrow{BC}.$$

So, given two vectors that we want to add, we locate their initial points at the same point, translate the initial point of one to the terminal point of the other and complete the parallelogram, as indicated in Figure 10.3b. The vector lying along the diagonal, with initial point at A and terminal point at C is the sum

$$\overrightarrow{AC} = \overrightarrow{AB} + \overrightarrow{AD}.$$

A second basic arithmetic operation for vectors is **scalar multiplication.** If we multiply a vector $\mathbf{u}$ by a scalar (a real number) $c > 0$, the resulting vector will have the same direction as $\mathbf{u}$, but will have magnitude $c\|\mathbf{u}\|$. On the other hand, multiplying a vector $\mathbf{u}$ by a scalar $c < 0$, will result in a vector with opposite direction from $\mathbf{u}$ and magnitude $|c|\|\mathbf{u}\|$ (see Figure 10.4).

Since the location of the initial point is irrelevant, we typically draw vectors with their initial point located at the origin. Such a vector is called a **position vector.** Notice that the terminal point of a position vector will completely determine the vector, so that specifying the terminal point will also specify the vector. For the position vector $\mathbf{a}$ with initial point at the origin and terminal point at the point $A(a_1, a_2)$ (see Figure 10.5), we denote the vector by

$$\mathbf{a} = \overrightarrow{OA} = \langle a_1, a_2 \rangle.$$

We call a_1 and a_2 the **components** of the vector $\mathbf{a}$; a_1 is the **first component** and a_2 is the **second component.** Be careful to distinguish between the *point* (a_1, a_2) and the position *vector* $\langle a_1, a_2 \rangle$. Note from Figure 10.5 that the magnitude of the position vector $\mathbf{a}$ follows directly from the Pythagorean Theorem. We have

$$\boxed{\|\mathbf{a}\| = \sqrt{a_1^2 + a_2^2}.}\qquad \text{Magnitude of a vector} \qquad (1.1)$$

Notice that it follows from the definition that for two position vectors $\mathbf{a} = \langle a_1, a_2 \rangle$ and $\mathbf{b} = \langle b_1, b_2 \rangle$, $\mathbf{a} = \mathbf{b}$ if and only if their terminal points are the same, that is if $a_1 = b_1$ and $a_2 = b_2$. In other words, two position vectors are equal only when their corresponding components are equal.

We said earlier that to add two vectors, you place their initial points at the same point, then locate the initial point of one of the vectors at the terminal point of the other vector and complete the parallelogram, as in Figure 10.3b. To see what this says about two position vectors, $\overrightarrow{OA} = \langle a_1, a_2 \rangle$ and $\overrightarrow{OB} = \langle b_1, b_2 \rangle$, we draw the position vectors in Figure 10.6 and complete the parallelogram, as before.

Notice from Figure 10.6 that

$$\overrightarrow{OA} + \overrightarrow{OB} = \overrightarrow{OC}.$$

Writing down the position vectors in their component form, we take this as our definition of vector addition:

$$\boxed{\langle a_1, a_2 \rangle + \langle b_1, b_2 \rangle = \langle a_1 + b_1, a_2 + b_2 \rangle.}\qquad \text{Vector addition} \qquad (1.2)$$

So, to add two vectors, we simply add the corresponding components. For this reason, we say that addition of vectors is done **componentwise.** Similarly, we define subtraction of vectors componentwise, so that

$$\langle a_1, a_2 \rangle - \langle b_1, b_2 \rangle = \langle a_1 - b_1, a_2 - b_2 \rangle. \tag{1.3}$$

We give a geometric interpretation of subtraction later in this section.

Next, consider the effect of scalar multiplication on a position vector. Recall that if we multiply a vector $\mathbf{a}$ by a scalar c, the result is a vector in the same direction as $\mathbf{a}$ (for $c > 0$) or the opposite direction as $\mathbf{a}$ (for $c < 0$), in each case with magnitude $|c|\|\mathbf{a}\|$. We indicate the case of a position vector $\mathbf{a} = \langle a_1, a_2 \rangle$ and scalar multiple $c > 1$ in Figure 10.7a and for $0 < c < 1$ in Figure 10.7b. The situation for negative values of the scalar multiple c is illustrated in Figures 10.7c and 10.7d.

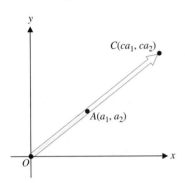

Figure 10.7a
Scalar multiplication ($c > 1$).

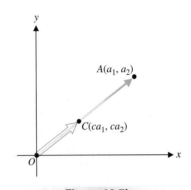

Figure 10.7b
Scalar multiplication ($0 < c < 1$).

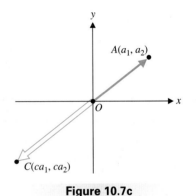

Figure 10.7c
Scalar multiplication ($c < -1$).

For the case where $c > 0$, notice that a vector in the same direction as $\mathbf{a}$, but with magnitude $|c|\|\mathbf{a}\|$ is the position vector $\langle ca_1, ca_2 \rangle$, since

$$\|\langle ca_1, ca_2 \rangle\| = \sqrt{(ca_1)^2 + (ca_2)^2} = \sqrt{c^2 a_1^2 + c^2 a_2^2}$$

$$= |c|\sqrt{a_1^2 + a_2^2} = |c|\|\mathbf{a}\|.$$

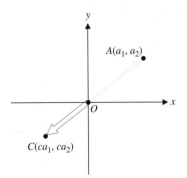

Figure 10.7d
Scalar multiplication ($-1 < c < 0$).

Similarly, if $c < 0$, you can show that $\langle ca_1, ca_2 \rangle$ is a vector in the **opposite** direction from $\mathbf{a}$, with magnitude $|c|\|\mathbf{a}\|$. For this reason, we define scalar multiplication of position vectors by

$$c \langle a_1, a_2 \rangle = \langle ca_1, ca_2 \rangle, \qquad \text{Scalar multiplication} \tag{1.4}$$

for any scalar c. Further, notice that this says that

$$\|c\mathbf{a}\| = |c|\|\mathbf{a}\|. \tag{1.5}$$

Example 1.1   Vector Arithmetic

For vectors $\mathbf{a} = \langle 2, 1 \rangle$ and $\mathbf{b} = \langle 3, -2 \rangle$, compute (a) $\mathbf{a} + \mathbf{b}$, (b) $2\mathbf{a}$, (c) $2\mathbf{a} + 3\mathbf{b}$, (d) $2\mathbf{a} - 3\mathbf{b}$ and (e) $\|2\mathbf{a} - 3\mathbf{b}\|$.

Solution   (a) From (1.2), we have

$$\mathbf{a} + \mathbf{b} = \langle 2, 1 \rangle + \langle 3, -2 \rangle = \langle 2 + 3, 1 - 2 \rangle = \langle 5, -1 \rangle.$$

(b) From (1.4), we have

$$2\mathbf{a} = 2\langle 2, 1 \rangle = \langle 2 \cdot 2, 2 \cdot 1 \rangle = \langle 4, 2 \rangle.$$

(c) From (1.2) and (1.4), we have

$$2\mathbf{a} + 3\mathbf{b} = 2\langle 2, 1 \rangle + 3\langle 3, -2 \rangle = \langle 4, 2 \rangle + \langle 9, -6 \rangle = \langle 13, -4 \rangle.$$

(d) From (1.3) and (1.4), we have

$$2\mathbf{a} - 3\mathbf{b} = 2\langle 2, 1 \rangle - 3\langle 3, -2 \rangle = \langle 4, 2 \rangle - \langle 9, -6 \rangle = \langle -5, 8 \rangle.$$

(e) Finally, from (1.1), we have

$$\|2\mathbf{a} - 3\mathbf{b}\| = \|\langle -5, 8 \rangle\| = \sqrt{25 + 64} = \sqrt{89}.$$

Observe that if we multiply any vector (with any direction) by the scalar $c = 0$, we get a vector with zero length, the **zero vector:**

$$\mathbf{0} = \langle 0, 0 \rangle.$$

Further, notice that this is the **only** vector with zero length. (Why is that?) The zero vector also has no particular direction. Finally, we define the **additive inverse** $-\mathbf{a}$ of a vector $\mathbf{a}$ in the expected way:

$$-\mathbf{a} = -\langle a_1, a_2 \rangle = (-1)\langle a_1, a_2 \rangle = \langle -a_1, -a_2 \rangle.$$

Notice that this says that the vector $-\mathbf{a}$ is a vector with the **opposite** direction as $\mathbf{a}$, and since

$$\|-\mathbf{a}\| = \|(-1)\langle a_1, a_2 \rangle\| = |-1|\|\mathbf{a}\| = \|\mathbf{a}\|,$$

$-\mathbf{a}$ has the same length as $\mathbf{a}$.

Definition 1.1

Two vectors having the same or opposite direction are called **parallel.**

Notice that this says that two (nonzero) position vectors $\mathbf{a}$ and $\mathbf{b}$ are parallel if and only if $\mathbf{b} = c\mathbf{a}$, for some scalar c.

Example 1.2   Determining When Two Vectors Are Parallel

Determine whether or not the given pair of vectors is parallel: (a) $\mathbf{a} = \langle 2, 3 \rangle$ and $\mathbf{b} = \langle 4, 5 \rangle$, (b) $\mathbf{a} = \langle 2, 3 \rangle$ and $\mathbf{b} = \langle -4, -6 \rangle$.

Solution   (a) Notice that from (1.4), we have that if $\mathbf{b} = c\mathbf{a}$, then

$$\langle 4, 5 \rangle = c\langle 2, 3 \rangle = \langle 2c, 3c \rangle.$$

For this to hold, we need for the corresponding components of the two vectors to be equal. That is, $4 = 2c$ (so that $c = 2$) **and** $5 = 3c$ (so that $c = 5/3$). This is a contradiction and so, $\mathbf{a}$ and $\mathbf{b}$ are not parallel.

(b) Again, from (1.4), we have

$$\langle -4, -6 \rangle = c\langle 2, 3 \rangle = \langle 2c, 3c \rangle.$$

In this case, we have $-4 = 2c$ (so that $c = -2$) and $-6 = 3c$ (which again leads us to $c = -2$). Notice that this says that $-2\mathbf{a} = \langle -4, -6 \rangle = \mathbf{b}$ and so, $\langle 2, 3 \rangle$ and $\langle -4, 6 \rangle$ are parallel.

We denote the set of all position vectors in two-dimensional space by

$$V_2 = \{\langle x, y \rangle \mid x, y \in \mathbb{R}\}.$$

You can easily show that the rules of algebra given in the following theorem hold for vectors in V_2.

Theorem 1.1

For any vectors **a**, **b** and **c** in V_2, and any scalars d and e in $\mathbb{R}$, the following hold:

   (i) $\mathbf{a} + \mathbf{b} = \mathbf{b} + \mathbf{a}$ (commutativity)
  (ii) $\mathbf{a} + (\mathbf{b} + \mathbf{c}) = (\mathbf{a} + \mathbf{b}) + \mathbf{c}$ (associativity)
 (iii) $\mathbf{a} + \mathbf{0} = \mathbf{a}$ (zero vector)
 (iv) $\mathbf{a} + (-\mathbf{a}) = \mathbf{0}$ (additive inverse)
  (v) $d(\mathbf{a} + \mathbf{b}) = d\mathbf{a} + d\mathbf{b}$ (distributive law)
 (vi) $(d + e)\mathbf{a} = d\mathbf{a} + e\mathbf{a}$ (distributive law)
 (vii) $(1)\mathbf{a} = \mathbf{a}$ (multiplication by 1) and
(viii) $(0)\mathbf{a} = \mathbf{0}$ (multiplication by 0).

Proof

We prove the first of these and leave the rest as exercises. By definition,

$$\mathbf{a} + \mathbf{b} = \langle a_1, a_2 \rangle + \langle b_1, b_2 \rangle = \langle a_1 + b_1, a_2 + b_2 \rangle \qquad \text{Since addition of real}$$
$$= \langle b_1 + a_1, b_2 + a_2 \rangle = \mathbf{b} + \mathbf{a}. \qquad \text{numbers is commutative.}$$

Notice that using the commutativity and associativity of vector addition, we have

$$\mathbf{b} + (\mathbf{a} - \mathbf{b}) = (\mathbf{a} - \mathbf{b}) + \mathbf{b} = \mathbf{a} + (-\mathbf{b} + \mathbf{b}) = \mathbf{a} + \mathbf{0} = \mathbf{a}.$$

From our graphical interpretation of vector addition, we get Figure 10.8. Notice that this now gives us a geometric interpretation of vector subtraction.

For any two points $A(x_1, y_1)$ and $B(x_2, y_2)$, observe from Figure 10.9 that the vector $\overrightarrow{AB}$ corresponds to the position vector $\langle x_2 - x_1, y_2 - y_1 \rangle$.

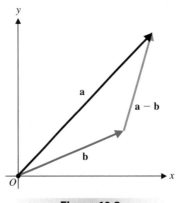

Figure 10.8
$\mathbf{b} + (\mathbf{a} - \mathbf{b}) = \mathbf{a}$.

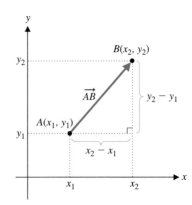

Figure 10.9
Vector from A to B.

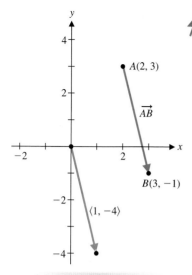

Figure 10.10a
$\overrightarrow{AB} = \langle 1, -4 \rangle.$

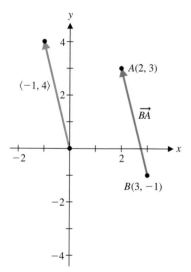

Figure 10.10b
$\overrightarrow{BA} = \langle -1, 4 \rangle.$

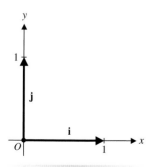

Figure 10.11
Standard basis.

Example 1.3   Finding a Position Vector

Find the vector with (a) initial point at $A(2, 3)$ and terminal point at $B(3, -1)$ and (b) initial point at B and terminal point at A.

Solution   (a) We show this graphically in Figure 10.10a. Notice that

$$\overrightarrow{AB} = \langle 3 - 2, -1 - 3 \rangle = \langle 1, -4 \rangle.$$

(b) Similarly, the vector with initial point at $B(3, -1)$ and terminal point at $A(2, 3)$ is given by

$$\overrightarrow{BA} = \langle 2 - 3, 3 - (-1) \rangle = \langle 2 - 3, 3 + 1 \rangle = \langle -1, 4 \rangle.$$

We indicate this graphically in Figure 10.10b.

We will occasionally find it convenient to write vectors in terms of some standard vectors. We define the **standard basis vectors i** and **j** by

$$\mathbf{i} = \langle 1, 0 \rangle \quad \text{and} \quad \mathbf{j} = \langle 0, 1 \rangle$$

(see Figure 10.11). Notice that $\|\mathbf{i}\| = \|\mathbf{j}\| = 1$. Any vector $\mathbf{a}$ with $\|\mathbf{a}\| = 1$ is called a **unit vector**. So, **i** and **j** are unit vectors.

Finally, we say that **i** and **j** form a **basis** for V_2, since we can write any vector $\mathbf{a} \in V_2$ in terms of **i** and **j**, as follows:

$$\mathbf{a} = \langle a_1, a_2 \rangle = a_1 \mathbf{i} + a_2 \mathbf{j}.$$

We call a_1 and a_2 the **horizontal** and **vertical components** of $\mathbf{a}$, respectively.

For any nonzero vector, we can always find a unit vector with the same direction, as in the following theorem.

Theorem 1.2

For any nonzero position vector $\mathbf{a} = \langle a_1, a_2 \rangle$, a unit vector having the same direction as $\mathbf{a}$ is given by

$$\mathbf{u} = \frac{1}{\|\mathbf{a}\|} \mathbf{a}. \quad \text{Unit vector}$$

The process of dividing a nonzero vector by its magnitude is sometimes called **normalization**. (A vector's magnitude is sometimes called its **norm**.) As we'll see, some problems are simplified by using normalized vectors.

Proof

First, notice that since $\mathbf{a} \neq \mathbf{0}$, $\|\mathbf{a}\| > 0$ and so, $\mathbf{u}$ is a **positive** scalar multiple of $\mathbf{a}$. This says that $\mathbf{u}$ and $\mathbf{a}$ have the same direction. To see that $\mathbf{u}$ is a unit vector, notice that since $\frac{1}{\|\mathbf{a}\|}$ is a positive scalar, we have from (1.5) that

$$\|\mathbf{u}\| = \left\| \frac{1}{\|\mathbf{a}\|} \mathbf{a} \right\| = \frac{1}{\|\mathbf{a}\|} \|\mathbf{a}\| = 1.$$

Example 1.4   Finding a Unit Vector

Find a unit vector in the same direction as $\mathbf{a} = \langle 3, -4 \rangle$.

Solution   First, note that

$$\|\mathbf{a}\| = \|\langle 3, -4 \rangle\| = \sqrt{3^2 + (-4)^2} = \sqrt{25} = 5.$$

A unit vector in the same direction as $\mathbf{a}$ is then

$$\mathbf{u} = \frac{1}{\|\mathbf{a}\|}\, \mathbf{a} = \frac{1}{5}\, \langle 3, -4 \rangle = \left\langle \frac{3}{5}, -\frac{4}{5} \right\rangle.$$

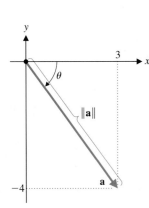

Figure 10.12
Polar form of a vector.

It is often convenient to write a vector explicitly in terms of its magnitude and direction. For instance, in example 1.4, we found that the magnitude of $\mathbf{a} = \langle 3, -4 \rangle$ is $\|\mathbf{a}\| = 5$, while its direction is indicated by the unit vector $\langle \frac{3}{5}, -\frac{4}{5} \rangle$. Notice that we can now write $\mathbf{a} = 5\langle \frac{3}{5}, -\frac{4}{5} \rangle$. Graphically, we can represent $\mathbf{a}$ as a position vector (see Figure 10.12). Notice also that if θ is the angle between the positive x-axis and $\mathbf{a}$, then

$$\mathbf{a} = 5\langle \cos\theta, \sin\theta \rangle,$$

where $\theta = \tan^{-1}\left(-\frac{4}{3}\right) \approx -0.93$. This representation is called the **polar form** of the vector $\mathbf{a}$. Note that this corresponds to writing the rectangular point $(3, -4)$ as the polar point (r, θ), where $r = \|\mathbf{a}\|$.

We close this section with two applications of vector arithmetic. Whenever two or more forces are acting on an object, the net force acting on the object (often referred to as the **resultant force**) is simply the sum of all of the force vectors. That is, the net effect of two or more forces acting on an object is the same as a single force (given by the sum) applied to the object.

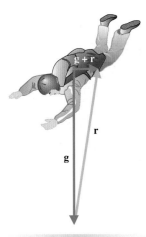

Figure 10.13
Forces on a skydiver.

Example 1.5   Finding the Net Force Acting on a Skydiver

At a certain point during a jump, there are two principal forces acting on a skydiver: gravity exerting a force of 180 pounds straight down, and air resistance exerting a force of 180 pounds up and 30 pounds to the right. What is the net force acting on the skydiver?

Solution   We write the gravity force vector as $\mathbf{g} = \langle 0, -180 \rangle$ and the air resistance force vector as $\mathbf{r} = \langle 30, 180 \rangle$. The net force on the skydiver is the sum of the two forces, $\mathbf{g} + \mathbf{r} = \langle 30, 0 \rangle$. We illustrate the forces in Figure 10.13. Notice that at this point, the vertical forces are balanced, producing a "free-fall" vertically, so that the skydiver is neither accelerating nor decelerating vertically. The net force is purely horizontal, combating the horizontal motion of the skydiver after jumping from the plane.

When flying an airplane, it's important to consider the velocity of the air in which you are flying. Observe that the effect of the velocity of the air can be quite significant. Think about it this way: if a plane flies at 200 mph (its air speed) and the air in which the plane is moving is itself moving at 35 mph in the same direction (i.e., there is a 35 mph tailwind), then the effective speed of the plane is 235 mph. Conversely, if the same 35 mph wind is moving in exactly the opposite direction (i.e., there is a 35 mph headwind), then the plane's effective speed is only 165 mph. A more complicated situation arises if the wind is blowing in some direction that's not parallel to the plane's direction of travel. In this case, we need to add the velocity vectors corresponding to the plane's air speed and the wind to get the effective velocity. We illustrate this with the following example.

Example 1.6 Steering an Aircraft in a Headwind and a Crosswind

An airplane has an air speed of 400 mph. Suppose that the wind velocity is given by the vector $\mathbf{w} = \langle 20, 30 \rangle$. In what direction should the airplane head in order to fly due west (i.e., in the direction of the unit vector $-\mathbf{i} = \langle -1, 0 \rangle$)?

Solution We illustrate the velocity vectors for the airplane and the wind in Figure 10.14. We let the airplane's velocity vector be $\mathbf{v} = \langle x, y \rangle$. The effective velocity of the plane is then $\mathbf{v} + \mathbf{w}$, which we set equal to $\langle c, 0 \rangle$, for some negative constant c. Since

$$\mathbf{v} + \mathbf{w} = \langle x + 20, y + 30 \rangle = \langle c, 0 \rangle,$$

we must have $x + 20 = c$ and $y + 30 = 0$, so that $y = -30$. Further, since the plane's air speed is 400 mph, we must have $\|\mathbf{v}\| = \sqrt{x^2 + y^2} = \sqrt{x^2 + 900} = 400$. Squaring this gives us $x^2 + 900 = 160,000$, so that $x = -\sqrt{159,100}$. (We take the negative square root so that the plane heads westward.) Consequently, the plane should head in the direction of $\mathbf{v} = \langle -\sqrt{159,100}, -30 \rangle$, which points left and down, or southwest, at an angle of $\tan^{-1}(30/\sqrt{159,100}) \approx 4°$ below due west.

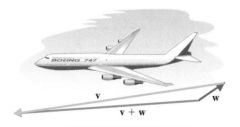

Figure 10.14
Forces on an airplane.

EXERCISES 10.1

1. Discuss whether each of the following is a vector or a scalar quantity: force, area, weight, height, temperature, wind velocity.

2. Some athletes are blessed with "good acceleration." In calculus, we define acceleration as the rate of change of velocity. Keeping in mind that the velocity vector has magnitude (i.e., speed) and direction, discuss why the ability to accelerate rapidly is beneficial.

3. We have emphasized that the location of the initial point of a vector is irrelevant. Using the example of a velocity vector, explain why we want to focus on the magnitude of the vector and its direction, but not on the initial point.

4. Describe the changes that occur when a vector is multiplied by a scalar $c \neq 0$. In your discussion, consider both positive and negative scalars, discuss changes both in the

components of the vector and in its graphical representation, and consider the specific case of a velocity vector.

In exercises 5–8, sketch the vectors 2a, − 3b, a + b and 2a − 3b.

5.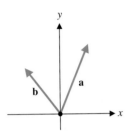

6.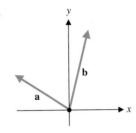

7.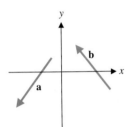

8.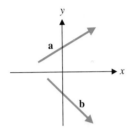

In exercises 9–12, compute a + b, a − 2b, 3a and ‖5b − 2a‖.

9. $\mathbf{a} = \langle 2, 4 \rangle, \mathbf{b} = \langle 3, -1 \rangle$ 10. $\mathbf{a} = \langle 3, -2 \rangle, \mathbf{b} = \langle 2, 0 \rangle$

11. $\mathbf{a} = \mathbf{i} + 2\mathbf{j}, \mathbf{b} = 3\mathbf{i} - \mathbf{j}$ 12. $\mathbf{a} = -2\mathbf{i} + \mathbf{j}, \mathbf{b} = 3\mathbf{i}$

In exercises 13–16, compute a − b, − 4b, 3a + b and ‖4a‖.

13. $\mathbf{a} = \langle -2, 3 \rangle, \mathbf{b} = \langle 1, 0 \rangle$ 14. $\mathbf{a} = \langle -1, -2 \rangle, \mathbf{b} = \langle 2, 3 \rangle$

15. $\mathbf{a} = \mathbf{i} + 2\mathbf{j}, \mathbf{b} = 3\mathbf{i} - \mathbf{j}$ 16. $\mathbf{a} = -2\mathbf{i} + \mathbf{j}, \mathbf{b} = 3\mathbf{i}$

17. For exercises 9 and 10, illustrate the sum $\mathbf{a} + \mathbf{b}$ graphically.

18. For exercises 13 and 14, illustrate the difference $\mathbf{a} - \mathbf{b}$ graphically.

In exercises 19–26, determine whether the vectors a and b are parallel.

19. $\mathbf{a} = \langle 2, 1 \rangle, \mathbf{b} = \langle -4, -2 \rangle$ 20. $\mathbf{a} = \langle 1, -2 \rangle, \mathbf{b} = \langle 2, 1 \rangle$

21. $\mathbf{a} = \langle -2, 3 \rangle, \mathbf{b} = \langle 4, 6 \rangle$ 22. $\mathbf{a} = \langle 1, -2 \rangle, \mathbf{b} = \langle -4, 8 \rangle$

23. $\mathbf{a} = \mathbf{i} + 2\mathbf{j}, \mathbf{b} = 3\mathbf{i} + 6\mathbf{j}$ 24. $\mathbf{a} = -2\mathbf{i} + \mathbf{j}, \mathbf{b} = 4\mathbf{i} + 2\mathbf{j}$

25. $\mathbf{a} = -5\mathbf{i}, \mathbf{b} = 10\mathbf{i} - 2\mathbf{j}$ 26. $\mathbf{a} = 2\mathbf{i} + \mathbf{j}, \mathbf{b} = -\mathbf{i} - \frac{1}{2}\mathbf{j}$

In exercises 27–34, find the vector with initial point A and terminal point B.

27. $A = (2, 3), B = (5, 4)$ 28. $A = (1, 2), B = (6, 4)$

29. $A = (4, 3), B = (1, 0)$ 30. $A = (4, 2), B = (2, 4)$

31. $A = (-1, 2), B = (1, -1)$

32. $A = (0, -2), B = (-3, 1)$

33. $A = (2, 0), B = (0, -3)$

34. $A = (1, 1), B = (-2, 4)$

In exercises 35–44, (a) find a unit vector in the same direction as the given vector and (b) write the given vector in polar form.

35. $\langle 4, -3 \rangle$ 36. $\langle 3, 6 \rangle$

37. $2\mathbf{i} - 4\mathbf{j}$ 38. $3\mathbf{i} + 4\mathbf{j}$

39. $4\mathbf{i}$ 40. $\sqrt{17}\mathbf{j}$

41. from $(2, 1)$ to $(5, 2)$ 42. from $(-2, 0)$ to $(1, 4)$

43. from $(5, -1)$ to $(2, 3)$ 44. from $(3, -2)$ to $(2, 0)$

In exercises 45–50, find a vector with the given magnitude in the same direction as the given vector.

45. magnitude 3, $\mathbf{v} = 3\mathbf{i} + 4\mathbf{j}$ 46. magnitude 4, $\mathbf{v} = 2\mathbf{i} - \mathbf{j}$

47. magnitude 29, $\mathbf{v} = \langle 2, 5 \rangle$ 48. magnitude 10, $\mathbf{v} = \langle 3, 1 \rangle$

49. magnitude 4, $\mathbf{v} = \langle 3, 0 \rangle$ 50. magnitude 5, $\mathbf{v} = \langle 0, -2 \rangle$

51. Suppose that there are two forces acting on a skydiver: gravity at 150 pounds down and air resistance at 140 pounds up and 20 pounds to the right. What is the net force acting on the skydiver?

52. Suppose that there are two forces acting on a skydiver: gravity at 200 pounds down and air resistance at 180 pounds up and 40 pounds to the right. What is the net force acting on the skydiver?

53. Suppose that there are two forces acting on a skydiver: gravity at 200 pounds down and air resistance. If the net force is 10 pounds down and 30 pounds to the right, what is the force of air resistance acting on the skydiver?

54. Suppose that there are two forces acting on a skydiver: gravity at 180 pounds down and air resistance. If the net force is 20 pounds down and 20 pounds to the left, what is the force of air resistance acting on the skydiver?

55. In the figure below, two ropes are attached to a large crate. Suppose that rope A exerts a force of $\langle -164, 115 \rangle$ pounds on the crate and rope B exerts a force of $\langle 177, 177 \rangle$ pounds on the crate. If the crate weighs 275 pounds, what is the net force acting on the crate? Based on your answer, which way will the crate move?

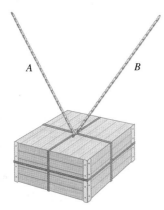

56. Repeat exercise 55 with forces of $\langle -131, 92 \rangle$ pounds from rope A and $\langle 92, 92 \rangle$ from rope B.

57. The thrust of an airplane's engines produces a speed of 300 mph in still air. The wind velocity is given by $\langle 30, -20 \rangle$. In what direction should the airplane head to fly due west?

58. The thrust of an airplane's engines produces a speed of 600 mph in still air. The wind velocity is given by $\langle -30, 60 \rangle$. In what direction should the airplane head to fly due west?

59. The thrust of an airplane's engines produces a speed of 400 mph in still air. The wind velocity is given by $\langle -20, 30 \rangle$. In what direction should the airplane head to fly due north?

60. The thrust of an airplane's engines produces a speed of 300 mph in still air. The wind velocity is given by $\langle 50, 0 \rangle$. In what direction should the airplane head to fly due north?

61. A paperboy is riding at 10 ft/s on a bicycle and tosses a paper over his left shoulder at 50 ft/s. If the porch is 50 ft off the road, how far up the street should the paperboy release the paper to hit the porch?

62. A papergirl is riding at 12 ft/s on a bicycle and tosses a paper over her left shoulder at 48 ft/s. If the porch is 40 ft off the road, how far up the street should the papergirl release the paper to hit the porch?

63. The water from a fire hose exerts a force of 200 pounds on the person holding the hose. The nozzle of the hose weighs 20 pounds. What force is required to hold the hose horizontal? At what angle to the horizontal is this force applied?

64. Repeat exercise 63 for holding the hose at a 45° angle to the horizontal.

65. If vector $\mathbf{a}$ has magnitude $\|\mathbf{a}\| = 3$ and vector $\mathbf{b}$ has magnitude $\|\mathbf{b}\| = 4$, what is the largest possible magnitude for the vector $\mathbf{a} + \mathbf{b}$? What is the smallest possible magnitude for the vector $\mathbf{a} + \mathbf{b}$? What will be the magnitude of $\mathbf{a} + \mathbf{b}$ if $\mathbf{a}$ and $\mathbf{b}$ are perpendicular?

66. Use vectors to show that the points $(1, 2)$, $(3, 1)$, $(4, 3)$ and $(2, 4)$ form a parallelogram.

67. Prove the associativity property of Theorem 1.1.

68. Prove the distributive laws of Theorem 1.1.

69. For vectors $\mathbf{a} = \langle 2, 3 \rangle$ and $\mathbf{b} = \langle 1, 4 \rangle$, compare $\|\mathbf{a} + \mathbf{b}\|$ and $\|\mathbf{a}\| + \|\mathbf{b}\|$. Repeat this comparison for two other choices of $\mathbf{a}$ and $\mathbf{b}$. Use the sketch in Figure 10.7 to explain why $\|\mathbf{a} + \mathbf{b}\| \leq \|\mathbf{a}\| + \|\mathbf{b}\|$ for any vectors $\mathbf{a}$ and $\mathbf{b}$.

70. To prove that $\|\mathbf{a} + \mathbf{b}\| \leq \|\mathbf{a}\| + \|\mathbf{b}\|$ for $\mathbf{a} = \langle a_1, a_2 \rangle$ and $\mathbf{b} = \langle b_1, b_2 \rangle$, start by showing that $2a_1 a_2 b_1 b_2 \leq a_1^2 b_2^2 + a_2^2 b_1^2$. [Hint: Compute $(a_1 b_2 - a_2 b_1)^2$.] Then, show that $a_1 b_1 + a_2 b_2 \leq \sqrt{a_1^2 + a_2^2} \sqrt{b_1^2 + b_2^2}$. (Hint: Square both sides and use the previous result.) Finally, compute $\|\mathbf{a} + \mathbf{b}\|^2 - (\|\mathbf{a}\| + \|\mathbf{b}\|)^2$ and use the previous inequality to show that this is less than or equal to 0.

71. In exercises 69 and 70, you explored the inequality $\|\mathbf{a} + \mathbf{b}\| \leq \|\mathbf{a}\| + \|\mathbf{b}\|$. Use the geometric interpretation of Figure 10.7 to conjecture the circumstances under which $\|\mathbf{a} + \mathbf{b}\| = \|\mathbf{a}\| + \|\mathbf{b}\|$. Similarly, use a geometric interpretation to determine circumstances under which $\|\mathbf{a} + \mathbf{b}\|^2 = \|\mathbf{a}\|^2 + \|\mathbf{b}\|^2$. In general, what is the relationship between $\|\mathbf{a} + \mathbf{b}\|^2$ and $\|\mathbf{a}\|^2 + \|\mathbf{b}\|^2$ (i.e., which is larger)?

72. The figure below shows a foot striking the ground, exerting a force of $\mathbf{F}$ pounds at an angle of θ from the vertical. The force is resolved into vertical and horizontal components $\mathbf{F}_v$ and $\mathbf{F}_h$, respectively. The friction force between floor and foot is $\mathbf{F}_f$, where $\|\mathbf{F}_f\| = \mu \|\mathbf{F}_v\|$ for a positive constant μ known as the **coefficient of friction.** Explain why the foot will slip if $\|\mathbf{F}_h\| > \|\mathbf{F}_f\|$ and show that this happens if and only if $\tan \theta > \mu$. Compare the angles θ at which slipping occurs for coefficients $\mu = 0.6$, $\mu = 0.4$ and $\mu = 0.2$.

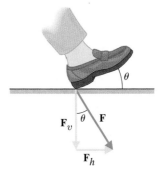

73.    The vectors **i** and **j** are not the only basis vectors that can be used. In fact, any two nonparallel vectors can be used as basis vectors for two-dimensional space. To see this, define $\mathbf{a} = \langle 1, 1 \rangle$ and $\mathbf{b} = \langle 1, -1 \rangle$. To write the vector $\langle 5, 1 \rangle$ in terms of these vectors, we want constants c_1 and c_2 such that $\langle 5, 1 \rangle = c_1 \mathbf{a} + c_2 \mathbf{b}$. Show that this requires that $c_1 + c_2 = 5$ and $c_1 - c_2 = 1$ and then solve for c_1 and c_2. Show that any vector $\langle x, y \rangle$ can be represented in terms of **a** and **b**.

HISTORICAL NOTES

William Rowan Hamilton (1805–1865) Irish mathematician who first defined and developed the theory of vectors. Hamilton was an outstanding student who was appointed Professor of Astronomy at Trinity College while still an undergraduate. After publishing several papers in the field of optics, Hamilton developed an innovative and highly influential approach to dynamics. He then became obsessed with the development of his theory of "quaternions" in which he also defined vectors. Hamilton thought that quaternions would revolutionize mathematical physics, but vectors have proved to be his most important contribution to mathematics.

10.2 VECTORS IN SPACE

One of the benefits of the way in which we defined vectors in the last section is the ease with which we can change the number of dimensions. In this section, we extend several ideas from the two-dimensional Euclidean space $\mathbb{R}^2$ to the three-dimensional Euclidean space, $\mathbb{R}^3$. In $\mathbb{R}^2$, we describe points by an ordered pair (a, b) of real numbers, where a represents (signed) distance from the origin along the x-axis and b represents (signed) distance from the origin along the y-axis. In much the same way, we can specify each point in three dimensions, by an ordered triple (a, b, c), where the coordinates a, b and c represent the (signed) distance from the origin along each of three coordinate axes (x, y and z). In this text, we adopt the convention that the positive x-axis is directed outward from the page, and to the left, the positive y-axis points outward and to the right and the positive z-axis points straight up, as indicated in Figure 10.15a. This orientation of the axes is an example of a **right-handed** coordinate system. That is, if you align the fingers of your right hand along the positive x-axis and then curl them toward the positive y-axis, your thumb will point in the direction of the positive z-axis (see Figure 10.15b). The formulas in this chapter are valid for right-handed coordinate systems (any right-handed system will do) but not all are valid for left-handed systems.

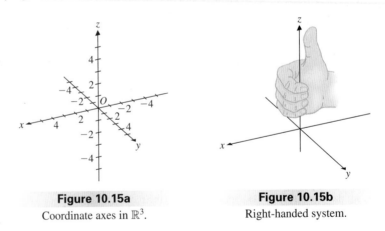

Figure 10.15a
Coordinate axes in $\mathbb{R}^3$.

Figure 10.15b
Right-handed system.

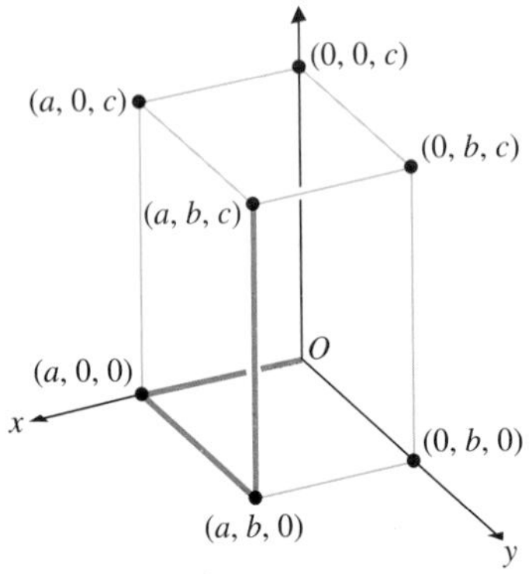

To locate the point $(a, b, c) \in \mathbb{R}^3$, where a, b and c are all positive, first move along the x-axis, a distance of a units from the origin. This will put you at the point $(a, 0, 0)$. Continuing from this point, move parallel to the y-axis a distance of b units from $(a, 0, 0)$. This leaves you at the point $(a, b, 0)$. Finally, continuing from this point, move c units parallel to the z-axis. This is the location of the point (a, b, c) (see Figure 10.16).

Figure 10.16
Locating the point (a, b, c).

Example 2.1   Plotting Points in Three Dimensions

Plot the points $(1, 2, 3)$, $(3, -2, 4)$ and $(-1, 3, -2)$.

Solution   Working as indicated above, we see the points plotted in Figures 10.17a, 10.17b and 10.17c, respectively.

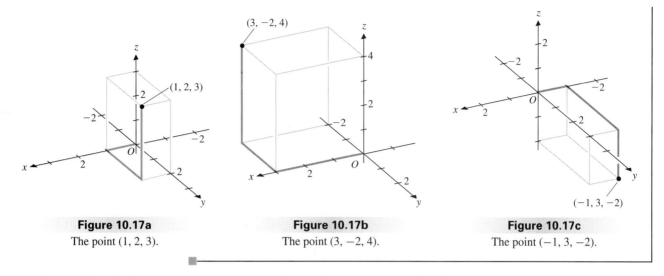

Figure 10.17a
The point $(1, 2, 3)$.

Figure 10.17b
The point $(3, -2, 4)$.

Figure 10.17c
The point $(-1, 3, -2)$.

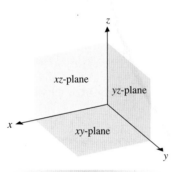

Figure 10.18
The coordinate planes.

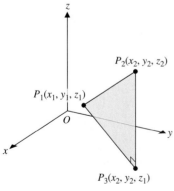

Figure 10.19
Distance in $\mathbb{R}^3$.

Distance in $\mathbb{R}^3$

Recall that in $\mathbb{R}^2$, the coordinate axes divide the xy-plane into four quadrants. In a similar fashion, the three coordinate planes in $\mathbb{R}^3$ (the xy-plane, the yz-plane and the xz-plane) divide space into **eight octants** (see Figure 10.18). The **first octant** is the one with $x > 0$, $y > 0$ and $z > 0$. We do not usually distinguish among the other seven octants.

One of the first questions we want to answer is how to compute the distance between two points in $\mathbb{R}^3$. As it turns out, we can resolve this by thinking of it as essentially a two-dimensional problem. For any two points $P_1(x_1, y_1, z_1)$ and $P_2(x_2, y_2, z_2)$ in $\mathbb{R}^3$, first locate a third point $P_3(x_2, y_2, z_1)$. There are several things to observe here. First, note that the three points determine a plane (in fact, any three noncolinear points determine a plane) and finding the distance between points in a plane is a familiar two-dimensional problem. Second, notice that the three points are the vertices of a right triangle, with the right angle at the point P_3 (see Figure 10.19). Since P_1, P_2 and P_3 form the vertices of a right triangle, the Pythagorean Theorem says that the distance between P_1 and P_2, denoted $d\{P_1, P_2\}$ satisfies

$$d\{P_1, P_2\}^2 = d\{P_1, P_3\}^2 + d\{P_2, P_3\}^2. \tag{2.1}$$

Notice that since P_2 lies directly above P_3 (or below, if $z_2 < z_1$), we have that

$$d\{P_2, P_3\} = d\{(x_2, y_2, z_2), (x_2, y_2, z_1)\} = |z_2 - z_1|.$$

Also notice that P_1 and P_3 both lie in the plane $z = z_1$. This says that we can ignore the third coordinates of these points (since they're the same!) and use the usual two-dimensional distance formula:

$$d\{P_1, P_3\} = d\{(x_1, y_1, z_1), (x_2, y_2, z_1)\} = \sqrt{(x_2 - x_1)^2 + (y_2 - y_1)^2}.$$

From (2.1), we now have

$$
\begin{aligned}
d\{P_1, P_2\}^2 &= d\{P_1, P_3\}^2 + d\{P_2, P_3\}^2 \\
&= \left[\sqrt{(x_2 - x_1)^2 + (y_2 - y_1)^2}\,\right]^2 + |z_2 - z_1|^2 \\
&= (x_2 - x_1)^2 + (y_2 - y_1)^2 + (z_2 - z_1)^2.
\end{aligned}
$$

Taking the square root of both sides gives us the **distance formula** for $\mathbb{R}^3$:

$$d\{(x_1, y_1, z_1), (x_2, y_2, z_2)\} = \sqrt{(x_2 - x_1)^2 + (y_2 - y_1)^2 + (z_2 - z_1)^2}. \tag{2.2}$$

Notice that (2.2) is a straightforward generalization of the familiar formula for the distance between two points in the plane.

Example 2.2   Computing Distance in $\mathbb{R}^3$

Find the distance between the points $(1, -3, 5)$ and $(5, 2, -3)$.

Solution   From (2.2), we have

$$d\{(1, -3, 5), (5, 2, -3)\} = \sqrt{(5-1)^2 + [2-(-3)]^2 + (-3-5)^2}$$
$$= \sqrt{4^2 + 5^2 + (-8)^2} = \sqrt{105}.$$

■

Vectors in $\mathbb{R}^3$

Now that we have a means of computing distances in $\mathbb{R}^3$, we can generalize the notion of vector to three-dimensional space. As in two dimensions, vectors have both direction and magnitude. We again visualize vectors as directed line segments, joining two points. A vector **v** is represented by any directed line segment with the appropriate magnitude and direction. The position vector **a** with terminal point at $A(a_1, a_2, a_3)$ (and initial point at the origin) is denoted by $\langle a_1, a_2, a_3 \rangle$ and is shown in Figure 10.20a.

We denote the set of all three-dimensional position vectors by

$$V_3 = \{\langle x, y, z \rangle \mid x, y, z \in \mathbb{R}\}.$$

The **magnitude** of the position vector $\mathbf{a} = \langle a_1, a_2, a_3 \rangle$ follows directly from the distance formula (2.2). We have

Magnitude of a vector

$$\|\mathbf{a}\| = \|\langle a_1, a_2, a_3 \rangle\| = \sqrt{a_1^2 + a_2^2 + a_3^2}. \tag{2.3}$$

Much as we had in $\mathbb{R}^2$, you can see from Figure 10.20b that the vector with initial point at $P(a_1, a_2, a_3)$ and terminal point at $Q(b_1, b_2, b_3)$ corresponds to the position vector

$$\overrightarrow{PQ} = \langle b_1 - a_1, b_2 - a_2, b_3 - a_3 \rangle.$$

We define vector addition in V_3 just as we did in V_2, by drawing a parallelogram, as seen in Figure 10.20c.

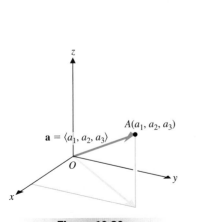

Figure 10.20a
Position vector in $\mathbb{R}^3$.

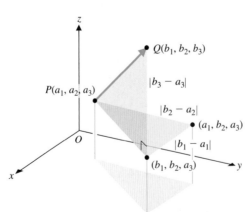

Figure 10.20b
Vector from P to Q.

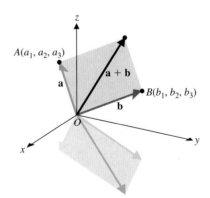

Figure 10.20c
Vector addition.

Notice that for vectors $\mathbf{a} = \langle a_1, a_2, a_3 \rangle$ and $\mathbf{b} = \langle b_1, b_2, b_3 \rangle$, we have

Vector addition

$$\mathbf{a} + \mathbf{b} = \langle a_1, a_2, a_3 \rangle + \langle b_1, b_2, b_3 \rangle = \langle a_1 + b_1, a_2 + b_2, a_3 + b_3 \rangle.$$

That is, as in V_2, addition of vectors in V_3 is done componentwise. Similarly, subtraction is done componentwise:

Vector subtraction

$$\mathbf{a} - \mathbf{b} = \langle a_1, a_2, a_3 \rangle - \langle b_1, b_2, b_3 \rangle = \langle a_1 - b_1, a_2 - b_2, a_3 - b_3 \rangle.$$

Again as in V_2, for any scalar $c \in \mathbb{R}$, $c\mathbf{a}$ is a vector in the same direction as $\mathbf{a}$ when $c > 0$ and the opposite direction as $\mathbf{a}$ when $c < 0$. We have

Scalar multiplication

$$c\mathbf{a} = c\langle a_1, a_2, a_3 \rangle = \langle ca_1, ca_2, ca_3 \rangle.$$

Further, it's easy to show using (2.3), that

$$\|c\mathbf{a}\| = |c| \, \|\mathbf{a}\|.$$

We define the **zero vector 0** to be the vector in V_3 of length 0:

$$\mathbf{0} = \langle 0, 0, 0 \rangle.$$

As in two dimensions, the zero vector has no particular direction. As we did in V_2, we define the **additive inverse** of a vector $\mathbf{a} \in V_3$ to be

$$-\mathbf{a} = -\langle a_1, a_2, a_3 \rangle = \langle -a_1, -a_2, -a_3 \rangle.$$

The rules of algebra established for vectors in V_2 hold verbatim in V_3, as seen in the following theorem.

Theorem 2.1

For any vectors $\mathbf{a}$, $\mathbf{b}$ and $\mathbf{c}$ in V_3, and any scalars d and e in $\mathbb{R}$, the following hold:

(i) $\mathbf{a} + \mathbf{b} = \mathbf{b} + \mathbf{a}$ (commutativity)
(ii) $\mathbf{a} + (\mathbf{b} + \mathbf{c}) = (\mathbf{a} + \mathbf{b}) + \mathbf{c}$ (associativity)
(iii) $\mathbf{a} + \mathbf{0} = \mathbf{a}$ (zero vector)
(iv) $\mathbf{a} + (-\mathbf{a}) = \mathbf{0}$ (additive inverse)
(v) $d(\mathbf{a} + \mathbf{b}) = d\mathbf{a} + d\mathbf{b}$ (distributive law)
(vi) $(d + e)\mathbf{a} = d\mathbf{a} + e\mathbf{a}$ (distributive law)
(vii) $(1)\mathbf{a} = \mathbf{a}$ (multiplication by 1) and
(viii) $(0)\mathbf{a} = \mathbf{0}$ (multiplication by 0)

We leave the proof of Theorem 2.1 as an exercise.

Since V_3 is three-dimensional, the standard basis consists of three unit vectors, each lying along one of the three coordinate axes. We define these as a straightforward generalization of the standard basis for V_2 by

$$\mathbf{i} = \langle 1, 0, 0 \rangle, \quad \mathbf{j} = \langle 0, 1, 0 \rangle \quad \text{and} \quad \mathbf{k} = \langle 0, 0, 1 \rangle,$$

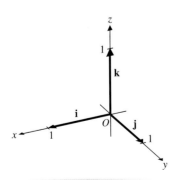

Figure 10.21
Standard basis for V_3.

as pictured in Figure 10.21. As in V_2, these basis vectors are unit vectors, since $\|\mathbf{i}\| = \|\mathbf{j}\| = \|\mathbf{k}\| = 1$. Also as in V_2, it is sometimes convenient to write position vectors in V_3 in terms of the standard basis. This is easily accomplished, as for any $\mathbf{a} \in V_3$, we can write

$$\mathbf{a} = \langle a_1, a_2, a_3 \rangle = a_1 \mathbf{i} + a_2 \mathbf{j} + a_3 \mathbf{k}.$$

If you're getting that déjà vu feeling that you've done all of this before, you're not imagining it. As we've seen, vectors in V_3 follow all of the same rules as vectors in V_2. We have simply been taking the time to develop them all. As a final note, observe that for any $\mathbf{a} = \langle a_1, a_2, a_3 \rangle \neq \mathbf{0}$, a unit vector in the same direction as $\mathbf{a}$ is given by

Unit vector

$$\mathbf{u} = \frac{1}{\|\mathbf{a}\|}\mathbf{a}. \qquad (2.4)$$

The proof of this result is identical to the proof of the corresponding result for vectors in V_2, found in Theorem 1.2. Once again, it is often convenient to normalize a vector (produce a vector in the same direction, but with length 1).

Example 2.3   Finding a Unit Vector

Find a unit vector in the same direction as $\langle 1, -2, 3 \rangle$ and write $\langle 1, -2, 3 \rangle$ as the product of its magnitude and a unit vector.

Solution   First, we find the magnitude of the vector:

$$\|\langle 1, -2, 3 \rangle\| = \sqrt{1^2 + (-2)^2 + 3^2} = \sqrt{14}.$$

From (2.4), we have that a unit vector having the same direction as $\langle 1, -2, 3 \rangle$ is given by

$$\mathbf{u} = \frac{1}{\sqrt{14}}\langle 1, -2, 3 \rangle = \left\langle \frac{1}{\sqrt{14}}, \frac{-2}{\sqrt{14}}, \frac{3}{\sqrt{14}} \right\rangle.$$

Further,

$$\langle 1, -2, 3 \rangle = \sqrt{14} \left\langle \frac{1}{\sqrt{14}}, \frac{-2}{\sqrt{14}}, \frac{3}{\sqrt{14}} \right\rangle.$$

Of course, going from two dimensions to three dimensions gives us a much richer geometry, with more interesting examples. For instance, we define a **sphere** to be the set of all points whose distance from a fixed point (the **center**) is constant.

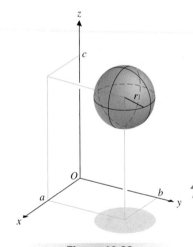

Figure 10.22
Sphere of radius r centered at (a, b, c).

Example 2.4   Finding the Equation of a Sphere

Find the equation of the sphere of radius r centered at the point (a, b, c).

Solution   The sphere consists of all points (x, y, z) whose distance from (a, b, c) is r, as illustrated in Figure 10.22. This says that

$$\sqrt{(x-a)^2 + (y-b)^2 + (z-c)^2} = d\{(x, y, z), (a, b, c)\} = r.$$

Squaring both sides gives us

Sphere of radius r, centered at (a, b, c)

$$(x-a)^2 + (y-b)^2 + (z-c)^2 = r^2,$$

the standard form of the equation of a sphere.

You will occasionally need to recognize when a given equation represents a common geometric shape, as in the following.

Example 2.5   Finding the Center and Radius of a Sphere

Find the geometric shape described by the following equation:

$$0 = x^2 + y^2 + z^2 - 4x + 8y - 10z + 36.$$

Solution   Completing the squares in each variable, we have

$$0 = (x^2 - 4x + 4) - 4 + (y^2 + 8y + 16) - 16 + (z^2 - 10z + 25) - 25 + 36$$
$$= (x - 2)^2 + (y + 4)^2 + (z - 5)^2 - 9.$$

Adding 9 to both sides gives us

$$3^2 = (x - 2)^2 + (y + 4)^2 + (z - 5)^2,$$

which is the equation of a sphere of radius 3 centered at the point $(2, -4, 5)$.

EXERCISES 10.2

1. Visualize the circle $x^2 + y^2 = 1$. With three-dimensional axes oriented as in Figure 10.15a, describe how to sketch this circle in the plane $z = 0$. Then, describe how to sketch the parabola $y = x^2$ in the plane $z = 0$. In general, explain how to translate a two-dimensional curve into a three-dimensional sketch.

2. It is difficult, if not impossible, for most people to visualize what points in four dimensions would look like. Nevertheless, it is easy to generalize the distance formula to four dimensions. Describe what the distance formula looks like in general dimension n, for $n \geq 4$.

3. It is very important to be able to quickly and accurately visualize three-dimensional relationships. In three dimensions, describe how many lines are perpendicular to the unit vector **i**. Describe all lines that are perpendicular to **i** and that pass through the origin. In three dimensions, describe how many planes are perpendicular to the unit vector **i**. Describe all planes that are perpendicular to **i** and that contain the origin.

4. In three dimensions, describe all planes that contain a given vector **a**. Describe all planes that contain two given vectors **a** and **b** (where **a** and **b** are not parallel). Describe all planes that contain a given vector **a** and pass through the origin. Describe all planes that contain two given (nonparallel) vectors **a** and **b** and pass through the origin.

In exercises 5 and 6, plot the indicated points.

5. (a) $(2, 1, 5)$ (b) $(3, 1, -2)$ (c) $(-1, 2, -4)$

6. (a) $(-2, 1, 2)$ (b) $(2, -3, -1)$ (c) $(3, -2, 2)$

In exercises 7 and 8, sketch the third axis to make xyz a right-handed system.

7.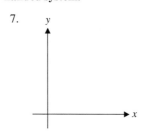

8.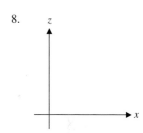

In exercises 9–14, find the distance between the given points.

9. $(2, 1, 2), (5, 5, 2)$ 10. $(1, 2, 0), (7, 10, 0)$

11. $(-1, 0, 2), (1, 2, 3)$ 12. $(3, 1, 0), (1, 3, -4)$

13. $(0, 0, 0), (2, 3, 5)$ 14. $(2, 1, -4), (0, 0, 0)$

In exercises 15–20, compute $a + b$, $a - 3b$ and $\|4a + 2b\|$.

15. $a = \langle 2, 1, -2 \rangle$, $b = \langle 1, 3, 0 \rangle$

16. $a = \langle 2, -1, 2 \rangle$, $b = \langle 1, 3, 0 \rangle$

17. $a = \langle -1, 0, 2 \rangle$, $b = \langle 4, 3, 2 \rangle$

18. $a = \langle 5, -1, 2 \rangle$, $b = \langle -1, 1, 2 \rangle$

19. $a = 3\mathbf{i} - \mathbf{j} + 4\mathbf{k}$, $b = 5\mathbf{i} + \mathbf{j}$

20. $a = \mathbf{i} - 4\mathbf{j} - 2\mathbf{k}$, $b = \mathbf{i} - 3\mathbf{j} + 4\mathbf{k}$

In exercises 21–28, (a) find two unit vectors parallel to the given vector and (b) write the given vector as the product of its magnitude and a unit vector.

21. $\langle 3, 1, 2 \rangle$ 22. $\langle 2, 3, 1 \rangle$

23. $\langle 2, -4, 6 \rangle$ 24. $\langle 4, -6, -2 \rangle$

25. $2\mathbf{i} - \mathbf{j} + 2\mathbf{k}$ 26. $4\mathbf{i} - 2\mathbf{j} + 4\mathbf{k}$

27. From $(1, 2, 3)$ to $(3, 2, 1)$ 28. From $(1, 4, 1)$ to $(3, 2, 2)$

In exercises 29–34, find a vector with the given magnitude and in the same direction as the given vector.

29. Magnitude 6, $\mathbf{v} = \langle 2, 2, -1 \rangle$

30. Magnitude 10, $\mathbf{v} = \langle 3, 0, -4 \rangle$

31. Magnitude 2, $\mathbf{v} = \langle 2, 0, -1 \rangle$

32. Magnitude 2, $\mathbf{v} = \langle 2, -3, 1 \rangle$

33. Magnitude 4, $\mathbf{v} = 2\mathbf{i} - \mathbf{j} + 3\mathbf{k}$

34. Magnitude 3, $\mathbf{v} = 3\mathbf{i} + 3\mathbf{j} - \mathbf{k}$

In exercises 35–40, find an equation of the sphere with radius r and center (a, b, c).

35. $r = 2$, $(a, b, c) = (3, 1, 4)$

36. $r = 3$, $(a, b, c) = (2, 0, 1)$

37. $r = 3$, $(a, b, c) = (2, 0, -3)$

38. $r = 4$, $(a, b, c) = (0, -3, -1)$

39. $r = \sqrt{5}$, $(a, b, c) = (\pi, 1, -3)$

40. $r = \sqrt{7}$, $(a, b, c) = (1, 3, 4)$

In exercises 41–52, identify the geometric shape described by the given equation.

41. $(x - 1)^2 + y^2 + (z + 2)^2 = 4$

42. $x^2 + (y - 1)^2 + (z - 4)^2 = 2$

43. $x^2 + y^2 - 2y + z^2 + 4z = 4$

44. $x^2 + 4x + y^2 - 6y + z^2 = 3$

45. $x^2 - 2x + y^2 + z^2 - 4z = 0$

46. $x^2 + x + y^2 - y + z^2 = \frac{7}{2}$

47. $y = 4$ 48. $x = -2$

49. $z = -1$ 50. $z = 3$

51. $x = 3$ 52. $y = -2$

In exercises 53–56, give an equation (e.g., $z = 0$) for the given figure.

53. xz-plane   54. xy-plane

55. yz-plane   56. x-axis

57. Prove the commutative property of Theorem 2.1.

58. Prove the associative property of Theorem 2.1.

59. Prove the distributive properties of Theorem 2.1.

60. Prove the multiplicative properties of Theorem 2.1.

61. Find the displacement vectors $\overrightarrow{PQ}$ and $\overrightarrow{QR}$ and determine whether the points $P = (2, 3, 1)$, $Q = (4, 2, 2)$ and $R = (8, 0, 4)$ are colinear (on the same line).

62. Find the displacement vectors $\overrightarrow{PQ}$ and $\overrightarrow{QR}$ and determine whether the points $P = (2, 3, 1)$, $Q = (0, 4, 2)$ and $R = (4, 1, 4)$ are colinear (on the same line).

63. Find the force needed to keep a helicopter hovering in place if the helicopter weighs 1000 pounds and a northeasterly wind exerts a force of 150 pounds.

64. Find the force needed to keep a helicopter hovering in place if the helicopter weighs 800 pounds and a northwesterly wind exerts a force of 100 pounds.

65. Find an equation describing all points equidistant from $A = (0, 1, 0)$ and $B = (2, 4, 4)$ and sketch a graph. Based on your graph, describe the relationship between the displacement vector $\overrightarrow{AB} = \langle 2, 3, 4 \rangle$ and your graph. Simplify your equation for the three-dimensional surface until 2, 3 and 4 appear as coefficients of x, y and z. Use what you have learned to quickly write down an equation for the set of all points equidistant from $A = (0, 1, 0)$ and $C = (5, 2, 3)$.

66. In this exercise, you will try to identify the three-dimensional surface defined by the equation $a(x - 1) + b(y - 2) + c(z - 3) = 0$ for nonzero constants a, b and c. First, show that $(1, 2, 3)$ is one point on the surface. Then, show that any point which is equidistant from the points $(1 + a, 2 + b, 3 + c)$ and $(1 - a, 2 - b, 3 - c)$ is on the surface. Use this geometric fact to identify the surface.

10.3   THE DOT PRODUCT

In the previous two sections, we defined vectors in $\mathbb{R}^2$ and $\mathbb{R}^3$ and examined many of the properties of vectors, including how to add and subtract two vectors. It turns out that two different kinds of products involving vectors have proved to be useful: the dot product (or scalar product) and the cross product (or vector product). We introduce the first of these two products in this section.

Definition 3.1

The **dot product** of two vectors $\mathbf{a} = \langle a_1, a_2, a_3 \rangle$ and $\mathbf{b} = \langle b_1, b_2, b_3 \rangle$ in V_3 is defined by

$$\mathbf{a} \cdot \mathbf{b} = \langle a_1, a_2, a_3 \rangle \cdot \langle b_1, b_2, b_3 \rangle = a_1 b_1 + a_2 b_2 + a_3 b_3. \tag{3.1}$$

Likewise, the dot product of two vectors in V_2 is defined by

$$\mathbf{a} \cdot \mathbf{b} = \langle a_1, a_2 \rangle \cdot \langle b_1, b_2 \rangle = a_1 b_1 + a_2 b_2.$$

Be sure to notice that the dot product of two vectors is a **scalar** (i.e., a number, not a vector).

Example 3.1   Computing a Dot Product in $\mathbb{R}^3$

Compute the dot product $\mathbf{a} \cdot \mathbf{b}$ for $\mathbf{a} = \langle 1, 2, 3 \rangle$ and $\mathbf{b} = \langle 5, -3, 4 \rangle$.

Solution   We have
$$\mathbf{a} \cdot \mathbf{b} = \langle 1, 2, 3 \rangle \cdot \langle 5, -3, 4 \rangle = (1)(5) + (2)(-3) + (3)(4) = 11.$$

You will no doubt agree that dot products are extremely simple to use, whether a vector is written in component form or written in terms of the standard basis vectors, as in the following example.

Example 3.2   Computing a Dot Product in $\mathbb{R}^2$

Find the dot product of the two vectors $\mathbf{a} = 2\mathbf{i} - 5\mathbf{j}$ and $\mathbf{b} = 3\mathbf{i} + 6\mathbf{j}$.

Solution   We have
$$\mathbf{a} \cdot \mathbf{b} = (2)(3) + (-5)(6) = 6 - 30 = -24.$$

Remark 3.1

From this point on, we will prove results about dot products for vectors in V_3 only, rather than separately for V_2 and V_3. Since vectors in V_2 can be thought of as a special case of vectors in V_3 (where the third component is zero), all of the results we prove for vectors in V_3 hold equally for vectors in V_2.

The dot product in V_2 or V_3 satisfies the following simple properties.

Theorem 3.1

For vectors $\mathbf{a}$, $\mathbf{b}$ and $\mathbf{c}$ and any scalar d, the following hold:
(i) $\mathbf{a} \cdot \mathbf{b} = \mathbf{b} \cdot \mathbf{a}$ (commutativity)
(ii) $\mathbf{a} \cdot (\mathbf{b} + \mathbf{c}) = \mathbf{a} \cdot \mathbf{b} + \mathbf{a} \cdot \mathbf{c}$ (distributive law)
(iii) $(d\mathbf{a}) \cdot \mathbf{b} = d(\mathbf{a} \cdot \mathbf{b}) = \mathbf{a} \cdot (d\mathbf{b})$
(iv) $\mathbf{0} \cdot \mathbf{a} = 0$ and
(v) $\mathbf{a} \cdot \mathbf{a} = \|\mathbf{a}\|^2.$

Proof

We prove (i) and (v). The remaining parts are left as exercises.

(i) For $\mathbf{a} = \langle a_1, a_2, a_3 \rangle$ and $\mathbf{b} = \langle b_1, b_2, b_3 \rangle$, we have from (3.1) that

$$\mathbf{a} \cdot \mathbf{b} = \langle a_1, a_2, a_3 \rangle \cdot \langle b_1, b_2, b_3 \rangle = a_1 b_1 + a_2 b_2 + a_3 b_3$$
$$= b_1 a_1 + b_2 a_2 + b_3 a_3 = \mathbf{b} \cdot \mathbf{a},$$

since multiplication of real numbers is commutative.

(v) For $\mathbf{a} = \langle a_1, a_2, a_3 \rangle$, we have

$$\mathbf{a} \cdot \mathbf{a} = \langle a_1, a_2, a_3 \rangle \cdot \langle a_1, a_2, a_3 \rangle = a_1^2 + a_2^2 + a_3^2 = \|\mathbf{a}\|^2.$$

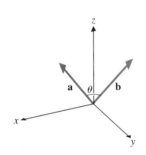

Figure 10.23a

The angle between two vectors.

Notice that properties (i)–(iv) of Theorem 3.1 are also properties of multiplication of real numbers. This is why we use the word *product* in dot product. However, there are some properties of multiplication of real numbers not shared by the dot product. For instance, we will see that $\mathbf{a} \cdot \mathbf{b} = 0$ does not imply that either $\mathbf{a} = \mathbf{0}$ or $\mathbf{b} = \mathbf{0}$.

One of the more valuable bits of information we can get from a dot product is the angle between two vectors. This is done as follows.

For any two **nonzero** vectors $\mathbf{a}$ and $\mathbf{b}$ in V_3, if we place their initial points at the same point, notice that the two vectors form an angle θ between them, where $0 \le \theta \le \pi$ (as illustrated in Figure 10.23a). We refer to this angle θ as the **angle between the vectors a and b.** That is, in the plane determined by $\mathbf{a}$ and $\mathbf{b}$, the angle θ between the two vectors is defined to be the smaller angle between $\mathbf{a}$ and $\mathbf{b}$.

Notice that if $\mathbf{a}$ and $\mathbf{b}$ have the **same** direction, then $\theta = 0$. If $\mathbf{a}$ and $\mathbf{b}$ have opposite directions, then $\theta = \pi$. We say that $\mathbf{a}$ and $\mathbf{b}$ are **orthogonal** or **perpendicular** if $\theta = \frac{\pi}{2}$. We consider the zero vector $\mathbf{0}$ to be orthogonal to every vector. The general case is stated in the following theorem.

Theorem 3.2

Let θ be the angle between nonzero vectors $\mathbf{a}$ and $\mathbf{b}$. Then,

$$\mathbf{a} \cdot \mathbf{b} = \|\mathbf{a}\| \, \|\mathbf{b}\| \cos \theta. \qquad (3.2)$$

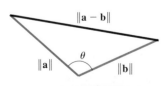

Figure 10.23b

The angle between two vectors.

Proof

We must prove the theorem for three separate cases.

(i) If $\mathbf{a}$ and $\mathbf{b}$ have the **same direction,** then $\mathbf{b} = c\mathbf{a}$, for some scalar $c > 0$ and the angle between $\mathbf{a}$ and $\mathbf{b}$ is $\theta = 0$. This says that

$$\mathbf{a} \cdot \mathbf{b} = \mathbf{a} \cdot (c\mathbf{a}) = c\mathbf{a} \cdot \mathbf{a} = c\|\mathbf{a}\|^2.$$

Further,

$$\|\mathbf{a}\| \, \|\mathbf{b}\| \cos \theta = \|\mathbf{a}\| \, |c| \, \|\mathbf{a}\| \cos 0 = c\|\mathbf{a}\|^2 = \mathbf{a} \cdot \mathbf{b},$$

since for $c > 0$, we have $|c| = c$.

(ii) If $\mathbf{a}$ and $\mathbf{b}$ have the **opposite direction,** the proof is nearly identical to case (i) above and we leave the details as an exercise.

(iii) If $\mathbf{a}$ and $\mathbf{b}$ are not parallel, then we have that $0 < \theta < \pi$, as shown in Figure 10.23b. Recall that the Law of Cosines allows us to relate the lengths of the sides of triangles like the one in Figure 10.23b. We have:

$$\|\mathbf{a} - \mathbf{b}\|^2 = \|\mathbf{a}\|^2 + \|\mathbf{b}\|^2 - 2\|\mathbf{a}\|\|\mathbf{b}\|\cos\theta. \tag{3.3}$$

Now, observe that

$$
\begin{aligned}
\|\mathbf{a} - \mathbf{b}\|^2 &= \|\langle a_1 - b_1, a_2 - b_2, a_3 - b_3\rangle\|^2 \\
&= (a_1 - b_1)^2 + (a_2 - b_2)^2 + (a_3 - b_3)^2 \\
&= \left(a_1^2 - 2a_1 b_1 + b_1^2\right) + \left(a_2^2 - 2a_2 b_2 + b_2^2\right) + \left(a_3^2 - 2a_3 b_3 + b_3^2\right) \\
&= \left(a_1^2 + a_2^2 + a_3^2\right) + \left(b_1^2 + b_2^2 + b_3^2\right) - 2(a_1 b_1 + a_2 b_2 + a_3 b_3) \\
&= \|\mathbf{a}\|^2 + \|\mathbf{b}\|^2 - 2\mathbf{a} \cdot \mathbf{b} \tag{3.4}
\end{aligned}
$$

Equating the right-hand sides of (3.3) and (3.4), we get (3.2), as desired.

We can use (3.2) to find the angle between two vectors, as in the following example.

Example 3.3 Finding the Angle between Two Vectors

Find the angle between the vectors $\mathbf{a} = \langle 2, 1, -3\rangle$ and $\mathbf{b} = \langle 1, 5, 6\rangle$.

Solution From (3.2), we have

$$\cos\theta = \frac{\mathbf{a} \cdot \mathbf{b}}{\|\mathbf{a}\|\|\mathbf{b}\|} = \frac{-11}{\sqrt{14}\sqrt{62}}.$$

It follows that

$$\theta = \cos^{-1}\left(\frac{-11}{\sqrt{14}\sqrt{62}}\right) \approx 1.953 \text{ (radians)}$$

(or about 112°), since $0 \le \theta \le \pi$ and the inverse cosine function returns an angle in this range.

The following result is an immediate and important consequence of Theorem 3.2.

Corollary 3.1

Two vectors $\mathbf{a}$ and $\mathbf{b}$ are orthogonal if and only if $\mathbf{a} \cdot \mathbf{b} = 0$.

Proof

First, observe that if either $\mathbf{a}$ or $\mathbf{b}$ is the zero vector, then $\mathbf{a} \cdot \mathbf{b} = 0$ and $\mathbf{a}$ and $\mathbf{b}$ are orthogonal, as the zero vector is considered orthogonal to every vector. If $\mathbf{a}$ and $\mathbf{b}$ are nonzero vectors and if θ is the angle between $\mathbf{a}$ and $\mathbf{b}$, we have from Theorem 3.2 that

$$\|\mathbf{a}\|\|\mathbf{b}\|\cos\theta = \mathbf{a} \cdot \mathbf{b} = 0$$

if and only if $\cos\theta = 0$ (since neither $\mathbf{a}$ nor $\mathbf{b}$ is the zero vector). This is equivalent to having $\mathbf{a}$ and $\mathbf{b}$ orthogonal and so, the result follows.

Example 3.4   Determining Whether Two Vectors Are Orthogonal

Determine whether or not the following pairs of vectors are orthogonal: (a) $\mathbf{a} = \langle 1, 3, -5 \rangle$ and $\mathbf{b} = \langle 2, 3, 10 \rangle$ and (b) $\mathbf{a} = \langle 4, 2, -1 \rangle$ and $\mathbf{b} = \langle 2, 3, 14 \rangle$.

Solution   For (a), we have:

$$\mathbf{a} \cdot \mathbf{b} = 2 + 9 - 50 = -39 \neq 0,$$

so that $\mathbf{a}$ and $\mathbf{b}$ are **not** orthogonal.

For (b), we have

$$\mathbf{a} \cdot \mathbf{b} = 8 + 6 - 14 = 0,$$

so that $\mathbf{a}$ and $\mathbf{b}$ are orthogonal, in this case.

The following two results provide us with some powerful tools for comparing the magnitudes of vectors.

Theorem 3.3 (Cauchy-Schwartz Inequality)
For any vectors $\mathbf{a}$ and $\mathbf{b}$,

$$|\mathbf{a} \cdot \mathbf{b}| \leq \|\mathbf{a}\| \|\mathbf{b}\|. \tag{3.5}$$

Proof

If either $\mathbf{a}$ or $\mathbf{b}$ is the zero vector, notice that (3.5) simply says that $0 \leq 0$, which is certainly true. On the other hand, if neither $\mathbf{a}$ nor $\mathbf{b}$ is the zero vector, we have from (3.2) that

$$|\mathbf{a} \cdot \mathbf{b}| = \|\mathbf{a}\| \|\mathbf{b}\| |\cos \theta| \leq \|\mathbf{a}\| \|\mathbf{b}\|,$$

since $|\cos \theta| \leq 1$ for all values of θ.

One benefit of the Cauchy-Schwartz Inequality is that it allows us to prove the following very useful result. If you were going to learn only one inequality in your lifetime, this is probably the one you would want to learn.

Theorem 3.4 (The Triangle Inequality)
For any vectors $\mathbf{a}$ and $\mathbf{b}$,

$$\|\mathbf{a} + \mathbf{b}\| \leq \|\mathbf{a}\| + \|\mathbf{b}\|. \tag{3.6}$$

Before we prove the theorem, consider the triangle formed by the vectors $\mathbf{a}$, $\mathbf{b}$ and $\mathbf{a} + \mathbf{b}$, shown in Figure 10.24. Notice that the Triangle Inequality says that the length of the vector $\mathbf{a} + \mathbf{b}$ never exceeds the sum of the individual lengths of $\mathbf{a}$ and $\mathbf{b}$.

Proof

From Theorem 3.1 (i), (ii) and (v), we have

$$\|\mathbf{a} + \mathbf{b}\|^2 = (\mathbf{a} + \mathbf{b}) \cdot (\mathbf{a} + \mathbf{b}) = \mathbf{a} \cdot \mathbf{a} + \mathbf{a} \cdot \mathbf{b} + \mathbf{b} \cdot \mathbf{a} + \mathbf{b} \cdot \mathbf{b}$$
$$= \|\mathbf{a}\|^2 + 2\mathbf{a} \cdot \mathbf{b} + \|\mathbf{b}\|^2.$$

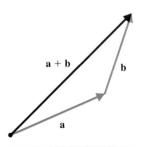

Figure 10.24
The Triangle Inequality.

From the Cauchy-Schwartz Inequality (3.5), we have $\mathbf{a} \cdot \mathbf{b} \leq |\mathbf{a} \cdot \mathbf{b}| \leq \|\mathbf{a}\|\|\mathbf{b}\|$ and so, we have

$$\|\mathbf{a} + \mathbf{b}\|^2 = \|\mathbf{a}\|^2 + 2\mathbf{a} \cdot \mathbf{b} + \|\mathbf{b}\|^2$$
$$\leq \|\mathbf{a}\|^2 + 2\|\mathbf{a}\|\|\mathbf{b}\| + \|\mathbf{b}\|^2 = (\|\mathbf{a}\| + \|\mathbf{b}\|)^2.$$

Taking square roots gives us (3.6).

■

Components and Projections

Think about the case where a vector represents a force. Very often, it's impossible or impractical to exert a force in the direction you'd like. For instance, in pulling a child's wagon, we exert a force in a convenient direction (in the direction determined by the position of the handle), instead of in the direction of motion (see Figure 10.25). An important question (particularly if you are pulling this wagon for a distance) is whether there is a force of smaller magnitude that can be exerted in a different direction and still produce the same effect on the wagon. Notice that it is the horizontal portion of the force that most directly contributes to the motion of the wagon. (The vertical portion of the force only acts to reduce friction.) We now consider how to compute such a component of a force.

Figure 10.25
Pulling a wagon.

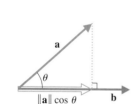

Figure 10.26a

$\text{comp}_\mathbf{b}\,\mathbf{a}$, for $0 < \theta < \frac{\pi}{2}$.

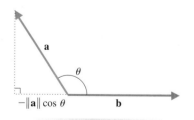

Figure 10.26b

$\text{comp}_\mathbf{b}\,\mathbf{a}$, for $\frac{\pi}{2} < \theta < \pi$.

Component of a along b

For any two nonzero position vectors $\mathbf{a}$ and $\mathbf{b}$, let the angle between the vectors be θ. If we drop a perpendicular line segment from the terminal point of $\mathbf{a}$ to the line containing the vector $\mathbf{b}$, then from elementary trigonometry, the base of the triangle (in the case where $0 < \theta < \frac{\pi}{2}$) has length given by $\|\mathbf{a}\| \cos \theta$ (see Figure 10.26a).

On the other hand, notice that if $\frac{\pi}{2} < \theta < \pi$, the length of the base is given by $-\|\mathbf{a}\| \cos \theta$ (see Figure 10.26b). In either case, we refer to $\|\mathbf{a}\| \cos \theta$ as the **component** of $\mathbf{a}$ along $\mathbf{b}$, denoted $\text{comp}_\mathbf{b}\,\mathbf{a}$. Using (3.2), observe that we can rewrite this as

$$\text{comp}_\mathbf{b}\,\mathbf{a} = \|\mathbf{a}\| \cos \theta = \|\mathbf{a}\| \frac{\|\mathbf{a}\|\|\mathbf{b}\|}{\|\mathbf{a}\|\|\mathbf{b}\|} \cos \theta$$

$$= \frac{\|\mathbf{a}\|}{\|\mathbf{a}\|\|\mathbf{b}\|} \|\mathbf{a}\|\|\mathbf{b}\| \cos \theta = \frac{1}{\|\mathbf{b}\|} \mathbf{a} \cdot \mathbf{b}.$$

or

$$\boxed{\text{comp}_\mathbf{b}\,\mathbf{a} = \frac{\mathbf{a} \cdot \mathbf{b}}{\|\mathbf{b}\|}.} \tag{3.7}$$

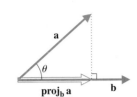

Figure 10.27a

$\text{proj}_b \mathbf{a}$, for $0 < \theta < \frac{\pi}{2}$.

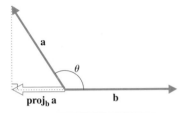

Figure 10.27b

$\text{proj}_b \mathbf{a}$, for $\frac{\pi}{2} < \theta < \pi$.

CAUTION

Be careful to distinguish between the **projection** of **a** onto **b** (a vector) and the **component** of **a** along **b** (a scalar). It is very common to confuse the two.

Be certain to notice that $\text{comp}_b \mathbf{a}$ is a scalar. Also, observe that we divide the dot product in (3.7) by $\|\mathbf{b}\|$ and not by $\|\mathbf{a}\|$. One way to keep this straight is to recognize that the components in Figures 10.26a and 10.26b depend on how long **a** is but not on how long **b** is. In (3.7), we have the dot product of the vector **a** and a unit vector in the direction of **b**, given by $\dfrac{\mathbf{b}}{\|\mathbf{b}\|}$.

Once again, consider the case where the vector **a** represents a force. Rather than the component of **a** along **b**, we are often interested in finding a force vector parallel to **b** having the same component along **b** as **a**. We call this vector the **projection** of **a** onto **b**, denoted $\text{proj}_b \mathbf{a}$, as indicated in Figures 10.27a and 10.27b. Since the projection has magnitude $|\text{comp}_b \mathbf{a}|$ and points in the direction of **b**, for $0 < \theta < \frac{\pi}{2}$ and opposite **b**, for $\frac{\pi}{2} < \theta < \pi$, we have from (3.7) that

$$\text{proj}_b \mathbf{a} = (\text{comp}_b \mathbf{a}) \frac{\mathbf{b}}{\|\mathbf{b}\|} = \left(\frac{\mathbf{a} \cdot \mathbf{b}}{\|\mathbf{b}\|} \right) \frac{\mathbf{b}}{\|\mathbf{b}\|},$$

or

$$\boxed{\text{proj}_b \mathbf{a} = \frac{\mathbf{a} \cdot \mathbf{b}}{\|\mathbf{b}\|^2} \mathbf{b},} \qquad \text{Projection of a onto b} \qquad (3.8)$$

where we have used $\dfrac{\mathbf{b}}{\|\mathbf{b}\|}$ to represent a unit vector in the direction of **b**.

In the following example, we illustrate the process of finding components and projections.

Example 3.5   Finding Components and Projections

For $\mathbf{a} = \langle 2, 3 \rangle$ and $\mathbf{b} = \langle -1, 5 \rangle$, find the component of **a** along **b** and the projection of **a** onto **b**.

Solution   From (3.7), we have

$$\text{comp}_b \mathbf{a} = \frac{\mathbf{a} \cdot \mathbf{b}}{\|\mathbf{b}\|} = \frac{\langle 2, 3 \rangle \cdot \langle -1, 5 \rangle}{\|\langle -1, 5 \rangle\|} = \frac{-2 + 15}{\sqrt{1 + 5^2}} = \frac{13}{\sqrt{26}}.$$

Similarly, from (3.8), we have

$$\text{proj}_b \mathbf{a} = \left(\frac{\mathbf{a} \cdot \mathbf{b}}{\|\mathbf{b}\|} \right) \frac{\mathbf{b}}{\|\mathbf{b}\|} = \left(\frac{13}{\sqrt{26}} \right) \frac{\langle -1, 5 \rangle}{\sqrt{26}}$$

$$= \frac{13}{26} \langle -1, 5 \rangle = \frac{1}{2} \langle -1, 5 \rangle = \left\langle -\frac{1}{2}, \frac{5}{2} \right\rangle.$$

We leave it as an exercise to show that in general, $\text{comp}_b \mathbf{a} \neq \text{comp}_a \mathbf{b}$ and $\text{proj}_b \mathbf{a} \neq \text{proj}_a \mathbf{b}$. One reason for needing to consider components of a vector in a given direction is to compute work, as we see in the following example.

Example 3.6   Calculating Work

You exert a constant force of 40 pounds in the direction of the handle of the wagon pictured in Figure 10.28. If the handle makes an angle of $\frac{\pi}{4}$ with the horizontal and you pull the wagon along a flat surface for 1 mile (5280 feet), find the work done.

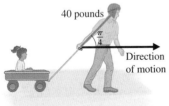

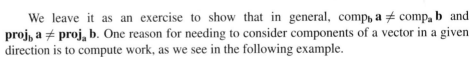

Figure 10.28

Pulling a wagon.

Solution First, recall from our discussion in Chapter 5 that if we apply a constant force F for a distance d, the work done is given by $W = Fd$. Unfortunately, the force exerted in the direction of motion is not given. Since the magnitude of the force is 40, the force vector must be

$$\mathbf{F} = 40 \left\langle \cos \frac{\pi}{4}, \sin \frac{\pi}{4} \right\rangle = 40 \left\langle \frac{\sqrt{2}}{2}, \frac{\sqrt{2}}{2} \right\rangle = \langle 20\sqrt{2}, 20\sqrt{2} \rangle.$$

The force exerted in the direction of motion is simply the component of the force along the vector $\mathbf{i}$ (that is, the horizontal component of $\mathbf{F}$) or $20\sqrt{2}$. The work done is then

$$W = Fd = 20\sqrt{2}\,(5280) \approx 149{,}341 \text{ foot-pounds.}$$

EXERCISES 10.3

1. Explain in words why the Triangle Inequality is true.

2. The dot product is called a "product" because the properties listed in Theorem 3.1 are true for multiplication of real numbers. Two other properties of multiplication of real numbers involve factoring: (1) if $ab = ac$ ($a \neq 0$) then $b = c$ and (2) if $ab = 0$ then $a = 0$ or $b = 0$. Discuss the extent to which these properties are true for the dot product.

3. We have asked you on several occasions to find unit vectors in given directions, but we have not discussed why unit vectors are important. Discuss how finding the component of a vector along a unit vector is easier than finding the component of the same vector along a nonunit vector. Explain why unit vectors are sometimes called **direction vectors** (indicating that the magnitude is irrelevant).

4. Some people are puzzled about why the vertical component of force does not contribute to the work computation of example 3.6. To understand this, explain why the vertical component of force does not contribute to the horizontal motion of the wagon. In practice, the vertical force is not useless. Explain why it is easier to pull a light wagon than a heavy wagon, and explain why a small vertical force essentially makes the wagon lighter.

In exercises 5–18, compute a · b.

5. $\mathbf{a} = \langle 3, 1 \rangle, \mathbf{b} = \langle 2, 4 \rangle$

6. $\mathbf{a} = \langle 2, 5 \rangle, \mathbf{b} = \langle 2, 1 \rangle$

7. $\mathbf{a} = \langle 0, -2 \rangle, \mathbf{b} = \langle -2, 4 \rangle$

8. $\mathbf{a} = \langle 2, 0 \rangle, \mathbf{b} = \langle -2, 4 \rangle$

9. $\mathbf{a} = 3\mathbf{i} + \mathbf{j}, \mathbf{b} = -2\mathbf{i} + 3\mathbf{j}$

10. $\mathbf{a} = -2\mathbf{i} - 3\mathbf{j}, \mathbf{b} = 2\mathbf{i} - 3\mathbf{j}$

11. $\mathbf{a} = \langle 2, -1, 3 \rangle, \mathbf{b} = \langle 0, 2, 4 \rangle$

12. $\mathbf{a} = \langle 3, 2, 0 \rangle, \mathbf{b} = \langle -2, 4, 3 \rangle$

13. $\mathbf{a} = 2\mathbf{i} - \mathbf{k}, \mathbf{b} = 4\mathbf{j} - \mathbf{k}$

14. $\mathbf{a} = 3\mathbf{i} + 3\mathbf{k}, \mathbf{b} = -2\mathbf{i} + \mathbf{j}$

In exercises 15–20, compute the angle between the vectors.

15. $\mathbf{a} = 3\mathbf{i} - 2\mathbf{j}, \mathbf{b} = \mathbf{i} + \mathbf{j}$

16. $\mathbf{a} = \langle 1, 4 \rangle, \mathbf{b} = \langle -2, 3 \rangle$

17. $\mathbf{a} = \langle 2, 0, -2 \rangle, \mathbf{b} = \langle 0, -2, 4 \rangle$

18. $\mathbf{a} = \langle 3, 2, 0 \rangle, \mathbf{b} = \langle -2, 4, 3 \rangle$

19. $\mathbf{a} = 3\mathbf{i} + \mathbf{j} - 4\mathbf{k}, \mathbf{b} = -2\mathbf{i} + 2\mathbf{j} + \mathbf{k}$

20. $\mathbf{a} = \mathbf{i} + 3\mathbf{j} - 2\mathbf{k}, \mathbf{b} = 2\mathbf{i} - 3\mathbf{k}$

In exercises 21–26, determine if the vectors are orthogonal.

21. $\mathbf{a} = \langle 2, -1 \rangle, \mathbf{b} = \langle 2, 4 \rangle$

22. $\mathbf{a} = \langle 4, 2 \rangle, \mathbf{b} = \langle -2, 3 \rangle$

23. $\mathbf{a} = \langle 4, -1, 1 \rangle, \mathbf{b} = \langle 2, 4, 4 \rangle$

24. $\mathbf{a} = \langle 1, 5, 2 \rangle, \mathbf{b} = \langle -4, 3, 5 \rangle$

25. $\mathbf{a} = 6\mathbf{i} + 2\mathbf{j}, \mathbf{b} = -\mathbf{i} + 3\mathbf{j}$

26. $\mathbf{a} = 3\mathbf{i}, \mathbf{b} = 6\mathbf{j} - 2\mathbf{k}$

In exercises 27–32, find a vector perpendicular to the given vector.

27. $\langle 2, -1 \rangle$

28. $\langle 4, 2 \rangle$

29. $\langle 4, -1, 1 \rangle$

30. $\langle -4, 3, 5 \rangle$

31. $6\mathbf{i} + 2\mathbf{j} - \mathbf{k}$

32. $2\mathbf{i} - 3\mathbf{k}$

In exercises 33–40, find $\text{comp}_b\, a$ and $\text{proj}_b\, a$.

33. $\mathbf{a} = \langle 2, 1 \rangle$, $\mathbf{b} = \langle 3, 4 \rangle$

34. $\mathbf{a} = \langle 4, 4 \rangle$, $\mathbf{b} = \langle -3, 4 \rangle$

35. $\mathbf{a} = 3\mathbf{i} + \mathbf{j}$, $\mathbf{b} = 4\mathbf{i} - 3\mathbf{j}$

36. $\mathbf{a} = -2\mathbf{i} - 3\mathbf{j}$, $\mathbf{b} = 2\mathbf{i} - 2\mathbf{j}$

37. $\mathbf{a} = \langle 2, -1, 3 \rangle$, $\mathbf{b} = \langle 1, 2, 2 \rangle$

38. $\mathbf{a} = \langle 1, 4, 5 \rangle$, $\mathbf{b} = \langle -2, 1, 2 \rangle$

39. $\mathbf{a} = \langle 2, 0, -2 \rangle$, $\mathbf{b} = \langle 0, -3, 4 \rangle$

40. $\mathbf{a} = \langle 3, 2, 0 \rangle$, $\mathbf{b} = \langle -2, 2, 1 \rangle$

41. Repeat example 3.6 with an angle of $\frac{\pi}{3}$ with the horizontal.

42. Repeat example 3.6 with an angle of $\frac{\pi}{6}$ with the horizontal.

43. Explain why the answers to exercises 41 and 42 aren't the same, even though the force exerted is the same. In this setting, explain why a larger amount of work corresponds to a more efficient use of the force.

44. Find the force needed in exercise 41 to produce the same amount of work as in example 3.6.

45. A constant force of $\langle 30, 20 \rangle$ pounds moves an object in a straight line from the point $(0, 0)$ to the point $(24, 10)$. Compute the work done.

46. A constant force of $\langle 60, -30 \rangle$ pounds moves an object in a straight line from the point $(0, 0)$ to the point $(10, -10)$. Compute the work done.

47. Label each statement as true or false. If it is true, briefly explain why; if it is false, give a counterexample.
    (a) If $\mathbf{a} \cdot \mathbf{b} = \mathbf{a} \cdot \mathbf{c}$, then $\mathbf{b} = \mathbf{c}$.
    (b) If $\mathbf{b} = \mathbf{c}$, then $\mathbf{a} \cdot \mathbf{b} = \mathbf{a} \cdot \mathbf{c}$.
    (c) $\mathbf{a} \cdot \mathbf{a} = \|\mathbf{a}\|^2$.
    (d) If $\|\mathbf{a}\| > \|\mathbf{b}\|$ then $\mathbf{a} \cdot \mathbf{c} > \mathbf{b} \cdot \mathbf{c}$.
    (e) If $\|\mathbf{a}\| = \|\mathbf{b}\|$ then $\mathbf{a} = \mathbf{b}$.

48. To compute $\mathbf{a} \cdot \mathbf{b}$, where $\mathbf{a} = \langle 2, 5 \rangle$ and $\mathbf{b} = \dfrac{\langle 4, 1 \rangle}{\sqrt{17}}$, you can first compute $\langle 2, 5 \rangle \cdot \langle 4, 1 \rangle$ and then divide the result (13) by $\sqrt{17}$. Which property of Theorem 3.1 is being used?

49. By the Cauchy-Schwartz inequality, $|\mathbf{a} \cdot \mathbf{b}| \le \|\mathbf{a}\|\|\mathbf{b}\|$. What relationship must exist between $\mathbf{a}$ and $\mathbf{b}$ to have $|\mathbf{a} \cdot \mathbf{b}| = \|\mathbf{a}\|\|\mathbf{b}\|$?

50. By the triangle inequality, $\|\mathbf{a} + \mathbf{b}\| \le \|\mathbf{a}\| + \|\mathbf{b}\|$. What relationship must exist between $\mathbf{a}$ and $\mathbf{b}$ to have $\|\mathbf{a} + \mathbf{b}\| = \|\mathbf{a}\| + \|\mathbf{b}\|$?

51. Use the triangle inequality to prove that $\|\mathbf{a} - \mathbf{b}\| \ge \|\mathbf{a}\| - \|\mathbf{b}\|$.

52. Prove parts (ii) and (iii) of Theorem 3.1.

53. In a methane molecule (CH_4), a carbon atom is surrounded by four hydrogen atoms. Assume that the hydrogen atoms are at $(0, 0, 0)$, $(1, 1, 0)$, $(1, 0, 1)$ and $(0, 1, 1)$ and the carbon atom is at $\left(\frac{1}{2}, \frac{1}{2}, \frac{1}{2}\right)$. Compute the **bond angle,** the angle from hydrogen atom to carbon atom to hydrogen atom.

54. Consider the parallelogram with vertices at $(0, 0)$, $(2, 0)$, $(3, 2)$ and $(1, 2)$. Find the angle at which the diagonals intersect.

55. Prove that $\text{comp}_c(\mathbf{a} + \mathbf{b}) = \text{comp}_c\,\mathbf{a} + \text{comp}_c\,\mathbf{b}$ for any nonzero vectors $\mathbf{a}$, $\mathbf{b}$ and $\mathbf{c}$.

56. The **orthogonal projection** of vector $\mathbf{a}$ along vector $\mathbf{b}$ is defined as $\text{orth}_b\,\mathbf{a} = \mathbf{a} - \text{proj}_b\,\mathbf{a}$. Sketch a picture showing vectors $\mathbf{a}$, $\mathbf{b}$, $\text{proj}_b\,\mathbf{a}$ and $\text{orth}_b\,\mathbf{a}$ and explain what is orthogonal about $\text{orth}_b\,\mathbf{a}$.

57. Suppose that a beam of an oil rig is installed in a direction parallel to $\langle 10, 1, 5 \rangle$. If a wave exerts a force of $\langle 0, -200, 0 \rangle$ Newtons, find the component of this force along the beam.

58. Repeat exercise 57 with a force of $\langle 13, -190, -61 \rangle$ Newtons. The forces here and in exercise 57 have nearly identical magnitudes. Explain why the force components are different.

59. A car makes a turn on a banked road. If the road is banked at $10°$ show that a vector parallel to the road is $\langle \cos 10°, \sin 10° \rangle$. If the car has weight 2000 pounds, find the component of the weight vector along the road vector. This component of weight provides a force that helps the car turn.

60. Find the component of the weight vector along the road vector for a 2500-pound car on a 15° bank.

61. In the diagram, a crate of weight w pounds is placed on a ramp inclined at angle θ above the horizontal. The vector **v** along the ramp is given by $\mathbf{v} = \langle \cos\theta, \sin\theta \rangle$ and the normal vector by $\mathbf{n} = \langle \sin\theta, -\cos\theta \rangle$. Show that **v** and **n** are perpendicular. Find the component of $\mathbf{w} = \langle 0, -w \rangle$ along **v** and the component of **w** along **n**.

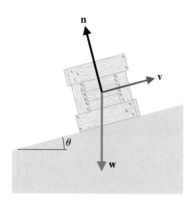

62. If the coefficient of static friction between the crate and ramp in exercise 61 equals μ_s, physics tells us that the crate will slide down the ramp if the component of **w** along **v** is greater than the product of μ_s and the component of **w** along **n**. Show that this occurs if the angle θ is steep enough that $\theta > \tan^{-1}\mu_s$.

63. A weight of 500 pounds is supported by two ropes that exert forces of $\mathbf{a} = \langle -100, 200 \rangle$ pounds and $\mathbf{b} = \langle 100, 300 \rangle$ pounds, respectively. Find the angle θ between the ropes.

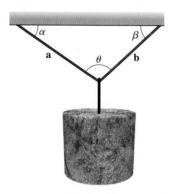

64. In the above diagram, find the angles α and β.

65. Suppose a small business sells three products. In a given month, if 3000 units of product A are sold, 2000 units of product B are sold and 4000 units of product C are sold, then the **sales vector** for that month is defined by $\mathbf{s} = \langle 3000, 2000, 4000 \rangle$. If the prices of products A, B and C are $20, $15 and $25, respectively, then the **price vector** is defined

by $\mathbf{p} = \langle 20, 15, 25 \rangle$. Compute $\mathbf{s} \cdot \mathbf{p}$ and discuss how it relates to monthly revenue.

66. Suppose that in a particular county ice cream sales (in thousands of gallons) for a year is given by the vector $\mathbf{s} = \langle 3, 5, 12, 40, 60, 100, 120, 160, 110, 50, 10, 2 \rangle$. That is, 3000 gallons were sold in January, 5000 gallons were sold in February, and so on. In the same county, suppose that murders for the year are given by the vector $\mathbf{m} = \langle 2, 0, 1, 6, 4, 8, 10, 13, 8, 2, 0, 6 \rangle$. Show that the average monthly ice cream sales is $\bar{s} = 56{,}000$ gallons and that the average monthly number of murders is $\bar{m} = 5$. Compute the vectors **a** and **b**, where the components of **a** equal the components of **s** with the mean 56 subtracted (so that $\mathbf{a} = \langle -53, -51, -44, \ldots \rangle$) and the components of **b** equal the components of **m** with the mean 5 subtracted. The correlation between ice cream sales and murders is defined as $\rho = \dfrac{\mathbf{a} \cdot \mathbf{b}}{\|\mathbf{a}\|\,\|\mathbf{b}\|}$. Often, a positive correlation is interpreted as meaning that **a** "causes" **b**. Explain why that conclusion would be invalid in this case.

67. One of the basic problems throughout calculus is computing distances. In this exercise, we will find the distance between a point (x_1, y_1) and a line $ax + by + d = 0$. To start with a concrete example, take the point $(5, 6)$ and the line $2x + 3y + 4 = 0$. First, show that the intercepts of the line are the points $\left(-\frac{4}{2}, 0\right)$ and $\left(0, -\frac{4}{3}\right)$. Show that the vector $\mathbf{b} = \langle 3, -2 \rangle$ is parallel to the displacement vector between these points and hence, also to the line. Sketch a picture showing the point $(5, 6)$, the line, the vector $\langle 3, -2 \rangle$ and the displacement vector **v** from $(-2, 0)$ to $(5, 6)$. Explain why the magnitude of the vector $\mathbf{v} - \mathbf{proj}_\mathbf{b}\,\mathbf{v}$ equals the desired distance between point and line. Compute this distance. Show that in general, the distance between the point (x_1, y_1) and the line $ax + by + d = 0$ equals $\dfrac{|ax_1 + by_1 + d|}{\sqrt{a^2 + b^2}}$.

68. In the figure below, the circle $x^2 + y^2 = r^2$ is shown. In this exercise, we will compute the time required for an object to travel the length of a chord from the top of the circle to another point on the circle at an angle of θ from the vertical, assuming that gravity (acting downward) is the only force.

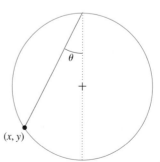

From our study of projectile motion in section 5.5, recall that an object traveling with a constant acceleration a covers a distance d in time $\sqrt{\frac{2d}{a}}$. Show that the component of gravity in the direction of the chord is $a = g\cos\theta$. If the chord ends at the point (x, y), show that the length of the chord is $d = \sqrt{2r^2 - 2ry}$. Also, show that $\cos\theta = \frac{r-y}{d}$. Putting this all together, compute the time it takes to travel the chord. Explain why it's surprising that the answer does not depend on the value of θ. Note that as θ increases, the distance d decreases but the effectiveness of gravity decreases. Discuss the balance between these two factors.

69. 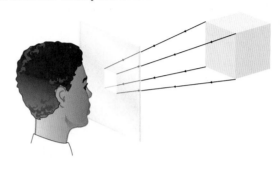 This exercise develops a basic principle used to create wireframe and other 3-D computer graphics. In the drawing, an artist traces the image of an object onto a pane of glass. Explain why the trace will be distorted unless the artist keeps the pane of glass perpendicular to the line of sight. The trace is thus a projection of the object onto the pane of glass. To make this precise, suppose that the artist is at the point $(100, 0, 0)$ and the point $P_1 = (2, 1, 3)$ is part of the object being traced. Find the projection $\mathbf{p}_1$ of the position vector $\langle 2, 1, 3 \rangle$ along the artist's position vector $\langle 100, 0, 0 \rangle$. Then find the vector $\mathbf{q}_1$ such that $\langle 2, 1, 3 \rangle = \mathbf{p}_1 + \mathbf{q}_1$. Which of the vectors $\mathbf{p}_1$ and $\mathbf{q}_1$ does the artist actually see and which one is hidden? Repeat this with the point $P_2 = (-2, 1, 3)$ and find vectors $\mathbf{p}_2$ and $\mathbf{q}_2$ such that $\langle -2, 1, 3 \rangle = \mathbf{p}_2 + \mathbf{q}_2$. The artist would plot both points P_1 and P_2 at the same point on the pane of glass. Identify which of the vectors $\mathbf{p}_1$, $\mathbf{q}_1$, $\mathbf{p}_2$ and $\mathbf{q}_2$ correspond to this point. From the artist's perspective, one of the points P_1 and P_2 is hidden behind the other. Identify which point is hidden and explain how the information in the vectors $\mathbf{p}_1$, $\mathbf{q}_1$, $\mathbf{p}_2$ and $\mathbf{q}_2$ can be used to determine which point is hidden.

10.4   THE CROSS PRODUCT

Notice that the dot product of two vectors results not in another vector, but in a scalar. In this section, we define another product of vectors, called the **cross product** or **vector product.** Unlike the dot product, the cross product of two vectors is another vector. The cross product has many important applications, from physics and engineering mechanics to space travel. Before we define the cross product, we need to define the notion of **determinant.**

Definition 4.1

The **determinant** of a 2×2 matrix of real numbers is defined by

2 × 2 determinant

$$\underbrace{\begin{vmatrix} a_1 & a_2 \\ b_1 & b_2 \end{vmatrix}}_{2 \times 2 \text{ matrix}} = a_1 b_2 - a_2 b_1. \tag{4.1}$$

Example 4.1   Computing a 2 × 2 Determinant

Evaluate the determinant $\begin{vmatrix} 1 & 2 \\ 3 & 4 \end{vmatrix}$.

Solution   From (4.1), we have

$$\begin{vmatrix} 1 & 2 \\ 3 & 4 \end{vmatrix} = (1)(4) - (2)(3) = -2.$$

Definition 4.2

The **determinant** of a 3×3 matrix of real numbers is defined as a combination of three 2×2 determinants, as follows:

3×3 determinant

$$\underbrace{\begin{vmatrix} a_1 & a_2 & a_3 \\ b_1 & b_2 & b_3 \\ c_1 & c_2 & c_3 \end{vmatrix}}_{3 \times 3 \text{ matrix}} = a_1 \begin{vmatrix} b_2 & b_3 \\ c_2 & c_3 \end{vmatrix} - a_2 \begin{vmatrix} b_1 & b_3 \\ c_1 & c_3 \end{vmatrix} + a_3 \begin{vmatrix} b_1 & b_2 \\ c_1 & c_2 \end{vmatrix}. \tag{4.2}$$

We can make some sense of the somewhat confusing form of this expression as follows. Equation (4.2) is referred to as an **expansion** of the determinant **along the first row.** Notice that the multipliers of each of the 2×2 determinants are the entries of the first row of the 3×3 matrix. Each 2×2 determinant is the determinant you get if you eliminate the row and column in which the corresponding multiplier lies. That is, for the *first* term, the multiplier is a_1 and the 2×2 determinant is found by eliminating the first row and *first* column from the 3×3 matrix:

$$\begin{vmatrix} \cancel{a_1} & a_2 & a_3 \\ \cancel{b_1} & b_2 & b_3 \\ \cancel{c_1} & c_2 & c_3 \end{vmatrix} = \begin{vmatrix} b_2 & b_3 \\ c_2 & c_3 \end{vmatrix}.$$

Likewise, the *second* 2×2 determinant is found by eliminating the first row and the *second* column from the 3×3 determinant:

$$\begin{vmatrix} a_1 & \cancel{a_2} & a_3 \\ b_1 & \cancel{b_2} & b_3 \\ c_1 & \cancel{c_2} & c_3 \end{vmatrix} = \begin{vmatrix} b_1 & b_3 \\ c_1 & c_3 \end{vmatrix}.$$

Be certain to notice the minus sign in front of this term. Finally, the *third* determinant is found by eliminating the first row and the *third* column from the 3×3 determinant:

$$\begin{vmatrix} a_1 & a_2 & \cancel{a_3} \\ b_1 & b_2 & \cancel{b_3} \\ c_1 & c_2 & \cancel{c_3} \end{vmatrix} = \begin{vmatrix} b_1 & b_2 \\ c_1 & c_2 \end{vmatrix}.$$

Example 4.2   Evaluating a 3×3 Determinant

Evaluate the determinant $\begin{vmatrix} 1 & 2 & 4 \\ -3 & 3 & 1 \\ 3 & -2 & 5 \end{vmatrix}$.

Solution   Expanding along the first row, we have:

$$\begin{vmatrix} 1 & 2 & 4 \\ -3 & 3 & 1 \\ 3 & -2 & 5 \end{vmatrix} = (1) \begin{vmatrix} 3 & 1 \\ -2 & 5 \end{vmatrix} - (2) \begin{vmatrix} -3 & 1 \\ 3 & 5 \end{vmatrix} + (4) \begin{vmatrix} -3 & 3 \\ 3 & -2 \end{vmatrix}$$

$$= (1)[(3)(5) - (1)(-2)] - (2)[(-3)(5) - (1)(3)]$$
$$+ (4)[(-3)(-2) - (3)(3)]$$
$$= 41.$$

We use determinant notation as a convenient device for defining the cross product, as follows.

Definition 4.3

For two vectors $\mathbf{a} = \langle a_1, a_2, a_3 \rangle$ and $\mathbf{b} = \langle b_1, b_2, b_3 \rangle$ in V_3, we define the **cross product** (or **vector product**) of $\mathbf{a}$ and $\mathbf{b}$ to be

Cross product

$$\mathbf{a} \times \mathbf{b} = \begin{vmatrix} \mathbf{i} & \mathbf{j} & \mathbf{k} \\ a_1 & a_2 & a_3 \\ b_1 & b_2 & b_3 \end{vmatrix} = \begin{vmatrix} a_2 & a_3 \\ b_2 & b_3 \end{vmatrix} \mathbf{i} - \begin{vmatrix} a_1 & a_3 \\ b_1 & b_3 \end{vmatrix} \mathbf{j} + \begin{vmatrix} a_1 & a_2 \\ b_1 & b_2 \end{vmatrix} \mathbf{k}. \quad (4.3)$$

Notice that by the way in which we defined it, $\mathbf{a} \times \mathbf{b}$ is also a vector in V_3. To compute $\mathbf{a} \times \mathbf{b}$, you must write the components of $\mathbf{a}$ in the second row and the components of $\mathbf{b}$ in the third row; **the order is important!** Also note that while we've used the determinant notation, the 3×3 determinant indicated in (4.3) is not really a determinant, in the sense in which we defined them, since the entries in the first row are vectors instead of scalars. Nonetheless, we find this slight abuse of notation convenient for computing cross products and we use it routinely.

Example 4.3   Computing a Cross Product

Compute $\langle 1, 2, 3 \rangle \times \langle 4, 5, 6 \rangle$.

Solution   From (4.3), we have

$$\langle 1, 2, 3 \rangle \times \langle 4, 5, 6 \rangle = \begin{vmatrix} \mathbf{i} & \mathbf{j} & \mathbf{k} \\ 1 & 2 & 3 \\ 4 & 5 & 6 \end{vmatrix} = \begin{vmatrix} 2 & 3 \\ 5 & 6 \end{vmatrix} \mathbf{i} - \begin{vmatrix} 1 & 3 \\ 4 & 6 \end{vmatrix} \mathbf{j} + \begin{vmatrix} 1 & 2 \\ 4 & 5 \end{vmatrix} \mathbf{k}$$

$$= -3\mathbf{i} + 6\mathbf{j} - 3\mathbf{k} = \langle -3, 6, -3 \rangle.$$

Remark 4.1

The cross product is defined only for vectors in V_3. There is no corresponding operation for vectors in V_2.

Theorem 4.1

For any vector $\mathbf{a} \in V_3$, $\mathbf{a} \times \mathbf{a} = \mathbf{0}$ and $\mathbf{a} \times \mathbf{0} = \mathbf{0}$.

Proof

We prove the first of these two results. The second we leave as an exercise. For $\mathbf{a} = \langle a_1, a_2, a_3 \rangle$, we have from (4.3) that

$$\mathbf{a} \times \mathbf{a} = \begin{vmatrix} \mathbf{i} & \mathbf{j} & \mathbf{k} \\ a_1 & a_2 & a_3 \\ a_1 & a_2 & a_3 \end{vmatrix} = \begin{vmatrix} a_2 & a_3 \\ a_2 & a_3 \end{vmatrix} \mathbf{i} - \begin{vmatrix} a_1 & a_3 \\ a_1 & a_3 \end{vmatrix} \mathbf{j} + \begin{vmatrix} a_1 & a_2 \\ a_1 & a_2 \end{vmatrix} \mathbf{k}$$

$$= (a_2 a_3 - a_3 a_2)\mathbf{i} - (a_1 a_3 - a_3 a_1)\mathbf{j} + (a_1 a_2 - a_2 a_1)\mathbf{k} = \mathbf{0}.$$

Let's take a brief look back at the result of example 4.3. There, we saw that

$$\langle 1, 2, 3 \rangle \times \langle 4, 5, 6 \rangle = \langle -3, 6, -3 \rangle.$$

HISTORICAL NOTES

**Josiah Willard Gibbs
(1839–1903)** American physicist
and mathematician who introduced
and named the dot product and the
cross product. A graduate of Yale,
Gibbs published important papers
in thermodynamics, statistical
mechanics and the electromagnetic
theory of light. Gibbs used vectors
to determine the orbit of a comet
from only three observations.
Originally produced as printed
notes for his students, Gibbs' vector
system greatly simplified the
original system developed by
Hamilton. Gibbs was well liked but
not famous in his lifetime. One
biographer wrote of Gibbs that,
"The greatness of his intellectual
achievements will never
overshadow the beauty and dignity
of his life."

You probably did not notice this at the time, but there is something rather interesting to observe here. Note that

$$\langle 1, 2, 3 \rangle \cdot \langle -3, 6, -3 \rangle = 0$$

and

$$\langle 4, 5, 6 \rangle \cdot \langle -3, 6, -3 \rangle = 0.$$

That is, both $\langle 1, 2, 3 \rangle$ and $\langle 4, 5, 6 \rangle$ are orthogonal to their cross product. As it turns out, this is true in general. We have the following result.

Theorem 4.2

For any vectors $\mathbf{a}$ and $\mathbf{b}$ in V_3, $\mathbf{a} \times \mathbf{b}$ is orthogonal to both $\mathbf{a}$ and $\mathbf{b}$.

Proof

Recall that two vectors are orthogonal if and only if their dot product is zero. Now, using (4.3), we have

$$\mathbf{a} \cdot (\mathbf{a} \times \mathbf{b}) = \langle a_1, a_2, a_3 \rangle \cdot \left[\begin{vmatrix} a_2 & a_3 \\ b_2 & b_3 \end{vmatrix} \mathbf{i} - \begin{vmatrix} a_1 & a_3 \\ b_1 & b_3 \end{vmatrix} \mathbf{j} + \begin{vmatrix} a_1 & a_2 \\ b_1 & b_2 \end{vmatrix} \mathbf{k} \right]$$

$$= a_1 \begin{vmatrix} a_2 & a_3 \\ b_2 & b_3 \end{vmatrix} - a_2 \begin{vmatrix} a_1 & a_3 \\ b_1 & b_3 \end{vmatrix} + a_3 \begin{vmatrix} a_1 & a_2 \\ b_1 & b_2 \end{vmatrix}$$

$$= a_1 [a_2 b_3 - a_3 b_2] - a_2 [a_1 b_3 - a_3 b_1] + a_3 [a_1 b_2 - a_2 b_1]$$

$$= a_1 a_2 b_3 - a_1 a_3 b_2 - a_1 a_2 b_3 + a_2 a_3 b_1 + a_1 a_3 b_2 - a_2 a_3 b_1$$

$$= 0,$$

so that $\mathbf{a}$ and $(\mathbf{a} \times \mathbf{b})$ are orthogonal. Similarly, you can show that $\mathbf{b} \cdot (\mathbf{a} \times \mathbf{b}) = 0$. We leave this as an exercise.

Notice that since $\mathbf{a} \times \mathbf{b}$ is orthogonal to both $\mathbf{a}$ and $\mathbf{b}$, it is also orthogonal to every vector lying in the plane containing $\mathbf{a}$ and $\mathbf{b}$. (We also say that $\mathbf{a} \times \mathbf{b}$ is orthogonal to the plane, in this case.) But, given a plane, out of which side of the plane does $\mathbf{a} \times \mathbf{b}$ point? We can get an idea by computing some simple cross products.

Notice that

$$\mathbf{i} \times \mathbf{j} = \begin{vmatrix} \mathbf{i} & \mathbf{j} & \mathbf{k} \\ 1 & 0 & 0 \\ 0 & 1 & 0 \end{vmatrix} = \begin{vmatrix} 0 & 0 \\ 1 & 0 \end{vmatrix} \mathbf{i} - \begin{vmatrix} 1 & 0 \\ 0 & 0 \end{vmatrix} \mathbf{j} + \begin{vmatrix} 1 & 0 \\ 0 & 1 \end{vmatrix} \mathbf{k} = \mathbf{k}.$$

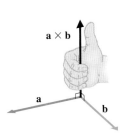

Figure 10.29a

$\mathbf{a} \times \mathbf{b}$.

Likewise,

$$\mathbf{j} \times \mathbf{k} = \mathbf{i}.$$

These are illustrations of the **right-hand rule:** If you align the fingers of your **right** hand along the vector $\mathbf{a}$ and bend your fingers around in the direction of rotation from $\mathbf{a}$ toward $\mathbf{b}$ (through an angle of less than $180°$), your thumb will point in the direction of $\mathbf{a} \times \mathbf{b}$ (see Figure 10.29a). Now, following the right-hand rule, $\mathbf{b} \times \mathbf{a}$ will point in the direction opposite $\mathbf{a} \times \mathbf{b}$ (see Figure 10.29b). In particular, notice that

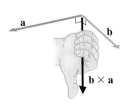

Figure 10.29b

$\mathbf{b} \times \mathbf{a}$.

$$\mathbf{j} \times \mathbf{i} = \begin{vmatrix} \mathbf{i} & \mathbf{j} & \mathbf{k} \\ 0 & 1 & 0 \\ 1 & 0 & 0 \end{vmatrix} = -\mathbf{k}.$$

We leave it as an exercise to show that

$$\mathbf{j} \times \mathbf{k} = \mathbf{i}, \qquad \mathbf{k} \times \mathbf{j} = -\mathbf{i},$$

$$\mathbf{k} \times \mathbf{i} = \mathbf{j} \quad \text{and} \quad \mathbf{i} \times \mathbf{k} = -\mathbf{j}.$$

Take the time to think through the right-hand rule for each of these cross products.
There are several other unusual things to observe here. Notice that

$$\mathbf{i} \times \mathbf{j} = \mathbf{k} \neq -\mathbf{k} = \mathbf{j} \times \mathbf{i}.$$

This says that the cross product is **not** commutative. Further, notice that

$$(\mathbf{i} \times \mathbf{j}) \times \mathbf{j} = \mathbf{k} \times \mathbf{j} = -\mathbf{i},$$

while

$$\mathbf{i} \times (\mathbf{j} \times \mathbf{j}) = \mathbf{i} \times \mathbf{0} = \mathbf{0},$$

so that the cross product is also **not** associative. That is, in general,

$$(\mathbf{a} \times \mathbf{b}) \times \mathbf{c} \neq \mathbf{a} \times (\mathbf{b} \times \mathbf{c}).$$

Since the cross product does not follow several of the rules you might expect a product to satisfy, you might ask what rules the cross product **does** satisfy. We summarize these in the following theorem.

Theorem 4.3

For any vectors $\mathbf{a}$, $\mathbf{b}$ and $\mathbf{c}$ in V_3 and any scalar d, the following hold:

  (i) $\mathbf{a} \times \mathbf{b} = -(\mathbf{b} \times \mathbf{a})$ (anticommutativity)

 (ii) $(d\mathbf{a}) \times \mathbf{b} = d(\mathbf{a} \times \mathbf{b}) = \mathbf{a} \times (d\mathbf{b})$

(iii) $\mathbf{a} \times (\mathbf{b} + \mathbf{c}) = \mathbf{a} \times \mathbf{b} + \mathbf{a} \times \mathbf{c}$ (distributive law)

(iv) $(\mathbf{a} + \mathbf{b}) \times \mathbf{c} = \mathbf{a} \times \mathbf{c} + \mathbf{b} \times \mathbf{c}$ (distributive law)

 (v) $\mathbf{a} \cdot (\mathbf{b} \times \mathbf{c}) = (\mathbf{a} \times \mathbf{b}) \cdot \mathbf{c}$ (scalar triple product) and

(vi) $\mathbf{a} \times (\mathbf{b} \times \mathbf{c}) = (\mathbf{a} \cdot \mathbf{c})\mathbf{b} - (\mathbf{a} \cdot \mathbf{b})\mathbf{c}$ (vector triple product)

Proof

We prove parts (i) and (iii) only. The remaining parts are left as exercises.

  (i) For $\mathbf{a} = \langle a_1, a_2, a_3 \rangle$ and $\mathbf{b} = \langle b_1, b_2, b_3 \rangle$, we have from (4.3) that

$$\mathbf{a} \times \mathbf{b} = \begin{vmatrix} \mathbf{i} & \mathbf{j} & \mathbf{k} \\ a_1 & a_2 & a_3 \\ b_1 & b_2 & b_3 \end{vmatrix} = \begin{vmatrix} a_2 & a_3 \\ b_2 & b_3 \end{vmatrix} \mathbf{i} - \begin{vmatrix} a_1 & a_3 \\ b_1 & b_3 \end{vmatrix} \mathbf{j} + \begin{vmatrix} a_1 & a_2 \\ b_1 & b_2 \end{vmatrix} \mathbf{k}$$

$$= -\begin{vmatrix} b_2 & b_3 \\ a_2 & a_3 \end{vmatrix} \mathbf{i} + \begin{vmatrix} b_1 & b_3 \\ a_1 & a_3 \end{vmatrix} \mathbf{j} - \begin{vmatrix} b_1 & b_2 \\ a_1 & a_2 \end{vmatrix} \mathbf{k} = -(\mathbf{b} \times \mathbf{a}),$$

since swapping two rows in a 2×2 matrix (or in a 3×3 matrix, for that matter) changes the sign of its determinant.

  (iii) For $\mathbf{c} = \langle c_1, c_2, c_3 \rangle$, we have

$$\mathbf{b} + \mathbf{c} = \langle b_1 + c_1, b_2 + c_2, b_3 + c_3 \rangle$$

and so,

$$\mathbf{a} \times (\mathbf{b} + \mathbf{c}) = \begin{vmatrix} \mathbf{i} & \mathbf{j} & \mathbf{k} \\ a_1 & a_2 & a_3 \\ b_1 + c_1 & b_2 + c_2 & b_3 + c_3 \end{vmatrix}.$$

Looking only at the **i** component of this, we have

$$\begin{vmatrix} a_2 & a_3 \\ b_2 + c_2 & b_3 + c_3 \end{vmatrix} = a_2(b_3 + c_3) - a_3(b_2 + c_2)$$

$$= (a_2 b_3 - a_3 b_2) + (a_2 c_3 - a_3 c_2)$$

$$= \begin{vmatrix} a_2 & a_3 \\ b_2 & b_3 \end{vmatrix} + \begin{vmatrix} a_2 & a_3 \\ c_2 & c_3 \end{vmatrix},$$

which you should note is also the **i** component of $\mathbf{a} \times \mathbf{b} + \mathbf{a} \times \mathbf{c}$. Likewise, you can easily show that the **j** and **k** components also match, which establishes the result.

You should always keep in mind that vectors are specified by two things: magnitude and direction. We have now shown that the direction of $\mathbf{a} \times \mathbf{b}$ is orthogonal to both **a** and **b**. But, what about its magnitude? Certainly, we could always compute it on a case-by-case basis, but there is something more general (and more useful) that we can say.

Theorem 4.4

For nonzero vectors **a** and **b** in V_3, if θ is the angle between **a** and **b** ($0 \leq \theta \leq \pi$), then

$$\|\mathbf{a} \times \mathbf{b}\| = \|\mathbf{a}\| \|\mathbf{b}\| \sin\theta. \tag{4.4}$$

Proof

From (4.3), we get

$$\|\mathbf{a} \times \mathbf{b}\|^2 = [a_2 b_3 - a_3 b_2]^2 + [a_1 b_3 - a_3 b_1]^2 + [a_1 b_2 - a_2 b_1]^2$$

$$= a_2^2 b_3^2 - 2a_2 a_3 b_2 b_3 + a_3^2 b_2^2 + a_1^2 b_3^2 - 2a_1 a_3 b_1 b_3 + a_3^2 b_1^2$$
$$\quad + a_1^2 b_2^2 - 2a_1 a_2 b_1 b_2 + a_2^2 b_1^2$$

$$= (a_1^2 + a_2^2 + a_3^2)(b_1^2 + b_2^2 + b_3^2) - (a_1 b_1 + a_2 b_2 + a_3 b_3)^2$$

$$= \|\mathbf{a}\|^2 \|\mathbf{b}\|^2 - (\mathbf{a} \cdot \mathbf{b})^2$$

$$= \|\mathbf{a}\|^2 \|\mathbf{b}\|^2 - \|\mathbf{a}\|^2 \|\mathbf{b}\|^2 \cos^2\theta \qquad \text{From Theorem 3.2.}$$

$$= \|\mathbf{a}\|^2 \|\mathbf{b}\|^2 (1 - \cos^2\theta)$$

$$= \|\mathbf{a}\|^2 \|\mathbf{b}\|^2 \sin^2\theta.$$

Taking square roots, we get

$$\|\mathbf{a} \times \mathbf{b}\| = \|\mathbf{a}\| \|\mathbf{b}\| \sin\theta,$$

since $\sin\theta \geq 0$, for $0 \leq \theta \leq \pi$.

The following characterization of parallel vectors is an immediate consequence of Theorem 4.4.

Corollary 4.1

Two nonzero vectors $\mathbf{a}, \mathbf{b} \in V_3$ are parallel if and only if $\mathbf{a} \times \mathbf{b} = \mathbf{0}$.

Proof

Recall that **a** and **b** are parallel if and only if the angle θ between them is either 0 or π. In either case, $\sin\theta = 0$ and so, by Theorem 4.4,

$$\|\mathbf{a} \times \mathbf{b}\| = \|\mathbf{a}\|\,\|\mathbf{b}\|\sin\theta = \|\mathbf{a}\|\,\|\mathbf{b}\|(0) = 0.$$

The result then follows from the fact that the only vector with zero magnitude is the zero vector.

Theorem 4.4 also provides us with the following interesting geometric interpretation of the cross product. For any two nonzero vectors **a** and **b**, as long as **a** and **b** are not parallel, they form two adjacent sides of a parallelogram, as seen in Figure 10.30. Notice that the area of the parallelogram is given by the product of the base and the altitude. We have

$$\text{Area} = (\text{base})(\text{altitude})$$
$$= \|\mathbf{b}\|\,\|\mathbf{a}\|\sin\theta = \|\mathbf{a} \times \mathbf{b}\|, \qquad (4.5)$$

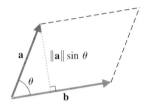

Figure 10.30

Parallelogram.

from Theorem 4.4. That is, the magnitude of the cross product of two vectors gives the area of the parallelogram with two adjacent sides formed by the vectors.

Example 4.4   Finding the Area of a Parallelogram Using the Cross Product

Find the area of the parallelogram with two adjacent sides formed by the vectors $\mathbf{a} = \langle 1, 2, 3\rangle$ and $\mathbf{b} = \langle 4, 5, 6\rangle$.

Solution   First notice that

$$\mathbf{a} \times \mathbf{b} = \begin{vmatrix} \mathbf{i} & \mathbf{j} & \mathbf{k} \\ 1 & 2 & 3 \\ 4 & 5 & 6 \end{vmatrix} = \mathbf{i}\begin{vmatrix} 2 & 3 \\ 5 & 6 \end{vmatrix} - \mathbf{j}\begin{vmatrix} 1 & 3 \\ 4 & 6 \end{vmatrix} + \mathbf{k}\begin{vmatrix} 1 & 2 \\ 4 & 5 \end{vmatrix} = \langle -3, 6, -3\rangle.$$

From (4.5), the area of the parallelogram is given by

$$\|\mathbf{a} \times \mathbf{b}\| = \|\langle -3, 6, -3\rangle\| = \sqrt{54} \approx 7.348.$$

We can also use Theorem 4.4 to find the distance from a point to a line in $\mathbb{R}^3$, as follows. Let d represent the distance from the point Q to the line through the points P and R. From elementary trigonometry, we have that

$$d = \|\overrightarrow{PQ}\|\sin\theta,$$

where θ is the angle between $\overrightarrow{PQ}$ and $\overrightarrow{PR}$ (see Figure 10.31). From (4.4), we have

$$\|\overrightarrow{PQ} \times \overrightarrow{PR}\| = \|\overrightarrow{PQ}\|\,\|\overrightarrow{PR}\|\sin\theta = \|\overrightarrow{PR}\|(d).$$

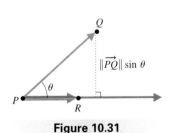

Figure 10.31

Distance from a point to a line.

Solving this for d, we get

$$d = \frac{\|\overrightarrow{PQ} \times \overrightarrow{PR}\|}{\|\overrightarrow{PR}\|}. \qquad (4.6)$$

Example 4.5 Finding the Distance from a Point to a Line

Find the distance from the point $Q(1, 2, 1)$ to the line through the points $P(2, 1, -3)$ and $R(2, -1, 3)$.

Solution First, we need position vectors corresponding to $\overrightarrow{PQ}$ and $\overrightarrow{PR}$. This is simple enough:

$$\overrightarrow{PQ} = \langle -1, 1, 4 \rangle \quad \text{and} \quad \overrightarrow{PR} = \langle 0, -2, 6 \rangle.$$

Notice that

$$\langle -1, 1, 4 \rangle \times \langle 0, -2, 6 \rangle = \begin{vmatrix} \mathbf{i} & \mathbf{j} & \mathbf{k} \\ -1 & 1 & 4 \\ 0 & -2 & 6 \end{vmatrix} = \langle 14, 6, 2 \rangle.$$

We then have from (4.6) that

$$d = \frac{\|\overrightarrow{PQ} \times \overrightarrow{PR}\|}{\|\overrightarrow{PR}\|} = \frac{\|\langle 14, 6, 2 \rangle\|}{\|\langle 0, -2, 6 \rangle\|} = \frac{\sqrt{236}}{\sqrt{40}} \approx 2.429.$$

For any three noncoplanar vectors $\mathbf{a}$, $\mathbf{b}$ and $\mathbf{c}$ (i.e., three vectors that do not lie in a single plane), consider the parallelepiped formed using the vectors as three adjacent edges (see Figure 10.32). Recall that the volume of such a solid is given by

$$\text{Volume} = (\text{Area of base})(\text{altitude}).$$

Further, since two adjacent sides of the base are formed by the vectors $\mathbf{a}$ and $\mathbf{b}$, we know that the area of the base is given by $\|\mathbf{a} \times \mathbf{b}\|$. Referring to Figure 10.32, notice that the altitude is given by

$$|\text{comp}_{\mathbf{a} \times \mathbf{b}}\, \mathbf{c}| = \frac{|\mathbf{c} \cdot (\mathbf{a} \times \mathbf{b})|}{\|\mathbf{a} \times \mathbf{b}\|},$$

from (3.7). The volume of the parallelepiped is then

$$\text{Volume} = \|\mathbf{a} \times \mathbf{b}\| \frac{|\mathbf{c} \cdot (\mathbf{a} \times \mathbf{b})|}{\|\mathbf{a} \times \mathbf{b}\|} = |\mathbf{c} \cdot (\mathbf{a} \times \mathbf{b})|.$$

The scalar $\mathbf{c} \cdot (\mathbf{a} \times \mathbf{b})$ is called the **scalar triple product** of the vectors $\mathbf{a}$, $\mathbf{b}$ and $\mathbf{c}$. As you can see from the following, you can evaluate the scalar triple product by computing a single

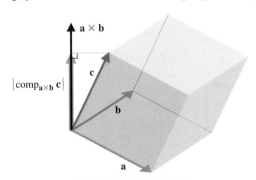

Figure 10.32
Parallelepiped formed by the
vectors $\mathbf{a}$, $\mathbf{b}$ and $\mathbf{c}$.

determinant. Note that for $\mathbf{a} = \langle a_1, a_2, a_3 \rangle$, $\mathbf{b} = \langle b_1, b_2, b_3 \rangle$ and $\mathbf{c} = \langle c_1, c_2, c_3 \rangle$, we have

$$\mathbf{c} \cdot (\mathbf{a} \times \mathbf{b}) = \mathbf{c} \cdot \begin{vmatrix} \mathbf{i} & \mathbf{j} & \mathbf{k} \\ a_1 & a_2 & a_3 \\ b_1 & b_2 & b_3 \end{vmatrix}$$

$$= \langle c_1, c_2, c_3 \rangle \cdot \left(\mathbf{i} \begin{vmatrix} a_2 & a_3 \\ b_2 & b_3 \end{vmatrix} - \mathbf{j} \begin{vmatrix} a_1 & a_3 \\ b_1 & b_3 \end{vmatrix} + \mathbf{k} \begin{vmatrix} a_1 & a_2 \\ b_1 & b_2 \end{vmatrix} \right)$$

$$= c_1 \begin{vmatrix} a_2 & a_3 \\ b_2 & b_3 \end{vmatrix} - c_2 \begin{vmatrix} a_1 & a_3 \\ b_1 & b_3 \end{vmatrix} + c_3 \begin{vmatrix} a_1 & a_2 \\ b_1 & b_2 \end{vmatrix}$$

$$= \begin{vmatrix} c_1 & c_2 & c_3 \\ a_1 & a_2 & a_3 \\ b_1 & b_2 & b_3 \end{vmatrix}. \tag{4.7}$$

Example 4.6 Finding the Volume of a Parallelepiped Using the Cross Product

Find the volume of the parallelepiped with three adjacent edges formed by the vectors $\mathbf{a} = \langle 1, 2, 3 \rangle$, $\mathbf{b} = \langle 4, 5, 6 \rangle$ and $\mathbf{c} = \langle 7, 8, 0 \rangle$.

Solution First, note that Volume $= |\mathbf{c} \cdot (\mathbf{a} \times \mathbf{b})|$. From (4.7), we have that

$$\mathbf{c} \cdot (\mathbf{a} \times \mathbf{b}) = \begin{vmatrix} 7 & 8 & 0 \\ 1 & 2 & 3 \\ 4 & 5 & 6 \end{vmatrix} = 7 \begin{vmatrix} 2 & 3 \\ 5 & 6 \end{vmatrix} - 8 \begin{vmatrix} 1 & 3 \\ 4 & 6 \end{vmatrix} + 0 \begin{vmatrix} 1 & 2 \\ 4 & 5 \end{vmatrix}$$

$$= 7(-3) - 8(-6) = 27.$$

So, the volume of the parallelepiped is Volume $= |\mathbf{c} \cdot (\mathbf{a} \times \mathbf{b})| = |27| = 27$.

Figure 10.33

Torque, τ.

Consider the action of a wrench on a bolt, as shown in Figure 10.33. In order to tighten the bolt, we apply a force $\mathbf{F}$ at the end of the handle, in the direction indicated in the figure. This force creates a **torque** τ acting along the axis of the bolt, drawing it in tight. Notice that the torque acts in the direction perpendicular to both $\mathbf{F}$ and the position vector $\mathbf{r}$ for the handle as indicated in Figure 10.33. In fact, using the right-hand rule, the torque acts in the same direction as $\mathbf{F} \times \mathbf{r}$ and physicists define the torque vector to be

$$\tau = \mathbf{F} \times \mathbf{r}.$$

In particular, this says that

$$\|\tau\| = \|\mathbf{F} \times \mathbf{r}\| = \|\mathbf{F}\| \, \|\mathbf{r}\| \sin \theta, \tag{4.8}$$

from (4.4). There are several observations we can make from this. First, this says that the farther away from the axis of the bolt we apply the force (i.e., the larger $\|\mathbf{r}\|$ is), the greater the magnitude of the torque. So, a longer wrench produces a greater torque, for a given amount of force applied. Second, to see which angle θ produces the largest torque, notice that $\sin \theta$ is maximized when $\theta = \frac{\pi}{2}$. We then have from (4.8) that the magnitude of the torque is maximized when $\theta = \frac{\pi}{2}$, that is, when the force vector $\mathbf{F}$ is orthogonal to the position vector $\mathbf{r}$. If you've ever spent any time using a wrench, this should fit well with your experience.

Example 4.7 Finding the Torque Applied by a Wrench

If you apply a force of magnitude 25 pounds at the end of a 15-inch-long wrench, at an angle of $\frac{\pi}{3}$ to the wrench, find the magnitude of the torque applied to the bolt. What is the maximum torque that a force of 25 pounds applied at that point can produce?

Solution   From (4.8), we have

$$\|\boldsymbol{\tau}\| = \|\mathbf{F}\| \|\mathbf{r}\| \sin\theta = 25\left(\frac{15}{12}\right)\sin\frac{\pi}{3}$$

$$= 25\left(\frac{15}{12}\right)\frac{\sqrt{3}}{2} \approx 27.1 \text{ foot-pounds.}$$

Further, the maximum torque is obtained when the angle between the wrench and the force vector is $\frac{\pi}{2}$. This would give us a maximum torque of

$$\|\boldsymbol{\tau}\| = \|\mathbf{F}\| \|\mathbf{r}\| \sin\theta = 25\left(\frac{15}{12}\right)(1) = 31.25 \text{ foot-pounds.}$$

Figure 10.34

Spinning ball.

Figure 10.35a

Backspin.

Figure 10.35b

Topspin.

In many sports, the action is at least partially influenced by the motion of a spinning ball. For instance, in baseball, batters must contend with pitchers' curveballs and in golf, players try to control their slice. In tennis, players hit shots with topspin, while in basketball, players improve their shooting by using backspin. The list goes on and on. These are all examples of the use of a force called the **Magnus force,** which we describe below.

Suppose that a ball is spinning with angular velocity ω, measured in radians per second (i.e., ω is the rate of change of the rotational angle). The ball spins about an axis, as shown in Figure 10.34. We define the spin vector $\mathbf{s}$ to have magnitude ω and direction parallel to the spin axis. You can use a right-hand rule to distinguish between the two directions parallel to the spin axis: curl the fingers of your right hand around the ball in the direction of the spin, and your thumb will point in the correct direction. Two examples are shown in Figures 10.35a and 10.35b. The motion of the ball disturbs the air through which it travels, creating a force acting on the ball called the **Magnus force.** The Magnus force $\mathbf{F}_m$ acting on a ball moving with velocity $\mathbf{v}$ and spin vector $\mathbf{s}$ is given by

$$\mathbf{F}_m = c(\mathbf{s} \times \mathbf{v}),$$

for some positive constant c. Suppose the balls in Figure 10.35a and Figure 10.35b are moving into the page and away from you. Using the usual sports terminology, the first ball has backspin and the second ball has topspin. Using the right-hand rule, we see that the Magnus force acting on the first ball acts in the upward direction, as shown in Figure 10.36a. This says that backspin (for example, on a basketball or golf shot) will produce an upward force which helps the ball land more softly than a ball with no spin. Again using the right-hand rule, you can see that the Magnus force acting on the second ball acts in the downward direction (see Figure 10.36b). This says that topspin (for example, on a tennis shot or baseball hit) will produce a downward force that causes the ball to drop to the ground more quickly than a ball with no spin.

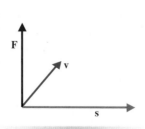

Figure 10.36a

Magnus force for a ball with backspin.

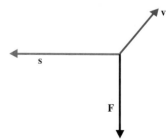

Figure 10.36b

Magnus force for a ball with topspin.

Figure 10.37a

Right-hand curveball.

Figure 10.37b

Right-hand golf shot.

Example 4.8   Finding the Direction of a Magnus Force

The balls shown in Figures 10.37a and 10.37b are moving into the page and away from you with spin as indicated. The first ball represents a right-handed baseball pitcher's curveball, while the second ball represents a right-handed golfer's shot. Determine the direction of the Magnus force and discuss the effects on the ball.

Solution   For the first ball, notice that the spin vector points up and to the left, so that $\mathbf{s} \times \mathbf{v}$ points down and to the left as shown in Figure 10.38a. Such a ball will curve to the left and drop faster than a ball that is not spinning, making it more difficult to hit. For the second ball, the spin vector points down and to the right, so $\mathbf{s} \times \mathbf{v}$ points up and to the right. Such a ball will move to the right (a "slice") and stay in the air longer than a ball that is not spinning (see Figure 10.38b).

Figure 10.38a

Magnus force for a right-handed curveball.

Figure 10.38b

Magnus force for a right-handed golf shot.

EXERCISES 10.4

1. In this chapter, we have developed several tests for geometric relationships. Briefly describe how to test if two vectors are (a) parallel; (b) perpendicular. Briefly describe how to test if (c) three points are colinear; (d) four points are coplanar.

2. The flip side of the problems in exercise 1 is to construct vectors with desired properties. Briefly describe how to construct a vector (a) parallel to a given vector; (b) perpendicular to a given vector. Given a vector, describe how to construct two other vectors such that the three vectors are mutually perpendicular.

3. Recall that torque is defined as $\boldsymbol{\tau} = \mathbf{F} \times \mathbf{r}$, where $\mathbf{F}$ is the force applied to the end of the handle and $\mathbf{r}$ is the position vector for the end of the handle. In example 4.7, how would the torque change if the force $\mathbf{F}$ were replaced with the force $-\mathbf{F}$? Answer both in mathematical terms and in physical terms.

4. Explain in geometric terms why $\mathbf{k} \times \mathbf{i} = \mathbf{j}$ and $\mathbf{k} \times \mathbf{j} = -\mathbf{i}$.

In exercises 5–8, compute the given determinant.

5. $\begin{vmatrix} 2 & 0 & -1 \\ 1 & 1 & 0 \\ -2 & -1 & 1 \end{vmatrix}$ 6. $\begin{vmatrix} 0 & 2 & -1 \\ 1 & -1 & 2 \\ 1 & 1 & 2 \end{vmatrix}$

7. $\begin{vmatrix} 2 & 3 & -1 \\ 0 & 1 & 0 \\ -2 & -1 & 3 \end{vmatrix}$ 8. $\begin{vmatrix} -2 & 2 & -1 \\ 0 & 3 & -2 \\ 0 & 1 & 2 \end{vmatrix}$

In exercises 9–16, compute the cross product $\mathbf{a} \times \mathbf{b}$.

9. $\mathbf{a} = \langle 1, 2, -1 \rangle$, $\mathbf{b} = \langle 1, 0, 2 \rangle$

10. $\mathbf{a} = \langle 3, 0, -1 \rangle$, $\mathbf{b} = \langle 1, 2, 2 \rangle$

11. $\mathbf{a} = \langle 0, 1, 4 \rangle$, $\mathbf{b} = \langle -1, 2, -1 \rangle$

12. $\mathbf{a} = \langle 2, -2, 0 \rangle$, $\mathbf{b} = \langle 3, 0, 1 \rangle$

13. $\mathbf{a} = \langle -2, -1, 4 \rangle$, $\mathbf{b} = \langle 1, 0, 0 \rangle$

14. $\mathbf{a} = \langle 2, 2, 1 \rangle$, $\mathbf{b} = \langle 1, 0, -3 \rangle$

15. $\mathbf{a} = 2\mathbf{i} - \mathbf{k}$, $\mathbf{b} = 4\mathbf{j} + \mathbf{k}$

16. $\mathbf{a} = -2\mathbf{i} + \mathbf{j} - 3\mathbf{k}$, $\mathbf{b} = 2\mathbf{j} - \mathbf{k}$

In exercises 17–22, find two unit vectors orthogonal to the two given vectors.

17. $\mathbf{a} = \langle 1, 0, 4 \rangle$, $\mathbf{b} = \langle 1, -4, 2 \rangle$

18. $\mathbf{a} = \langle 2, -2, 1 \rangle$, $\mathbf{b} = \langle 0, 0, -2 \rangle$

19. $\mathbf{a} = \langle 2, -1, 0 \rangle$, $\mathbf{b} = \langle 1, 0, 3 \rangle$

20. $\mathbf{a} = \langle 0, 2, 1 \rangle$, $\mathbf{b} = \langle 1, 0, -1 \rangle$

21. $\mathbf{a} = 3\mathbf{i} - \mathbf{j}$, $\mathbf{b} = 4\mathbf{j} + \mathbf{k}$

22. $\mathbf{a} = -2\mathbf{i} + 3\mathbf{j} - 3\mathbf{k}$, $\mathbf{b} = 2\mathbf{i} - \mathbf{k}$

In exercises 23–26, use the cross product to determine the angle between the vectors.

23. $\mathbf{a} = \langle 1, 0, 4 \rangle$, $\mathbf{b} = \langle 2, 0, 1 \rangle$

24. $\mathbf{a} = \langle 2, 2, 1 \rangle$, $\mathbf{b} = \langle 0, 0, 2 \rangle$

25. $\mathbf{a} = 3\mathbf{i} + \mathbf{k}$, $\mathbf{b} = 4\mathbf{j} + \mathbf{k}$

26. $\mathbf{a} = \mathbf{i} + 3\mathbf{j} + 3\mathbf{k}$, $\mathbf{b} = 2\mathbf{i} + \mathbf{j}$

In exercises 27–30, find the distance from the point Q to the given line.

27. $Q = (1, 2, 0)$, line through $(0, 1, 2)$ and $(3, 1, 1)$

28. $Q = (2, 0, 1)$, line through $(1, -2, 2)$ and $(3, 0, 2)$

29. $Q = (3, -2, 1)$, line through $(2, 1, -1)$ and $(1, 1, 1)$

30. $Q = (1, 3, 1)$, line through $(1, 3, -2)$ and $(1, 0, -2)$

31. If you apply a force of magnitude 20 pounds at the end of an 8-inch-long wrench at an angle of $\frac{\pi}{4}$ to the wrench, find the magnitude of the torque applied to the bolt.

32. If you apply a force of magnitude 40 pounds at the end of an 18-inch-long wrench at an angle of $\frac{\pi}{3}$ to the wrench, find the magnitude of the torque applied to the bolt.

33. If you apply a force of magnitude 30 pounds at the end of an 8-inch-long wrench at an angle of $\frac{\pi}{6}$ to the wrench, find the magnitude of the torque applied to the bolt.

34. If you apply a force of magnitude 30 pounds at the end of an 8-inch-long wrench at an angle of $\frac{\pi}{3}$ to the wrench, find the magnitude of the torque applied to the bolt.

In exercises 35–42, assume that the balls are moving into the page (and away from you) with the indicated spin. Determine the direction of the Magnus force.

35. 36.

37. 38.

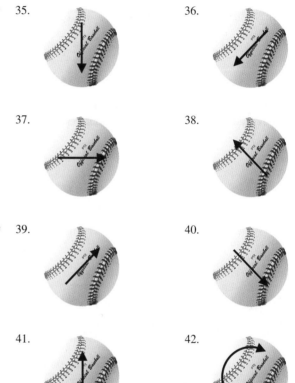

39. 40.

41. 42.

In exercises 43–56, a sports situation is described, with the typical ball spin shown in the indicated exercise above. Discuss the effects on the ball and how the game is affected.

43. Baseball overhand fastball, spin in exercise 35

44. Baseball overhand curveball, spin in exercise 41

45. Baseball right-handed "three-quarters" curveball, spin in exercise 39

46. Baseball right-handed "sidearm" fastball, spin in exercise 37

47. Tennis topspin groundstroke, spin in exercise 41

48. Tennis backspin groundstroke, spin in exercise 35

49. Football spiral pass, spin in exercise 42

50. Tennis right-handed slice serve, spin in exercise 39

51. Golf "pure" hit, spin in exercise 35

52. Golf right-handed "hook" shot, spin in exercise 40

53. Basketball free throw, spin in exercise 35

54. Basketball "finger-roll" shot, spin in exercise 41

55. Soccer left-footed "curl" kick, spin in exercise 36

56. Soccer right-footed "curl" kick, spin in exercise 40

In exercises 57–62, label each statement as true or false. If it is true, briefly explain why. If it is false, give a counterexample.

57. If $\mathbf{a} \times \mathbf{b} = \mathbf{a} \times \mathbf{c}$, then $\mathbf{b} = \mathbf{c}$.

58. $\mathbf{a} \times \mathbf{b} = -\mathbf{b} \times \mathbf{a}$

59. $\mathbf{a} \times \mathbf{a} = \|\mathbf{a}\|^2$

60. $\mathbf{a} \cdot (\mathbf{b} \times \mathbf{c}) = (\mathbf{a} \cdot \mathbf{b}) \times \mathbf{c}$

61. If the force is doubled, the torque doubles.

62. If the spin rate is doubled, the Magnus force is doubled.

In exercises 63–68, find the indicated area or volume.

63. Area of the parallelogram with two adjacent sides formed by $\langle 2, 3 \rangle$ and $\langle 1, 4 \rangle$

64. Area of the parallelogram with two adjacent sides formed by $\langle -2, 1 \rangle$ and $\langle 1, -3 \rangle$

65. Area of the triangle with vertices $(0, 0, 0)$, $(2, 3, -1)$ and $(3, -1, 4)$

66. Area of the triangle with vertices $(0, 0, 0)$, $(0, -2, 1)$ and $(1, -3, 0)$

67. Volume of the parallelepiped with three adjacent edges formed by $\langle 2, 1, 0 \rangle$, $\langle -1, 2, 0 \rangle$ and $\langle 1, 1, 2 \rangle$

68. Volume of the parallelepiped with three adjacent edges formed by $\langle 0, -1, 0 \rangle$, $\langle 0, 2, -1 \rangle$ and $\langle 1, 0, 2 \rangle$

In exercises 69–74, use geometry to identify the cross product (do not compute!).

69. $\mathbf{i} \times (\mathbf{j} \times \mathbf{k})$

70. $\mathbf{j} \times (\mathbf{j} \times \mathbf{k})$

71. $\mathbf{j} \times (\mathbf{j} \times \mathbf{i})$

72. $(\mathbf{j} \times \mathbf{i}) \times \mathbf{k}$

73. $\mathbf{i} \times (3\mathbf{k})$

74. $\mathbf{k} \times (2\mathbf{i})$

In exercises 75–78, use the parallelepiped volume formula to determine if the vectors are coplanar.

75. $\langle 2, 3, 1 \rangle$, $\langle 1, 0, 2 \rangle$ and $\langle 0, 3, -3 \rangle$

76. $\langle 1, -3, 1 \rangle$, $\langle 2, -1, 0 \rangle$ and $\langle 0, -5, 2 \rangle$

77. $\langle 1, 0, -2 \rangle$, $\langle 3, 0, 1 \rangle$ and $\langle 2, 1, 0 \rangle$

78. $\langle 1, 1, 2 \rangle$, $\langle 0, -1, 0 \rangle$ and $\langle 3, 2, 4 \rangle$

79. Show that $\|\mathbf{a} \times \mathbf{b}\|^2 = \|\mathbf{a}\|^2 \|\mathbf{b}\|^2 - (\mathbf{a} \cdot \mathbf{b})^2$.

80. Prove parts (ii), (iv), (v) and (vi) of Theorem 4.3.

81. Use the torque formula $\boldsymbol{\tau} = \mathbf{F} \times \mathbf{r}$ to explain the positioning of doorknobs. In particular, explain why the knob is placed as far as possible from the hinges and at a height that makes it possible for most people to push or pull on the door at a right angle to the door.

82. In the diagram, a foot applies a force $\mathbf{F}$ vertically to a bicycle pedal. Compute the torque on the sprocket in terms of θ and $\mathbf{F}$. Determine the angle θ at which the torque is maximized. When helping a young person to learn to ride a bicycle, most people rotate the sprocket so that the pedal sticks straight out to the front. Explain why this is helpful.

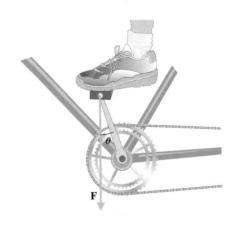

10.5 LINES AND PLANES IN SPACE

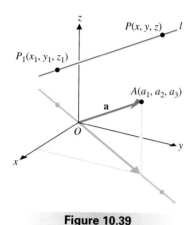

Figure 10.39
Line in space.

Think for a moment about how you specify a line in the xy-plane. Normally, you specify two points on the line (**any** two points will do) or you specify a single point on the line and its direction. Of course, in two dimensions, the direction is indicated by the **slope** of the line. The situation in three dimensions is not all that different. As you can imagine, specifying two points on a line will still determine the line. An alternative is to specify a single point on the line and its *direction*. But, how can you specify the direction of a line in three dimensions? Hopefully, you will think about vectors right away.

Let's look for the line that passes through the point $P_1(x_1, y_1, z_1)$ and that is parallel to the position vector $\mathbf{a} = \langle a_1, a_2, a_3 \rangle$ (see Figure 10.39). Notice that for any other point $P(x, y, z)$ on the line, the vector $\overrightarrow{P_1 P}$ must be parallel to $\mathbf{a}$. Further, recall that two vectors are parallel if and only if one is a scalar multiple of the other. This says that

$$\overrightarrow{P_1 P} = t\mathbf{a}, \tag{5.1}$$

for some scalar t. The line then consists of all points $P(x, y, z)$ for which (5.1) holds. Now, since

$$\overrightarrow{P_1 P} = \langle x - x_1, y - y_1, z - z_1 \rangle,$$

we have from (5.1) that

$$\langle x - x_1, y - y_1, z - z_1 \rangle = t\mathbf{a} = t\langle a_1, a_2, a_3 \rangle.$$

Finally, since two vectors are equal if and only if all of their components are equal, we get

Parametric equations of a line

$$\boxed{x - x_1 = a_1 t, \quad y - y_1 = a_2 t, \quad \text{and} \quad z - z_1 = a_3 t.} \tag{5.2}$$

We call (5.2) **parametric equations** for the line, where t is the **parameter.** As in the two-dimensional case, you should recognize that a line in space can be represented by many different sets of parametric equations. Provided none of a_1, a_2 or a_3 are zero, we can solve for the parameter in each of the three equations, to get

Symmetric equations of a line

$$\boxed{\frac{x - x_1}{a_1} = \frac{y - y_1}{a_2} = \frac{z - z_1}{a_3}.} \tag{5.3}$$

We refer to (5.3) as the **symmetric equations** of the line.

Example 5.1 Finding Equations of a Line Given a Point and a Vector

Find an equation of the line through the point $(1, 5, 2)$ and parallel to the vector $\langle 4, 3, 7 \rangle$. Also, determine where the line intersects the yz-plane.

Solution Notice that we can write down the equations immediately. From (5.2), parametric equations for the line are

$$x - 1 = 4t, \quad y - 5 = 3t \quad \text{and} \quad z - 2 = 7t.$$

From (5.3), the symmetric equations of the line are

$$\frac{x - 1}{4} = \frac{y - 5}{3} = \frac{z - 2}{7}. \tag{5.4}$$

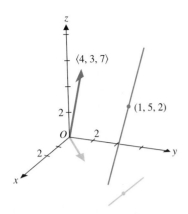

Figure 10.40

The line $x = 1 + 4t$, $y = 5 + 3t$, $z = 2 + 7t$.

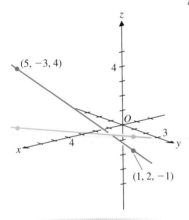

Figure 10.41

The line $\dfrac{x-1}{4} = \dfrac{y-2}{-5} = \dfrac{z+1}{5}$.

We show the graph of the line in Figure 10.40. Note that the line intersects the yz-plane where $x = 0$. Setting $x = 0$ in (5.4), we solve for y and z to obtain

$$y = \frac{17}{4} \quad \text{and} \quad z = \frac{1}{4}.$$

Alternatively, observe that we could solve $x - 1 = 4t$ for t (again where $x = 0$) and substitute this into the parametric equations for y and z. So, the line intersects the yz-plane at the point $\left(0, \frac{17}{4}, \frac{1}{4}\right)$.

■

Given two points, we can easily find the equations of the line passing through them, as in the following example.

Example 5.2   Finding Equations of a Line Given Two Points

Find an equation of the line passing through the points $P(1, 2, -1)$ and $Q(5, -3, 4)$.

Solution   First, we need to find a vector that is parallel to the given line. The obvious choice is

$$\overrightarrow{PQ} = \langle 5 - 1, -3 - 2, 4 - (-1) \rangle = \langle 4, -5, 5 \rangle.$$

Picking either point will give us equations for the line. Here, we use P, so that parametric equations for the line are

$$x - 1 = 4t, \quad y - 2 = -5t \quad \text{and} \quad z + 1 = 5t.$$

Similarly, the symmetric equations of the line are

$$\frac{x-1}{4} = \frac{y-2}{-5} = \frac{z+1}{5}.$$

We show the graph of the line in Figure 10.41.

■

Since we have specified a line by choosing a point on the line and a vector with the same direction, the following definition should be transparent.

Definition 5.1

Let l_1 and l_2 be two lines in $\mathbb{R}^3$, with parallel vectors $\mathbf{a}$ and $\mathbf{b}$, respectively, and let θ be the angle between $\mathbf{a}$ and $\mathbf{b}$.

(i)   The lines l_1 and l_2 are **parallel** whenever $\mathbf{a}$ and $\mathbf{b}$ are parallel.

(ii)   If l_1 and l_2 intersect, then

(a) the angle between l_1 and l_2 is θ and

(b) the lines l_1 and l_2 are **orthogonal** whenever $\mathbf{a}$ and $\mathbf{b}$ are orthogonal.

Recall that in two dimensions, two lines are either parallel or they must intersect. This is not true in three dimensions, as we see in the following example.

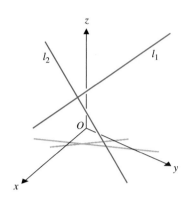

Figure 10.42
Skew lines.

| **Example 5.3** | Showing Two Lines Are Not Parallel but Do Not Intersect |

Show that the lines

$$l_1: x - 2 = -t, \quad y - 1 = 2t \quad \text{and} \quad z - 5 = 2t$$

and

$$l_2: x - 1 = s, \quad y - 2 = -s \quad \text{and} \quad z - 1 = 3s$$

are not parallel, yet do not intersect.

Solution   You will notice immediately that we have used different letters (t and s) as parameters for the two lines. In this setting, the parameter is a dummy variable, so the letter used is not significant. However, solving the first parametric equation of each line for the parameter in terms of x, we get

$$t = 2 - x \quad \text{and} \quad s = x - 1,$$

respectively. This says that the parameter represents something different in each line, so we must use different letters. Notice from the graph in Figure 10.42 that the lines are most certainly not parallel, but it is unclear whether or not they intersect. (Remember, the graph is a two-dimensional rendering of lines in three dimensions and so, while the two-dimensional lines drawn do intersect, it's unclear whether or not the three-dimensional lines that they represent intersect.)

You can read from the parametric equations that a vector parallel to l_1 is $\mathbf{a}_1 = \langle -1, 2, 2 \rangle$, while a vector parallel to l_2 is $\mathbf{a}_2 = \langle 1, -1, 3 \rangle$. Since $\mathbf{a}_1$ is not a scalar multiple of $\mathbf{a}_2$ (check this!), the vectors are not parallel and so, the lines l_1 and l_2 are not parallel. You might expect that the lines must then intersect, but let's take a careful look at this. The lines will intersect if there's a choice of the parameters s and t that produces the same point, that is, that produces the same values for all of x, y and z. Setting the x-values equal, we get

$$2 - t = 1 + s,$$

so that $s = 1 - t$. Setting the y-values equal and setting $s = 1 - t$, we get

$$1 + 2t = 2 - s = 2 - (1 - t) = 1 + t.$$

Solving this for t yields $t = 0$, which further implies that $s = 1$. Setting the z-components equal gives

$$5 + 2t = 3s + 1,$$

but this is not satisfied when $t = 0$ and $s = 1$. So, l_1 and l_2 are not parallel, yet do not intersect.

Definition 5.2
Nonparallel, nonintersecting lines are called **skew** lines.

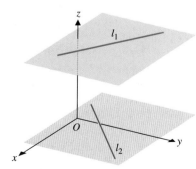

Figure 10.43
Skew lines.

Note that it's fairly easy to visualize skew lines. Draw two planes that are parallel and draw a line in each plane (so that it's completely contained in the plane). As long as the two lines are not parallel, these are skew lines (see Figure 10.43).

Planes in $\mathbb{R}^3$

Now, think about how you might specify a plane in space. What kind of information will be adequate to do this? As a simple example, how can you describe the yz-plane? You might observe that the yz-plane is the set of all points whose x-coordinate is zero. Unfortunately, this kind of description will only work for planes parallel to one of the three coordinate planes. (Why?) Alternatively, we can think of the yz-plane as a set of points in space such that every vector connecting two points in the set is orthogonal to **i**. But notice that there are many such planes. In fact, any plane parallel to the yz-plane satisfies this criterion (see Figure 10.44). In order to select the one that corresponds to the yz-plane, you need to select a point that it passes through (any one will do).

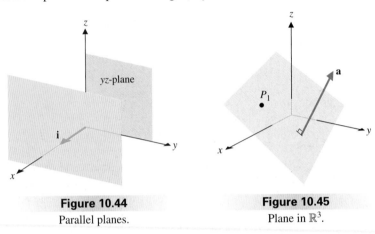

Figure 10.44
Parallel planes.

Figure 10.45
Plane in $\mathbb{R}^3$.

In general, a plane in space is determined by specifying a vector $\mathbf{a} = \langle a_1, a_2, a_3 \rangle$ that is **normal** to the plane (i.e., orthogonal to every vector lying in the plane) and a point $P_1(x_1, y_1, z_1)$ lying in the plane (see Figure 10.45). In order to find an equation of the plane, let $P(x, y, z)$ represent any point in the plane. Then, since P and P_1 are both points in the plane, the vector $\overrightarrow{P_1 P} = \langle x - x_1, y - y_1, z - z_1 \rangle$ lies in the plane and so, must be orthogonal to **a**. By Corollary 3.1, we have that

$$0 = \mathbf{a} \cdot \overrightarrow{P_1 P} = \langle a_1, a_2, a_3 \rangle \cdot \langle x - x_1, y - y_1, z - z_1 \rangle$$

or

Equation of a plane

$$0 = a_1(x - x_1) + a_2(y - y_1) + a_3(z - z_1). \qquad (5.5)$$

Equation (5.5) is an equation for the plane passing through the point (x_1, y_1, z_1) with normal vector $\langle a_1, a_2, a_3 \rangle$. As you can imagine, it's a simple matter to use this to find the equation of any particular plane. We illustrate this in the following example.

Example 5.4   The Equation of a Plane Given a Point and a Normal Vector

Find an equation of the plane containing the point $(1, 2, 3)$ with normal vector $\langle 4, 5, 6 \rangle$.

Solution   From (5.5), we have the equation

$$0 = 4(x - 1) + 5(y - 2) + 6(z - 3). \qquad (5.6)$$

To draw the plane, we locate three points lying in the plane. In this case, the simplest way to do this is to look at the intersections of the plane with each of the coordinate

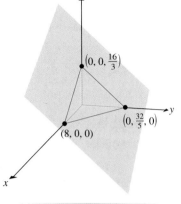

Figure 10.46

The plane through $(8, 0, 0)$, $\left(0, \frac{32}{5}, 0\right)$ and $\left(0, 0, \frac{16}{3}\right)$.

axes. When $y = z = 0$, we get from (5.6) that

$$0 = 4(x - 1) + 5(0 - 2) + 6(0 - 3) = 4x - 4 - 10 - 18,$$

so that $4x = 32$ or $x = 8$. The intersection of the plane with the x-axis is then the point $(8, 0, 0)$. Similarly, you can find the intersections of the plane with the y- and z-axes: $\left(0, \frac{32}{5}, 0\right)$ and $\left(0, 0, \frac{16}{3}\right)$, respectively. Using these three points, we can draw the plane seen in Figure 10.46. We start by drawing the triangle with vertices at the three points; the plane we want is the one containing this triangle. Notice that since the plane intersects all three of the coordinate axes, the portion of the plane in the first octant is the indicated triangle.

Note that if we expand out the expression in (5.5), we get

$$0 = a_1(x - x_1) + a_2(y - y_1) + a_3(z - z_1)$$
$$= a_1 x + a_2 y + a_3 z + \underbrace{(-a_1 x_1 - a_2 y_1 - a_3 z_1)}_{\text{constant}}.$$

We refer to this last equation as a **linear equation** in the three variables x, y and z. In particular, this says that every linear equation of the form

$$0 = ax + by + cz + d,$$

where a, b, c and d are constants is the equation of a plane with normal vector $\langle a, b, c \rangle$.

We observed earlier that three points determine a plane. But, how can you find an equation of a plane given only three points? If you are to use (5.5), you'll first need to find a normal vector. We can easily resolve this, as in the following example.

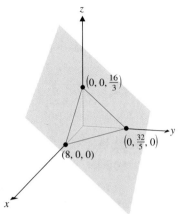

Figure 10.47

Plane containing three points.

Example 5.5 *Finding the Equation of a Plane Given Three Points*

Find the plane containing the three points $P(1, 2, 2)$, $Q(2, -1, 4)$ and $R(3, 5, -2)$.

Solution First, we'll need to find a vector normal to the plane (**any** one will do). Notice that two vectors lying in the plane are

$$\overrightarrow{PQ} = \langle 1, -3, 2 \rangle \quad \text{and} \quad \overrightarrow{QR} = \langle 1, 6, -6 \rangle.$$

Consequently, one vector orthogonal to both of $\overrightarrow{PQ}$ and $\overrightarrow{QR}$ is the cross product

$$\overrightarrow{PQ} \times \overrightarrow{QR} = \begin{vmatrix} \mathbf{i} & \mathbf{j} & \mathbf{k} \\ 1 & -3 & 2 \\ 1 & 6 & -6 \end{vmatrix} = \langle 6, 8, 9 \rangle.$$

Since $\overrightarrow{PQ}$ and $\overrightarrow{QR}$ are not parallel, $\overrightarrow{PQ} \times \overrightarrow{QR}$ must be orthogonal to the plane, as well. (Why is that?) From (5.5), an equation for the plane is then

$$0 = 6(x - 1) + 8(y - 2) + 9(z - 2).$$

In Figure 10.47, we show the triangle with vertices at the three points. The plane in question is the one containing the indicated triangle.

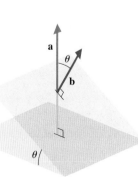

Figure 10.48

Angle between planes.

In three dimensions, two planes are either parallel or they intersect in a straight line. (Think about this some.) Suppose that two planes having normal vectors **a** and **b**, respectively, intersect. Then the angle between the planes is the same as the angle between **a** and **b** (see Figure 10.48). With this in mind, we say that the two planes are **parallel** whenever their normal vectors are parallel and the planes are **orthogonal** whenever their normal vectors are orthogonal.

| **Example 5.6** | The Equation of a Plane Given a Point and a Parallel Plane |

Find an equation for the plane through the point $(1, 4, -5)$ and parallel to the plane defined by $2x - 5y + 7z = 12$.

Solution   First, notice that a normal vector to the given plane is $\langle 2, -5, 7 \rangle$. Since the two planes are to be parallel, this vector is also normal to the new plane. From (5.5), we can write down the equation of the plane:

$$0 = 2(x - 1) - 5(y - 4) + 7(z + 5).$$

It's particularly easy to see that some planes are parallel to the coordinate planes.

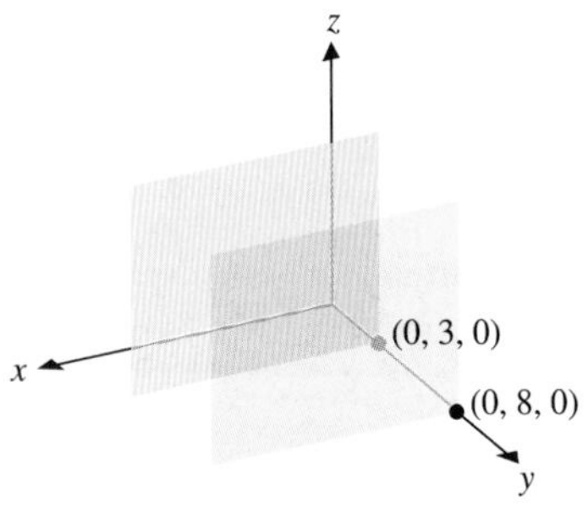

| **Example 5.7** | Drawing Some Simple Planes |

Draw the plane $y = 3$; draw the plane $y = 8$.

Solution   First, notice that both equations represent planes with the same normal vector, $\langle 0, 1, 0 \rangle = \mathbf{j}$. This says that the planes are both parallel to the xz-plane, the first one passing through the point $(0, 3, 0)$ and the second one passing through $(0, 8, 0)$, as seen in Figure 10.49.

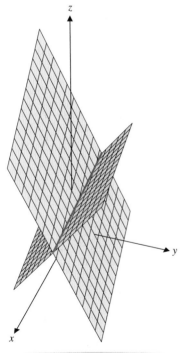

Figure 10.49
The planes $y = 3$ and $y = 8$.

You should recognize that the intersection of two nonparallel planes will be a line. (Think about this some!) In the following example, we see how to find an equation of the line of intersection.

| **Example 5.8** | Finding the Intersection of Two Planes |

Find the line of intersection of the planes: $x + 2y + z = 3$ and $x - 4y + 3z = 5$.

Solution   Solving both equations for x, we get

$$x = 3 - 2y - z \quad \text{and} \quad x = 5 + 4y - 3z. \qquad (5.7)$$

Setting these expressions for x equal gives us

$$3 - 2y - z = 5 + 4y - 3z.$$

Solving this for z gives us

$$2z = 6y + 2 \quad \text{or} \quad z = 3y + 1.$$

Returning to either equation in (5.7), we can solve for x (also in terms of y). We have

$$x = 3 - 2y - z = 3 - 2y - (3y + 1) = -5y + 2.$$

If we take y as our parameter (if you wish, let $y = t$), we have parametric equations for the line of intersection:

$$x = -5t + 2, \quad y = t \quad \text{and} \quad z = 3t + 1.$$

You can see the line of intersection in the computer-generated graph of the two planes seen in Figure 10.50.

Figure 10.50
Intersection of planes.

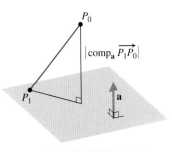

Figure 10.51

Distance from a point to a plane.

Suppose that we wanted to find the distance from the plane $ax + by + cz + d = 0$ to a point $P_0(x_0, y_0, z_0)$ not on the plane. Notice that the distance is measured along a line segment connecting the point to the plane that is orthogonal to the plane (see Figure 10.51). To compute this distance, pick any point $P_1(x_1, y_1, z_1)$ lying in the plane and let $\mathbf{a} = \langle a, b, c \rangle$ denote a vector normal to the plane. Next, notice from Figure 10.51 that the distance from P_0 to the plane is simply $|\text{comp}_{\mathbf{a}} \overrightarrow{P_1 P_0}|$. Also notice that

$$\overrightarrow{P_1 P_0} = \langle x_0 - x_1, y_0 - y_1, z_0 - z_1 \rangle.$$

From (3.7), we can now write the distance as

$$|\text{comp}_{\mathbf{a}} \overrightarrow{P_1 P_0}| = \left| \overrightarrow{P_1 P_0} \cdot \frac{\mathbf{a}}{\|\mathbf{a}\|} \right|$$

$$= \left| \langle x_0 - x_1, y_0 - y_1, z_0 - z_1 \rangle \cdot \frac{\langle a, b, c \rangle}{\|\langle a, b, c \rangle\|} \right|$$

$$= \frac{|a(x_0 - x_1) + b(y_0 - y_1) + c(z_0 - z_1)|}{\sqrt{a^2 + b^2 + c^2}}$$

$$= \frac{|ax_0 + by_0 + cz_0 - (ax_1 + by_1 + cz_1)|}{\sqrt{a^2 + b^2 + c^2}}$$

$$= \frac{|ax_0 + by_0 + cz_0 + d|}{\sqrt{a^2 + b^2 + c^2}}, \qquad (5.8)$$

since (x_1, y_1, z_1) lies in the plane and $ax + by + cz = -d$, for every point (x, y, z) in the plane.

Example 5.9 Finding the Distance between Parallel Planes

Find the distance between the parallel planes:

$$P_1 : 2x - 3y + z = 6$$

and

$$P_2 : 4x - 6y + 2z = 8.$$

Solution First, notice that the planes are parallel, since their normal vectors $\langle 2, -3, 1 \rangle$ and $\langle 4, -6, 2 \rangle$ are parallel. Next, notice that since the planes are parallel, the distance from the plane P_1 to every point in the plane P_2 is the same. So, pick any point in P_2, say $(0, 0, 4)$. (This is certainly convenient.) The distance d from the point $(0, 0, 4)$ to the plane P_1 is then given by (5.8) to be

$$d = \frac{|(2)(0) - (3)(0) + (1)(4) - 6|}{\sqrt{2^2 + 3^2 + 1^2}} = \frac{2}{\sqrt{14}}.$$

EXERCISES 10.5

1. Explain how to shift back and forth between the parametric and symmetric equations of a line. Describe one situation in which you would prefer to have parametric equations to work with, and one situation in which symmetric equations would be more convenient.

2. Lines and planes can both be specified with a point and a vector. Discuss the differences in the vectors used, and explain why the normal vector of the plane specifies a full plane while the direction vector of the line merely specifies a line.

3. Notice that if $c = 0$ in the general equation $ax + by + cz + d = 0$ of a plane, you have an equation that would describe a line in the xy-plane. Describe how this line relates to the plane.

4. Our hint about visualizing skew lines was to place the lines in parallel planes. Discuss whether every pair of skew lines must necessarily lie in parallel planes. (Hint: Discuss how the cross product of the direction vectors of the lines would relate to the parallel planes.)

In exercises 5–16, find (a) parametric equations and (b) symmetric equations of the line.

5. The line through $(1, 2, -3)$ and parallel to $\langle 2, -1, 4 \rangle$

6. The line through $(3, -2, 4)$ and parallel to $\langle 3, 2, -1 \rangle$

7. The line through $(2, 1, 3)$ and $(4, 0, 4)$

8. The line through $(0, 2, 1)$ and $(2, 0, 2)$

9. The line through $(1, 4, 1)$ and parallel to the line $x = 2 - 3t$, $y = 4$, $z = 6 + t$

10. The line through $(2, 0, -1)$ and parallel to the line $x = -3t$, $y = 2 - 4t$, $z = -6$

11. The line through $(3, 1, -1)$ and parallel to the line $\dfrac{x - 2}{3} = \dfrac{y + 1}{-4} = \dfrac{z}{2}$

12. The line through $(-1, 0, 0)$ and parallel to the line $\dfrac{x + 1}{-2} = \dfrac{y}{3} = z - 2$

13. The line through $(2, 0, 1)$ and perpendicular to both $\langle 1, 0, 2 \rangle$ and $\langle 0, 2, 1 \rangle$

14. The line through $(-3, 1, 0)$ and perpendicular to both $\langle 0, -3, 1 \rangle$ and $\langle 4, 2, -1 \rangle$

15. The line through $(1, 2, -1)$ and normal to the plane $2x - y + 3z = 12$

16. The line through $(0, -2, 1)$ and normal to the plane $y + 3z = 4$

In exercises 17–22, state if the lines are parallel or perpendicular or find the angle between the lines.

17. $\begin{cases} x = 1 - 3t \\ y = 2 + 4t \\ z = -6 + t \end{cases}$ and $\begin{cases} x = 1 + 2s \\ y = 2 - 2s \\ z = -6 + s \end{cases}$

18. $\begin{cases} x = 4 - 2t \\ y = 3t \\ z = -1 + 2t \end{cases}$ and $\begin{cases} x = 4 + s \\ y = -2s \\ z = -1 + 3s \end{cases}$

19. $\begin{cases} x = 1 + 2t \\ y = 3 \\ z = -1 + t \end{cases}$ and $\begin{cases} x = 2 - s \\ y = 10 + 5s \\ z = 3 + 2s \end{cases}$

20. $\begin{cases} x = 1 - 2t \\ y = 2t \\ z = 5 - t \end{cases}$ and $\begin{cases} x = 3 + 2s \\ y = -2 - 2s \\ z = 6 + s \end{cases}$

21. $\begin{cases} x = -1 + 2t \\ y = 3 + 4t \\ z = -6t \end{cases}$ and $\begin{cases} x = 3 - s \\ y = 1 - 2s \\ z = 3s \end{cases}$

22. $\begin{cases} x = 3 - t \\ y = 4 \\ z = -2 + 2t \end{cases}$ and $\begin{cases} x = 1 + 2s \\ y = 7 - 3s \\ z = -3 + s \end{cases}$

In exercises 23–26, determine if the lines are parallel, skew or intersect.

23. $\begin{cases} x = 4 + t \\ y = 2 \\ z = 3 + 2t \end{cases}$ and $\begin{cases} x = 2 + 2s \\ y = 2s \\ z = -1 + 4s \end{cases}$

24. $\begin{cases} x = 3 + t \\ y = 3 + 3t \\ z = 4 - t \end{cases}$ and $\begin{cases} x = 2 - s \\ y = 1 - 2s \\ z = 6 + 2s \end{cases}$

25. $\begin{cases} x = 1 + 2t \\ y = 3 \\ z = -1 - 4t \end{cases}$ and $\begin{cases} x = 2 - s \\ y = 2 \\ z = 3 + 2s \end{cases}$

26. $\begin{cases} x = 1 - 2t \\ y = 2t \\ z = 5 - t \end{cases}$ and $\begin{cases} x = 3 + 2s \\ y = -2 \\ z = 3 + 2s \end{cases}$

In exercises 27–40, find an equation of the given plane.

27. The plane containing the point $(1, 3, 2)$ with normal vector $\langle 2, -1, 5 \rangle$

28. The plane containing the point $(-2, 0, 3)$ with normal vector $\langle 4, 3, -2 \rangle$

29. The plane containing the point $(-2, 1, 0)$ with normal vector $\langle -3, 0, 2 \rangle$

30. The plane containing the point $(0, -4, 1)$ with normal vector $\langle 0, 4, 2 \rangle$

31. The plane containing the points $(2, 0, 3)$, $(1, 1, 0)$ and $(3, 2, -1)$

32. The plane containing the points $(1, -2, 1)$, $(2, -1, 0)$ and $(3, -2, 2)$

33. The plane containing the points $(-2, 2, 0)$, $(-2, 3, 2)$ and $(1, 2, 2)$

34. The plane containing the points $(3, -2, 1)$, $(1, -2, 2)$ and $(3, 2, 2)$

35. The plane containing the point $(2, 1, -1)$ and parallel to the plane $3x - y + 2z = 1$

36. The plane containing the point $(3, -2, 1)$ and parallel to the plane $x + 3y - 4z = 2$

37. The plane containing the point $(0, -2, -1)$ and parallel to the plane $-2x + 4y = 3$

38. The plane containing the point $(3, 1, 0)$ and parallel to the plane $-3x - 3y + 2z = 4$

39. The plane containing the point $(1, 2, 1)$ and perpendicular to the planes $x + y = 2$ and $2x + y - z = 1$

40. The plane containing the point $(3, 0, -1)$ and perpendicular to the planes $x + 2y - z = 2$ and $2x - z = 1$

In exercises 41–52, sketch the given plane.

41. $x + y + z = 4$ 42. $2x - y + 4z = 4$

43. $3x + 6y - z = 6$ 44. $2x + y + 3z = 6$

45. $x = 4$ 46. $y = 3$

47. $z = 2$ 48. $y = -3$

49. $x + y = 1$ 50. $2x - z = 2$

51. $y = x + 2$ 52. $z = -2x + 1$

In exercises 53–56, find the intersection of the planes.

53. $2x - y - z = 4$ and $3x - 2y + z = 0$

54. $3x + y - z = 2$ and $2x - 3y + z = -1$

55. $3x + 4y = 1$ and $x + y - z = 3$

56. $x - 2y + z = 2$ and $x + 3y - 2z = 0$

In exercises 57–62, find the distance between the given objects.

57. The point $(2, 0, 1)$ and the plane $2x - y + 2z = 4$

58. The point $(1, 3, 0)$ and the plane $3x + y - 5z = 2$

59. The point $(2, -1, -1)$ and the plane $x - y + z = 4$

60. The point $(0, -1, 1)$ and the plane $2x - 3y = 2$

61. The planes $2x - y - z = 1$ and $2x - y - z = 4$

62. The planes $x + 3y - 2z = 3$ and $x + 3y - 2z = 1$

63. Show that the distance between planes $ax + by + cz = d_1$ and $ax + by + cz = d_2$ is given by $\dfrac{|d_2 - d_1|}{\sqrt{a^2 + b^2 + c^2}}$.

64. Suppose that $\langle 2, 1, 3 \rangle$ is a normal vector for a plane containing the point $(2, -3, 4)$. Show that an equation of the plane is $2x + y + 3z = 13$. Explain why another normal vector for this plane is $\langle -4, -2, -6 \rangle$. Use this normal vector to find an equation of the plane and show that the equation reduces to the same equation, $2x + y + 3z = 13$.

65. Find an equation of the plane containing the lines
$$\begin{cases} x = 4 + t \\ y = 2 \\ z = 3 + 2t \end{cases} \text{ and } \begin{cases} x = 2 + 2s \\ y = 2s \\ z = -1 + 4s \end{cases}.$$

66. Find an equation of the plane containing the lines
$$\begin{cases} x = 1 - t \\ y = 2 + 3t \\ z = 2t \end{cases} \text{ and } \begin{cases} x = 1 - s \\ y = 5 \\ z = 4 - 2s \end{cases}.$$

67. Suppose two airplanes fly paths described by the parametric equations $P_1 : \begin{cases} x = 3 \\ y = 6 - 2t \\ z = 3t + 1 \end{cases}$ and $P_2 : \begin{cases} x = 1 + 2s \\ y = 3 + s \\ z = 2 + 2s \end{cases}$. Describe the shape of the flight paths. Determine whether the paths intersect. Determine if the planes collide.

68. Compare the equations that we have developed for the distance between a (two-dimensional) point and a line and for a (three-dimensional) point and a plane. Based on these equations, hypothesize a formula for the distance between the (four-dimensional) point (x_1, y_1, z_1, w_1) and the hyperplane $ax + by + cz + dw + e = 0$.

69. In this exercise, we will explore the geometrical object determined by the parametric equations $\begin{cases} x = 2s + 3t \\ y = 3s + 2t \\ z = s + t \end{cases}$. Given that there are two parameters, what dimension do you expect the object to have? Given that the individual parametric equations are linear, what do you expect the object to be? Show that the points $(0, 0, 0)$, $(2, 3, 1)$ and $(3, 2, 1)$ are on the object. Find an equation of the plane containing these three points. Substitute in the equations for x, y and z and show that the object lies in the plane. Argue that the object is, in fact, the entire plane.

10.6 SURFACES IN SPACE

In Chapter 3, you expended considerable effort learning how to use the calculus to draw graphs of a wide range of functions in two dimensions. Now that we have discussed lines and planes in $\mathbb{R}^3$, we continue our graphical development by drawing more complicated objects in three dimensions. Don't expect a general theory like we developed for two-dimensional graphs. Drawing curves and surfaces in three dimensions by hand or correctly interpreting computer-generated graphics is something of an art. After all, you must draw a two-dimensional image that somehow represents an object in three dimensions. Our goal here is not to produce artists, but rather to leave you with the ability to deal with a small group of surfaces in three dimensions. For our presentation over the next several chapters, you will want to have at your disposal a small number of familiar surfaces. You will need to recognize these when you see them and have a reasonable facility for drawing a picture by hand. We also urge you to learn to produce and interpret computer-generated graphs. Follow our hints carefully and work **lots** of problems. In numerous exercises in the chapters that follow, taking a few extra minutes to draw a better graph will often result in a huge savings of time and effort.

Cylindrical Surfaces

We begin with a simple type of three-dimensional surface. When you see the word *cylinder*, you probably think of a right circular cylinder. For instance, consider the graph of the equation $x^2 + y^2 = 9$ in **three** dimensions. Your first reaction might be to say that this is the equation for a circle, but you'd only be partly correct. The graph of $x^2 + y^2 = 9$ in **two** dimensions is the circle of radius 3, centered at the origin, but what about its graph in **three** dimensions? Consider the intersection of the surface with the plane $z = k$, for some constant k. Since the equation has no z's in it, the intersection with every such plane (called the **trace** of the surface in the plane $z = k$) is the same: a circle of radius 3, centered at the origin. Think about it: whatever this three-dimensional surface is, its intersection with every plane parallel to the xy-plane is a circle of radius 3, centered at the origin. This describes a right circular cylinder, in this case one of radius 3, whose axis is the z-axis (see Figure 10.52).

More generally, the term **cylinder** is used to refer to any surface whose traces in every plane parallel to a given plane are the same. With this definition, many surfaces qualify as cylinders.

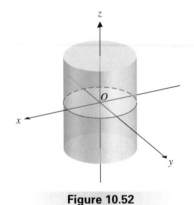

Figure 10.52
Right circular cylinder.

Example 6.1 Sketching a Surface

Draw a graph of the surface $z = y^2$ in $\mathbb{R}^3$.

Solution Notice that since there are no x's in the equation, the trace of the graph in the plane $x = k$ is the same for every k. This is then a cylinder whose trace in every plane parallel to the yz-plane is the parabola $z = y^2$. To draw this, we first draw the trace in the yz-plane (see Figure 10.53a) and then make several copies of the trace, locating the vertices at various points along the x-axis and finally, connect the traces with lines parallel to the x-axis to give the drawing its three-dimensional look (see Figure 10.53b). A computer-generated wireframe graph of the same surface is seen in Figure 10.53c. Notice that the wireframe consists of numerous traces for fixed values of x or y.

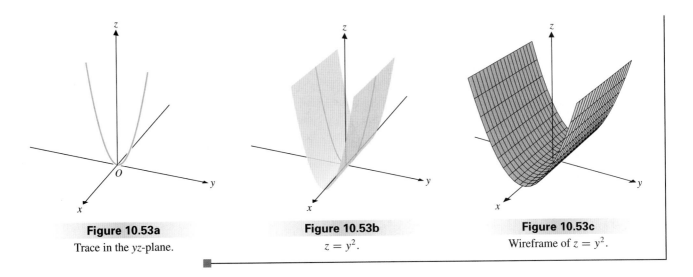

Figure 10.53a	**Figure 10.53b**	**Figure 10.53c**
Trace in the yz-plane.	$z = y^2$.	Wireframe of $z = y^2$.

Example 6.2   Sketching an Unusual Cylinder

Draw a graph of the surface $z = \sin x$ in $\mathbb{R}^3$.

Solution   Notice that once again, one of the variables is missing. In this case, there are no y's and so, traces of the surface in any plane parallel to the xz-plane are the same. They all look like the two-dimensional graph of $z = \sin x$. We draw one of these in the xz-plane and then make copies in planes parallel to the xz-plane, finally connecting the endpoints with lines parallel to the y-axis (see Figure 10.54a). In Figure 10.54b, we show a computer-generated wireframe plot of the same surface. In this case, the cylinder looks like a plane with ripples in it.

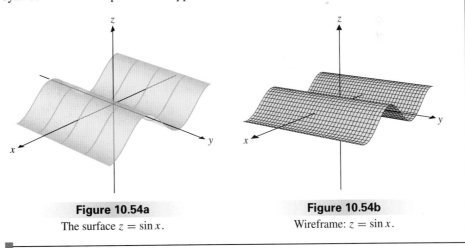

Figure 10.54a	**Figure 10.54b**
The surface $z = \sin x$.	Wireframe: $z = \sin x$.

Quadric Surfaces

The graph of the equation

$$ax^2 + by^2 + cz^2 + dxy + eyz + fxz + gx + hy + jz + k = 0$$

in three-dimensional space (where $a, b, c, d, e, f, g, h, j$ and k are all constants and at least one of a, b, c, d, e or f is nonzero) is referred to as a **quadric surface.**

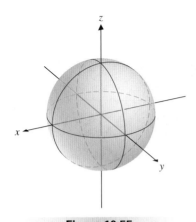

Figure 10.55

Sphere.

The most familiar quadric surface is the **sphere:**

$$(x - a)^2 + (y - b)^2 + (z - c)^2 = r^2$$

of radius r centered at the point (a, b, c). To draw the sphere centered at $(0, 0, 0)$, first draw a circle of radius r, centered at the origin in the yz-plane. Then, to give the surface its three-dimensional look, draw circles of radius r centered at the origin, in both the xz- and xy-planes, as in Figure 10.55. Note that due to the perspective, these circles will look like ellipses and will only be partially visible (we indicate the hidden parts of the circles with dashed lines).

A generalization of the sphere is the **ellipsoid:**

$$\frac{(x - a)^2}{d^2} + \frac{(y - b)^2}{e^2} + \frac{(z - c)^2}{f^2} = 1.$$

(Notice that when $d = e = f$, the surface is a sphere.)

Example 6.3   Sketching an Ellipsoid

Graph the ellipsoid

$$\frac{x^2}{1} + \frac{y^2}{4} + \frac{z^2}{9} = 1.$$

Solution   To get an idea of what the graph looks like, first draw its traces in the three coordinate planes. (In general, you may need to look at the traces in planes parallel to the three coordinate planes, but the traces in the three coordinate planes will suffice, here.) In the yz-plane, $x = 0$, so we have the ellipse

$$\frac{y^2}{4} + \frac{z^2}{9} = 1,$$

which we graph in Figure 10.56a. Next, add to Figure 10.56a the traces in the xy- and xz-planes. These are

$$\frac{x^2}{1} + \frac{y^2}{4} = 1 \quad \text{and} \quad \frac{x^2}{1} + \frac{z^2}{9} = 1,$$

respectively, and are both ellipses (see Figure 10.56b).

CASs have the capability of plotting functions of several variables in three dimensions. Many graphing calculators with three-dimensional plotting capabilities only

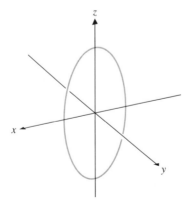

Figure 10.56a

Ellipse in yz-plane.

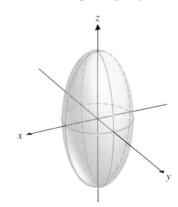

Figure 10.56b

Ellipsoid.

produce three-dimensional plots when given z as a function of x and y. For the problem at hand, notice that we can solve for z and plot the two functions $z = 3\sqrt{1 - x^2 - \frac{y^2}{4}}$ and $z = -3\sqrt{1 - x^2 - \frac{y^2}{4}}$ to obtain the graph of the surface. Observe that the wireframe graph in Figure 10.56c is not particularly smooth and appears to have some gaps. To correctly interpret such a graph, you must mentally fill in the gaps. This requires an understanding of how the graph should look, which we obtained drawing Figure 10.56b.

As an alternative, many CASs enable you to graph the equation $x^2 + \dfrac{y^2}{4} + \dfrac{z^2}{9} = 1$ using **implicit plot** mode. In this mode, the CAS numerically solves the equation for the value of z corresponding to each one of a large number of sample values of x and y and plots the resulting points. The graph obtained in Figure 10.56d is an improvement over Figure 10.56c, but doesn't show the traces that we used to construct Figure 10.56b.

The best option, when available, is often a **parametric plot.** In three dimensions, this involves writing each of the three variables x, y and z in terms of two parameters, with the resulting surface plotted by plotting points corresponding to a sample of values of the two parameters. (A more extensive discussion of the mathematics of parametric surfaces is given in section 14.6.) As we develop in the exercises, parametric equations for the ellipsoid are $x = \sin s \cos t$, $y = 2 \sin s \sin t$ and $z = 3 \cos s$, with the parameters taken to be in the intervals $0 \le s \le 2\pi$ and $0 \le t \le 2\pi$. Notice how Figure 10.56e shows a nice smooth plot and clearly shows the elliptical traces.

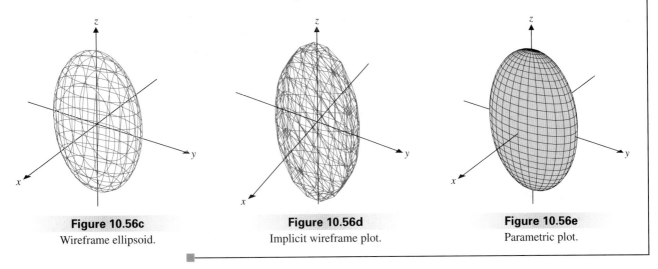

Figure 10.56c
Wireframe ellipsoid.

Figure 10.56d
Implicit wireframe plot.

Figure 10.56e
Parametric plot.

Figure 10.57a
Traces.

Example 6.4   Sketching a Paraboloid

Draw a graph of the quadric surface

$$x^2 + y^2 = z.$$

Solution   To get an idea of what the graph looks like, first draw its traces in the three coordinate planes. In the yz-plane, we have $x = 0$ and so $y^2 = z$ (a parabola). In the xz-plane, we have $y = 0$ and so, $x^2 = z$ (a parabola). In the xy-plane, we have $z = 0$ and so, $x^2 + y^2 = 0$ (a point—the origin). We sketch the traces in Figure 10.57a. Finally, since the trace in the xy-plane is just a point, we consider the traces in the planes $z = k$ (for $k > 0$). Notice that these are the circles $x^2 + y^2 = k$, where for larger values of z (i.e., larger values of k), we get circles of larger radius. We sketch the surface in

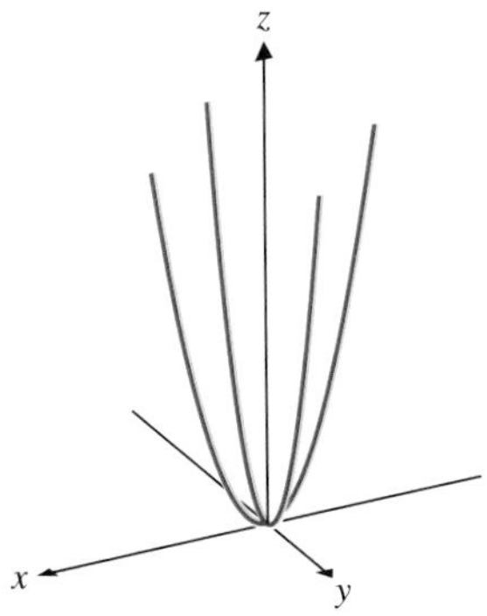

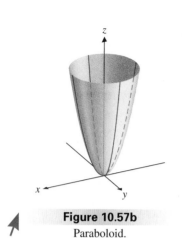

Figure 10.57b

Paraboloid.

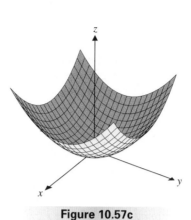

Figure 10.57c

Wireframe paraboloid.

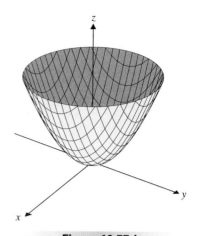

Figure 10.57d

Wireframe paraboloid for
$0 \le z \le 15$.

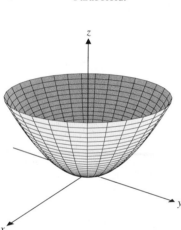

Figure 10.57e

Parametric plot paraboloid.

Figure 10.57b. Such surfaces are called **paraboloids** and since the traces in planes parallel to the xy-plane are circles, this is called a **circular paraboloid.**

Graphing utilities with three-dimensional capabilities generally produce a graph like Figure 10.57c for $z = x^2 + y^2$. Notice that the parabolic traces are visible, but not the circular cross sections we drew in Figure 10.57b. The four peaks visible in Figure 10.57c are due to the rectangular domain used for the plot (in this case, $-5 \le x \le 5$ and $-5 \le y \le 5$).

An improvement to this can be made by restricting the range of z-values. With $0 \le z \le 15$, you can clearly see the circular cross section in the plane $z = 15$ in Figure 10.57d.

As in example 6.3, a parametric surface plot is even better. Here, we have $x = s \cos t$, $y = s \sin t$ and $z = s^2$ with $0 \le s \le 5$ and $0 \le t \le 2\pi$. Figure 10.57e clearly shows the circular cross sections in the planes $z = k$, for $k > 0$. ■

Notice that in each of the last several examples, we have had to use some thought to produce computer-generated graphs that adequately show the important features of the given quadric surface. We want to encourage you to use your graphing calculator or CAS for drawing three-dimensional plots, because computer graphics are powerful tools for visualization and problem solving. However, be aware that you will need a basic understanding of the geometry of quadric surfaces to effectively produce and interpret computer-generated graphs.

Example 6.5   **Sketching an Elliptic Cone**

Draw a graph of the quadric surface

$$x^2 + \frac{y^2}{4} = z^2.$$

Solution   Be careful not to jump to conclusions. While this equation may look a lot like that of an ellipsoid, there is a significant difference. (Look where the z^2 term is!) Again, we start by looking at the traces in the coordinate planes. For the yz-plane, we have $x = 0$ and so, $\frac{y^2}{4} = z^2$ or $y^2 = 4z^2$, so that $y = \pm 2z$. That is, the trace is a pair of

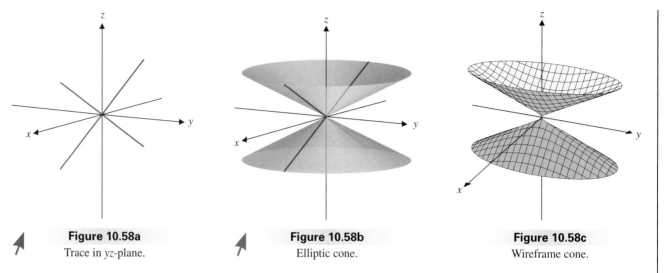

Figure 10.58a
Trace in yz-plane.

Figure 10.58b
Elliptic cone.

Figure 10.58c
Wireframe cone.

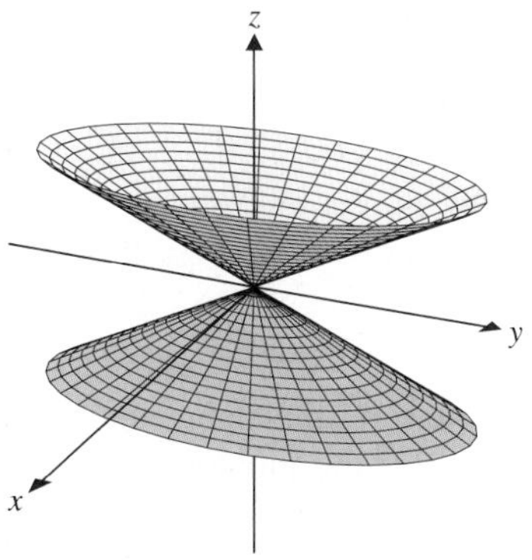

Figure 10.58d
Parametric plot.

lines: $y = 2z$ and $y = -2z$. We show these in Figure 10.58a. Likewise, the trace in the xz-plane is a pair of lines: $x = \pm z$. The trace in the xy-plane is simply the origin. (Why?) Finally, the traces in the planes $z = k$ $(k \neq 0)$, parallel to the xy-plane are the ellipses: $x^2 + \frac{y^2}{4} = k^2$. Adding these to the drawing gives us the double-cone seen in Figure 10.58b.

Since the traces in planes parallel to the xy-plane are ellipses, we refer to this as an **elliptic cone**.

Notice that one way to plot this with a CAS is to graph the two functions $z = \sqrt{x^2 + \frac{y^2}{4}}$ and $z = -\sqrt{x^2 + \frac{y^2}{4}}$. In Figure 10.58c, we restrict the z-range to $-10 \leq z \leq 10$ to show the circular cross sections. Notice that this plot shows a gap between the two halves of the cone. If you have drawn Figure 10.58b yourself, this plotting deficiency won't fool you. Alternatively, the parametric plot shown in Figure 10.58d, with $x = \sqrt{s^2} \cos t$, $y = 2\sqrt{s^2} \sin t$ and $z = s$ with $-5 \leq s \leq 5$ and $0 \leq t \leq 2\pi$, shows the full cone with its circular and linear traces.

Example 6.6 Sketching a Hyperboloid of One Sheet

Draw a graph of the quadric surface

$$\frac{x^2}{4} + y^2 - \frac{z^2}{2} = 1.$$

Solution   The traces in the coordinate planes are as follows:

$$yz\text{-plane } (x = 0): y^2 - \frac{z^2}{2} = 1 \text{ (hyperbola)}$$

(see Figure 10.59a on the following page),

$$xy\text{-plane } (z = 0): \frac{x^2}{4} + y^2 = 1 \text{ (ellipse)}$$

and

$$xz\text{-plane } (y = 0): \frac{x^2}{4} - \frac{z^2}{2} = 1 \text{ (hyperbola)}.$$

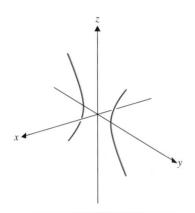

Figure 10.59a

Trace in yz-plane.

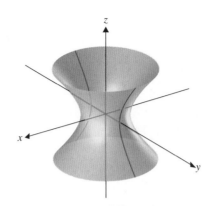

Figure 10.59b

Hyperboloid of one sheet.

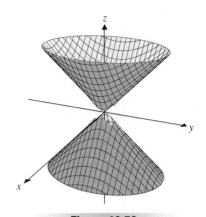

Figure 10.59c

Wireframe hyperboloid.

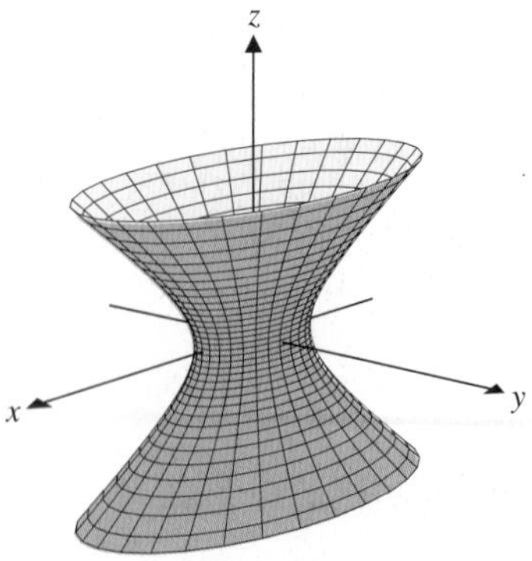

Figure 10.59d

Parametric plot.

Further, notice that the trace of the surface in each plane $z = k$ (parallel to the xy-plane) is also an ellipse:

$$\frac{x^2}{4} + y^2 = \frac{k^2}{2} + 1.$$

Finally, observe that the larger k is, the larger the axes of the ellipses are. Adding this information to Figure 10.59a, we draw the surface seen in Figure 10.59b. We call this surface a **hyperboloid of one sheet.**

To plot this with a CAS, you could graph the two functions $z = \sqrt{2\left(\frac{x^2}{4} + y^2 - 1\right)}$ and $z = -\sqrt{2\left(\frac{x^2}{4} + y^2 - 1\right)}$. (See Figure 10.59c, where we have restricted the z-range to $-10 \le z \le 10$, to show the circular cross sections.) Notice that this plot looks more like a cone than the hyperboloid in Figure 10.59b. If you have drawn Figure 10.59b yourself, this plotting problem won't fool you.

Alternatively, the parametric plot seen in Figure 10.59d, with $x = 2\cos s \cosh t$, $y = \sin s \cosh t$ and $z = \sqrt{2} \sinh t$, with $0 \le s \le 2\pi$ and $-5 \le t \le 5$, shows the full hyperboloid with its circular and hyperbolic traces.

Example 6.7   Sketching a Hyperboloid of Two Sheets

Draw a graph of the quadric surface

$$\frac{x^2}{4} - y^2 - \frac{z^2}{2} = 1.$$

Solution   First, notice that this is the same equation as in example 6.6, except for the sign of the y-term. As we have done before, we first look at the traces in the three coordinate planes. The trace in the yz-plane ($x = 0$) is defined by

$$-y^2 - \frac{z^2}{2} = 1.$$

Since it is clearly impossible for two negative numbers to add up to something positive, this is a contradiction and there is no trace in the yz-plane. That is, the surface does not intersect the yz-plane. The traces in the other two coordinate planes are as follows:

$$xy\text{-plane } (z = 0): \frac{x^2}{4} - y^2 = 1 \text{ (hyperbola)}$$

and

$$xz\text{-plane } (y = 0): \frac{x^2}{4} - \frac{z^2}{2} = 1 \text{ (hyperbola)}.$$

We show these traces in Figure 10.60a. Finally, notice that for $x = k$, we have that

$$y^2 + \frac{z^2}{2} = \frac{k^2}{4} - 1,$$

so that the traces in the plane $x = k$ are ellipses for $k^2 > 4$. It is important to notice here that if $k^2 < 4$, the equation $y^2 + \frac{z^2}{9} = \frac{k^2}{4} - 1$ has no solution. (Why is that?) So, if $-2 < k < 2$, the surface has no trace at all in the plane $x = k$, leaving a gap which separates the hyperbola into two *sheets*. Putting this all together, we have the surface seen in Figure 10.60b. We call this surface a **hyperboloid of two sheets.**

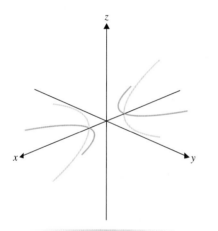

Figure 10.60a

Traces in xy- and xz-planes.

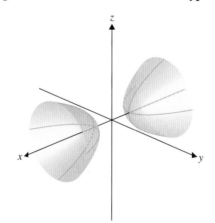

Figure 10.60b

Hyperboloid of two sheets.

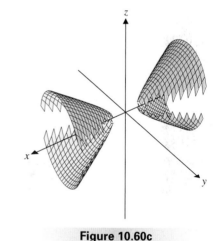

Figure 10.60c

Wireframe hyperboloid.

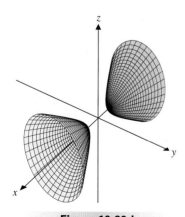

Figure 10.60d

Parametric plot.

We can plot this on a CAS by graphing the two functions $z = \sqrt{2\left(\frac{x^2}{4} - y^2 - 1\right)}$ and $z = -\sqrt{2\left(\frac{x^2}{4} - y^2 - 1\right)}$. (See Figure 10.60c, where we have restricted the z-range to $-10 \le z \le 10$, to show the circular cross sections.) Notice that this plot shows large gaps between the two halves of the hyperboloid. If you have drawn Figure 10.60b yourself, this plotting deficiency won't fool you.

Alternatively, the parametric plot with $x = 2\cosh s$, $y = \sinh s \cos t$ and $z = \sqrt{2} \sinh s \sin t$, for $-4 \le s \le 4$ and $0 \le t \le 2\pi$, produces the left half of the hyperboloid with its circular and hyperbolic traces. The right half of the hyperboloid has parametric equations $x = -2\cosh s$, $y = \sinh s \cos t$ and $z = \sqrt{2} \sinh s \sin t$ with $-4 \le s \le 4$ and $0 \le t \le 2\pi$. We show both halves in Figure 10.60d.

As our final example, we offer one of the more interesting quadric surfaces. It is also one of the more difficult surfaces to sketch.

Example 6.8   Sketching a Hyperbolic Paraboloid

Sketch the graph of the quadric surface defined by the equation

$$z = 2y^2 - x^2.$$

Solution   We first consider the traces in planes parallel to each of the coordinate planes:

parallel to xy-plane $(z = k)$: $2y^2 - x^2 = k$ (hyperbola, for $k \neq 0$),

parallel to xz-plane $(y = k)$: $z = -x^2 + 2k^2$ (parabola opening down)

and

parallel to yz-plane $(x = k)$: $z = 2y^2 - k^2$ (parabola opening up).

We begin by drawing the traces in the xz- and yz-planes, as seen in Figure 10.61a. Since the trace in the xy-plane is the degenerate hyperbola $2y^2 = x^2$ (two lines: $x = \pm 2y$), we instead draw the trace in several of the planes $z = k$. Notice that for $k > 0$, these are hyperbolas opening toward the positive and negative y-direction and for $k < 0$, these are hyperbolas opening toward the positive and negative x-direction. We indicate one of these for $k > 0$ and one for $k < 0$ in Figure 10.61b, where we show a sketch of the surface. We refer to this surface as a **hyperbolic paraboloid.** More than anything else, the surface resembles a saddle. In fact, we refer to the origin as a **saddle point** for this graph. (We'll discuss the significance of saddle points in Chapter 12.)

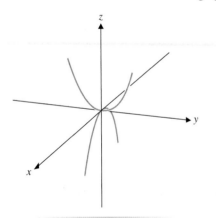

Figure 10.61a
Traces in the xz- and yz-planes.

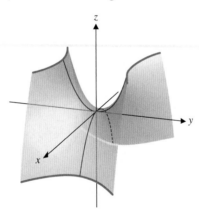

Figure 10.61b
The surface $z = 2y^2 - x^2$.

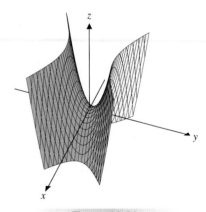

Figure 10.61c
Wireframe plot of $z = 2y^2 - x^2$.

A wireframe graph of $z = 2y^2 - x^2$ is shown in Figure 10.61c (with $-5 \leq x \leq 5$ and $-5 \leq y \leq 5$ and where we limited the z-range to $-8 \leq z \leq 12$). Note that only the parabolic cross sections are drawn, but the graph shows all the features of Figure 10.61b. Plotting this surface parametrically is fairly tedious (requiring four different sets of equations) and doesn't improve the graph noticeably.

An Application

You may have noticed the large number of paraboloids around you. For instance, radio-telescopes and even home television satellite dishes have the shape of a portion of a paraboloid. Reflecting telescopes have parabolic mirrors that again, are a portion of a paraboloid. There is a very good reason for this. It turns out that in all of these cases, light waves and radio waves striking **any** point on the parabolic dish or mirror are reflected toward **one** point, the focus of each parabolic cross section through the vertex of the paraboloid. This remarkable fact means that all light waves and radio waves end up being concentrated at just one point. In the case of a radiotelescope, placing a small receiver just in front of the focus can take a very faint signal and increase its effective strength immensely (see Figure 10.62). The same principle is used in optical telescopes to concentrate the light from a faint source (e.g., a distant star). In this case, a small mirror is mounted in a line from the parabolic mirror to the focus. The small mirror then reflects the concentrated light to an eyepiece for viewing (see Figure 10.63).

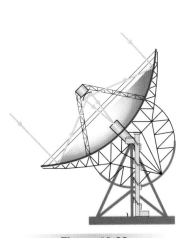

Figure 10.62
Radiotelescope.

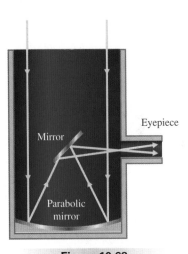

Figure 10.63
Reflecting telescope.

EXERCISES 10.6

1. In the text, different hints were given for graphing cylinders as opposed to quadric surfaces. Explain how to tell from the equation whether you have a cylinder, a quadric surface, a plane or some other surface.

2. The first step in graphing a quadric surface is identifying traces. Given the traces, explain how to tell whether you have an ellipsoid, elliptical cone, paraboloid or hyperboloid. (Hint: For a paraboloid, how many traces are parabolas?)

3. Suppose you have identified that a given equation represents a hyperboloid. Explain how to determine whether the hyperboloid has one sheet or two sheets.

4. Circular paraboloids have a bowl-like shape. However, the paraboloids $z = x^2 + y^2$, $z = 4 - x^2 - y^2$, $y = x^2 + z^2$ and $x = y^2 + z^2$ all open up in different directions. Explain why these paraboloids are different and how to determine in which direction a paraboloid opens.

In exercises 5–44, sketch the appropriate traces and then sketch and identify the surface.

5. $z = x^2$

6. $z = 4 - y^2$

7. $x^2 + \dfrac{y^2}{9} + \dfrac{z^2}{4} = 1$

8. $\dfrac{x^2}{4} + \dfrac{y^2}{4} + \dfrac{z^2}{9} = 1$

9. $z = 4x^2 + 4y^2$

10. $z = x^2 + 4y^2$

11. $z^2 = 4x^2 + y^2$

12. $z^2 = \dfrac{x^2}{4} + \dfrac{y^2}{9}$

13. $z = x^2 - y^2$

14. $z = y^2 - x^2$

15. $x^2 - y^2 + z^2 = 1$

16. $x^2 + \dfrac{y^2}{4} - z^2 = 1$

17. $x^2 - \dfrac{y^2}{9} - z^2 = 1$

18. $x^2 - y^2 - \dfrac{z^2}{4} = 1$

19. $z = \cos x$

20. $z = \sqrt{x^2 + 4y^2}$

21. $z = 4 - x^2 - y^2$

22. $x = y^2 + z^2$

23. $z = x^3$

24. $z = 4 - y^2$

25. $z = \sqrt{x^2 + y^2}$

26. $z = \sin y$

27. $y = x^2$

28. $x = 2 - y^2$

29. $y = x^2 + z^2$

30. $z = 9 - x^2 - y^2$

31. $x^2 + 4y^2 + 16z^2 = 16$

32. $2x - z = 4$

33. $4x^2 - y^2 - z = 0$

34. $-x^2 - y^2 + 9z^2 = 9$

35. $4x^2 + y^2 - z^2 = 4$

36. $x^2 - y^2 + 9z^2 = 9$

37. $-4x^2 + y^2 - z^2 = 4$

38. $x^2 - 4y^2 + z = 0$

39. $x + y = 1$

40. $9x^2 + y^2 + 9z^2 = 9$

41. $x^2 + y^2 = 4$

42. $9x^2 + z^2 = 9$

43. $x^2 + y^2 - z = 4$

44. $x + y^2 + z^2 = 2$

In exercises 45–48, sketch the given traces on a single three-dimensional coordinate system.

45. $z = x^2 + y^2; x = 0, x = 1, x = 2$

46. $z = x^2 + y^2; y = 0, y = 1, y = 2$

47. $z = x^2 - y^2; x = 0, x = 1, x = 2$

48. $z = x^2 - y^2; y = 0, y = 1, y = 2$

49. Hyperbolic paraboloids are sometimes called "saddle" graphs. The architect of the Saddle Dome in the Canadian city of Calgary used this shape to create an attractive and symbolically meaningful structure.

One issue in using this shape is water drainage from the roof. If the Saddle Dome roof is described by $z = x^2 - y^2$,

$-1 \le x \le 1, -1 \le y \le 1$, in which direction would the water drain? First, consider traces for which y is constant. Show that the trace has a minimum at $x = 0$. Identify the plane $x = 0$ in the picture. Next, show that the trace at $x = 0$ has an absolute maximum at $y = 0$. Use this information to identify the two primary points at which the water would drain.

50. Cooling towers for nuclear reactors are often constructed as hyperboloids of one sheet because of the structural stability of that surface. (See the photo below.) Suppose all horizontal cross sections are circular, with a minimum radius of 200 feet. The tower is to be 800 feet tall with a maximum cross-sectional radius of 300 feet. Find an equation for the structure.

51. If $x = a \sin s \cos t$, $y = b \sin s \sin t$ and $z = c \cos s$, show that (x, y, z) lies on the ellipsoid $\dfrac{x^2}{a^2} + \dfrac{y^2}{b^2} + \dfrac{z^2}{c^2} = 1$.

52. If $x = as \cos t$, $y = bs \sin t$ and $z = s^2$, show that (x, y, z) lies on the paraboloid $z = \dfrac{x^2}{a^2} + \dfrac{y^2}{b^2}$.

53. If $x = a\sqrt{s^2} \cos t$, $y = b\sqrt{s^2} \sin t$ and $z = s$, show that (x, y, z) lies on the cone $z^2 = \dfrac{x^2}{a^2} + \dfrac{y^2}{b^2}$.

54. If $x = a \cos s \cosh t$, $y = b \sin s \cosh t$ and $z = c \sinh t$, show that (x, y, z) lies on the hyperboloid of one sheet $\dfrac{x^2}{a^2} + \dfrac{y^2}{b^2} - \dfrac{z^2}{c^2} = 1$.

55. If $a > 0$ and $x = a \cosh s$, $y = b \sinh s \cos t$ and $z = c \sinh s \sin t$, show that (x, y, z) lies on the right half of the hyperboloid of two sheets $\dfrac{x^2}{a^2} - \dfrac{y^2}{b^2} - \dfrac{z^2}{c^2} = 1$.

56. If $a < 0$ and $x = a \cosh s$, $y = b \sinh s \cos t$ and $z = c \sinh s \sin t$, show that (x, y, z) lies on the left half of the hyperboloid of two sheets $\dfrac{x^2}{a^2} - \dfrac{y^2}{b^2} - \dfrac{z^2}{c^2} = 1$.

57. Find parametric equations as in exercises 51–56 for the surfaces in exercises 7, 9 and 11. Use a CAS to graph the parametric surfaces.

58. Find parametric equations as in exercises 51–56 for the surfaces in exercises 15 and 17. Use a CAS to graph the parametric surfaces.

59. Find parametric equations for the surface in exercise 21.

60. Find parametric equations for the surface in exercise 37.

61. You can improve the appearance of a wireframe graph by carefully choosing the viewing window. We commented on the curved edge in Figure 10.57c. Graph this function with domain $-5 \le x \le 5$ and $-5 \le y \le 5$ but limit the z-range to $-1 \le z \le 20$. Does this look more like Figure 10.57b?

62. Golf club manufacturers use ellipsoids (called **inertia ellipsoids**) to visualize important characteristics of golf clubs. A three-dimensional coordinate system is set up as shown in the figure. The (second) moments of inertia are then computed for the clubhead about each coordinate axis. The inertia ellipsoid is defined as $I_{xx}x^2 + I_{yy}y^2 + I_{zz}z^2 + 2I_{xy}xy + 2I_{yz}yz + 2I_{xz}xz = 1$. The graph of this ellipsoid provides important information to the club designer. For comparison purposes, a homogeneous spherical shell would have a perfect sphere as its inertia ellipsoid. In *Science and Golf II*, the following data are provided for a 6-iron and driver, respectively. Graph the ellipsoids and compare the shapes (recall that the larger the moment of inertia of an object, the harder it is to rotate). For the 6-iron, $89.4x^2 + 195.8y^2 + 124.9z^2 - 48.6xy - 111.8xz + 0.4yz = 1,000,000$ and for the driver, $119.3x^2 + 243.9y^2 + 139.4z^2 - 1.2xy - 71.4xz - 25.8yz = 1,000,000$.

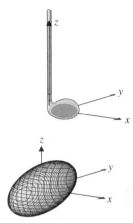

63. Sketch the graphs of $x^2 + cy^2 - z^2 = 1$ for a variety of positive and negative constants c. If your CAS allows you to animate a sequence of graphs, set up an animation that shows a sequence of hyperboloids of one sheet morphing into hyperboloids of two sheets.

CHAPTER REVIEW EXERCISES

In exercises 1–4, compute a + b, 4b and ‖2b − a‖.

1. $\mathbf{a} = \langle -2, 3 \rangle, \mathbf{b} = \langle 1, 0 \rangle$

2. $\mathbf{a} = \langle -1, -2 \rangle, \mathbf{b} = \langle 2, 3 \rangle$

3. $\mathbf{a} = 10\mathbf{i} + 2\mathbf{j} - 2\mathbf{k}, \mathbf{b} = -4\mathbf{i} + 3\mathbf{j} + 2\mathbf{k}$

4. $\mathbf{a} = -\mathbf{i} - \mathbf{j} + 2\mathbf{k}, \mathbf{b} = -\mathbf{i} + \mathbf{j} - 2\mathbf{k}$

In exercises 5–8, determine whether a and b are parallel, orthogonal or neither.

5. $\mathbf{a} = \langle 2, 3 \rangle, \mathbf{b} = \langle 4, 5 \rangle$ 6. $\mathbf{a} = \mathbf{i} - 2\mathbf{j}, \mathbf{b} = 2\mathbf{i} - \mathbf{j}$

7. $\mathbf{a} = \langle -2, 3, 1 \rangle, \mathbf{b} = \langle 4, -6, -2 \rangle$

8. $\mathbf{a} = 2\mathbf{i} - \mathbf{j} + 2\mathbf{k}, \mathbf{b} = 4\mathbf{i} - 2\mathbf{j} + \mathbf{k}$

In exercises 9 and 10, find the displacement vector $\overrightarrow{PQ}$.

9. $P = (3, 1, -2), Q = (2, -1, 1)$

10. $P = (3, 1), Q = (1, 4)$

In exercises 11–16, find a unit vector in the same direction as the given vector.

11. $\langle 3, 6 \rangle$ 12. $\langle -2, 3 \rangle$

13. $10\mathbf{i} + 2\mathbf{j} - 2\mathbf{k}$ 14. $-\mathbf{i} - \mathbf{j} + 2\mathbf{k}$

15. from $(4, 1, 2)$ to $(1, 1, 6)$

16. from $(2, -1, 0)$ to $(0, 3, -2)$

In exercises 17 and 18, find the distance between the given points.

17. $(0, -2, 2), (3, 4, 1)$

18. $(3, 1, 0), (1, 4, 1)$

In exercises 19 and 20, find a vector with the given magnitude and in the same direction as the given vector.

19. magnitude 2, $\mathbf{v} = 2\mathbf{i} - 2\mathbf{j} + 2\mathbf{k}$

20. magnitude $\frac{1}{2}$, $\mathbf{v} = -\mathbf{i} - \mathbf{j} + \mathbf{k}$

21. The thrust of an airplane's engine produces a speed of 500 mph in still air. The wind velocity is given by $\langle 20, -80 \rangle$. In what direction should the plane head to fly due east?

22. Two ropes are attached to a crate. The ropes exert forces of $\langle -160, 120 \rangle$ and $\langle 160, 160 \rangle$, respectively. If the crate weighs 300 pounds, what is the net force on the crate?

In exercises 23 and 24, find an equation of the sphere with radius r and center (a, b, c).

23. $r = 6, (a, b, c) = (0, -2, 0)$

24. $r = \sqrt{3}, (a, b, c) = (-3, 1, 2)$

In exercises 25–28, compute a · b.

25. $\mathbf{a} = \langle 2, -1 \rangle, \mathbf{b} = \langle 2, 4 \rangle$

26. $\mathbf{a} = \mathbf{i} - 2\mathbf{j}, \mathbf{b} = 4\mathbf{i} + 2\mathbf{j}$

27. $\mathbf{a} = 3\mathbf{i} + \mathbf{j} - 4\mathbf{k}, \mathbf{b} = -2\mathbf{i} + 2\mathbf{j} + \mathbf{k}$

28. $\mathbf{a} = \mathbf{i} + 3\mathbf{j} - 2\mathbf{k}, \mathbf{b} = 2\mathbf{i} - 3\mathbf{k}$

In exercises 29 and 30, find the angle between the vectors.

29. $\langle 3, 2, 1 \rangle$ and $\langle -1, 1, 2 \rangle$

30. $\langle 3, 4 \rangle$ and $\langle 2, -1 \rangle$

In exercises 31 and 32, find comp$_b$ a and proj$_b$ a.

31. $\mathbf{a} = 3\mathbf{i} + \mathbf{j} - 4\mathbf{k}, \mathbf{b} = \mathbf{i} + 2\mathbf{j} + \mathbf{k}$

32. $\mathbf{a} = \mathbf{i} + 3\mathbf{j} - 2\mathbf{k}, \mathbf{b} = 2\mathbf{i} - 3\mathbf{k}$

In exercises 33–36, compute the cross product a x b.

33. $\mathbf{a} = \langle 1, -2, 1 \rangle, \mathbf{b} = \langle 2, 0, 1 \rangle$

34. $\mathbf{a} = \langle 1, -2, 0 \rangle, \mathbf{b} = \langle 1, 0, -2 \rangle$

35. $\mathbf{a} = 2\mathbf{j} + \mathbf{k}, \mathbf{b} = 4\mathbf{i} + 2\mathbf{j} - \mathbf{k}$

36. $\mathbf{a} = \mathbf{i} - 2\mathbf{j} - 3\mathbf{k}, \mathbf{b} = 2\mathbf{i} - \mathbf{j}$

In exercises 37 and 38, find two unit vectors orthogonal to both given vectors.

37. $\mathbf{a} = 2\mathbf{i} + \mathbf{k}, \mathbf{b} = -\mathbf{i} + 2\mathbf{j} - \mathbf{k}$

38. $\mathbf{a} = 3\mathbf{i} + \mathbf{j} - 2\mathbf{k}, \mathbf{b} = 2\mathbf{i} - \mathbf{j}$

39. A force of $\langle 40, -30 \rangle$ pounds moves an object in a straight line from $(1, 0)$ to $(60, 22)$. Compute the work done.

40. Use vectors to find the angles in the triangle with vertices $(0, 0), (3, 1)$ and $(1, 4)$.

In exercises 41 and 42, find the distance from the point Q to the given line.

41. $Q = (1, -1, 0)$, line $\begin{cases} x = t + 1 \\ y = 2t - 1 \\ z = 3 \end{cases}$

42. $Q = (0, 1, 0)$, line $\begin{cases} x = 2t - 1 \\ y = 4t \\ z = 3t + 2 \end{cases}$

In exercises 43 and 44, find the indicated area or volume.

43. Area of the parallelogram with adjacent edges formed by $\langle 2, 0, 1 \rangle$ and $\langle 0, 1, -3 \rangle$

44. Volume of the parallelepiped with three adjacent edges formed by $\langle 1, -1, 2 \rangle, \langle 0, 0, 4 \rangle$ and $\langle 3, 0, 1 \rangle$

45. A force of magnitude 50 pounds is applied at the end of a 6-inch-long wrench at an angle of $\frac{\pi}{6}$ to the wrench. Find the magnitude of the torque applied to the bolt.

46. A ball is struck with backspin. Find the direction of the Magnus force and describe the effect on the ball.

In exercises 47–50, find (a) parametric equations and (b) symmetric equations of the line.

47. The line through $(2, -1, -3)$ and $(0, 2, -3)$

48. The line through $(-1, 0, 2)$ and $(-3, 0, -2)$

49. The line through $(2, -1, 1)$ and parallel to $\frac{x-1}{2} = 2y = \frac{z+2}{-3}$

50. The line through $(0, 2, 1)$ and normal to the plane $2x - 3y + z = 4$

In exercises 51 and 52, find the angle between the lines.

51. $\begin{cases} x = 4+t \\ y = 2 \\ z = 3+2t \end{cases}$ and $\begin{cases} x = 4+2s \\ y = 2+2s \\ z = 3+4s \end{cases}$

52. $\begin{cases} x = 3+t \\ y = 3+3t \\ z = 4-t \end{cases}$ and $\begin{cases} x = 3-s \\ y = 3-2s \\ z = 4+2s \end{cases}$

In exercises 53 and 54, determine if the lines are parallel, skew or intersect.

53. $\begin{cases} x = 2t \\ y = 3+t \\ z = -1+4t \end{cases}$ and $\begin{cases} x = 4 \\ y = 4+s \\ z = 3+s \end{cases}$

54. $\begin{cases} x = 1-t \\ y = 2t \\ z = 5-t \end{cases}$ and $\begin{cases} x = 3+3s \\ y = 2 \\ z = 1-3s \end{cases}$

In exercises 55–58, find an equation of the given plane.

55. The plane containing the point $(-5, 0, 1)$ with normal vector $\langle 4, 1, -2 \rangle$

56. The plane containing the point $(2, -1, 2)$ with normal vector $\langle 3, -1, 0 \rangle$

57. The plane containing the points $(2, 1, 3)$, $(2, -1, 2)$ and $(3, 3, 2)$

58. The plane containing the points $(2, -1, 2)$, $(1, -1, 4)$ and $(3, -1, 2)$

In exercises 59–72, sketch and identify the surface.

59. $9x^2 + y^2 + z = 9$

60. $x^2 + y + z^2 = 1$

61. $y^2 + z^2 = 1$

62. $x^2 + 4y^2 = 4$

63. $x^2 - 2x + y^2 + z^2 = 3$

64. $x^2 + (y+2)^2 + z^2 = 6$

65. $y = 2$

66. $z = 5$

67. $2x - y + z = 4$

68. $3x + 2y - z = 6$

69. $x^2 - y^2 + 4z^2 = 4$

70. $x^2 - y^2 - z = 1$

71. $x^2 - y^2 - 4z^2 = 4$

72. $x^2 + y^2 - z = 1$

73. As we focus on three-dimensional geometry in the next several chapters, some projections will be difficult but important to visualize. In this exercise, we contrast the curves C_1 and C_2 defined parametrically by $\begin{cases} x = \cos t \\ y = \cos t \\ z = \sin t \end{cases}$ and $\begin{cases} x = \cos t \\ y = \cos t \\ z = \sqrt{2}\sin t \end{cases}$, respectively. If you have access to three-dimensional graphics, try sketching each curve from a variety of perspectives. Our question will be whether either curve is a circle. For both curves, note that $x = y$. Describe in words and sketch a graph of the plane $x = y$. Next, note that the projection of C_1 back into the yz-plane is a circle ($y = \cos t$, $z = \sin t$). If C_1 is actually a circle in the plane $x = y$, discuss what its projection (shadow) in the yz-plane would look like. Given this, explain whether C_1 is actually a circle or an ellipse. Compare your description of the projection of a circle into the yz-plane to the projection of C_2 into the yz-plane. To make this more quantitative, we can use the general rule that for a two-dimensional region, the area of its projection onto a plane equals the area of the region multiplied by $\cos\theta$, where θ is the angle between the plane in which the region lies and the plane into which it is being projected. Given this, compute the radius of the circle C_2.

CHAPTER 11

VECTOR-VALUED FUNCTIONS

Modern fighter jet pilots are routinely required to perform tasks that are literally super-human. While traveling at supersonic speeds, pilots need to track missiles and other aircraft and identify which are friendly and which are not. In less than the time it takes to blink an eye, a pilot must decide whether to speed up or slow down and which way to turn. These decisions must be made even if the jet is already in a turn so tight that most people would pass out from the "g" forces acting on them. As you may have discovered playing computer-simulated flying games, the vision and reaction times required are beyond human abilities. Fighter pilots must rely on sophisticated computer equipment to accurately determine the positions and velocities of all objects in the air around them.

Airline pilots face similar challenges. Although they do not typically need to perform intricate rolls and turns to avoid danger, airline pilots need information about numerous airborne objects, and this information must be given on their terms. That is, instead of hearing that another plane is also 340 miles east of Los Angeles heading toward St. Louis, a pilot needs to know that another plane is less than 1 mile ahead and heading toward him. Pilots need positions and velocities relative to their own position, not relative to some arbitrary fixed point on the ground. One of our goals in this chapter is the construction of a moving three-dimensional coordinate system that can provide this kind of information.

In this chapter, we take another step toward a fully three-dimensional calculus. Our first major objective is to describe the position, velocity and acceleration of a moving object. However, since an object's location in space is a three-dimensional quantity, we cannot describe its location with a single function. Instead, we will extend the notion of function from the familiar scalar real-valued functions (ones whose values are a single real number) to **vector-valued functions,** whose values are vectors. In the first two sections of this chapter, we develop the calculus of vector-valued functions. Much of the material is a straightforward extension to three dimensions of the familiar one-dimensional calculus. In fact, many of the definitions, theorems and calculations will seem familiar to you. However, as you have already seen, the geometry of three dimensions can be much more interesting than two dimensions. Because of this, we devote section 11.3 specifically to motion in three-dimensional space. In particular, we will see how centripetal force relates to speed and turning radius. This relationship is critical to aeronautical engineers, who must understand the forces that can act on the aircraft they design.

The calculations in the first three sections are appropriate for tracking an object from a fixed reference point. As we've already observed, this information is much less useful to pilots than information given relative to their own position and velocity. In this sense, pilots view their cockpit as the origin, the point from which measurements should be made. Further, an airplane's velocity determines a specific orientation of space, in terms of which relative directions such as "behind" and "above" are defined. In section 11.4, we define the unit tangent vector as a unit vector in the direction of velocity. Then, in section 11.5, we devise a three-dimensional coordinate system with one axis pointing in the direction of the unit tangent vector. The acceleration of an airplane can then be resolved into a tangential component that affects the speed of the airplane and a normal component that causes the airplane to turn. As you will see with the many problems and applications we present, the analysis of vector-valued functions is of interest even if you don't fly jet aircraft.

11.1 VECTOR-VALUED FUNCTIONS

Consider the circuitous path of the airplane indicated in Figure 11.1a. How could you describe its location at any given time? You might consider using a point (x, y, z) in three dimensions, but it turns out to be more convenient if we describe its location at any given time by the endpoint of a vector. (We call such a vector a **position vector.**) You should realize that this means that we need a different vector for every time. (See Figure 11.1b for vectors indicating the location of the plane at a number of times.) Notice that a **function** that gives us a vector in V_3 for each time t would do the job nicely. This is the concept of a vector-valued function. More precisely, we have the following definition.

Definition 1.1

A **vector-valued function** $\mathbf{r}(t)$ is a mapping from its domain $D \subset \mathbb{R}$ to its range $R \subset V_3$, so that for each t in D, $\mathbf{r}(t) = \mathbf{v}$ for only one vector $\mathbf{v} \in V_3$. We can always write a vector-valued function as

$$\mathbf{r}(t) = f(t)\mathbf{i} + g(t)\mathbf{j} + h(t)\mathbf{k}, \tag{1.1}$$

for some scalar functions f, g and h (called the **component functions** of $\mathbf{r}$).

Figure 11.1a

Airplane's flight path.

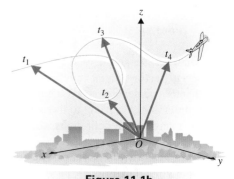

Figure 11.1b

Vectors indicating plane's position
at several times.

Notice that if $\mathbf{r}(t)$ is defined by (1.1), its value is a (different) vector in V_3, for each t in D. We can likewise define a vector-valued function $\mathbf{r}(t)$ in V_2 by

$$\mathbf{r}(t) = f(t)\mathbf{i} + g(t)\mathbf{j},$$

for some scalar functions f and g.

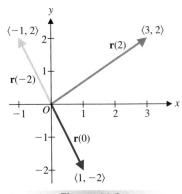

Figure 11.2a

Some values of
$\mathbf{r}(t) = (t + 1)\mathbf{i} + (t^2 - 2)\mathbf{j}$.

Remark 1.1

We routinely use the variable t to represent the independent variable for vector-valued functions, since in many applications t represents **time.**

Example 1.1

Sketching the Curve Defined by a Vector-Valued Function

Sketch a graph of the curve traced out by the endpoint of the two-dimensional vector-valued function

$$\mathbf{r}(t) = (t + 1)\mathbf{i} + (t^2 - 2)\mathbf{j}.$$

Solution Substituting some values for t, we have $\mathbf{r}(0) = \mathbf{i} - 2\mathbf{j} = \langle 1, -2 \rangle$ and $\mathbf{r}(2) = 3\mathbf{i} + 2\mathbf{j} = \langle 3, 2 \rangle$ and $\mathbf{r}(-2) = \langle -1, 2 \rangle$. We plot these in Figure 11.2a. The endpoints of all position vectors $\mathbf{r}(t)$ lie on the curve C, described parametrically by

$$C : x = t + 1, \ y = t^2 - 2, \ t \in \mathbb{R}.$$

We can eliminate the parameter by solving for t in terms of x:

$$t = x - 1.$$

The curve is then given by

$$y = t^2 - 2 = (x - 1)^2 - 2.$$

Notice that the graph of this is a parabola opening up, with vertex at the point $(1, -2)$, as seen in Figure 11.2b. The small arrows marked on the graph indicate the **orientation,** that is, the direction of increasing values of t. If the curve describes the path of an object, then the orientation indicates the direction in which the object traverses the path. In this case, we can easily determine the orientation from the parametric representation of the curve. Since $x = t + 1$, observe that x increases as t increases.

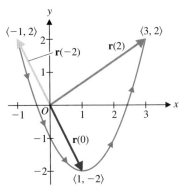

Figure 11.2b

Curve defined by
$\mathbf{r}(t) = (t + 1)\mathbf{i} + (t^2 - 2)\mathbf{j}$.

Notice from our work in example 1.1 that the curve traced out by the endpoint of the vector-valued function $\mathbf{r}(t) = f(t)\mathbf{i} + g(t)\mathbf{j}$ is identical to the curve described by the parametric equations $x = f(t), y = g(t)$, as presented in section 9.1. The ideas presented here are not new; only the notation and terminology are new.

You may recall from your experience with parametric equations in Chapter 9 that eliminating the parameter from the parametric representation of a curve is not always so easy as it was in example 1.1. We illustrate this in the following example.

Example 1.2

A Vector-Valued Function Defining an Ellipse

Sketch a graph of the curve traced out by the endpoint of the vector-valued function $\mathbf{r}(t) = 4 \cos t\,\mathbf{i} - 3 \sin t\,\mathbf{j}, t \in \mathbb{R}$.

Solution In this case, the curve can be written parametrically as

$$x = 4\cos t, \quad y = -3\sin t, \quad t \in \mathbb{R}.$$

In order to eliminate the parameter here, you do not want to solve for the parameter t. Instead, look for some relationship between the variables. You should notice that

$$\left(\frac{x}{4}\right)^2 + \left(\frac{y}{3}\right)^2 = \cos^2 t + \sin^2 t = 1$$

or

$$\left(\frac{x}{4}\right)^2 + \left(\frac{y}{3}\right)^2 = 1,$$

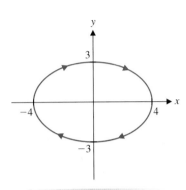

Figure 11.3

Curve defined by
$\mathbf{r}(t) = 4\cos t\,\mathbf{i} - 3\sin t\,\mathbf{j}.$

which is the equation of an ellipse (see Figure 11.3). Determining the orientation of the curve here is just a bit more tricky than in example 1.1. In this case, you'll need to look carefully at both parametric equations. First, fix a starting place on the curve, for convenience, say $(4, 0)$. This corresponds to $t = 0, \pm 2\pi, \pm 4\pi, \ldots$. As t increases, notice that $\cos t$ (and hence, x) decreases initially, while $\sin t$ increases, so that $y = -3\sin t$ decreases (initially). With both x and y decreasing initially, we get the clockwise orientation indicated in Figure 11.3.

Just as the endpoint of a vector-valued function in two dimensions traces out a curve, if we were to plot the value of $\mathbf{r}(t) = f(t)\mathbf{i} + g(t)\mathbf{j} + h(t)\mathbf{k}$ for every value of t, the endpoints of the vectors also trace out a curve in three dimensions.

Example 1.3 A Vector-Valued Function Defining an Elliptical Helix

Plot the curve traced out by the vector-valued function $\mathbf{r}(t) = \sin t\,\mathbf{i} - 3\cos t\,\mathbf{j} + 2t\mathbf{k}$, $t \geq 0$.

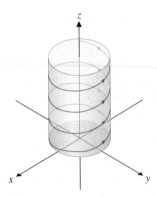

Figure 11.4a

Elliptical helix,
$\mathbf{r}(t) = \sin t\,\mathbf{i} - 3\cos t\,\mathbf{j} + 2t\mathbf{k}.$

Solution The curve is given parametrically by

$$x = \sin t, \quad y = -3\cos t, \quad z = 2t, \quad t \geq 0.$$

While most curves in three dimensions are difficult to recognize, there is something that you should see here. Notice that there is a relationship between x and y, namely,

$$x^2 + \left(\frac{y}{3}\right)^2 = \sin^2 t + \cos^2 t = 1. \tag{1.2}$$

In two dimensions, this is the equation of an ellipse. In three dimensions, since the equation does not involve z, (1.2) is the equation of an elliptic cylinder whose axis is the z-axis. This says that every point on the curve defined by $\mathbf{r}(t)$ lies on the cylinder. Notice from the parametric equations for x and y that (in two dimensions) the ellipse is traversed in the counterclockwise direction. This says that the curve will wrap itself around the cylinder (counterclockwise, as you look down the positive z-axis toward the origin), as t increases. Finally, since $z = 2t$, z will increase as t increases and so, the curve will wind its way up the cylinder, as t increases. We show the curve and the elliptical cylinder in Figure 11.4a. We call this curve an **elliptical helix.** In Figure 11.4b, we display a computer-generated graph of the same helix. There, rather than the usual x-, y- and z-axes, we show a framed graph, where the values of x, y and z are indicated on three adjacent edges of a box containing the graph.

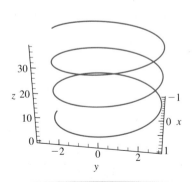

Figure 11.4b

Computer sketch:
$\mathbf{r}(t) = \sin t\,\mathbf{i} - 3\cos t\,\mathbf{j} + 2t\mathbf{k}.$

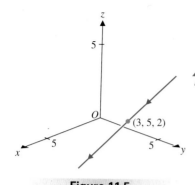

Figure 11.5
Straight line:
$\mathbf{r}(t) = \langle 3 + 2t, 5 - 3t, 2 - 4t \rangle$.

We can use vector-valued functions as a convenient representation of some very familiar curves, as we see in the following example.

Example 1.4 A Vector-Valued Function Defining a Line

Plot the curve traced out by the vector-valued function

$$\mathbf{r}(t) = \langle 3 + 2t, 5 - 3t, 2 - 4t \rangle, \quad t \in \mathbb{R}.$$

Solution Notice that the curve is given parametrically by

$$x = 3 + 2t, \quad y = 5 - 3t, \quad z = 2 - 4t, \quad t \in \mathbb{R}.$$

You should recognize these equations as parametric equations for the straight line parallel to the vector $\langle 2, -3, -4 \rangle$ and passing through the point $(3, 5, 2)$, as seen in Figure 11.5.

Most three-dimensional graphs are very challenging to sketch by hand. You will probably want to use computer-generated graphics for most sketches. Even so, you will need to be knowledgeable enough to know when to zoom in or out or rotate a graph to uncover a hidden feature. You should be able to draw several basic curves by hand, like those in examples 1.3 and 1.4. More importantly, you should be able to recognize the effects various components have on the graph of a three-dimensional curve. In the following example, we walk you through matching four vector-valued functions with their computer-generated graphs.

Example 1.5 Matching a Vector-Valued Function to Its Graph

Match each of the vector-valued functions $\mathbf{f}_1(t) = \langle \cos t, \ln t, \sin t \rangle$, $\mathbf{f}_2(t) = \langle t \cos t, t \sin t, t \rangle$, $\mathbf{f}_3(t) = \langle 3 \sin 2t, t, t \rangle$ and $\mathbf{f}_4(t) = \langle 5 \sin^3 t, 5 \cos^3 t, t \rangle$ with the corresponding computer-generated graph.

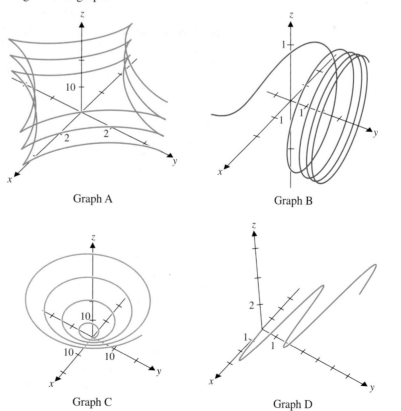

Graph A

Graph B

Graph C

Graph D

Solution   First, realize that there is no single, correct procedure for solving this problem. You need to match familiar functions with familiar graphical properties. From example 1.3, recall that certain combinations of sines and cosines will produce curves that lie on a cylinder. Notice that for the function $\mathbf{f}_1(t)$, $x = \cos t$ and $z = \sin t$, so that

$$x^2 + z^2 = \cos^2 t + \sin^2 t = 1.$$

This says that every point on the curve lies on the cylinder $x^2 + z^2 = 1$. (This is a right circular cylinder of radius 1 whose axis is the y-axis.) Further, the function $y = \ln t$ tends rapidly to $-\infty$ as $t \to 0$ and increases slowly as t increases beyond $t = 1$. Notice that the curve in Graph B appears to lie on a right circular cylinder and that the spirals get closer together as you move to the right (as $y \to \infty$) and move very far apart as you move to the left (as $y \to -\infty$). At first glance, you might expect the curve traced out by $\mathbf{f}_2(t)$ to also lie on a right circular cylinder, but look more closely. Here, we have $x = t \cos t$, $y = t \sin t$ and $z = t$, so that

$$x^2 + y^2 = t^2 \cos^2 t + t^2 \sin^2 t = t^2 = z^2.$$

This says that the curve lies on the surface defined by $x^2 + y^2 = z^2$ (a right circular cone with axis along the z-axis). Notice that only the curve shown in Graph C fits this description. Next, notice that for $\mathbf{f}_3(t)$, the y and z components are identical and so, the curve must lie in the plane $y = z$. Replacing t by y, we have $x = 3 \sin 2t = 3 \sin 2y$, a sine curve lying in the plane $y = z$. Clearly, the curve in Graph D fits this description. This leaves us with Graph A for function $\mathbf{f}_4(t)$. Beyond Graph A being the only curve remaining from which to choose, notice that if the cosine and sine terms weren't cubed, we'd simply have the helix of example 1.3. The cubes square off the smooth curve of the helix. Since $z = t$, each point on the curve is a point on the cylinder defined parametrically by $x = 5 \sin^3 t$ and $y = 5 \cos^3 t$. You need only look at the graph of the cross section of the cylinder (found by graphing the parametric equations $x = 5 \sin^3 t$ and $y = 5 \cos^3 t$ in two dimensions) to decide that Graph A is the obvious choice. We show this cross section in Figure 11.6.

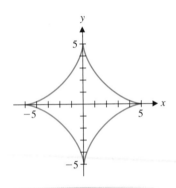

Figure 11.6

The cross section $x = 5 \sin^3 t$, $y = 5 \cos^3 t$.

Arc Length in $\mathbb{R}^3$

A natural question to ask about a curve is, "how long is it?" Indeed, we have already answered this question in the case of curves in two dimensions. Recall from section 5.4 that if f and f' are continuous on the interval $[a, b]$, then the arc length of the curve $y = f(x)$ on that interval is given by

$$s = \int_a^b \sqrt{1 + [f'(x)]^2}\, dx.$$

In section 9.3, we extended this to the case of a curve defined parametrically by $x = f(t)$, $y = g(t)$, where f, f', g and g' are all continuous for $t \in [a, b]$. In this case, we showed that if the curve is traversed exactly once as t increases from a to b, then the arc length is given by

$$s = \int_a^b \sqrt{[f'(t)]^2 + [g'(t)]^2}\, dt. \tag{1.3}$$

In both cases, recall that we developed the arc length formula by first breaking the curve into small pieces (i.e., we **partitioned** the interval $[a, b]$) and then approximating the

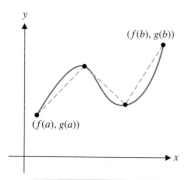

Figure 11.7a

Approximate arc length in $\mathbb{R}^2$.

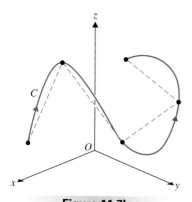

Figure 11.7b

Approximate arc length in $\mathbb{R}^3$.

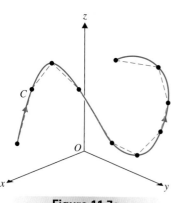

Figure 11.7c

Improved arc length approximation.

length with the sum of the lengths of small line segments connecting successive points (see Figure 11.7a). Finally, we made the approximation exact by taking a limit as the number of points in the partition tended to infinity. This says that if the curve C in $\mathbb{R}^2$ is traced out by the endpoint of the vector-valued function $\mathbf{r}(t) = \langle f(t), g(t) \rangle$, then the arc length is given by (1.3).

The situation in three dimensions is a straightforward extension of the two-dimensional case. Suppose that a curve is traced out by the endpoint of the vector-valued function $\mathbf{r}(t) = \langle f(t), g(t), h(t) \rangle$, where f, f', g, g', h and h' are all continuous for $t \in [a, b]$ and where the curve is traversed exactly once as t increases from a to b. As we have done countless times now, we begin by approximating the quantity of interest, in this case, the arc length. To do this, we partition the interval $[a, b]$ into n subintervals of equal size: $a = t_0 < t_1 < \cdots < t_n = b$, where $t_i - t_{i-1} = \Delta t = \frac{b-a}{n}$, for all $i = 1, 2, \ldots, n$. Next, for each $i = 1, 2, \ldots, n$, we approximate the arc length s_i of that portion of the curve joining the points $(f(t_{i-1}), g(t_{i-1}), h(t_{i-1}))$ and $(f(t_i), g(t_i), h(t_i))$ by the straight-line distance between the points (see Figure 11.7b for an illustration of the case where $n = 4$). From the distance formula, we have

$$s_i \approx d\{(f(t_{i-1}), g(t_{i-1}), h(t_{i-1})), (f(t_i), g(t_i), h(t_i))\}$$
$$= \sqrt{[f(t_i) - f(t_{i-1})]^2 + [g(t_i) - g(t_{i-1})]^2 + [h(t_i) - h(t_{i-1})]^2}.$$

Applying the Mean Value Theorem three times (why can we do this?), we get

$$f(t_i) - f(t_{i-1}) = f'(c_i)(t_i - t_{i-1}) = f'(c_i) \, \Delta t,$$
$$g(t_i) - g(t_{i-1}) = g'(d_i)(t_i - t_{i-1}) = g'(d_i) \, \Delta t$$

and

$$h(t_i) - h(t_{i-1}) = h'(e_i)(t_i - t_{i-1}) = h'(e_i) \, \Delta t,$$

for some points c_i, d_i and e_i in the interval (t_{i-1}, t_i). This gives us

$$s_i \approx \sqrt{[f(t_i) - f(t_{i-1})]^2 + [g(t_i) - g(t_{i-1})]^2 + [h(t_i) - h(t_{i-1})]^2}$$
$$= \sqrt{[f'(c_i) \, \Delta t]^2 + [g'(d_i) \, \Delta t]^2 + [h'(e_i) \, \Delta t]^2}$$
$$= \sqrt{[f'(c_i)]^2 + [g'(d_i)]^2 + [h'(e_i)]^2} \, \Delta t.$$

Notice that if Δt is small, then all of c_i, d_i and e_i are very close and we can make the further approximation

$$s_i \approx \sqrt{[f'(c_i)]^2 + [g'(c_i)]^2 + [h'(c_i)]^2} \, \Delta t,$$

for each $i = 1, 2, \ldots, n$. The total arc length is then approximately

$$s \approx \sum_{i=1}^{n} \sqrt{[f'(c_i)]^2 + [g'(c_i)]^2 + [h'(c_i)]^2} \, \Delta t,$$

where the total error in the approximation of arc length tends to 0, as $\Delta t \to 0$. (Carefully consider Figures 11.7b and 11.7c to see why.)

Taking the limit as $n \to \infty$ gives the exact arc length:

$$s = \lim_{n \to \infty} \sum_{i=1}^{n} \sqrt{[f'(c_i)]^2 + [g'(c_i)]^2 + [h'(c_i)]^2} \, \Delta t,$$

provided the limit exists. You should recognize this as the definite integral

Arc length

$$s = \int_a^b \sqrt{[f'(t)]^2 + [g'(t)]^2 + [h'(t)]^2}\, dt.$$ (1.4)

You should now observe that the arc length formula for a curve in $\mathbb{R}^2$ (1.3) is a special case of (1.4). Unfortunately, the integral in (1.4) can only rarely be computed exactly and we must typically be satisfied with a numerical approximation. The following example illustrates one of the very few arc lengths in $\mathbb{R}^3$ that can be computed exactly.

Example 1.6   Computing Arc Length in $\mathbb{R}^3$

Find the arc length of the curve traced out by the endpoint of the vector-valued function $\mathbf{r}(t) = \langle 2t, \ln t, t^2 \rangle$, for $1 \leq t \leq e$.

Solution   First, notice that $\mathbf{r}'(t) = \langle 2, \dfrac{1}{t}, 2t \rangle$. From (1.4), we now have

$$s = \int_1^e \sqrt{2^2 + \left(\frac{1}{t}\right)^2 + (2t)^2}\, dt = \int_1^e \sqrt{4 + \frac{1}{t^2} + 4t^2}\, dt$$

$$= \int_1^e \sqrt{\frac{1 + 4t^2 + 4t^4}{t^2}}\, dt = \int_1^e \sqrt{\frac{(1 + 2t^2)^2}{t^2}}\, dt$$

$$= \int_1^e \frac{1 + 2t^2}{t}\, dt = \int_1^e \left(\frac{1}{t} + 2t\right) dt$$

$$= \left(\ln|t| + 2\frac{t^2}{2}\right)\Bigg|_1^e = (\ln e + e^2) - (\ln 1 + 1) = e^2.$$

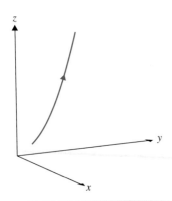

Figure 11.8
The curve defined by
$\mathbf{r}(t) = \langle 2t, \ln t, t^2 \rangle$.

We show a graph of the curve for $1 \leq t \leq e$ in Figure 11.8.

Don't worry if you don't see how to calculate an arc length integral exactly; **most can't** be done exactly! This is why we have developed numerical methods of integration. In any case, it's more important that you know what arc length is and how it is calculated, than it is for you to be able to carry out any particular calculations.

Example 1.7   Approximating Arc Length in $\mathbb{R}^3$

Find the arc length of the curve traced out by the endpoint of the vector-valued function $\mathbf{r}(t) = \langle e^{2t}, \sin t, t \rangle$, for $0 \leq t \leq 2$.

Solution   First, note that $\mathbf{r}'(t) = \langle 2e^{2t}, \cos t, 1 \rangle$. From (1.4), we now have

$$s = \int_0^2 \sqrt{(2e^{2t})^2 + (\cos t)^2 + 1^2}\, dt = \int_0^2 \sqrt{4e^{4t} + \cos^2 t + 1}\, dt.$$

Since you don't know how to evaluate this integral exactly (this is usually the case), you can approximate the integral using Simpson's Rule or the numerical integration routine built into your calculator or computer algebra system to find that the arc length is approximately $s \approx 53.8$.

EXERCISES 11.1

1. Discuss the differences, if any, between the curve traced out by the terminal point of the vector-valued function $\mathbf{r}(t) = \langle f(t), g(t) \rangle$ and the curve defined parametrically by $x = f(t), y = g(t)$.

2. In example 1.3, describe the "shadow" of the helix in the xy-plane (the shadow created by shining a light down from the "top" of the z-axis). Equivalently, if the helix is collapsed down into the xy-plane, describe the resulting curve. Compare this curve to the ellipse defined parametrically by $x = \sin t, y = -3 \cos t$.

3. Discuss how you would compute the arc length of a curve in four or more dimensions. Specifically, for the curve traced out by the terminal point of the n-dimensional vector-valued function $\mathbf{r}(t) = \langle f_1(t), f_2(t), \ldots, f_n(t) \rangle$ for $n \geq 4$, state the arc length formula and discuss how it relates to the n-dimensional distance formula.

4. The helix in Figure 11.4a is shown from a standard viewpoint (above the xy-plane, in between the x- and y-axes). Describe what an observer at the point $(0, 0, -1000)$ would see. Also, describe what observers at the points $(1000, 0, 0)$ and $(0, 1000, 0)$ would see.

In exercises 5–8, plot the values of the vector-valued function.

5. $\mathbf{r}(t) = \langle 3t, t^2, 2t - 1 \rangle, t = 0, t = 1, t = 2$

6. $\mathbf{r}(t) = (4 - t)\mathbf{i} + (1 - t^2)\mathbf{j} + (t^3 - 1)\mathbf{k}, t = -2, t = 0, t = 2$

7. $\mathbf{r}(t) = \langle \cos 3t, 2, \sin 2t - 1 \rangle, t = -\frac{\pi}{2}, t = 0, t = \frac{\pi}{2}$

8. $\mathbf{r}(t) = \langle e^{2-t}, 1 - t, 3 \rangle, t = -1, t = 0, t = 1$

In exercises 9–34, sketch the curve traced out by the given vector-valued function.

9. $\mathbf{r}(t) = \langle 2 \cos t, \sin t - 1 \rangle$ 　　10. $\mathbf{r}(t) = \langle \sin t - 2, 4 \cos t \rangle$

11. $\mathbf{r}(t) = \langle 2 \cos t + \sin 2t, 2 \sin t + \cos 2t \rangle$

12. $\mathbf{r}(t) = \langle 2 \cos 3t + \sin 5t, 2 \sin 3t + \cos 5t \rangle$

13. $\mathbf{r}(t) = \langle 4 \cos 4t - 6 \cos t, 4 \sin 4t - 6 \sin t \rangle$

14. $\mathbf{r}(t) = \langle 8 \cos t + 2 \cos 7t, 8 \sin t + 2 \sin 7t \rangle$

15. $\mathbf{r}(t) = \langle 2 \cos t, 2 \sin t, 3 \rangle$ 　　16. $\mathbf{r}(t) = \langle \cos 2t, \sin 2t, 1 \rangle$

17. $\mathbf{r}(t) = \langle t, t^2 + 1, -1 \rangle$ 　　18. $\mathbf{r}(t) = \langle 3, t, t^2 - 1 \rangle$

19. $\mathbf{r}(t) = \langle t, 1, 3t^2 \rangle$

20. $\mathbf{r}(t) = \langle t + 2, 2t - 1, t + 2 \rangle$

21. $\mathbf{r}(t) = \langle 4t - 1, 2t + 1, -6t \rangle$

22. $\mathbf{r}(t) = \langle -2t, 2t, 3 - t \rangle$ 　　23. $\mathbf{r}(t) = \langle 3 \cos t, 3 \sin t, t \rangle$

24. $\mathbf{r}(t) = \langle 2 \cos t, \sin t, 3t \rangle$ 　　25. $\mathbf{r}(t) = \langle 2 \cos t, 3 \sin t, 2t \rangle$

26. $\mathbf{r}(t) = \langle -1, 2 \cos t, 2 \sin t \rangle$ 　　27. $\mathbf{r}(t) = \langle t \cos 2t, t \sin 2t, 2t \rangle$

28. $\mathbf{r}(t) = \langle t \cos t, 2t, t \sin t \rangle$ 　　29. $\mathbf{r}(t) = \langle \cos 5t, \sin t, \sin 6t \rangle$

30. $\mathbf{r}(t) = \langle 3 \cos 2t, \sin t, \cos 3t \rangle$

31. $\mathbf{r}(t) = \langle t, t, 2t^2 - 1 \rangle$ 　　32. $\mathbf{r}(t) = \langle t^3 - t, t^2, 2t - 4 \rangle$

33. $\mathbf{r}(t) = \langle \tan t, \sin t^2, \cos t \rangle$ 　　34. $\mathbf{r}(t) = \langle \sin t, -\csc t, \cot t \rangle$

35. In exercises a–f, match the vector-valued function with its graph. Give reasons for your choices.

(a) $\mathbf{r}(t) = \langle \cos t^2, t, t \rangle$

(b) $\mathbf{r}(t) = \langle \cos t, \sin t, \sin t^2 \rangle$

(c) $\mathbf{r}(t) = \langle \sin 16\sqrt{t}, \cos 16\sqrt{t}, t \rangle$

(d) $\mathbf{r}(t) = \langle \sin t^2, \cos t^2, t \rangle$

(e) $\mathbf{r}(t) = \langle t, t, 6 - 4t^2 \rangle$

(f) $\mathbf{r}(t) = \langle t^3 - t, 0.5t^2, 2t - 4 \rangle$

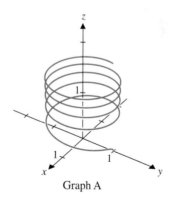

Graph A

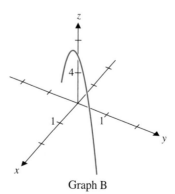

Graph B

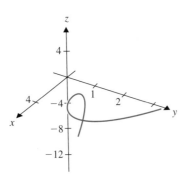

Graph C

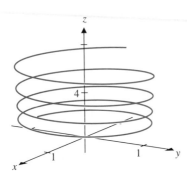

Graph D

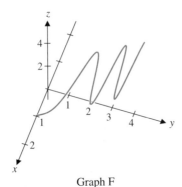

Graph E

Graph F

36. Of the functions in exercise 35, which are periodic? Which are bounded?

In exercises 37–42, use a CAS to sketch the curve and find its arc length.

37. $\mathbf{r}(t) = \langle \cos t, \sin t, \cos 2t \rangle, 0 \leq t \leq 2\pi$

38. $\mathbf{r}(t) = \langle \cos t, \sin t, \sin t + \cos t \rangle, 0 \leq t \leq 2\pi$

39. $\mathbf{r}(t) = \langle \cos \pi t, \sin \pi t, \cos 16t \rangle, 0 \leq t \leq 2$

40. $\mathbf{r}(t) = \langle \cos \pi t, \sin \pi t, \cos 16t \rangle, 0 \leq t \leq 4$

41. $\mathbf{r}(t) = \langle t, t^2 - 1, t^3 \rangle, 0 \leq t \leq 2$

42. $\mathbf{r}(t) = \langle t^2 + 1, 2t, t^2 - 1 \rangle, 0 \leq t \leq 2$

43. Show that the curve in exercise 37 lies on the hyperbolic paraboloid $z = x^2 - y^2$. Use a CAS to sketch both the surface and the curve.

44. Show that the curve in exercise 38 lies on the plane $z = x + y$. Use a CAS to sketch both the plane and the curve.

45. Show that the curve $\mathbf{r}(t) = \langle 2t, 4t^2 - 1, 8t^3 \rangle, 0 \leq t \leq 1$ has the same arc length as the curve in exercise 41.

46. Show that the curve $\mathbf{r}(t) = \langle t + 1, 2\sqrt{t}, t - 1 \rangle, 0 \leq t \leq 4$ has the same arc length as the curve in exercise 42.

47. Compare the graphs of $\mathbf{r}(t) = \langle t, t^2, t^2 \rangle$, $\mathbf{g}(t) = \langle \cos t, \cos^2 t, \cos^2 t \rangle$ and $\mathbf{h}(t) = \langle \sqrt{t}, t, t \rangle$. Explain the similarities and the differences.

48. Compare the graphs of $\mathbf{r}(t) = \langle 2t - 1, t^2, t \rangle$, $\mathbf{g}(t) = \langle 2 \sin t - 1, \sin^2 t, \sin t \rangle$ and $\mathbf{h}(t) = \langle 2e^t - 1, e^{2t}, e^t \rangle$. Explain the similarities and the differences.

49. Use a graphing utility to sketch the graph of $\mathbf{r}(t) = \langle \cos t, \cos t, \sin t \rangle$ with $0 \leq t \leq 2\pi$. Explain why the graph should be the same with $0 \leq t \leq T$ for any $T \geq 2\pi$. Try several larger domains ($0 \leq t \leq 2\pi$, $0 \leq t \leq 10\pi$, $0 \leq t \leq 50\pi$ etc.) with your graphing utility. Eventually, the ellipse should start looking thicker and for large enough domains you will see a mess of jagged lines. Explain what has gone wrong with the graphing utility.

50. It may surprise you that the curve in exercise 49 is not a circle. Show that the traces in the xz-plane and yz-plane are circles. Show that the curve lies in the plane $x = y$. Sketch a graph showing the plane $x = y$ and a circular trace in the yz-plane. To draw a curve in the plane $x = y$ with the circular trace, explain why the curve must be wider in the xy-direction than in the z-direction. In other words, the curve is not circular.

51. In contrast to exercises 49 and 50, the graph of $\mathbf{r}(t) = \langle \cos t, \cos t, \sqrt{2} \sin t \rangle$ **is** a circle. To verify this, start by showing that $\|\mathbf{r}(t)\| = \sqrt{2}$ for all t. Then observe that the

curve lies in the plane $x = y$. Explain why this proves that the graph is a (portion of a) circle. A little more insight can be gained by looking at basis vectors. The circle lies in the plane $x = y$, which contains the vector $\mathbf{u} = \frac{1}{\sqrt{2}}\langle 1, 1, 0 \rangle$. The plane $x = y$ also contains the vector $\mathbf{v} = \langle 0, 0, 1 \rangle$. Show that **any** vector $\mathbf{w}$ in the plane $x = y$ can be written as $\mathbf{w} = c_1\mathbf{u} + c_2\mathbf{v}$ for some constants c_1 and c_2. Also, show that $\mathbf{r}(t) = (\sqrt{2}\cos t)\mathbf{u} + (\sqrt{2}\sin t)\mathbf{v}$. Recall that in two dimensions, a circle of radius r centered at the origin can be written parametrically as $(r\cos t)\mathbf{i} + (r\sin t)\mathbf{j}$. In general, suppose that $\mathbf{u}$ and $\mathbf{v}$ are any orthogonal unit vectors. If $\mathbf{r}(t) = (r\cos t)\mathbf{u} + (r\sin t)\mathbf{v}$, show that $\mathbf{r}(t) \cdot \mathbf{r}(t) = r^2$.

52. Referring back to exercises 13 and 14, examine the graphs of several vector-valued functions of the form $\mathbf{r}(t) = \langle a\cos ct + b\cos dt, a\sin ct + b\sin dt \rangle$ for constants a, b, c and d. Determine the values of these constants that produce graphs of different types. For example, starting with the graph of $\langle 4\cos 4t - 6\cos t, 4\sin 4t - 6\sin t \rangle$, change $c = 4$ to $c = 3$, $c = 5$, $c = 2$, etc. Conjecture a relationship between the number of loops and the difference between c and d. Test this conjecture on other vector-valued functions. Returning to $\langle 4\cos 4t - 6\cos t, 4\sin 4t - 6\sin t \rangle$, change $a = 4$ to other values. Conjecture a relationship between the size of the loops and the value of a.

11.2 THE CALCULUS OF VECTOR-VALUED FUNCTIONS

Now that we have defined vector-valued functions, we need some tools for examining them. In this section, we begin to explore the calculus of vector-valued functions. As with scalar functions, we begin with the notion of limit and progress to continuity, derivatives and finally, integrals. Take careful note of how our presentation parallels that from Chapters 1, 2 and 4. We follow this same kind of progression again when we examine functions of several variables in Chapter 12. We define everything in this section in terms of vector-valued functions in three dimensions. The definitions can be interpreted for vector-valued functions in two dimensions in the obvious way, by simply dropping the third component everywhere.

For a vector-valued function $\mathbf{r}(t) = \langle f(t), g(t), h(t) \rangle$, if we write

$$\lim_{t \to a} \mathbf{r}(t) = \mathbf{u},$$

we mean that as t gets closer and closer to a, the vector $\mathbf{r}(t)$ is getting closer and closer to the vector $\mathbf{u}$. If we write $\mathbf{u} = \langle u_1, u_2, u_3 \rangle$, this means that

$$\lim_{t \to a} \mathbf{r}(t) = \lim_{t \to a} \langle f(t), g(t), h(t) \rangle = \mathbf{u} = \langle u_1, u_2, u_3 \rangle.$$

Notice that for this to occur, we must have that $f(t)$ is approaching u_1, $g(t)$ is approaching u_2 and $h(t)$ is approaching u_3. In view of this, we make the following definition.

Definition 2.1

For a vector-valued function $\mathbf{r}(t) = \langle f(t), g(t), h(t) \rangle$, the **limit** of $\mathbf{r}(t)$ as t approaches a is given by

Limit of a vector-valued function

$$\lim_{t \to a} \mathbf{r}(t) = \lim_{t \to a} \langle f(t), g(t), h(t) \rangle = \left\langle \lim_{t \to a} f(t), \lim_{t \to a} g(t), \lim_{t \to a} h(t) \right\rangle, \qquad (2.1)$$

provided **all** of the indicated limits exist. If any of the limits indicated on the right-hand side of (2.1) fail to exist, then $\lim_{t \to a} \mathbf{r}(t)$ **does not exist.**

In the following example, we see that calculating a limit of a vector-valued function simply consists of calculating three separate limits of scalar functions.

Example 2.1   Finding the Limit of a Vector-Valued Function

Find $\lim_{t \to 0}\langle t^2 + 1, 5\cos t, \sin t\rangle$.

Solution   Recall that each of the component functions is continuous (for all t) and so, we can calculate their limits simply by substituting the values for t. We have

$$\lim_{t \to 0}\langle t^2 + 1, 5\cos t, \sin t\rangle = \left\langle \lim_{t \to 0}(t^2 + 1), 5\lim_{t \to 0}\cos t, \lim_{t \to 0}\sin t\right\rangle$$

$$= \langle 1, 5, 0\rangle.$$

Example 2.2   A Limit That Does Not Exist

Find $\lim_{t \to 0}\langle e^{2t} + 5, t^2 + 2t - 3, 1/t\rangle$.

Solution   Notice that the limit of the third component is $\lim_{t \to 0}\dfrac{1}{t}$, which does not exist. So, even though the limits of the first two components exist, the limit of the vector-valued function does not exist.

Recall that for a scalar function f, we say that f is **continuous** at a if and only if

$$\lim_{t \to a} f(t) = f(a).$$

That is, a scalar function is continuous at a point whenever the limit and the value of the function are the same. We define the continuity of vector-valued functions in the same way.

Definition 2.2

The vector-valued function $\mathbf{r}(t) = \langle f(t), g(t), h(t)\rangle$ is **continuous** at $t = a$ whenever

$$\lim_{t \to a}\mathbf{r}(t) = \mathbf{r}(a)$$

Continuity of a vector-valued function

(i.e., whenever the limit and the value of the vector-valued function are the same).

Notice that in terms of the components of $\mathbf{r}$, this says that $\mathbf{r}(t)$ is continuous at $t = a$ whenever

$$\lim_{t \to a}\langle f(t), g(t), h(t)\rangle = \langle f(a), g(a), h(a)\rangle.$$

Further, since

$$\lim_{t \to a}\langle f(t), g(t), h(t)\rangle = \left\langle \lim_{t \to a} f(t), \lim_{t \to a} g(t), \lim_{t \to a} h(t)\right\rangle,$$

it follows that $\mathbf{r}$ is continuous at $t = a$ if and only if

$$\left\langle \lim_{t \to a} f(t), \lim_{t \to a} g(t), \lim_{t \to a} h(t)\right\rangle = \langle f(a), g(a), h(a)\rangle.$$

Finally, note that this occurs if and only if

$$\lim_{t \to a} f(t) = f(a), \quad \lim_{t \to a} g(t) = g(a) \quad \text{and} \quad \lim_{t \to a} h(t) = h(a).$$

If you look carefully at what we have just said, you'll notice that we just proved the following theorem.

Theorem 2.1

A vector-valued function $\mathbf{r}(t) = \langle f(t), g(t), h(t) \rangle$ is continuous at $t = a$ if and only if **all** of f, g and h are continuous at $t = a$.

Notice that Theorem 2.1 says that if you want to determine whether or not a vector-valued function is continuous, you need only check the continuity of each component function (something you already know how to do!). We demonstrate this in the following two examples.

Example 2.3 Determining Where a Vector-Valued Function is Continuous

Determine for what values of t the vector-valued function $\mathbf{r}(t) = \langle e^{5t}, \ln(t + 1), \cos t \rangle$ is continuous.

Solution From Theorem 2.1, we need only consider the continuity of the component functions; $\mathbf{r}(t)$ will be continuous wherever **all** its components are continuous. We have: e^{5t} is continuous for all t, $\ln(t + 1)$ is continuous for $t > -1$ and $\cos t$ is continuous for all t. This says that $\mathbf{r}(t)$ is continuous for $t > -1$.

■ ────────

Example 2.4 A Vector-Valued Function with Infinitely Many Discontinuities

Determine for what values of t the vector-valued function $\mathbf{r}(t) = \langle \tan t, |t + 3|, \frac{1}{t-2} \rangle$ is continuous.

Solution First, note that $\tan t$ is continuous, except at $t = \dfrac{(2n + 1)\pi}{2}$, for $n = 0$, ± 1, ± 2, ... (i.e., $\tan t$ is continuous except at odd multiples of $\frac{\pi}{2}$). The second component $|t + 3|$ is continuous for all t (although it's not differentiable at $t = -3$). Finally, the third component $\dfrac{1}{t - 2}$ is continuous except at $t = 2$. Since all three components must be continuous in order for $\mathbf{r}(t)$ to be continuous, we have that $\mathbf{r}(t)$ is continuous, except at $t = 2$ and $t = \dfrac{(2n + 1)\pi}{2}$, for $n = 0, \pm 1, \pm 2, \ldots$.

■ ────────

Recall that in Chapter 2, we defined the derivative of a scalar function f to be

$$f'(t) = \lim_{h \to 0} \frac{f(t + h) - f(t)}{h}.$$

Replacing h by Δt, we can rewrite this as

$$f'(t) = \lim_{\Delta t \to 0} \frac{f(t + \Delta t) - f(t)}{\Delta t}.$$

You may be wondering why we want to change from a perfectly nice variable like h to something more unusual like Δt. The only reason is that we want to use the notation to emphasize that Δt is an **increment** of the variable t. In Chapter 12, we'll be defining partial derivatives of functions of more than one variable, where we'll use this type of notation to make it clear which variable is being incremented.

We now define the derivative of a vector-valued function in the expected way.

Definition 2.3
The **derivative $\mathbf{r}'(t)$** of the vector-valued function $\mathbf{r}(t)$ is defined by

$$\mathbf{r}'(t) = \lim_{\Delta t \to 0} \frac{\mathbf{r}(t + \Delta t) - \mathbf{r}(t)}{\Delta t}, \qquad (2.2)$$

for any values of t for which the limit exists. When the limit exists for $t = a$, we say that $\mathbf{r}$ is **differentiable** at $t = a$.

Fortunately, you will not need to learn any new differentiation rules, as the derivative of a vector-valued function is found directly from the derivatives of the individual components, as we see in the following result.

Theorem 2.2
Let $\mathbf{r}(t) = \langle f(t), g(t), h(t) \rangle$ and suppose that the components f, g and h are all differentiable for some value of t. Then $\mathbf{r}$ is also differentiable at that value of t and its derivative is given by

$$\mathbf{r}'(t) = \langle f'(t), g'(t), h'(t) \rangle. \qquad (2.3)$$

Proof

From the definition of derivative of a vector-valued function (2.2), we have

$$\mathbf{r}'(t) = \lim_{\Delta t \to 0} \frac{\mathbf{r}(t + \Delta t) - \mathbf{r}(t)}{\Delta t}$$

$$= \lim_{\Delta t \to 0} \frac{1}{\Delta t} [\langle f(t + \Delta t), g(t + \Delta t), h(t + \Delta t) \rangle - \langle f(t), g(t), h(t) \rangle]$$

$$= \lim_{\Delta t \to 0} \frac{1}{\Delta t} \langle f(t + \Delta t) - f(t), g(t + \Delta t) - g(t), h(t + \Delta t) - h(t) \rangle,$$

where we have used the definition of vector subtraction. Distributing the scalar $\dfrac{1}{\Delta t}$ into each component and using the definition of limit of a vector-valued function (2.1), we have

$$\mathbf{r}'(t) = \lim_{\Delta t \to 0} \frac{1}{\Delta t} \langle f(t + \Delta t) - f(t), g(t + \Delta t) - g(t), h(t + \Delta t) - h(t) \rangle$$

$$= \lim_{\Delta t \to 0} \left\langle \frac{f(t + \Delta t) - f(t)}{\Delta t}, \frac{g(t + \Delta t) - g(t)}{\Delta t}, \frac{h(t + \Delta t) - h(t)}{\Delta t} \right\rangle$$

$$= \left\langle \lim_{\Delta t \to 0} \frac{f(t + \Delta t) - f(t)}{\Delta t}, \lim_{\Delta t \to 0} \frac{g(t + \Delta t) - g(t)}{\Delta t}, \lim_{\Delta t \to 0} \frac{h(t + \Delta t) - h(t)}{\Delta t} \right\rangle$$

$$= \langle f'(t), g'(t), h'(t) \rangle,$$

where in the last step we recognized the definition of the derivatives of each of the component functions f, g and h.

Notice that thanks to Theorem 2.2, in order to differentiate a vector-valued function, we need only differentiate the individual component functions, using the usual rules of differentiation. We illustrate this in the following example.

Example 2.5 Finding the Derivative of a Vector-Valued Function

Find the derivative of $\mathbf{r}(t) = \langle \sin(t^2), e^{\cos t}, t \ln t \rangle$.

Solution Applying the chain rule to the first two components and the product rule to the third, we have (for $t > 0$):

$$\mathbf{r}'(t) = \left\langle \frac{d}{dt}[\sin(t^2)], \frac{d}{dt}(e^{\cos t}), \frac{d}{dt}(t \ln t) \right\rangle$$

$$= \left\langle \cos(t^2) \frac{d}{dt}(t^2), e^{\cos t} \frac{d}{dt}(\cos t), \frac{d}{dt}(t) \ln t + t \frac{d}{dt}(\ln t) \right\rangle$$

$$= \left\langle \cos(t^2)(2t), e^{\cos t}(-\sin t), (1) \ln t + t \frac{1}{t} \right\rangle$$

$$= \langle 2t \cos(t^2), -\sin t \, e^{\cos t}, \ln t + 1 \rangle.$$

For the most part, to compute derivatives of vector-valued functions, we only need to use the already familiar rules for differentiation of scalar functions. There are several special derivative rules, however, which we state in the following theorem.

Theorem 2.3

Suppose that $\mathbf{r}(t)$ and $\mathbf{s}(t)$ are differentiable vector-valued functions, $f(t)$ is a differentiable scalar function and c is any scalar constant. Then

(i) $\dfrac{d}{dt}[\mathbf{r}(t) + \mathbf{s}(t)] = \mathbf{r}'(t) + \mathbf{s}'(t)$

(ii) $\dfrac{d}{dt}[c\mathbf{r}(t)] = c\mathbf{r}'(t)$

(iii) $\dfrac{d}{dt}[f(t)\mathbf{r}(t)] = f'(t)\mathbf{r}(t) + f(t)\mathbf{r}'(t)$

(iv) $\dfrac{d}{dt}[\mathbf{r}(t) \cdot \mathbf{s}(t)] = \mathbf{r}'(t) \cdot \mathbf{s}(t) + \mathbf{r}(t) \cdot \mathbf{s}'(t)$ and

(v) $\dfrac{d}{dt}[\mathbf{r}(t) \times \mathbf{s}(t)] = \mathbf{r}'(t) \times \mathbf{s}(t) + \mathbf{r}(t) \times \mathbf{s}'(t)$.

Notice that parts (iii), (iv) and (v) are the product rules for the various kinds of products we can define. In (iii), we have the derivative of a product of a scalar function and a vector-valued function; in (iv) we have the derivative of a dot product and in (v), we have the derivative of a cross product. In each of these three cases, it's important for you to recognize that these follow the same pattern as the product rule for the derivative of the product of two scalar functions.

Proof

(i) For $\mathbf{r}(t) = \langle f_1(t), g_1(t), h_1(t) \rangle$ and $\mathbf{s}(t) = \langle f_2(t), g_2(t), h_2(t) \rangle$, we have from (2.3) and the rules for vector addition that

$$\frac{d}{dt}[\mathbf{r}(t) + \mathbf{s}(t)] = \frac{d}{dt}[\langle f_1(t), g_1(t), h_1(t) \rangle + \langle f_2(t), g_2(t), h_2(t) \rangle]$$

$$= \frac{d}{dt}\langle f_1(t) + f_2(t), g_1(t) + g_2(t), h_1(t) + h_2(t) \rangle$$

$$= \langle f_1'(t) + f_2'(t), g_1'(t) + g_2'(t), h_1'(t) + h_2'(t) \rangle$$

$$= \langle f_1'(t), g_1'(t), h_1'(t) \rangle + \langle f_2'(t), g_2'(t), h_2'(t) \rangle$$

$$= \mathbf{r}'(t) + \mathbf{s}'(t).$$

(iv) From the definition of dot product and the usual product rule for the product of two scalar functions, we have

$$\frac{d}{dt}[\mathbf{r}(t) \cdot \mathbf{s}(t)] = \frac{d}{dt}[\langle f_1(t), g_1(t), h_1(t) \rangle \cdot \langle f_2(t), g_2(t), h_2(t) \rangle]$$

$$= \frac{d}{dt}[f_1(t)f_2(t) + g_1(t)g_2(t) + h_1(t)h_2(t)]$$

$$= f_1'(t)f_2(t) + f_1(t)f_2'(t) + g_1'(t)g_2(t) + g_1(t)g_2'(t)$$
$$+ h_1'(t)h_2(t) + h_1(t)h_2'(t)$$

$$= [f_1'(t)f_2(t) + g_1'(t)g_2(t) + h_1'(t)h_2(t)]$$
$$+ [f_1(t)f_2'(t) + g_1(t)g_2'(t) + h_1(t)h_2'(t)]$$

$$= \mathbf{r}'(t) \cdot \mathbf{s}(t) + \mathbf{r}(t) \cdot \mathbf{s}'(t).$$

We leave the proofs of (ii), (iii) and (v) as exercises.

We next explore an important graphical interpretation of the derivative of a vector-valued function. First, recall that one interpretation of the derivative of a scalar function is that the value of the derivative at a point gives the slope of the tangent line to the curve at that point. For the case of the vector-valued function $\mathbf{r}(t)$, notice that from (2.2), the derivative of $\mathbf{r}(t)$ at $t = a$ is given by

$$\mathbf{r}'(a) = \lim_{\Delta t \to 0} \frac{\mathbf{r}(a + \Delta t) - \mathbf{r}(a)}{\Delta t}.$$

Again, recall that the endpoint of the vector-valued function $\mathbf{r}(t)$ traces out a curve C in $\mathbb{R}^3$. In Figure 11.9a, we show the position vectors $\mathbf{r}(a)$, $\mathbf{r}(a + \Delta t)$ and $\mathbf{r}(a + \Delta t) - \mathbf{r}(a)$, for some fixed $\Delta t > 0$, using our graphical interpretation of vector subtraction, developed in Chapter 10. (How does the picture differ if $\Delta t < 0$?) Notice that for $\Delta t > 0$, the vector $\frac{\mathbf{r}(a + \Delta t) - \mathbf{r}(a)}{\Delta t}$ points in the same direction as $\mathbf{r}(a + \Delta t) - \mathbf{r}(a)$.

If we take smaller and smaller values of Δt, $\frac{\mathbf{r}(a + \Delta t) - \mathbf{r}(a)}{\Delta t}$ will approach $\mathbf{r}'(a)$. We illustrate this graphically in Figures 11.9b and 11.9c.

As $\Delta t \to 0$, notice that the vector $\frac{\mathbf{r}(a + \Delta t) - \mathbf{r}(a)}{\Delta t}$ approaches a vector that is tangent to the curve C at the terminal point of $\mathbf{r}(a)$, as seen in Figure 11.9d. We refer to

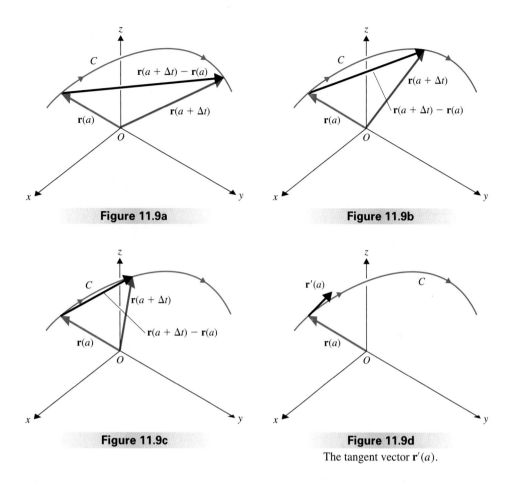

Figure 11.9a

Figure 11.9b

Figure 11.9c

Figure 11.9d
The tangent vector $\mathbf{r}'(a)$.

$\mathbf{r}'(a)$ as the **tangent vector** to the curve C at the point corresponding to $t = a$. Be sure to observe that $\mathbf{r}'(a)$ lies along the tangent line to the curve at $t = a$ and points in the direction of the orientation of C. (Recognize that Figures 11.9a, 11.9b and 11.9c are all drawn so that $\Delta t > 0$. What changes in each of the figures if $\Delta t < 0$?)

We illustrate this notion for a simple curve in $\mathbb{R}^2$ in the following example.

Example 2.6   Drawing Position and Tangent Vectors

For $\mathbf{r}(t) = \langle -\cos 2t, \sin 2t \rangle$, plot the curve traced out by the endpoint of $\mathbf{r}(t)$ and draw the position vector and tangent vector at $t = \frac{\pi}{4}$.

Solution   First, notice that

$$\mathbf{r}'(t) = \langle 2 \sin 2t, 2 \cos 2t \rangle.$$

Also, the curve traced out by $\mathbf{r}(t)$ is given parametrically by

$$C : x = -\cos 2t, \quad y = \sin 2t, \ t \in \mathbb{R}.$$

You should observe that, in this case,

$$x^2 + y^2 = \cos^2 2t + \sin^2 2t = 1,$$

so that the curve is the circle of radius 1, centered at the origin. Further, from the parameterization, you can see that the orientation is clockwise. The position and tangent

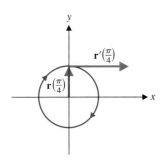

y

$\mathbf{r}'\!\left(\frac{\pi}{4}\right)$

$\mathbf{r}\!\left(\frac{\pi}{4}\right)$

x

Figure 11.10

Position and tangent vectors.

vectors at $t = \frac{\pi}{4}$ are given by

$$\mathbf{r}\!\left(\frac{\pi}{4}\right) = \left\langle -\cos\frac{\pi}{2},\ \sin\frac{\pi}{2} \right\rangle = \langle 0, 1\rangle$$

and

$$\mathbf{r}'\!\left(\frac{\pi}{4}\right) = \left\langle 2\sin\frac{\pi}{2},\ 2\cos\frac{\pi}{2} \right\rangle = \langle 2, 0\rangle,$$

respectively. We show the curve, along with the vectors $\mathbf{r}\!\left(\frac{\pi}{4}\right)$ and $\mathbf{r}'\!\left(\frac{\pi}{4}\right)$ in Figure 11.10. In particular, you might note that

$$\mathbf{r}\!\left(\frac{\pi}{4}\right) \cdot \mathbf{r}'\!\left(\frac{\pi}{4}\right) = 0,$$

so that $\mathbf{r}\!\left(\frac{\pi}{4}\right)$ and $\mathbf{r}'\!\left(\frac{\pi}{4}\right)$ are orthogonal. In fact, $\mathbf{r}(t)$ and $\mathbf{r}'(t)$ are orthogonal for every t, as follows:

$$\mathbf{r}(t) \cdot \mathbf{r}'(t) = \langle -\cos 2t,\ \sin 2t\rangle \cdot \langle 2\sin 2t,\ 2\cos 2t\rangle$$
$$= -2\cos 2t\,\sin 2t + 2\sin 2t\,\cos 2t = 0.$$

Were you surprised to find in example 2.6 that the position vector and the tangent vector were orthogonal at every point? As it turns out, this is a special case of a more general result, which we state in the following theorem.

Theorem 2.4

$\|\mathbf{r}(t)\| = $ constant if and only if $\mathbf{r}(t)$ and $\mathbf{r}'(t)$ are orthogonal, for all t.

Proof

(i) Suppose that $\|\mathbf{r}(t)\| = c$, for some constant c. Recall that

$$\mathbf{r}(t) \cdot \mathbf{r}(t) = \|\mathbf{r}(t)\|^2 = c^2. \qquad (2.4)$$

Differentiating both sides of (2.4), we get

$$\frac{d}{dt}[\mathbf{r}(t) \cdot \mathbf{r}(t)] = \frac{d}{dt}c^2 = 0.$$

From Theorem 2.3 (iv), we now have

$$0 = \frac{d}{dt}[\mathbf{r}(t) \cdot \mathbf{r}(t)] = \mathbf{r}'(t) \cdot \mathbf{r}(t) + \mathbf{r}(t) \cdot \mathbf{r}'(t) = 2\,\mathbf{r}(t) \cdot \mathbf{r}'(t),$$

so that $\mathbf{r}(t) \cdot \mathbf{r}'(t) = 0$, as desired.

(ii) We leave the proof of the converse as an exercise.

You should observe that Theorem 2.4 has some geometric significance. First, note that in two dimensions, if $\|\mathbf{r}(t)\| = c$ (where c is a constant), then the curve traced out by the position vector $\mathbf{r}(t)$ must lie on the circle of radius c, centered at the origin. (Think about this some!) We can then interpret Theorem 2.4 to say that the path traced out by $\mathbf{r}(t)$ lies on a circle centered at the origin if and only if the tangent vector is orthogonal to the position vector at every point on the curve. Likewise, in three dimensions, if $\|\mathbf{r}(t)\| = c$ (where c is a constant), the curve traced out by $\mathbf{r}(t)$ lies on the sphere of radius c centered at the origin. In this case, we can interpret Theorem 2.4 to say that the curve traced out by $\mathbf{r}(t)$ lies on a sphere centered at the origin if and only if the tangent vector is orthogonal to the position vector at every point on the curve.

We conclude this section by making a few straightforward definitions. Recall that when we say that the scalar function $F(t)$ is an antiderivative of the scalar function $f(t)$, we mean that F is any function such that $F'(t) = f(t)$. We extend this notion to vector-valued functions in the following definition.

Definition 2.4

The vector-valued function $\mathbf{R}(t)$ is an **antiderivative** of the vector-valued function $\mathbf{r}(t)$ whenever $\mathbf{R}'(t) = \mathbf{r}(t)$.

Notice that if $\mathbf{r}(t) = \langle f(t), g(t), h(t) \rangle$ and f, g and h have antiderivatives F, G and H, respectively, then

$$\frac{d}{dt} \langle F(t), G(t), H(t) \rangle = \langle F'(t), G'(t), H'(t) \rangle = \langle f(t), g(t), h(t) \rangle.$$

That is, $\langle F(t), G(t), H(t) \rangle$ is an antiderivative of $\mathbf{r}(t)$. In fact, $\langle F(t) + c_1, G(t) + c_2, H(t) + c_3 \rangle$ is also an antiderivative of $\mathbf{r}(t)$, for any choice of constants c_1, c_2 and c_3. This leads us to the following definition.

Definition 2.5

If $\mathbf{R}(t)$ is any antiderivative of $\mathbf{r}(t)$, the **indefinite integral** of $\mathbf{r}(t)$ is defined to be

$$\int \mathbf{r}(t)\, dt = \mathbf{R}(t) + \mathbf{c},$$

where $\mathbf{c}$ is an arbitrary constant vector.

As in the scalar case, $\mathbf{R}(t) + \mathbf{c}$ is the most general antiderivative of $\mathbf{r}(t)$. (Why is that?) Notice that this says that

Indefinite integral of a vector-valued function

$$\int \mathbf{r}(t)\, dt = \int \langle f(t), g(t), h(t) \rangle\, dt = \left\langle \int f(t)\, dt, \int g(t)\, dt, \int h(t)\, dt \right\rangle. \quad (2.5)$$

That is, you integrate a vector-valued function by integrating each of the individual components.

Example 2.7   Evaluating the Indefinite Integral of a Vector-Valued Function

Evaluate the indefinite integral $\int \left\langle t^2 + 2, \sin 2t, 4t e^{t^2} \right\rangle dt$.

Solution   From (2.5), we have

$$\int \left\langle t^2 + 2, \sin 2t, 4t e^{t^2} \right\rangle dt = \left\langle \int (t^2 + 2)\, dt, \int \sin 2t\, dt, \int 4t e^{t^2}\, dt \right\rangle$$

$$= \left\langle \frac{1}{3} t^3 + 2t + c_1, -\frac{1}{2} \cos 2t + c_2, 2e^{t^2} + c_3 \right\rangle$$

$$= \left\langle \frac{1}{3} t^3 + 2t, -\frac{1}{2} \cos 2t, 2e^{t^2} \right\rangle + \mathbf{c},$$

where $\mathbf{c} = \langle c_1, c_2, c_3 \rangle$ is an arbitrary constant vector.

Similarly, we define the **definite integral** of a vector-valued function in the obvious way.

Definition 2.6

For the vector-valued function $\mathbf{r}(t) = \langle f(t), g(t), h(t) \rangle$, we define the **definite integral** of $\mathbf{r}(t)$ by

Definite integral
of a vector-valued function

$$\int_a^b \mathbf{r}(t)\, dt = \int_a^b \langle f(t), g(t), h(t) \rangle\, dt = \left\langle \int_a^b f(t)\, dt, \int_a^b g(t)\, dt, \int_a^b h(t)\, dt \right\rangle. \quad (2.6)$$

Notice that this says that the definite integral of a vector-valued function $\mathbf{r}(t)$ is simply the vector whose components are the definite integrals of the corresponding components of $\mathbf{r}(t)$. With this in mind, we now extend the Fundamental Theorem of Calculus to vector-valued functions.

Theorem 2.5

Suppose that $\mathbf{R}(t)$ is an antiderivative of $\mathbf{r}(t)$ on the interval $[a, b]$. Then,

$$\int_a^b \mathbf{r}(t)\, dt = \mathbf{R}(b) - \mathbf{R}(a).$$

Proof

The proof is straightforward and we leave this as an exercise.

Example 2.8   Evaluating the Definite Integral
of a Vector-Valued Function

Evaluate $\int_0^1 \langle \sin \pi t, 6t^2 + 4t \rangle\, dt$.

Solution   Notice that an antiderivative for the integrand is

$$\left\langle -\frac{1}{\pi} \cos \pi t, \frac{6t^3}{3} + 4\frac{t^2}{2} \right\rangle = \left\langle -\frac{1}{\pi} \cos \pi t, 2t^3 + 2t^2 \right\rangle$$

From Theorem 2.5, we have that

$$\int_0^1 \langle \sin \pi t, 6t^2 + 4t \rangle\, dt = \left\langle -\frac{1}{\pi} \cos \pi t, 2t^3 + 2t^2 \right\rangle \bigg|_0^1$$

$$= \left\langle -\frac{1}{\pi} \cos \pi, 2 + 2 \right\rangle - \left\langle -\frac{1}{\pi} \cos 0, 0 \right\rangle$$

$$= \left\langle \frac{1}{\pi} + \frac{1}{\pi}, 4 - 0 \right\rangle = \left\langle \frac{2}{\pi}, 4 \right\rangle.$$

EXERCISES 11.2

1. Suppose that $\mathbf{r}(t) = \langle f(t), g(t), h(t) \rangle$, where $\lim_{t\to 0} f(t) = \lim_{t\to 0} g(t) = 0$ and $\lim_{t\to 0} h(t) = \infty$. Describe what is happening graphically as $t \to 0$ and explain why (even though the limits of two of the component functions exist) the limit of $\mathbf{r}(t)$ as $t \to 0$ does not exist.

2. In example 2.3, describe what is happening graphically for $t \le -1$. Explain why we don't say that $\mathbf{r}(t)$ is continuous for $t \le -1$.

3. Suppose that $\mathbf{r}(t)$ is a vector-valued function such that $\mathbf{r}(0) = \langle a, b, c \rangle$ and $\mathbf{r}'(0)$ exists. Imagine zooming in on the curve traced out by $\mathbf{r}(t)$ near the point (a, b, c). Describe what the curve will look like and how it relates to the tangent vector $\mathbf{r}'(0)$.

4. There is a quotient rule corresponding to the product rule in Theorem 2.3, part (iii). State this rule and describe in words how you would prove it. Explain why there isn't a quotient rule corresponding to the product rules in parts (iv) and (v) of Theorem 2.3.

In exercises 5–10, find the limit if it exists.

5. $\lim_{t\to 0} \langle t^2 - 1, e^{2t}, \sin t \rangle$

6. $\lim_{t\to 1} \langle t^2, e^{2t}, \sqrt{t^2 + 2t} \rangle$

7. $\lim_{t\to 0} \left\langle \frac{\sin t}{t}, \cos t, \frac{t + 1}{t - 1} \right\rangle$

8. $\lim_{t\to 1} \left\langle \sqrt{t - 1}, t^2 + 3, \frac{t + 1}{t - 1} \right\rangle$

9. $\lim_{t\to 0} \langle \ln t, \sqrt{t^2 + 1}, t - 3 \rangle$

10. $\lim_{t\to \pi/2} \langle \cos t, t^2 + 3, \tan t \rangle$

In exercises 11–16, determine all values of t at which the given vector-valued function is continuous.

11. $\mathbf{r}(t) = \left\langle \frac{t + 1}{t - 1}, t^2, 2t \right\rangle$

12. $\mathbf{r}(t) = \left\langle \sin t, \cos t, \frac{3}{t} \right\rangle$

13. $\mathbf{r}(t) = \langle \tan t, \sin t^2, \cos t \rangle$

14. $\mathbf{r}(t) = \langle \cos 5t, \tan t, 6\sin t \rangle$

15. $\mathbf{r}(t) = \langle 4\cos t, \sqrt{t}, 4\sin t \rangle$

16. $\mathbf{r}(t) = \langle \sin t, -\csc t, \cot t \rangle$

In exercises 17–22, find the derivative of the given vector-valued function.

17. $\mathbf{r}(t) = \left\langle t^4, \sqrt{t + 1}, \frac{3}{t^2} \right\rangle$

18. $\mathbf{r}(t) = \left\langle \frac{t - 3}{t + 1}, te^{2t}, t^3 \right\rangle$

19. $\mathbf{r}(t) = \langle \sin t, \sin t^2, \cos t \rangle$

20. $\mathbf{r}(t) = \langle \cos 5t, \tan t, 6\sin t \rangle$

21. $\mathbf{r}(t) = \left\langle e^{t^2}, t^2, \sec 2t \right\rangle$

22. $\mathbf{r}(t) = \left\langle \sqrt{t^2 + 1}, \cos t, e^{-3t} \right\rangle$

In exercises 23–26, sketch the curve traced out by the endpoint of the given vector-valued function and plot the position and tangent vectors at the indicated points.

23. $\mathbf{r}(t) = \langle \cos t, \sin t \rangle, t = 0, t = \frac{\pi}{2}, t = \pi$

24. $\mathbf{r}(t) = \langle t, t^2 - 1 \rangle, t = 0, t = 1, t = 2$

25. $\mathbf{r}(t) = \langle \cos t, t, \sin t \rangle, t = 0, t = \frac{\pi}{2}, t = \pi$

26. $\mathbf{r}(t) = \langle t, t, t^2 - 1 \rangle, t = 0, t = 1, t = 2$

In exercises 27–36, evaluate the given indefinite or definite integral.

27. $\int \langle 3t - 1, \sqrt{t} \rangle \, dt$

28. $\int \left\langle \frac{3}{t^2}, \frac{4}{t} \right\rangle \, dt$

29. $\int \langle \cos 3t, \sin t, e^{4t} \rangle \, dt$

30. $\int \langle e^{-3t}, \sin 5t, t^{3/2} \rangle \, dt$

31. $\int \left\langle te^{t^2}, 3t\sin t, \frac{3t}{t^2 + 1} \right\rangle \, dt$

32. $\int \langle e^{-3t}, t^2\cos t^3, t\cos t \rangle \, dt$

33. $\int_0^1 \langle t^2 - 1, 3t \rangle \, dt$

34. $\int_1^4 \langle \sqrt{t}, 5 \rangle \, dt$

35. $\int_0^2 \left\langle \frac{4}{t + 1}, e^{t-2}, te^t \right\rangle \, dt$

36. $\int_0^4 \left\langle 2te^{4t}, t^2 - 1, \frac{4t}{t^2 + 1} \right\rangle \, dt$

In exercises 37–40, find t such that $\mathbf{r}(t)$ and $\mathbf{r}'(t)$ are perpendicular.

37. $\mathbf{r}(t) = \langle \cos t, \sin t \rangle$

38. $\mathbf{r}(t) = \langle 2\cos t, \sin t \rangle$

39. $\mathbf{r}(t) = \langle t, t, t^2 - 1 \rangle$

40. $\mathbf{r}(t) = \langle t^2, t, t^2 - 5 \rangle$

41. In each of exercises 37 and 38, show that there are no values of t such that $\mathbf{r}(t)$ and $\mathbf{r}'(t)$ are parallel.

42. In each of exercises 39 and 40, show that there are no values of t such that $\mathbf{r}(t)$ and $\mathbf{r}'(t)$ are parallel.

In exercises 43–46, find all values of t such that $\mathbf{r}'(t)$ lies in the xy-plane.

43. $\mathbf{r}(t) = \langle t, t, t^3 - 3 \rangle$
44. $\mathbf{r}(t) = \langle t^2, t, \sin t^2 \rangle$

45. $\mathbf{r}(t) = \langle \cos t, \sin t, \sin 2t \rangle$

46. $\mathbf{r}(t) = \left\langle \sqrt{t+1}, \cos t, t^4 - 8t^2 \right\rangle$

47. Prove Theorem 2.3, part (ii).

48. In Theorem 2.3, part (ii), replace the scalar product $c\mathbf{r}(t)$ with the dot product $\mathbf{c} \cdot \mathbf{r}(t)$, for a constant vector $\mathbf{c}$ and prove the result.

49. Prove Theorem 2.3, part (iii).

50. Prove Theorem 2.3, part (v).

51. Label as true or false and explain why. If $\mathbf{u}(t) = \dfrac{1}{\|\mathbf{r}(t)\|} \mathbf{r}(t)$ and $\mathbf{u}(t) \cdot \mathbf{u}'(t) = 0$ then $\mathbf{r}(t) \cdot \mathbf{r}'(t) = 0$.

52. Label as true or false and explain why. If $\mathbf{r}(t_0) \cdot \mathbf{r}'(t_0) = 0$ for some t_0, then $\|\mathbf{r}(t)\|$ is constant.

53. Prove that if $\mathbf{r}(t)$ and $\mathbf{r}'(t)$ are orthogonal for all t, then $\|\mathbf{r}(t)\| = $ constant [Theorem 2.4, part (ii)].

54. Prove Theorem 2.5.

55. Find all values of t such that $\mathbf{r}'(t) = \mathbf{0}$ for each function: (a) $\mathbf{r}(t) = \langle t, t^2 - 1 \rangle$, (b) $\mathbf{r}(t) = \langle 2\cos t + \sin 2t, 2\sin t + \cos 2t \rangle$, (c) $\mathbf{r}(t) = \langle 2\cos 3t + \sin 5t, 2\sin 3t + \cos 5t \rangle$

and (d) $\mathbf{r}(t) = \langle t^2, t^4 - 1 \rangle$. Based on your results, conjecture the graphical significance of having the derivative of a vector-valued function equal the zero vector. If $\mathbf{r}(t)$ is the position function of some object in motion, explain the physical significance of having a zero derivative. Explain your geometric interpretation in light of your physical interpretation.

56. You may recall that a scalar function has either a discontinuity, a "sharp corner" or a cusp at places where the derivative doesn't exist. In this exercise, we look at the analogous **smoothness** of graphs of vector-valued functions. A curve C is said to be **smooth** if it is traced out by a vector-valued function $\mathbf{r}(t)$, where $\mathbf{r}'(t)$ is continuous and $\mathbf{r}'(t) \neq \mathbf{0}$ for all values of t. Sketch the graph of $\mathbf{r}(t) = \langle t, \sqrt[3]{t^2} \rangle$ and explain why we include the requirement that $\mathbf{r}'(t)$ be continuous. Sketch the graph of $\mathbf{r}(t) = \langle 2\cos t + \sin 2t, 2\sin t + \cos 2t \rangle$ and show that $\mathbf{r}'(0) = \mathbf{0}$. Explain why we include the requirement that $\mathbf{r}'(t)$ be nonzero. Sketch the graph of $\mathbf{r}(t) = \langle 2\cos 3t + \sin 5t, 2\sin 3t + \cos 5t \rangle$ and show that $\mathbf{r}'(t)$ never equals the zero vector. By zooming in on the edges of the graph, show that this curve is accurately described as smooth. Sketch the graphs of $\mathbf{r}(t) = \langle t, t^2 - 1 \rangle$ and $\mathbf{g}(t) = \langle t^2, t^4 - 1 \rangle$ for $t \geq 0$ and observe that they trace out the same curve. Show that $\mathbf{g}'(0) = \mathbf{0}$, but that the curve is smooth at $t = 0$. Explain why this says that the requirement that $\mathbf{r}'(t) \neq \mathbf{0}$ need not hold for **every** $\mathbf{r}(t)$ tracing out the curve. [This requirement only needs to hold for one such $\mathbf{r}(t)$.] Determine which of the following curves are smooth. If the curve is not smooth, identify the graphical characteristic that is "unsmooth": $\mathbf{r}(t) = \langle \cos t, \sin t, t \rangle$, $\mathbf{r}(t) = \langle \cos t, \sin t, \sqrt[3]{t^2} \rangle$, $\mathbf{r}(t) = \langle \tan t, \sin t^2, \cos t \rangle$, $\mathbf{r}(t) = \langle 5\sin^3 t, 5\cos^3 t, t \rangle$ and $\mathbf{r}(t) = \langle \cos t, t^2 e^{-t}, \cos^2 t \rangle$.

11.3 MOTION IN SPACE

We are finally at a point where we have sufficient mathematical machinery to describe the motion of an object in a three-dimensional setting. Problems such as this were among the earliest and most important applications of the calculus and continue to be of great importance today. For instance, if you launch a rocket (e.g., a space shuttle), you probably want to know where it will go. Problems such as this, dealing with motion, were one of the primary focuses of Newton and many of his contemporaries. While Newton certainly wasn't thinking about launching space shuttles into orbit, he used his newly invented calculus to explain all kinds of motion, from the motion of a projectile (such as a ball) hurled through the air, to the motion of the planets. His stunning achievements in this field unlocked mysteries that had eluded the greatest minds for centuries and form the basis of our understanding of mechanics today. In this section, we use vector-valued functions to describe objects in motion.

Suppose that an object moves along a curve described parametrically by

$$C : x = f(t), \ y = g(t), \ z = h(t),$$

where t represents time and where $t \in [a, b]$. Recall that we can think of the curve as being traced out by the endpoint of the vector-valued function

$$\mathbf{r}(t) = \langle f(t), g(t), h(t) \rangle.$$

We observed in section 11.2 that the value of $\mathbf{r}'(t)$ for any given value of t is a tangent vector pointing in the direction of the orientation of the curve. We can now give another interpretation of this. From (2.3), we have

$$\mathbf{r}'(t) = \langle f'(t), g'(t), h'(t) \rangle$$

and the magnitude of this vector-valued function is

$$\|\mathbf{r}'(t)\| = \sqrt{[f'(t)]^2 + [g'(t)]^2 + [h'(t)]^2}.$$

(Where have you seen this expression before?) Notice that from (1.4), given any number $t_0 \in [a, b]$, the arc length of the portion of the curve from $u = t_0$ up to $u = t$ is given by

$$s(t) = \int_{t_0}^{t} \sqrt{[f'(u)]^2 + [g'(u)]^2 + [h'(u)]^2} \, du. \qquad (3.1)$$

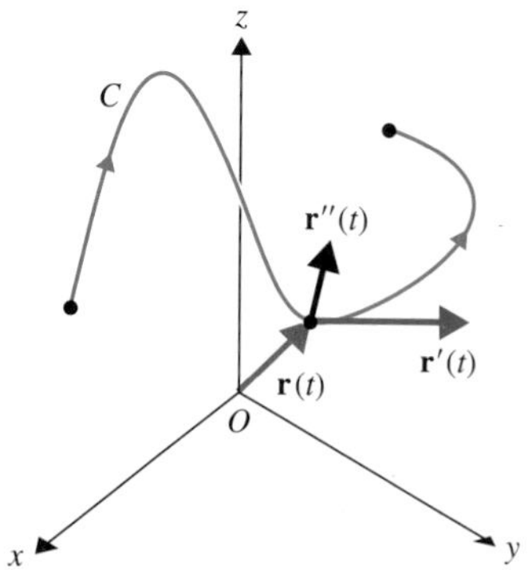

Part II of the Fundamental Theorem of Calculus says that if we differentiate both sides of (3.1), we get

$$s'(t) = \sqrt{[f'(t)]^2 + [g'(t)]^2 + [h'(t)]^2} = \|\mathbf{r}'(t)\|.$$

Think about the physical interpretation of $s'(t)$. Since $s(t)$ represents arc length, $s'(t)$ gives the instantaneous rate of change of arc length with respect to time, that is, the **speed** of the object as it moves along the curve. So, for any given value of t, $\mathbf{r}'(t)$ is a tangent vector pointing in the direction of the orientation of C (i.e., the direction followed by the object) and whose magnitude gives the speed of the object. What would you call $\mathbf{r}'(t)$? We call it the **velocity** vector, usually denoted $\mathbf{v}(t)$. As in the case of one-dimensional motion, we refer to the derivative of the velocity vector $\mathbf{v}'(t) = \mathbf{r}''(t)$ as the **acceleration** vector, denoted $\mathbf{a}(t)$. When drawing the velocity and acceleration vectors, we locate both of their initial points at the terminal point of $\mathbf{r}(t)$ (i.e., at the point on the curve), as shown in Figure 11.11.

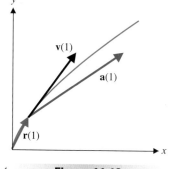

Figure 11.11

Position, velocity and acceleration vectors.

Figure 11.12

Position, velocity and acceleration vectors.

Example 3.1   Finding Velocity and Acceleration Vectors

Find the velocity and acceleration vectors if the position of an object moving in the xy-plane is given by $\mathbf{r}(t) = \langle t^3, 2t^2 \rangle$.

Solution   We have

$$\mathbf{v}(t) = \mathbf{r}'(t) = \langle 3t^2, 4t \rangle \quad \text{and} \quad \mathbf{a}(t) = \mathbf{r}''(t) = \langle 6t, 4 \rangle.$$

In particular, this says that at $t = 1$, we have $\mathbf{r}(1) = \langle 1, 2 \rangle$, $\mathbf{v}(1) = \mathbf{r}'(1) = \langle 3, 4 \rangle$ and $\mathbf{a}(1) = \mathbf{r}''(1) = \langle 6, 4 \rangle$. We plot the curve and these vectors in Figure 11.12.

Just as in the case of one-dimensional motion, the relationship between position, velocity and acceleration vectors goes both ways. That is, given the acceleration vector, we can determine the velocity and position vectors, provided we have some additional information.

Example 3.2   Finding Velocity and Position from Acceleration

Find the velocity and position of an object at any time t, given that its acceleration is $\mathbf{a}(t) = \langle 6t, 12t + 2, e^t \rangle$, its initial velocity is $\mathbf{v}(0) = \langle 2, 0, 1 \rangle$ and its initial position is $\mathbf{r}(0) = \langle 0, 3, 5 \rangle$.

Solution   Since $\mathbf{a}(t) = \mathbf{v}'(t)$, we can integrate once to obtain

$$\mathbf{v}(t) = \int \mathbf{a}(t)\, dt = \int [6t\mathbf{i} + (12t + 2)\mathbf{j} + e^t\mathbf{k}]\, dt$$

$$= 3t^2\mathbf{i} + (6t^2 + 2t)\mathbf{j} + e^t\mathbf{k} + \mathbf{c}_1,$$

where $\mathbf{c}_1$ is an arbitrary constant vector. To determine the value of $\mathbf{c}_1$, we use the initial velocity:

$$\langle 2, 0, 1 \rangle = \mathbf{v}(0) = (0)\mathbf{i} + (0)\mathbf{j} + (1)\mathbf{k} + \mathbf{c}_1,$$

so that $\mathbf{c}_1 = \langle 2, 0, 0 \rangle$. This gives us the velocity

$$\mathbf{v}(t) = (3t^2 + 2)\mathbf{i} + (6t^2 + 2t)\mathbf{j} + e^t\mathbf{k}.$$

Since $\mathbf{v}(t) = \mathbf{r}'(t)$, we integrate once again, to obtain

$$\mathbf{r}(t) = \int \mathbf{v}(t)\, dt = \int [(3t^2 + 2)\mathbf{i} + (6t^2 + 2t)\mathbf{j} + e^t\mathbf{k}]\, dt$$

$$= (t^3 + 2t)\mathbf{i} + (2t^3 + t^2)\mathbf{j} + e^t\mathbf{k} + \mathbf{c}_2,$$

where $\mathbf{c}_2$ is an arbitrary constant vector. We can use the initial position to determine the value of $\mathbf{c}_2$, as follows:

$$\langle 0, 3, 5 \rangle = \mathbf{r}(0) = (0)\mathbf{i} + (0)\mathbf{j} + (1)\mathbf{k} + \mathbf{c}_2,$$

so that $\mathbf{c}_2 = \langle 0, 3, 4 \rangle$. This gives us the position vector

$$\mathbf{r}(t) = (t^3 + 2t)\mathbf{i} + (2t^3 + t^2 + 3)\mathbf{j} + (e^t + 4)\mathbf{k}.$$

We show the curve and indicate sample vectors for $\mathbf{r}(t)$, $\mathbf{v}(t)$ and $\mathbf{a}(t)$ in Figure 11.13.

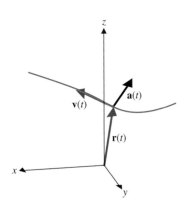

Figure 11.13

Position, velocity and acceleration vectors.

We have already seen **Newton's second law of motion** several times now. In the case of one-dimensional motion, we had that the net force acting on an object equals the product of the mass and the acceleration ($F = ma$). In the case of motion in two or more dimensions, we have the vector form of Newton's second law:

$$\mathbf{F} = m\mathbf{a}.$$

Here, m is the mass, $\mathbf{a}$ is the acceleration vector and $\mathbf{F}$ is the vector representing the net force acting on the object.

Example 3.3   Finding the Force Acting on an Object

Find the force acting on an object moving along a circular path of radius b centered at the origin, with constant angular speed.

Solution   Here, by constant **angular speed,** we mean that if θ is the angle made by the position vector and the positive x-axis and t is time (see Figure 11.14a, where the indicated orientation is for the case where $\omega > 0$), then we have that

$$\frac{d\theta}{dt} = \omega \text{ (constant)}.$$

Notice that this says that $\theta = \omega t + c$, for some constant c. Further, we can think of the circular path as the curve traced out by the endpoint of the vector-valued function

$$\mathbf{r}(t) = \langle b\cos\theta, b\sin\theta \rangle = \langle b\cos(\omega t + c), b\sin(\omega t + c) \rangle.$$

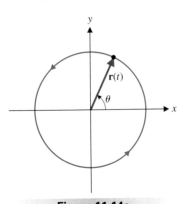

Figure 11.14a

Motion along a circle.

Notice that the path is the same for every value of c. (Think about what the value of c affects.) For simplicity, we take $\theta = 0$ when $t = 0$, so that $\theta = \omega t$ and

$$\mathbf{r}(t) = \langle b \cos \omega t, b \sin \omega t \rangle.$$

Now that we know the position at any time t, we can differentiate to find the velocity and acceleration. We have

$$\mathbf{v}(t) = \mathbf{r}'(t) = \langle -b\omega \sin \omega t, b\omega \cos \omega t \rangle$$

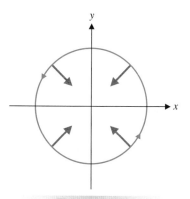

Figure 11.14b
Centripetal force.

and

$$\mathbf{a}(t) = \mathbf{v}'(t) = \mathbf{r}''(t) = \langle -b\omega^2 \cos \omega t, -b\omega^2 \sin \omega t \rangle$$
$$= -\omega^2 \langle b \cos \omega t, b \sin \omega t \rangle = -\omega^2 \mathbf{r}(t).$$

From Newton's second law of motion, we now have

$$\mathbf{F}(t) = m\mathbf{a}(t) = -m\omega^2 \mathbf{r}(t).$$

Notice that since $m\omega^2 > 0$, this says that the force acting on the object points in the direction opposite the position vector. That is, at any point on the path, it points in toward the origin (see Figure 11.14b). We call such a force a **centripetal** (center-seeking) force. Finally, observe that on this circular path, $\|\mathbf{r}(t)\| = b$, so that at every point on the path, the force vector has constant magnitude:

$$\|\mathbf{F}(t)\| = \|-m\omega^2 \mathbf{r}(t)\| = m\omega^2 \|\mathbf{r}(t)\| = m\omega^2 b.$$

Notice that one consequence of the result $\mathbf{F}(t) = -m\omega^2 \mathbf{r}(t)$ from example 3.3 is that the magnitude of the force increases as the rotation rate ω increases. You have experienced this if you have been on a roller coaster with tight turns or loops. The faster you are going, the stronger the force that your seat exerts on you.

Just as we did in the one-dimensional case, we can use Newton's second law of motion to do more than simply identify the force acting on an object with a given position function. It's much more important to be able to determine the position of an object given only a knowledge of the forces acting on it. For instance, one of the most significant problems faced by the military is how to aim a projectile (e.g., a missile) so that it will end up hitting its intended target. This problem is harder than it sounds, even when the target is standing still. When the target is an aircraft moving faster than the speed of sound, the problem presents significant challenges. In any case, the calculus can be used to arrive at a meaningful solution. We present the simplest possible case (where neither the target nor the source of the projectile are moving) in the following example.

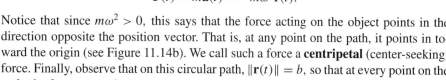

Example 3.4   Analyzing the Motion of a Projectile

A projectile is launched with an initial speed of 140 feet per second from ground level at an angle of $\frac{\pi}{4}$ to the horizontal. Assuming that the only force acting on the object is gravity (i.e., there is no air resistance, etc.), find the maximum altitude, the horizontal range and the speed at impact of the projectile.

Solution   From Newton's second law of motion, we have

$$\mathbf{F}(t) = m\mathbf{a}(t).$$

Notice that here, the motion is in a single plane (so that we need only consider two dimensions) and the only force acting on the object is the force of gravity, which acts straight down. In this case, we have that $\mathbf{a}(t)$ is simply the acceleration due to gravity.

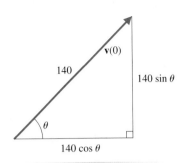

Figure 11.15a
Initial velocity vector.

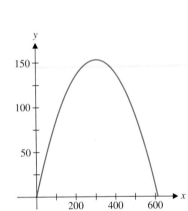

Figure 11.15b
Path of a projectile.

Although this is not constant, it is nearly so at altitudes reasonably close to sea level. We will assume that

$$\mathbf{a}(t) = -g\mathbf{j},$$

where g is the constant acceleration due to gravity, $g \approx 32$ feet/second^2. So, we have

$$\mathbf{v}'(t) = \mathbf{a}(t) = -32\mathbf{j}.$$

Integrating this once gives us

$$\mathbf{v}(t) = \int \mathbf{a}(t)\, dt = -32t\mathbf{j} + \mathbf{c}_1, \tag{3.2}$$

where $\mathbf{c}_1$ is an arbitrary constant vector. If we knew the initial velocity vector $\mathbf{v}(0)$, we could use this to solve for $\mathbf{c}_1$, but we only know the initial speed (i.e., the magnitude of the velocity vector). Referring to Figure 11.15a, notice that you can read off the components of $\mathbf{v}(0)$. From the definitions of the sine and cosine functions, we have

$$\mathbf{v}(0) = \left\langle 140\cos\frac{\pi}{4}, 140\sin\frac{\pi}{4} \right\rangle = \langle 70\sqrt{2}, 70\sqrt{2} \rangle.$$

From (3.2), we now have

$$\langle 70\sqrt{2}, 70\sqrt{2} \rangle = \mathbf{v}(0) = (-32)(0)\mathbf{j} + \mathbf{c}_1 = \mathbf{c}_1.$$

Substituting this back into (3.2), we have

$$\mathbf{v}(t) = -32t\mathbf{j} + \langle 70\sqrt{2}, 70\sqrt{2} \rangle = \langle 70\sqrt{2}, 70\sqrt{2} - 32t \rangle. \tag{3.3}$$

Integrating (3.3) will give us the position vector

$$\mathbf{r}(t) = \int \mathbf{v}(t)\, dt = \langle 70\sqrt{2}t, 70\sqrt{2}t - 16t^2 \rangle + \mathbf{c}_2,$$

where $\mathbf{c}_2$ is an arbitrary constant vector. Since the initial location was not specified, we choose it to be the origin. (This is usually most convenient.) This gives us

$$\mathbf{0} = \mathbf{r}(0) = \mathbf{c}_2,$$

so that

$$\mathbf{r}(t) = \langle 70\sqrt{2}t, 70\sqrt{2}t - 16t^2 \rangle. \tag{3.4}$$

We show a graph of the path of the projectile in Figure 11.15b. Now that we have found expressions for the position and velocity vectors for any time, we can answer the physical questions. Notice that the maximum altitude occurs at the instant when the object stops moving up (just before it starts to fall). This says that the vertical ($\mathbf{j}$) component of velocity must be zero. From (3.3), we get

$$0 = 70\sqrt{2} - 32t,$$

so that the time at the maximum altitude is

$$t = \frac{70\sqrt{2}}{32}.$$

The maximum altitude is then found from the vertical component of the position vector at this time:

$$\text{Maximum altitude} = 70\sqrt{2}t - 16t^2 \Big|_{t=\frac{70\sqrt{2}}{32}} = 70\sqrt{2}\left(\frac{70\sqrt{2}}{32}\right) - 16\left(\frac{70\sqrt{2}}{32}\right)^2$$

$$= \frac{1225}{8} = 153.125 \text{ feet.}$$

To determine the horizontal range, we first need to determine the instant at which the object strikes the ground. Notice that this occurs when the vertical component of the position vector is zero (i.e., when the height above the ground is zero). From (3.4), we see that this occurs when

$$0 = 70\sqrt{2}\,t - 16t^2 = 2t\left(35\sqrt{2} - 8t\right).$$

There are two solutions of this equation: $t = 0$ (the time at which the projectile is launched) and $t = \dfrac{35\sqrt{2}}{8}$ (the time of impact). The horizontal range is then the horizontal (**i**) component of position at this time:

$$\text{Range} = 70\sqrt{2}\,t\Big|_{t=\frac{35\sqrt{2}}{8}} = \left(70\sqrt{2}\right)\left(\frac{35\sqrt{2}}{8}\right) = \frac{1225}{2} = 612.5 \text{ feet}.$$

Finally, the speed at impact is the magnitude of the velocity vector at the time of impact:

$$\left\|\mathbf{v}\left(\frac{35\sqrt{2}}{8}\right)\right\| = \left\|\left\langle 70\sqrt{2}, 70\sqrt{2} - 32\left(\frac{35\sqrt{2}}{8}\right)\right\rangle\right\|$$

$$= \left\|\left\langle 70\sqrt{2}, -70\sqrt{2}\right\rangle\right\| = 140 \text{ ft/sec}.$$

■

You might have noticed in example 3.4 that the speed at impact was the same as the initial speed. Don't expect this to always be the case. Generally, this will only be true for a projectile of constant mass that is fired from ground level and returns to ground level and that is not subject to air resistance or other forces.

Equations of Motion

We now derive the equations of motion for a projectile in a slightly more general setting than that described in example 3.4. Consider a projectile fired from an altitude h above the ground at an angle θ to the horizontal and with initial speed v_0. We can use Newton's second law of motion to determine the position of the projectile at any time t and once we have this, we can answer any questions about the motion.

We again start with Newton's second law and assume that the only force acting on the object is gravity. We have

$$\mathbf{F}(t) = m\mathbf{a}(t),$$

where $\mathbf{F}(t) = -mg\mathbf{j}$. This gives us (as in example 3.4)

$$\mathbf{v}'(t) = \mathbf{a}(t) = -g\mathbf{j}. \qquad (3.5)$$

Integrating (3.5) gives us

$$\mathbf{v}(t) = \int \mathbf{a}(t)\,dt = -gt\mathbf{j} + \mathbf{c}_1, \qquad (3.6)$$

where $\mathbf{c}_1$ is an arbitrary constant vector. In order to solve for $\mathbf{c}_1$, we need the value of $\mathbf{v}(t)$ for some t, but we are given only the initial speed v_0 and the angle at which the projectile is fired. Notice that from the definitions of sine and cosine, we can read off the components of $\mathbf{v}(0)$ from Figure 11.16a. From this and (3.6), we have

$$\langle v_0 \cos\theta, v_0 \sin\theta \rangle = \mathbf{v}(0) = \mathbf{c}_1.$$

This gives us the velocity vector

$$\mathbf{v}(t) = \langle v_0 \cos\theta, v_0 \sin\theta - gt \rangle. \qquad (3.7)$$

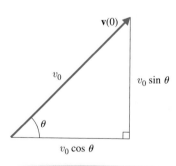

Figure 11.16a

Initial velocity.

Since $\mathbf{r}'(t) = \mathbf{v}(t)$, we integrate (3.7) to get the position:

$$\mathbf{r}(t) = \int \mathbf{v}(t)\,dt = \left\langle (v_0\cos\theta)t, (v_0\sin\theta)t - \frac{gt^2}{2}\right\rangle + \mathbf{c}_2.$$

To solve for $\mathbf{c}_2$, we want to use the initial position $\mathbf{r}(0)$, but we're not given it. We're only told that the projectile starts from an altitude of h feet above the ground. If we select the origin to be the point on the ground directly below the launching point, we have

$$\langle 0, h\rangle = \mathbf{r}(0) = \mathbf{c}_2,$$

so that

$$\mathbf{r}(t) = \left\langle (v_0\cos\theta)t, (v_0\sin\theta)t - \frac{gt^2}{2}\right\rangle + \langle 0, h\rangle$$

$$= \left\langle (v_0\cos\theta)t, h + (v_0\sin\theta)t - \frac{gt^2}{2}\right\rangle. \qquad (3.8)$$

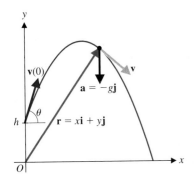

Figure 11.16b

Path of the projectile.

Notice that the path traced out by $\mathbf{r}(t)$ (from $t = 0$ until impact) is a portion of a parabola (see Figure 11.16b).

Now that we have derived (3.7) and (3.8), we have all we need to answer any further questions about the motion. For instance, if we need to know the maximum altitude, this occurs at the time at which the vertical ($\mathbf{j}$) component of velocity is zero (i.e., at the time when the projectile stops rising). From (3.7), we solve

$$0 = v_0\sin\theta - gt,$$

so that the time at which the maximum altitude is reached is given by

Time to reach maximum altitude

$$t_{\max} = \frac{v_0\sin\theta}{g}.$$

The maximum altitude itself is the vertical component of the position vector at this time. From (3.8), we have

$$\text{Maximum altitude} = h + (v_0\sin\theta)t - \frac{gt^2}{2}\bigg|_{t=t_{\max}}$$

$$= h + (v_0\sin\theta)\left(\frac{v_0\sin\theta}{g}\right) - \frac{g}{2}\left(\frac{v_0\sin\theta}{g}\right)^2$$

Maximum altitude

$$= h + \frac{1}{2}\frac{v_0^2\sin^2\theta}{g}.$$

To find the horizontal range or the speed at impact, we must first find the time of impact. To get this, we set the vertical component of position to zero. From (3.8), we have

$$0 = h + (v_0\sin\theta)t - \frac{gt^2}{2}.$$

Notice that this is simply a quadratic equation for t. Given v_0, θ and h, we can solve for the time t using the quadratic formula.

In all of the foregoing analysis, we left the constant acceleration due to gravity as g. You will usually use one of the two approximations:

$$g \approx 32 \text{ ft/sec}^2 \quad \text{or} \quad g \approx 9.8 \text{ m/sec}^2.$$

When using any other units, simply adjust the units to feet or meters and the time scale to seconds or make the corresponding adjustments to the value of g.

In the following example, we examine a fully three-dimensional projectile motion problem for the first time.

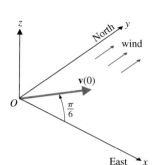

Figure 11.17a

The initial velocity and wind velocity vectors.

| **Example 3.5** | Analyzing the Motion of a Projectile in Three Dimensions |

A projectile of mass 1 kg is launched from ground level toward the east at 200 meters/second, at an angle of $\frac{\pi}{6}$ to the horizontal. If a gusting northerly wind applies a steady force of 2 newtons to the projectile, find the landing location of the projectile and its speed at impact.

Solution   Notice that because of the cross wind, the motion is fully three-dimensional. We orient the x-, y- and z-axes so that the positive y-axis points north, the positive x-axis points east and the positive z-axis points up, as in Figure 11.17a, where we also show the initial velocity vector and vectors indicating the force due to the wind. The two forces acting on the projectile are gravity (in the negative z-direction with magnitude $9.8m = 9.8$ newtons) and wind (in the y-direction with magnitude 2 newtons). Newton's second law is $\mathbf{F} = m\mathbf{a} = \mathbf{a}$. We have

$$\mathbf{a}(t) = \mathbf{v}'(t) = \langle 0, 2, -9.8 \rangle$$

Integrating gives us the velocity function

$$\mathbf{v}(t) = \langle 0, 2t, -9.8t \rangle + \mathbf{c}_1, \qquad (3.9)$$

where $\mathbf{c}_1$ is an arbitrary constant vector. Note that the initial velocity is

$$\mathbf{v}(0) = \left\langle 200 \cos \frac{\pi}{6}, 0, 200 \sin \frac{\pi}{6} \right\rangle = \langle 100\sqrt{3}, 0, 100 \rangle.$$

From (3.9), we now have

$$\langle 100\sqrt{3}, 0, 100 \rangle = \mathbf{v}(0) = \mathbf{c}_1.$$

This gives us

$$\mathbf{v}(t) = \langle 100\sqrt{3}, 2t, 100 - 9.8t \rangle.$$

We integrate this to get the position vector:

$$\mathbf{r}(t) = \langle 100\sqrt{3}t, t^2, 100t - 4.9t^2 \rangle + \mathbf{c}_2,$$

for a constant vector $\mathbf{c}_2$. Taking the initial position to be the origin, we get

$$\mathbf{0} = \mathbf{r}(0) = \mathbf{c}_2,$$

so that

$$\mathbf{r}(t) = \langle 100\sqrt{3}t, t^2, 100t - 4.9t^2 \rangle. \qquad (3.10)$$

Note that the projectile strikes the ground when the $\mathbf{k}$ component of position is zero. From (3.10), we have that this occurs when

$$0 = 100t - 4.9t^2 = t(100 - 4.9t).$$

So, the projectile is on the ground when $t = 0$ (time of launch) and when $t = \frac{100}{4.9} \approx 20.4$ seconds (the time of impact). The location of impact is then the endpoint of the vector $\mathbf{r}\left(\frac{100}{4.9}\right) \approx \langle 3534.8, 416.5, 0 \rangle$ and the speed at impact is

$$\left\| \mathbf{v}\left(\frac{100}{4.9}\right) \right\| \approx 204 \text{ m/s.}$$

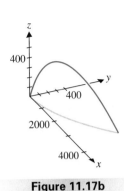

Figure 11.17b

Path of the projectile.

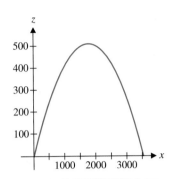

Figure 11.17c

Projection of path onto the xz-plane.

We show a computer-generated graph of the path of the projectile in Figure 11.17b. In this figure, we also indicate the shadow made by the path of the projectile on the ground. In Figure 11.17c, we show the projection of the projectile's path onto the xz-plane. Observe that this parabola is analogous to the parabola shown in Figure 11.15b.

EXERCISES 11.3

1. Explain why it makes sense in example 3.4 that the speed at impact equals the initial speed. (Hint: What force would slow the object down?) If the projectile were launched from above ground, discuss how the speed at impact would compare to the initial speed.

2. For an actual projectile, taking into account air resistance, explain why the speed at impact would be less than the initial speed.

3. In this section, we assumed that the acceleration due to gravity is constant. By contrast, air resistance is a function of velocity (the faster the object goes, the more air resistance there is). Explain why including air resistance in our Newton's law model of projectile motion would make the mathematics **much** more complicated.

4. In example 3.5, use the x- and y-components of the position function to explain why the projection of the projectile's path onto the xy-plane would be a parabola. The projection onto the xz-plane is also a parabola. Discuss whether or not the path in Figure 11.17b is a parabola. If you were watching the projectile, would the path appear to be parabolic?

In exercises 5–10, find the velocity and acceleration functions for the given position function.

5. $\mathbf{r}(t) = \langle 5\cos 2t, 5\sin 2t \rangle$

6. $\mathbf{r}(t) = \langle 2\cos t + \sin 2t, 2\sin t + \cos 2t \rangle$

7. $\mathbf{r}(t) = \langle 25t, -16t^2 + 15t + 5 \rangle$

8. $\mathbf{r}(t) = \langle 25te^{-2t}, -16t^2 + 10t + 20 \rangle$

9. $\mathbf{r}(t) = \langle 4te^{-2t}, 2e^{-2t}, -16t^2 \rangle$

10. $\mathbf{r}(t) = \langle 3e^{-3t}, \sin 2t, t^3 - 3t \rangle$

In exercises 11–18, find the position function from the given velocity or acceleration function.

11. $\mathbf{v}(t) = \langle 10, -32t + 4 \rangle, \mathbf{r}(0) = \langle 3, 8 \rangle$

12. $\mathbf{v}(t) = \langle 4t, t^2 - 1 \rangle, \mathbf{r}(0) = \langle 10, -2 \rangle$

13. $\mathbf{a}(t) = \langle 0, -32 \rangle, \mathbf{v}(0) = \langle 5, 0 \rangle, \mathbf{r}(0) = \langle 0, 16 \rangle$

14. $\mathbf{a}(t) = \langle t, \sin t \rangle, \mathbf{v}(0) = \langle 2, -6 \rangle, \mathbf{r}(0) = \langle 10, 4 \rangle$

15. $\mathbf{v}(t) = \langle 10, 3e^{-t}, -32t + 4 \rangle, \mathbf{r}(0) = \langle 0, -6, 20 \rangle$

16. $\mathbf{v}(t) = \langle t + 2, t^2, e^{-t/3} \rangle, \mathbf{r}(0) = \langle 4, 0, -3 \rangle$

17. $\mathbf{a}(t) = \langle t, 0, -16 \rangle, \mathbf{v}(0) = \langle 12, -4, 0 \rangle, \mathbf{r}(0) = \langle 5, 0, 2 \rangle$

18. $\mathbf{a}(t) = \langle e^{-3t}, t, \sin t \rangle, \mathbf{v}(0) = \langle 4, -2, 4 \rangle, \mathbf{r}(0) = \langle 0, 4, -2 \rangle$

In exercises 19–22, find the centripetal force on an object of mass 10 kg with the given position function (in units of meters and seconds).

19. $\mathbf{r}(t) = \langle 4\cos 2t, 4\sin 2t \rangle$ 20. $\mathbf{r}(t) = \langle 3\cos 5t, 3\sin 5t \rangle$

21. $\mathbf{r}(t) = \langle 6\cos 4t, 6\sin 4t \rangle$ 22. $\mathbf{r}(t) = \langle 2\cos 3t, 2\sin 3t \rangle$

In exercises 23–26, find the force acting on an object of mass 10 kg with the given position function (in units of meters and seconds).

23. $\mathbf{r}(t) = \langle 3\cos 2t, 5\sin 2t \rangle$

24. $\mathbf{r}(t) = \langle 3\cos 4t, 2\sin 5t \rangle$

25. $\mathbf{r}(t) = \langle 3t^2 + t, 3t - 1 \rangle$

26. $\mathbf{r}(t) = \langle 20t - 3, -16t^2 + 2t + 30 \rangle$

In exercises 27–32, a projectile is fired with initial speed v_0 feet per second from a height of h feet at an angle of θ above the horizontal. Assuming that the only force acting on the object is gravity, find the maximum altitude, horizontal range and speed at impact.

27. $v_0 = 100, h = 0, \theta = \frac{\pi}{3}$ 28. $v_0 = 100, h = 0, \theta = \frac{\pi}{6}$

29. $v_0 = 160, h = 10, \theta = \frac{\pi}{4}$ 30. $v_0 = 120, h = 10, \theta = \frac{\pi}{3}$

31. $v_0 = 320, h = 10, \theta = \frac{\pi}{4}$ 32. $v_0 = 240, h = 10, \theta = \frac{\pi}{3}$

33. Based on your answers to exercises 29 and 31, what effect does doubling the initial speed have on the horizontal range?

34. The angles $\frac{\pi}{3}$ and $\frac{\pi}{6}$ are symmetric about $\frac{\pi}{4}$; that is, $\frac{\pi}{4} - \frac{\pi}{6} = \frac{\pi}{3} - \frac{\pi}{4}$. Based on your answers to exercises 27 and 28, how do horizontal ranges for symmetric angles compare?

35. For the general projectile of equation (3.8) with $h = 0$, show that the horizontal range is $\dfrac{v_0^2 \sin 2\theta}{g}$.

36. Given the result of exercise 35, find the angle that produces the maximum horizontal range.

In exercises 37–44, neglect all forces except gravity. In all these situations, the effect of air resistance is actually significant, but your calculations will give a good first approximation.

37. A baseball is hit from a height of 3 feet with initial speed 120 feet per second and at an angle of 30 degrees above the horizontal. Find a vector-valued function describing the position of the ball t seconds after it is hit. To be a home run, the ball must clear a wall that is 385 feet away and 6 feet tall. Determine if this is a home run.

38. Repeat exercise 37 if the ball is launched with an initial angle of 31 degrees.

39. A baseball pitcher throws a pitch horizontally from a height of 6 feet with an initial speed of 130 feet per second. Find a vector-valued function describing the position of the ball t seconds after release. If home plate is 60 feet away, how high is the ball when it crosses home plate?

40. If a person drops a ball from height 6 feet, how high will the ball be when the pitch of exercise 39 crosses home plate?

41. A tennis serve is struck horizontally from a height of 8 feet with initial speed 120 feet per second. For the serve to count (be "in"), it must clear a net which is 39 feet away and 3 feet high and must land before the service line 60 feet away. Find a vector function for the position of the ball and determine if this serve is in or out.

42. Repeat exercise 41 if the ball is struck with an initial speed of (a) 80 ft/s or (b) 65 ft/s.

43. A football punt is launched at an angle of 50 degrees with an initial speed of 55 mph. Assuming the punt is launched from ground level, compute the "hang time" (the amount of time in the air) for the punt.

44. Compute the extra hang time if the punt in exercise 43 has an initial speed of 60 mph.

45. Find the landing point in exercise 27 if the object has mass 1 slug, is launched due east and there is a northerly wind force of 8 pounds.

46. Find the landing point in exercise 28 if the object has mass 1 slug, is launched due east and there is a southerly wind force of 4 pounds.

47. Suppose an airplane is acted on by three forces: gravity, wind and engine thrust. Assume that the force vector for gravity is $m\mathbf{g} = m\langle 0, 0, -32\rangle$, the force vector for wind is $\mathbf{w} = \langle 0, 1, 0\rangle$ for $0 \le t \le 1$ and $\mathbf{w} = \langle 0, 2, 0\rangle$ for $t > 1$, and the force vector for engine thrust is $\mathbf{e} = \langle 2t, 0, 24\rangle$. Newton's second law of motion gives us $m\mathbf{a} = m\mathbf{g} + \mathbf{w} + \mathbf{e}$. Assume that $m = 1$ and the initial velocity vector is $\mathbf{v}(0) = \langle 100, 0, 10\rangle$. Show that the velocity vector for $0 \le t \le 1$ is $\mathbf{v}(t) = \langle t^2 + 100, t, 10 - 8t\rangle$. For $t > 1$, integrate the equation $\mathbf{a} = \mathbf{g} + \mathbf{w} + \mathbf{e}$ to get $\mathbf{v}(t) = \langle t^2 + a, 2t + b, -8t + c\rangle$ for constants a, b and c. Explain (on physical grounds) why the function $\mathbf{v}(t)$ should be continuous and find the values of the constants that make it so. Show that $\mathbf{v}(t)$ is not differentiable. Given the nature of the force function, why does this make sense?

48. Find the position function for the airplane in exercise 47.

49. A roller coaster is designed to travel a circular loop of radius 100 feet. If the riders feel weightless at the top of the loop, what is the speed of the roller coaster?

50. For a satellite in earth orbit, the speed v in miles per second is related to the height h miles above the surface of the earth by $v = \sqrt{\frac{95{,}600}{4000 + h}}$. Suppose a satellite is in orbit 15,000 miles above the surface of the earth. How much does the speed need to increase to raise the orbit to a height of 20,000 miles?

51. A jet pilot executing a circular turn experiences an acceleration of "5 g's" (that is, $\|\mathbf{a}\| = 5g$). If the jet's speed is 900 km/hr, what is the radius of the turn?

52. For the jet pilot of exercise 51, how many g's would be experienced if the speed were 1800 km/hr?

53. **Angular momentum** is defined by $\mathbf{L}(t) = m\mathbf{r}(t) \times \mathbf{v}(t)$. Show that $\mathbf{L}'(t) = m\mathbf{r}(t) \times \mathbf{a}(t)$.

54. Prove the law of conservation of angular momentum: in the absence of torque, $\mathbf{L}(t)$ is constant. [Hint: The torque is given by $m\mathbf{r}(t) \times \mathbf{a}(t)$.]

55. A ball rolls off a table of height 3 feet. Its initial velocity is horizontal with speed v_0. Determine where the ball hits the ground and the velocity vector of the ball at the moment of impact. Find the angle between the horizontal and the impact velocity vector. Next, assume that the next bounce of the ball starts with the ball being launched from the ground with initial conditions determined by the impact velocity. The launch speed equals 0.6 times the impact speed (so the ball won't bounce forever) and the launch angle equals the (positive) angle between the horizontal and the impact velocity vector. Using these conditions, determine where the ball next hits the ground. Continue on to find the third point at which the ball bounces.

56. In many sports such as golf and ski jumping, it is important to determine the range of a projectile on a slope. Suppose that the ground passes through the origin and slopes at an angle of α to the horizontal. Show that an equation of the ground is $y = -(\tan\alpha)x$. An object is launched at height $h = 0$ with initial speed v_0 at an angle of θ from the horizontal. Note that equation (3.8) still holds, but the landing condition is now $y = -(\tan\alpha)x$. Find the x-coordinate of the landing point and show that the range (the distance along the ground) is given by $R = \frac{2}{g}v_0^2 \sec\alpha\cos\theta(\sin\theta + \tan\alpha\cos\theta)$. Use trigonometric identities to rewrite this as $R = \frac{1}{g}v_0^2 \sec^2\alpha[\sin\alpha + \sin(\alpha + 2\theta)]$. Use this formula to find the value of θ that maximizes the range. For flat ground ($\alpha = 0$), the optimal angle is $45°$. State an easy way of taking the value of α (say, $\alpha = 10°$ or $\alpha = -8°$) and adjusting from $45°$ to the optimal angle.

11.4   CURVATURE

Imagine that you are designing a new highway. Nearly all roads have curves, to avoid both natural and human-made obstacles. So that cars are able to maintain a reasonable speed on your new road, you will need to design the road in such a way so as to avoid curves that are too sharp. To do this, you'll first need to have some concept of how sharp a given curve is. In this section, we develop a measure of how much a curve is twisting and turning at any given point. First, realize that any given curve has infinitely many different parameterizations. For instance, the parametric equations $x = t^2$ and $y = t$ describe a parabola that opens to the right. In fact, for any real number $a > 0$, the equations $x = (at)^2$ and $y = at$ describe the same parabola. So, any measure of how sharp a curve is should be independent of the parameterization. The simplest choice of a parameter (for conceptual purposes, but not for computational purposes) is arc length. Further, observe that this is the correct parameter to use, as we measure how sharp a curve is by seeing how much it twists and turns per unit length. (Think about it this way: a turn of 90° over a quarter mile is not particularly sharp in comparison with a turn of 90° over a distance of 30 feet.)

For the curve traced out by the endpoint of the vector-valued function $\mathbf{r}(t) = \langle f(t), g(t), h(t) \rangle$, for $a \leq t \leq b$, we define the arc length parameter $s(t)$ to be the arc length of that portion of the curve from $u = a$ up to $u = t$. That is, from (1.4),

$$s(t) = \int_a^t \sqrt{[f'(u)]^2 + [g'(u)]^2 + [h'(u)]^2} \, du.$$

Recognizing that $\sqrt{[f'(u)]^2 + [g'(u)]^2 + [h'(u)]^2} = \|\mathbf{r}'(u)\|$, we can write this more simply as

$$s(t) = \int_a^t \|\mathbf{r}'(u)\| \, du. \tag{4.1}$$

Although explicitly finding an arc length parameterization of a curve is not the central thrust of our discussion here, we briefly pause to construct such a parameterization now, for the purpose of illustration.

Example 4.1   Parameterizing a Curve in Terms of Arc Length

Find an arc length parameterization of the circle of radius 4 centered at the origin.

Solution   First, note that one parameterization of this circle is

$$C : x = f(t) = 4 \cos t, \quad y = g(t) = 4 \sin t, \quad 0 \leq t \leq 2\pi.$$

In this case, the arc length from $u = 0$ to $u = t$ is given by

$$s(t) = \int_0^t \sqrt{[f'(u)]^2 + [g'(u)]^2} \, du$$

$$= \int_0^t \sqrt{[-4 \sin u]^2 + [4 \cos u]^2} \, du = 4 \int_0^t 1 \, du = 4t.$$

That is, $t = s/4$, so that an arc length parameterization for C is

$$C : x = 4 \cos \left(\frac{s}{4} \right), \quad y = 4 \sin \left(\frac{s}{4} \right), \quad 0 \leq s \leq 8\pi.$$

Consider the curve C traced out by the endpoint of the vector-valued function $\mathbf{r}(t)$. Recall that for each t, $\mathbf{v}(t) = \mathbf{r}'(t)$ can be thought of as both the velocity vector and a tangent vector, pointing in the direction of motion (i.e., the orientation of C). Notice that

Unit tangent vector

$$\boxed{\mathbf{T}(t) = \frac{\mathbf{r}'(t)}{\|\mathbf{r}'(t)\|}} \tag{4.2}$$

is also a tangent vector, but has length one ($\|\mathbf{T}(t)\| = 1$). We call $\mathbf{T}(t)$ the **unit tangent vector** to the curve C. That is, for each t, $\mathbf{T}(t)$ is a tangent vector of length one pointing in the direction of the orientation of C.

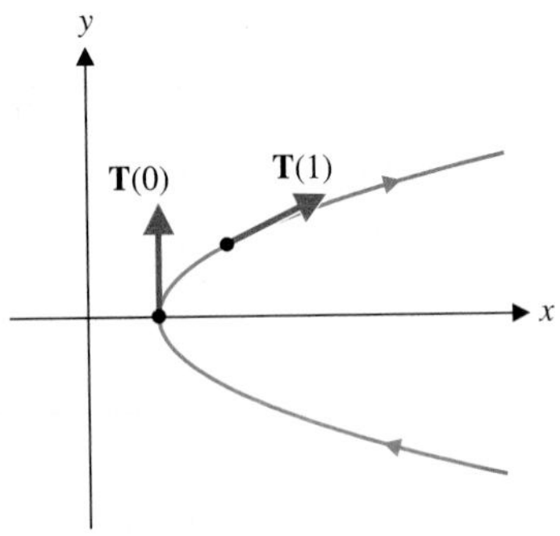

Example 4.2 Finding a Unit Tangent Vector

Find the unit tangent vector to the curve determined by $\mathbf{r}(t) = \langle t^2 + 1, t \rangle$.

Solution We have

$$\mathbf{r}'(t) = \langle 2t, 1 \rangle,$$

so that

$$\|\mathbf{r}'(t)\| = \sqrt{(2t)^2 + 1} = \sqrt{4t^2 + 1}.$$

From (4.2), the unit tangent vector is given by

$$\mathbf{T}(t) = \frac{\mathbf{r}'(t)}{\|\mathbf{r}'(t)\|} = \frac{\langle 2t, 1 \rangle}{\sqrt{4t^2 + 1}} = \left\langle \frac{2t}{\sqrt{4t^2 + 1}}, \frac{1}{\sqrt{4t^2 + 1}} \right\rangle.$$

In particular, we have $\mathbf{T}(0) = \langle 0, 1 \rangle$ and $\mathbf{T}(1) = \left\langle \frac{2}{\sqrt{5}}, \frac{1}{\sqrt{5}} \right\rangle$. We indicate both of these in Figure 11.18.

Figure 11.18
Unit tangent vectors.

In Figures 11.19a and 11.19b, we show two curves, both connecting the points A and B. Think about driving a car along roads in the shape of these two curves. The curve in Figure 11.19b indicates a much sharper turn than the curve in Figure 11.19a. The question before us is to see how to mathematically describe this degree of "sharpness." You should get an idea of this from Figures 11.19c and 11.19d. These are the same curves as those

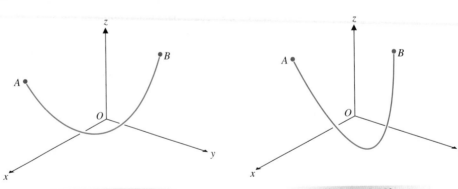

Figure 11.19a
Gentle curve.

Figure 11.19b
Sharp curve.

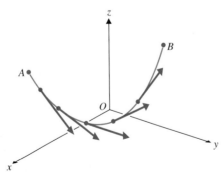

Figure 11.19c
Unit tangent vectors.

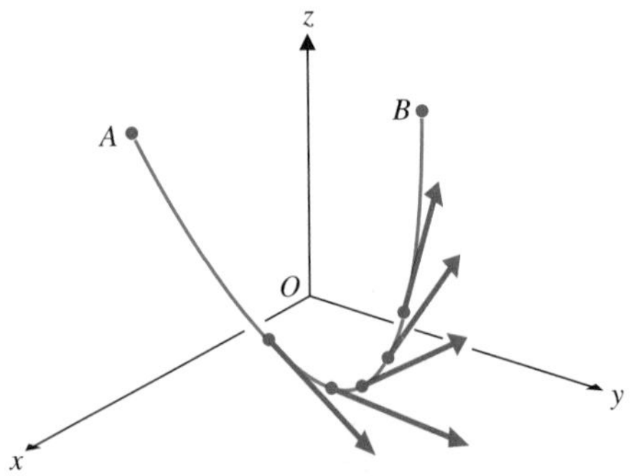

Figure 11.19d
Unit tangent vectors.

shown in Figures 11.19a and 11.19b, respectively, but we have drawn in a number of unit tangent vectors at equally spaced points on the curves. Notice that the unit tangent vectors change very slowly along the gentle curve in Figure 11.19c, but twist and turn quite rapidly in the vicinity of the sharp curve in Figure 11.19d. Based on our analysis of Figures 11.19c and 11.19d, notice that the rate of change of the unit tangent vectors with respect to arc length along the curve will give us a measure of sharpness. To this end, we make the following definition.

Definition 4.1
The **curvature** κ of a curve is the scalar quantity

$$\kappa = \left\| \frac{d\mathbf{T}}{ds} \right\|. \tag{4.3}$$

Note that, while the definition of curvature makes sense intuitively, it is not a simple matter to compute κ directly from (4.3). To do so, we would need to first find the arc length parameter and the unit tangent vector $\mathbf{T}(t)$, rewrite $\mathbf{T}(t)$ in terms of the arc length parameter s and then differentiate with respect to s. This is not usually done. Instead, observe that by the chain rule,

$$\mathbf{T}'(t) = \frac{d\mathbf{T}}{dt} = \frac{d\mathbf{T}}{ds}\frac{ds}{dt},$$

so that

$$\kappa = \left\| \frac{d\mathbf{T}}{ds} \right\| = \frac{\|\mathbf{T}'(t)\|}{\left| \dfrac{ds}{dt} \right|}. \tag{4.4}$$

Now, from (4.1), we had

$$s(t) = \int_a^t \|\mathbf{r}'(u)\|\,du,$$

so that by part II of the Fundamental Theorem of Calculus,

$$\frac{ds}{dt} = \|\mathbf{r}'(t)\|. \tag{4.5}$$

From (4.4) and (4.5), we now have

Curvature
$$\boxed{\kappa = \frac{\|\mathbf{T}'(t)\|}{\|\mathbf{r}'(t)\|}.} \tag{4.6}$$

Notice that it should be comparatively simple to use (4.6) to compute the curvature. We illustrate this with the following simple example.

Example 4.3 Finding the Curvature of a Straight Line

Find the curvature of a straight line.

Solution First, think about what we're asking. Straight lines are, well, straight, so their curvature should be zero at every point. Let's see. Suppose that the line is traced out by the vector-valued function $\mathbf{r}(t) = \langle at + b, ct + d, et + f \rangle$, for some constants a, b, c, d, e and f. Then,

$$\mathbf{r}'(t) = \langle a, c, e \rangle$$

and so,

$$\|\mathbf{r}'(t)\| = \sqrt{a^2 + c^2 + e^2} = \text{constant.}$$

The unit tangent vector is then

$$\mathbf{T}(t) = \frac{\mathbf{r}'(t)}{\|\mathbf{r}'(t)\|} = \frac{\langle a, c, e \rangle}{\sqrt{a^2 + c^2 + e^2}},$$

which is a constant vector. This gives us $\mathbf{T}'(t) = \mathbf{0}$, for all t. From (4.6), we now have

$$\kappa = \frac{\|\mathbf{T}'(t)\|}{\|\mathbf{r}'(t)\|} = \frac{\|\mathbf{0}\|}{\sqrt{a^2 + c^2 + e^2}} = 0,$$

as expected.

Well, if a line has zero curvature, can you think of a geometrical object with lots of curvature? The first one to come to mind is likely the circle, which we discuss next.

Example 4.4   Finding the Curvature of a Circle

Find the curvature for a circle of radius a.

Solution   We leave it as an exercise to show that the curvature does not depend on the location of the center of the circle. (Intuitively, it certainly should not.) So, for simplicity, we assume that the circle is centered at the origin. Notice that the circle of radius a centered at the origin is traced out by the vector-valued function $\mathbf{r}(t) = \langle a \cos t, a \sin t \rangle$. Differentiating, we get

$$\mathbf{r}'(t) = \langle -a \sin t, a \cos t \rangle$$

and

$$\|\mathbf{r}'(t)\| = \sqrt{(-a \sin t)^2 + (a \cos t)^2} = a\sqrt{\sin^2 t + \cos^2 t} = a.$$

The unit tangent vector is then given by

$$\mathbf{T}(t) = \frac{\mathbf{r}'(t)}{\|\mathbf{r}'(t)\|} = \frac{\langle -a \sin t, a \cos t \rangle}{a} = \langle -\sin t, \cos t \rangle.$$

Differentiating this gives us

$$\mathbf{T}'(t) = \langle -\cos t, -\sin t \rangle$$

and from (4.6), we have

$$\kappa = \frac{\|\mathbf{T}'(t)\|}{\|\mathbf{r}'(t)\|} = \frac{\|\langle -\cos t, -\sin t \rangle\|}{a} = \frac{\sqrt{(-\cos t)^2 + (-\sin t)^2}}{a} = \frac{1}{a}.$$

Notice that the result of example 4.4 is consistent with your intuition. First, you should be able to drive a car around a circular track while holding the steering wheel in a fixed position. (That is, the curvature should be constant.) Further, the smaller that the radius of a circular track is, the sharper you will need to turn (that is, the larger the curvature). On the other hand, on a circular track of very large radius, it would seem as is if you were driving fairly straight (i.e., the curvature will be close to 0).

You probably noticed that computing the curvature of the curves in examples 4.3 and 4.4 was just the slightest bit tedious. We can simplify this process somewhat with the following result.

Theorem 4.1

The curvature of the curve traced out by the vector-valued function $\mathbf{r}(t)$ is given by

$$\kappa = \frac{\|\mathbf{r}'(t) \times \mathbf{r}''(t)\|}{\|\mathbf{r}'(t)\|^3}. \tag{4.7}$$

The proof of Theorem 4.1 is rather long and involved and so, we omit it at this time, in the interest of brevity. We return to this in section 11.5, where the proof becomes a simple consequence of another significant result.

Notice that it is a relatively simple matter to use (4.7) to compute the curvature for nearly any three-dimensional curve.

Example 4.5 Finding the Curvature of a Helix

Find the curvature of the helix traced out by $\mathbf{r}(t) = \langle 2\sin t, 2\cos t, 4t \rangle$.

Solution A graph of the helix is indicated in Figure 11.20. We have

$$\mathbf{r}'(t) = \langle 2\cos t, -2\sin t, 4 \rangle$$

and

$$\mathbf{r}''(t) = \langle -2\sin t, -2\cos t, 0 \rangle.$$

Now,

$$\mathbf{r}'(t) \times \mathbf{r}''(t) = \langle 2\cos t, -2\sin t, 4 \rangle \times \langle -2\sin t, -2\cos t, 0 \rangle$$

$$= \begin{vmatrix} \mathbf{i} & \mathbf{j} & \mathbf{k} \\ 2\cos t & -2\sin t & 4 \\ -2\sin t & -2\cos t & 0 \end{vmatrix}$$

$$= \langle 8\cos t, -8\sin t, -4\cos^2 t - 4\sin^2 t \rangle$$

$$= \langle 8\cos t, -8\sin t, -4 \rangle.$$

From (4.7), we get that the curvature is

$$\kappa = \frac{\|\mathbf{r}'(t) \times \mathbf{r}''(t)\|}{\|\mathbf{r}'(t)\|^3}$$

$$= \frac{\|\langle 8\cos t, -8\sin t, -4 \rangle\|}{\|\langle 2\cos t, -2\sin t, 4 \rangle\|^3} = \frac{\sqrt{80}}{(\sqrt{20})^3} = \frac{1}{10}.$$

Note that this says that the helix has a constant curvature, as you should suspect from the graph in Figure 11.20.

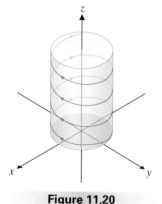

Figure 11.20
Circular helix.

In the case of a plane curve that is the graph of a function, $y = f(x)$, we can derive a particularly simple formula for the curvature. Notice that such a curve is traced out by the vector-valued function $\mathbf{r}(t) = \langle t, f(t), 0 \rangle$, where the third component is 0, since the curve lies completely in the xy-plane. Further, $\mathbf{r}'(t) = \langle 1, f'(t), 0 \rangle$ and $\mathbf{r}''(t) = \langle 0, f''(t), 0 \rangle$. From (4.7), we have

$$\kappa = \frac{\|\mathbf{r}'(t) \times \mathbf{r}''(t)\|}{\|\mathbf{r}'(t)\|^3} = \frac{\|\langle 1, f'(t), 0 \rangle \times \langle 0, f''(t), 0 \rangle\|}{\|\langle 1, f'(t), 0 \rangle\|^3}$$

$$= \frac{|f''(t)|}{\{1 + [f'(t)]^2\}^{3/2}},$$

where we have left the calculation of the cross product as a simple exercise. Since the parameter $t = x$, we can write the curvature as

Curvature for the plane curve $y = f(x)$

$$\kappa = \frac{|f''(x)|}{\{1 + [f'(x)]^2\}^{3/2}}. \tag{4.8}$$

Example 4.6   Finding the Curvature of a Parabola

Find the curvature of the parabola $y = ax^2 + bx + c$. Also, find the limiting value of the curvature as $x \to \infty$.

Solution   Taking $f(x) = ax^2 + bx + c$, we have that $f'(x) = 2ax + b$ and $f''(x) = 2a$. From (4.8), we have that

$$\kappa = \frac{|2a|}{[1 + (2ax + b)^2]^{3/2}}.$$

Taking the limit as $x \to \infty$, we have

$$\lim_{x \to \infty} \kappa = \lim_{x \to \infty} \frac{|2a|}{[1 + (2ax + b)^2]^{3/2}} = 0.$$

In other words, as $x \to \infty$, the parabola straightens out. You've certainly observed this in the graphs of parabolas for some time. Now, we have verified that this is not some sort of optical illusion; it's reality. It is a straightforward exercise to show that the maximum curvature occurs at the vertex of the parabola ($x = -b/2a$).

EXERCISES 11.4

1. Explain what it means for a curve to have zero curvature (a) at a point and (b) on an interval of t-values.

2. Throughout our study of calculus, we have looked at tangent line approximations to curves. Some tangent lines approximate a curve well over a fairly lengthy interval while some stay close to a curve for only very short intervals. If the curvature at $x = a$ is large, would you expect the tangent line at $x = a$ to approximate the curve well over a lengthy interval or a short interval? What if the curvature is small? Explain.

3. Discuss the relationship between curvature and concavity for a function $y = f(x)$.

4. Explain why the curvature $\kappa = \frac{1}{10}$ of the helix in example 4.5 is less than the curvature of the circle $\langle 2 \sin t, 2 \cos t \rangle$ in two dimensions.

In exercises 5–8, find an arc length parameterization of the given two-dimensional curve.

5. The circle of radius 2 centered at the origin

6. The circle of radius 5 centered at the origin

7. The line segment from the origin to the point $(3, 4)$

8. The line segment from $(1, 2)$ to the point $(5, -2)$

In exercises 9–14, find the unit tangent vector to the curve at the indicated points.

9. $\mathbf{r}(t) = \langle 3t, t^2 \rangle, t = 0, t = -1, t = 1$

10. $\mathbf{r}(t) = \langle 2t^3, \sqrt{t} \rangle, t = 1, t = 2, t = 3$

11. $\mathbf{r}(t) = \langle 3\cos t, 2\sin t \rangle, t = 0, t = -\frac{\pi}{2}, t = \frac{\pi}{2}$

12. $\mathbf{r}(t) = \langle 4\sin t, 2\cos t \rangle, t = -\pi, t = 0, t = \pi$

13. $\mathbf{r}(t) = \langle 3t, \cos 2t, \sin 2t \rangle, t = 0, t = -\pi, t = \pi$

14. $\mathbf{r}(t) = \langle 4t, 2t, t^2 \rangle, t = -1, t = 0, t = 1$

15.   Sketch the curve in exercise 11 along with the vectors $\mathbf{r}(0)$, $\mathbf{T}(0)$, $\mathbf{r}\left(\frac{\pi}{2}\right)$ and $\mathbf{T}\left(\frac{\pi}{2}\right)$.

16.   Sketch the curve in exercise 12 along with the vectors $\mathbf{r}(0)$, $\mathbf{T}(0)$, $\mathbf{r}\left(\frac{\pi}{2}\right)$ and $\mathbf{T}\left(\frac{\pi}{2}\right)$.

17.   Sketch the curve in exercise 13 along with the vectors $\mathbf{r}(0)$, $\mathbf{T}(0)$, $\mathbf{r}(\pi)$ and $\mathbf{T}(\pi)$.

18.   Sketch the curve in exercise 14 along with the vectors $\mathbf{r}(0)$, $\mathbf{T}(0)$, $\mathbf{r}(1)$ and $\mathbf{T}(1)$.

In exercises 19–26, find the curvature at the given point.

19. $\mathbf{r}(t) = \langle e^{-2t}, 2t, 4 \rangle, t = 0$

20. $\mathbf{r}(t) = \langle 2, \sin \pi t, \ln t \rangle, t = 1$

21. $\mathbf{r}(t) = \langle t, \sin 2t, 3t \rangle, t = 0$

22. $\mathbf{r}(t) = \langle t, t^2 + t - 1, t \rangle, t = 0$

23. $f(x) = 3x^2 - 1, x = 1$

24. $f(x) = x^3 + 2x - 1, x = 2$

25. $f(x) = \sin x, x = \pi/2$

26. $f(x) = e^{-3x}, x = 0$

27.   For $f(x) = \sin x$ (see exercise 25), show that the curvature is the same at $x = \frac{\pi}{2}$ and $x = \frac{3\pi}{2}$. Use the graph of $y = \sin x$ to predict whether the curvature would be larger or smaller at $x = \pi$.

28.   For $f(x) = e^{-3x}$ (see exercise 26), show that the curvature is larger at $x = 0$ than at $x = 2$. Use the graph of $y = e^{-3x}$ to predict whether the curvature would be larger or smaller at $x = 4$.

In exercises 29–32, sketch the curve and compute the curvature at the indicated points.

29. $\mathbf{r}(t) = \langle 2\cos 2t, 2\sin 2t, 3t \rangle, t = 0, t = \frac{\pi}{2}$

30. $\mathbf{r}(t) = \langle \cos 2t, 2\sin 2t, 4t \rangle, t = 0, t = \frac{\pi}{2}$

31. $\mathbf{r}(t) = \langle t, t, t^2 - 1 \rangle, t = 0, t = 2$

32. $\mathbf{r}(t) = \langle 2t - 1, t + 2, t - 3 \rangle, t = 0, t = 2$

In exercises 33–36, sketch the curve and find any points of maximum or minimum curvature.

33. $\mathbf{r}(t) = \langle 2\cos t, 3\sin t \rangle$ 34. $\mathbf{r}(t) = \langle 4\cos t, 3\sin t \rangle$

35. $y = 4x^2 - 3$ 36. $y = \sin x$

In exercises 37–40, graph the curvature function $\kappa(x)$ and find the limit of the curvature as $x \to \infty$.

37. $y = e^{2x}$ 38. $y = e^{-2x}$

39. $y = x^3$ 40. $y = \sqrt{x}$

41.   Explain how the answers to exercises 37–40 relate to the graphs.

42.   Find the curvature of the circular helix $\langle a\cos t, a\sin t, bt \rangle$.

43.   Label as True or False and explain: at a relative extremum of $y = f(x)$, the curvature is either a minimum or maximum.

44.   Label as True or False and explain: at an inflection point of $y = f(x)$, the curvature is zero.

45.   Label as True or False and explain: the curvature of the two-dimensional curve $y = f(x)$ is the same as the curvature of the three-dimensional curve $\mathbf{r}(t) = \langle t, f(t), c \rangle$ for any constant c.

46.   Label as True or False and explain: the curvature of the two-dimensional curve $y = f(x)$ is the same as the curvature of the three-dimensional curve $\mathbf{r}(t) = \langle t, f(t), t \rangle$.

47.   Show that the curvature of the polar curve $r = f(\theta)$ is given by
$$\kappa = \frac{|2[f'(\theta)]^2 - f(\theta)f''(\theta) + [f(\theta)]^2|}{\{[f'(\theta)]^2 + [f(\theta)]^2\}^{3/2}}.$$

48.   If $f(0) = 0$, show that the curvature of the polar curve $r = f(\theta)$ at $\theta = 0$ is given by $\kappa = \dfrac{2}{|f'(0)|}$.

In exercises 49–52, use exercises 47 and 48 to find the curvature of the polar curve at the indicated points.

49. $r = \sin 3\theta, \theta = 0, \theta = \frac{\pi}{6}$

50. $r = 3 + 2\cos \theta, \theta = 0, \theta = \frac{\pi}{2}$

51. $r = 3e^{2\theta}, \theta = 0, \theta = 1$

52. $r = 1 - 2\sin\theta, \theta = 0, \theta = \frac{\pi}{2}$

53. Find the curvature of the helix traced out by $\mathbf{r}(t) = \langle 2\sin t, 2\cos t, 0.4t \rangle$ and compare to the result of example 4.5.

54. Find the limit as $n \to 0$ of the curvature of $\mathbf{r}(t) = \langle 2\sin t, 2\cos t, nt \rangle$ for $n > 0$. Explain this result graphically.

55. In this exercise, we explore an unusual two-dimensional parametric curve sometimes known as the **Cornu spiral.** Define the vector-valued function $\mathbf{r}(t) = \left\langle \int_0^t \cos\left(\frac{\pi u^2}{2}\right) du, \int_0^t \sin\left(\frac{\pi u^2}{2}\right) du \right\rangle$. Use a graphing utility to sketch the graph of $\mathbf{r}(t)$ for $-\pi \le t \le \pi$. Compute the arc length of the curve from $t = 0$ to $t = c$ and compute the curvature at $t = c$. What is the remarkable property that you find?

11.5 TANGENT AND NORMAL VECTORS

Up to this point, we have used a single frame of reference for all of our work with vectors. That is, we have written all vectors in terms of the standard unit basis vectors $\mathbf{i}$, $\mathbf{j}$ and $\mathbf{k}$. However, this is not always the most convenient framework for describing vectors. For instance, imagine that you need to investigate the forces acting on an aircraft as it flies across the sky. A fixed frame of reference would be particularly inconvenient here. A much better frame of reference would be one that moves along with the aircraft. As it turns out, such a moving frame of reference sheds light on a wide variety of problems. In this section, we will construct this moving reference frame and see how this immediately provides useful information regarding the forces acting on an object in motion.

Consider an object moving along the curve traced out by the vector-valued function $\mathbf{r}(t) = \langle f(t), g(t), h(t) \rangle$. If we are to define a reference frame that moves with the object, we will need to have (at each point on the curve) three mutually orthogonal unit vectors. One of these should point in the direction of motion (i.e., in the direction of the orientation of the curve). In section 11.4, we defined the unit tangent vector $\mathbf{T}(t)$ by

$$\mathbf{T}(t) = \frac{\mathbf{r}'(t)}{\|\mathbf{r}'(t)\|}.$$

Further, recall from Theorem 2.4 that since $\mathbf{T}(t)$ is a unit vector (and consequently has the constant magnitude of 1), $\mathbf{T}(t)$ must be orthogonal to $\mathbf{T}'(t)$ for each t. This gives us a second unit vector in our moving frame of reference, as follows.

Definition 5.1
The **principal unit normal vector** $\mathbf{N}(t)$ is a unit vector having the same direction as $\mathbf{T}'(t)$ and is defined by

$$\mathbf{N}(t) = \frac{\mathbf{T}'(t)}{\|\mathbf{T}'(t)\|}. \tag{5.1}$$

You might wonder about the direction in which $\mathbf{N}(t)$ points. Simply saying that it's orthogonal to $\mathbf{T}(t)$ is not quite enough. After all, in three dimensions, there are infinitely many directions that are orthogonal to $\mathbf{T}(t)$. (In two dimensions, there are only two possible directions.) We can clarify this with the following observation.

Recall that from (4.5), we have that $\frac{ds}{dt} = \|\mathbf{r}'(t)\| > 0$. (This followed from the definition of the arc length parameter in (4.1).) In particular, this says that $\left|\frac{ds}{dt}\right| = \frac{ds}{dt}$. From the chain rule, we have

$$\mathbf{T}'(t) = \frac{d\mathbf{T}}{dt} = \frac{d\mathbf{T}}{ds}\frac{ds}{dt}.$$

This gives us

$$\mathbf{N}(t) = \frac{\mathbf{T}'(t)}{\|\mathbf{T}'(t)\|} = \frac{\frac{d\mathbf{T}}{ds}\frac{ds}{dt}}{\left\|\frac{d\mathbf{T}}{ds}\right\|\left|\frac{ds}{dt}\right|} = \frac{\frac{d\mathbf{T}}{ds}}{\left\|\frac{d\mathbf{T}}{ds}\right\|}$$

or equivalently,

$$\mathbf{N}(t) = \frac{1}{\kappa}\frac{d\mathbf{T}}{ds}, \tag{5.2}$$

where we have used the definition of curvature in (4.3), $\kappa = \left\|\frac{d\mathbf{T}}{ds}\right\|$.

You should note that (5.2) is not particularly useful as a formula for computing $\mathbf{N}(t)$. (Why not?) However, we can use it to interpret the meaning of $\mathbf{N}(t)$. Since $\kappa > 0$ in order for (5.2) to make sense, $\mathbf{N}(t)$ will have the same direction as $\frac{d\mathbf{T}}{ds}$. Note that $\frac{d\mathbf{T}}{ds}$ is the instantaneous rate of change of the unit tangent vector with respect to arc length. This says that $\frac{d\mathbf{T}}{ds}$ (and consequently also, $\mathbf{N}$) points in the direction in which $\mathbf{T}$ is turning as arc length increases. That is, $\mathbf{N}(t)$ will always point to the **concave** side of the curve (see Figure 11.21).

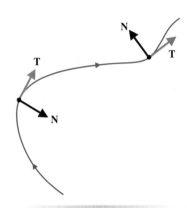

Figure 11.21

Principal unit normal vectors.

Example 5.1 Finding Unit Tangent and Principal Unit Normal Vectors

Find the unit tangent and principal unit normal vectors to the curve defined by $\mathbf{r}(t) = \langle t^2, t \rangle$.

Solution Notice that $\mathbf{r}'(t) = \langle 2t, 1 \rangle$ and so from (4.2), we have

$$\mathbf{T}(t) = \frac{\mathbf{r}'(t)}{\|\mathbf{r}'(t)\|} = \frac{\langle 2t, 1 \rangle}{\|\langle 2t, 1 \rangle\|} = \frac{\langle 2t, 1 \rangle}{\sqrt{4t^2 + 1}}$$

$$= \frac{2t}{\sqrt{4t^2 + 1}}\mathbf{i} + \frac{1}{\sqrt{4t^2 + 1}}\mathbf{j}.$$

Using the quotient rule, we have

$$\mathbf{T}'(t) = \frac{2\sqrt{4t^2 + 1} - 2t\left(\frac{1}{2}\right)(4t^2 + 1)^{-1/2}(8t)}{4t^2 + 1}\mathbf{i} - \frac{1}{2}(4t^2 + 1)^{-3/2}(8t)\mathbf{j}$$

$$= 2(4t^2 + 1)^{-1/2}\frac{(4t^2 + 1) - 4t^2}{4t^2 + 1}\mathbf{i} - (4t^2 + 1)^{-3/2}(4t)\mathbf{j}$$

$$= 2(4t^2 + 1)^{-3/2}\langle 1, -2t \rangle.$$

Further,

$$\|\mathbf{T}'(t)\| = 2(4t^2 + 1)^{-3/2}\|\langle 1, -2t \rangle\|$$

$$= 2(4t^2 + 1)^{-3/2}\sqrt{1 + 4t^2} = 2(4t^2 + 1)^{-1}.$$

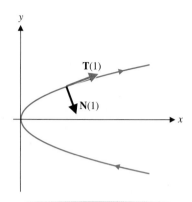

Figure 11.22
Unit tangent and principal unit
normal vectors.

From (5.1), the principal unit normal is then

$$\mathbf{N}(t) = \frac{\mathbf{T}'(t)}{\|\mathbf{T}'(t)\|} = \frac{2(4t^2 + 1)^{-3/2}\langle 1, -2t \rangle}{2(4t^2 + 1)^{-1}}$$
$$= (4t^2 + 1)^{-1/2}\langle 1, -2t \rangle.$$

In particular, for $t = 1$, we get $\mathbf{T}(1) = \left\langle \frac{2}{\sqrt{5}}, \frac{1}{\sqrt{5}} \right\rangle$ and $\mathbf{N}(1) = \left\langle \frac{1}{\sqrt{5}}, -\frac{2}{\sqrt{5}} \right\rangle$. We sketch the curve and these two sample vectors in Figure 11.22.

The calculations are similar in three dimensions, as you can see from the following example.

Example 5.2   Finding Unit Tangent and Principal
Unit Normal Vectors

Find the unit tangent and principal unit normal vectors to the curve determined by $\mathbf{r}(t) = \langle \sin 2t, \cos 2t, t \rangle$.

Solution   First, observe that $\mathbf{r}'(t) = \langle 2\cos 2t, -2\sin 2t, 1 \rangle$ and so we have from (4.2) that

$$\mathbf{T}(t) = \frac{\mathbf{r}'(t)}{\|\mathbf{r}'(t)\|} = \frac{\langle 2\cos 2t, -2\sin 2t, 1 \rangle}{\|\langle 2\cos 2t, -2\sin 2t, 1 \rangle\|} = \frac{1}{\sqrt{5}}\langle 2\cos 2t, -2\sin 2t, 1 \rangle.$$

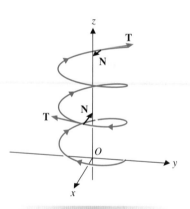

Figure 11.23
Unit tangent and principal unit
normal vectors.

This gives us

$$\mathbf{T}'(t) = \frac{1}{\sqrt{5}}\langle -4\sin 2t, -4\cos 2t, 0 \rangle$$

and so, from (5.1), the principal unit normal is

$$\mathbf{N}(t) = \frac{\mathbf{T}'(t)}{\|\mathbf{T}'(t)\|} = \frac{1}{4}\langle -4\sin 2t, -4\cos 2t, 0 \rangle = \langle -\sin 2t, -\cos 2t, 0 \rangle.$$

Notice that the curve here is a circular helix and that at each point, $\mathbf{N}(t)$ points straight back toward the z-axis (see Figure 11.23).

To get a third unit vector orthogonal to both $\mathbf{T}(t)$ and $\mathbf{N}(t)$, we simply take their cross product.

Definition 5.2
We define the **binormal** vector $\mathbf{B}(t)$ to be

$$\mathbf{B}(t) = \mathbf{T}(t) \times \mathbf{N}(t).$$

Notice that by definition, $\mathbf{B}(t)$ is orthogonal to both $\mathbf{T}(t)$ and $\mathbf{N}(t)$ and by Theorem 4.4 in Chapter 10, its magnitude is given by

$$\|\mathbf{B}(t)\| = \|\mathbf{T}(t) \times \mathbf{N}(t)\| = \|\mathbf{T}(t)\| \, \|\mathbf{N}(t)\| \sin\theta,$$

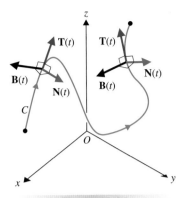

Figure 11.24

The **TNB** frame.

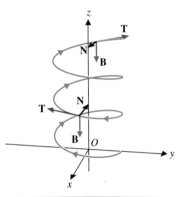

Figure 11.25

The **TNB** frame for
$\mathbf{r}(t) = \langle \sin 2t, \cos 2t, t \rangle$.

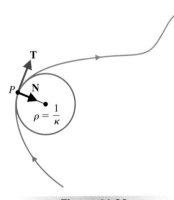

Figure 11.26

Osculating circle.

where θ is the angle between $\mathbf{T}(t)$ and $\mathbf{N}(t)$. However, since $\mathbf{T}(t)$ and $\mathbf{N}(t)$ are both unit vectors, $\|\mathbf{T}(t)\| = \|\mathbf{N}(t)\| = 1$. Further, $\mathbf{T}(t)$ and $\mathbf{N}(t)$ are orthogonal, so that $\sin \theta = 1$ and consequently, $\|\mathbf{B}(t)\| = 1$, too. This triple of three unit vectors $\mathbf{T}(t)$, $\mathbf{N}(t)$ and $\mathbf{B}(t)$ forms a frame of reference, called the **TNB frame** (or the **moving trihedral**), that moves along the curve defined by $\mathbf{r}(t)$ (see Figure 11.24). This moving frame of reference has particular importance in a branch of mathematics called **differential geometry** and is used in the navigation of spacecraft.

As you can see, the definition of the binormal vector is certainly straightforward. We illustrate this now for the curve from example 5.2.

Example 5.3   Finding the Binormal Vector

Find the binormal vector $\mathbf{B}(t)$ for the curve traced out by $\mathbf{r}(t) = \langle \sin 2t, \cos 2t, t \rangle$.

Solution   Recall from example 5.2 that the unit tangent vector is given by $\mathbf{T}(t) = \frac{1}{\sqrt{5}} \langle 2\cos 2t, -2\sin 2t, 1 \rangle$ and the principal unit normal vector is given by $\mathbf{N}(t) = \langle -\sin 2t, -\cos 2t, 0 \rangle$. From Definition 5.2, we now have that the binormal vector is given by

$$\mathbf{B}(t) = \mathbf{T}(t) \times \mathbf{N}(t) = \frac{1}{\sqrt{5}} \langle 2\cos 2t, -2\sin 2t, 1 \rangle \times \langle -\sin 2t, -\cos 2t, 0 \rangle$$

$$= \frac{1}{\sqrt{5}} \begin{vmatrix} \mathbf{i} & \mathbf{j} & \mathbf{k} \\ 2\cos 2t & -2\sin 2t & 1 \\ -\sin 2t & -\cos 2t & 0 \end{vmatrix}$$

$$= \frac{1}{\sqrt{5}} [\mathbf{i}(\cos 2t) - \mathbf{j}(\sin 2t) + \mathbf{k}(-2\cos^2 2t - 2\sin^2 2t)]$$

$$= \frac{1}{\sqrt{5}} \langle \cos 2t, -\sin 2t, -2 \rangle.$$

We illustrate the **TNB** frame for this curve in Figure 11.25.

For each value of t, the plane determined by $\mathbf{N}(t)$ and $\mathbf{B}(t)$ is called the **normal plane**. By definition, the normal plane to a curve at a point contains all of the lines that are orthogonal to the tangent vector at the given point on the curve. For each value of t, the plane determined by $\mathbf{T}(t)$ and $\mathbf{N}(t)$ is called the **osculating plane**. For a two-dimensional curve, the osculating plane is simply the xy-plane.

For a given value of t, say $t = t_0$, if the curvature κ of the curve at the point P corresponding to t_0 is nonzero, then the circle of radius $\rho = \frac{1}{\kappa}$ lying completely in the osculating plane and whose center lies a distance of $\frac{1}{\kappa}$ from P along the normal $\mathbf{N}(t)$ is called the **osculating circle** (or the **circle of curvature**). Recall from example 4.4 that the curvature of a circle is the reciprocal of its radius. Notice that this says that the osculating circle has the same tangent and curvature at P as the curve. Further, since the normal vector always points to the concave side of the curve, the osculating circle lies on the concave side of the curve. In this sense, then, the osculating circle is the circle that "best fits" the curve at the point P (see Figure 11.26). The radius of the osculating circle is called the **radius of curvature** and the center of the circle is called the **center of curvature**.

Example 5.4 Finding the Osculating Circle

Find the osculating circle for the parabola defined by $\mathbf{r}(t) = \langle t^2, t \rangle$ at $t = 0$.

Solution In example 5.1, we found that the unit tangent vector is

$$\mathbf{T}(t) = (4t^2 + 1)^{-1/2} \langle 2t, 1 \rangle,$$

$$\mathbf{T}'(t) = 2(4t^2 + 1)^{-3/2} \langle 1, -2t \rangle$$

and the principal unit normal is

$$\mathbf{N}(t) = (4t^2 + 1)^{-1/2} \langle 1, -2t \rangle.$$

So, from (4.6), the curvature is given by

$$\kappa(t) = \frac{\|\mathbf{T}'(t)\|}{\|\mathbf{r}'(t)\|}$$

$$= \frac{2(4t^2 + 1)^{-3/2}(1 + 4t^2)^{1/2}}{(4t^2 + 1)^{1/2}} = 2(4t^2 + 1)^{-3/2}.$$

We now have $\kappa(0) = 2$, so that the radius of curvature for $t = 0$ is $\rho = \dfrac{1}{\kappa} = \dfrac{1}{2}$. Further, $\mathbf{N}(0) = \langle 1, 0 \rangle$ and $\mathbf{r}(0) = \langle 0, 0 \rangle$, so that the center of curvature is located $\rho = \frac{1}{2}$ unit from the origin in the direction of $\mathbf{N}(0)$ (i.e., along the positive x-axis). We draw the curve and the osculating circle in Figure 11.27.

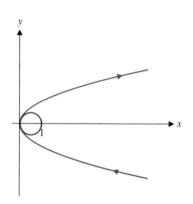

Figure 11.27
Osculating circle.

Tangential and Normal Components of Acceleration

Now that we have defined the unit tangent and principal unit normal vectors, we can make a remarkable observation about the motion of an object. In particular, we'll see how this observation helps to explain the behavior of an automobile as it travels along a curved stretch of road.

Suppose that the position of an object at time t is given by the terminal point of the vector-valued function $\mathbf{r}(t)$. Recall from the definition of the unit tangent vector that $\mathbf{T}(t) = \dfrac{\mathbf{r}'(t)}{\|\mathbf{r}'(t)\|}$ and from (4.5), $\|\mathbf{r}'(t)\| = \dfrac{ds}{dt}$, where s represents arc length. Then, the velocity of the object is given by

$$\mathbf{v}(t) = \mathbf{r}'(t) = \|\mathbf{r}'(t)\| \, \mathbf{T}(t) = \frac{ds}{dt} \, \mathbf{T}(t).$$

Using the product rule [Theorem 2.3 (iii)], we have that the acceleration is given by

$$\mathbf{a}(t) = \mathbf{v}'(t) = \frac{d}{dt}\left(\frac{ds}{dt} \, \mathbf{T}(t) \right) = \frac{d^2s}{dt^2} \, \mathbf{T}(t) + \frac{ds}{dt} \, \mathbf{T}'(t). \tag{5.3}$$

Recall that we had defined the principal unit normal by $\mathbf{N}(t) = \dfrac{\mathbf{T}'(t)}{\|\mathbf{T}'(t)\|}$, so that

$$\mathbf{T}'(t) = \|\mathbf{T}'(t)\| \, \mathbf{N}(t). \tag{5.4}$$

Further, by the chain rule,

$$\|\mathbf{T}'(t)\| = \left\| \frac{d\mathbf{T}}{dt} \right\| = \left\| \frac{d\mathbf{T}}{ds} \frac{ds}{dt} \right\|$$

$$= \left| \frac{ds}{dt} \right| \left\| \frac{d\mathbf{T}}{ds} \right\| = \kappa \frac{ds}{dt}, \tag{5.5}$$

where we have also used the definition of the curvature κ given in (4.3) and the fact that $\dfrac{ds}{dt} > 0$. Putting together (5.4) and (5.5), we now have that

$$\mathbf{T}'(t) = \|\mathbf{T}'(t)\|\, \mathbf{N}(t) = \kappa\, \frac{ds}{dt}\, \mathbf{N}(t).$$

Using this together with (5.3), we now get

$$\mathbf{a}(t) = \frac{d^2 s}{dt^2}\, \mathbf{T}(t) + \kappa \left(\frac{ds}{dt}\right)^2 \mathbf{N}(t). \qquad (5.6)$$

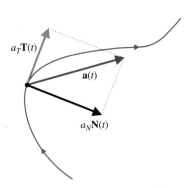

$a_T\mathbf{T}(t)$

$\mathbf{a}(t)$

$a_N\mathbf{N}(t)$

Figure 11.28
Tangential and normal components
of acceleration.

Equation (5.6) provides us with a surprising wealth of insight into the motion of an object. First, notice that since $\mathbf{a}(t)$ is written as a sum of a vector parallel to $\mathbf{T}(t)$ and a vector parallel to $\mathbf{N}(t)$, the acceleration vector always lies in the plane determined by $\mathbf{T}(t)$ and $\mathbf{N}(t)$ (i.e., the osculating plane). In particular, this says that the acceleration is always orthogonal to the binormal $\mathbf{B}(t)$. We call the coefficient of $\mathbf{T}(t)$ in (5.6) the **tangential component of acceleration** a_T and the coefficient of $\mathbf{N}(t)$ the **normal component of acceleration** a_N. That is,

$$a_T = \frac{d^2 s}{dt^2} \quad \text{and} \quad a_N = \kappa \left(\frac{ds}{dt}\right)^2. \qquad (5.7)$$

See Figure 11.28 for a graphical depiction of this decomposition of $\mathbf{a}(t)$ into tangential and normal components.

No doubt you have been in a car as it negotiates a curve in the road. We can use (5.6) to explore a strategy for keeping the car on the road (see Figure 11.29). From Newton's second law of motion, the net force acting on your car at any given time t is $\mathbf{F}(t) = m\mathbf{a}(t)$, where m is the mass of the car. From (5.6), we have

$$\mathbf{F}(t) = m\mathbf{a}(t) = m\, \frac{d^2 s}{dt^2}\, \mathbf{T}(t) + m\kappa \left(\frac{ds}{dt}\right)^2 \mathbf{N}(t).$$

$\mathbf{F}(t)$

$a_T\mathbf{T}(t)$

$a_N\mathbf{N}(t)$

Figure 11.29
Driving around a curve.

Since $\mathbf{T}(t)$ points in the direction of the path of motion, you want the component of the force acting in the direction of $\mathbf{T}(t)$ to be as large as possible compared to the component of the force acting in the direction of the normal $\mathbf{N}(t)$. (Observe that if the normal component of the force is too large, it may exceed the normal component of the force of friction between the tires and the highway, causing the car to skid off the road.) If the curve is sharp (i.e., the curvature κ is large), notice that the only way to minimize the force applied in this direction is to make $\left(\dfrac{ds}{dt}\right)^2$ small. Recall that $\dfrac{ds}{dt}$ is the speed and so, reducing speed is the only way to reduce the normal component of the force. In order to maximize the tangential component of the force, you want to make $\dfrac{d^2 s}{dt^2}$ as large as possible. Certainly, if $\dfrac{ds}{dt}$ is the speed, then $\dfrac{d^2 s}{dt^2}$ is the instantaneous rate of change of speed with respect to time. So, to maximize the tangential component of the force, you need to be accelerating while in the curve. In particular, you'll need $\dfrac{d^2 s}{dt^2} > 0$. You have probably noticed advisory signs on the highway as you approach a sharp curve. They advise you to slow down **before** you enter the curve. You should now see that this is in fact the optimal strategy for keeping your car on the road. If you wait until you're in the curve to slow down, then $\dfrac{d^2 s}{dt^2} < 0$ in the curve and so, the tangential component of the force is negative and this force is, in fact, making it harder to get through the curve. Notice that reducing speed $\left(\text{i.e., reducing } \dfrac{ds}{dt}\right)$ before the curve and then gently accelerating $\left(\text{keeping } \dfrac{d^2 s}{dt^2} > 0\right)$ once

Figure 11.30

Net force: $\dfrac{d^2s}{dt^2} < 0$.

you're in the curve will help you to negotiate the curve, by keeping the resultant force $\mathbf{F}(t)$ pointing in the general direction you wish to go. Alternatively, keeping $\dfrac{d^2s}{dt^2} < 0$ makes $\dfrac{d^2s}{dt^2}\mathbf{T}(t)$ point in the **opposite** direction as $\mathbf{T}(t)$. The net force $\mathbf{F}(t)$ will then point away from the direction of motion, working against driving through the curve (see Figure 11.30).

Example 5.5   Finding Tangential and Normal Components of Acceleration

Find the tangential and normal components of acceleration for an object with position vector $\mathbf{r}(t) = \langle 2\sin t, 2\cos t, 4t \rangle$.

Solution   In example 4.5, we found that the curvature of this curve is $\kappa = \frac{1}{10}$. We also have $\mathbf{r}'(t) = \langle 2\cos t, -2\sin t, 4 \rangle$, so that

$$\frac{ds}{dt} = \|\mathbf{r}'(t)\| = \sqrt{20}$$

and so, $\dfrac{d^2s}{dt^2} = 0$, for all t. From (5.6), we have that the acceleration is

$$\mathbf{a}(t) = \frac{d^2s}{dt^2}\,\mathbf{T}(t) + \kappa \left(\frac{ds}{dt}\right)^2 \mathbf{N}(t)$$

$$= (0)\,\mathbf{T}(t) + \frac{1}{10}\left(\sqrt{20}\right)^2 \mathbf{N}(t) = 2\mathbf{N}(t).$$

So, here we have $a_T = 0$ and $a_N = 2$.

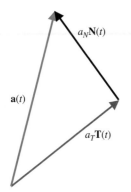

Figure 11.31

Components of $\mathbf{a}(t)$.

Notice that it's reasonably simple to compute $a_T = \dfrac{d^2s}{dt^2}$. You must only calculate $\dfrac{ds}{dt} = \|\mathbf{r}'(t)\|$ and then differentiate the result. On the other hand, computing a_N is a bit more complicated, since it requires you to first compute the curvature κ. We can simplify the calculation of a_N with the following observation. From (5.6), we have

$$\mathbf{a}(t) = \frac{d^2s}{dt^2}\,\mathbf{T}(t) + \kappa \left(\frac{ds}{dt}\right)^2 \mathbf{N}(t) = a_T\mathbf{T}(t) + a_N\mathbf{N}(t).$$

This says that $\mathbf{a}(t)$ is the vector resulting from adding the **orthogonal** vectors $a_T\mathbf{T}(t)$ and $a_N\mathbf{N}(t)$. (See Figure 11.31, where we have drawn the vectors so that the initial point of $a_N\mathbf{N}(t)$ is located at the terminal point of $a_T\mathbf{T}(t)$.) From the Pythagorean Theorem, we have that

$$\|\mathbf{a}(t)\|^2 = \|a_T\mathbf{T}(t)\|^2 + \|a_N\mathbf{N}(t)\|^2$$

$$= a_T^2 + a_N^2, \tag{5.8}$$

since $\mathbf{T}(t)$ and $\mathbf{N}(t)$ are unit vectors (i.e., $\|\mathbf{T}(t)\| = \|\mathbf{N}(t)\| = 1$). Solving (5.8) for a_N, we get

$$a_N = \sqrt{\|\mathbf{a}(t)\|^2 - a_T^2}, \tag{5.9}$$

where we have taken the positive root since $a_N = \kappa\left(\dfrac{ds}{dt}\right)^2 \geq 0$. Once you know $\mathbf{a}(t)$ and a_T, you can use (5.9) to quickly calculate a_N, without first computing the curvature. As an alternative, observe that a_T is the component of $\mathbf{a}(t)$ along the velocity vector $\mathbf{v}(t)$. Further, from (5.7) and (5.9), we can compute a_N and κ. This allows us to compute a_T, a_N and κ without first computing the speed.

Example 5.6   Finding Tangential and Normal Components of Acceleration

Find the tangential and normal components of acceleration for an object whose path is defined by $\mathbf{r}(t) = \langle t, 2t, t^2 \rangle$. In particular, find these components at $t = 1$. Also, find the curvature.

Solution   First, we compute the velocity $\mathbf{v}(t) = \mathbf{r}'(t) = \langle 1, 2, 2t \rangle$ and the acceleration $\mathbf{a}(t) = \langle 0, 0, 2 \rangle$. This gives us

$$\frac{ds}{dt} = \|\mathbf{r}'(t)\| = \|\langle 1, 2, 2t \rangle\| = \sqrt{1^2 + 2^2 + (2t)^2} = \sqrt{5 + 4t^2}.$$

The tangential component of acceleration is then

$$a_T = \frac{d^2 s}{dt^2} = \frac{d}{dt}\sqrt{5 + 4t^2}$$

$$= \frac{1}{2}(5 + 4t^2)^{-1/2}(8t) = \frac{4t}{\sqrt{5 + 4t^2}}.$$

Equivalently, we can compute a_T as the component of $\mathbf{a}(t) = \langle 0, 0, 2 \rangle$ along $\mathbf{v}(t) = \langle 1, 2, 2t \rangle$:

$$a_T = \langle 0, 0, 2 \rangle \cdot \frac{\langle 1, 2, 2t \rangle}{\sqrt{5 + 4t^2}} = \frac{4t}{\sqrt{5 + 4t^2}}.$$

From (5.9), we have that the normal component of acceleration is

$$a_N = \sqrt{\|\mathbf{a}(t)\|^2 - a_T^2} = \sqrt{2^2 - \frac{16t^2}{5 + 4t^2}}$$

$$= \sqrt{\frac{4(5 + 4t^2) - 16t^2}{5 + 4t^2}} = \frac{\sqrt{20}}{\sqrt{5 + 4t^2}}.$$

Think about computing a_N from its definition in (5.7) and notice how much simpler it was to use (5.9). Further, at $t = 1$, we have

$$a_T = \frac{4}{3} \quad \text{and} \quad a_N = \frac{\sqrt{20}}{3}.$$

Finally, from (5.7), the curvature is

$$\kappa = \frac{a_N}{\left(\dfrac{ds}{dt}\right)^2} = \frac{\sqrt{20}}{\sqrt{5 + 4t^2}} \frac{1}{\left(\sqrt{5 + 4t^2}\right)^2}$$

$$= \frac{\sqrt{20}}{\left(5 + 4t^2\right)^{3/2}}.$$

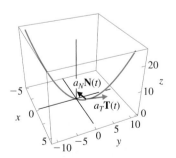

Figure 11.32

Tangential and normal components of acceleration at $t = 1$.

Notice how easy it was to compute the curvature in this way. In Figure 11.32, we show a plot of the curve traced out by $\mathbf{r}(t)$, along with the tangential and normal components of acceleration at $t = 1$.

Equation (5.6) has a wealth of applications. Among many others, it provides us with a relatively simple proof of Theorem 4.1, which we had deferred until now. You may recall that the result says that the curvature of a path traced out by the vector-valued function $\mathbf{r}(t)$ is given by

$$\kappa = \frac{\|\mathbf{r}'(t) \times \mathbf{r}''(t)\|}{\|\mathbf{r}'(t)\|^3}. \tag{5.10}$$

Proof

From (5.6), we have

$$\mathbf{a}(t) = \frac{d^2 s}{dt^2} \mathbf{T}(t) + \kappa \left(\frac{ds}{dt} \right)^2 \mathbf{N}(t).$$

Taking the cross product of both sides of this equation with $\mathbf{T}(t)$ gives us

$$\mathbf{T}(t) \times \mathbf{a}(t) = \frac{d^2 s}{dt^2} \mathbf{T}(t) \times \mathbf{T}(t) + \kappa \left(\frac{ds}{dt} \right)^2 \mathbf{T}(t) \times \mathbf{N}(t)$$

$$= \kappa \left(\frac{ds}{dt} \right)^2 \mathbf{T}(t) \times \mathbf{N}(t),$$

since the cross product of any vector with itself is the zero vector. Taking the magnitude of both sides and recognizing that $\mathbf{T}(t) \times \mathbf{N}(t) = \mathbf{B}(t)$, we get

$$\|\mathbf{T}(t) \times \mathbf{a}(t)\| = \kappa \left(\frac{ds}{dt} \right)^2 \|\mathbf{T}(t) \times \mathbf{N}(t)\|$$

$$= \kappa \left(\frac{ds}{dt} \right)^2 \|\mathbf{B}(t)\|. \tag{5.11}$$

Since the binormal vector $\mathbf{B}(t)$ is a unit vector, we have that $\|\mathbf{B}(t)\| = 1$ and equation (5.11) then simplifies to

$$\|\mathbf{T}(t) \times \mathbf{a}(t)\| = \kappa \left(\frac{ds}{dt} \right)^2.$$

Recalling that $\mathbf{T}(t) = \dfrac{\mathbf{r}'(t)}{\|\mathbf{r}'(t)\|}$, $\mathbf{a}(t) = \mathbf{r}''(t)$ and $\dfrac{ds}{dt} = \|\mathbf{r}'(t)\|$ gives us

$$\frac{\|\mathbf{r}'(t) \times \mathbf{r}''(t)\|}{\|\mathbf{r}'(t)\|} = \kappa \|\mathbf{r}'(t)\|^2.$$

Finally, solving for κ leaves us with (5.10), as desired.

Kepler's Laws

We are now in a position to present one of the most profound discoveries ever made by humankind. For hundreds of years, people believed that the sun, the other stars and the planets all revolved around the earth. The year 1543 saw the publication of the astronomer Copernicus' theory that the earth and other planets in fact, revolved around the sun. Sixty years later, based on a very careful analysis of a massive number of astronomical observations, the German astronomer Johannes Kepler formulated three laws that he reasoned must be followed by every planet. We present these now.

Kepler's Laws of Planetary Motion

1. Each planet follows an elliptical orbit, with the sun at one focus.
2. The line segment joining the sun to a planet sweeps out equal areas in equal times.
3. If T is the time required for a given planet to make one orbit of the sun and if the length of the major axis of its elliptical orbit is $2a$, then $T^2 = ka^3$, for some constant k (i.e., T^2 is proportional to a^3).

HISTORICAL NOTES

Johannes Kepler (1571–1630)
German astronomer and mathematician whose discoveries revolutionized Western science. Kepler's remarkable mathematical ability and energy produced connections among many areas of research. A study of observations of the moon led to work in optics that included the first fundamentally correct description of the operation of the human eye. Kepler's model of the solar system used an ingenious nesting of the five Platonic solids to describe the orbits of the six known planets. (Kepler invented the term "satellite" to describe the moon, which his system demoted from planetary status.) A study of the wine casks opened at his wedding led Kepler to compute volumes of solids of revolution, inventing techniques that were vital to the subsequent development of calculus.

Kepler's exhaustive analysis of the data changed our perception of our place in the universe. While Kepler's work was empirical in nature, Newton's approach to the same problem was not. In 1687, in his book *Principia Mathematica,* Newton showed how to use his calculus to derive Kepler's three laws from two of Newton's laws: his second law of motion and his law of universal gravitation. You should not underestimate the significance of this achievement. With this work, Newton shed light on some of the fundamental physical laws that govern our universe.

In order to simplify our analysis, we assume that we are looking at a solar system consisting of one sun and one planet. This is a reasonable assumption, since the gravitational attraction of the sun is far greater than that of any other body (planet, moon, comet, etc.), owing to the sun's far greater mass. (As it turns out, the gravitational attraction of other bodies does have an effect. In fact, it was an observation of the irregularities in the orbit of Uranus that led astronomers to hypothesize the existence of Neptune before it had ever been observed in a telescope.)

We assume that the center of mass of the sun is located at the origin and that the center of mass of the planet is located at the terminal point of the vector-valued function $\mathbf{r}(t)$. The velocity vector for the planet is then $\mathbf{v}(t) = \mathbf{r}'(t)$, with the acceleration given by $\mathbf{a}(t) = \mathbf{r}''(t)$. From Newton's second law of motion, we have that the net (gravitational) force $\mathbf{F}(t)$ acting on the planet is

$$\mathbf{F}(t) = m\mathbf{a}(t),$$

where m is the mass of the planet. From Newton's law of universal gravitation, we have that if M is the mass of the sun, then the gravitational attraction between the two bodies satisfies

$$\mathbf{F}(t) = -\frac{GmM}{\|\mathbf{r}(t)\|^2}\frac{\mathbf{r}(t)}{\|\mathbf{r}(t)\|},$$

where G is the **universal gravitational constant.**[1] We have written $\mathbf{F}(t)$ in this form so that you can see that at each point, the gravitational attraction acts in the direction **opposite** the position vector $\mathbf{r}(t)$. Further, the gravitational attraction is jointly proportional to the masses of the sun and the planet and inversely proportional to the square of the distance between the sun and the planet. For simplicity, we will let $r = \|\mathbf{r}\|$ and not explicitly indicate the t-variable. Taking $\mathbf{u}(t) = \frac{\mathbf{r}(t)}{\|\mathbf{r}(t)\|}$ (a unit vector in the direction of $\mathbf{r}(t)$), we can then write Newton's laws as simply

$$\mathbf{F} = m\mathbf{a} \quad \text{and} \quad \mathbf{F} = -\frac{GmM}{r^2}\mathbf{u}.$$

---

[1]If we measure mass in kilograms, force in newtons and distance in meters, G is given approximately by $G \approx 6.672 \times 10^{-11}$ N m^2/kg^2.

We begin by demonstrating that the orbit of a planet lies in a plane. Equating the two expressions above for **F** and canceling out the common factor of m, we have

$$\mathbf{a} = -\frac{GM}{r^2}\mathbf{u}. \tag{5.12}$$

Notice that this says that the acceleration **a** always points in the **opposite** direction from **r**, so that the force of gravity accelerates the planet toward the sun at all times. Since **a** and **r** are parallel, we have that

$$\mathbf{r} \times \mathbf{a} = \mathbf{0}. \tag{5.13}$$

Next, notice from the product rule [Theorem 2.3 (v)], that we have

$$\frac{d}{dt}(\mathbf{r} \times \mathbf{v}) = \frac{d\mathbf{r}}{dt} \times \mathbf{v} + \mathbf{r} \times \frac{d\mathbf{v}}{dt}$$

$$= \mathbf{v} \times \mathbf{v} + \mathbf{r} \times \mathbf{a} = \mathbf{0},$$

in view of (5.13) and since $\mathbf{v} \times \mathbf{v} = \mathbf{0}$. Integrating both sides of this expression gives us

$$\mathbf{r} \times \mathbf{v} = \mathbf{c}, \tag{5.14}$$

for some constant vector **c**. Notice that this says that for each t, $\mathbf{r}(t)$ is orthogonal to the constant vector **c**. In particular, this says that the terminal point of $\mathbf{r}(t)$ (and consequently, the orbit of the planet) lies in the plane orthogonal to the vector **c** and containing the origin.

Now that we have established that a planet's orbit lies in a plane, we are in a position to prove Kepler's first law. For the sake of simplicity, we assume that the plane containing the orbit is the xy-plane, so that **c** is parallel to the z-axis (see Figure 11.33). Now, observe that since $\mathbf{r} = r\mathbf{u}$, we have by the product rule [Theorem 2.3 (iii)] that

$$\mathbf{v} = \frac{d\mathbf{r}}{dt} = \frac{d}{dt}(r\mathbf{u}) = \frac{dr}{dt}\mathbf{u} + r\frac{d\mathbf{u}}{dt}.$$

Substituting this into (5.14), and replacing **r** by $r\mathbf{u}$, we have

$$\mathbf{c} = \mathbf{r} \times \mathbf{v} = r\mathbf{u} \times \left(\frac{dr}{dt}\mathbf{u} + r\frac{d\mathbf{u}}{dt}\right)$$

$$= r\frac{dr}{dt}(\mathbf{u} \times \mathbf{u}) + r^2\left(\mathbf{u} \times \frac{d\mathbf{u}}{dt}\right)$$

$$= r^2\left(\mathbf{u} \times \frac{d\mathbf{u}}{dt}\right),$$

since $\mathbf{u} \times \mathbf{u} = \mathbf{0}$. Together with (5.12), this gives us

$$\mathbf{a} \times \mathbf{c} = -\frac{GM}{r^2}\mathbf{u} \times r^2\left(\mathbf{u} \times \frac{d\mathbf{u}}{dt}\right)$$

$$= -GM\mathbf{u} \times \left(\mathbf{u} \times \frac{d\mathbf{u}}{dt}\right)$$

$$= -GM\left[\left(\mathbf{u} \cdot \frac{d\mathbf{u}}{dt}\right)\mathbf{u} - (\mathbf{u} \cdot \mathbf{u})\frac{d\mathbf{u}}{dt}\right], \tag{5.15}$$

where we have rewritten the vector triple product using Theorem 4.3 (vi) in Chapter 10. There are two other things to note here. First, since **u** is a unit vector, $\mathbf{u} \cdot \mathbf{u} = \|\mathbf{u}\|^2 = 1$. Further, from Theorem 2.4, since **u** is a vector-valued function of constant magnitude, $\mathbf{u} \cdot \frac{d\mathbf{u}}{dt} = 0$. Consequently, (5.15) simplifies to

$$\mathbf{a} \times \mathbf{c} = GM\frac{d\mathbf{u}}{dt} = \frac{d}{dt}(GM\mathbf{u}),$$

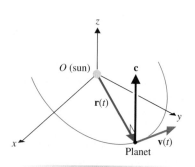

Figure 11.33

Position and velocity vectors for planetary motion.

since G and M are constants. Observe that using the definition of $\mathbf{a}$, we can write

$$\mathbf{a} \times \mathbf{c} = \frac{d\mathbf{v}}{dt} \times \mathbf{c} = \frac{d}{dt}(\mathbf{v} \times \mathbf{c}),$$

since $\mathbf{c}$ is a constant vector. Equating these last two expressions for $\mathbf{a} \times \mathbf{c}$ gives us

$$\frac{d}{dt}(\mathbf{v} \times \mathbf{c}) = \frac{d}{dt}(GM\mathbf{u}).$$

Integrating both sides gives us

$$\mathbf{v} \times \mathbf{c} = GM\mathbf{u} + \mathbf{b}, \tag{5.16}$$

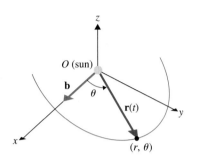

Figure 11.34

Polar coordinates for the position of the planet.

for some constant vector $\mathbf{b}$. Now, note that $\mathbf{v} \times \mathbf{c}$ must be orthogonal to $\mathbf{c}$ and so, $\mathbf{v} \times \mathbf{c}$ must lie in the xy-plane. (Recall that we had chosen the orientation of the xy-plane so that $\mathbf{c}$ was a vector orthogonal to the plane. This says further that every vector orthogonal to $\mathbf{c}$ must lie in the xy-plane.) From (5.16), since $\mathbf{u}$ and $\mathbf{v} \times \mathbf{c}$ lie in the xy-plane, $\mathbf{b}$ must also lie in the same plane. (Think about why this must be so.) Next, align the x-axis so that the positive x-axis points in the same direction as $\mathbf{b}$ (see Figure 11.34). Also, let θ be the angle from the positive x-axis to $\mathbf{r}(t)$, so that (r, θ) are polar coordinates for the endpoint of the position vector $\mathbf{r}(t)$, as indicated in Figure 11.34.

Next, let $b = \|\mathbf{b}\|$ and $c = \|\mathbf{c}\|$. Then, from (5.14), we have

$$c^2 = \mathbf{c} \cdot \mathbf{c} = (\mathbf{r} \times \mathbf{v}) \cdot \mathbf{c} = \mathbf{r} \cdot (\mathbf{v} \times \mathbf{c}),$$

where we have rewritten the scalar triple product using Theorem 4.3 (v) in Chapter 10. Putting this together with (5.16), and writing $\mathbf{r} = r\mathbf{u}$, we get

$$
\begin{aligned}
c^2 = \mathbf{r} \cdot (\mathbf{v} \times \mathbf{c}) &= r\mathbf{u} \cdot (GM\mathbf{u} + \mathbf{b}) \\
&= rGM\mathbf{u} \cdot \mathbf{u} + r\mathbf{u} \cdot \mathbf{b}. \tag{5.17}
\end{aligned}
$$

Since $\mathbf{u}$ is a unit vector, $\mathbf{u} \cdot \mathbf{u} = \|\mathbf{u}\|^2 = 1$ and by Theorem 3.2 in Chapter 10,

$$\mathbf{u} \cdot \mathbf{b} = \|\mathbf{u}\| \|\mathbf{b}\| \cos\theta = b\cos\theta,$$

where θ is the angle between $\mathbf{b}$ and $\mathbf{u}$ (i.e., the angle between the positive x-axis and $\mathbf{r}$). Together with (5.17), this gives us

$$c^2 = rGM + rb\cos\theta = r(GM + b\cos\theta).$$

Solving this for r gives us

$$r = \frac{c^2}{GM + b\cos\theta}.$$

Dividing numerator and denominator by GM reduces this to

$$r = \frac{ed}{1 + e\cos\theta}, \tag{5.18}$$

where $e = \dfrac{b}{GM}$ and $d = \dfrac{c^2}{b}$. Recall from Theorem 7.2 in Chapter 9 that (5.18) is a polar equation for a conic section with focus at the origin and eccentricity e. Finally, since the orbit of a planet is a closed curve, this must be the equation of an ellipse, since the other conic sections (parabolas and hyperbolas) are not closed curves. We have now proved that (assuming one sun and one planet and no other celestial bodies), the orbit of a planet is an ellipse with one focus located at the center of mass of the sun.

You may be thinking what a long derivation this was. (We haven't been keeping score, but it was probably one of the longest derivations of anything in this book.) Take a moment, though, to realize the enormity of what we have done. Thanks to the genius of Newton and his second law of motion and his law of universal gravitation, we have used the calculus to settle one of the most profound questions of our existence: How does the mechanics of a solar system work? Through the power of reason and the use of considerable calculus, we have found an answer that is consistent with the observed motion of the planets, first postulated by Kepler. This magnificent achievement came more than 300 years ago and was one of the earliest (and most profound) success stories for the calculus. Since that time, the calculus has proven to be an invaluable tool for countless engineers, physicists, mathematicians and others.

EXERCISES 11.5

1. Suppose that you are driving a car, going slightly uphill as the road curves to the left. Describe the directions of the unit tangent, principal unit normal and binormal vectors. What changes if the road curves to the right?

2. If the components of $\mathbf{r}(t)$ are linear functions, explain why you can't compute the principal unit normal vector. Describe graphically why it is impossible to define a single direction for the principal unit normal.

3. Previously in your study of calculus, you have approximated curves with lines and the graphs of other polynomials (Taylor polynomials). Discuss possible circumstances in which the osculating circle would be a better or worse approximation of a curve than the graph of a polynomial.

4. Suppose that you are flying in a fighter jet and an enemy jet is headed straight at you with velocity vector parallel to your principal unit normal vector. Discuss how much danger you are in and what maneuver(s) you might want to make to avoid danger.

In exercises 5–12, find the unit tangent and principal unit normal vectors at the given points.

5. $\mathbf{r}(t) = \langle t, t^2 \rangle$ at $t = 0, t = 1$

6. $\mathbf{r}(t) = \langle t, t^3 \rangle$ at $t = 0, t = 1$

7. $\mathbf{r}(t) = \langle \cos 2t, \sin 2t \rangle$ at $t = 0, t = \frac{\pi}{4}$

8. $\mathbf{r}(t) = \langle 2 \cos t, 3 \sin t \rangle$ at $t = 0, t = \frac{\pi}{4}$

9. $\mathbf{r}(t) = \langle \cos 2t, t, \sin 2t \rangle$ at $t = 0, t = \frac{\pi}{2}$

10. $\mathbf{r}(t) = \langle \cos t, \sin t, \sin t \rangle$ at $t = 0, t = \frac{\pi}{2}$

11. $\mathbf{r}(t) = \langle t, t^2 - 1, t \rangle$ at $t = 0, t = 1$

12. $\mathbf{r}(t) = \langle t, t, 3 \sin 2t \rangle$ at $t = 0, t = -\pi$

In exercises 13–16, find the osculating circle at the given points.

13. $\mathbf{r}(t) = \langle t, t^2 \rangle$ at $t = 0$ 14. $\mathbf{r}(t) = \langle t, t^3 \rangle$ at $t = 0$

15. $\mathbf{r}(t) = \langle \cos 2t, \sin 2t \rangle$ at $t = \frac{\pi}{4}$

16. $\mathbf{r}(t) = \langle 2 \cos t, 3 \sin t \rangle$ at $t = \frac{\pi}{4}$

In exercises 17–20, find the tangential and normal components of acceleration for the given position functions at the given points.

17. $\mathbf{r}(t) = \langle 8t, 16t - 16t^2 \rangle$ at $t = 0, t = 1$

18. $\mathbf{r}(t) = \langle \cos 2t, \sin 2t \rangle$ at $t = 0, t = 2$

19. $\mathbf{r}(t) = \langle \cos 2t, t^2, \sin 2t \rangle$ at $t = 0, t = \frac{\pi}{4}$

20. $\mathbf{r}(t) = \langle 2 \cos t, 3 \sin t, t^2 \rangle$ at $t = 0, t = \frac{\pi}{4}$

21. In exercise 19, determine whether the speed of the object is increasing or decreasing at the given points.

22. In exercise 20, determine whether the speed of the object is increasing or decreasing at the given points.

23. For the circular helix traced out by $\mathbf{r}(t) = \langle a \cos t, a \sin t, bt \rangle$, find the tangential and normal components of acceleration.

24. For the linear path traced out by $\mathbf{r}(t) = \langle a + bt, c + dt, e + ft \rangle$, find the tangential and normal components of acceleration.

In exercises 25–28, find the binormal vector $\mathbf{B}(t) = \mathbf{T}(t) \times \mathbf{N}(t)$ at $t = 0$ and $t = 1$. Also, sketch the curve traced out by $\mathbf{r}(t)$ and the vectors $\mathbf{T}, \mathbf{N}$ and $\mathbf{B}$ at these points.

25. $\mathbf{r}(t) = \langle t, 2t, t^2 \rangle$ 26. $\mathbf{r}(t) = \langle t, 2t, t^3 \rangle$

27. $\mathbf{r}(t) = \langle 4 \cos \pi t, 4 \sin \pi t, t \rangle$

28. $\mathbf{r}(t) = \langle 3 \cos 2\pi t, t, \sin 2\pi t \rangle$

In exercises 29–32, label the statement as true (i.e., always true) or false and explain your answer.

29. $\mathbf{T} \cdot \dfrac{d\mathbf{T}}{ds} = 0$ 30. $\mathbf{T} \cdot \mathbf{B} = 0$

31. $\dfrac{d}{ds}(\mathbf{T} \cdot \mathbf{T}) = 0$ 32. $\mathbf{T} \cdot (\mathbf{N} \times \mathbf{B}) = 1$

The friction force required to keep a car from skidding on a curve is given by $\mathbf{F}_s(t) = m a_N \mathbf{N}(t)$. In exercises 33–36, find the friction force needed to keep a car of mass $m = 100$ (slugs) from skidding.

33. $\mathbf{r}(t) = \langle 100 \cos \pi t, 100 \sin \pi t \rangle$

34. $\mathbf{r}(t) = \langle 200 \cos \pi t, 200 \sin \pi t \rangle$

35. $\mathbf{r}(t) = \langle 100 \cos 2\pi t, 100 \sin 2\pi t \rangle$

36. $\mathbf{r}(t) = \langle 300 \cos 2t, 300 \sin 2t \rangle$

37. Based on your answers to exercises 33 and 34, how does the required friction force change when the radius of a turn is doubled?

38. Based on your answers to exercises 33 and 35, how does the required friction force change when the speed of a car on a curve is doubled?

39. Compare the radii of the osculating circles for $y = \cos x$ at $x = 0$ and $x = \frac{\pi}{4}$. Compute the concavity of the curve at these points, and use this information to explain why one circle is larger than the other.

40. Compare the osculating circles for $y = \cos x$ at $x = 0$ and $x = \pi$. Compute the concavity of the curve at these points, and use this information to help explain why the circles have the same radius.

41. For $y = x^2$, show that each center of curvature lies on the curve traced out by $\mathbf{r}(t) = \langle 2t + 4t^3, \frac{1}{2} + 3t^2 \rangle$. Graph this curve.

42. For $r = e^{a\theta}, a > 0$, show that the radius of curvature is $e^{a\theta}\sqrt{a^2 + 1}$. Show that each center of curvature lies on the curve traced out by $ae^{at}\langle -\sin t, \cos t \rangle$ and graph the curve.

43. In this exercise, we prove Kepler's second law. Denote the (two-dimensional) path of the planet in polar coordinates by $\mathbf{r} = (r \cos \theta)\mathbf{i} + (r \sin \theta)\mathbf{j}$. Show that $\mathbf{r} \times \mathbf{v} = r^2 \dfrac{d\theta}{dt}\mathbf{k}$. Conclude that $r^2 \dfrac{d\theta}{dt} = \|\mathbf{r} \times \mathbf{v}\|$. Recall that in polar coordinates, the area swept out by the curve $r = r(\theta)$ is given by $A = \displaystyle\int_a^b \frac{1}{2}r^2 d\theta$, and show that $\dfrac{dA}{dt} = \frac{1}{2}r^2\dfrac{d\theta}{dt}$. From $\dfrac{dA}{dt} = \frac{1}{2}\|\mathbf{r} \times \mathbf{v}\|$, conclude that equal areas are swept out in equal times.

44. In this exercise, we prove Kepler's third law. Recall that the area of the ellipse $\dfrac{x^2}{a^2} + \dfrac{y^2}{b^2} = 1$ is πab. From exercise 43, the rate at which area is swept out is given by $\dfrac{dA}{dt} = \frac{1}{2}\|\mathbf{r} \times \mathbf{v}\|$. Conclude that the period of the orbit is $T = \dfrac{\pi ab}{\frac{1}{2}\|\mathbf{r} \times \mathbf{v}\|}$ and so, $T^2 = \dfrac{4\pi^2 a^2 b^2}{\|\mathbf{r} \times \mathbf{v}\|^2}$. Use (5.18) to show that the minimum value of r is $r_{\min} = \dfrac{ed}{1+e}$ and that the maximum value of r is $r_{\max} = \dfrac{ed}{1-e}$. Explain why $2a = r_{\min} + r_{\max}$ and use this to show that $a = \dfrac{ed}{1-e^2}$. Given that $1 - e^2 = \dfrac{b^2}{a^2}$, show that $\dfrac{b^2}{a} = ed$. From $e = \dfrac{b}{GM}$ and $d = \dfrac{c^2}{b}$, show that $ed = \dfrac{\|\mathbf{r} \times \mathbf{v}\|^2}{GM}$. It then follows that $\dfrac{b^2}{a} = \dfrac{\|\mathbf{r} \times \mathbf{v}\|^2}{GM}$. Finally, show that $T^2 = ka^3$, where the constant $k = \dfrac{4\pi^2}{GM}$ does not depend on the specific orbit of the planet.

45. (a) Show that $\dfrac{d\mathbf{B}}{ds}$ is orthogonal to $\mathbf{T}$. (b) Show that $\dfrac{d\mathbf{B}}{ds}$ is orthogonal to $\mathbf{B}$.

46. Use the result of exercise 45 to show that $\dfrac{d\mathbf{B}}{ds} = -\tau \mathbf{N}$, for some scalar τ. (τ is called the **torsion,** which measures how much a curve twists.) Also, show that $\tau = -\dfrac{d\mathbf{B}}{ds} \cdot \mathbf{N}$.

47. Show that the torsion for the curve traced out by $\mathbf{r}(t) = \langle f(t), g(t), k \rangle$ is zero for any constant k. (In general, the torsion is zero for any curve that lies in a single plane.)

48. The following three formulas (called the **Frenet-Serret formulas**) are of great significance in the field of differential geometry:

   (a) $\dfrac{d\mathbf{T}}{ds} = \kappa \mathbf{N}$ [equation (5.2)]

   (b) $\dfrac{d\mathbf{B}}{ds} = -\tau \mathbf{N}$ (see exercise 46)

   (c) $\dfrac{d\mathbf{N}}{ds} = -\kappa \mathbf{T} + \tau \mathbf{B}$

   Use the fact that $\mathbf{N} = \mathbf{B} \times \mathbf{T}$ and the product rule [Theorem 2.3 (v)] to establish (c).

49. Use the Frenet-Serret formulas (see exercise 48) to establish each of the following formulas:

   (a) $\mathbf{r}''(t) = s''(t)\mathbf{T} + \kappa[s'(t)]^2\mathbf{N}$

   (b) $\mathbf{r}'(t) \times \mathbf{r}''(t) = \kappa[s'(t)]^3\mathbf{B}$

   (c) $\mathbf{r}'''(t) = \{s'''(t) - \kappa^2[s'(t)]^3\}\mathbf{T} + \{3\kappa s'(t)s''(t) + \kappa'(t)[s'(t)]^2\}\mathbf{N} + \kappa\tau[s'(t)]^3\mathbf{B}$

   (d) $\tau = \dfrac{[\mathbf{r}'(t) \times \mathbf{r}''(t)] \cdot \mathbf{r}'''(t)}{|\mathbf{r}'(t) \times \mathbf{r}''(t)|^2}$

50. Show that the torsion for the helix traced out by $\mathbf{r}(t) = \langle a \cos t, a \sin t, bt \rangle$ is given by $\tau = \dfrac{b}{a^2 + b^2}$. [Hint: See exercise 49 (d).]

51. In this exercise, we explore some ramifications of the precise form of Newton's law of universal gravitation. Suppose that the gravitational force between objects is $\mathbf{F} = -\dfrac{GMm}{r^n}\mathbf{u}$, for some positive integer $n \geq 1$ (the actual law has $n = 2$). Show that the path of the planet would still be planar, and that Kepler's second law still holds. Also, show that the circular orbit $\mathbf{r} = \langle r \cos kt, r \sin kt \rangle$ (where r is a constant) satisfies the equation $\mathbf{F} = m\mathbf{a}$ and hence is a potential path for the orbit. For this path, find the relationship between the period of the orbit and the radius of the orbit.

CHAPTER REVIEW EXERCISES

In exercises 1 and 2, sketch the curve and plot the values of the vector-valued function.

1. $\mathbf{r}(t) = \langle t^2, 2 - t^2, 1 \rangle, t = 0, t = 1, t = 2$

2. $\mathbf{r}(t) = \langle \sin t, 2 \cos t, 3 \rangle, t = -\pi, t = 0, t = \pi$

In exercises 3–12, sketch the curve traced out by the given vector-valued function.

3. $\mathbf{r}(t) = \langle 3 \cos t + 1, \sin t \rangle$

4. $\mathbf{r}(t) = \langle 2 \sin t, \cos t + 2 \rangle$

5. $\mathbf{r}(t) = \langle 3 \cos t + 2 \sin 3t, 3 \sin t + 2 \cos 3t \rangle$

6. $\mathbf{r}(t) = \langle 3 \cos t + \sin 3t, 3 \sin t + \cos 3t \rangle$

7. $\mathbf{r}(t) = \langle 2 \cos t, 3, 3 \sin t \rangle$

8. $\mathbf{r}(t) = \langle 3 \cos t, -2, 2 \sin t \rangle$

9. $\mathbf{r}(t) = \langle 4 \cos 3t + 6 \cos t, 6 \sin t, 4 \sin 3t \rangle$

10. $\mathbf{r}(t) = \langle \sin \pi t, \sqrt{t^2 + t^3}, \cos \pi t \rangle$

11. $\mathbf{r}(t) = \langle \tan t, 4 \cos t, 4 \sin t \rangle$

12. $\mathbf{r}(t) = \langle \cos 5t, \tan t, 6 \sin t \rangle$

13. In parts (a)–(f), match the vector-valued function with its graph.
    (a) $\mathbf{r}(t) = \langle \sin t, t, \sin 2t \rangle$
    (b) $\mathbf{r}(t) = \langle t, \sin t, \sin 2t \rangle$
    (c) $\mathbf{r}(t) = \langle 6 \sin \pi t, t, 6 \cos \pi t \rangle$
    (d) $\mathbf{r}(t) = \langle \sin^5 t, \sin^2 t, \cos t \rangle$
    (e) $\mathbf{r}(t) = \langle \cos t, 1 - \cos^2 t, \cos t \rangle$
    (f) $\mathbf{r}(t) = \langle t^2 + 1, t^2 + 2, t - 1 \rangle$

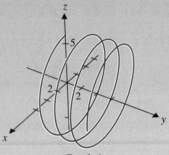

Graph A

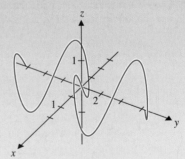

Graph B

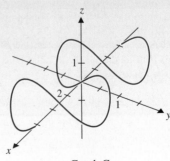

Graph C

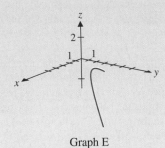

Graph D

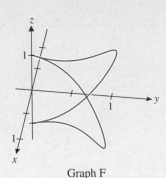

Graph E

z

1

1

y

1

x

Graph F

In exercises 14–16, sketch the curve and find its arc length.

14. $\mathbf{r}(t) = \langle \cos \pi t, \sin \pi t, \cos 4\pi t \rangle$, $0 \le t \le 2$

15. $\mathbf{r}(t) = \langle \cos t, \sin t, 6t \rangle$, $0 \le t \le 2\pi$

16. $\mathbf{r}(t) = \langle t, 4t - 1, 2 - 6t \rangle$, $0 \le t \le 2$

In exercises 17 and 18, find the limit if it exists.

17. $\lim\limits_{t \to 1} \langle t^2 - 1, e^{2t}, \cos \pi t \rangle$ 18. $\lim\limits_{t \to 1} \langle e^{-2t}, \csc \pi t, t^3 - 5t \rangle$

In exercises 19 and 20, determine all values of t at which the given vector-valued function is continuous.

19. $\mathbf{r}(t) = \langle e^{4t}, \ln t^2, 2t \rangle$

20. $\mathbf{r}(t) = \left\langle \sin t, \tan 2t, \dfrac{3}{t^2 - 1} \right\rangle$

In exercises 21 and 22, find the derivative of the given vector-valued function.

21. $\mathbf{r}(t) = \left\langle \sqrt{t^2 + 1}, \sin 4t, \ln 4t \right\rangle$

22. $\mathbf{r}(t) = \langle te^{-2t}, t^3, 5 \rangle$

In exercises 23–26, evaluate the given indefinite or definite integral.

23. $\displaystyle \int \left\langle e^{-4t}, \frac{2}{t^3}, 4t - 1 \right\rangle dt$ 24. $\displaystyle \int \left\langle \frac{2t^2}{t^3 + 2}, \sqrt{t + 1} \right\rangle dt$

25. $\displaystyle \int_0^1 \langle \cos \pi t, 4t, 2 \rangle \, dt$ 26. $\displaystyle \int_0^2 \langle e^{-3t}, 6t^2 \rangle \, dt$

In exercises 27 and 28, find the velocity and acceleration vectors for the given position vector.

27. $\mathbf{r}(t) = \langle 4 \cos 2t, 4 \sin 2t, 4t \rangle$

28. $\mathbf{r}(t) = \langle t^2 + 2, 4, t^3 \rangle$

In exercises 29–32, find the position vector from the given velocity or acceleration vector.

29. $\mathbf{v}(t) = \langle 2t + 4, -32t \rangle$, $\mathbf{r}(0) = \langle 2, 1 \rangle$

30. $\mathbf{v}(t) = \langle 4, t^2 - 1 \rangle$, $\mathbf{r}(0) = \langle -4, 2 \rangle$

31. $\mathbf{a}(t) = \langle 0, -32 \rangle$, $\mathbf{v}(0) = \langle 4, 3 \rangle$, $\mathbf{r}(0) = \langle 2, 6 \rangle$

32. $\mathbf{a}(t) = \langle t, e^{2t} \rangle$, $\mathbf{v}(0) = \langle 2, 0 \rangle$, $\mathbf{r}(0) = \langle 4, 0 \rangle$

In exercises 33 and 34, find the force acting on an object of mass 4 with the given position vector.

33. $\mathbf{r}(t) = \langle 12t, 12 - 16t^2 \rangle$

34. $\mathbf{r}(t) = \langle 3 \cos 2t, 2 \sin 2t \rangle$

In exercises 35 and 36, a projectile is fired with initial speed v_0 feet per second from a height of h feet at an angle of θ above the horizontal. Assuming that the only force acting on the object is gravity, find the maximum altitude, horizontal range and speed at impact.

35. $v_0 = 80, h = 0, \theta = \frac{\pi}{12}$ 36. $v_0 = 80, h = 6, \theta = \frac{\pi}{4}$

In exercises 37 and 38, find the unit tangent vector to the curve at the indicated points.

37. $\mathbf{r}(t) = \langle e^{-2t}, 2t, 4 \rangle$, $t = 0, t = 1$

38. $\mathbf{r}(t) = \langle 2, \sin \pi t^2, \ln t \rangle$, $t = 1, t = 2$

In exercises 39–42, find the curvature of the curve at the indicated points.

39. $\mathbf{r}(t) = \langle \cos t, \sin t, \sin t \rangle, t = 0, t = \frac{\pi}{4}$

40. $\mathbf{r}(t) = \langle 4\cos 2t, 3\sin 2t \rangle, t = 0, t = \frac{\pi}{4}$

41. $\mathbf{r}(t) = \langle 4, 3t \rangle, t = 0, t = 1$

42. $\mathbf{r}(t) = \langle t^2, t^3, t^4 \rangle, t = 0, t = 2$

In exercises 43 and 44, find the unit tangent and principal unit normal vectors at the given points.

43. $\mathbf{r}(t) = \langle \cos t, \sin t, \sin t \rangle$ at $t = 0$

44. $\mathbf{r}(t) = \langle \cos t, \sin t, \sin t \rangle$ at $t = \frac{\pi}{2}$

In exercises 45 and 46, find the tangential and normal components of acceleration at the given points.

45. $\mathbf{r}(t) = \langle 2t, t^2, 2 \rangle$ at $t = 0, t = 1$

46. $\mathbf{r}(t) = \langle t^2, 3, 2t \rangle$ at $t = 0, t = 2$

In exercises 47 and 48, the friction force required to keep a car from skidding on a curve is given by $\mathbf{F}_s(t) = ma_N \mathbf{N}(t)$. Find the friction force needed to keep a car of mass $m = 120$ (slugs) from skidding.

47. $\mathbf{r}(t) = \langle 80\cos 6t, 80\sin 6t \rangle$

48. $\mathbf{r}(t) = \langle 80\cos 4t, 80\sin 4t \rangle$

49. A tennis serve is struck at an angle θ below the horizontal from a height of 8 feet and with initial speed 120 feet per second. For the serve to count (be "in"), it must clear a net that is 39 feet away and 3 feet high and must land before the service line 60 feet away. Find the range of angles for which the serve is in.

50. A baseball pitcher throws a pitch at an angle θ below the horizontal from a height of 6 feet with an initial speed of 130 feet per second. Home plate is 60 feet away. For the pitch to be a strike, the ball must cross home plate between $20''$ and $42''$ above the ground. Find the range of angles for which the pitch will be a strike.

CHAPTER 12

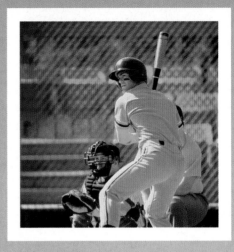

FUNCTIONS OF SEVERAL VARIABLES AND PARTIAL DIFFERENTIATION

Few things in baseball are as exciting as a home run. In the summer of 1998, the entire baseball world got caught up in the excitement, as sluggers Mark McGwire of the St. Louis Cardinals and Sammy Sosa of the Chicago Cubs both approached and then passed Roger Maris' record for the most home runs in a single season. With each crack of the bat, spectators and players alike watched the flight of the ball, wondering if it would carry far enough to clear the fence for another home run or fall short and be caught by a waiting fielder. It's usually difficult to tell whether a ball will stay in the park and get caught or fly over the fence for a home run. It's reasonable then, to ask what factors determine which outcome will occur.

In sections 5.5 and 11.3, we developed the equations for the flight of such a projectile, although always under the (unrealistic) assumption of no air resistance. There, we discovered that a ball hit at angle θ above the horizontal with initial velocity v will have a horizontal range of

$$R = \frac{1}{32} v^2 \sin 2\theta.$$

Using the properties of the sine function, we have been able to draw some interesting conclusions from this formula. However, notice that the equation for R differs from most of the functions we have studied as yet, in that R depends on **two** independent (unrelated) variables, v and θ. So far, we have only developed the calculus for functions of **one** variable.

It may have occurred to you that the situation is even more complicated than we've described. The range certainly depends on the initial velocity and launch angle, but it also depends on air effects in the form of air drag, wind velocity and the Magnus force. Air drag, in turn, varies with temperature, humidity and altitude, among other factors. It should now be clear that the range of a fly ball is not particularly easy to predict.

In this chapter, we introduce functions of several variables and extend the ideas of calculus to those functions. Often, we will gain valuable insight by considering the effect of one variable at a time. For instance, from the formula $R = \frac{1}{32} v^2 \sin 2\theta$, we can conclude that for a given initial velocity, the maximum range is obtained with $\theta = \pi/4$ (so that $\sin 2\theta = 1$). For other functions, the interplay among variables may be more subtle, so we will learn to combine information from the individual variables into information about the entire function. In section 12.3, we introduce partial derivatives to analyze one aspect of the relationship among variables. The chain rule (introduced in section 12.5) and directional derivatives (discussed in section 12.6) extend our ability to analyze functions of two or more variables.

After studying the basic calculus for functions of several variables, you should be able to find extrema of relatively simple functions. Perhaps more importantly, you should understand enough about such functions to be able to approximate extrema of more complicated functions. Of course, in real applications problems, you are rarely given a convenient formula. Even so, the understanding of multivariable calculus that you develop here will help you to make sense of a broad range of complex phenomena.

12.1 FUNCTIONS OF SEVERAL VARIABLES

The first ten chapters of this book focused on functions $f(x)$ whose domain and range were subsets of the real numbers. In Chapter 11, we studied vector-valued functions $\mathbf{F}(t)$ whose domain was a subset of the real numbers, but whose range was a set of vectors in two or more dimensions. In this section, we expand our concept of function to include functions that depend on more than one variable, that is, functions whose *domain* is multi-dimensional.

A **function of two variables** is a rule that assigns a real number $f(x, y)$ to each ordered pair of real numbers (x, y) in the domain of the function. For a function f defined on the domain $D \subset \mathbb{R}^2$, we sometimes write $f : D \subset \mathbb{R}^2 \to \mathbb{R}$ to indicate that f maps points in two dimensions to real numbers. You may think of such a function as a rule whose input is a pair of real numbers and whose output is a single real number. For instance, $f(x, y) = xy^2$ and $g(x, y) = x^2 - e^y$ are both functions of the two variables x and y.

Likewise, a **function of three variables** is a rule that assigns a real number $f(x, y, z)$ to each ordered triple of real numbers (x, y, z) in the domain $D \subset \mathbb{R}^3$ of the function. We sometimes write $f : D \subset \mathbb{R}^3 \to \mathbb{R}$ to indicate that f maps points in three dimensions to real numbers. For instance, $f(x, y, z) = xy^2 \cos z$ and $g(x, y, z) = 3zx^2 - e^y$ are both functions of the three variables x, y and z.

In principle, there is no difficulty defining functions of four (or five or more) variables. In practice, the notation gets a bit awkward and graphs become problematic. (How would you graph a four-dimensional point?) We focus here on functions of two and three variables, although most of our results can be easily extended to higher dimensions.

Unless specifically stated otherwise, the domain of a function of several variables is taken to be the set of all values of the variables for which the function is defined. One complication is that the resulting domain is of higher dimension than you are used to.

Example 1.1   Finding the Domain of a Function of Two Variables

Find and sketch the domain for (a) $f(x, y) = x \ln y$ and (b) $g(x, y) = \dfrac{2x}{y - x^2}$.

Solution   (a) For $f(x, y) = x \ln y$, recall that $\ln y$ is defined only for $y > 0$. The domain of f is then the set $D = \{(x, y) | y > 0\}$, that is, the half-plane lying above the x-axis (see Figure 12.1a).

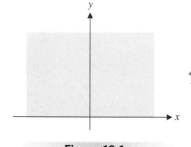

Figure 12.1a
The domain of $f(x, y) = x \ln y$.

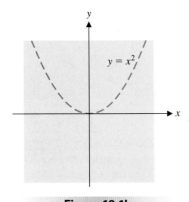

Figure 12.1b

The domain of $g(x, y) = \dfrac{2x}{y - x^2}$.

(b) For $g(x, y) = \dfrac{2x}{y - x^2}$, note that g is defined unless there is a division by zero, which occurs when $y - x^2 = 0$. The domain of g is then $\{(x, y) | y \neq x^2\}$, which is the entire xy-plane with the parabola $y = x^2$ removed (see Figure 12.1b).

Example 1.2 Finding the Domain of a Function of Three Variables

Find and describe in graphical terms the domains of (a) $f(x, y, z) = \dfrac{\cos(x + z)}{xy}$ and (b) $g(x, y, z) = \sqrt{9 - x^2 - y^2 - z^2}$.

Solution (a) For $f(x, y, z) = \dfrac{\cos(x + z)}{xy}$, there is a division by zero if $xy = 0$, which occurs if $x = 0$ or $y = 0$. The domain is then $\{(x, y, z) | x \neq 0 \text{ and } y \neq 0\}$, which is all of three-dimensional space excluding the yz-plane ($x = 0$) and the xz-plane ($y = 0$).

(b) Notice that for $g(x, y, z) = \sqrt{9 - x^2 - y^2 - z^2}$ to be defined, you'll need to have $9 - x^2 - y^2 - z^2 \geq 0$, or $x^2 + y^2 + z^2 \leq 9$. The domain of g is then the sphere of radius 3 centered at the origin and its interior.

In many applications, you won't have a formula representing a function of interest. Rather, you may only know values of the function at a relatively small number of points, as in the following example.

Example 1.3 A Function Defined by a Table of Data

A computer simulation of the flight of a baseball provided the data displayed in the following table for the range in feet of a ball hit with initial velocity v ft/s and backspin rate of ω rpm. Each ball is struck at an angle of 30° above the horizontal.

v \ ω	0	1000	2000	3000	4000
150	294	312	333	350	367
160	314	334	354	373	391
170	335	356	375	395	414
180	355	376	397	417	436

Thinking of the range as a function $R(v, \omega)$, find $R(180, 0)$, $R(160, 0)$, $R(160, 4000)$ and $R(160, 2000)$. Discuss the results in baseball terms.

Solution The function values are found by looking in the row with the given value of v and the column with the given value of ω. Thus, $R(180, 0) = 355$, $R(160, 0) = 314$, $R(160, 4000) = 391$ and $R(160, 2000) = 354$. This says that a ball with no backspin and initial velocity 180 ft/s flies 41 ft farther than one with initial velocity 160 ft/s (no surprise there). However, observe that if a 160 ft/s ball also has backspin of 4000 rpm, it actually flies 36 ft farther than the 180 ft/s ball with no backspin. (The backspin gives the ball a lift force that keeps it in the air longer.) The combination of 160 ft/s and 2000 rpm produces almost exactly the same distance as 180 ft/s with no spin. (Watts and Bahill estimate that hitting the ball $\frac{1}{4}''$ below center produces 2000 rpm.) Thus, both initial velocity and spin have significant effects on the distance the ball flies.

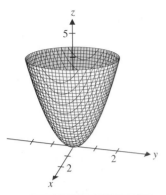

Figure 12.2a
$z = x^2 + y^2$.

The **graph of a function of two variables** is the graph of the equation $z = f(x, y)$. This is not new, as you have already graphed a number of quadric surfaces that represent functions of two variables.

Example 1.4 Graphing Functions of Two Variables

Graph (a) $f(x, y) = x^2 + y^2$ and (b) $g(x, y) = \sqrt{4 - x^2 + y^2}$.

Solution (a) For $f(x, y) = x^2 + y^2$, you may recognize the surface $z = x^2 + y^2$ as a circular paraboloid. Notice that the traces in the planes $z = k > 0$ are circles, while the traces in the planes $x = k$ and $y = k$ are parabolas. A graph is shown in Figure 12.2a.

(b) For $g(x, y) = \sqrt{4 - x^2 + y^2}$, note that the surface $z = \sqrt{4 - x^2 + y^2}$ is the top half of the surface $z^2 = 4 - x^2 + y^2$ or $x^2 - y^2 + z^2 = 4$. Here, observe that the traces in the planes $x = k$ and $z = k$ are hyperbolas, while the traces in the planes $y = k$ are circles. This gives us a hyperboloid of one sheet, wrapped around the y-axis. The graph of $z = g(x, y)$ is the top half of the hyperboloid, as shown in Figure 12.2b.

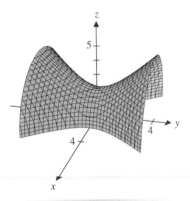

Figure 12.2b
$z = \sqrt{4 - x^2 + y^2}$.

Recall from your earlier experience drawing surfaces in three dimensions that an analysis of traces is helpful in sketching many graphs.

Example 1.5 Graphing Functions in Three Dimensions

Graph (a) $f(x, y) = \sin x \cos y$ and (b) $g(x, y) = e^{-x^2}(y^2 + 1)$.

Solution (a) For $f(x, y) = \sin x \cos y$, notice that the traces in the planes $y = k$ are the sine curves $z = \sin x \cos k$, while its traces in the planes $x = k$ are the cosine curves $z = \sin k \cos y$. The traces in the planes $z = k$ are the curves $k = \sin x \cos y$. These are a bit more unusual, as seen in Figure 12.3a (which is computer-generated) for $k = 0.5$. The surface should look like a sine wave in all directions, as shown in the computer-generated plot in Figure 12.3b.

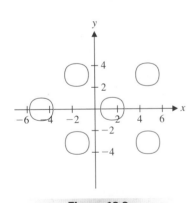

Figure 12.3a
The traces of the surface in the plane $z = 0.5$.

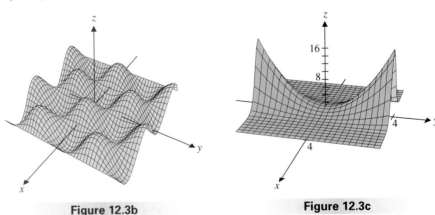

Figure 12.3b
$z = \sin x \cos y$.

Figure 12.3c
$z = e^{-x^2}(y^2 + 1)$.

(b) For $g(x, y) = e^{-x^2}(y^2 + 1)$, observe that the traces of the surface in the planes $x = k$ are parabolic, while the traces in the planes $y = k$ are proportional to $z = e^{-x^2}$, which are bell-shaped curves. The traces in the planes $z = k$ are not particularly helpful here. A sketch of the surface is shown in Figure 12.3c.

Graphing functions of more than one variable is not a simple business. For most functions of two variables, you must take hints from the expressions and try to piece together the clues to identify the surface. Your knowledge of functions of one variable is critical here. You will need to make frequent reference to your knowledge of the graphs of basic functions of a single variable.

Example 1.6   Matching a Function of Two Variables to Its Graph

Match the functions $f_1(x, y) = \cos(x^2 + y^2)$, $f_2(x, y) = \cos(e^x + e^y)$, $f_3(x, y) = \ln(x^2 + y^2)$ and $f_4(x, y) = e^{-xy}$ to the surfaces shown in Figures 12.4a–12.4d.

Solution   There are two properties of $f_1(x, y)$ that you should immediately notice. First, since the cosine of any angle lies between -1 and 1, $z = f_1(x, y)$ must always lie between -1 and 1. Second, the expression $x^2 + y^2$ is significant. Given any value of r, and any point (x, y) on the circle $x^2 + y^2 = r^2$, the height of the surface at the point (x, y) is a constant, given by $z = f_1(x, y) = \cos(r^2)$. Look for a surface that is bounded (this rules out Figure 12.4a) and has circular cross sections parallel to the xy-plane (ruling out Figures 12.4b and 12.4d). That leaves Figure 12.4c for the graph of $z = f_1(x, y)$.

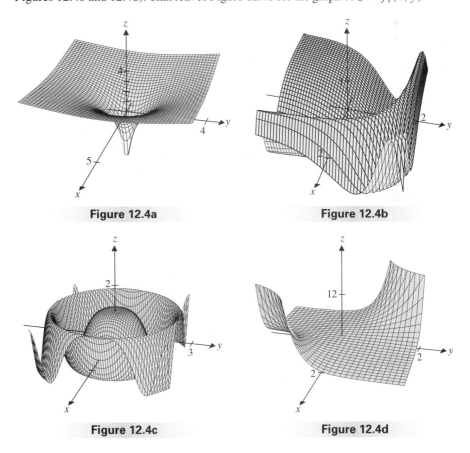

Figure 12.4a	**Figure 12.4b**
Figure 12.4c	**Figure 12.4d**

You should notice that $y = f_3(x, y)$ also has circular cross sections parallel to the xy-plane, again because of the expression $x^2 + y^2$ (think of polar coordinates). Another important property of $f_3(x, y)$ for you to recognize is that the logarithm tends to $-\infty$ as its argument (in this case, $x^2 + y^2$) approaches 0. This appears to be what is indicated in Figure 12.4a, with the surface dropping sharply at the center of the sketch. So, $z = f_3(x, y)$ corresponds to Figure 12.4a.

The remaining two functions involve exponentials. The most important distinction between them is that $f_2(x, y)$ lies between -1 and 1, due to the cosine term. This suggests that the graph of $f_2(x, y)$ is given in Figure 12.4b. To avoid jumping to a decision prematurely (after all, the domains used to produce these figures are all slightly different and could be misleading), make sure that the properties of $f_4(x, y)$ correspond to Figure 12.4d. Note that $e^{-xy} \to 0$ as $xy \to \infty$ and $e^{-xy} \to \infty$ as $xy \to -\infty$. As you move away from the origin in regions where x and y have the same sign, the surface should approach the xy-plane ($z = 0$). In regions where x and y have opposite signs, the surface should rise sharply. Notice that this behavior is exactly what you are seeing in Figure 12.4d.

Remark 1.1

The analysis we went through in example 1.6 may seem a bit slow, but we urge you to practice this on your own. The more you think (carefully) about how the properties of functions correspond to the structures of surfaces in three dimensions, the easier this chapter will be.

As with any use of technology, the creation of informative three-dimensional graphs can require a significant amount of knowledge and trial-and-error exploration. Even when you have an idea of what a graph should look like (and most often you won't!), you may need to change the viewing window several times before you can clearly see a particular feature. The wireframe graph in Figure 12.5a is a poor representation of $f(x, y) = x^2 + y^2$. Notice that this graph shows numerous traces in the planes $x = c$ and $y = c$ for $-5 \le c \le 5$. However, no traces are drawn in planes parallel to the xy-plane, so you get no sense that the figure has circular cross sections. One way to improve this is to limit the range of z-values to $0 \le z \le 20$ as in Figure 12.5b. Observe that cutting off the graph here (i.e., not displaying all values of z for the displayed values of x and y) reveals the circular cross section at $z = 20$.

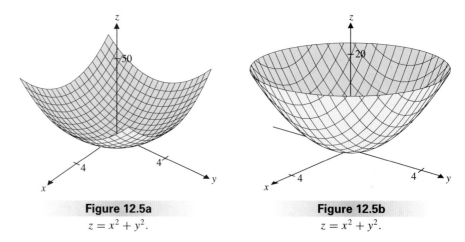

Figure 12.5a
$z = x^2 + y^2$.

Figure 12.5b
$z = x^2 + y^2$.

An important feature of three-dimensional graphs that is not present in two-dimensional graphs is the **viewpoint** from which the graph is drawn. In Figures 12.5a and 12.5b, we are looking at the paraboloid from a viewpoint that is above the xy-plane and between the positive x- and y-axes. This is the default viewpoint for many graphing utilities and is very similar to the way we have drawn graphs by hand. Different graphing utilities have different ways of changing the viewpoint. With some utilities, you can specify the coordinates of the point from which you want to look. Other utilities let you control the amount of tilt (the angle measured downward from the positive z-axis) and turn (the angle in the xy-plane measured from the positive x-axis). Figure 12.3c shows the default viewpoint of $f(x, y) = e^{-x^2}(y^2 + 1)$. In Figure 12.6a, we switch the viewpoint to the positive y-axis, from which we can see the bell-shaped profile of the graph. This viewpoint shows us several traces with $y = c$, so that we see a number of curves of the form $z = ke^{-x^2}$. In Figure 12.6b, the viewpoint is the positive x-axis, so that we see parabolic traces of the form $z = k(y^2 + 1)$. Figure 12.6c shows the view from high above the x-axis.

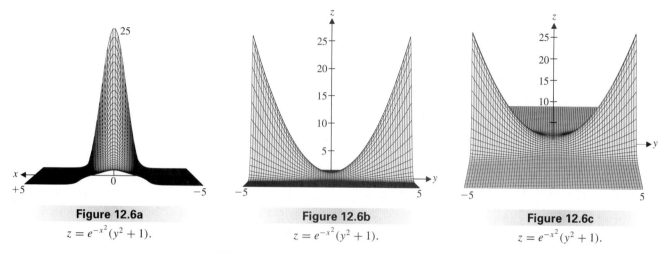

Figure 12.6a

$z = e^{-x^2}(y^2 + 1)$.

Figure 12.6b

$z = e^{-x^2}(y^2 + 1)$.

Figure 12.6c

$z = e^{-x^2}(y^2 + 1)$.

Many graphing utilities offer alternatives to wireframe graphs. One deficiency of wireframe graphs is the lack of traces parallel to the xy-plane. This is not a problem in Figures 12.6a to 12.6c, where traces in the planes $z = c$ are too complicated to be helpful. However, in Figures 12.5a and 12.5b, the circular cross sections provide valuable information about the structure of the graph. To see such traces, many graphing utilities provide a "contour mode" or "parametric surface" option. These are shown in Figures 12.7a and 12.7b for $f(x, y) = x^2 + y^2$, and are explored further in the exercises.

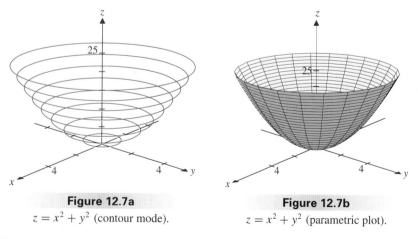

Figure 12.7a

$z = x^2 + y^2$ (contour mode).

Figure 12.7b

$z = x^2 + y^2$ (parametric plot).

Two other types of graphs, the **contour plot** and the **density plot,** provide the same information condensed into a two-dimensional picture. Recall that for two of the surfaces in example 1.6, it was important to recognize that the surface had circular cross sections, since x and y appeared only in the combination $x^2 + y^2$. The contour plot and the density plot will aid in identifying features such as this.

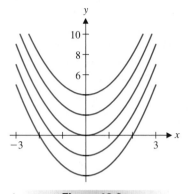

Contour plot of
$f(x, y) = -x^2 + y$.

A **level curve** of the function $f(x, y)$ is the (two-dimensional) graph of the equation $f(x, y) = c$, for some constant c. (So, the level curve $f(x, y) = c$ is a two-dimensional graph of the trace of the surface $z = f(x, y)$ in the plane $z = c$.) A **contour plot** of $f(x, y)$ is a graph of numerous level curves $f(x, y) = c$ for representative values of c.

Example 1.7   Sketching Contour Plots

Sketch contour plots for (a) $f(x, y) = -x^2 + y$ and (b) $g(x, y) = x^2 + y^2$.

Solution   (a) First, note that the level curves of $f(x, y)$ are defined by $-x^2 + y = c$, where c is a constant. Solving for y, you can identify the level curves as the parabolas $y = x^2 + c$. A contour plot with $c = -4, -2, 0, 2$ and 4 is shown in Figure 12.8a.

(b) The level curves for $g(x, y)$ are the circles $x^2 + y^2 = c$. In this case, note that there are level curves **only** for $c \geq 0$. A contour plot with $c = 1, 4, 7$ and 10 is shown in Figure 12.8b.

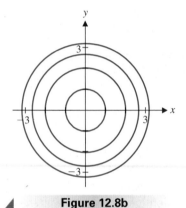

Contour plot of
$g(x, y) = x^2 + y^2$.

Note that in example 1.7, we used values for c that were equally spaced. There is no requirement that you do so, but it can help you to get a sense for how the level curves would "stack up" to produce the three-dimensional graph. We show a more extensive contour plot for $g(x, y) = x^2 + y^2$ in Figure 12.9a. In Figure 12.9b, we show a plot of the surface, with a number of traces drawn (in planes parallel to the xy-plane). Notice that the projections of these traces onto the xy-plane correspond to the contour plot in Figure 12.9a.

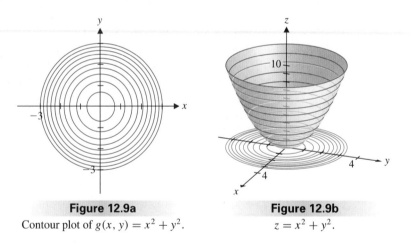

Figure 12.9a
Contour plot of $g(x, y) = x^2 + y^2$.

Figure 12.9b
$z = x^2 + y^2$.

Look carefully at Figure 12.9a and observe that the contour plot indicates that the increase in the radii of the circles is not constant as you increase z.

As you might expect, for more complicated functions, the process of matching contour plots with surfaces becomes more challenging.

Example 1.8   Matching Surfaces to Contour Plots

Match the surfaces of example 1.6 to the contour plots shown in Figures 12.10a–12.10d.

Solution   In Figures 12.4a and 12.4c, the level curves are circular, so these surfaces correspond to the contour plots in Figures 12.10a and 12.10b. But, which is which? The principle feature of the surface in Figure 12.4a is the vertical asymptote at the origin. Because of the rapid change in the function near the origin, there will be a large number

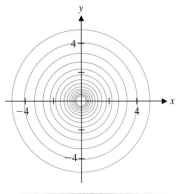

Figure 12.10a

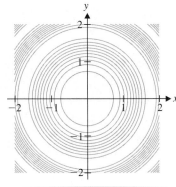

Figure 12.10b

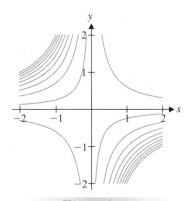

Figure 12.10c

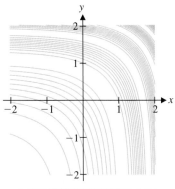

Figure 12.10d

of level curves near the origin (think about this!). By contrast, the oscillations in Figure 12.4c would produce level curves that alternately get closer together and farther apart. We can conclude that Figure 12.4a matches with Figure 12.10a, while Figure 12.4c matches with Figure 12.10b. Now, consider the two remaining surfaces and level curves. Imagine intersecting the surface in Figure 12.4d with the plane $z = 4$. You would get two separate curves that open in opposite directions (to the lower left and upper right of Figure 12.4d). These correspond to the hyperbolas seen in Figure 12.10c. The final match of Figure 12.10d to Figure 12.4b is hard to see, but notice how the curves of Figure 12.10d correspond to the curve of the peaks in Figure 12.4b. (To see this, you will need to adjust for the y-axis pointing up in Figure 12.10d and to the right in Figure 12.4b.) As an additional means of distinguishing the last two graphs, notice that Figure 12.4d is very flat near the origin. This corresponds to the lack of level curves near the origin in Figure 12.10c. By contrast, Figure 12.4b shows oscillation near the origin and there are several level curves near the origin in Figure 12.10d.

Remark 1.2

If the level curves in a contour plot are plotted for equally spaced values of z, observe that a tightly packed region of the contour plot will correspond to a region of rapid change in the function. Alternatively, blank space in the contour plot corresponds to a region of slow change in the function.

A **density plot** is closely related to a contour plot, in that they are both two-dimensional representations of a surface in three dimensions. For a density plot, the (x-y) graphing window is divided into rectangles (like pixels, although usually much larger). Each rectangle is shaded according to the size of the function value of a representative point in the rectangle, ranging from light blue (maximum function value) to black (minimum function value). In a density plot, notice that level curves can be seen as curves formed by a specific shade of gray.

Example 1.9   Matching Functions and Density Plots

Match the density plots in Figures 12.11a–12.11c with the functions $f_1(x, y) = \dfrac{1}{y^2 - x^2}$, $f_2(x, y) = \dfrac{2x}{y - x^2}$ and $f_3(x, y) = \cos(x^2 + y^2)$.

Solution   As we did with contour plots, we start with the most obvious properties of the functions and try to identify the corresponding properties in the density plots.

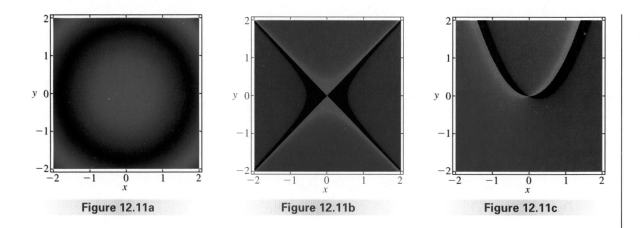

Figure 12.11a   **Figure 12.11b**   **Figure 12.11c**

Both $f_1(x, y)$ and $f_2(x, y)$ have gaps in their domains due to divisions by zero. Near the discontinuities, you should expect large functions values. Notice that Figure 12.11b shows a lighter color band in the shape of a hyperbola (like $y^2 - x^2 = c$ for a small number c) and Figure 12.11c shows a lighter color band in the shape of a parabola (like $y - x^2 = 0$). This tells you that the density plot for $f_1(x, y)$ is Figure 12.11b and the density plot for $f_2(x, y)$ is Figure 12.11c. That leaves Figure 12.11a for $f_3(x, y)$. You should be able to see the circular bands in the density plot arising from the $x^2 + y^2$ term in $f_3(x, y)$.

There are many examples of contour plots and density plots that you see every day. Weather maps often show level curves of atmospheric pressure (see Figure 12.12a). In this setting, the level curves are called **isobars** (that is, curves along which the barometric pressure is constant). Other weather maps represent temperature or degree of wetness with color coding (see Figure 12.12b), which are essentially density plots.

Scientists also use density plots while studying other climatic phenomena. For instance, in Figures 12.12c and 12.12d, we show two density plots indicating sea-surface height (which correlates with ocean heat content) indicating changes in the El Niño phenomenon over a period of several weeks.

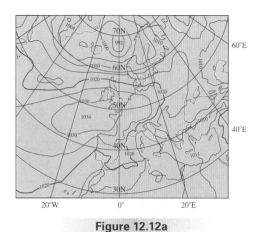

Figure 12.12a
Weather map showing
barometric pressure.

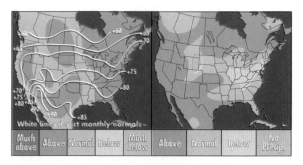

Figure 12.12b
Weather maps showing bands of
temperature and precipitation.

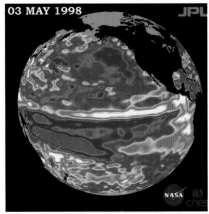

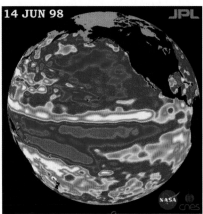

Figure 12.12c
Ocean heat content.

Figure 12.12d
Ocean heat content.

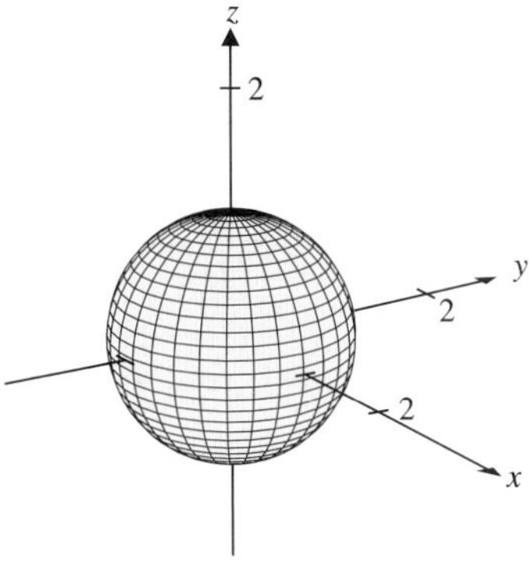

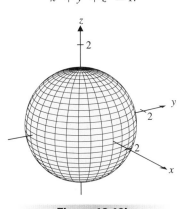

Figure 12.13a
$x^2 + y^2 + z^2 = 1$.

Figure 12.13b
$x^2 + y^2 + z^2 = 2$.

We close this section by briefly looking at the graphs of functions of three variables, $f(x, y, z)$. We won't actually graph any such functions, since a true graph would require four dimensions (three independent variables plus one dependent variable), which we cannot represent in two dimensions. We can, however, gain important information from looking at graphs of the **level surfaces** of a function f. These are the graphs of the equation $f(x, y, z) = c$, for different choices of the constant c. Much as level curves do for functions of two variables, level surfaces can help you identify symmetries and regions of rapid or slow change in a function of three variables.

Example 1.10   **Sketching Level Surfaces**

Sketch several level surfaces of $f(x, y, z) = x^2 + y^2 + z^2$.

Solution   The level surfaces are described by the equation $x^2 + y^2 + z^2 = c$. Of course, these are spheres of radius $\sqrt{c}$ for $c > 0$. Surfaces with $c = 1$ and $c = 2$ are shown in Figures 12.13a and 12.13b, respectively.

Notice that the function in example 1.10 measures the square of the distance from the origin. If you didn't recognize this at first, the level surfaces would clearly show you the symmetry and gradual increase of the function.

EXERCISES 12.1

1.   In example 1.4, we sketched a paraboloid and the top half of a hyperboloid as examples of graphs of functions of two variables. Explain why neither a full hyperboloid nor an ellipsoid would be the graph of a function of two variables. Develop a "vertical line test" for determining whether a given surface is the graph of a function of two variables.

2.   In example 1.4, we used traces to help sketch the surface, but in example 1.5 the traces were less helpful. Discuss the differences in the functions involved and how you can tell whether traces will be helpful or not.

3.   In examples 1.7 and 1.8, we discussed how to identify a contour plot given the formula for a function. In this exercise, you will discuss the inverse problem. That is, given a contour plot, what can be said about the function? For example, explain why a contour plot without labels (identifying the value of z) could correspond to more than one function. If the contour plot shows a set of concentric circles around a point, explain why you would expect that point to be the location of a local extremum. Explain why, without labels, you could not distinguish a local maximum from a local minimum.

4.   For this exercise, imagine a contour plot that shows level curves for equally spaced z-values (e.g., $z = 0$, $z = 2$ and $z = 4$). Near point A, the level curves are very close together, but near point B, there are no level curves showing at all. Discuss the behavior of the function near points A and B, especially commenting on whether the function is changing rapidly or slowly.

In exercises 5–8, describe and sketch the domain of the function.

5. $f(x, y) = \dfrac{1}{x + y}$ 6. $f(x, y) = \dfrac{3xy}{y - x^2}$

7. $f(x, y) = \ln(2 + x + y)$ 8. $f(x, y) = \sqrt{1 - x^2 - y^2}$

In exercises 9–14, describe the range of the function.

9. $f(x, y) = \sqrt{2 + x - y}$ 10. $f(x, y) = \sqrt{9 - x^2 - y^2}$

11. $f(x, y) = \cos(x^2 + y^2)$ 12. $f(x, y) = e^{x - y}$

13. $f(x, y) = x^2 + y^2 - 1$ 14. $f(x, y) = 4 - x^2 - y^2$

In exercises 15 and 16, compute the indicated function values.

15. $f(x, y) = x^2 + y$; $f(1, 2)$, $f(0, 3)$

16. $f(x, y) = \sqrt{x + y^2}$; $f(3, 1)$, $f(4, 0)$

In exercises 17 and 18, use the table in example 1.3.

17.   Find (a) $R(150, 1000)$, (b) $R(150, 2000)$ and (c) $R(150, 3000)$. (d) Based on your answers, how much extra distance is gained from an additional 1000 rpm of backspin?

18.   Find (a) $R(150, 2000)$, (b) $R(160, 2000)$ and (c) $R(170, 2000)$. (d) Based on your answers, how much extra distance is gained from an additional 10 ft/s of initial velocity?

In exercises 19–22, sketch the indicated traces and graph $z = f(x, y)$.

19. $f(x, y) = x^2 + y^2$; $z = 1, z = 4, z = 9, x = 0$

20. $f(x, y) = x^2 - y^2$; $z = 0, z = 1, y = 0, y = 2$

21. $f(x, y) = \sqrt{x^2 + y^2}$; $z = 1, z = 2, z = 3, y = 0$

22. $f(x, y) = x - 2y$; $z = 0, z = 1, x = 0, y = 0$

In exercises 23–38, use a graphing utility to sketch graphs of $z = f(x, y)$ from two different viewpoints, showing different features of the graphs.

23. $f(x, y) = x^2 + y^3$ 24. $f(x, y) = x^2 + y^4$

25. $f(x, y) = x^2 + y^2 - x^4$ 26. $f(x, y) = \dfrac{x^2}{x^2 + y^2 + 1}$

27. $f(x, y) = \dfrac{4}{(x^2 + y^2)^2 - 1}$ 28. $f(x, y) = \dfrac{x - y}{(x - y)^2 - 1}$

29. $f(x, y) = \sin x + \sin y$ 30. $f(x, y) = \cos \sqrt{x^2 + y^2}$

31. $f(x, y) = \sin x^3 - \tan y^2$ 32. $f(x, y) = \sin^2 x + \cos^2 y$

33. $f(x, y) = xye^{-x^2 - y^2}$ 34. $f(x, y) = e^{y - x^2}$

35. $f(x, y) = \ln(x^2 + y^2 - 1)$ 36. $f(x, y) = \ln(x^2 + y^2 + 1)$

37. $f(x, y) = 2x \sin xy \ln y$ 38. $f(x, y) = x^2 \sec y \cos 5xy$

In exercises 39–46, sketch a contour plot.

39. $f(x, y) = x^2 + 4y^2$ 40. $f(x, y) = 4 - x^2 - y^2$

41. $f(x, y) = \cos \sqrt{x^2 + y^2}$ 42. $f(x, y) = e^{-x^2 - y^2}$

43. $f(x, y) = y - 4x^2$ 44. $f(x, y) = y^3 - 2x$

45. $f(x, y) = e^{y - x^3}$ 46. $f(x, y) = ye^x$

In exercises 47–50, use a CAS to sketch a contour plot.

47. $f(x, y) = xye^{-x^2 - y^2}$ 48. $f(x, y) = x^3 - 3xy + y^2$

49. $f(x, y) = \sin x \sin y$ 50. $f(x, y) = \sin(y - x^2)$

51.   In parts a–f, match the functions to the surfaces.
 (a) $f(x, y) = x^2 + 3x^7$
 (b) $f(x, y) = x^2 - y^3$
 (c) $f(x, y) = \cos^2 x + y^2$
 (d) $f(x, y) = \cos(x^2 + y^2)$
 (e) $f(x, y) = \sin(x^2 + y^2)$
 (f) $f(x, y) = e^{-x^2 - y^2}$

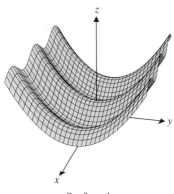

Surface A

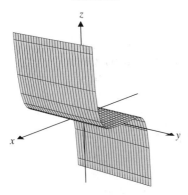

Surface B

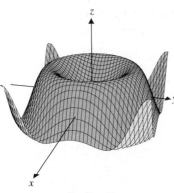

Surface C

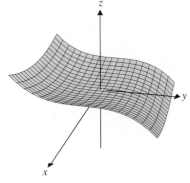

Surface D

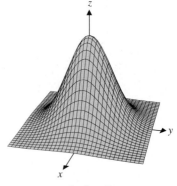

Surface E

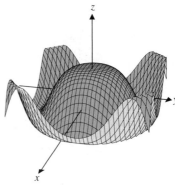

Surface F

52. In parts a–d, match the surfaces to the contour plots.

(a)

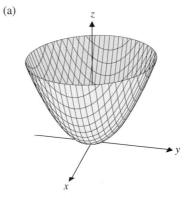

(b)

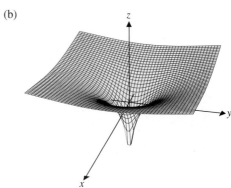

(c)

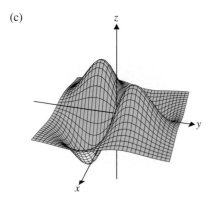

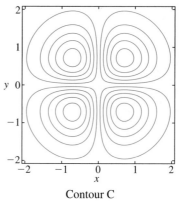

Contour C

(d)

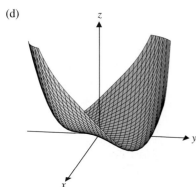

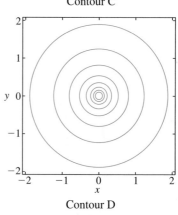

Contour D

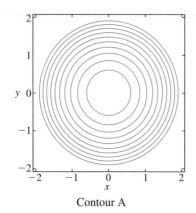

Contour A

53. In parts a–d, match the density plots to the contour plots of exercise 52.

(a)

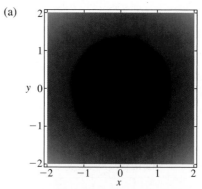

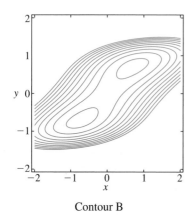

Contour B

(b)

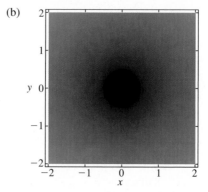

(c)

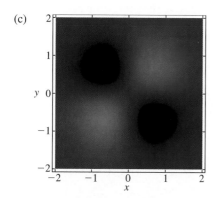

(d)

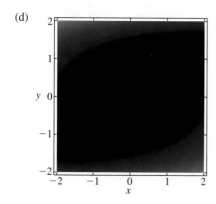

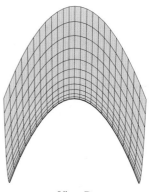

View B

58. The graph of $f(x, y) = x^2 y^2 - y^4 + x^3$ is shown from two different viewpoints. Identify which is viewed from (a) the positive x-axis and (b) the positive y-axis.

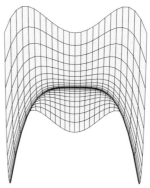

View A

In exercises 54–56, sketch several level surfaces of the given function.

54. $f(x, y, z) = x^2 - y^2 + z^2$

55. $f(x, y, z) = x^2 + y^2 - z$

56. $f(x, y, z) = z - \sqrt{x^2 + y^2}$

57. The graph of $f(x, y) = x^2 - y^2$ is shown from two different viewpoints. Identify which is viewed from (a) the positive x-axis and (b) the positive y-axis.

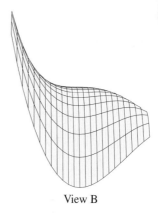

View B

59. For the graphs in exercises 57 and 58, most software that produces wireframe graphs will show the view from the z-axis as a square grid. Explain why this is an accurate (although not very helpful) representation.

60. Suppose that you are shining a flashlight down at a surface from the positive z-axis. Explain why the result will be similar to a density plot.

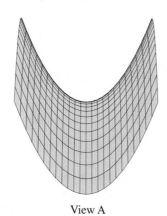

View A

61. Describe in words the graph of $z = \sin(x + y)$. In which direction is the "wave" traveling? Explain why the wireframe graph as viewed from the point (100, 100, 0) appears to be rectangular.

62. Find a viewpoint from which a wireframe graph of $z = \sin(x + y)$ shows only a single sine wave.

63. Find a viewpoint from which a wireframe graph of $z = (y - \sqrt{3}x)^2$ shows only a single parabola.

64. Find all viewpoints from which a wireframe graph of $z = e^{-x^2 - y^2}$ shows a bell-shaped curve.

65. Suppose that the following contour plot represents the population density in a city at a particular time in the evening. If there is a large rock concert that evening, locate the stadium. Speculate on what might account for other circular level curves and the linear level curves.

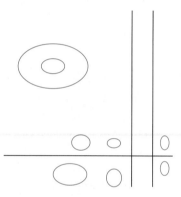

66. Suppose that the following contour plot represents the temperature in a room. If it is winter, identify likely positions for a heating vent and a window. Speculate on what the circular level curves might represent.

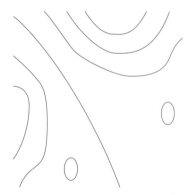

67. Suppose that the following contour plot represents the coefficient of restitution (the "bounciness") at various locations on a tennis racket. Locate the point of maximum power for the racket, and explain why you know it's **maximum** power and not minimum power. Racket manufacturers sometimes call one of the level curves the "sweet spot" of the racket. Explain why this is reasonable.

68. Suppose that the following contour plot represents the elevation on a golf putting green. Assume that the elevation increases as you move up the contour plot. If the hole is at point H, describe what putts from points A, B and C would be like.

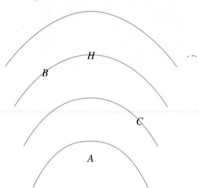

69. A well-known college uses the following formula to predict the grade average of prospective students:

$$PGA = 0.708 * HS + 0.0018 * SATV + 0.001 * SATM - 1.13$$

Here, PGA is the predicted grade average, HS is the student's high school grade average (in core academic courses, on a 4-point scale), SATV is the student's SAT verbal score and SATM is the student's SAT math score. Use your scores to compute your own predicted grade average. Determine if it is possible to have a predicted average of 4.0, or a negative predicted grade average. In this formula, the predicted grade average is a function of three variables. State which variable you think is the most important and explain why you think so.

70. In *The Hidden Game of Football,* Carroll, Palmer and Thorn give the following formula for the probability p of the team with the ball winning a game:

$$\ln\!\left(\frac{p}{1-p}\right) = 0.6s + 0.084\frac{s}{\sqrt{t/60}} - 0.0073(y - 74).$$

Here, s is the current score differential (+ if you're winning, − if you're losing), t is the number of minutes remaining and y is the number of yards to the goal line. For the function

$p(s, t, y)$, compute $p(2, 10, 40)$, $p(3, 10, 40)$, $p(3, 10, 80)$ and $p(3, 20, 40)$ and interpret the differences in football terms.

71. Suppose that you drive x mph for d miles and then y mph for d miles. Show that your average speed S is given by $S(x, y) = \dfrac{2xy}{x + y}$ mph. On a 40-mile trip, if you average 30 mph for the first 20 miles, how fast must you go to average 40 mph for the entire trip? How fast must you go to average 60 mph for the entire trip?

72. The price-to-earnings ratio of a stock is defined by $R = \frac{P}{E}$, where P is the price per share of the stock and E is the earnings. The yield of the stock is defined by $Y = \frac{d}{P}$, where d is the dividends per share. Find the yield as a function of R, d and E.

73. If your graphing utility can draw three-dimensional parametric graphs, compare the wireframe graph of $z = x^2 + y^2$ with the parametric graph of $x(r, t) = r \cos t$, $y(r, t) = r \sin t$ and $z(r, t) = r^2$. (Change parameter letters from r and t to whichever letters your utility uses.)

74. If your graphing utility can draw three-dimensional parametric graphs, compare the wireframe graph of $z = \ln(x^2 + y^2)$ with the parametric graph of $x(r, t) = r \cos t$, $y(r, t) = r \sin t$ and $z(r, t) = \ln(r^2)$.

75. If your graphing utility can draw three-dimensional parametric graphs, find parametric equations for $z = \cos(x^2 + y^2)$ and compare the wireframe and parametric graphs.

76. If your graphing utility can draw three-dimensional parametric graphs, compare the wireframe graphs of $z = \pm\sqrt{1 - x^2 - y^2}$ with the parametric graph of $x(u, v) = \cos u \sin v$, $y(u, v) = \sin u \sin v$ and $z(u, v) = \cos v$.

77. Explore the results graphically of the transformations $g_1(x, y) = f(x, y) + c$, $g_2(x, y) = f(x, y + c)$ and $g_3(x, y) = f(x + c, y)$. [Hint: Take a specific function like $f(x, y) = x^2 + y^2$ and look at the graphs of the transformed functions $x^2 + y^2 + 2$, $x^2 + (y + 2)^2$ and $(x + 2)^2 + y^2$.] Determine what changes occur when the constant is added. Test your hypothesis for other constants (be sure to try negative constants, too). Then, explore the transformations $g_4(x, y) = cf(x, y)$ and $g_5(x, y) = f(c_1 x, c_2 y)$.

78. One common use of functions of two or more variables is in image processing. For instance, to digitize a black-and-white photograph, you can superimpose a rectangular grid and label each subrectangle with a number representing the brightness of that portion of the photograph. The grid defines the x- and y-values and the brightness numbers are the func-

tion values. Briefly describe how this function differs from other functions in this section. (Hint: How many x- and y-values are there?) Near the soccer jersey in the photograph shown, describe how the brightness function behaves. To "sharpen" the picture by increasing the contrast, should you transform the function values to make them closer together or further apart?

B&W photo

Photo with grid

7	8	8	9
5	2	5	6
6	7	7	6
6	6	4	6

Digitized photo

12.2   LIMITS AND CONTINUITY

The progression of topics that we have followed several times now (at the beginning of our study of the calculus and again when we introduced vector-valued functions) has been to first look at graphs of functions. We then developed limits, continuity, derivatives and integrals. We follow this same progression now by first extending the concept of limit to functions of two (and then three) variables. As you will see, the increase in dimension causes some interesting complications.

First, we remind you that the concept of limit is fairly simple. For a function of a single variable, if we write $\lim_{x \to a} f(x) = L$, we mean that as x gets closer and closer to a, $f(x)$ gets closer and closer to the number L. Recall that when we say that x gets closer and closer to a, we mean that x gets arbitrarily close to a and can approach a from either side of a ($x < a$ or $x > a$). Further, the limit must be the same as x approaches a from either side. For functions of several variables, the idea is very similar. When we write

$$\lim_{(x,y) \to (a,b)} f(x, y) = L,$$

we mean that as (x, y) gets closer and closer to (a, b), $f(x, y)$ is getting closer and closer to the number L. In this case, (x, y) may approach (a, b) along any path through (a, b). Note that unlike the case for functions of a single variable, there are many (in fact, infinitely many) different paths passing through any given point (a, b).

For instance, $\lim_{(x,y) \to (2,3)} (xy - 2)$ asks us to identify what happens to the function $xy - 2$ as x approaches 2 and y approaches 3. Clearly, $xy - 2$ approaches $2(3) - 2 = 4$ and we write

$$\lim_{(x,y) \to (2,3)} (xy - 2) = 4.$$

Similarly, you can reason that

$$\lim_{(x,y) \to (-1,\pi)} (\sin xy - x^2 y) = \sin(-\pi) - \pi = -\pi.$$

In other words, for many (nice) functions, we can compute limits simply by substituting into the function.

Unfortunately, as with functions of a single variable, the limits we're most interested in cannot be computed by simply substituting values for x and y. For instance, for

$$\lim_{(x,y) \to (1,0)} \frac{y}{x + y - 1},$$

substituting in $x = 1$ and $y = 0$ gives the indeterminate form $\frac{0}{0}$. To evaluate this limit, we must investigate further.

You may recall from our discussion in section 1.5 that for a function f of a single variable defined on an open interval containing a (but not necessarily at a), we say that $\lim_{x \to a} f(x) = L$ if given any number $\varepsilon > 0$, there is another number $\delta > 0$ such that $|f(x) - L| < \varepsilon$ whenever $0 < |x - a| < \delta$. In other words, no matter how close you wish to make $f(x)$ to L (we represent this distance by ε), you can make it that close, just by making x sufficiently close to a (i.e., within a distance δ of a).

The definition of the limit of a function of two variables is completely analogous to the definition for a function of a single variable. We say that $\lim_{(x,y) \to (a,b)} f(x, y) = L$, if we can make $f(x, y)$ as close as desired to L by making the point (x, y) sufficiently close to (a, b). We make this more precise in the following definition.

Definition 2.1 (Formal Definition of Limit)

Let f be defined on the interior of a circle centered at the point (a, b), except possibly at (a, b) itself. We say that $\lim_{(x,y)\to(a,b)} f(x, y) = L$ if for every $\varepsilon > 0$ there exists a $\delta > 0$ such that $|f(x, y) - L| < \varepsilon$ whenever $0 < \sqrt{(x - a)^2 + (y - b)^2} < \delta$.

We illustrate the definition in Figure 12.14.

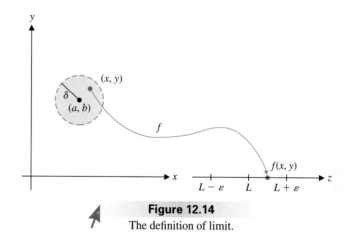

Figure 12.14

The definition of limit.

Notice that the definition says that given any desired degree of closeness $\varepsilon > 0$, you must be able to find another number $\delta > 0$, so that **all** points lying within a distance δ of (a, b) are mapped by f to points within distance ε of L on the real line.

Example 2.1   Using the Definition of Limit

Verify that $\lim_{(x,y)\to(a,b)} x = a$ and $\lim_{(x,y)\to(a,b)} y = b$.

Solution   Certainly, both of these limits are intuitively quite clear. We can use Definition 2.1 to verify them, however. Given any number $\varepsilon > 0$, we must find another number $\delta > 0$ so that $|x - a| < \varepsilon$ whenever $0 < \sqrt{(x - a)^2 + (y - b)^2} < \delta$. Notice that

$$\sqrt{(x - a)^2 + (y - b)^2} > \sqrt{(x - a)^2} = |x - a|,$$

and so, taking $\delta = \varepsilon$, we have that

$$|x - a| = \sqrt{(x - a)^2} < \sqrt{(x - a)^2 + (y - b)^2} < \varepsilon,$$

whenever $0 < \sqrt{(x - a)^2 + (y - b)^2} < \delta$. Likewise, we can show that $\lim_{(x,y)\to(a,b)} y = b$.

With this definition of limit, we can prove the usual results for limits of sums, products and quotients. That is, if $f(x, y)$ and $g(x, y)$ both have limits as (x, y) approaches (a, b), we have

$$\lim_{(x,y)\to(a,b)} [f(x, y) \pm g(x, y)] = \lim_{(x,y)\to(a,b)} f(x, y) \pm \lim_{(x,y)\to(a,b)} g(x, y)$$

(i.e., the limit of a sum or difference is the sum or difference of the limits),

$$\lim_{(x,y)\to(a,b)} [f(x,y)g(x,y)] = \left[\lim_{(x,y)\to(a,b)} f(x,y)\right]\left[\lim_{(x,y)\to(a,b)} g(x,y)\right]$$

(i.e., the limit of a product is the product of the limits) and

$$\lim_{(x,y)\to(a,b)} \frac{f(x,y)}{g(x,y)} = \frac{\displaystyle\lim_{(x,y)\to(a,b)} f(x,y)}{\displaystyle\lim_{(x,y)\to(a,b)} g(x,y)}$$

(i.e., the limit of a quotient is the quotient of the limits), **provided** $\displaystyle\lim_{(x,y)\to(a,b)} g(x,y) \neq 0$.

A **polynomial** in the two variables x and y is any sum of terms of the form $cx^n y^m$, where c is a constant and n and m are nonnegative integers. Using the preceding results and example 2.1, we can show that the limit of any polynomial always exists and is found simply by substitution.

Example 2.2   Finding a Simple Limit

Evaluate $\displaystyle\lim_{(x,y)\to(2,1)} \frac{2x^2 y + 3xy}{5xy^2 + 3y}$.

Solution   First, note that this is the limit of a rational function (i.e., the quotient of two polynomials). Since the limit in the denominator is

$$\lim_{(x,y)\to(2,1)} (5xy^2 + 3y) = 10 + 3 = 13 \neq 0,$$

we have

$$\lim_{(x,y)\to(2,1)} \frac{2x^2 y + 3xy}{5xy^2 + 3y} = \frac{\displaystyle\lim_{(x,y)\to(2,1)} (2x^2 y + 3xy)}{\displaystyle\lim_{(x,y)\to(2,1)} (5xy^2 + 3y)} = \frac{14}{13}.$$

Think about the implications of Definition 2.1 (even if you are a little unsure of the role of ε and δ). If there is **any** way to approach the point (a, b) without the function values approaching the value L (e.g., by virtue of the function values blowing up, oscillating or by approaching some other value), then the limit will not equal L. Of course, since we're in two dimensions, there are an infinite number of paths along which we can approach any given point (a, b), including lines of various slopes and curves of different types. For the limit to equal L, the function has to approach L along **every** possible path. This gives us a simple method for determining that a limit does not exist.

Remark 2.1

If $f(x, y)$ approaches L_1 as (x, y) approaches (a, b) along a path P_1 and $f(x, y)$ approaches $L_2 \neq L_1$ as (x, y) approaches (a, b) along a path P_2, then $\displaystyle\lim_{(x,y)\to(a,b)} f(x, y)$ **does not exist.**

Don't worry; this may look complicated at first, but with practice it becomes easy. This is analogous to checking the left- and right-hand limits for functions of a single variable. The major difference is that in two dimensions, instead of just two paths approaching a given point, but there are infinitely many (and you obviously can't check each one individually). In practice, when you suspect that a limit does not exist, you should check the limit along the simplest paths first. We will use the following guidelines.

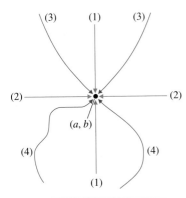

Figure 12.15
Various paths to (a, b).

Remark 2.2

The simplest paths to try are (1) $x = a$, $y \to b$ (vertical lines); (2) $y = b$, $x \to a$ (horizontal lines); (3) $y = g(x)$, $x \to a$ [where $b = g(a)$] and (4) $x = g(y)$, $y \to b$ [where $a = g(b)$].

Several of these paths are illustrated in Figure 12.15.

Example 2.3 A Limit That Does Not Exist

Evaluate $\displaystyle\lim_{(x,y) \to (1,0)} \frac{y}{x + y - 1}$.

Solution First, we consider the vertical line path along the line $x = 1$ and compute the limit as y approaches 0. If $(x, y) \to (1, 0)$ along the line $x = 1$, we have

$$\lim_{(1,y) \to (1,0)} \frac{y}{1 + y - 1} = \lim_{y \to 0} 1 = 1.$$

We next consider the horizontal line $y = 0$ and compute the limit as x approaches 1. Here, we have

$$\lim_{(x,0) \to (1,0)} \frac{0}{x + 0 - 1} = \lim_{x \to 0} 0 = 0.$$

Since the function is approaching two different values along two different paths to the point $(1, 0)$, the limit does not exist.

Many of our examples and exercises have (x, y) approaching $(0, 0)$. In this case, notice that another simple path passing through $(0, 0)$ is the line $y = x$.

Example 2.4 A Limit That Is the Same Along Two Paths but Does Not Exist

Evaluate $\displaystyle\lim_{(x,y) \to (0,0)} \frac{xy}{x^2 + y^2}$.

Solution First, we consider the limit along the path $x = 0$. We have

$$\lim_{(0,y) \to (0,0)} \frac{0}{0 + y^2} = \lim_{y \to 0} 0 = 0.$$

Similarly, for the path $y = 0$, we have

$$\lim_{(x,0) \to (0,0)} \frac{0}{x^2 + 0} = \lim_{x \to 0} 0 = 0.$$

Be careful; just because the limits along the first two paths you try are the same does **not** mean that the limit exists. Keep in mind that for a limit to exist, we'll need the limit to be the same along **all** paths through $(0, 0)$ (not just along two). We may simply need to look at more paths. Notice that for the path $y = x$, we have

$$\lim_{(x,x) \to (0,0)} \frac{x(x)}{x^2 + x^2} = \lim_{x \to 0} \frac{x^2}{2x^2} = \frac{1}{2}.$$

Since the limit along this path doesn't match the limit along the first two paths, the limit does not exist.

As you've seen in examples 2.3 and 2.4, substitutions for particular paths often result in the function reducing to a constant. When choosing paths, you should look for substitutions that will simplify the function dramatically.

Example 2.5

A Limit Problem Requiring a More Complicated Choice of Path

Evaluate $\lim\limits_{(x,y)\to(0,0)} \dfrac{xy^2}{x^2+y^4}$.

Solution   First, we consider the path $x = 0$ and get

$$\lim_{(0,y)\to(0,0)} \frac{0}{0+y^4} = \lim_{y\to0} 0 = 0.$$

Next, following the path $y = 0$, we get

$$\lim_{(x,0)\to(0,0)} \frac{0}{x^2+0} = \lim_{x\to0} 0 = 0.$$

Since the limits along the first two paths are the same, we try another path. As in example 2.4, the next most obvious choice of path through $(0, 0)$ is the line $y = x$. As it turns out, this limit is

$$\lim_{(x,x)\to(0,0)} \frac{x^3}{x^2+x^4} = \lim_{x\to0} \frac{x}{1+x^2} = 0,$$

also, so we'll need to try yet another path. You should quickly get tired of looking for the limit along straight-line paths through the origin. We leave it as an exercise to show that the limit along **every** straight line through the origin is 0. (Let $y = kx$ and let $x \to 0$ to verify this.) However, we still cannot conclude that the limit is 0. For this to happen, the limit along **all** paths (not just along all straight-line paths) must be 0. At this point, there are two possibilities: either the limit exists (and equals 0) or the limit does not exist, in which case, we must find some path through $(0, 0)$ along which the path is not 0. (Yes, there are a lot of things to consider here; just be patient.) Finally, notice that along the path $x = y^2$, the terms x^2 and y^4 will be equal. We then have

$$\lim_{(y^2,y)\to(0,0)} \frac{y^2(y^2)}{(y^2)^2+y^4} = \lim_{y\to0} \frac{y^4}{2y^4} = \frac{1}{2}.$$

Since this limit does not agree with the limits along the earlier paths, the original limit does not exist.

Before discussing how to show that a limit *does* exist, we pause to explore example 2.5 graphically. First, try to imagine what the graph of $f(x, y) = \dfrac{xy^2}{x^2+y^4}$ might look like. The function is defined except at the origin, it approaches 0 along the x-axis, y-axis and along any line $y = kx$ through the origin. Yet, $f(x, y)$ approaches $\frac{1}{2}$ along the parabola $x = y^2$. What does the surface look like? A standard sketch of the surface $z = f(x, y)$ with $-5 \le x \le 5$ and $-5 \le y \le 5$ gives us some good clues, but you need to know what you're looking for. (You can see part of the ridge at $z = 0.5$, as well as a trough at $z = -0.5$ corresponding to $x = -y^2$, in Figure 12.16a.) A density plot clearly shows the parabola of large function values in light blue and a parabola of small function values in black (see Figure 12.16b). Near the origin, the surface has a ridge at $x = y^2$, $z = \frac{1}{2}$, dropping off quickly to a smooth surface that approaches the origin. The ridge is in two pieces ($y > 0$ and $y < 0$) separated by the origin.

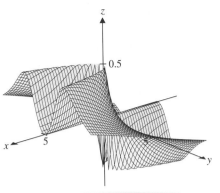

Figure 12.16a

$z = \dfrac{xy^2}{x^2 + y^4}$, for $-5 \le x \le 5$,
$-5 \le y \le 5$.

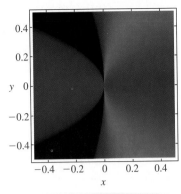

Figure 12.16b

Density plot of $f(x, y) = \dfrac{xy^2}{x^2 + y^4}$.

We want to caution you that the procedure followed in the last three examples was used to show that a limit does **not** exist. What if a limit does exist? Realize that you'll never be able to establish that a limit exists by taking limits along specific paths. There are infinitely many paths through any given point and you'll never be able to exhaust all of the possibilities. However, after following a number of paths and getting the same limit along each of them, you should begin to suspect that the limit just might exist. As yet, we haven't developed any strategy for evaluating limits that do exist. One tool we can use is the following generalization of the Squeeze Theorem presented in section 1.2.

Theorem 2.1

Suppose that $|f(x, y) - L| \le g(x, y)$ for all (x, y) in the interior of some circle centered at (a, b), except possibly at (a, b). If $\displaystyle\lim_{(x,y)\to(a,b)} g(x, y) = 0$, then $\displaystyle\lim_{(x,y)\to(a,b)} f(x, y) = L$.

Proof

For any given $\varepsilon > 0$, we know from the definition of $\displaystyle\lim_{(x,y)\to(a,b)} g(x, y) = 0$, that there is a number $\delta > 0$ such that $0 < \sqrt{(x - a)^2 + (y - b)^2} < \delta$ guarantees that $|g(x, y) - 0| < \varepsilon$. For any such points (x, y), we have

$$|f(x, y) - L| \le g(x, y) < \varepsilon.$$

It now follows from the definition of limit that $\displaystyle\lim_{(x,y)\to(a,b)} f(x, y) = L$.

In other words, the theorem simply states that if $|f(x, y) - L|$ is trapped between 0 (the absolute value is never negative) and a function (g) that approaches 0, then $|f(x, y) - L|$ must also have a 0 limit.

In practice, you start with a conjecture for the limit L (obtained for instance, by calculating the limit along several simple paths). Then, look for a simpler function that is larger than $|f(x, y) - L|$. You can sometimes find such a function by discarding terms in $|f(x, y) - L|$ to get a "larger" function $g(x, y)$. As long as $g(x, y)$ tends to zero as (x, y) approaches (a, b), you can use Theorem 2.1.

Example 2.6 Proving That a Limit Exists

Evaluate $\displaystyle\lim_{(x,y)\to(0,0)} \frac{x^2 y}{x^2 + y^2}$.

Solution As we did in earlier examples, we start by looking at the limit along several paths through $(0, 0)$. (If these limits don't agree, then the limit does not exist and we are done! If they do agree, then they provide us with a conjecture for the value of the limit.) Along the path $x = 0$, we have

$$\lim_{(0,y)\to(0,0)} \frac{0}{0 + y^2} = 0.$$

Along the path $y = 0$, we have

$$\lim_{(x,0)\to(0,0)} \frac{0}{x^2 + 0} = 0.$$

Further, along the path $y = x$, we have

$$\lim_{(x,x)\to(0,0)} \frac{x^3}{x^2 + x^2} = \lim_{x\to 0} \frac{x}{2} = 0.$$

At this stage, we know that if the limit exists, it must equal 0. We could try other paths, but our last calculation gives an important clue that the limit does exist. In that calculation, after simplifying the expression, there remained an extra power of x in the numerator forcing the limit to 0. To show that the limit equals $L = 0$, consider

$$|f(x, y) - L| = |f(x, y) - 0| = \left| \frac{x^2 y}{x^2 + y^2} \right|.$$

Notice that if there was no y^2 term in the denominator, we could cancel the x^2 terms. Since $x^2 + y^2 \geq x^2$, we have for $x \neq 0$

$$|f(x, y) - L| = \left| \frac{x^2 y}{x^2 + y^2} \right| \leq \left| \frac{x^2 y}{x^2} \right| = |y|.$$

Certainly, $\displaystyle\lim_{(x,y)\to(0,0)} |y| = 0$ and so, Theorem 2.1 gives us $\displaystyle\lim_{(x,y)\to(0,0)} \frac{x^2 y}{x^2 + y^2} = 0$, also.

■

When (x, y) approaches a point other than $(0, 0)$, the idea is the same as in example 2.6, but the algebra may get messier, as we see in the following example.

Example 2.7 Finding a Limit of a Function of Two Variables

Evaluate $\displaystyle\lim_{(x,y)\to(1,0)} \frac{(x - 1)^2 \ln x}{(x - 1)^2 + y^2}$.

Solution Along the path $x = 1$, we have

$$\lim_{(1,y)\to(1,0)} \frac{0}{y^2} = 0.$$

Along the path $y = 0$, we have

$$\lim_{(x,0)\to(1,0)} \frac{(x-1)^2 \ln x}{(x-1)^2} = \lim_{x\to 1} \ln x = 0.$$

A third path through $(1, 0)$ is the line $y = x - 1$ (note that in this case, we must have $y \to 0$ as $x \to 1$). We have

$$\lim_{(x,x-1)\to(1,0)} \frac{(x-1)^2 \ln x}{(x-1)^2 + (x-1)^2} = \lim_{x\to 1} \frac{(x-1)^2 \ln x}{2(x-1)^2} = \lim_{x\to 1} \frac{\ln x}{2} = 0.$$

At this point, you should begin to suspect that the limit just might be 0. You never know, though, until you find another path along which the limit is different or until you prove that the limit actually is 0. To show this, we consider

$$|f(x, y) - L| = \left| \frac{(x-1)^2 \ln x}{(x-1)^2 + y^2} \right|.$$

Notice that if the y^2 term were not present in the denominator, then we could cancel the $(x-1)^2$ terms. We have

$$|f(x, y) - L| = \left| \frac{(x-1)^2 \ln x}{(x-1)^2 + y^2} \right| \le \left| \frac{(x-1)^2 \ln x}{(x-1)^2} \right| = |\ln x|$$

Since $\lim\limits_{(x,y)\to(1,0)} |\ln x| = 0$, it follows from Theorem 2.1 that $\lim\limits_{(x,y)\to(1,0)} \dfrac{(x-1)^2 \ln x}{(x-1)^2 + y^2} = 0$, also.

As with functions of one variable and (more recently) vector-valued functions, the concept of continuity is closely connected to limits. Recall that in these cases, a function (or vector-valued function) is continuous at a point whenever the limit and the value of the function are the same. In particular, note that this says that limits of continuous functions of a single variable (or vector-valued functions) are found by simply substituting in. This same characterization applies to continuous functions of several variables, as we see in the following definition.

Definition 2.2

Suppose that $f(x, y)$ is defined in the interior of a circle centered at the point (a, b). We say that f is **continuous** at (a, b) if $\lim\limits_{(x,y)\to(a,b)} f(x, y) = f(a, b)$.

If $f(x, y)$ is not continuous at (a, b), then we call (a, b) a **discontinuity** of f.

Notice that this definition is completely analogous to our previous definitions of continuity for the cases of functions of one variable and vector-valued functions. The graphical interpretation is similar, although we have already seen that three-dimensional graphs can be more complicated. Still, the idea is that for a continuous function $f(x, y)$, if (x, y) changes slightly, then $f(x, y)$ also changes slightly.

Before we define the concept of continuity on a region $R \subset \mathbb{R}^2$, we first need to define open and closed regions in two dimensions. We refer to the interior of a circle (i.e., the set of all points inside but not on the circle) as an **open disk** (see Figure 12.17a). A **closed disk** consists of the circle and its interior (see Figure 12.17b). These are the two-dimensional analogs of open and closed intervals, respectively, of the real line. For a given two-dimensional region R, a point (a, b) in R is called an **interior point** of R if there is an open disk centered at

Figure 12.17a
Open disk.

Figure 12.17b
Closed disk.

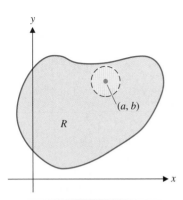

Figure 12.18a

Interior point.

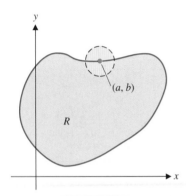

Figure 12.18b

Boundary point.

(a, b) that lies **completely** inside of R (see Figure 12.18a). A point (a, b) in R is called a **boundary point** of R if **every** open disk centered at (a, b) contains points in R **and** points outside R (see Figure 12.18b). A set R is **closed** if it contains **all** of its boundary points. Alternatively, R is **open** if it contains **none** of its boundary points. Note that these are analogous to closed and open intervals of the real line: closed intervals include all (both) of their boundary points (endpoints), while open intervals include none (neither) of their boundary points.

If the domain of a function contains any of its boundary points, we will need to modify our definition of continuity slightly, to ensure that the limit is calculated over paths that lie inside the domain only. (Recall that this is essentially what we did to define continuity of a function of a single variable on a closed interval.) If (a, b) is a boundary point of the domain D of a function f, we say that f is continuous at (a, b) if

$$\lim_{\substack{(x,y)\to(a,b)\\(x,y)\in D}} f(x, y) = f(a, b).$$

This notation indicates that the limit is taken only along paths lying completely inside D. (This corresponds to a one-sided limit for functions of a single variable.) Note that this limit requires a slight modification of Definition 2.1, as follows.

We say that

$$\lim_{\substack{(x,y)\to(a,b)\\(x,y)\in D}} f(x, y) = L$$

if for every $\varepsilon > 0$ there exists a $\delta > 0$ such that $|f(x) - L| < \varepsilon$ whenever $(x, y) \in D$ and $0 < \sqrt{(x - a)^2 + (y - b)^2} < \delta$.

We say that a function $f(x, y)$ is **continuous on a region** R if it is continuous at each point in R.

Notice that because we define continuity in terms of limits, we immediately have the following results, which follow directly from the corresponding results for limits. If $f(x, y)$ and $g(x, y)$ are continuous at (a, b), then $f + g$, $f - g$ and $f \cdot g$ are all continuous at (a, b). Further, f/g is continuous at (a, b), if in addition $g(a, b) \neq 0$. We leave the proof of these statements as (easy) exercises. Notice right away that this says that all polynomials are continuous everywhere. (Think about why this must be true.)

In many cases, determining where a function is continuous involves identifying where the function isn't defined and using our continuity results for functions of a single variable.

Example 2.8   Determining Where a Function of Two Variables Is Continuous

Find all points where the given function is continuous: (a) $f(x, y) = \dfrac{x}{x^2 - y}$ and (b) $g(x, y) = \begin{cases} \dfrac{x^4}{x(x^2 + y^2)}, & \text{if } (x, y) \neq (0, 0) \\ 0, & \text{if } (x, y) = (0, 0) \end{cases}$.

Solution   For (a), notice that $f(x, y)$ is a quotient of two polynomials (i.e., a rational function) and so, it is continuous at any point where we don't divide by 0. Since division by zero occurs only when $y = x^2$, we have that f is continuous at all points (x, y) with $y \neq x^2$. For (b), the function g is also a quotient of polynomials, except at the origin. Notice that there is a division by 0 whenever $x = 0$. We must consider the point $(0, 0)$ separately, however, since the function is not defined by the rational expression there. We can verify that $\lim_{(x,y)\to(0,0)} g(x, y) = 0$ using the following string of inequalities. Notice that for $(x, y) \neq (0, 0)$,

$$|g(x, y)| = \left| \frac{x^4}{x(x^2 + y^2)} \right| \leq \left| \frac{x^4}{x(x^2)} \right| = |x|$$

and $|x| \to 0$ as $(x, y) \to (0, 0)$. By Theorem 2.1, we have that

$$\lim_{(x,y) \to (0,0)} g(x, y) = 0 = g(0, 0),$$

so that g is continuous at $(0, 0)$. Putting this all together, we get that g is continuous at the origin and also at all points (x, y) with $x \neq 0$.

$\blacksquare$

The following theorem shows that we can use all of our established continuity results for functions of a single variable when considering functions of several variables.

Theorem 2.2

Suppose that $f(x, y)$ is continuous at (a, b) and $g(x)$ is continuous at the point $f(a, b)$. Then

$$h(x, y) = (g \circ f)(x, y) = g(f(x, y))$$

is continuous at (a, b).

Sketch of the Proof

We leave the proof as an exercise, but it goes something like this. Notice that if (x, y) is close to (a, b), then by the continuity of f at (a, b), $f(x, y)$ will be close to $f(a, b)$. By the continuity of g at the point $f(a, b)$, it follows that $g(f(x, y))$ will be close to $g(f(a, b))$, so that $g \circ f$ is also continuous at (a, b).

$\blacksquare$

Example 2.9 Determining Where a Composition of Functions Is Continuous

Determine where $f(x, y) = e^{x^2 y}$ is continuous.

Solution Notice that $f(x, y) = g(h(x, y))$, where $g(t) = e^t$ and $h(x, y) = x^2 y$. Since g is continuous for all values of t and h is a polynomial in x and y (and hence continuous for all x and y), it follows from Theorem 2.2 that f is continuous for all x and y.

$\blacksquare$

Remark 2.3

All of the foregoing analysis is extended to functions of three (or more) variables in the obvious fashion.

Definition 2.3

Let the function $f(x, y, z)$ be defined on the interior of a sphere, centered at the point (a, b, c), except possibly at (a, b, c) itself. We say that $\lim\limits_{(x,y,z) \to (a,b,c)} f(x, y, z) = L$ if for every $\varepsilon > 0$ there exists a $\delta > 0$ such that $|f(x, y, z) - L| < \varepsilon$ whenever

$$0 < \sqrt{(x - a)^2 + (y - b)^2 + (z - c)^2} < \delta.$$

Observe that, as with limits of functions of two variables, Definition 2.3 says that in order to have $\lim\limits_{(x,y,z)\to(a,b,c)} f(x, y, z) = L$, we must have that $f(x, y, z)$ approaches L along every possible path through the point (a, b, c). Just as with functions of two variables, notice that if a function of three variables approaches different limits along two particular paths, then the limit does not exist.

Example 2.10   A Limit in Three Dimensions That Does Not Exist

Evaluate $\lim\limits_{(x,y,z)\to(0,0,0)} \dfrac{x^2 + y^2 - z^2}{x^2 + y^2 + z^2}$.

Solution   First, we consider the path $x = y = 0$ (the z-axis). There, we have

$$\lim_{(0,0,z)\to(0,0,0)} \frac{0^2 + 0^2 - z^2}{0^2 + 0^2 + z^2} = \lim_{z\to0} \frac{-z^2}{z^2} = -1.$$

Along the path $x = z = 0$ (the y-axis), we have

$$\lim_{(0,y,0)\to(0,0,0)} \frac{0^2 + y^2 - 0^2}{0^2 + y^2 + 0^2} = \lim_{y\to0} \frac{y^2}{y^2} = 1.$$

Since the limits along these two specific paths do not agree, the limit does not exist.

■

We extend the definition of continuity to functions of three variables in the obvious way, as follows.

Definition 2.4

Suppose that $f(x, y, z)$ is defined in the interior of a sphere centered at (a, b, c). We say that f is **continuous** at (a, b, c) if $\lim\limits_{(x,y,z)\to(a,b,c)} f(x, y, z) = f(a, b, c)$.

If $f(x, y, z)$ is not continuous at (a, b, c), then we call (a, b, c) a **discontinuity** of f.

As you can see, limits and continuity for functions of three variables work essentially the same as they do for functions of two variables. You will examine these in more detail in the exercises.

EXERCISES 12.2

1. Choosing between the paths $y = x$ and $x = y^2$, explain why $y = x$ is a better choice in example 2.4 but $x = y^2$ is a better choice in example 2.5.

2. In terms of Definition 2.1, explain why the limit in example 2.5 does not exist. That is, explain why making (x, y) close to $(0, 0)$ doesn't guarantee that $f(x, y)$ is close to 0.

3. A friend claims that a limit equals 0, but you found that it does not exist. Looking over your friend's work, you see that the path with $x = 0$ and the path with $y = 0$ both produce a limit of 0. No other work is shown. Explain to your friend why other paths must be checked.

4. Explain why the path $y = x$ is not a valid path for the limit in example 2.7.

In exercises 5–10, compute the indicated limit.

5. $\lim\limits_{(x,y)\to(1,3)} \dfrac{x^2 y}{4x^2 - y}$

6. $\lim\limits_{(x,y)\to(2,-1)} \dfrac{x + y}{x^2 - 2xy}$

7. $\lim\limits_{(x,y)\to(\pi,1)} \dfrac{\cos xy}{y^2+1}$

8. $\lim\limits_{(x,y)\to(-3,0)} \dfrac{e^{xy}}{x^2+y^2}$

9. $\lim\limits_{(x,y,z)\to(1,0,2)} \dfrac{4xz}{y^2+z^2}$

10. $\lim\limits_{(x,y,z)\to(1,1,2)} \dfrac{e^{x+y-z}}{x-z}$

34. $\lim\limits_{(x,y,z)\to(0,0,0)} \dfrac{x^2y^2z^2}{x^2+y^2+z^2}$

In exercises 11–24, show that the indicated limit does not exist.

11. $\lim\limits_{(x,y)\to(0,0)} \dfrac{3x^2}{x^2+y^2}$

12. $\lim\limits_{(x,y)\to(0,0)} \dfrac{2y^2}{2x^2-y^2}$

13. $\lim\limits_{(x,y)\to(0,0)} \dfrac{4xy}{3y^2-x^2}$

14. $\lim\limits_{(x,y)\to(0,0)} \dfrac{2xy}{x^2+2y^2}$

15. $\lim\limits_{(x,y)\to(0,0)} \dfrac{2xy^2}{x^2+y^4}$

16. $\lim\limits_{(x,y)\to(0,0)} \dfrac{3x^3\sqrt{y}}{x^4+y^2}$

17. $\lim\limits_{(x,y)\to(0,0)} \dfrac{\sqrt[3]{x}\,y^2}{x+y^3}$

18. $\lim\limits_{(x,y)\to(0,0)} \dfrac{2xy^3}{x^2+8y^6}$

19. $\lim\limits_{(x,y)\to(0,0)} \dfrac{y\sin x}{x^2+y^2}$

20. $\lim\limits_{(x,y)\to(0,0)} \dfrac{x(\cos y-1)}{x^3+y^3}$

21. $\lim\limits_{(x,y)\to(1,2)} \dfrac{xy-2x-y+2}{x^2-2x+y^2-4y+5}$

22. $\lim\limits_{(x,y)\to(2,0)} \dfrac{2y^2}{(x-2)^2+y^2}$

23. $\lim\limits_{(x,y,z)\to(0,0,0)} \dfrac{3x^2}{x^2+y^2+z^2}$

24. $\lim\limits_{(x,y,z)\to(0,0,0)} \dfrac{x^2+y^2+z^2}{x^2-y^2+z^2}$

25. $\lim\limits_{(x,y,z)\to(0,0,0)} \dfrac{xyz}{x^3+y^3+z^3}$

26. $\lim\limits_{(x,y,z)\to(0,0,0)} \dfrac{x^2yz}{x^4+y^4+z^4}$

In exercises 27–34, show that the indicated limit exists.

27. $\lim\limits_{(x,y)\to(0,0)} \dfrac{xy^2}{x^2+y^2}$

28. $\lim\limits_{(x,y)\to(0,0)} \dfrac{x^2y}{x^2+y^2}$

29. $\lim\limits_{(x,y)\to(0,0)} \dfrac{2x^2\sin y}{2x^2+y^2}$

30. $\lim\limits_{(x,y)\to(0,0)} \dfrac{x^3y+x^2y^3}{x^2+y^2}$

31. $\lim\limits_{(x,y)\to(0,0)} \dfrac{x^3+4x^2+2y^2}{2x^2+y^2}$

32. $\lim\limits_{(x,y)\to(0,0)} \dfrac{x^2y-x^2-y^2}{x^2+y^2}$

33. $\lim\limits_{(x,y,z)\to(0,0,0)} \dfrac{3x^3}{x^2+y^2+z^2}$

In exercises 35–38, use graphs and density plots to explain why the limit in the indicated exercise does not exist.

35. Exercise 11

36. Exercise 12

37. Exercise 13

38. Exercise 14

In exercises 39–48, determine all points at which the given function is continuous.

39. $f(x,y)=4xy+\sin 3x^2y$

40. $f(x,y)=e^{3x-4y}+x^2-y$

41. $f(x,y)=\sqrt{9-x^2-y^2}$

42. $f(x,y)=\sqrt{x^2-y^2-1}$

43. $f(x,y)=\ln(3-x^2+y)$

44. $f(x,y)=\tan(x+y)$

45. $f(x,y,z)=\dfrac{x^3}{y}+\sin z$

46. $f(x,y,z)=4xe^{y-z}$

47. $f(x,y,z)=\sqrt{x^2+y^2+z^2-4}$

48. $f(x,y,z)=\sqrt{z-x^2-y^2}$

In exercises 49 and 50, estimate the indicated limit numerically.

49. $\lim\limits_{(x,y)\to(0,0)} \dfrac{1-\cos xy}{x^2y^2+x^2y^3}$

50. $\lim\limits_{(x,y)\to(0,0)} \dfrac{3\sin xy^2}{x^2y^2+xy^2}$

In exercises 51–54, label the statement as true or false and explain.

51. If $\lim\limits_{(x,y)\to(a,b)} f(x,y)=L$, then $\lim\limits_{x\to a} f(x,b)=L$.

52. If $\lim\limits_{x\to a} f(x,b)=L$, then $\lim\limits_{(x,y)\to(a,b)} f(x,y)=L$.

53. If $\lim\limits_{x\to a} f(x,b)=\lim\limits_{y\to b} f(a,y)=L$, then $\lim\limits_{(x,y)\to(a,b)} f(x,y)=L$.

54. If $\lim\limits_{(x,y)\to(0,0)} f(x,y)=0$, then $\lim\limits_{(x,y)\to(0,0)} f(cx,y)=0$ for any constant c.

55. Show that the function
$$f(x,y)=\begin{cases} \dfrac{xy^2}{x^2+y^4}, & \text{if } (x,y)\neq(0,0) \\ 0, & \text{if } (x,y)=(0,0) \end{cases}$$
is not continuous at $(0,0)$. Notice that this function is closely related to that of example 2.5.

56. Show that the function in exercise 55 "acts" continuous at the origin along any straight line through the origin in the sense that for any such line L with the limit restricted to points (x, y) on L, $\displaystyle\lim_{(x,y)\to(0,0)} f(x, y) = f(0, 0)$.

In exercises 57–60, use polar coordinates to find the indicated limit, if it exists. Note that $(x, y) \to (0, 0)$ is equivalent to $r \to 0$.

57. $\displaystyle\lim_{(x,y)\to(0,0)} \frac{\sqrt{x^2+y^2}}{\sin\sqrt{x^2+y^2}}$

58. $\displaystyle\lim_{(x,y)\to(0,0)} \frac{e^{x^2+y^2}-1}{x^2+y^2}$

59. $\displaystyle\lim_{(x,y)\to(0,0)} \frac{xy^2}{x^2+y^2}$

60. $\displaystyle\lim_{(x,y)\to(0,0)} \frac{x^2y}{x^2+y^2}$

61. 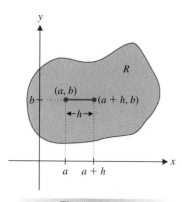 In this exercise, you will explore how the patterns of contour plots relate to the existence of limits.

Start by showing that $\displaystyle\lim_{(x,y)\to(0,0)} \frac{x^2}{x^2+y^2}$ doesn't exist and $\displaystyle\lim_{(x,y)\to(0,0)} \frac{x^2y}{x^2+y^2} = 0$. Then sketch several contour plots for each function while zooming in on the point $(0, 0)$. For a function whose limit exists as (x, y) approaches (a, b), what should be happening to the range of function values as you zoom in on the point (a, b)? Describe the appearance of each contour plot for $\dfrac{x^2y}{x^2+y^2}$ near $(0, 0)$. By contrast, what should be happening to the range of function values as you zoom in on a point at which the limit doesn't exist? Explain how this appears in the contour plots for $\dfrac{x^2}{x^2+y^2}$. Use contour plots to conjecture whether or not the following limits exist: $\displaystyle\lim_{(x,y)\to(0,0)} \frac{xy}{x^2+y}$ and $\displaystyle\lim_{(x,y)\to(0,0)} \frac{x\sin y}{x^2+y^2}$.

12.3   PARTIAL DERIVATIVES

Recall that for a function f of a single variable, we define the derivative function as

$$f'(x) = \lim_{h\to 0} \frac{f(x+h) - f(x)}{h},$$

for any values of x for which the limit exists. At any particular value $x = a$, we interpret $f'(a)$ as the instantaneous rate of change of the function with respect to x at that point. For instance, if $f(t)$ represents the temperature of an object at time t, then $f'(t)$ gives the rate of change of temperature with respect to time. In this section, we generalize the notion of derivative to functions of more than one variable.

Consider a flat metal plate in the shape of the region $R \subset \mathbb{R}^2$. Suppose that the temperature at any point $(x, y) \in R$ is given by $f(x, y)$. [Since the temperature only depends on the location (x, y), recognize that this says that the temperature is independent of time.] A reasonable question might be to ask what the rate of change of f is in the x-direction at a point $(a, b) \in R$. Think about it this way: if you move along the horizontal line segment from (a, b) to $(a + h, b)$, what is the average rate of change of the temperature with respect to the horizontal distance x (see Figure 12.19)? Notice that on this line segment, y is a constant ($y = b$). So, the average rate of change on this line segment is given by

$$\frac{f(a+h, b) - f(a, b)}{h}.$$

To get the instantaneous rate of change of f in the x-direction at the point (a, b), we take the limit as $h \to 0$:

$$\lim_{h\to 0} \frac{f(a+h, b) - f(a, b)}{h}.$$

You should recognize this limit as a derivative. Since f is a function of two variables and we have held the one variable fixed ($y = b$), we call this the **partial derivative of f with**

Figure 12.19
Average temperature on a horizontal line segment.

respect to x at the point (a, b), denoted

$$\frac{\partial f}{\partial x}(a, b) = \lim_{h \to 0} \frac{f(a + h, b) - f(a, b)}{h}.$$

This says that $\frac{\partial f}{\partial x}(a, b)$ gives the instantaneous rate of change of f with respect to x (i.e., in the x-direction) at the point (a, b). Graphically, observe that in defining $\frac{\partial f}{\partial x}(a, b)$, we are looking only at points in the plane $y = b$. The intersection of $z = f(x, y)$ and $y = b$ is a curve, as shown in Figures 12.20a and 12.20b. Notice that the partial derivative $\frac{\partial f}{\partial x}(a, b)$ then gives the slope of the tangent line to this curve at $x = a$, as indicated in Figure 12.20b.

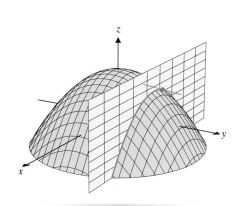

Figure 12.20a

Intersection of the surface
$z = f(x, y)$ with the plane $y = b$.

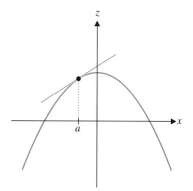

Figure 12.20b

The curve $z = f(x, b)$.

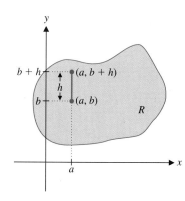

Figure 12.21

Average temperature on a vertical
line segment.

Likewise, if we move along a vertical line segment from (a, b) to $(a, b + h)$ (see Figure 12.21), the average rate of change of f along this segment is given by

$$\frac{f(a, b + h) - f(a, b)}{h}.$$

The instantaneous rate of change of f in the y-direction at the point (a, b) is then given by

$$\lim_{h \to 0} \frac{f(a, b + h) - f(a, b)}{h},$$

which you should again recognize as a derivative. In this case, however, we have held the value of x fixed ($x = a$) and refer to this as the **partial derivative of f with respect to y** at the point (a, b), denoted

$$\frac{\partial f}{\partial y}(a, b) = \lim_{h \to 0} \frac{f(a, b + h) - f(a, b)}{h}.$$

Graphically, observe that in defining $\frac{\partial f}{\partial y}(a, b)$, we are looking only at points in the plane $x = a$. The intersection of $z = f(x, y)$ and $x = a$ is a curve, as shown in Figures 12.22a and 12.22b (on the following page). In this case, notice that the partial derivative $\frac{\partial f}{\partial y}(a, b)$ gives the slope of the tangent line to the curve at $y = b$, as shown in Figure 12.22b.

Graphical interpretations of second-order partial derivatives will be explored in the exercises.

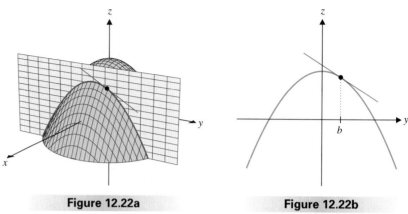

Figure 12.22a	**Figure 12.22b**
The intersection of the surface $z = f(x, y)$ with the plane $x = a$.	The curve $z = f(a, y)$.

More generally, we define the partial derivative functions as follows.

Definition 3.1

The **partial derivative of $f(x, y)$ with respect to x**, written $\dfrac{\partial f}{\partial x}$, is defined by

$$\frac{\partial f}{\partial x}(x, y) = \lim_{h \to 0} \frac{f(x + h, y) - f(x, y)}{h},$$

for any values of x and y for which the limit exists. The **partial derivative of $f(x, y)$ with respect to y**, written $\dfrac{\partial f}{\partial y}$, is defined by

$$\frac{\partial f}{\partial y}(x, y) = \lim_{h \to 0} \frac{f(x, y + h) - f(x, y)}{h},$$

for any values of x and y for which the limit exists.

Notation: Notice that since we are now dealing with functions of several variables, we can no longer use the same old prime notation for denoting partial derivatives. [Which partial derivative would $f'(x, y)$ denote?] We introduce several convenient types of notation here. For $z = f(x, y)$, we write

$$\frac{\partial f}{\partial x}(x, y) = f_x(x, y) = \frac{\partial z}{\partial x}(x, y) = \frac{\partial}{\partial x}[f(x, y)].$$

The expression $\dfrac{\partial}{\partial x}$ is a **partial differential operator.** It tells you to take the partial derivative (with respect to x) of whatever expression follows it. Similarly, we have

$$\frac{\partial f}{\partial y}(x, y) = f_y(x, y) = \frac{\partial z}{\partial y}(x, y) = \frac{\partial}{\partial y}[f(x, y)].$$

Look carefully at how we defined these derivatives and you'll see that we can compute partial derivatives using all of our usual rules for computing ordinary derivatives. Notice that in the definition of $\dfrac{\partial f}{\partial x}$, the value of y is held constant, say at $y = b$. If we define $g(x) = f(x, b)$, then

$$\frac{\partial f}{\partial x}(x, b) = \lim_{h \to 0} \frac{f(x + h, b) - f(x, b)}{h} = \lim_{h \to 0} \frac{g(x + h) - g(x)}{h} = g'(x).$$

That is, to compute the partial derivative $\dfrac{\partial f}{\partial x}$, you simply take an ordinary derivative with respect to x, while treating y as a constant. Similarly, you can compute $\dfrac{\partial f}{\partial y}$ by taking an ordinary derivative with respect to y, while treating x as a constant.

Example 3.1 — Computing Partial Derivatives

For $f(x, y) = 3x^2 + x^3 y + 4y^2$, compute $\dfrac{\partial f}{\partial x}(x, y)$, $\dfrac{\partial f}{\partial y}(x, y)$, $f_x(1, 0)$ and $f_y(2, -1)$.

Solution   Compute $\dfrac{\partial f}{\partial x}$ by treating y as a constant (just like the coefficients 3 and 4). We have

$$\frac{\partial f}{\partial x} = \frac{\partial}{\partial x}(3x^2 + x^3 y + 4y^2) = 6x + (3x^2)y + 0 = 6x + 3x^2 y.$$

Notice here that the partial derivative of $4y^2$ with respect to x is 0, since $4y^2$ is treated as if it were a constant when differentiating with respect to x. Next, we compute $\dfrac{\partial f}{\partial y}$ by treating x as a constant. We have

$$\frac{\partial f}{\partial y} = \frac{\partial}{\partial y}(3x^2 + x^3 y + 4y^2) = 0 + x^3(1) + 8y = x^3 + 8y.$$

Substituting values in for x and y, we get

$$f_x(1, 0) = \frac{\partial f}{\partial x}(1, 0) = 6 + 0 = 6$$

and

$$f_y(2, -1) = \frac{\partial f}{\partial y}(2, -1) = 8 - 8 = 0.$$

Again, since we are holding one of the variables fixed when we compute a partial derivative, we can use all of our familiar rules for computing derivatives. For instance, we have the product rules:

$$\frac{\partial}{\partial x}(uv) = \frac{\partial u}{\partial x}v + u\frac{\partial v}{\partial x}$$

and

$$\frac{\partial}{\partial y}(uv) = \frac{\partial u}{\partial y}v + u\frac{\partial v}{\partial y}$$

and the quotient rule:

$$\frac{\partial}{\partial x}\left(\frac{u}{v}\right) = \frac{\dfrac{\partial u}{\partial x}v - u\dfrac{\partial v}{\partial x}}{v^2},$$

with a corresponding quotient rule holding for $\dfrac{\partial}{\partial y}\left(\dfrac{u}{v}\right)$.

Example 3.2 — Computing Partial Derivatives

For $f(x, y) = e^{xy} + \dfrac{x}{y}$, compute $\dfrac{\partial f}{\partial x}$ and $\dfrac{\partial f}{\partial y}$.

Solution   Recall that if $g(x) = e^{4x} + \dfrac{x}{4}$, then $g'(x) = 4e^{4x} + \dfrac{1}{4}$, from the chain rule. Replacing the 4 with y and treating it as we would any other constant, we have

$$\frac{\partial f}{\partial x} = \frac{\partial}{\partial x}\left(e^{xy} + \frac{x}{y}\right) = ye^{xy} + \frac{1}{y}.$$

For the y-partial derivative, recall that if $h(y) = \frac{4}{y}$, then $h'(y) = -\frac{4}{y^2}$. Replacing the 4 with x and treating it as you would any other constant, we have

$$\frac{\partial f}{\partial y} = \frac{\partial}{\partial y}\left(e^{xy} + \frac{x}{y}\right) = xe^{xy} - \frac{x}{y^2}.$$

■

We interpret partial derivatives as rates of change, in the same way as we interpret ordinary derivatives of functions of a single variable.

Example 3.3   An Application of Partial Derivatives to Thermodynamics

For a real gas, van der Waals' equation states that

$$\left(P + \frac{n^2 a}{V^2}\right)(V - nb) = nRT.$$

Here, P is the pressure of the gas, V is the volume of the gas, T is the temperature (in degrees Kelvin), n is the number of moles of gas, R is the universal gas constant and a and b are constants. Compute and interpret $\frac{\partial P}{\partial V}$ and $\frac{\partial T}{\partial P}$.

Solution   We first solve for P to get

$$P = \frac{nRT}{V - nb} - \frac{n^2 a}{V^2}$$

and compute

$$\frac{\partial P}{\partial V} = \frac{\partial}{\partial V}\left(\frac{nRT}{V - nb} - \frac{n^2 a}{V^2}\right) = -\frac{nRT}{(V - nb)^2} + 2\frac{n^2 a}{V^3}.$$

Notice that this gives the rate of change of pressure relative to a change in volume (with temperature held constant). Next, solving van der Waals' equation for T, we get

$$T = \frac{1}{nR}\left(P + \frac{n^2 a}{V^2}\right)(V - nb)$$

and compute

$$\frac{\partial T}{\partial P} = \frac{\partial}{\partial P}\left[\frac{1}{nR}\left(P + \frac{n^2 a}{V^2}\right)(V - nb)\right] = \frac{1}{nR}(V - nb).$$

This gives the rate of change of temperature relative to a change in pressure (with volume held constant). You will discover an interesting fact about these partial derivatives in the exercises.

■

Notice that the partial derivatives found in the preceding examples are themselves functions of two variables. We have seen that second- and higher-order derivatives of functions of a single variable provide much significant information. Not surprisingly, **higher-order partial derivatives** are also very important in applications.

For functions of two variables, there are four different second-order partial derivatives. The partial derivative with respect to x of $\frac{\partial f}{\partial x}$ is $\frac{\partial}{\partial x}\left(\frac{\partial f}{\partial x}\right)$, usually abbreviated as $\frac{\partial^2 f}{\partial x^2}$ or f_{xx}. Similarly, taking two successive partial derivatives with respect to y gives us $\frac{\partial}{\partial y}\left(\frac{\partial f}{\partial y}\right) = \frac{\partial^2 f}{\partial y^2} = f_{yy}$. For **mixed second-order partial derivatives,** one derivative is taken with respect to each variable. If the first partial derivative is taken with respect to x, we have $\frac{\partial}{\partial y}\left(\frac{\partial f}{\partial x}\right)$, abbreviated as $\frac{\partial^2 f}{\partial y \partial x}$, or $(f_x)_y = f_{xy}$. If the first partial derivative is taken with respect to y, we have $\frac{\partial}{\partial x}\left(\frac{\partial f}{\partial y}\right)$, abbreviated as $\frac{\partial^2 f}{\partial x \partial y}$, or $(f_y)_x = f_{yx}$.

Example 3.4   Computing Second-Order Partial Derivatives

Find all second-order partial derivatives of $f(x, y) = x^2 y - y^3 + \ln x$.

Solution   We start by computing the first-order partial derivatives: $\frac{\partial f}{\partial x} = 2xy + \frac{1}{x}$ and $\frac{\partial f}{\partial y} = x^2 - 3y^2$. We then have

$$\frac{\partial^2 f}{\partial x^2} = \frac{\partial}{\partial x}\left(\frac{\partial f}{\partial x}\right) = \frac{\partial}{\partial x}\left(2xy + \frac{1}{x}\right) = 2y - \frac{1}{x^2},$$

$$\frac{\partial^2 f}{\partial y \partial x} = \frac{\partial}{\partial y}\left(\frac{\partial f}{\partial x}\right) = \frac{\partial}{\partial y}\left(2xy + \frac{1}{x}\right) = 2x,$$

$$\frac{\partial^2 f}{\partial x \partial y} = \frac{\partial}{\partial x}\left(\frac{\partial f}{\partial y}\right) = \frac{\partial}{\partial x}(x^2 - 3y^2) = 2x$$

and finally,

$$\frac{\partial^2 f}{\partial y^2} = \frac{\partial}{\partial y}\left(\frac{\partial f}{\partial y}\right) = \frac{\partial}{\partial y}(x^2 - 3y^2) = -6y.$$

Notice in example 3.4 that $\frac{\partial^2 f}{\partial y \partial x} = \frac{\partial^2 f}{\partial x \partial y}$. It turns out that this is true for most, but **not all,** of the functions that you will encounter. The proof of the following result can be found in most texts on advanced calculus.

Theorem 3.1

If $f_{xy}(x, y)$ and $f_{yx}(x, y)$ are continuous on an open set containing (a, b), then $f_{xy}(a, b) = f_{yx}(a, b)$.

We can, of course, continue taking derivatives, computing third-, fourth- or even higher-order partial derivatives. Theorem 3.1 can be extended to show that as long as the partial derivatives are all continuous in an open set, the order of differentiation doesn't matter. With higher-order partial derivatives, notations such as $\frac{\partial^3 f}{\partial x \partial y \partial x}$ become quite awkward and so, we usually use f_{xyx} instead.

Example 3.5   Computing Higher-Order Partial Derivatives

For $f(x, y) = \cos(xy) - x^3 + y^4$, compute f_{xyy} and f_{xyyy}.

Solution   We have

$$f_x = \frac{\partial}{\partial x}\left[\cos(xy) - x^3 + y^4\right] = -y \sin(xy) - 3x^2.$$

Differentiating f_x with respect to y gives us

$$f_{xy} = \frac{\partial}{\partial y}[-y\sin(xy) - 3x^2] = -\sin(xy) - xy\cos(xy)$$

and

$$f_{xyy} = \frac{\partial}{\partial y}[-\sin(xy) - xy\cos(xy)] = -2x\cos(xy) + x^2 y\sin(xy).$$

Finally, we have

$$f_{xyyy} = \frac{\partial}{\partial y}[-2x\cos(xy) + x^2 y\sin(xy)]$$

$$= 2x^2\sin(xy) + x^2\sin(xy) + x^3 y\cos(xy) = 3x^2\sin(xy) + x^3 y\cos(xy).$$

Thus far, we have worked with partial derivatives of functions of two variables. The extensions to functions of three or more variables are completely analogous to what we have discussed here. In the exercises, you will be asked to write out definitions for partial derivatives of functions of three variables. In the following example, you can see that the calculations proceed just as you would expect.

Example 3.6   Partial Derivatives of Functions of Three Variables

For $f(x, y, z) = \sqrt{xy^3 z} + 4x^2 y$, defined for $x, y, z \geq 0$, compute f_x, f_{xy} and f_{xyz}.

Solution   To keep x, y and z as separate as possible, we first rewrite f as

$$f(x, y, z) = x^{1/2} y^{3/2} z^{1/2} + 4x^2 y.$$

To compute the partial derivative with respect to x, we treat y and z as constants and obtain

$$f_x = \frac{\partial}{\partial x}\left(x^{1/2} y^{3/2} z^{1/2} + 4x^2 y\right) = \left(\frac{1}{2}x^{-1/2}\right) y^{3/2} z^{1/2} + 8xy.$$

Next, treating x and z as constants, we get

$$f_{xy} = \frac{\partial}{\partial y}\left(\frac{1}{2}x^{-1/2} y^{3/2} z^{1/2} + 8xy\right) = \left(\frac{1}{2}x^{-1/2}\right)\left(\frac{3}{2}y^{1/2}\right) z^{1/2} + 8x.$$

Finally, treating x and y as constants, we get

$$f_{xyz} = \frac{\partial}{\partial z}\left[\left(\frac{1}{2}x^{-1/2}\right)\left(\frac{3}{2}y^{1/2}\right) z^{1/2} + 8x\right] = \left(\frac{1}{2}x^{-1/2}\right)\left(\frac{3}{2}y^{1/2}\right)\left(\frac{1}{2}z^{-1/2}\right).$$

Notice that this derivative is defined for $x, z > 0$ and $y \geq 0$. Further, you can show that all first-, second- and third-order partial derivatives are continuous for $x, y, z > 0$, so the order in which we take the partial derivatives is irrelevant in this case.

Example 3.7   An Application of Partial Derivatives to a Sagging Beam

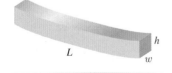

Figure 12.23

A horizontal beam.

The sag in a beam of length L, width w and height h (see Figure 12.23) is given by $S(L, w, h) = c\dfrac{L^4}{wh^3}$ for some constant c. Show that $\dfrac{\partial S}{\partial L} = \dfrac{4}{L}S$, $\dfrac{\partial S}{\partial w} = -\dfrac{1}{w}S$ and $\dfrac{\partial S}{\partial h} = -\dfrac{3}{h}S$. Use this result to determine which variable has the greatest proportional effect on the sag.

Solution   We start by computing

$$\frac{\partial S}{\partial L} = \frac{\partial}{\partial L}\left(c\frac{L^4}{wh^3}\right) = c\frac{4L^3}{wh^3}.$$

We need to manipulate this expression to rewrite it in terms of S. Multiplying top and bottom by L, we get

$$\frac{\partial S}{\partial L} = c\frac{4L^3}{wh^3} = c\frac{4L^4}{wh^3 L} = \frac{4}{L}c\frac{L^4}{wh^3} = \frac{4}{L}S.$$

The other calculations are similar and are left as exercises. To interpret the results, suppose that a small change ΔL in length produces a small change ΔS in the sag. We now have that $\frac{\Delta S}{\Delta L} \approx \frac{\partial S}{\partial L} = \frac{4}{L}S$. Rearranging the terms, we have

$$\frac{\Delta S}{S} \approx 4\frac{\Delta L}{L}.$$

That is, the proportional change in S is approximately four times the proportional change in L. Similarly, we have that in absolute value, the proportional change in S is approximately the proportional change in w and three times the proportional change in h. Proportionally then, a change in the length has the greatest effect on the amount of sag. In this sense, length is the most *important* of the three dimensions.

In many applications, no formula for the function is available and we can only estimate the value of the partial derivatives from a small collection of data points.

Example 3.8   Estimating Partial Derivatives from a Table of Data

A computer simulation of the flight of a baseball provided the data displayed in the following table for the range $f(v, \omega)$ in feet of a ball hit with initial velocity v ft/s and backspin rate of ω rpm. Each ball is struck at an angle of $30°$ above the horizontal.

v ω	0	1000	2000	3000	4000
150	294	312	333	350	367
160	314	334	354	373	391
170	335	356	375	395	414
180	355	376	397	417	436

Use the data to estimate $\frac{\partial f}{\partial v}(160, 2000)$ and $\frac{\partial f}{\partial \omega}(160, 2000)$. Interpret both quantities in baseball terms.

Solution   From the definition of partial derivative, we know that

$$\frac{\partial f}{\partial v}(160, 2000) = \lim_{h \to 0}\frac{f(160 + h, 2000) - f(160, 2000)}{h},$$

so we can approximate the value of the partial derivative by computing the difference quotient $\frac{f(160 + h, 2000) - f(160, 2000)}{h}$ for as small a value of h as possible. Since data points are provided for $v = 150$, we can compute the difference quotient for

$h = -10$, to get

$$\frac{\partial f}{\partial v}(160, 2000) \approx \frac{f(150, 2000) - f(160, 2000)}{150 - 160} = \frac{333 - 354}{150 - 160} = 2.1.$$

We can also use the data point for $v = 170$, to get

$$\frac{\partial f}{\partial v}(160, 2000) \approx \frac{f(170, 2000) - f(160, 2000)}{170 - 160} = \frac{375 - 354}{170 - 160} = 2.1$$

Since both estimates equal 2.1, we make the estimate $\frac{\partial f}{\partial v}(160, 2000) \approx 2.1$. The data point $f(160, 2000) = 354$ tells us that a ball struck with initial velocity 160 ft/s and backspin 2000 rpm will fly 354 feet. The partial derivative tells us that increasing the initial velocity by 1 ft/s will add approximately 2.1 feet to the distance.

We do similar computations to estimate $\frac{\partial f}{\partial \omega}(160, 2000)$, noting that the closest data values to $\omega = 2000$ are $\omega = 1000$ and $\omega = 3000$. We get

$$\frac{\partial f}{\partial \omega}(160, 2000) \approx \frac{f(160, 1000) - f(160, 2000)}{1000 - 2000} = \frac{334 - 354}{1000 - 2000} = 0.02$$

and

$$\frac{\partial f}{\partial \omega}(160, 2000) \approx \frac{f(160, 3000) - f(160, 2000)}{3000 - 2000} = \frac{373 - 354}{3000 - 2000} = 0.019$$

Reasonable estimates for $\frac{\partial f}{\partial \omega}(160, 2000)$ are then 0.02, 0.019 or 0.0195 (the average of the two calculations). Using 0.02 as our approximation, we can interpret this to mean that an increase in backspin of 1 rpm will add approximately 0.02 ft to the distance. A simpler way to interpret this is to say that an increase of 100 rpm will add approximately 2 ft to the distance.

EXERCISES 12.3

1. Suppose that the function $f(x, y)$ is a sum of terms where each term contains x or y but not both. Explain why $f_{xy} = 0$.

2. In Definition 3.1, explain how to remember which partial derivative involves the term $f(x + h, y)$ and which involves the term $f(x, y + h)$.

3. In section 2.8, we computed derivatives implicitly, by using the chain rule and differentiating both sides of an equation with respect to x. In the process of doing so, we made calculations such as $(x^2 y^2)' = 2xy^2 + 2x^2 yy'$. Explain why this derivative is computed differently than the partial derivatives of this section.

4. For $f(x, y, z) = x^3 e^{4x \sin y} + y^2 \sin xy + 4xyz$, you could compute f_{xyz} in a variety of orders. Discuss how many different orders are possible and which order(s) would be the easiest.

In exercises 5–16, find all first-order partial derivatives.

5. $f(x, y) = x^3 - 4xy^2 + y^4$

6. $f(x, y) = x^2 y^3 - 3x$

7. $f(x, y) = x^2 e^y - 4y$

8. $f(x, y) = y \sin x^2 + x^3$

9. $f(x, y) = x^2 \sin xy - 3y^3$

10. $f(x, y) = 3e^{x^2 y} - \sqrt{x - 1}$

11. $f(x, y) = 4e^{x/y} - \frac{y}{x}$

12. $f(x, y) = \frac{x - 3}{y} + x^2 \tan y$

13. $f(x, y, z) = 3x \sin y + 4x^3 y^2 z$

14. $f(x, y, z) = 4y \sin z - 3x^2 z$

15. $f(x, y, z) = \frac{2}{\sqrt{x^2 + y^2 + z^2}}$

16. $f(x, y, z) = \dfrac{y}{\sqrt{x^2 + y^2 + z^2}}$

In exercises 17–26, find the indicated partial derivatives.

17. $f(x, y) = x^3 - 4xy^2 + 3y;\ \dfrac{\partial^2 f}{\partial x^2}, \dfrac{\partial^2 f}{\partial y^2}, \dfrac{\partial^2 f}{\partial y \partial x}$

18. $f(x, y) = x^2 y - 4x + 3\sin y;\ \dfrac{\partial^2 f}{\partial x^2}, \dfrac{\partial^2 f}{\partial y^2}, \dfrac{\partial^2 f}{\partial y \partial x}$

19. $f(x, y) = x^4 - 3x^2 y^3 + 5y;\ f_{xx}, f_{xy}, f_{xyy}$

20. $f(x, y) = e^{4x} - \sin y^2 - \sqrt{xy};\ f_{xx}, f_{xy}, f_{yyx}$

21. $f(x, y, z) = x^3 y^2 - \sin yz;\ f_{xx}, f_{yz}, f_{xyz}$

22. $f(x, y, z) = xy^2 z^3 - e^{xyz};\ f_{yy}, f_{zz}, f_{xyy}$

23. $f(x, y, z) = e^{2xy} - \dfrac{z^2}{y} + xz \sin y;\ f_{xx}, f_{yy}, f_{yyzz}$

24. $f(x, y, z) = \ln(xyz^2);\ f_{xx}, f_{yyz}, f_{xxyzz}$

25. $f(w, x, y, z) = w^2 xy - e^{wz};\ f_{ww}, f_{wxy}, f_{wwxyz}$

26. $f(w, x, y, z) = \sqrt{wyz} - x^3 \sin w;\ f_{xx}, f_{yy}, f_{wxyz}$

In exercises 27–32, (a) sketch the graph of $z = f(x, y)$ and (b) on this graph, highlight the appropriate two-dimensional trace and interpret the partial derivative as a slope.

27. $f(x, y) = 4 - x^2 - y^2,\ \dfrac{\partial f}{\partial x}(1, 1)$

28. $f(x, y) = \sqrt{x^2 + y^2},\ \dfrac{\partial f}{\partial x}(1, 0)$

29. $f(x, y) = 4 - x^2 - y^2,\ \dfrac{\partial f}{\partial y}(1, 1)$

30. $f(x, y) = \sqrt{x^2 + y^2},\ \dfrac{\partial f}{\partial y}(1, 0)$

31. $f(x, y) = 4 - x^2 - y^2,\ \dfrac{\partial f}{\partial y}(2, 0)$

32. $f(x, y) = \sqrt{x^2 + y^2},\ \dfrac{\partial f}{\partial y}(0, 2)$

33. Compute and interpret $\dfrac{\partial V}{\partial T}$ for van der Waals' equation (see example 3.3).

34. For van der Waals' equation, show that $\dfrac{\partial T}{\partial P} \dfrac{\partial P}{\partial V} \dfrac{\partial V}{\partial T} = -1$. If you misunderstood the chain rule, why might you expect this product to equal to 1?

35. In example 3.7, show that $\dfrac{\partial S}{\partial w} = -\dfrac{1}{w} S$.

36. In example 3.7, show that $\dfrac{\partial S}{\partial h} = -\dfrac{3}{h} S$.

37. If the sag in the beam of example 3.7 were given by $S(L, w, h) = c \dfrac{L^3}{wh^4}$, determine which variable would have the greatest proportional effect.

38. Based on example 3.7 and your result in exercise 37, state a simple rule for determining which variable has the greatest proportional effect.

In exercises 39–42, find all points at which $\dfrac{\partial f}{\partial x} = \dfrac{\partial f}{\partial y} = 0$ and interpret the significance of the points graphically.

39. $f(x, y) = x^2 + y^2$

40. $f(x, y) = x^2 + y^2 - x^4$

41. $f(x, y) = \sin x \sin y$

42. $f(x, y) = e^{-x^2 - y^2}$

In exercises 43–46, use the contour plot to estimate $\dfrac{\partial f}{\partial x}$ and $\dfrac{\partial f}{\partial y}$ at the origin.

43.

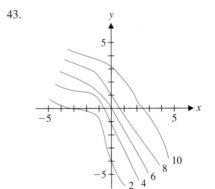

44.

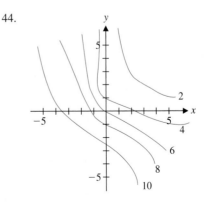

45.

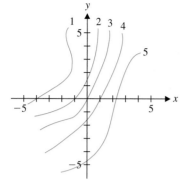

46.

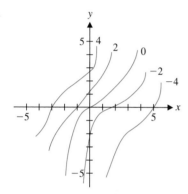

47. The table shows wind chill (how cold it "feels" outside) as a function of temperature (degrees Fahrenheit) and wind speed (mph). We can think of this as a function $C(t, s)$. Estimate the partial derivatives $\frac{\partial C}{\partial t}(10, 10)$ and $\frac{\partial C}{\partial s}(10, 10)$. Interpret each partial derivative and explain why it is surprising that $\frac{\partial C}{\partial t}(10, 10) \neq 1$.

Speed \ Temp	30	20	10	0	−10
0	30	20	10	0	−10
5	27	16	6	−5	−15
10	16	4	−9	−24	−33
15	9	−5	−18	−32	−45
20	4	−10	−25	−39	−53
25	0	−15	−29	−44	−59
30	−2	−18	−33	−48	−63

48. Rework exercise 47 using the point (10, 20). Explain the significance of the inequality $\left|\frac{\partial C}{\partial s}(10, 10)\right| > \left|\frac{\partial C}{\partial s}(10, 20)\right|$.

49. Using the baseball data in example 3.8, estimate and interpret $\frac{\partial f}{\partial v}(170, 3000)$ and $\frac{\partial f}{\partial w}(170, 3000)$.

50. According to the data in example 3.8, a baseball with initial velocity 170 ft/s and backspin 3000 rpm flies 395 ft. Suppose that the ball must go 400 ft to clear the fence for a home run. Based on your answers to exercise 49, how much extra backspin is needed for a home run?

51. Carefully write down a definition for the three first-order partial derivatives of a function of three variables $f(x, y, z)$.

52. Determine how many second-order partial derivatives there are of $f(x, y, z)$. Assuming a result analogous to Theorem 3.1, how many of these second-order partial derivatives are actually different?

53. Show that the functions $f_n(x, t) = \sin n\pi x \cos n\pi ct$ satisfy the **wave equation** $c^2 \frac{\partial^2 f}{\partial x^2} = \frac{\partial^2 f}{\partial t^2}$ for any positive integer n and any constant c.

54. Show that if $f(x)$ is a function with a continuous second derivative, then $f(x - ct)$ is a solution of the wave equation of exercise 53. If x represents position and t represents time, explain why c can be interpreted as the velocity of the wave.

55. The value of an investment of \$1000 invested at a constant 10% rate for 5 years is $V = 1000\left[\frac{1 + 0.1(1 - T)}{1 + I}\right]^5$, where T is the tax rate and I is the inflation rate. Compute $\frac{\partial V}{\partial I}$ and $\frac{\partial V}{\partial T}$ and discuss whether the tax rate or the inflation rate has a greater influence on the value of the investment.

56. The value of an investment of \$1000 invested at a rate r for 5 years with a tax rate of 28% is $V = 1000\left[\frac{1 + 0.72r}{1 + I}\right]^5$, where I is the inflation rate. Compute $\frac{\partial V}{\partial r}$ and $\frac{\partial V}{\partial I}$ and discuss whether the investment rate or the inflation rate has a greater influence on the value of the investment.

57. Suppose that the position of a guitar string of length L varies according to $p(x, t) = \sin x \cos t$, where x represents the distance along the string, $0 \leq x \leq L$, and t represents time. Compute and interpret $\frac{\partial p}{\partial x}$ and $\frac{\partial p}{\partial t}$.

58. Suppose that the concentration of some pollutant in a river as a function of position x and time t is given by $p(x, t) = p_0(x - ct)e^{-\mu t}$ for constants p_0, c and μ. Show that $\frac{\partial p}{\partial t} = -c\frac{\partial p}{\partial x} - \mu p$. Interpret both $\frac{\partial p}{\partial t}$ and $\frac{\partial p}{\partial x}$ and explain how this equation relates the change in pollution at a specific location to the current of the river and the rate at which the pollutant decays.

59. In a chemical reaction, the temperature T, entropy S, Gibbs free energy G and enthalpy H are related by $G = H - TS$. Show that $\frac{\partial(G/T)}{\partial T} = -\frac{H}{T^2}$.

60. For the chemical reaction of exercise 59, show that $\frac{\partial(G/T)}{\partial(1/T)} = H$. Chemists measure the enthalpy of a reaction by measuring this rate of change.

61. Suppose that three resistors are in parallel in an electrical circuit. If the resistances are R_1, R_2 and R_3 ohms, respectively, then the net resistance in the circuit equals $R = \frac{R_1 R_2 R_3}{R_1 R_2 + R_1 R_3 + R_2 R_3}$. Compute and interpret the partial derivative $\frac{\partial R}{\partial R_1}$. Given this partial derivative, explain how to quickly write down the partial derivatives $\frac{\partial R}{\partial R_2}$ and $\frac{\partial R}{\partial R_3}$.

62. The ideal gas law relating pressure, temperature and volume is $P = \frac{cT}{V}$ for some constant c. Show that $T \frac{\partial P}{\partial T} \frac{\partial V}{\partial T} = c$.

63. A process called **tag-and-recapture** is used to estimate populations of animals in the wild. First, some number T of the animals are captured, tagged and released into the wild. Later, a number S of the animals are captured, of which t are tagged. The estimate of the total population is then $P(T, S, t) = \frac{TS}{t}$. Compute $P(100, 60, 15)$; the proportion of tagged animals in the recapture is $\frac{15}{60} = \frac{1}{4}$. Based on your estimate of the total population, what proportion of the total population has been tagged? Now compute $\frac{\partial P}{\partial t}(100, 60, 15)$ and use it to estimate how much your population estimate would change if one more recaptured animal was tagged.

64. Suppose that L hours of labor and K dollars of investment by a company results in a productivity of $P = L^{0.75} K^{0.25}$. Compute the marginal productivity of labor, defined by $\frac{\partial P}{\partial L}$, and the marginal productivity of capital, defined by $\frac{\partial P}{\partial K}$.

65. For the function
$$f(x, y) = \begin{cases} \frac{xy(x^2 - y^2)}{x^2 + y^2}, & \text{if } (x, y) \neq (0, 0) \\ 0, & \text{if } (x, y) = (0, 0) \end{cases}$$
use the limit definitions of partial derivatives to show that $f_{xy}(0, 0) = -1$ but $f_{yx}(0, 0) = 1$. Determine which assumption in Theorem 3.1 is not true.

66. For $f(x, y) = \begin{cases} \frac{xy^2}{x^2 + y^4}, & \text{if } (x, y) \neq (0, 0) \\ 0, & \text{if } (x, y) = (0, 0) \end{cases}$, show that $\frac{\partial f}{\partial x}(0, 0) = \frac{\partial f}{\partial y}(0, 0) = 0$. [Note that we have previously shown that this function is not continuous at $(0, 0)$.]

67. Suppose that $f(x, y)$ is a function with continuous second-order partial derivatives. Consider the curve obtained by intersecting the surface $z = f(x, y)$ with the plane $y = y_0$. Explain

how the slope of this curve at the point $x = x_0$ relates to $\frac{\partial f}{\partial x}(x_0, y_0)$. Relate the concavity of this curve at the point $x = x_0$ to $\frac{\partial^2 f}{\partial x^2}(x_0, y_0)$.

68. As in exercise 67, develop a graphical interpretation of $\frac{\partial^2 f}{\partial y^2}(x_0, y_0)$.

69. In exercises 67 and 68, you interpreted the second-order partial derivatives f_{xx} and f_{yy} as concavity. In this exercise, you will develop a geometric interpretation of the mixed partial derivative f_{xy}. (More information can be found in the article "What is f_{xy}?" by Brian McCartin in the March 1998 issue of the journal *PRIMUS*.) Start by using Taylor's Theorem (see section 8.7) to show that

$$\lim_{k \to 0} \lim_{h \to 0} \frac{f(x, y) - f(x + h, y) - f(x, y + k) + f(x + h, y + k)}{hk} = f_{xy}(x, y).$$

[Hint: Treating y as a constant, you have $f(x + h, y) = f(x, y) + hf_x(x, y) + h^2 g(x, y)$ for some function $g(x, y)$. Similarly expand the other terms in the numerator.] Therefore, for small h and k, $f_{xy}(x, y) \approx \frac{f_0 - f_1 - f_2 + f_3}{hk}$, where $f_0 = f(x, y)$, $f_1 = f(x + h, y)$, $f_2 = f(x, y + k)$ and $f_3 = f(x + h, y + k)$. The four points $P_0 = (x, y, f_0)$, $P_1 = (x + h, y, f_1)$, $P_2 = (x, y + k, f_2)$ and $P_3 = (x + h, y + k, f_3)$ determine a parallelepiped as shown in the figure.

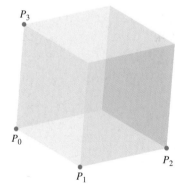

Recalling that the volume of a parallelepiped formed by vectors **a**, **b** and **c** is given by $\mathbf{a} \cdot (\mathbf{b} \times \mathbf{c})$, show that the volume of this box equals $|(f_0 - f_1 - f_2 + f_3)hk|$. That is, the volume is approximately equal to $|f_{xy}(x, y)|(hk)^2$. Conclude that the larger $|f_{xy}(x, y)|$ is, the greater the volume of the box and hence, the further the point P_3 is from the plane determined by the points P_0, P_1 and P_2. To see what this means graphically, start with the function $f(x, y) = x^2 + y^2$ at the point $(1, 1, 2)$. With $h = k = 0.1$, show that the points $(1, 1, 2)$, $(1.1, 1, 2.21)$, $(1, 1.1, 2.21)$ and $(1.1, 1.1, 2.42)$ all lie in the same plane. The derivative $f_{xy}(1, 1) = 0$ indicates that at the point $(1.1, 1.1, 2.42)$, the graph does not curve away from the plane of the points $(1, 1, 2)$, $(1.1, 1, 2.21)$ and $(1, 1.1, 2.21)$. Contrast this to the behavior of the function $f(x, y) = x^2 + xy$ at the point $(1, 1, 2)$. This says that f_{xy} measures the amount of curving of the surface as you sequentially change x and y by small amounts.

70. 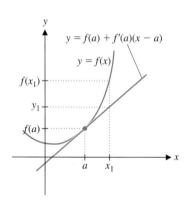 A ball, such as a baseball, flying through the air encounters air resistance in the form of **air drag.** The magnitude of the drag force is typically the product of a number (called the drag coefficient) and the square of the velocity. The drag coefficient is not actually a constant. The figure (reprinted from *Keep Your Eye on the Ball* by Watts and Bahill) shows experimental data for the drag coefficient as a function of the roughness of the ball (measured by ε/D, where ε is the size of the bumps on the ball and D is the diameter of the ball) and the Reynolds number (Re, which is proportional to velocity). We'll call the drag coefficient f, rename $u = \varepsilon/D$ and $v = \text{Re}$ and consider $f(u, v)$. Use the graph to estimate $\dfrac{\partial f}{\partial u}(0.005, 1.5 \times 10^5)$ and $\dfrac{\partial f}{\partial v}(0.005, 1.5 \times 10^5)$ and interpret each partial derivative. All golf balls have "dimples" that make the surface of the golf ball rougher. Explain why a golf ball with dimples, traveling at a velocity corresponding to a Reynolds number of about 0.9×10^5, will fly much farther than a ball with no dimples.

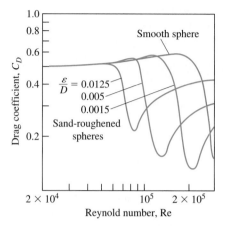

12.4 TANGENT PLANES AND LINEAR APPROXIMATIONS

Recall that one of the features of the tangent line to the curve $y = f(x)$ at $x = a$ is that it stays close to the curve near the point of tangency. This enables us to use the tangent line to approximate values of the function close to the point of tangency (see Figure 12.24a). Recall that the equation of the tangent line is given by

$$y = f(a) + f'(a)(x - a). \tag{4.1}$$

Notice that the function on the right side of (4.1) is a linear function of x. When we discussed this in section 3.1, we called this the **linear approximation** to $f(x)$ at $x = a$.

In much the same way, information about the behavior of a function of two variables near a given point can be obtained from the tangent *plane* to the surface at that point. For instance, the graph of $z = 6 - x^2 - y^2$ and its tangent plane at the point $(1, 2, 1)$ are shown in Figure 12.24b. Notice that near the point $(1, 2, 1)$, the surface and the tangent plane are very close together. We will develop and exploit this approximation in this section.

Our development of the tangent plane will parallel the development of the tangent line in section 2.1. For functions of two variables, we will see that the tangent plane is determined by two slopes, given by the partial derivatives.

We want to find a general equation for the tangent plane to $z = f(x, y)$ at the point $(a, b, f(a, b))$. You should refer to Figures 12.25a and 12.25b to visualize the process. Starting from a standard graphing window (Figure 12.25a shows $z = 6 - x^2 - y^2$ with $-3 \leq x \leq 3$ and $-3 \leq y \leq 3$), zoom in on the point $(a, b, f(a, b))$, as in Figure 12.25b (showing $z = 6 - x^2 - y^2$ with $0.9 \leq x \leq 1.1$ and $1.9 \leq y \leq 2.1$). The surface in Figure 12.25b looks like a plane. What has happened is that we have zoomed in far enough that the surface and its tangent plane are difficult to distinguish visually. This suggests that for points (x, y) close to the point of tangency, we can use the corresponding z-value on the tangent plane as an approximation to the value of the function at that point. First, we must

Figure 12.24a

Linear approximation.

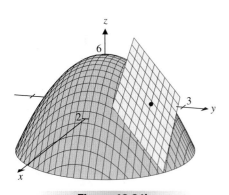

Figure 12.24b

$z = 6 - x^2 - y^2$ and the tangent plane at $(1, 2, 1)$.

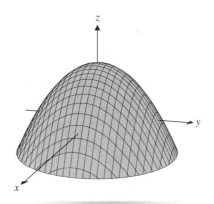

Figure 12.25a

$z = 6 - x^2 - y^2$, with $-3 \le x \le 3$ and $-3 \le y \le 3$.

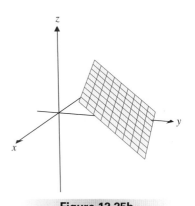

Figure 12.25b

$z = 6 - x^2 - y^2$, with $0.9 \le x \le 1.1$ and $1.9 \le y \le 2.1$.

find an equation of the tangent plane. Recall that an equation of a plane can be constructed from a point in the plane and any vector normal to the plane. One point lying in the tangent plane is, of course, the point of tangency $(a, b, f(a, b))$. To find a normal vector, we will find two vectors lying in the plane and then take their cross product to find a vector orthogonal to both (and thus, orthogonal to the plane).

Imagine intersecting the surface $z = f(x, y)$ with the plane $y = b$, as shown in Figure 12.26a. As we observed in section 12.3, the result is a curve in the plane $y = b$ whose slope at $x = a$ is given by $f_x(a, b)$. Along the tangent line at $x = a$, a change of 1 unit in x corresponds to a change of $f_x(a, b)$ in z. Since we're looking at a curve that lies in the plane $y = b$, the value of y doesn't change at all along the curve. A vector with the same direction as the tangent line is then $\langle 1, 0, f_x(a, b) \rangle$. Because of the way in which we constructed it, this vector must lie in the tangent plane (think about this some). Now, intersecting the surface $z = f(x, y)$ with the plane $x = a$, as shown in Figure 12.26b, we get a curve lying in the plane $x = a$, whose slope at $y = b$ is given by $f_y(a, b)$. A vector with the same direction as the tangent line at $y = b$ is then $\langle 0, 1, f_y(a, b) \rangle$.

We have now found two vectors in the tangent plane: $\langle 1, 0, f_x(a, b) \rangle$ and $\langle 0, 1, f_y(a, b) \rangle$. A vector normal to the plane is then given by the cross product:

$$\langle 0, 1, f_y(a, b) \rangle \times \langle 1, 0, f_x(a, b) \rangle = \langle f_x(a, b), f_y(a, b), -1 \rangle,$$

where we have left the details of the calculation as an exercise. We indicate the tangent plane and normal vector at a point in Figure 12.26c. We have now derived the following result.

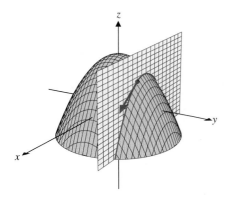

Figure 12.26a

The intersection of the surface $z = f(x, y)$ with the plane $y = b$.

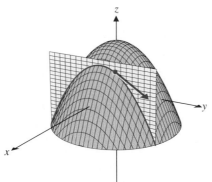

Figure 12.26b

The intersection of the surface $z = f(x, y)$ with the plane $x = a$.

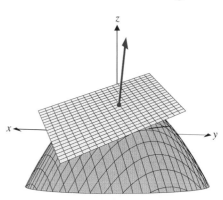

Figure 12.26c

Tangent plane and normal vector.

Remark 4.1

Notice the similarity between the equation of the tangent plane given in (4.2) and the equation of the tangent line to $y = f(x)$ given in (4.1).

Theorem 4.1

Suppose that $f(x, y)$ has continuous first partial derivatives at (a, b). A normal vector to the tangent plane to $z = f(x, y)$ at (a, b) is then $\langle f_x(a, b), f_y(a, b), -1 \rangle$. Further, an equation of the tangent plane is given by

$$f_x(a, b)(x - a) + f_y(a, b)(y - b) - [z - f(a, b)] = 0$$

or

$$z = f(a, b) + f_x(a, b)(x - a) + f_y(a, b)(y - b). \qquad (4.2)$$

Observe that since we now know a normal vector to the tangent plane, a line orthogonal to the tangent plane is given by

$$x = a + f_x(a, b)t, \quad y = b + f_y(a, b)t, \quad z = f(a, b) - t. \qquad (4.3)$$

This line is called the **normal line** to the surface at the point $(a, b, f(a, b))$.

It's now a simple matter to use Theorem 4.1 to construct the equations of a tangent plane and normal line to nearly any surface, as we illustrate in the following examples.

Example 4.1   Finding Equations of the Tangent Plane and the Normal Line

Find equations of the tangent plane and the normal line to $z = 6 - x^2 - y^2$ at the point $(1, 2, 1)$.

Solution   For $f(x, y) = 6 - x^2 - y^2$, we have $f_x = -2x$ and $f_y = -2y$. This gives us $f_x(1, 2) = -2$ and $f_y(1, 2) = -4$. A normal vector is then $\langle -2, -4, -1 \rangle$ and from (4.2), an equation of the tangent plane is

$$z - 1 = -2(x - 1) - 4(y - 2).$$

From (4.3), equations of the normal line are

$$x = 1 - 2t, \quad y = 2 - 4t, \quad z = 1 - t.$$

A sketch of the surface, the tangent plane and the normal line is shown in Figure 12.27.

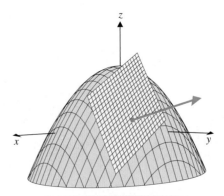

Figure 12.27
Surface, tangent plane and normal line at the point $(1, 2, 1)$.

Finding Equations of the Tangent Plane
▮ **Example 4.2**   and the Normal Line

Find equations of the tangent plane and the normal line to $z = x^3 + y^3 + \dfrac{x^2}{y}$ at (2, 1, 13).

Solution   First, notice that here, $f_x = 3x^2 + \dfrac{2x}{y}$ and $f_y = 3y^2 - \dfrac{x^2}{y^2}$, so that $f_x(2, 1) = 12 + 4 = 16$ and $f_y(2, 1) = 3 - 4 = -1$. A normal vector is then $\langle 16, -1, -1 \rangle$ and from (4.2), an equation of the tangent plane is

$$z - 13 = 16(x - 2) - (y - 1).$$

From (4.3), equations of the normal line are

$$x = 2 + 16t, \quad y = 1 - t, \quad z = 13 - t.$$

A sketch of the surface, the tangent plane and the normal line is shown in Figure 12.28.

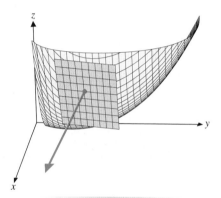

Figure 12.28
Surface, tangent plane and normal
line at the point (2, 1, 13).

      Notice that in each of Figures 12.27 and 12.28, the tangent plane appears to stay close to the surface near the point of tangency. This says that the z-values on the tangent plane should be close to the corresponding z-values on the surface, which are given by the function values $f(x, y)$, at least for (x, y) close to the point of tangency. Further, the simple form of the equation for the tangent plane makes it ideal for approximating the value of complicated functions. From Theorem 4.1, we have that the tangent plane to $z = f(x, y)$ at the point (a, b) is given by

$$z = f(a, b) + f_x(a, b)(x - a) + f_y(a, b)(y - b).$$

We define the **linear approximation** $L(x, y)$ of $f(x, y)$ at the point (a, b) to be the function defining the z-values on the tangent plane, namely,

$$L(x, y) = f(a, b) + f_x(a, b)(x - a) + f_y(a, b)(y - b). \qquad (4.4)$$

We illustrate this with the following example.

▮ **Example 4.3**   Finding a Linear Approximation

Compute the linear approximation of $f(x, y) = 2x + e^{x^2 - y}$ at (0, 0). Compare the linear approximation to the actual function values for (a) $x = 0$ and y near 0; (b) $y = 0$ and x near 0; (c) $y = x$, with both x and y near 0 and (d) $y = 2x$, with both x and y near 0.

Solution   First, notice that $f_x = 2 + 2xe^{x^2 - y}$ and $f_y = -e^{x^2 - y}$, so that $f_x(0, 0) = 2$ and $f_y(0, 0) = -1$. Also, $f(0, 0) = 1$. From (4.4), the linear approximation is then given by

$$L(x, y) = 1 + 2(x - 0) - (y - 0) = 1 + 2x - y.$$

The following table compares values of $L(x, y)$ and $f(x, y)$ for a number of points of the form $(0, y)$, $(x, 0)$, (x, x) and $(x, 2x)$.

(x, y)	$f(x, y)$	$L(x, y)$	(x, y)	$f(x, y)$	$L(x, y)$
$(0, 0.1)$	0.905	0.9	$(0.1, 0.1)$	1.11393	1.1
$(0, 0.01)$	0.99005	0.99	$(0.01, 0.01)$	1.01015	1.01
$(0, -0.1)$	1.105	1.1	$(-0.1, -0.1)$	0.91628	0.9
$(0, -0.01)$	1.01005	1.01	$(-0.01, -0.01)$	0.99015	0.99
$(0.1, 0)$	1.21005	1.2	$(0.1, 0.2)$	1.02696	1.0
$(0.01, 0)$	1.02010	1.02	$(0.01, 0.02)$	1.00030	1.0
$(-0.1, 0)$	0.81005	0.8	$(-0.1, -0.2)$	1.03368	1.0
$(-0.01, 0)$	0.98010	0.98	$(-0.01, -0.02)$	1.00030	1.0

Notice that the closer a given point is to the point of tangency, the more accurate the linear approximation is at that point. This is typical of this type of approximation and was also the case for linear approximations to functions of a single variable. We will explore this further in the exercises.

■

Increments and Differentials

Now that we have examined linear approximations from a graphical perspective, we will examine these in a symbolic fashion. In the course of doing so, we will gain some insight into the behavior of functions of several variables. First, we remind you of the notation and some alternative language that we have already used in section 3.1 for functions of a single variable. We defined the **increment** Δy of the function $f(x)$ at $x = a$ to be

$$\Delta y = f(a + \Delta x) - f(a).$$

Referring to Figure 12.29, notice that for Δx small,

$$\Delta y \approx dy = f'(a) \, \Delta x,$$

where we referred to dy as the **differential** of y. Further, observe that if f is differentiable at $x = a$ and $\varepsilon = \dfrac{\Delta y - dy}{\Delta x}$, then we have

$$\varepsilon = \frac{\Delta y - dy}{\Delta x} = \frac{f(a + \Delta x) - f(a) - f'(a) \, \Delta x}{\Delta x}$$

$$= \frac{f(a + \Delta x) - f(a)}{\Delta x} - f'(a) \to 0,$$

as $\Delta x \to 0$. (You'll need to recognize the definition of derivative here!) Finally, solving

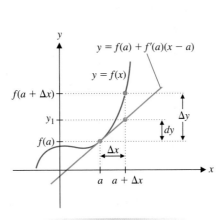

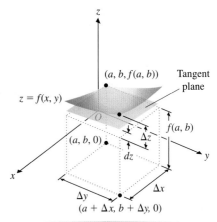

Figure 12.29

Increments and differentials for a
function of one variable.

Figure 12.30

Linear approximation.

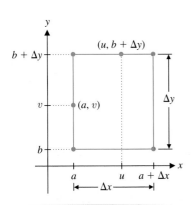

Figure 12.31

Intermediate points from the Mean
Value Theorem.

for Δy in terms of ε, we have

$$\Delta y = dy + \varepsilon \, \Delta x,$$

where $\varepsilon \to 0$, as $\Delta x \to 0$. We can make a similar observation for functions of several variables, as follows.

For $z = f(x, y)$, we define the **increment** of f to be

$$\Delta z = f(a + \Delta x, b + \Delta y) - f(a, b).$$

That is, Δz is the change in z that occurs when a is incremented by Δx and b is incremented by Δy, as illustrated in Figure 12.30. Notice that as long as f is continuous in some open region containing (a, b) and f has first partial derivatives on that region, we can write

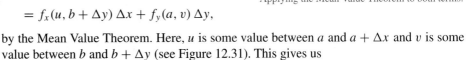

$$\Delta z = f(a + \Delta x, b + \Delta y) - f(a, b)$$

$$= [f(a + \Delta x, b + \Delta y) - f(a, b + \Delta y)] + [f(a, b + \Delta y) - f(a, b)]$$

Adding and subtracting $f(a, b + \Delta y)$.

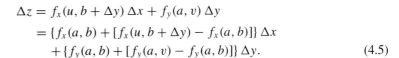

$$= f_x(u, b + \Delta y)[(a + \Delta x) - a] + f_y(a, v)[(b + \Delta y) - b]$$

Applying the Mean Value Theorem to both terms.

$$= f_x(u, b + \Delta y) \, \Delta x + f_y(a, v) \, \Delta y,$$

by the Mean Value Theorem. Here, u is some value between a and $a + \Delta x$ and v is some value between b and $b + \Delta y$ (see Figure 12.31). This gives us

$$\Delta z = f_x(u, b + \Delta y) \, \Delta x + f_y(a, v) \, \Delta y$$
$$= \{f_x(a, b) + [f_x(u, b + \Delta y) - f_x(a, b)]\} \, \Delta x$$
$$+ \{f_y(a, b) + [f_y(a, v) - f_y(a, b)]\} \, \Delta y. \qquad (4.5)$$

Notice that we can rewrite (4.5) as

$$\Delta z = f_x(a, b) \, \Delta x + f_y(a, b) \, \Delta y + \varepsilon_1 \, \Delta x + \varepsilon_2 \, \Delta y,$$

where

$$\varepsilon_1 = f_x(u, b + \Delta y) - f_x(a, b) \quad \text{and} \quad \varepsilon_2 = f_y(a, v) - f_y(a, b).$$

Finally, observe that if f_x and f_y are both continuous in some open region containing (a, b), then ε_1 and ε_2 will both tend to 0, as $(\Delta x, \Delta y) \to (0, 0)$. In fact, you should recognize that since $\varepsilon_1, \varepsilon_2 \to 0$, as $(\Delta x, \Delta y) \to (0, 0)$, the products $\varepsilon_1 \, \Delta x$ and $\varepsilon_2 \, \Delta y$ both tend to 0 even faster than do $\varepsilon_1, \varepsilon_2, \Delta x$ or Δy individually. (Think about this!)

We have now established the following result.

Theorem 4.2 Suppose that $z = f(x, y)$ is defined on the rectangle $R = \{(x, y)|x_0 < x < x_1, y_0 < y < y_1\}$ and f_x and f_y are defined on R and are continuous at $(a, b) \in R$. Then for $(a + \Delta x, b + \Delta y) \in R$,

$$\Delta z = f_x(a, b)\,\Delta x + f_y(a, b)\,\Delta y + \varepsilon_1\,\Delta x + \varepsilon_2\,\Delta y, \qquad (4.6)$$

where ε_1 and ε_2 are functions of Δx and Δy that both tend to zero, as $(\Delta x, \Delta y) \to (0, 0)$.

For some very simple functions, we can compute Δz by hand, as illustrated in the following example.

Example 4.4 Computing the Increment Δz

For $z = f(x, y) = x^2 - 5xy$, find Δz and write it in the form indicated in Theorem 4.2.

Solution We have

$$\begin{aligned}
\Delta z &= f(x + \Delta x, y + \Delta y) - f(x, y) \\
&= [(x + \Delta x)^2 - 5(x + \Delta x)(y + \Delta y)] - (x^2 - 5xy) \\
&= x^2 + 2x\,\Delta x + (\Delta x)^2 - 5(xy + x\,\Delta y + y\,\Delta x + \Delta x\,\Delta y) - x^2 + 5xy \\
&= \underbrace{(2x - 5y)}_{f_x}\,\Delta x + \underbrace{(-5x)}_{f_y}\,\Delta y + \underbrace{(\Delta x)}_{\varepsilon_1}\,\Delta x + \underbrace{(-5\Delta x)}_{\varepsilon_2}\,\Delta y \\
&= f_x(x, y)\,\Delta x + f_y(x, y)\,\Delta y + \varepsilon_1\,\Delta x + \varepsilon_2\,\Delta y,
\end{aligned}$$

where $\varepsilon_1 = \Delta x$ and $\varepsilon_2 = -5\Delta x$ both tend to zero, as $(\Delta x, \Delta y) \to (0, 0)$, as indicated in Theorem 4.2.

Look closely at the first two terms in the expansion of the increment Δz given in (4.6). If we take $\Delta x = x - a$ and $\Delta y = y - b$, then they correspond to the linear approximation of $f(x, y)$. In this context, we give this a special name. If we increment x by the amount $dx = \Delta x$ and increment y by $dy = \Delta y$, then we define the **differential** of z to be

$$\boxed{dz = f_x(x, y)\,dx + f_y(x, y)\,dy.}$$

This is sometimes referred to as a **total differential.** Notice that for dx and dy small, we have from (4.6) that

$$\Delta z \approx dz.$$

You should recognize that this is the same approximation as the linear approximation developed in the beginning of this section. In this case though, we have developed this from an analytical perspective, rather than the geometrical one used in the beginning of the section.

Functions that can be approximated linearly in the above fashion we give a special name, as in the following definition.

Definition 4.1

Let $z = f(x, y)$. We say that f is **differentiable** at (a, b) if we can write

$$\Delta z = f_x(a, b)\, \Delta x + f_y(a, b)\, \Delta y + \varepsilon_1\, \Delta x + \varepsilon_2\, \Delta y,$$

where ε_1 and ε_2 are both functions of Δx and Δy and $\varepsilon_1, \varepsilon_2 \to 0$, as $(\Delta x, \Delta y) \to (0, 0)$. We say that f is differentiable on a region $R \subset \mathbb{R}^2$ whenever f is differentiable at every point in R.

Note that from Theorem 4.2, if f_x and f_y are defined on some open rectangle R containing the point (a, b) and if f_x and f_y are continuous at (a, b), then f will be differentiable at (a, b). Just as with functions of a single variable, it can be shown that if f is differentiable at a point (a, b), then it is also continuous at (a, b). Further, owing to Theorem 4.2, if a function is differentiable at a point, then the linear approximation (differential) at that point provides a good approximation to the function near that point. Be very careful of what this does **not** say, however. If a function has partial derivatives at a point, it need **not** be differentiable or even continuous at that point. (In the exercises, you will see an example of a function with partial derivatives defined everywhere, but that is not continuous at a point.)

The idea of the linear approximation extends easily to three or more dimensions. We lose the graphical interpretation of a tangent plane approximating a surface, but the definition should make sense.

Definition 4.2

The **linear approximation** to $f(x, y, z)$ at the point (a, b, c) is given by

$$L(x, y, z) = f(a, b, c) + \frac{\partial f}{\partial x}(a, b, c)(x - a)$$

$$+ \frac{\partial f}{\partial y}(a, b, c)(y - b) + \frac{\partial f}{\partial z}(a, b, c)(z - c).$$

We can write the linear approximation in the context of increments and differentials, as follows. If we increment x by Δx, y by Δy and z by Δz, then the increment of $w = f(x, y, z)$ is given by

$$\Delta w = f(x + \Delta x, y + \Delta y, z + \Delta z) - f(x, y, z)$$
$$\approx dw = f_x(x, y, z)\, \Delta x + f_y(x, y, z)\, \Delta y + f_z(x, y, z)\, \Delta z.$$

A good way to interpret (and remember!) the linear approximation is that each partial derivative represents the change in the function relative to the change in that variable. The linear approximation starts with the function value at the known point and adds in the approximate changes corresponding to each of the independent variables.

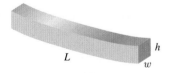

Figure 12.32

A typical beam.

Example 4.5   Approximating the Sag in a Beam

Suppose that the sag in a beam of length L, width w and height h is given by $S(L, w, h) = 0.0004\dfrac{L^4}{wh^3}$, with all lengths measured in inches. We illustrate the beam in Figure 12.32. A beam is supposed to measure $L = 36$, $w = 2$ and $h = 6$ with a corresponding sag of 1.5552 inches. Due to weathering and other factors, the manufacturer only guarantees measurements with error tolerances $L = 36 \pm 1$, $w = 2 \pm 0.4$ and $h = 6 \pm 0.8$. Use a linear approximation to estimate the possible range of sags in the beam.

Solution   We first compute $\dfrac{\partial S}{\partial L} = 0.0016\dfrac{L^3}{wh^3}$, $\dfrac{\partial S}{\partial w} = -0.0004\dfrac{L^4}{w^2h^3}$ and $\dfrac{\partial S}{\partial h} = -0.0012\dfrac{L^4}{wh^4}$. At the point $(36, 2, 6)$, we then have $\dfrac{\partial S}{\partial L}(36, 2, 6) = 0.1728$, $\dfrac{\partial S}{\partial w}(36, 2, 6) = -0.7776$ and $\dfrac{\partial S}{\partial h}(36, 2, 6) = -0.7776$. From Definition 4.2, the linear approximation of the sag is then given by

$$S \approx 1.5552 + 0.1728(L - 36) - 0.7776(w - 2) - 0.7776(h - 6).$$

From the stated tolerances, $L - 36$ must be between -1 and 1, $w - 2$ must be between -0.4 and 0.4 and $h - 6$ must be between -0.8 and 0.8. Notice that the maximum sag then occurs with $L - 36 = 1$, $w - 2 = -0.4$ and $h - 6 = -0.8$. The linear approximation predicts that

$$S - 1.5552 \approx 0.1728 + 0.31104 + 0.62208 = 1.10592.$$

Similarly, the minimum sag occurs with $L - 36 = -1$, $w - 2 = 0.4$ and $h - 6 = 0.8$. The linear approximation predicts that

$$S - 1.5552 \approx -0.1728 - 0.31104 - 0.62208 = -1.10592.$$

Based on the linear approximation, the sag is 1.5552 ± 1.10592, or between 0.44928 and 2.66112. As you can see, in this case, the uncertainty in the sag is substantial.

In many real-world situations, we do not have a formula for the quantity we are interested in computing. Even so, given sufficient information, we can still use linear approximations to estimate the desired quantity.

Example 4.6   Estimating the Gauge of a Sheet of Metal

Manufacturing plants create rolls of metal of a desired gauge (thickness) by feeding the metal through very large rollers. The thickness of the resulting metal depends on the gap between the working rollers, the speed at which the rollers turn and the temperature of the metal. Suppose that for a certain metal, a gauge of 4 mm is produced by a gap of 4 mm, a speed of 10 m/s and a temperature of $900°$. Experiments show that an increase in speed of 0.2 m/s increases the gauge by 0.06 mm and an increase in temperature of $10°$ decreases the gauge by 0.04 mm. Use a linear approximation to estimate the gauge at 10.1 m/s and $880°$.

Solution   With no change in gap, we assume that the gauge is a function $g(s, t)$ of the speed s and the temperature t. Based on our data, $\dfrac{\partial g}{\partial s} \approx \dfrac{0.06}{0.2} = 0.3$ and $\dfrac{\partial g}{\partial t} \approx \dfrac{-0.04}{10} = -0.004$. From Definition 4.2, the linear approximation of $g(s, t)$ is given by

$$g(s, t) \approx 4 + 0.3(s - 10) - 0.004(t - 900).$$

With $s = 10.1$ and $t = 880$, we get the estimate

$$g(10.1, 880) \approx 4 + 0.3(0.1) - 0.004(-20) = 4.11.$$

EXERCISES 12.4

1. Describe which graphical properties of the surface $z = f(x, y)$ would cause the linear approximation of $f(x, y)$ at (a, b) to be particularly accurate or inaccurate.

2. Temperature varies with longitude (x), latitude (y) and altitude (z). Speculate whether or not the temperature function would be differentiable and what significance the answer would have for weather prediction.

3. Imagine a surface $z = f(x, y)$ with a ridge of discontinuities along the line $y = x$. Explain in graphical terms why $f(x, y)$ would not be differentiable at $(0, 0)$ or any other point on the line $y = x$.

4. The function in exercise 3 might have first partial derivatives $f_x(0, 0)$ and $f_y(0, 0)$. Explain why the slopes along $x = 0$ and $y = 0$ could have limits as x and y approach 0. If *differentiable* is intended to describe functions with smooth graphs, explain why differentiability is not defined in terms of the existence of partial derivatives.

In exercises 5–16, find equations of the tangent plane and normal line to the surface at the given point.

5. $z = x^2 + y^2 - 1$ at $(2, 1, 4)$

6. $z = x^2 + y^2 - 1$ at $(0, 2, 3)$

7. $z = e^{-x^2-y^2}$ at $(0, 0, 1)$

8. $z = e^{-x^2-y^2}$ at $(1, 1, e^{-2})$

9. $z = \sin x \cos y$ at $(0, \pi, 0)$

10. $z = \sin x \cos y$ at $\left(\frac{\pi}{2}, \pi, -1\right)$

11. $z = x^3 - 2xy$ at $(-2, 3, 4)$

12. $z = x^3 - 2xy$ at $(1, -1, 3)$

13. $z = \sqrt{x^2 + y^2}$ at $(-3, 4, 5)$

14. $z = \sqrt{x^2 + y^2}$ at $(8, -6, 10)$

15. $z = \dfrac{4x}{y}$ at $(1, 2, 2)$

16. $z = \dfrac{4x}{y}$ at $(-1, 4, -1)$

In exercises 17–26, compute the linear approximation of the function at the given point.

17. $f(x, y) = \sqrt{x^2 + y^2}$ at $(3, 0)$

18. $f(x, y) = \sqrt{x^2 + y^2}$ at $(0, -3)$

19. $f(x, y) = \sin x \cos y$ at $(0, \pi)$

20. $f(x, y) = \sin x \cos y$ at $\left(\frac{\pi}{2}, \pi\right)$

21. $f(x, y) = xe^{y^2} - 4x$ at $(2, 0)$

22. $f(x, y) = xe^{xy^2} + 3y^2$ at $(0, 1)$

23. $f(x, y, z) = \sin yz^2 + x^3z$ at $(-2, 0, 1)$

24. $f(x, y, z) = xe^{yz} - \sqrt{x - y^2}$ at $(4, 1, 0)$

25. $f(w, x, y, z) = w^2xy - e^{wyz}$ at $(-2, 3, 1, 0)$

26. $f(w, x, y, z) = \cos xyz - w^3x^2$ at $(2, -1, 4, 0)$

In exercises 27–30, compare the linear approximation from the indicated exercise to the exact function value at the given points.

27. Exercise 17 at $(3, -0.1)$, $(3.1, 0)$, $(3.1, -0.1)$

28. Exercise 18 at $(0.1, -3)$, $(0, -3.1)$, $(0.1, -3.1)$

29. Exercise 19 at $(0, 3)$, $(0.1, \pi)$, $(0.1, 3)$

30. Exercise 21 at $(2.1, 0)$, $(2, 0.2)$, $(1, -1)$

31. Use a linear approximation to estimate the range of sags in the beam of example 4.5 if the error tolerances are $L = 36 \pm 0.5$, $w = 2 \pm 0.2$ and $h = 6 \pm 0.5$.

32. Use a linear approximation to estimate the range of sags in the beam of example 4.5 if the error tolerances are $L = 32 \pm 0.4$, $w = 2 \pm 0.3$ and $h = 8 \pm 0.4$.

33. Use a linear approximation to estimate the gauge of the metal in example 4.6 at 9.9 m/s and 930°.

34. Use a linear approximation to estimate the gauge of the metal in example 4.6 at 10.2 m/s and 910°.

35. Suppose that for a metal similar to that of example 4.6, an increase in speed of 0.3 m/s increases the gauge by 0.03 mm

and an increase in temperature of $20°$ decreases the gauge by 0.02 mm. Use a linear approximation to estimate the gauge at 10.2 m/s and $890°$.

36. Suppose that for the metal in example 4.6, a decrease of 0.05 mm in the gap between the working rolls decreases the gauge by 0.04 mm. Use a linear approximation in three variables to estimate the gauge at 10.15 m/s, $905°$ and a gap of 3.98 mm.

In exercises 37–40, find the increment Δz and write it in the form given in Theorem 4.2.

37. $f(x, y) = 2xy + y^2$

38. $f(x, y) = (x + y)^2$

39. $f(x, y) = x^2 + y^2$

40. $f(x, y) = x^3 - 3xy$

41. Determine whether or not $f(x, y) = x^2 + 3xy$ is differentiable.

42. Determine whether or not $f(x, y) = xy^2$ is differentiable.

In exercises 43 and 44, find the total differential of $f(x, y)$.

43. $f(x, y) = ye^x + \sin x$

44. $f(x, y) = \sqrt{x + y}$

In exercises 45 and 46, show that the partial derivatives $f_x(0, 0)$ and $f_y(0, 0)$ both exist, but the function $f(x, y)$ is not continuous at $(0, 0)$. Show that f is not differentiable at $(0, 0)$.

45. $f(x, y) = \begin{cases} \dfrac{2xy}{x^2 + y^2}, & \text{if } (x, y) \neq (0, 0) \\ 0, & \text{if } (x, y) = (0, 0) \end{cases}$

46. $f(x, y) = \begin{cases} \dfrac{xy^2}{x^2 + y^2}, & \text{if } (x, y) \neq (0, 0) \\ 0, & \text{if } (x, y) = (0, 0) \end{cases}$

47. In this exercise, we visualize the linear approximation of example 4.3. Start with a contour plot of $f(x, y) = 2x + e^{x^2 - y}$ with $-1 \leq x \leq 1$ and $-1 \leq y \leq 1$. Then zoom in on the point $(0, 0)$ of the contour plot until the level curves appear straight and equally spaced. (Level curves for z-values between 0.9 and 1.1 with a graphing window of $-0.1 \leq x \leq 0.1$ and $-0.1 \leq y \leq 0.1$ should work.) You will need the z-values for the level curves. Notice that to move from the $z = 1$ level curve to the $z = 1.05$ level curve you move 0.025 units to the right. Then $\dfrac{\partial f}{\partial x}(0, 0) \approx \dfrac{\Delta z}{\Delta x} = \dfrac{0.05}{0.025} = 2$. Verify graphically that $\dfrac{\partial f}{\partial y}(0, 0) \approx -1$. Explain how to use the contour plot to reproduce the linear approximation $1 + 2x - y$.

48. Use the graphical method of exercise 47 to find the linear approximation of $f(x, y) = \sin(x^2 + 2xy)$ at the point $(1, 3)$.

In exercises 49–52, use the graphical method of exercises 47–48 to estimate the linear approximation of $f(x, y)$ at $(0, 0)$.

49.

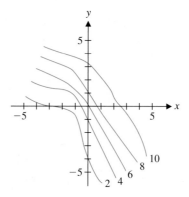

50.

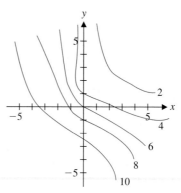

51.

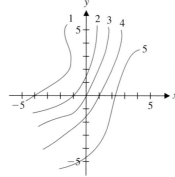

52.

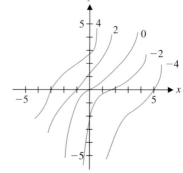

53. The table below shows wind chill (how cold it "feels" outside) as a function of temperature (degrees Fahrenheit) and wind speed (mph). We can think of this as a function $w(t, s)$. Estimate the partial derivatives $\frac{\partial w}{\partial t}(10, 10)$ and $\frac{\partial w}{\partial s}(10, 10)$ and the linear approximation of $w(t, s)$ at $(10, 10)$. Use the linear approximation to estimate the wind chill at $(12, 13)$.

Speed Temp	30	20	10	0	−10
0	30	20	10	0	−10
5	27	16	6	−5	−15
10	16	4	−9	−24	−33
15	9	−5	−18	−32	−45
20	4	−10	−25	−39	−53
25	0	−15	−29	−44	−59
30	−2	−18	−33	−48	−63

54. Estimate the linear approximation of wind chill at $(10, 15)$ and use it to estimate the wind chill at $(12, 13)$. Explain any differences between this answer and that of exercise 53.

55. In exercise 47, we specified that you zoom in on the contour plot until the level curves appear linear **and** equally spaced. To see why the second condition is necessary, sketch a contour plot of $f(x, y) = e^{x-y}$ with $-1 \leq x \leq 1$ and $-1 \leq y \leq 1$. Use this plot to estimate $\frac{\partial f}{\partial y}(0, 0)$ and $\frac{\partial f}{\partial y}(0, 0)$ and compare to the exact values. Zoom in until the level curves are equally spaced and estimate again. Explain why this estimate is much better.

56. Show that $\left\langle 0, 1, \frac{\partial f}{\partial y}(a, b) \right\rangle \times \left\langle 1, 0, \frac{\partial f}{\partial x}(a, b) \right\rangle = \left\langle \frac{\partial f}{\partial x}(a, b), \frac{\partial f}{\partial y}(a, b), -1 \right\rangle.$

For exercises 57 and 58, we need to use the notation of matrix algebra. First, we define the 2×2 matrix A to be a two-dimensional array of real numbers, $A = \begin{bmatrix} a & b \\ c & d \end{bmatrix}$, we define a column vector x to be a one-dimensional array of real numbers, $x = \begin{bmatrix} x_1 \\ x_2 \end{bmatrix}$ and define the product of a matrix and column vector to be the (column) vector $\begin{bmatrix} a & b \\ c & d \end{bmatrix} \begin{bmatrix} x_1 \\ x_2 \end{bmatrix} = \begin{bmatrix} ax_1 + bx_2 \\ cx_1 + dx_2 \end{bmatrix}$. You can learn much more about matrix algebra by looking at one of the many textbooks on the subject, including those bearing the title *Linear Algebra*.

57. We can extend the linear approximation of this section to quadratic approximations. Define the **Hessian**

matrix $H = \begin{bmatrix} f_{xx} & f_{xy} \\ f_{yx} & f_{yy} \end{bmatrix}$, the **gradient vector** $\nabla f(x_0, y_0) = \langle f_x(x_0, y_0), f_y(x_0, y_0) \rangle$, the column vector $\mathbf{x} = \begin{bmatrix} x \\ y \end{bmatrix}$, the vector $\mathbf{x}_0 = \begin{bmatrix} x_0 \\ y_0 \end{bmatrix}$ and the transpose vector $\mathbf{x}^T = [x \ y]$. The **quadratic approximation** of $f(x, y)$ at the point (x_0, y_0) is defined by

$$Q(x, y) = f(x_0, y_0) + \nabla f(x_0, y_0) \cdot (\mathbf{x} - \mathbf{x}_0)$$
$$+ \frac{1}{2}(\mathbf{x} - \mathbf{x}_0)^T H(x_0, y_0)(\mathbf{x} - \mathbf{x}_0)$$

Find the quadratic approximation of $f(x, y) = 2x + e^{x^2 - y}$ and compute $Q(x, y)$ for the points in the table of example 4.3.

58. An important application of linear approximations of functions $f(x)$ is Newton's method for finding solutions of equations of the form $f(x) = 0$. In this exercise, we extend Newton's method to functions of several variables. To be specific, suppose that $f_1(x, y)$ and $f_2(x, y)$ are functions of two variables with continuous partial derivatives. To solve the equations $f_1(x, y) = 0$ and $f_2(x, y) = 0$ simultaneously, start with a guess $x = x_0$ and $y = y_0$. The idea is to replace $f_1(x, y)$ and $f_2(x, y)$ with their linear approximations $L_1(x, y) = 0$ and $L_2(x, y) = 0$ and solve these (simpler) equations simultaneously. Write out the linear approximations and show that we want

$$\frac{\partial f_1}{\partial x}(x_0, y_0)(x - x_0) + \frac{\partial f_1}{\partial y}(x_0, y_0)(y - y_0) = -f_1(x_0, y_0)$$
$$\frac{\partial f_2}{\partial x}(x_0, y_0)(x - x_0) + \frac{\partial f_2}{\partial y}(x_0, y_0)(y - y_0) = -f_2(x_0, y_0)$$

Recall that there are several ways (substitution and elimination are popular) to solve two linear equations in two unknowns. The simplest way symbolically is to use matrices. If we define the **Jacobian matrix**

$$J(\mathbf{x}_0) = \begin{bmatrix} \frac{\partial f_1}{\partial x}(\mathbf{x}_0) & \frac{\partial f_1}{\partial y}(\mathbf{x}_0) \\ \frac{\partial f_2}{\partial x}(\mathbf{x}_0) & \frac{\partial f_2}{\partial y}(\mathbf{x}_0) \end{bmatrix}$$

where $\mathbf{x}_0$ represents the point (x_0, y_0), the above equations can be written as $J(\mathbf{x}_0)(\mathbf{x} - \mathbf{x}_0) = -\mathbf{f}(\mathbf{x}_0)$, which has solution $\mathbf{x} - \mathbf{x}_0 = -J^{-1}(\mathbf{x}_0)\mathbf{f}(\mathbf{x}_0)$ or $\mathbf{x} = \mathbf{x}_0 - J^{-1}(\mathbf{x}_0)\mathbf{f}(\mathbf{x}_0)$. Here, the matrix $J^{-1}(\mathbf{x}_0)$ is called the **inverse** of the matrix $J(\mathbf{x}_0)$. The inverse A^{-1} of a matrix A (when it is defined) is the matrix for which $\mathbf{a} = A\mathbf{b}$ if and only if $\mathbf{b} = A^{-1}\mathbf{a}$, for all column vectors $\mathbf{a}$ and $\mathbf{b}$. In general, Newton's method is defined by the iteration

$$\mathbf{x}_{n+1} = \mathbf{x}_n - J^{-1}(\mathbf{x}_n)\mathbf{f}(\mathbf{x}_n)$$

Use Newton's method with an initial guess of $\mathbf{x}_0 = (-1, 0.5)$ to approximate a solution of the equations $x^2 - 2y = 0$ and $x^2 y - \sin y = 0$.

12.5   THE CHAIN RULE

You already are quite familiar with the chain rule for functions of a single variable. Even if you don't remember the name, you certainly remember how it's used. For instance, to differentiate the function $e^{\sin(x^2)}$, we have

$$\frac{d}{dx}\left[e^{\sin(x^2)}\right] = e^{\sin(x^2)} \underbrace{\frac{d}{dx}[\sin(x^2)]}_{\text{the derivative of the }inside}$$

$$= e^{\sin(x^2)} \cos(x^2) \underbrace{\frac{d}{dx}(x^2)}_{\text{the derivative of the }inside}$$

$$= e^{\sin(x^2)} \cos(x^2)(2x).$$

You have used this rule hundreds of times, but you may be less familiar with the general form of the chain rule. For differentiable functions f and g, we have

$$\frac{d}{dx}[f(g(x))] = f'(g(x))\underbrace{g'(x)}_{\text{the derivative of the }inside}.$$

We now pause to extend the chain rule to functions of several variables. This takes several slightly different forms, depending on the number of independent variables. Keep in mind that these are variations of the already familiar chain rule for functions of a single variable.

First, consider a differentiable function $f(x, y)$ where the arguments x and y are both in turn, differentiable functions of a single variable t. If we wish to find the derivative of $f(x, y)$ with respect to t, we can first write $g(t) = f(x(t), y(t))$. Then, from the definition of (an ordinary) derivative, we have

$$\frac{d}{dt}[f(x(t), y(t))] = g'(t) = \lim_{\Delta t \to 0} \frac{g(t + \Delta t) - g(t)}{\Delta t}$$

$$= \lim_{\Delta t \to 0} \frac{f(x(t + \Delta t), y(t + \Delta t)) - f(x(t), y(t))}{\Delta t}.$$

For simplicity, we write $\Delta x = x(t + \Delta t) - x(t)$, $\Delta y = y(t + \Delta t) - y(t)$ and $\Delta z = f(x(t + \Delta t), y(t + \Delta t)) - f(x(t), y(t))$. We now have

$$\frac{d}{dt}[f(x(t), y(t))] = \lim_{\Delta t \to 0} \frac{\Delta z}{\Delta t}.$$

Since f is a differentiable function of x and y, we have (from the definition of differentiability) that

$$\Delta z = \frac{\partial f}{\partial x}\Delta x + \frac{\partial f}{\partial y}\Delta y + \varepsilon_1 \Delta x + \varepsilon_2 \Delta y,$$

where ε_1 and ε_2 both tend to 0, as $(\Delta x, \Delta y) \to (0, 0)$. Dividing through by Δt gives us

$$\frac{\Delta z}{\Delta t} = \frac{\partial f}{\partial x}\frac{\Delta x}{\Delta t} + \frac{\partial f}{\partial y}\frac{\Delta y}{\Delta t} + \varepsilon_1 \frac{\Delta x}{\Delta t} + \varepsilon_2 \frac{\Delta y}{\Delta t}.$$

Taking the limit as $\Delta t \to 0$ now gives us

$$\frac{d}{dt}[f(x(t), y(t))] = \lim_{\Delta t \to 0} \frac{\Delta z}{\Delta t}$$

$$= \frac{\partial f}{\partial x} \lim_{\Delta t \to 0} \frac{\Delta x}{\Delta t} + \frac{\partial f}{\partial y} \lim_{\Delta t \to 0} \frac{\Delta y}{\Delta t}$$

$$+ \lim_{\Delta t \to 0} \varepsilon_1 \lim_{\Delta t \to 0} \frac{\Delta x}{\Delta t} + \lim_{\Delta t \to 0} \varepsilon_2 \lim_{\Delta t \to 0} \frac{\Delta y}{\Delta t}. \qquad (5.1)$$

Notice that

$$\lim_{\Delta t \to 0} \frac{\Delta x}{\Delta t} = \lim_{\Delta t \to 0} \frac{x(t + \Delta t) - x(t)}{\Delta t} = \frac{dx}{dt}$$

and

$$\lim_{\Delta t \to 0} \frac{\Delta y}{\Delta t} = \lim_{\Delta t \to 0} \frac{y(t + \Delta t) - y(t)}{\Delta t} = \frac{dy}{dt}.$$

Further, notice that since $x(t)$ and $y(t)$ are differentiable, they are also continuous and so,

$$\lim_{\Delta t \to 0} \Delta x = \lim_{\Delta t \to 0} [x(t + \Delta t) - x(t)] = 0.$$

Likewise, $\lim_{\Delta t \to 0} \Delta y = 0$, also. Consequently, since $(\Delta x, \Delta y) \to (0, 0)$, as $\Delta t \to 0$, we have

$$\lim_{\Delta t \to 0} \varepsilon_1 = \lim_{\Delta t \to 0} \varepsilon_2 = 0.$$

From (5.1), we now have

$$\frac{d}{dt}[f(x(t), y(t))] = \frac{\partial f}{\partial x} \lim_{\Delta t \to 0} \frac{\Delta x}{\Delta t} + \frac{\partial f}{\partial y} \lim_{\Delta t \to 0} \frac{\Delta y}{\Delta t}$$

$$+ \lim_{\Delta t \to 0} \varepsilon_1 \lim_{\Delta t \to 0} \frac{\Delta x}{\Delta t} + \lim_{\Delta t \to 0} \varepsilon_2 \lim_{\Delta t \to 0} \frac{\Delta y}{\Delta t}$$

$$= \frac{\partial f}{\partial x} \frac{dx}{dt} + \frac{\partial f}{\partial y} \frac{dy}{dt}.$$

We summarize the chain rule for the derivative of $f(x(t), y(t))$ with the following result.

Theorem 5.1 (Chain Rule)
If $z = f(x(t), y(t))$, where $x(t)$ and $y(t)$ are differentiable and $f(x, y)$ is a differentiable function of x and y, then

$$\frac{dz}{dt} = \frac{d}{dt}[f(x(t), y(t))] = \frac{\partial f}{\partial x}(x(t), y(t)) \frac{dx}{dt} + \frac{\partial f}{\partial y}(x(t), y(t)) \frac{dy}{dt}.$$

As a convenient device for remembering the chain rule, we often use a **tree diagram** like the one shown in the margin of the following page. Notice that if $z = f(x, y)$ and x and y are both functions of the variable t, then t is the independent variable. We consider x and y to be **intermediate variables,** since they both depend on t. In the tree diagram, we list the dependent variable z at the top, followed by each of the intermediate variables x and

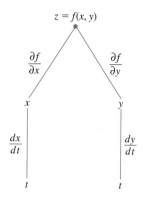

$z = f(x, y)$

$\frac{\partial f}{\partial x}$ $\frac{\partial f}{\partial y}$

x y

$\frac{dx}{dt}$ $\frac{dy}{dt}$

t t

y, with the dependent variable t at the bottom level, with each of the variables connected by a path. Next to each of the paths, we indicate the corresponding derivative $\left(\text{i.e., between } z \text{ and } x, \text{ we indicate } \frac{\partial z}{\partial x}\right)$. The chain rule then gives $\frac{dz}{dt}$ as the sum of all of the products of the derivatives along each path to t. That is,

$$\frac{dz}{dt} = \frac{\partial z}{\partial x}\frac{dx}{dt} + \frac{\partial z}{\partial y}\frac{dy}{dt}.$$

This device is especially useful for functions of several variables that are in turn functions of several other variables, as we will see shortly.

We illustrate the use of this new chain rule in the following example.

Example 5.1   Using the Chain Rule

For $f(x, y) = x^2 e^y$, $x(t) = t^2 - 1$ and $y(t) = \sin t$, find the derivative of $g(t) = f(x(t), y(t))$.

Solution   We first compute the derivatives $\frac{\partial f}{\partial x} = 2xe^y$, $\frac{\partial f}{\partial y} = x^2 e^y$, $x'(t) = 2t$ and $y'(t) = \cos t$. The chain rule (Theorem 5.1) then gives us

$$g'(t) = \frac{\partial f}{\partial x}\frac{dx}{dt} + \frac{\partial f}{\partial y}\frac{dy}{dt} = 2xe^y(2t) + x^2 e^y \cos t$$

$$= 2(t^2 - 1)e^{\sin t}(2t) + (t^2 - 1)^2 e^{\sin t} \cos t.$$

In example 5.1, notice that you could have first substituted for x and y and then computed the derivative of $g(t) = (t^2 - 1)^2 e^{\sin t}$, using the usual rules of differentiation. In fact, when direct substitution is possible, it is usually preferable. In the following example, you don't have any alternative but to use the chain rule.

Example 5.2   A Case Where the Chain Rule Is Needed

Suppose the production of a firm is modeled by the **Cobb-Douglas production** function $P(k, l) = 20k^{1/4}l^{3/4}$, where k measures capital (in millions of dollars) and l measures the labor force (in thousands of workers). Suppose that $l = 2$ and $k = 6$, the labor force is decreasing at the rate of 20 workers per year and capital is growing at the rate of $\$400,000$ per year. Determine the rate of change of production.

Solution   Suppose that $g(t) = P(k(t), l(t))$. From the chain rule, we have

$$g'(t) = \frac{\partial P}{\partial k}k'(t) + \frac{\partial P}{\partial l}l'(t).$$

Notice that $\frac{\partial P}{\partial k} = 5k^{-3/4}l^{3/4}$ and $\frac{\partial P}{\partial l} = 15k^{1/4}l^{-1/4}$. With $l = 2$ and $k = 6$, this gives us $\frac{\partial P}{\partial k}(6, 2) \approx 2.1935$ and $\frac{\partial P}{\partial l}(6, 2) \approx 19.7411$. Since k is measured in millions of dollars and l is measured in thousands of workers, we have $k'(t) = 0.4$ and $l'(t) = -0.02$. From the chain rule, we now have

$$g'(t) = \frac{\partial P}{\partial k}k'(t) + \frac{\partial P}{\partial l}l'(t)$$

$$\approx 2.1935(0.4) + 19.7411(-0.02) = 0.48258.$$

We can easily extend Theorem 5.1 to the case of a function $f(x, y)$, where x and y are both functions of the two independent variables s and t, $x = x(s, t)$ and $y = y(s, t)$. Notice that if we differentiate with respect to s, we treat t as a constant. Applying Theorem 5.1 (while holding t fixed), we have

$$\frac{\partial}{\partial s}[f(x, y)] = \frac{\partial f}{\partial x}\frac{\partial x}{\partial s} + \frac{\partial f}{\partial y}\frac{\partial y}{\partial s}.$$

Similarly, we can find a chain rule for $\frac{\partial}{\partial t}[f(x, y)]$. This gives us the following more general form of the chain rule.

Theorem 5.2 (Chain Rule)

Suppose that $z = f(x, y)$, where f is a differentiable function of x and y and where $x = x(s, t)$ and $y = y(s, t)$ both have first-order partial derivatives. Then we have the chain rules:

$$\frac{\partial z}{\partial s} = \frac{\partial z}{\partial x}\frac{\partial x}{\partial s} + \frac{\partial z}{\partial y}\frac{\partial y}{\partial s}$$

and

$$\frac{\partial z}{\partial t} = \frac{\partial z}{\partial x}\frac{\partial x}{\partial t} + \frac{\partial z}{\partial y}\frac{\partial y}{\partial t}.$$

Observe that the tree diagram shown in the margin serves as a convenient reminder of the chain rules indicated in Theorem 5.2, again by summing the products of the indicated partial derivatives along each path from z to s or t, respectively.

The chain rule is easily extended to functions of three or more variables. You will explore this in the exercises.

Example 5.3 Using the Chain Rule

Suppose that $f(x, y) = e^{xy}$, $x(u, v) = 3u \sin v$ and $y(u, v) = 4v^2 u$. For $g(u, v) = f(x(u, v), y(u, v))$, find the partial derivatives $\frac{\partial g}{\partial u}$ and $\frac{\partial g}{\partial v}$.

Solution We first compute the partial derivatives $\frac{\partial f}{\partial x} = ye^{xy}$, $\frac{\partial f}{\partial y} = xe^{xy}$, $\frac{\partial x}{\partial u} = 3 \sin v$ and $\frac{\partial y}{\partial u} = 4v^2$. The chain rule (Theorem 5.2) gives us

$$\frac{\partial g}{\partial u} = \frac{\partial f}{\partial x}\frac{\partial x}{\partial u} + \frac{\partial f}{\partial y}\frac{\partial y}{\partial u} = ye^{xy}(3 \sin v) + xe^{xy}(4v^2).$$

Substituting for x and y, we get

$$\frac{\partial g}{\partial u} = 4v^2 u e^{12u^2 v^2 \sin v}(3 \sin v) + 3u \sin v e^{12u^2 v^2 \sin v}(4v^2).$$

For the partial derivative of g with respect to v, we compute $\frac{\partial x}{\partial v} = 3u \cos v$ and $\frac{\partial y}{\partial v} = 8vu$. Here, the chain rule gives us

$$\frac{\partial g}{\partial v} = ye^{xy}(3u \cos v) + xe^{xy}(8vu).$$

Substituting for x and y, we have

$$\frac{\partial g}{\partial v} = 4v^2 u e^{12u^2 v^2 \sin v}(3u \cos v) + 3u \sin v e^{12u^2 v^2 \sin v}(8vu).$$

Once again, it is often simpler to first substitute in the expressions for x and y. We leave it as an exercise to show that you get the same derivatives either way. On the other hand, there are plenty of times where the chain rules seen in Theorems 5.1 and 5.2 are indispensable. You will see some of these in the exercises, while we present several important uses next.

Example 5.4   Converting from Rectangular to Polar Coordinates

For a differentiable function $f(x, y)$ with $x = r \cos\theta$ and $y = r \sin\theta$, show that $f_r = f_x \cos\theta + f_y \sin\theta$ and $f_{rr} = f_{xx} \cos^2\theta + 2f_{xy} \cos\theta \sin\theta + f_{yy} \sin^2\theta$.

Solution   First, notice that $\dfrac{\partial x}{\partial r} = \cos\theta$ and $\dfrac{\partial y}{\partial r} = \sin\theta$. From Theorem 5.2, we now have

$$f_r = \frac{\partial f}{\partial r} = \frac{\partial f}{\partial x}\frac{\partial x}{\partial r} + \frac{\partial f}{\partial y}\frac{\partial y}{\partial r} = f_x \cos\theta + f_y \sin\theta.$$

Be very careful when computing the second partial derivative. Using the expression we have already found for f_r and Theorem 5.2, we have

$$
\begin{aligned}
f_{rr} &= \frac{\partial(f_r)}{\partial r} = \frac{\partial}{\partial r}(f_x \cos\theta + f_y \sin\theta) = \frac{\partial}{\partial r}(f_x)\cos\theta + \frac{\partial}{\partial r}(f_y)\sin\theta \\
&= \left[\frac{\partial}{\partial x}(f_x)\frac{\partial x}{\partial r} + \frac{\partial}{\partial y}(f_x)\frac{\partial y}{\partial r}\right]\cos\theta + \left[\frac{\partial}{\partial x}(f_y)\frac{\partial x}{\partial r} + \frac{\partial}{\partial y}(f_y)\frac{\partial y}{\partial r}\right]\sin\theta \\
&= (f_{xx}\cos\theta + f_{xy}\sin\theta)\cos\theta + (f_{yx}\cos\theta + f_{yy}\sin\theta)\sin\theta \\
&= f_{xx}\cos^2\theta + 2f_{xy}\cos\theta\sin\theta + f_{yy}\sin^2\theta,
\end{aligned}
$$

as desired.

Implicit Differentiation

Suppose that the equation $F(x, y) = 0$ defines y implicitly as a function of x, say $y = f(x)$. In section 2.8, we saw how to calculate $\dfrac{dy}{dx}$ in such a case. We can use the chain rule for functions of several variables to obtain an alternative method for calculating this. Moreover, this will provide us with new insights into when this can be done and, more important yet, this will generalize to functions of several variables defined implicitly by an equation.

We let $z = F(x, y)$, where $x = t$ and $y = f(t)$. From Theorem 5.1, we have

$$\frac{dz}{dt} = F_x \frac{dx}{dt} + F_y \frac{dy}{dt}.$$

But, since $z = F(x, y) = 0$, we have $\dfrac{dz}{dt} = 0$, too. Further, since $x = t$, we have $\dfrac{dx}{dt} = 1$ and $\dfrac{dy}{dt} = \dfrac{dy}{dx}$. This leaves us with

$$0 = F_x + F_y \frac{dy}{dx}.$$

Notice that we can solve this for $\dfrac{dy}{dx}$, provided $F_y \neq 0$. In this case, we have

$$\frac{dy}{dx} = -\frac{F_x}{F_y}.$$

Recognize that we already know how to calculate $\dfrac{dy}{dx}$ implicitly, so this doesn't appear to give us anything new. However, it turns out that the **Implicit Function Theorem** (proved in a course in advanced calculus) says that if F_x and F_y are continuous on an open disk containing the point (a, b) where $F(a, b) = 0$ and $F_y(a, b) \neq 0$, then the equation $F(x, y) = 0$ implicitly defines y as a function of x nearby the point (a, b).

More significantly, we can extend this notion to functions of several variables defined implicitly, as follows. Suppose that the equation $F(x, y, z) = 0$ implicitly defines a function $z = f(x, y)$, where f is differentiable. Then, we can find the partial derivatives f_x and f_y using the chain rule, as follows. We first let $w = F(x, y, z)$. From the chain rule, we have

$$\frac{\partial w}{\partial x} = F_x \frac{\partial x}{\partial x} + F_y \frac{\partial y}{\partial x} + F_z \frac{\partial z}{\partial x}.$$

Notice that since $w = F(x, y, z) = 0$, $\dfrac{\partial w}{\partial x} = 0$. Also, $\dfrac{\partial x}{\partial x} = 1$ and $\dfrac{\partial y}{\partial x} = 0$, since x and y are independent variables. This gives us

$$0 = F_x + F_z \frac{\partial z}{\partial x}.$$

We can solve this for $\dfrac{\partial z}{\partial x}$, as long as $F_z \neq 0$, to obtain

$$\frac{\partial z}{\partial x} = -\frac{F_x}{F_z}. \tag{5.2}$$

Likewise, differentiating w with respect to y leads us to

$$\frac{\partial z}{\partial y} = -\frac{F_y}{F_z}, \tag{5.3}$$

again, as long as $F_z \neq 0$. Much as in the two variables case, the Implicit Function Theorem for functions of three variables says that if F_x, F_y and F_z are continuous inside a sphere containing the point (a, b, c), where $F(a, b, c) = 0$ and $F_z(a, b, c) \neq 0$, then the equation $F(x, y, z) = 0$ implicitly defines z as a function of x and y nearby the point (a, b, c).

Example 5.5 Finding Partial Derivatives Implicitly

Find $\dfrac{\partial z}{\partial x}$ and $\dfrac{\partial z}{\partial y}$, given that $F(x, y, z) = xy^2 + z^3 + \sin(xyz) = 0$.

Solution   First, note that using the usual chain rule, we have

$$F_x = y^2 + yz \cos(xyz),$$
$$F_y = 2xy + xz \cos(xyz)$$

and

$$F_z = 3z^2 + xy \cos(xyz).$$

From (5.2), we now have

$$\frac{\partial z}{\partial x} = -\frac{F_x}{F_z} = -\frac{y^2 + yz \cos(xyz)}{3z^2 + xy \cos(xyz)}.$$

Likewise, from (5.3), we have

$$\frac{\partial z}{\partial y} = -\frac{F_y}{F_z} = -\frac{2xy + xz \cos(xyz)}{3z^2 + xy \cos(xyz)}.$$

Notice that, much like implicit differentiation with two variables, implicit differentiation with three variables yields expressions for the derivatives that depend on all three variables.

EXERCISES 12.5

1. In example 5.1, we mentioned that direct substitution followed by differentiation was an option (see exercises 3 and 4 below) and is often preferable. Discuss the advantages and disadvantages of direct substitution versus the method of example 5.1.

2. In example 5.4, we treated z as a function of x and y. Explain how to modify our results from the Implicit Function Theorem for treating x as a function of y and z.

3. Repeat example 5.1 by first substituting $x = t^2 - 1$ and $y = \sin t$ and then computing $g'(t)$.

4. Repeat example 5.3 by first substituting $x = 3u \sin v$ and $y = 4v^2 u$ and then computing $\dfrac{\partial g}{\partial u}$ and $\dfrac{\partial g}{\partial v}$.

In exercises 5–8, use the chain rule to find the indicated derivative(s).

5. $g'(t)$ where $g(t) = f(x(t), y(t))$, $f(x, y) = x^2 y - \sin y$, $x(t) = \sqrt{t^2 + 1}$, $y(t) = e^t$

6. $g'(t)$ where $g(t) = f(x(t), y(t))$, $f(x, y) = \sqrt{x^2 + y^2}$, $x(t) = \sin t$, $y(t) = t^2 + 2$

7. $\dfrac{\partial g}{\partial u}$ and $\dfrac{\partial g}{\partial v}$ where $g(u, v) = f(x(u, v), y(u, v))$, $f(x, y) = 4x^2 y^3$, $x(u, v) = u^3 - v \sin u$, $y(u, v) = 4u^2$

8. $\dfrac{\partial g}{\partial u}$ and $\dfrac{\partial g}{\partial v}$ where $g(u, v) = f(x(u, v), y(u, v))$, $f(x, y) = xy^3 - 4x^2$, $x(u, v) = e^{u^2}$, $y(u, v) = \sqrt{v^2 + 1} \sin u$

In exercises 9–12, state the chain rule for the general composite function.

9. $g(t) = f(x(t), y(t), z(t))$

10. $g(u, v) = f(x(u, v), y(u, v), z(u, v))$

11. $g(u, v, w) = f(x(u, v, w), y(u, v, w))$

12. $g(u, v, w) = f(x(u, v, w), y(u, v, w), z(u, v, w))$

13. In example 5.2, suppose that $l = 4$ and $k = 6$, the labor force is decreasing at the rate of 60 workers per year and capital is growing at the rate of \$100,000 per year. Determine the rate of change of production.

14. In example 5.2, suppose that $l = 3$ and $k = 4$, the labor force is increasing at the rate of 80 workers per year and capital is decreasing at the rate of \$200,000 per year. Determine the rate of change of production.

15. Suppose the production of a firm is modeled by $P(k, l) = 16k^{1/3} l^{2/3}$, with k and l defined as in example 5.2. Suppose that $l = 3$ and $k = 4$, the labor force is increasing at the rate of 80 workers per year and capital is decreasing at the rate of \$200,000 per year. Determine the rate of change of production.

16. Suppose the production of a firm is modeled by $P(k, l) = 16k^{1/3} l^{2/3}$, with k and l defined as in example 5.2. Suppose that $l = 2$ and $k = 5$, the labor force is decreasing at the rate of 40 workers per year and capital is decreasing at the rate of \$100,000 per year. Determine the rate of change of production.

17. For a business product, income is the product of the quantity sold and the price, which we can write as $I = qp$. If the quantity sold increases at a rate of 5% and the price increases at a rate of 3%, show that income increases at a rate of 8%.

18. Assume that $I = qp$ as in exercise 17. If the quantity sold decreases at a rate of 3% and price increases at a rate of 5%, determine the rate of increase or decrease in income.

In exercises 19–22, use the chain rule twice to find the indicated derivative.

19. $g(t) = f(x(t), y(t))$, find $g''(t)$

20. $g(t) = f(x(t), y(t), z(t))$, find $g''(t)$

21. $g(u, v) = f(x(u, v), y(u, v))$, find $\dfrac{\partial^2 g}{\partial u^2}$

22. $g(u, v) = f(x(u, v), y(u, v))$, find $\dfrac{\partial^2 g}{\partial u \partial v}$

In exercises 23–26, use implicit differentiation to find $\dfrac{\partial z}{\partial x}$ and $\dfrac{\partial z}{\partial y}$.

23. $3x^2 z + 2z^3 - 3yz = 0$ 24. $xyz - 4y^2 z^2 + \cos xy = 0$

25. $3e^{xyz} - 4xz^2 + x\cos y = 2$

26. $3yz^2 - e^{4x}\cos 4z - 3y^2 = 4$

27. For a differentiable function $f(x, y)$ with $x = r\cos\theta$ and $y = r\sin\theta$, show that $f_\theta = -f_x r \sin\theta + f_y r \cos\theta$.

28. For a differentiable function $f(x, y)$ with $x = r\cos\theta$ and $y = r\sin\theta$, show that $f_{\theta\theta} = f_{xx}r^2\sin^2\theta - 2f_{xy}r^2\cos\theta\sin\theta + f_{yy}r^2\cos^2\theta - f_x r\cos\theta - f_y r\sin\theta$.

29. For a differentiable function $f(x, y)$ with $x = r\cos\theta$ and $y = r\sin\theta$, use the results of exercises 27 and 28 and example 5.4 to show that $f_{xx} + f_{yy} = f_{rr} + \frac{1}{r}f_r + \frac{1}{r^2}f_{\theta\theta}$. This expression is called the **Laplacian** of f.

30. Given that $r = \sqrt{x^2 + y^2}$, show that $\dfrac{\partial r}{\partial x} = \dfrac{x}{\sqrt{x^2 + y^2}} = \dfrac{x}{r} = \cos\theta$. Starting from $r = \dfrac{x}{\cos\theta}$, does it follow that $\dfrac{\partial r}{\partial x} = \dfrac{1}{\cos\theta}$? Explain why it's not possible for both calculations to be correct. Find all mistakes.

31. A baseball player who has h hits in b at bats has a batting average of $a = \dfrac{h}{b}$. For example, 100 hits in 400 at bats would be an average of 0.250. It is traditional to carry three decimal places and to describe this average as being "250 points." To use the chain rule to estimate the change in batting average after a player gets a hit, assume that h and b are functions of time and that getting a hit means $h' = b' = 1$. Show that $a' = \dfrac{b - h}{b^2}$. Early in a season, a typical batter might have 50 hits in 200 at bats. Show that getting a hit will increase batting average by about 4 points. Find the approximate increase in batting average later in the season for a player with 100 hits in 400 at bats. In general, if b and h are both doubled, how does a' change?

32. For the baseball players of exercise 31, approximate the number of points that the batting average will decrease by making an out.

33. We have previously done calculations of the amount of work done by some force. Recall that if a scalar force $F(x)$ is applied as x increases from $x = a$ to $x = b$, then the work done equals $W = \int_a^b F(x)\,dx$. If the position x is a differentiable function of time, then we can write $W = \int_0^T F(x(t))x'(t)\,dt$, where $x(0) = a$ and $x(T) = b$. **Power** is defined as the time derivative of work. Work is sometimes measured in foot-pounds, so power could be measured in foot-pounds per second (ft-lb/s). One horsepower is equal to 550 ft-lb/s. Show that if force and velocity are constant, then power is the product of force and velocity. Determine how many pounds of force are required to maintain 400 hp at 80 mph. For a variable force and velocity, use the chain rule to compute power.

34. Engineers and physicists (and therefore mathematicians) spend countless hours studying the properties of forced oscillators. Two physical situations that are well modeled by the same mathematical equations are a spring oscillating due to some force and a basic electrical circuit with a voltage source. A general solution of a forced oscillator looks like $u(t) = g(t) - \int_0^t g(u)e^{-(t-u)/2}\left[\cos\frac{\sqrt{3}}{2}(t - u) + \frac{2}{3}\sin\frac{\sqrt{3}}{2}(t - u)\right]du$. If $g(0) = 1$ and $g'(0) = 2$, compute $u(0)$ and $u'(0)$.

12.6   THE GRADIENT AND DIRECTIONAL DERIVATIVES

Suppose that you are hiking in rugged terrain. You can think of your altitude at the point given by longitude x and latitude y as defining a function $f(x, y)$. This is obviously not a function for which you would be likely to have a handy formula, but you can learn more about this function than you might expect. Observe that if you face due east, the slope of the terrain (which you could measure) is given by the partial derivative $\dfrac{\partial f}{\partial x}(x, y)$. Similarly, facing due north, the slope of the terrain is given by $\dfrac{\partial f}{\partial y}(x, y)$. However, in terms of $f(x, y)$, how would you compute the slope in some other direction, say north-by-northwest? In this section, we develop the notion of **directional derivative,** which will answer this question.

Suppose we want to find the instantaneous rate of change of $f(x, y)$ at the point $P(a, b)$ and in the direction given by the unit vector $\mathbf{u} = \langle u_1, u_2 \rangle$. Let $Q(x, y)$ be any point on the line through $P(a, b)$ in the direction of $\mathbf{u}$. Notice that the vector $\overrightarrow{PQ}$ is then parallel

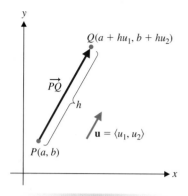

Figure 12.33

The vector $\overrightarrow{PQ}$.

to **u**. We know that two vectors are parallel if and only if one is a scalar multiple of the other, in which case, $\overrightarrow{PQ} = h\mathbf{u}$, for some scalar h. This says that

$$\overrightarrow{PQ} = \langle x - a, y - b \rangle = h\mathbf{u} = h\langle u_1, u_2 \rangle = \langle hu_1, hu_2 \rangle.$$

Remember that two vectors are equal only when all of their components are the same. We must then have

$$x = a + hu_1 \quad \text{and} \quad y = b + hu_2,$$

so that the point Q is described by $(a + hu_1, b + hu_2)$, as indicated in Figure 12.33. Notice that we can write the average rate of change of $z = f(x, y)$ along the line from P to Q as

$$\Delta z = \frac{f(a + hu_1, b + hu_2) - f(a, b)}{h}.$$

The instantaneous rate of change of $f(x, y)$ at the point $P(a, b)$ and in the direction of the unit vector **u** is then found by taking the limit as $h \to 0$. We give this limit a special name in the following definition.

Definition 6.1

The **directional derivative of** $f(x, y)$ at the point (a, b) and in the direction of the unit vector $\mathbf{u} = \langle u_1, u_2 \rangle$ is given by

$$D_\mathbf{u} f(a, b) = \lim_{h \to 0} \frac{f(a + hu_1, b + hu_2) - f(a, b)}{h},$$

provided the limit exists.

Notice that this limit resembles the definition of partial derivative, except that in this case, both variables may change. Further, you should observe that the directional derivative in the direction of the positive x-axis (i.e., in the direction of the unit vector $\mathbf{u} = \langle 1, 0 \rangle$) is

$$D_\mathbf{u} f(a, b) = \lim_{h \to 0} \frac{f(a + h, b) - f(a, b)}{h},$$

which you should recognize as the partial derivative $\dfrac{\partial f}{\partial x}$. Likewise, the directional derivative in the direction of the positive y-axis (i.e., in the direction of the unit vector $\mathbf{u} = \langle 0, 1 \rangle$) is $\dfrac{\partial f}{\partial y}$. In fact, it turns out that any directional derivative can be calculated simply, in terms of the first partial derivatives, as we see in the following theorem.

Theorem 6.1

Suppose that f is differentiable at (a, b) and $\mathbf{u} = \langle u_1, u_2 \rangle$ is any unit vector. Then, we can write

$$D_\mathbf{u} f(a, b) = f_x(a, b)u_1 + f_y(a, b)u_2.$$

Proof

Let $g(h) = f(a + hu_1, b + hu_2)$. Then, $g(0) = f(a, b)$ and so, from Definition 6.1, we have

$$D_\mathbf{u} f(a, b) = \lim_{h \to 0} \frac{f(a + hu_1, b + hu_2) - f(a, b)}{h} = \lim_{h \to 0} \frac{g(h) - g(0)}{h} = g'(0).$$

If we define $x = a + hu_1$ and $y = b + hu_2$, we have $g(h) = f(x, y)$. From the chain rule (Theorem 5.1), we have

$$g'(h) = \frac{\partial f}{\partial x}\frac{dx}{dh} + \frac{\partial f}{\partial y}\frac{dy}{dh} = \frac{\partial f}{\partial x}u_1 + \frac{\partial f}{\partial y}u_2.$$

Finally, taking $h = 0$ gives us

$$D_{\mathbf{u}}f(a, b) = g'(0) = \frac{\partial f}{\partial x}(a, b)u_1 + \frac{\partial f}{\partial y}(a, b)u_2,$$

as desired.

Example 6.1 Computing Directional Derivatives

For $f(x, y) = x^2 y - 4y^3$, compute $D_{\mathbf{u}}f(2, 1)$ for the directions (a) $\mathbf{u} = \langle \frac{\sqrt{3}}{2}, \frac{1}{2} \rangle$ and (b) $\mathbf{u}$ in the direction from $(2, 1)$ to $(4, 0)$.

Solution Regardless of the direction, we first need to compute the first partial derivatives $\frac{\partial f}{\partial x} = 2xy$ and $\frac{\partial f}{\partial y} = x^2 - 12y^2$. Then, $f_x(2, 1) = 4$ and $f_y(2, 1) = -8$. For (a), the unit vector is given as $\mathbf{u} = \langle \frac{\sqrt{3}}{2}, \frac{1}{2} \rangle$ and so, from Theorem 6.1 we have

$$D_{\mathbf{u}}f(2, 1) = f_x(2, 1)u_1 + f_y(2, 1)u_2 = 4\frac{\sqrt{3}}{2} - 8\left(\frac{1}{2}\right) = 2\sqrt{3} - 4 \approx -0.5.$$

In particular, notice that this says that the function is decreasing in this direction.

For (b), we must first find the unit vector $\mathbf{u}$ in the indicated direction. Observe that the vector from $(2, 1)$ to $(4, 0)$ corresponds to the position vector $\langle 2, -1 \rangle$ and so, the unit vector in that direction is $\mathbf{u} = \langle \frac{2}{\sqrt{5}}, -\frac{1}{\sqrt{5}} \rangle$. We then have from Theorem 6.1 that

$$D_{\mathbf{u}}f(2, 1) = f_x(2, 1)u_1 + f_y(2, 1)u_2 = 4\frac{2}{\sqrt{5}} - 8\left(-\frac{1}{\sqrt{5}}\right) = \frac{16}{\sqrt{5}}.$$

Notice that this says that the function is increasing rapidly in this direction.

For convenience, we define the **gradient** of a function to be the vector-valued function whose components are the first-order partial derivatives of f. We denote the gradient of a function f by **grad** f or ∇f (read "del f").

Definition 6.2

The **gradient** of $f(x, y)$ is the vector-valued function

$$\nabla f(x, y) = \left\langle \frac{\partial f}{\partial x}, \frac{\partial f}{\partial y} \right\rangle = \frac{\partial f}{\partial x}\mathbf{i} + \frac{\partial f}{\partial y}\mathbf{j},$$

provided both partial derivatives exist.

Observe that, using the gradient, we can write a directional derivative as the dot product of the gradient and the unit vector in the direction of interest, as follows. For any unit vector $\mathbf{u} = \langle u_1, u_2 \rangle$,

$$
\begin{aligned}
D_{\mathbf{u}} f(x, y) &= f_x(x, y)u_1 + f_y(x, y)u_2 \\
&= \langle f_x(x, y), f_y(x, y) \rangle \cdot \langle u_1, u_2 \rangle \\
&= \nabla f(x, y) \cdot \mathbf{u}.
\end{aligned}
$$

We state this result in the following theorem.

Theorem 6.2

If f is a differentiable function of x and y and $\mathbf{u}$ is any unit vector, then

$$
D_{\mathbf{u}} f(x, y) = \nabla f(x, y) \cdot \mathbf{u}.
$$

Theorem 6.2 makes it easy to compute directional derivatives. Further, writing directional derivatives as a dot product has many important consequences, one of which we see in the following example.

Example 6.2   Finding Directional Derivatives

For $f(x, y) = x^2 + y^2$, find $D_{\mathbf{u}} f(1, -1)$ for (a) $\mathbf{u}$ in the direction of $\mathbf{v} = \langle -3, 4 \rangle$ and (b) $\mathbf{u}$ in the direction of $\mathbf{v} = \langle 3, -4 \rangle$.

Solution   First, note that

$$
\nabla f = \left\langle \frac{\partial f}{\partial x}, \frac{\partial f}{\partial y} \right\rangle = \langle 2x, 2y \rangle.
$$

At the point $(1, -1)$, we have $\nabla f(1, -1) = \langle 2, -2 \rangle$. For (a), a unit vector in the same direction as $\mathbf{v}$ is $\mathbf{u} = \langle -\frac{3}{5}, \frac{4}{5} \rangle$. The directional derivative in this direction is then

$$
D_{\mathbf{u}} f(1, -1) = \langle 2, -2 \rangle \cdot \left\langle -\frac{3}{5}, \frac{4}{5} \right\rangle = \frac{-6 - 8}{5} = -\frac{14}{5}.
$$

For (b), the unit vector is $\mathbf{u} = \langle \frac{3}{5}, -\frac{4}{5} \rangle$ and so, the directional derivative in this direction is

$$
D_{\mathbf{u}} f(1, -1) = \langle 2, -2 \rangle \cdot \left\langle \frac{3}{5}, -\frac{4}{5} \right\rangle = \frac{6 + 8}{5} = \frac{14}{5}.
$$

A graphical interpretation of the directional derivatives in example 6.2 is given on the following page. Suppose we intersect the surface $z = f(x, y)$ with a plane passing through the point $(1, -1, 2)$, which is perpendicular to the xy-plane and parallel to the vector $\mathbf{u}$ (see Figure 12.34a). Notice that the intersection is a curve in two dimensions. Sketch this curve on a new set of coordinate axes, chosen so that the new origin corresponds to the point $(1, -1, 2)$, the new vertical axis is in the z-direction and the new positive horizontal axis points in the direction of the vector $\mathbf{u}$. In Figure 12.34b, we show the case for $\mathbf{u} = \langle -\frac{3}{5}, \frac{4}{5} \rangle$ and in Figure 12.34c, we show the case for $\mathbf{u} = \langle \frac{3}{5}, -\frac{4}{5} \rangle$. In each case, the directional derivative gives the slope of the curve at the origin (in the new coordinate system). Notice that the direction vectors in example 6.2 parts (a) and (b) differ only by sign and the resulting curves in Figures 12.34b and 12.34c are exact mirror images of each other.

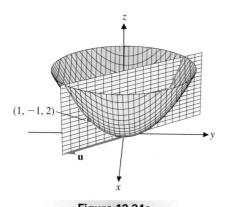

Figure 12.34a

Intersection of surface with plane.

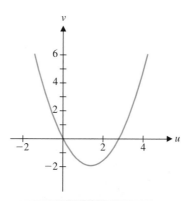

Figure 12.34b

$\mathbf{u} = \left\langle -\frac{3}{5}, \frac{4}{5} \right\rangle.$

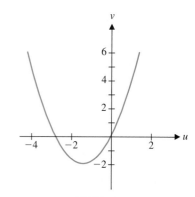

Figure 12.34c

$\mathbf{u} = \left\langle \frac{3}{5}, -\frac{4}{5} \right\rangle.$

Another way of viewing the directional derivative graphically is with level curves.

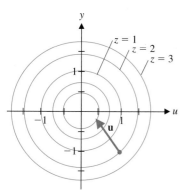

Figure 12.35

Contour plot of $z = x^2 + y^2$.

Example 6.3   *Directional Derivatives and Level Curves*

Use a contour plot of $z = x^2 + y^2$ to estimate $D_{\mathbf{u}} f(1, -1)$ for $\mathbf{u} = \left\langle -\frac{3}{5}, \frac{4}{5} \right\rangle$.

Solution   A contour plot of $z = x^2 + y^2$ is shown in Figure 12.35 with the direction vector $\mathbf{u} = \left\langle -\frac{3}{5}, \frac{4}{5} \right\rangle$ sketched in with its initial point located at the point $(1, -1)$. The level curves shown correspond to $z = 0.2, 0.5, 1, 2$ and 3. From the graph, you can approximate the directional derivative by estimating $\dfrac{\Delta z}{\Delta u}$, where Δu is the distance traveled along the unit vector $\mathbf{u}$. For the unit vector shown, $\Delta u = 1$. Further, the vector appears to extend from the $z = 2$ level curve to the $z = 0.2$ level curve. In this case, $\Delta z = 0.2 - 2 = -1.8$ and our estimate of the directional derivative is $\dfrac{\Delta z}{\Delta u} = -1.8$. Compared to the actual directional derivative of $-\frac{14}{5} = -2.8$ (found in example 6.2), this is not very accurate. A better estimate could be obtained with a smaller Δu. For example, to get from the $z = 2$ level curve to the $z = 1$ level curve, it appears that we travel along about 40% of the unit vector. Then $\dfrac{\Delta z}{\Delta u} \approx \dfrac{1 - 2}{0.4} = -2.5$. You could continue this process by drawing more level curves, corresponding to values of z closer to $z = 2$.

Keep in mind that the directional derivative gives the rate of change of a function in a given direction. In this case, it's reasonable to ask in what direction a given function has its maximum or minimum rate of increase. In order to answer such questions, you must first recall from Theorem 3.2 in Chapter 10 that for any two vectors $\mathbf{a}$ and $\mathbf{b}$, we have $\mathbf{a} \cdot \mathbf{b} = \|\mathbf{a}\| \|\mathbf{b}\| \cos \theta$, where θ is the angle between the vectors $\mathbf{a}$ and $\mathbf{b}$. Applying this to the form of the directional derivative given in Theorem 6.2, we have

$$D_{\mathbf{u}} f(a, b) = \nabla f(a, b) \cdot \mathbf{u}$$
$$= \|\nabla f(a, b)\| \|\mathbf{u}\| \cos \theta = \|\nabla f(a, b)\| \cos \theta,$$

where θ is the angle between the gradient vector at (a, b) and the direction vector $\mathbf{u}$.

Notice now that the maximum value of $\|\nabla f(a, b)\| \cos \theta$ occurs when $\theta = 0$, so that $\cos \theta = 1$. The directional derivative is then $\|\nabla f(a, b)\|$. Further, observe that the angle $\theta = 0$ when $\nabla f(a, b)$ and $\mathbf{u}$ are in the **same** direction, so that $\mathbf{u} = \dfrac{\nabla f(a, b)}{\|\nabla f(a, b)\|}$. Similarly, the minimum value of the directional derivative occurs when $\theta = \pi$, so that $\cos \theta = -1$.

In this case, $\nabla f(a, b)$ and $\mathbf{u}$ have **opposite** directions, so that $\mathbf{u} = -\dfrac{\nabla f(a, b)}{\|\nabla f(a, b)\|}$. Finally, observe that when $\theta = \frac{\pi}{2}$, $\mathbf{u}$ is perpendicular to $\nabla f(a, b)$ and the directional derivative in this direction is zero. Since the level curves are curves in the xy-plane on which f is constant, notice that a zero directional derivative at a point indicates that $\mathbf{u}$ is tangent to a level curve. We summarize these observations in the following theorem.

Theorem 6.3

Suppose that f is a differentiable function of x and y at the point (a, b). Then
  (i)   the maximum rate of change of f at (a, b) is $\|\nabla f(a, b)\|$ and occurs in the direction of the gradient, $\mathbf{u} = \dfrac{\nabla f(a, b)}{\|\nabla f(a, b)\|}$;
  (ii) the minimum rate of change of f at (a, b) is $-\|\nabla f(a, b)\|$ and occurs in the direction opposite the gradient, $\mathbf{u} = -\dfrac{\nabla f(a, b)}{\|\nabla f(a, b)\|}$ and
  (iii) the gradient $\nabla f(a, b)$ is orthogonal to the level curve $f(x, y) = c$ at the point (a, b), where $c = f(a, b)$.

In using Theorem 6.3, remember that the directional derivative corresponds to the rate of change of the function $f(x, y)$ in the given direction.

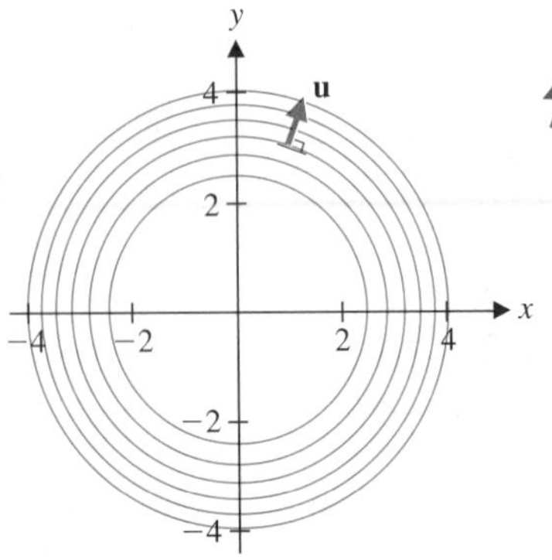

Example 6.4   Finding Maximum and Minimum Rates of Change

Find the maximum and minimum rates of change of the function $f(x, y) = x^2 + y^2$ at the point $(1, 3)$.

Solution   We first compute the gradient $\nabla f = \langle 2x, 2y \rangle$ and evaluate it at the point $(1, 3)$: $\nabla f(1, 3) = \langle 2, 6 \rangle$. From Theorem 6.3, the maximum rate of change of f at $(1, 3)$ is $\|\nabla f(1, 3)\| = \|\langle 2, 6 \rangle\| = \sqrt{40}$, and occurs in the direction

$$\mathbf{u} = \frac{\nabla f(1, 3)}{\|\nabla f(1, 3)\|} = \frac{\langle 2, 6 \rangle}{\sqrt{40}}.$$

Similarly, the minimum rate of change is $-\|\nabla f(1, 3)\| = -\|\langle 2, 6 \rangle\| = -\sqrt{40}$, which occurs in the direction

$$\mathbf{u} = -\frac{\nabla f(1, 3)}{\|\nabla f(1, 3)\|} = \frac{-\langle 2, 6 \rangle}{\sqrt{40}}.$$

Figure 12.36

Contour plot of $z = x^2 + y^2$.

Notice that the direction of maximum increase in example 6.3 points away from the origin, since the displacement vector from $(0, 0)$ to $(1, 3)$ is parallel to $\mathbf{u} = \langle 2, 6 \rangle / \sqrt{40}$. This should make sense given the familiar shape of the paraboloid. The contour plot of $f(x, y)$ shown in Figure 12.36 indicates that the gradient is perpendicular to the level curves. We expand on this idea in the following example.

Example 6.5   Finding the Direction of Steepest Ascent

The contour plot of $f(x, y) = 3x - x^3 - 3xy^2$ shown in Figure 12.37 indicates several level curves near a relative maximum at $(1, 0)$. Find the direction of maximum increase from the point $A(0.6, -0.7)$ and sketch in the path of steepest ascent.

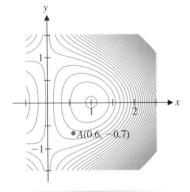

Figure 12.37

Contour plot of $z = 3x - x^3 - 3xy^2$.

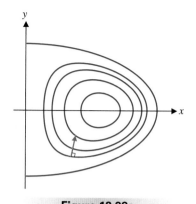

Figure 12.38a

Direction of steepest ascent at
$(0.6, -0.7)$.

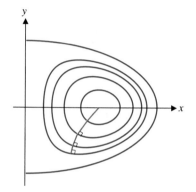

Figure 12.38b

Path of steepest ascent.

Solution   From Theorem 6.3, the direction of maximum increase at $(0.6, -0.7)$ is given by the gradient $\nabla f(0.6, -0.7)$. We have $\nabla f = \langle 3 - 3x^2 - 3y^2, -6xy \rangle$ and so, $\nabla f(0.6, -0.7) = \langle 0.45, 2.52 \rangle$. The unit vector in this direction is then $\mathbf{u} = \langle 0.176, 0.984 \rangle$. A vector in this direction (not drawn to scale) at the point $(0.6, -0.7)$ is shown in Figure 12.38a. Notice that this vector **does not** point to the maximum at $(1, 0)$. (By analogy, on a mountain, the steepest path from a given point will not always point to the actual peak.) The **path of steepest ascent** is a curve that remains perpendicular to each level curve it passes through. Notice that at the tip of the vector drawn in Figure 12.38a, the vector is no longer perpendicular to the level curve. Finding an equation for the path of steepest ascent is difficult. In Figure 12.38b, we sketch in a plausible steepest curve.

■

Most of the results of this section extend easily to functions of any number of variables.

Definition 6.3

The **directional derivative of** $f(x, y, z)$ at the point (a, b, c) in the direction of the unit vector $\mathbf{u} = \langle u_1, u_2, u_3 \rangle$ is given by

$$D_{\mathbf{u}} f(a, b, c) = \lim_{h \to 0} \frac{f(a + hu_1, b + hu_2, c + hu_3)}{h},$$

provided the limit exists.

The **gradient** of $f(x, y, z)$ is the vector-valued function

$$\nabla f(x, y, z) = \left\langle \frac{\partial f}{\partial x}, \frac{\partial f}{\partial y}, \frac{\partial f}{\partial z} \right\rangle = \frac{\partial f}{\partial x}\mathbf{i} + \frac{\partial f}{\partial y}\mathbf{j} + \frac{\partial f}{\partial z}\mathbf{k},$$

provided all the partial derivatives are defined.

As was the case for functions of two variables, the gradient gives us a simple representation of directional derivatives in three dimensions.

Theorem 6.4

If f is a differentiable function of x, y and z and $\mathbf{u}$ is any unit vector, then

$$D_{\mathbf{u}} f(x, y, z) = \nabla f(x, y, z) \cdot \mathbf{u}. \qquad (6.1)$$

As in two dimensions, we have that

$$D_{\mathbf{u}} f(x, y, z) = \nabla f(x, y, z) \cdot \mathbf{u} = \|\nabla f(x, y, z)\| \|\mathbf{u}\| \cos\theta$$
$$= \|\nabla f(x, y, z)\| \cos\theta,$$

where θ is the angle between the vectors $\nabla f(x, y, z)$ and $\mathbf{u}$. For precisely the same reasons as in two dimensions, you can now see that the direction of maximum increase at any given point is given by the gradient at that point.

Example 6.6   Finding the Direction of Maximum Increase

If the temperature at point (x, y, z) is given by $T(x, y, z) = 85 + (1 - z/100)e^{-(x^2+y^2)}$, find the direction from the point $(2, 0, 99)$ in which the temperature increases most rapidly.

Solution   We first compute the gradient

$$\nabla f = \left\langle \frac{\partial f}{\partial x}, \frac{\partial f}{\partial y}, \frac{\partial f}{\partial z} \right\rangle$$

$$= \left\langle -2x \left(1 - \frac{z}{100}\right) e^{-(x^2+y^2)}, \; -2y \left(1 - \frac{z}{100}\right) e^{-(x^2+y^2)}, \; -\left(\frac{1}{100}\right) e^{-(x^2+y^2)} \right\rangle$$

and $\nabla f(2, 0, 99) = \left\langle -\frac{1}{25} e^{-4}, 0, -\frac{1}{100} e^{-4} \right\rangle$. To find a unit vector in this direction, you can simplify the algebra by canceling the common factor of e^{-4} (think about why this makes sense) and multiplying by 100. A unit vector in the direction of $\langle -4, 0, -1 \rangle$, and also in the direction of $\nabla f(2, 0, 99)$, is then $\dfrac{\langle -4, 0, -1 \rangle}{\sqrt{17}}$.

Recall that for any constant k, the equation $f(x, y, z) = k$ defines a level surface of the function $f(x, y, z)$. Now, suppose that **u** is any unit vector lying in the tangent plane to the level surface $f(x, y, z) = k$ at a point (a, b, c) on the level surface. Then, it follows that the rate of change of f in the direction of **u** at (a, b, c) [given by the directional derivative $D_{\mathbf{u}} f(a, b, c)$] is zero, since f is constant on a level surface. From (6.1), we now have that

$$0 = D_{\mathbf{u}} f(a, b, c) = \nabla f(a, b, c) \cdot \mathbf{u}.$$

This occurs only when the vectors $\nabla f(a, b, c)$ and **u** are orthogonal. Since **u** was taken to be any vector lying in the tangent plane, we now have that $\nabla f(a, b, c)$ is orthogonal to every vector lying in the tangent plane at the point (a, b, c). Observe that this says that $\nabla f(a, b, c)$ is a normal vector to the tangent plane to the surface $f(x, y, z) = k$ at the point (a, b, c). This proves the following theorem.

Theorem 6.5

Suppose that $f(x, y, z)$ has continuous partial derivatives at the point (a, b, c). Then, $\nabla f(a, b, c)$ is a normal vector to the tangent plane to the surface $f(x, y, z) = k$, at the point (a, b, c). Further, the equation of the tangent plane is

$$0 = f_x(a, b, c)(x - a) + f_y(a, b, c)(y - b) + f_z(a, b, c)(z - c).$$

We refer to the line through (a, b, c) in the direction of $\nabla f(a, b, c)$ as the **normal line** to the surface at the point (a, b, c). Observe that this has equations

$$x = a + f_x(a, b, c)t, \quad y = b + f_y(a, b, c)t, \quad z = c + f_z(a, b, c)t.$$

In the following example, we illustrate the use of the gradient at a point to find the tangent plane to a surface at that point.

Example 6.7   Using a Gradient to Find a Tangent Plane and Normal Line to a Surface

Find equations of the tangent plane and the normal line to $x^3 y - y^2 + z^2 = 7$ at the point $(1, 2, 3)$.

Solution   If we interpret the surface as a level surface of the function $f(x, y, z) = x^3 y - y^2 + z^2$, a normal vector to the tangent plane at the point $(1, 2, 3)$ is given by $\nabla f(1, 2, 3)$. We have $\nabla f = \langle 3x^2 y, x^3 - 2y, 2z \rangle$ and $\nabla f(1, 2, 3) = \langle 6, -3, 6 \rangle$. Given the normal vector $\langle 6, -3, 6 \rangle$ and point $(1, 2, 3)$, an equation of the tangent plane is

$$6(x - 1) - 3(y - 2) + 6(z - 3) = 0.$$

The normal line has equations

$$x = 1 + 6t, \quad y = 2 - 3t, \quad z = 3 + 6t.$$

Recall that in section 12.4, we found that a normal vector to the tangent plane to the surface $z = f(x, y)$ at the point $(a, b, f(a, b))$ is given by $\left\langle \dfrac{\partial f}{\partial x}(a, b), \dfrac{\partial f}{\partial y}(a, b), -1 \right\rangle$. Note that this is simply a special case of the gradient formula of Theorem 6.5, as follows. First, observe that we can rewrite the equation $z = f(x, y)$ as $f(x, y) - z = 0$. We can then think of this surface as a level surface of the function $g(x, y, z) = f(x, y) - z$, which at the point $(a, b, f(a, b))$ has normal vector

$$\nabla g(a, b, f(a, b)) = \left\langle \frac{\partial f}{\partial x}(a, b), \frac{\partial f}{\partial y}(a, b), -1 \right\rangle.$$

Just as it is important to constantly think of ordinary derivatives as slopes of tangent lines and as instantaneous rates of change, it is crucial to keep in mind at all times the interpretations of gradients. **Always** think of gradients as vector-valued functions whose values specify the direction of maximum increase of a function and whose values provide normal vectors (to the level curves in two dimensions and to the level surfaces in three dimensions).

Example 6.8   Using a Gradient to Find a Tangent Plane to a Surface

Find an equation of the tangent plane to $z = \sin(x + y)$ at the point $(\pi, \pi, 0)$.

Solution   We rewrite the equation of the surface as $g(x, y, z) = \sin(x + y) - z = 0$ and compute $\nabla g(x, y, z) = \langle \cos(x + y), \cos(x + y), -1 \rangle$. At the point $(\pi, \pi, 0)$, the normal to the surface is given by $\nabla g(\pi, \pi, 0) = \langle 1, 1, -1 \rangle$. An equation of the tangent plane is then

$$(x - \pi) + (y - \pi) - z = 0.$$

EXERCISES 12.6

1. Pick an area outside your classroom that has a small hill. Starting at the bottom of the hill, describe how to follow the gradient path to the top. In particular, describe how to determine the direction in which the gradient points at a given point on the hill. In general, should you be looking ahead or down at the ground? Should individual blades of grass count? What should you do if you encounter a wall?

2. Discuss whether the gradient path described in exercise 1 is guaranteed to get you to the top of the hill. Discuss whether the gradient path is the shortest path, the quickest path or the easiest path.

3. Use the sketch in Figure 12.34a to explain why the curves in Figures 12.34b and 12.34c are different.

4. Suppose the function $f(x, y)$ represents the altitude at various points on a ski slope. Explain in physical terms why the direction of maximum increase is $180°$ opposite the direction of maximum decrease, with the direction of zero change halfway in between. If $f(x, y)$ represents altitude on a rugged mountain instead of a ski slope, explain why the results (which are still true) are harder to visualize.

In exercises 5–8, find the gradient of the given function.

5. $f(x, y) = x^2 + 4xy^2 - y^5$ 6. $f(x, y) = x^3 e^{3y} - y^4$

7. $f(x, y) = xe^{xy^2} + \cos y^2$

8. $f(x, y) = e^{3y/x} - x^2 y^3$

In exercises 9–14, find the gradient of the given function at the indicated point.

9. $f(x, y) = 2e^{4x/y} - 2x$, $(2, -1)$

10. $f(x, y) = \sin 3xy + y^2$, $(\pi, 1)$

11. $f(x, y) = \sqrt{x^2 + y^2}$, $(4, -3)$

12. $f(x, y) = x^2\sqrt{y^2 + 1}$, $(2, 0)$

13. $f(x, y, z) = 3x^2y - z\cos x$, $(0, 2, -1)$

14. $f(x, y, z) = z^2e^{2x-y} - 4xz^2$, $(1, 2, 2)$

In exercises 15–30, compute the directional derivative of f at the given point in the direction of the indicated vector.

15. $f(x, y) = x^2y + 4y^2$, $(2, 1)$, $\mathbf{u} = \left\langle \frac{1}{2}, \frac{\sqrt{3}}{2} \right\rangle$

16. $f(x, y) = x^3y - 4y^2$, $(2, -2)$, $\mathbf{u} = \left\langle \frac{1}{2}, \frac{\sqrt{3}}{2} \right\rangle$

17. $f(x, y) = x^2y + 4y^2$, $(2, 1)$, $\mathbf{u} = \left\langle \frac{1}{2}, -\frac{\sqrt{3}}{2} \right\rangle$

18. $f(x, y) = x^3y - 4y^2$, $(2, -1)$, $\mathbf{u} = \left\langle \frac{1}{\sqrt{2}}, \frac{1}{\sqrt{2}} \right\rangle$

19. $f(x, y) = \sqrt{x^2 + y^2}$, $(3, -4)$, $\mathbf{u}$ in the direction of $\langle 3, -2 \rangle$

20. $f(x, y) = \sqrt{xy} - y^2$, $(1, 4)$, $\mathbf{u}$ in the direction of $\langle 1, 4 \rangle$

21. $f(x, y) = e^{4x^2 - y}$, $(1, 4)$, $\mathbf{u}$ in the direction of $\langle -2, -1 \rangle$

22. $f(x, y) = y^2e^{4x}$, $(0, -2)$, $\mathbf{u}$ in the direction of $\langle 3, -1 \rangle$

23. $f(x, y) = \cos(2x - y)$, $(\pi, 0)$, $\mathbf{u}$ in the direction from $(\pi, 0)$ to $(2\pi, \pi)$

24. $f(x, y) = x^2\sin 4y$, $\left(-2, \frac{\pi}{8}\right)$, $\mathbf{u}$ in the direction from $\left(-2, \frac{\pi}{8}\right)$ to $(0, 0)$

25. $f(x, y) = x^2 - 2xy + y^2$, $(-2, -1)$, $\mathbf{u}$ in the direction from $(-2, -1)$ to $(2, -3)$

26. $f(x, y) = y^2 + 2ye^{4x}$, $(0, -2)$, $\mathbf{u}$ in the direction from $(0, -2)$ to $(-4, 4)$

27. $f(x, y, z) = x^3yz^2 - 4xy$, $(1, -1, 2)$, $\mathbf{u}$ in the direction of $\langle 2, 0, -1 \rangle$

28. $f(x, y, z) = \sqrt{x^2 + y^2 + z^2}$, $(1, -4, 8)$, $\mathbf{u}$ in the direction of $\langle 1, 1, -2 \rangle$

29. $f(x, y, z) = e^{xy+z}$, $(1, -1, 1)$, $\mathbf{u}$ in the direction of $\langle 4, -2, 3 \rangle$

30. $f(x, y, z) = \cos xy + z$, $(0, -2, 4)$, $\mathbf{u}$ in the direction of $\langle 0, 3, -4 \rangle$

In exercises 31–40, find the directions of maximum and minimum change of f at the given point, and the values of the maximum and minimum rates of change.

31. $f(x, y) = x^2 - y^3$, $(2, 1)$

32. $f(x, y) = x^2 - y^3$, $(-1, -2)$

33. $f(x, y) = y^2e^{4x}$, $(0, -2)$

34. $f(x, y) = y^2e^{4x}$, $(3, -1)$

35. $f(x, y) = x\cos 3y$, $(2, 0)$

36. $f(x, y) = x\cos 3y$, $(-2, \pi)$

37. $f(x, y) = \sqrt{2x^2 - y}$, $(3, 2)$

38. $f(x, y) = \sqrt{x^2 + y^2}$, $(3, -4)$

39. $f(x, y, z) = 4x^2yz^3$, $(1, 2, 1)$

40. $f(x, y, z) = \sqrt{x^2 + y^2 + z^2}$, $(1, 2, -2)$

41. In exercises 38 and 40, compare the gradient direction to the position vector from the origin to the given point. Explain in terms of the graph of f why this relationship should hold.

42. Suppose that $g(x)$ is a differentiable function and $f(x, y) = g(x^2 + y^2)$. Show that $\nabla f(a, b)$ is parallel to $\langle a, b \rangle$. Explain this in graphical terms.

43. Graph $z = \sin(x + y)$. Compute $\nabla \sin(x + y)$ and explain why the gradient gives you the direction that the sine wave travels. In which direction would the sine wave travel for $z = \sin(2x - y)$?

44. Show that the vector $\langle 100, -100 \rangle$ is perpendicular to $\nabla \sin(x + y)$. Explain why the directional derivative of $\sin(x + y)$ in the direction of $\langle 100, -100 \rangle$ must be zero. Sketch a wireframe graph of $z = \sin(x + y)$ from the viewpoint $(100, -100, 0)$. Explain why you only see one trace. Find a viewpoint from which $z = \sin(2x - y)$ only shows one trace.

In exercises 45–48, find equations of the tangent plane and normal line to the surface at the given point.

45. $z = x^2 + y^3$ at $(1, -1, 0)$ 46. $z = \sqrt{x^2 + y^2}$ at $(3, -4, 5)$

47. $x^2 + y^2 + z^2 = 6$ at $(-1, 2, 1)$

48. $z^2 = x^2 - y^2$ at $(5, -3, -4)$

In exercises 49 and 50, find all points at which the tangent plane to the surface is parallel to the xy-plane. Discuss the graphical significance of each point.

49. $z = 2x^2 - 4xy + y^4$ 50. $z = \sin x \cos y$

In exercises 51–54, sketch in the path of steepest ascent from the indicated point.

51.
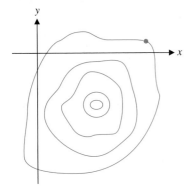

52.
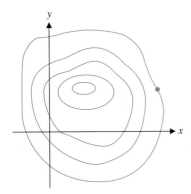

53.
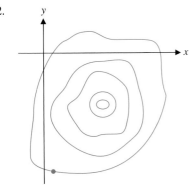

54.
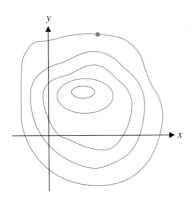

In exercises 55 and 56, use the contour plot to estimate $\nabla f(0,0)$.

55.
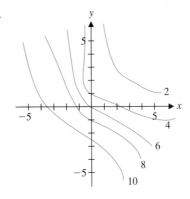

56.
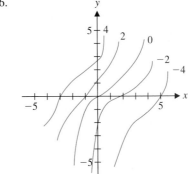

In exercises 57 and 58, use the table to estimate $\nabla f(0,0)$.

57.

y \ x	−0.2	−0.1	0	0.1	0.2
−0.4	2.1	2.5	2.8	3.1	3.4
−0.2	1.9	2.2	2.4	2.6	2.9
0	1.6	1.8	2.0	2.2	2.5
0.2	1.3	1.4	1.6	1.8	2.1
0.4	1.1	1.2	1.1	1.4	1.7

58.

y \ x	−0.4	−0.2	0	0.2	0.4
−0.6	2.4	2.1	1.8	1.3	1.0
−0.3	2.6	2.2	1.9	1.5	1.2
0	2.7	2.4	2.0	1.6	1.3
0.3	2.9	2.5	2.1	1.7	1.5
0.6	3.1	2.7	2.3	1.9	1.7

59. At a certain point on a mountain, a surveyor sights due east and measures a 10° drop-off, then sights due north and measures a 6° rise. Find the direction of steepest ascent and compute the degree rise in that direction.

60. At a certain point on a mountain, a surveyor sights due west and measures a 4° rise, then sights due north and measures a 3° rise. Find the direction of steepest ascent and compute the degree rise in that direction.

61. Suppose that the elevation on a hill is given by $f(x, y) = 200 - y^2 - 4x^2$. From the site at $(1, 2)$, in which direction will the rain run off?

62. For the mountain of exercise 61, if a level road is to be built at elevation 100, find the shape of the road.

63. If the temperature at the point (x, y, z) is given by $T(x, y, z) = 80 + 5e^{-z}(x^{-2} + y^{-1})$, find the direction from the point $(1, 4, 8)$ in which the temperature decreases most rapidly.

64. If the temperature at the point (x, y, z) is given by $T(x, y, z) = 80 + 5e^{-z}(x^{-2} + y^{-1})$, find the direction from the point $(1, 4, 8)$ in which the temperature increases most rapidly.

65. In example 4.6 of this chapter, we looked at a manufacturing process. Suppose that a gauge of 4 mm results from a gap of 4 mm, a speed of 10 m/s and a temperature of 900°. Further, suppose that an increase in gap of 0.05 mm increases the gauge by 0.04 mm, an increase in speed of 0.2 m/s increases the gauge by 0.06 mm and an increase in temperature of 10° decreases the gauge by 0.04 mm. Thinking of gauge as a function of gap, speed and temperature, find the direction of maximum increase of gauge.

66. In exercise 65, find the direction of maximum decrease in gauge.

67. The **Laplacian** of a function $f(x, y)$ is defined by $\nabla^2 f(x, y) = f_{xx}(x, y) + f_{yy}(x, y)$. Compute $\nabla^2 f(x, y)$ for $f(x, y) = x^3 - 2xy + y^2$.

68. For $f(x, y) = \begin{cases} \dfrac{xy^2}{x^2 + y^4}, & \text{if } (x, y) \neq (0, 0) \\ 0, & \text{if } (x, y) = (0, 0) \end{cases}$, show that the directional derivative $D_{\mathbf{u}} f(0, 0)$ exists for all directions $\mathbf{u}$, even though f is not continuous at $(0, 0)$.

69. The horizontal range of a baseball that has been hit depends on its launch angle and the rate of backspin on the ball. The following figure (reprinted from *Keep Your Eye on the Ball* by Watts and Bahill) shows level curves for the range as a function of angle and spin rate for an initial speed of 110 mph. Watts and Bahill suggest using the dashed line to find the best launch angle for a given spin rate. For example, start at $\omega = 2000$, move horizontally to the dashed line and then vertically down to $\theta = 30$. For a spin rate of 2000 rpm, the

greatest range is achieved with a launch angle of 30°. To understand why, note that the dashed line intersects level curves at points where the level curves have horizontal tangents. Start at a point where the dashed line intersects a level curve and explain why you can conclude from the graph that changing the angle would decrease the range. Therefore, the dashed line indicates optimal angles. As ω increases, does the optimal angle increase or decrease? Explain in physical terms why this makes sense. Explain why you know that the dashed line does not follow a gradient path, and explain what a gradient path would represent.

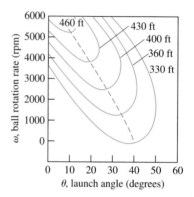

70. With the computer revolution of the 1990s came a new need to generate realistic-looking graphics. In this exercise, we look at one of the basic principles of three-dimensional graphics. We have often used wireframe graphs such as Figure A below to visualize surfaces in three dimensions. Certainly, the graphic in Figure A is crude, but even this sketch is quite informative to us, as we can clearly see a local maximum. By having the computer plot more points, as in Figure B, some of the rough edges get smoothed out. Still, there is something missing, isn't there? Almost everything we see in real life is shaded by a light source from above. This shading gives us very important clues about the three-dimensional structure of the surface. In Figure C, we have simply added some shading to Figure B. There is more work to be done in smoothing out Figure C, but for now we want to understand how the shading works. In particular, we'll

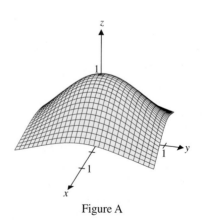

Figure A

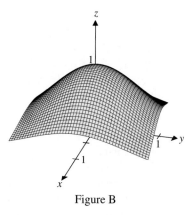

Figure B

Figure C

discuss a basic type of shading called **Lambert shading.** The idea is to shade a portion of the picture based on the size of the angle between the normal to the surface and the line to the light source. The larger the angle, the darker the portion of the picture should be. Explain why this works. For the surface $z = e^{-x^2-y^2}$ (shown in Figures A–C with $-1 \le x \le 1$ and $-1 \le y \le 1$) and

a light source at $(0, 0, 100)$, compute the angle at the points $(0, 0, 1)$, $(0, 1, e^{-1})$ and $(1, 0, e^{-1})$. Show that all points with $x^2 + y^2 = 1$ have the same angle and explain why in terms of the symmetry of the surface. If the position of the light source is changed, will these points remain equally well lit? Based on Figure C, try to determine where the light source is located.

12.7   EXTREMA OF FUNCTIONS OF SEVERAL VARIABLES

One of the most important uses of calculus is for optimization. Indeed, you have seen optimization problems reappear in a number of places, since we first introduced the idea in section 3.7. Undoubtedly, you recognize that in this age, manufacturing plants need to maximize the quality of their products by minimizing defects. In order to design more and more efficient spacecraft, engineers need to minimize the weight of a structure while maximizing its strength. The list goes on and on. In this section, we introduce the mathematical basis for optimizing functions of several variables.

Carefully examine the surface $z = xe^{-x^2/2-y^3/3+y}$, shown in Figure 12.39a for $-2 \le x \le 4$ and $-1 \le y \le 4$. From the graph, notice that you can identify both a peak and a trough. We can zoom in to get a better view of the peak (see Figure 12.39b for $0.9 \le x \le 1.1$ and $0.9 \le y \le 1.1$). Referring to Figure 12.39a, we can zoom in to get a better view of the trough (see Figure 12.39c for $-1.1 \le x \le -0.9$ and $0.9 \le y \le 1.1$). Such points are referred to as local extrema, which we define as follows.

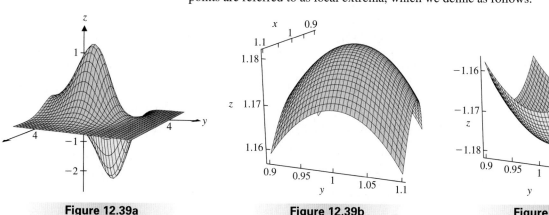

Figure 12.39a
$z = xe^{-x^2/2-y^3/3+y}$.

Figure 12.39b
Local maximum.

Figure 12.39c
Local minimum.

Definition 7.1

We call $f(a, b)$ a **local maximum** of f if there is an open disk R centered at (a, b), for which $f(a, b) \geq f(x, y)$ for all $(x, y) \in R$. Similarly, $f(a, b)$ is called a **local minimum** of f if there is an open disk R centered at (a, b), for which $f(a, b) \leq f(x, y)$ for all $(x, y) \in R$. In either case $f(a, b)$ is called a **local extremum** of f.

Note the similarity between Definition 7.1 and the definition of local extrema given in section 3.3. The idea here is the same as it was in Chapter 3. That is, if $f(a, b) \geq f(x, y)$ for all (x, y) "near" (a, b), we call $f(a, b)$ a local maximum.

Look carefully at Figures 12.39b and 12.39c; it appears that at both local extrema, the tangent plane is horizontal. Think about this for a moment and convince yourself that if the tangent plane was tilted, the function would be increasing in one direction and decreasing in another direction, which can't happen at a local extremum (maximum or minimum). Much as with functions of one variable, it turns out that local extrema can only occur where the first (partial) derivatives are zero or do not exist. However, as we'll see, having zero first partial derivatives at a point does not by itself guarantee a local extremum.

Definition 7.2

The point (a, b) is a **critical point** of the function $f(x, y)$ if (a, b) is in the domain of f and either $\dfrac{\partial f}{\partial x}(a, b) = \dfrac{\partial f}{\partial y}(a, b) = 0$ or one or both of $\dfrac{\partial f}{\partial x}$ and $\dfrac{\partial f}{\partial y}$ do not exist at (a, b).

Recall that for a function $f(x)$ of a single variable, if f has a local extremum at $x = a$, then a must be a critical number of f [i.e., $f'(a) = 0$ or $f'(a)$ is undefined]. As we show now, the same kind of behavior is seen in functions of several variables. In this case, we have that if $f(a, b)$ is a local extremum (local maximum or local minimum), then (a, b) must be a critical point of f. Be careful, though; although local extrema can only occur at critical points, every critical point need **not** correspond to a local extremum. For this reason, we refer to critical points as **candidates** for local extrema.

Theorem 7.1

If $f(x, y)$ has a local extremum at (a, b), then (a, b) must be a critical point of f.

Proof

Suppose that $f(x, y)$ has a local maximum at (a, b). Holding y constant at $y = b$, notice that the function $g(x) = f(x, b)$ has a local maximum at $x = a$. By Fermat's Theorem (Theorem 3.2 in Chapter 3), either $g'(a) = 0$ or $g'(a)$ doesn't exist. Note that $g'(a) = \dfrac{\partial f}{\partial x}(a, b)$. Likewise, holding x constant at $x = a$, observe that the function $h(y) = f(a, y)$ has a local maximum at $y = b$. It follows that $h'(b) = 0$ or $h'(b)$ doesn't exist. Note that $h'(b) = \dfrac{\partial f}{\partial y}(a, b)$. Combining these two observations, we have that each of $\dfrac{\partial f}{\partial x}(a, b)$ and $\dfrac{\partial f}{\partial y}(a, b)$ equals 0 or doesn't exist. We can then conclude that (a, b) must be a critical point of f. A nearly identical argument shows that if $f(x, y)$ has a local minimum at (a, b) then (a, b) must be a critical point of f.

When looking for local extrema, you must first find all critical points, since local extrema can only occur at critical points. Then, analyze each critical point to determine if it is the location of a local maximum, local minimum or neither. We now return to the function $f(x, y) = xe^{-x^2/2-y^3/3+y}$ discussed in the introduction to the section.

Example 7.1   Finding Local Extrema Graphically

Find all critical points of $f(x, y) = xe^{-x^2/2-y^3/3+y}$ and analyze each critical point graphically.

Solution   First, we compute the first partial derivatives:

$$\frac{\partial f}{\partial x} = e^{-x^2/2-y^3/3+y} + x(-x)e^{-x^2/2-y^3/3+y} = (1 - x^2)e^{-x^2/2-y^3/3+y}$$

and

$$\frac{\partial f}{\partial y} = x(-y^2 + 1)e^{-x^2/2-y^3/3+y}.$$

Since exponentials are always positive, we have $\frac{\partial f}{\partial x} = 0$ if and only if $1 - x^2 = 0$, that is, when $x = \pm 1$. We have $\frac{\partial f}{\partial y} = 0$ if and only if $x(-y^2 + 1) = 0$, that is, when $x = 0$ or $y = \pm 1$. Notice that both partial derivatives exist for all (x, y) and so, the only critical points are solutions of $\frac{\partial f}{\partial x} = \frac{\partial f}{\partial y} = 0$. For this to occur, notice that we need $x = \pm 1$ and either $x = 0$ or $y = \pm 1$. However, if $x = 0$, then $\frac{\partial f}{\partial x} \neq 0$ so there are no critical points with $x = 0$. This leaves all combinations of $x = \pm 1$ and $y = \pm 1$ as critical points: $(1, 1), (-1, 1), (1, -1)$ and $(-1, -1)$. Keep in mind that the critical points are only candidates for local extrema; we must look further to determine whether they correspond to extrema. We zoom in on each critical point in turn, to graphically identify any local extrema. We have already seen (see Figures 12.39b and 12.39c) that $f(x, y)$ has a local maximum at $(1, 1)$ and a local minimum at $(-1, 1)$. Figures 12.40a and 12.40b show $z = f(x, y)$ zoomed in on $(1, -1)$ and $(-1, -1)$, respectively. In Figure 12.40a, notice that in the plane $x = 1$ (extending left to right), the point at $(1, -1)$ is a local minimum. However, in the plane $y = -1$ (extending back to front), the point at $(1, -1)$ is a local maximum. This point is therefore not a local extremum. We refer to such a point as a

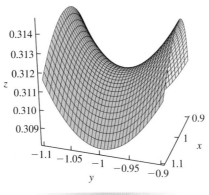

Figure 12.40a

Saddle point at $(1, -1)$.

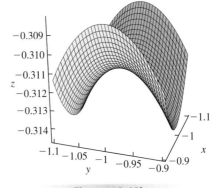

Figure 12.40b

Saddle point at $(-1, -1)$.

saddle point. (It looks like a saddle, doesn't it?) Similarly, in Figure 12.40b, notice that in the plane $x = -1$ (extending left to right), the point at $(-1, -1)$ is a local maximum. However, in the plane $y = -1$ (extending back to front), the point at $(-1, -1)$ is a local minimum. Again, at $(-1, -1)$ we have a saddle point.

We now pause to carefully define saddle points.

Definition 7.3

The point $P(a, b, f(a, b))$ is a **saddle point** of $z = f(x, y)$ if (a, b) is a critical point of f and if every open disk centered at (a, b) contains points (x, y) in the domain of f for which $f(x, y) < f(a, b)$ and points (x, y) in the domain of f for which $f(x, y) > f(a, b)$.

To further explore example 7.1 graphically, we show a contour plot of $f(x, y) = xe^{-x^2/2 - y^3/3 + y}$ in Figure 12.41. Notice that near the local maximum at $(1, 1)$ and the local minimum at $(-1, 1)$ the level curves look like concentric circles. This corresponds to the paraboloid-like shape of the surface near these points (see Figures 12.39b and 12.39c). Concentric ovals are characteristic of local extrema. Notice that, without the level curves labeled, there is no way to tell from the contour plot which is the maximum and which is the minimum. Saddle points are typically characterized by the hyperbolic-looking curves seen near $(-1, -1)$ and $(1, -1)$.

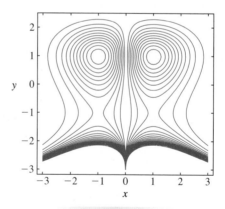

Figure 12.41
Contour plot of
$f(x, y) = xe^{-x^2/2 - y^3/3 + y}$.

Of course, we can't rely on interpreting three-dimensional graphs for finding local extrema. Recall that for functions of a single variable, we developed two tests (the first derivative test and the second derivative test) that give various criteria for determining when a given critical number corresponds to a local maximum or a local minimum or neither. The following result, which we prove at the end of the section, is surprisingly simple and is a generalization of the second derivative test for functions of a single variable.

> **Theorem 7.2 (Second Derivatives Test)**
>
> Suppose that $f(x, y)$ has continuous second-order partial derivatives in some open disk containing the point (a, b) and that $f_x(a, b) = f_y(a, b) = 0$. Define the **discriminant** D for the point (a, b) by
>
> $$D(a, b) = f_{xx}(a, b)f_{yy}(a, b) - [f_{xy}(a, b)]^2.$$
>
> (i)   If $D(a, b) > 0$ and $f_{xx}(a, b) > 0$, then f has a local minimum at (a, b).
> (ii) If $D(a, b) > 0$ and $f_{xx}(a, b) < 0$, then f has a local maximum at (a, b).
> (iii) If $D(a, b) < 0$, then f has a saddle point at (a, b).
> (iv) If $D(a, b) = 0$, then no conclusion can be drawn.

It's important to make some sense of this result (in other words, to understand it and not just memorize it). Note that to have $D(a, b) > 0$, we must have **both** $f_{xx}(a, b) > 0$ and $f_{yy}(a, b) > 0$ **or** both $f_{xx}(a, b) < 0$ and $f_{yy}(a, b) < 0$. In the first case, notice that the surface $z = f(x, y)$ will be concave up in the plane $y = b$ and concave up in the plane $x = a$. In this case, the surface will look like an upward-shaped paraboloid near the point (a, b). Consequently, the critical point (a, b) must be the location of a local minimum. In the second case, both $f_{xx}(a, b) < 0$ and $f_{yy}(a, b) < 0$. This says that the surface $z = f(x, y)$ will be concave down in the plane $y = b$ and concave down in the plane $x = a$. So, in this case, the surface looks like a downward-shaped paraboloid near the point (a, b), which must then be the location of a local maximum. Observe that one way to get $D(a, b) < 0$ is for $f_{xx}(a, b)$ and $f_{yy}(a, b)$ to have opposite signs (one positive and one negative). To have opposite concavities in the planes $x = a$ and $y = b$ means that there is a saddle point at (a, b), as in Figures 12.40a and 12.40b.

> **Example 7.2**   Using the Discriminant to Find Local Extrema
>
> Locate and classify all critical points for $f(x, y) = 2x^2 - y^3 - 2xy$.
>
> **Solution**   We first compute the first partial derivatives: $f_x = 4x - 2y$ and $f_y = -3y^2 - 2x$. Since both partial derivatives are defined for all (x, y), the critical points are solutions of the two equations:
>
> $$f_x = 4x - 2y = 0$$
>
> and
>
> $$f_y = -3y^2 - 2x = 0.$$
>
> Solving the first equation for y, we get $y = 2x$. Substituting this into the second equation, we have
>
> $$0 = -3(4x^2) - 2x = -12x^2 - 2x$$
> $$= -2x(6x + 1),$$
>
> so that $x = 0$ or $x = -\frac{1}{6}$. The corresponding y-values are $y = 0$ and $y = -\frac{1}{3}$. The only two critical points are then $(0, 0)$ and $\left(-\frac{1}{6}, -\frac{1}{3}\right)$. To classify these points, we first compute the second partial derivatives: $f_{xx} = 4$, $f_{yy} = -6y$ and $f_{xy} = -2$ and then test the discriminant. We have
>
> $$D(0, 0) = (4)(0) - (-2)^2 = -4$$
>
> and
>
> $$D\left(-\tfrac{1}{6}, -\tfrac{1}{3}\right) = (4)(2) - (-2)^2 = 4.$$

From Theorem 7.2, we conclude that there is a saddle point of f at $(0, 0)$, since $D(0, 0) < 0$. Further, there is a local minimum at $\left(-\frac{1}{6}, -\frac{1}{3}\right)$ since $D\left(-\frac{1}{6}, -\frac{1}{3}\right) > 0$ and $f_{xx}\left(-\frac{1}{6}, -\frac{1}{3}\right) > 0$. The surface is shown in Figure 12.42.

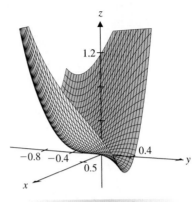

Figure 12.42

$z = 2x^2 - y^3 - 2xy$.

As we see in the following example, the second derivatives test does not always help us to classify a critical point.

Example 7.3   Classifying Critical Points

Locate and classify all critical points for $f(x, y) = x^3 - 2y^2 - 2y^4 + 3x^2y$.

Solution   Here, we have $f_x = 3x^2 + 6xy$ and $f_y = -4y - 8y^3 + 3x^2$. Since both f_x and f_y exist for all (x, y), the critical points are solutions of the two equations:

$$f_x = 3x^2 + 6xy = 0$$

and

$$f_y = -4y - 8y^3 + 3x^2 = 0.$$

From the first equation, we have

$$0 = 3x^2 + 6xy = 3x(x + 2y),$$

so that at a critical point, $x = 0$ or $x = -2y$. Substituting $x = 0$ into the second equation, we have

$$0 = -4y - 8y^3 = -4y(1 + 2y^2).$$

The only (real) solution of this equation is $y = 0$. This says that for $x = 0$, we have only one critical point: $(0, 0)$. Substituting $x = -2y$ into the second equation, we have

$$0 = -4y - 8y^3 + 3(4y^2) = -4y(1 + 2y^2 - 3y) = -4y(2y - 1)(y - 1).$$

The solutions of this equation are $y = 0$, $y = \frac{1}{2}$ and $y = 1$, with corresponding critical points $(0, 0)$, $\left(-1, \frac{1}{2}\right)$ and $(-2, 1)$. To classify the critical points, we compute the second partial derivatives: $f_{xx} = 6x + 6y$, $f_{yy} = -4 - 24y^2$ and $f_{xy} = 6x$ and evaluate the discriminant at each critical point. We have

$$D(0, 0) = (0)(-4) - (0)^2 = 0,$$

$$D\left(-1, \frac{1}{2}\right) = (-3)(-10) - (-6)^2 = -6$$

and

$$D(-2, 1) = (-6)(-28) - (-12)^2 = 24.$$

From Theorem 7.2, we conclude that f has a saddle point at $\left(-1, \frac{1}{2}\right)$, since $D\left(-1, \frac{1}{2}\right) < 0$. Further, f has a local maximum at $(-2, 1)$ since $D(-2, 1) > 0$ and $f_{xx}(-2, 1) < 0$. Unfortunately, Theorem 7.2 gives us no information about the critical point $(0, 0)$, since $D(0, 0) = 0$. However, notice that in the plane $y = 0$ we have $f(x, y) = x^3$. In two dimensions, the curve $z = x^3$ has an inflection point at $x = 0$. This shows that there is no local extremum at $(0, 0)$. The surface near $(0, 0)$ is shown in Figure 12.43a. The surface near $(-2, 1)$ and $\left(-1, \frac{1}{2}\right)$ is shown in Figures 12.43b and 12.43c, respectively. Since the graphs are not especially clear, it's good that we have done the analysis! You will show in the exercises that there is a saddle point at $(0, 0)$.

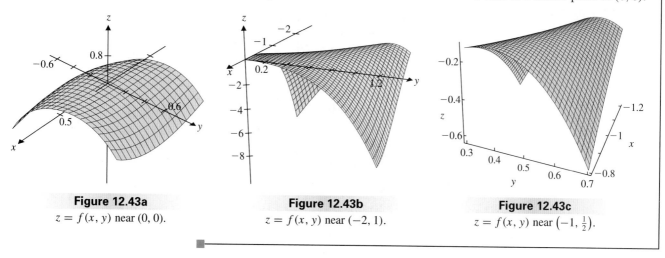

Figure 12.43a	**Figure 12.43b**	**Figure 12.43c**
$z = f(x, y)$ near $(0, 0)$.	$z = f(x, y)$ near $(-2, 1)$.	$z = f(x, y)$ near $\left(-1, \frac{1}{2}\right)$.

One commonly used application of the theory of local extrema is the statistical technique of **least squares.** This technique (or, more accurately, this criterion) is essential to many commonly used curve-fitting and data analysis procedures. The following example illustrates the use of least squares in **linear regression.**

Example 7.4   Linear Regression

Population data from the U.S. census is shown in the following table.

Year	Population
1960	179,323,175
1970	203,302,031
1980	226,542,203
1990	248,709,873

Find the straight line that "best" fits the data.

Solution   To make the data more manageable, we first transform the raw data into variables x (the number of decades since 1960) and y (population, in millions of people, rounded off to the nearest whole number). We display this data in the table in the margin.

x	y
0	179
1	203
2	227
3	249

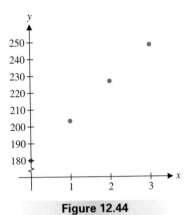

Figure 12.44

U.S. population since 1960
(in millions).

A plot of x and y is shown in Figure 12.44. From the plot, it would appear that the population data is nearly linear. Our goal is to find the line that "best" fits the data. (This is called the **regression line**.) The criterion for "best" fit is the least-squares criterion, as defined below. The equation of our line will be $y = ax + b$, with constants a and b to be determined. For a value of x represented in the data, the error (or **residual**) is given by the difference between the actual y-value and the predicted value $ax + b$. The least-squares criterion is to choose a and b to minimize the sum of the squares of all the residuals. (In a sense, this minimizes the total error.) For our data, the residuals are given in the following table.

x	$ax + b$	y	residual
0	b	179	$b - 179$
1	$a + b$	203	$a + b - 203$
2	$2a + b$	227	$2a + b - 227$
3	$3a + b$	249	$3a + b - 249$

The sum of the squares of the residuals is then given by the function

$$f(a, b) = (b - 179)^2 + (a + b - 203)^2 + (2a + b - 227)^2 + (3a + b - 249)^2.$$

From Theorem 7.1, we must have $\dfrac{\partial f}{\partial a} = \dfrac{\partial f}{\partial b} = 0$ at the minimum point, since f_a and f_b are defined everywhere. We have

$$0 = \frac{\partial f}{\partial a} = 2(a + b - 203) + 4(2a + b - 227) + 6(3a + b - 249)$$

and

$$0 = \frac{\partial f}{\partial b} = 2(b - 179) + 2(a + b - 203) + 2(2a + b - 227) + 2(3a + b - 249).$$

After multiplying out all terms, we have

$$28a + 12b = 2808$$

and

$$12a + 8b = 1716.$$

The second equation reduces to $3a + 2b = 429$, so that $a = 143 - \frac{2}{3}b$. Substituting this into the first equation, we have

$$28\left(143 - \frac{2}{3}b\right) + 12b = 2808,$$

or

$$4004 - 2808 = \left(\frac{56}{3} - 12\right)b.$$

This gives us $b = \frac{897}{5} = 179.4$, so that

$$a = 143 - \frac{2}{3}\left(\frac{897}{5}\right) = \frac{117}{5} = 23.4.$$

The regression line with these coefficients is

$$y = 23.4x + 179.4.$$

Realize that all we have determined so far is that (a, b) is a critical point, a candidate for a local extremum. To verify that our choice of a and b give the **minimum** function value, note that the surface $z = f(x, y)$ is a paraboloid opening toward the positive

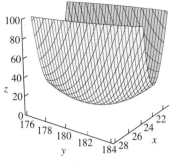

Figure 12.45

$z = f(x, y)$.

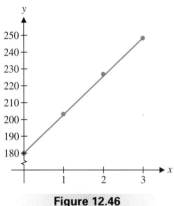

Figure 12.46

The regression line.

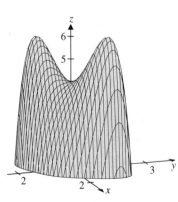

Figure 12.47

$z = 4xy - x^4 - y^4 + 4$.

z-axis (see Figure 12.45) and the only critical point of an upward-curving paraboloid is an absolute minimum. A plot of the regression line $y = 23.4x + 179.4$ with the data points is shown in Figure 12.46. Look carefully and notice that the line matches the data quite well. This also gives us confidence that we have found the minimum sum of the squared residuals.

As you will quickly discover in the exercises, finding critical points of even simple-looking functions of several variables can be quite a challenge. For the complicated functions that often arise in applications, finding critical points by hand can be nearly impossible. Because of this, numerical procedures for estimating maxima and minima are essential. We briefly introduce one such method here.

Given a function $f(x, y)$, make your best guess (x_0, y_0) of the location of a local maximum (or minimum). We call this your **initial guess** and want to use this to obtain a more precise estimate of the location of the maximum (or minimum). How might we do that? Well, recall that the direction of maximum increase of the function from the point (x_0, y_0) is given by the gradient $\nabla f(x_0, y_0)$. So, starting at (x_0, y_0), if we move in the direction of $\nabla f(x_0, y_0)$, f should be increasing, but how far should we go in this direction? One strategy (the method of **steepest ascent**), is to continue moving in the direction of the gradient until the function stops increasing. We call this stopping point (x_1, y_1). Starting anew from (x_1, y_1), we repeat the process, by computing a new gradient $\nabla f(x_1, y_1)$ and following it until $f(x, y)$ stops increasing, at some point (x_2, y_2). We then continue this process until the change in function values from $f(x_n, y_n)$ to $f(x_{n+1}, y_{n+1})$ is insignificant. Likewise, to find a local minimum, follow the path of **steepest descent,** by moving in the direction opposite the gradient, $-\nabla f(x_0, y_0)$ (the direction of maximum decrease of the function). We illustrate the steepest ascent algorithm in the following example.

Example 7.5 Method of Steepest Ascent

Use the steepest ascent algorithm to estimate the maximum of $f(x, y) = 4xy - x^4 - y^4 + 4$.

Solution A sketch of the surface is shown in Figure 12.47. Notice that it appears from the graph that the function has two local maxima, but it's hard to tell which (if either) is actually higher. We will estimate the maximum on the right by starting with an initial guess of $(2, 3)$, where $f(2, 3) = -69$. (Note that this is obviously not the maximum, but it will suffice as a crude initial guess.) From this point, we want to follow the path of steepest ascent and move in the direction of $\nabla f(2, 3)$. We have

$$\nabla f(x, y) = \langle 4y - 4x^3, 4x - 4y^3 \rangle$$

and so, $\nabla f(2, 3) = \langle -20, -100 \rangle$. Note that every point lying on the line through $(2, 3)$ in the direction of $\langle -20, -100 \rangle$ will have the form $(2 - 20h, 3 - 100h)$, for some value of $h > 0$. (Think about this!) Our goal is to move in this direction until $f(x, y)$ stops increasing. Notice that this puts us at a critical point for function values on the line of points $(2 - 20h, 3 - 100h)$. Since the function values along this line are given by $g(h) = f(2 - 20h, 3 - 100h)$, we find the smallest positive h such that $g'(h) = 0$. From the chain rule, we have

$$g'(h) = -20 \frac{\partial f}{\partial x}(2 - 20h, 3 - 100h) - 100 \frac{\partial f}{\partial y}(2 - 20h, 3 - 100h)$$

$$= -20[4(3 - 100h) - 4(2 - 20h)^3] - 100[4(2 - 20h) - 4(3 - 100h)^3].$$

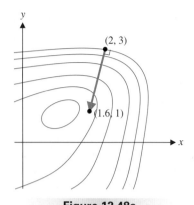

Figure 12.48a
First step of steepest ascent.

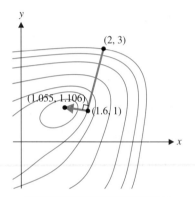

Figure 12.48b
Second step of steepest ascent.

Solving the equation $g'(h) = 0$ (we did it numerically), we get $h \approx 0.02$. This moves us to the point $(x_1, y_1) = (2 - 20h, 3 - 100h) = (1.6, 1)$, with function value $f(x_1, y_1) = 2.846$. A contour plot of $f(x, y)$ with this first step is shown in Figure 12.48a. Notice that since $f(x_1, y_1) > f(x_0, y_0)$, we have found an improved approximation of the local maximum. To improve this further, we repeat the process starting with the new point. In this case, we have $\nabla f(1.6, 1) = \langle -12.384, 2.4 \rangle$ and we look for a critical point for the new function $g(h) = f(1.6 - 12.384h, 1 + 2.4h)$, for $h > 0$. Again, from the chain rule, we have

$$g'(h) = -12.384 \frac{\partial f}{\partial x}(1.6 - 12.384h, 1 + 2.4h) + 2.4 \frac{\partial f}{\partial y}(1.6 - 12.384h, 1 + 2.4h).$$

Solving $g'(h) = 0$ numerically gives us $h \approx 0.044$. This moves us to the point $(x_2, y_2) = (1.6 - 12.384h, 1 + 2.4h) = (1.055, 1.106)$, with function value $f(x_2, y_2) = 5.932$. Notice that we have again improved our approximation of the local maximum. A contour plot of $f(x, y)$ with the first two steps is shown in Figure 12.48b. From the contour plot, it appears that we are now very near a local maximum. In practice, you continue this process until you are no longer improving the approximation significantly. (This is easily implemented on a computer.) In the accompanying table, we show the first seven steps of steepest ascent. We leave it as an exercise to show that the local maximum is actually at $(1, 1)$ with function value $f(1, 1) = 6$.

n	x_n	y_n	$f(x_n, y_n)$
0	2	3	−69
1	1.6	1	2.846
2	1.055	1.106	5.932
3	1.0315	1.0035	5.994
4	1.0049	1.0094	5.9995
5	1.0029	1.0003	5.99995
6	1.0005	1.0009	5.999995
7	1.0003	1.0003	5.9999993

We define absolute extrema in a similar fashion to local extrema.

Definition 7.4

We call $f(a, b)$ the **absolute maximum** of f on the region R if $f(a, b) \geq f(x, y)$ for all $(x, y) \in R$. Similarly, $f(a, b)$ is called the **absolute minimum** of f on R if $f(a, b) \leq f(x, y)$ for all $(x, y) \in R$. In either case, $f(a, b)$ is called an **absolute extremum** of f.

Recall that for a function f of a single variable, we observed that whenever f is continuous on the closed interval $[a, b]$, it will assume a maximum and a minimum value on $[a, b]$. Further, we proved that absolute extrema must occur at either critical numbers of f or at the endpoints of the interval $[a, b]$. The situation for absolute extrema of functions of two variables is very similar. First, we need some terminology. We say that a region $R \subset \mathbb{R}^2$ is **bounded** if there is a disk that completely contains R. We now have the following result (whose proof can be found in more advanced texts).

Theorem 7.3 (Extreme Value Theorem)
Suppose that $f(x, y)$ is continuous on the closed and bounded region $R \subset \mathbb{R}^2$. Then f has both an absolute maximum and an absolute minimum on R. Further, the absolute extrema must occur at either a critical point in R or on the boundary of R.

Note that if $f(a, b)$ is an absolute extremum of f in R and (a, b) is in the interior of R, then (a, b) is also a local extremum of f, in which case, (a, b) must be a critical point. This says that all of the absolute extrema of a function f in a region R occur either at critical points (and we already know how to find these) or on the boundary of the region. Observe that this also provides us with a method for locating absolute extrema of continuous functions on closed and bounded regions. That is, we find the extrema on the boundary and compare these against the local extrema. We examine this in the following example. Notice that the basic steps in example 7.6 are as follows:

- Find all critical points of f in the region R.
- Find the maximum and minimum values of f on the boundary of R.
- Compare the values of f at the critical points with the maximum and minimum values of f on the boundary of R.

Example 7.6   Finding Absolute Extrema

Find the absolute extrema of $f(x, y) = 5 + 4x - 2x^2 + 3y - y^2$ on the region R bounded by the lines $y = 2$, $y = x$ and $y = -x$.

Solution   We show a sketch of the surface in Figure 12.49a and a sketch of the region R in Figure 12.49b. From the sketch of the surface, notice that the absolute minimum appears to occur on the line $x = -2$ and the absolute maximum occurs somewhere near the line $x = 1$. (Be careful; as we've already observed, it's difficult to accurately determine the location of extrema from a three-dimensional graph. You should use this as a guide for what to expect, only.) Keep in mind that the extrema can only occur at a critical point or at a point on the boundary of R. So, we first check to see if there are any interior critical points. We have $f_x = 4 - 4x = 0$ for $x = 1$ and $f_y = 3 - 2y = 0$ for $y = \frac{3}{2}$. So, there is only one critical point $\left(1, \frac{3}{2}\right)$ and it is located in the interior of R. Next, we look for the maximum and minimum values of f on the boundary of R. In this case, R consists of three separate pieces: the portion of the line $y = 2$ for $-2 \leq x \leq 2$, the portion of the line $y = x$, for $0 \leq x \leq 2$ and the portion of the line $y = -x$, for $-2 \leq x \leq 0$. We look for the maximum value of f on each of these separately. On the portion of the line $y = 2$ for $-2 \leq x \leq 2$, we have

$$f(x, y) = f(x, 2) = 5 + 4x - 2x^2 + 6 - 4 = 7 + 4x - 2x^2 = g(x).$$

To find the maximum and minimum values of f on this portion of the boundary, we need only find the maximum and minimum values of g on the interval $[-2, 2]$. We have $g'(x) = 4 - 4x = 0$ only for $x = 1$. Comparing the value of g at the endpoints and the only critical number in the interval, we have: $g(-2) = -9$, $g(2) = 7$ and $g(1) = 9$. So, the maximum value of f on this portion of the boundary is 9 and the minimum value is -9.

On the portion of the line $y = x$, for $0 \leq x \leq 2$, we have

$$f(x, y) = f(x, x) = 5 + 7x - 3x^2 = h(x).$$

We have $h'(x) = 7 - 6x = 0$, only for $x = \frac{7}{6}$, which is in the interval. Comparing the values of h at the endpoints and the critical number, we have: $h(0) = 5$, $h(2) = 7$ and

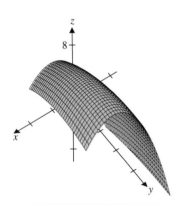

z
8

Figure 12.49a
The surface
$z = 5 + 4x - 2x^2 + 3y - y^2$.

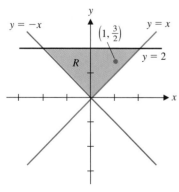

$y = -x$ y $y = x$
$\left(1, \frac{3}{2}\right)$
R $y = 2$
x

Figure 12.49b
The region R.

$h\left(\frac{7}{6}\right) \approx 9.08$. So, the maximum value of f on this portion of the boundary is approximately 9.08 and its minimum value is 5.

On the portion of the line $y = -x$, for $-2 \leq x \leq 0$, we have

$$f(x, y) = f(x, -x) = 5 + x - 3x^2 = k(x).$$

We have $k'(x) = 1 - 6x = 0$, only for $x = \frac{1}{6}$, which is **not** in the interval $[-2, 0]$ under consideration. Comparing the values of k at the endpoints, we have $k(-2) = -9$ and $k(0) = 5$, so that the maximum value of f on this portion of the boundary is 5 and its minimum value is -9.

Finally, we compute the value of f at the lone critical point in $R : f\left(1, \frac{3}{2}\right) = \frac{37}{4} = 9.25$. The largest of all these values we have computed is the absolute maximum in R and the smallest is the absolute minimum. So, the absolute maximum is $f\left(1, \frac{3}{2}\right) = 9.25$ and the absolute minimum is $f(-2, 2) = -9$. Note that these are also consistent with what we observed in Figure 12.49a.

We close this section with a proof of the Second Derivatives Test (Theorem 7.2). To keep the notation to a minimum, we will assume that the critical point to be tested is $(0, 0)$. The proof can be extended to any critical point by a change of variables.

Proof of the Second Derivatives Test

Suppose that $(0, 0)$ is a critical point of $f(x, y)$ with $f_x(0, 0) = f_y(0, 0) = 0$. We will look at the change in $f(x, y)$ from $(0, 0)$ in the direction of the unit vector $\mathbf{u} = \dfrac{\langle 1, k \rangle}{\sqrt{1 + k^2}}$ for some constant k. In this direction, notice that $y = kx$. If we define $g(x) = f(x, kx)$, then by the chain rule, we have

$$g'(x) = f_x(x, kx) + kf_y(x, kx) \tag{7.1}$$

and

$$g''(x) = f_{xx}(x, kx) + kf_{xy}(x, kx) + kf_{yx}(x, kx) + k^2 f_{yy}(x, kx).$$

At $x = 0$, this gives us

$$g''(0) = f_{xx}(0, 0) + 2kf_{xy}(0, 0) + k^2 f_{yy}(0, 0), \tag{7.2}$$

where we have $f_{xy} = f_{yx}$, since f was assumed to have continuous second partial derivatives. Since $f_x(0, 0) = f_y(0, 0) = 0$, we have from (7.1) that

$$g'(0) = f_x(0, 0) + kf_y(0, 0) = 0.$$

From the second derivative test for functions of a single variable, the sign of $g''(0)$ can tell us whether there is a local maximum or a local minimum of g at $x = 0$. To determine whether $g''(0)$ is positive or negative, observe that using (7.2), we can write $g''(0)$ as

$$g''(0) = ak^2 + 2bk + c = p(k),$$

where a, b and c are the constants $a = f_{yy}(0, 0)$, $b = f_{xy}(0, 0)$ and $c = f_{xx}(0, 0)$. Of course, the graph of $p(k)$ is a parabola. Recall that for any parabola, if $a > 0$, then $p(k)$ has a minimum at $k = -\frac{b}{a}$, given by $p\left(-\frac{b}{a}\right) = -\frac{b^2}{a} + c$. (Hint: Complete the square.) In case (i) of the theorem, we assume that the discriminant satisfies

$$0 < D(0, 0) = f_{xx}(0, 0)f_{yy}(0, 0) - [f_{xy}(0, 0)]^2 = ac - b^2,$$

so that $-\frac{b^2}{a} + c > 0$. In this case,

$$p(k) \geq p\left(-\frac{b}{a}\right) = -\frac{b^2}{a} + c > 0.$$

We have shown that, in case (i), when $D(0, 0) > 0$ and $f_{xx}(0, 0) > 0$, $g''(0) = p(k) > 0$ for all k. So, g has a local minimum at 0 and consequently, in all directions, the point at $(0, 0)$ is a local minimum of f. For case (ii), where $D(0, 0) > 0$ and $f_{xx}(0, 0) < 0$, we consider $p(k)$ with $a < 0$. In a similar fashion, we can show that here, $p(k) \leq -\frac{b^2}{a} + c < 0$. Given that we have $g''(0) = p(k) < 0$ for all k, we conclude that the point at $(0, 0)$ is a local maximum of f. For case (iii), where the discriminant $D(0, 0) < 0$, the parabola $p(k)$ will assume both positive and negative values. For some values of k, we have $g''(0) > 0$ and the point $(0, 0)$ is a local minimum along the path $y = kx$, while for other values of k, we have $g''(0) < 0$ and the point $(0, 0)$ is a local maximum along the path $y = kx$. Taken together, this says that the point at $(0, 0)$ must be a saddle point of f.

EXERCISES 12.7

1. If $f(x, y)$ has a local minimum at (a, b), explain why the point $(a, b, f(a, b))$ is a local minimum in the intersection of $z = f(x, y)$ with any vertical plane. Explain why the condition $f_x(a, b) = f_y(a, b) = 0$ guarantees that (a, b) is a critical point in any such plane.

2. Suppose that $f_x(a, b) \neq 0$. Explain why the tangent plane to $z = f(x, y)$ at (a, b) must be "tilted" so that there is not a local extremum at (a, b).

3. Suppose that $f_x(a, b) = f_y(a, b) = 0$ and $f_{xx}(a, b) f_{yy}(a, b) < 0$. Explain why there must be a saddle point at (a, b).

4. Explain why the center of a set of concentric circles in a contour plot will often represent a local extremum.

In exercises 5–12, locate all critical points and classify them using Theorem 7.2.

5. $f(x, y) = e^{-x^2}(y^2 + 1)$

6. $f(x, y) = \cos^2 x + y^2$

7. $f(x, y) = x^3 - 3xy + y^3$

8. $f(x, y) = 4xy - x^4 - y^4 + 4$

9. $f(x, y) = y^2 + x^2y + x^2 - 2y$

10. $f(x, y) = 2x^2 + y^3 - x^2y - 3y$

11. $f(x, y) = e^{-x^2 - y^2}$ 12. $f(x, y) = x \sin y$

In exercises 13–18, locate all critical points and analyze each graphically. If you have a CAS, use Theorem 7.2 to classify each point.

13. $f(x, y) = x^2 - \dfrac{4xy}{y^2 + 1}$ 14. $f(x, y) = \dfrac{x + y}{x^2 + y^2 + 1}$

15. $f(x, y) = xe^{-x^2 - y^2}$ 16. $f(x, y) = x^2 e^{-x^2 - y^2}$

17. $f(x, y) = xye^{-x^2 - y^2}$ 18. $f(x, y) = xye^{-x^2 - y^4}$

In exercises 19–22, numerically approximate all critical points. Classify each point graphically or with Theorem 7.2.

19. $f(x, y) = xy^2 - x^2 - y + \frac{1}{16}x^4$

20. $f(x, y) = 2y(x + 2) - x^2 + y^4 - 9y^2$

21. $f(x, y) = (x^2 - y^3)e^{-x^2 - y^2}$

22. $f(x, y) = (x^2 - 3x)e^{-x^2 - y^2}$

In exercises 23–29, use least squares as in example 7.4 to find a linear model of the data.

23. A famous mental calculation prodigy named Jacques Inaudi was timed in 1894 at various mental arithmetic problems. His times are shown below. (Think about what your times might be!) The data is taken from *The Number Sense* by Stanislas Dehaene. Treat the number of operations as the independent variable (x) and time as the dependent variable (y).

Number of operations	1	4	9	16
Time (sec)	0.6	2.0	6.4	21
Example	$3 \cdot 7$	$63 \cdot 58$	$638 \cdot 823$	$7286 \cdot 5397$

24. Repeat exercise 23 with the following data point added: 36 operations in 240 seconds (an example is $729, 856 \cdot 297, 143$). How much effect does this last point have on the linear model?

25. The Dow Jones Industrial averages for several days starting in June 1998 are shown. Use your linear model to predict the

average on day 12. Linear models of similar data can be found in information supplied by financial consulting firms, typically with the warning to not use the linear model for forecasting. Explain why this warning is appropriate.

Date (number of days)	0	2	4	6	8
Dow Jones average	8910	8800	9040	9040	9050

26. The following data shows the average price of a gallon of regular gasoline in California. Use the linear model to predict the price in 1990 and 1995. The actual prices were $1.09 and $1.23. Explain why your forecasts were not accurate.

Year	1970	1975	1980	1985
Price	$0.34	$0.59	$1.23	$1.11

27. The following data shows the height and weight of a small number of people. Use the linear model to predict the weight of a $6'8''$ person and a $5'0''$ person. Comment on how accurate you think the model is.

Height (inches)	68	70	70	71
Weight (pounds)	160	172	184	180

28. The following data shows the age and income for a small number of people. Use the linear model to predict the income of a 45-year-old and of an 80-year-old. Comment on how accurate you think the model is.

Age (years)	24	32	40	56
Income ($)	30,000	34,000	52,000	82,000

29. The following data shows the average number of points professional football teams score when starting different distances from the opponents' goal line. (For more information, see Hal Stern's "A Statistician Reads the Sports Pages" in *Chance*, Summer 1998. The number of points is determined by the next score, so that if the opponent scores next, the number of points is negative.) Use the linear model to predict the average number of points starting (a) 60 yards from the goal line and (b) 40 yards from the goal line.

Yards from goal	15	35	55	75	95
Average points	4.57	3.17	1.54	0.24	−1.25

30. In *The Hidden Game of Pro Football*, authors Carroll, Palmer and Thorn claim that the above data supports the conclusion

that when a team loses a fumble they lose an average of 4 points *regardless of where they are on the field*. That is, a fumble at the 50-yard line costs the same number of points as a fumble at the opponents' 10-yard line. Use your result from exercise 29 to verify this claim.

In exercises 31–34, calculate the first two steps of the steepest ascent algorithm from the given starting point.

31. $f(x, y) = 2xy - 2x^2 + y^3$, $(0, -1)$

32. $f(x, y) = 3xy - x^3 - y^2$, $(1, 1)$

33. $f(x, y) = x - x^2y^4 + y^2$, $(1, 1)$

34. $f(x, y) = xy^2 - x^2 - y$, $(1, 0)$

35. Calculate one step of the steepest ascent algorithm for $f(x, y) = 2xy - 2x^2 + y^3$ starting at $(0, 0)$. Explain in graphical terms what goes wrong.

36. Define a **steepest descent algorithm** for finding local minima.

In exercises 37–40, find the absolute extrema of the function on the region.

37. $f(x, y) = x^2 + 3y - 3xy$, region bounded by $y = x$, $y = 0$ and $x = 2$

38. $f(x, y) = x^2 + y^2 - 4xy$, region bounded by $y = x$, $y = -3$ and $x = 3$

39. $f(x, y) = x^2 + y^2$, region bounded by $(x - 1)^2 + y^2 = 4$

40. $f(x, y) = x^2 + y^2 - 2x - 4y$, region bounded by $y = x$, $y = 3$ and $x = 0$

41. Find all critical points of $f(x, y) = x^2y^2$ and show that Theorem 7.2 fails to identify any of them. Use the form of the function to determine what each critical point represents.

42. Repeat exercise 41 for $f(x, y) = x^{2/3}y^2$.

43. Complete the square to identify all local extrema of $f(x, y) = x^2 + 2x + y^2 - 4y + 1$.

44. Complete the square to identify all local extrema of $f(x, y) = x^4 - 6x^2 + y^4 + 2y^2 - 1$.

45. In exercise 7, there is a saddle point at $(0, 0)$. This means that there is (at least) one trace of $z = x^3 - 3xy + y^3$ with a local minimum at $(0, 0)$ and (at least) one trace with a local maximum at $(0, 0)$. To analyze traces in the planes $y = kx$ (for some constant k), substitute $y = kx$ and show that

$z = (1 + k^3)x^3 - 3kx^2$. Show that $f(x) = (1 + k^3)x^3 - 3kx^2$ has a local minimum at $x = 0$ if $k < 0$ and a local maximum at $x = 0$ if $k > 0$. (Hint: Use the Second Derivative Test from section 3.5.)

46. In exercise 8, there is a saddle point at $(0, 0)$. As in exercise 45, find traces such that there is a local maximum at $(0, 0)$ and traces such that there is a local minimum at $(0, 0)$.

47. In example 7.3, $(0, 0)$ is a critical point but is not classified by Theorem 7.2. Use the technique of exercise 45 to analyze this saddle point.

48. Repeat exercise 47 for $f(x, y) = x^2 - 3xy^2 + 4x^3y$.

49. For $f(x, y, z) = xz - x + y^3 - 3y$, show that $(0, 1, 1)$ is a critical point. To classify this critical point, show that $f(0 + \Delta x, 1 + \Delta y, 1 + \Delta z) = \Delta x \Delta z + 3\Delta y^2 + \Delta y^3 + f(0, 1, 1)$. Setting $\Delta y = 0$ and $\Delta x \Delta z > 0$, conclude that $f(0, 1, 1)$ is not a local maximum. Setting $\Delta y = 0$ and $\Delta x \Delta z < 0$, conclude that $f(0, 1, 1)$ is not a local minimum.

50. Repeat exercise 49 for the point $(0, -1, 1)$.

In exercises 51–54, label the statement as true or false and explain why.

51. If $f(x, y)$ has a local maximum at (a, b), then $\dfrac{\partial f}{\partial x}(a, b) = \dfrac{\partial f}{\partial y}(a, b) = 0$.

52. If $\dfrac{\partial f}{\partial x}(a, b) = \dfrac{\partial f}{\partial y}(a, b) = 0$, then $f(x, y)$ has a local maximum at (a, b).

53. In between any two local maxima of $f(x, y)$ there must be at least one local minimum of $f(x, y)$.

54. If $f(x, y)$ has exactly two critical points, they can't both be local maxima.

55. In the contour plot, the locations of four local extrema and nine saddle points are visible. Identify these critical points.

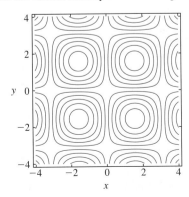

56. In the contour plot, the locations of one local extremum and one saddle point are visible. Identify each critical point.

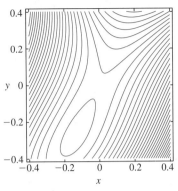

In exercises 57–60, use the contour plot to conjecture the locations of all local extrema and saddle points.

57.

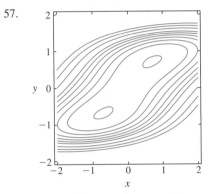

58.

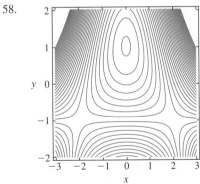

59.

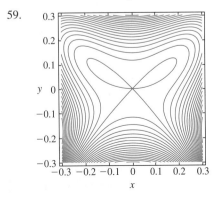

60.

61. Construct the function $d(x, y)$ giving the distance from a point (x, y, z) on the paraboloid $z = 4 - x^2 - y^2$ to the point $(3, -2, 1)$. Then determine the point that minimizes $d(x, y)$.

62. Use the method of exercise 61 to find the closest point on the cone $z = \sqrt{x^2 + y^2}$ to the point $(2, -3, 0)$.

63. Use the method of exercise 61 to find the closest point on the sphere $x^2 + y^2 + z^2 = 9$ to the point $(2, 1, -3)$.

64. Use the method of exercise 61 to find the closest point on the plane $3x - 4y + 3x = 12$ to the origin.

65. Show that the function $f(x, y) = 5xe^y - x^5 - e^{5y}$ has exactly one critical point, which is a local maximum but not an absolute maximum.

66. Show that the function $f(x, y) = 2x^4 + e^{4y} - 4x^2e^y$ has exactly two critical points, both of which are local minima.

67. Prove that the situation of exercise 66 (two local minima without a local maximum) can never occur for differentiable functions of one variable.

68. The **Hardy-Weinberg law** of genetics describes the relationship between the proportions of different genes in populations. Suppose that a certain gene has three types (e.g., blood types of A, B and O). If the three types have proportions p, q and r, respectively, in the population, then the Hardy-Weinberg law states that the proportion of people who carry two different types of genes equals $f(p, q, r) = 2pq + 2pr + 2qr$. Explain why $p + q + r = 1$ and then show that the maximum value of $f(p, q, r)$ is $\frac{2}{3}$.

69. In example 7.4, we found the "best" linear fit to population data using the least-squares criterion. Use the least-squares criterion to find the best quadratic fit to the data. That is, for functions of the form $ax^2 + bx + c$, find the values of the constants a, b and c that minimize the sum of the squares of the residuals. For the given data, show that the sum of the squares of the residuals for the quadratic model is less than for the linear model. Explain why this has to be true mathematically. In spite of this, explain why the linear model might be preferable to the quadratic model. (Hint: Use both models to predict 100 years into the future and backtrack 100 years into the past.)

70. Use the least-squares criterion to find the best exponential fit $(y = ae^{bx})$ to the data of example 7.4. Compare the actual residuals of this model to the actual residuals of the linear model. Explain why there is some theoretical justification for using an exponential model, and then discuss the advantages and disadvantages of the exponential and linear models.

71. A practical flaw with the method of steepest ascent presented in example 7.5 is that the equation $g'(h) = 0$ may be difficult to solve. An alternative is to use Newton's method to approximate a solution. A method commonly used in practice is to approximate h using one iteration of Newton's method with initial guess $h = 0$. We derive the resulting formula here. Recall that $g(h) = f(x_k + ah, y_k + bh)$ where (x_k, y_k) is the current point and $\langle a, b \rangle = \nabla f(x_k, y_k)$. Newton's method applied to $g'(h) = 0$ with $h_0 = 0$ is given by $h_1 = -\frac{g'(0)}{g''(0)}$. Show that $g'(0) = af_x(x_k, y_k) + bf_y(x_k, y_k) = a^2 + b^2 = \nabla f(x_k, y_k) \cdot \nabla f(x_k, y_k)$. Also, show that $g''(0) = a^2 f_{xx}(x_k, y_k) + 2ab f_{xy}(x_k, y_k) + b^2 f_{yy}(x_k, y_k) = \nabla f(x_k, y_k) \cdot H(x_k, y_k) \nabla f(x_k, y_k)$, where the Hessian matrix is defined by $H = \begin{bmatrix} f_{xx} & f_{xy} \\ f_{yx} & f_{yy} \end{bmatrix}$. Putting this together with the work in example 7.5, the method of steepest ascent becomes $\mathbf{v}_{k+1} = \mathbf{v}_k - \frac{\nabla f(\mathbf{v}_k) \cdot \nabla f(\mathbf{v}_k)}{\nabla f(\mathbf{v}_k) \cdot H(\mathbf{v}_k) \nabla f(\mathbf{v}_k)} \nabla f(\mathbf{v}_k)$, where $\mathbf{v}_k = \begin{bmatrix} x_k \\ y_k \end{bmatrix}$.

12.8 CONSTRAINED OPTIMIZATION AND LAGRANGE MULTIPLIERS

The local extrema we found in section 12.7 form one piece of the optimization puzzle. In many applications, the goal is not to identify theoretical maximum or minimum values, but to achieve the absolute best possible product given a large set of constraints such as limited resources or technology. For example, an automotive engineer's objective might be to minimize wind drag in the design of a car. However, new designs are severely limited by customer demands of luxury and attractiveness and particularly by manufacturing constraints such as cost. In this section, we develop a technique for finding the maximum or minimum of a function given some constraint on the function's domain.

We first consider the two-dimensional geometric problem of finding the point on the line $y = 3 - 2x$ that is closest to the origin. As is often the case, there are several ways to solve this problem. We use a highly geometric approach, which we then extend to an important general result. We show a graph of the line in Figure 12.50a. Notice that the set of points that are 1 unit from the origin form the circle $x^2 + y^2 = 1$. In Figure 12.50b, you can see that the line $y = 3 - 2x$ lies entirely outside this circle. This tells us that every point on the line $y = 3 - 2x$ lies more than 1 unit from the origin. Looking at the circle $x^2 + y^2 = 4$ in Figure 12.50c, you can clearly see that there are infinitely many points on the line $y = 3 - 2x$ that are less than 2 units from the origin. If we shrink the circle in Figure 12.50c (or enlarge the circle in Figure 12.50b), it will eventually reach a size at which the line is tangent to the circle (see Figure 12.50d). You should recognize that the point of tangency is the closest point on the line to the origin, since all other points on the line are outside the circle and hence, are farther away from the origin.

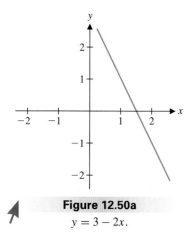

Figure 12.50a

$y = 3 - 2x$.

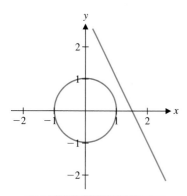

Figure 12.50b

$y = 3 - 2x$ and the circle of radius 1 centered at $(0, 0)$.

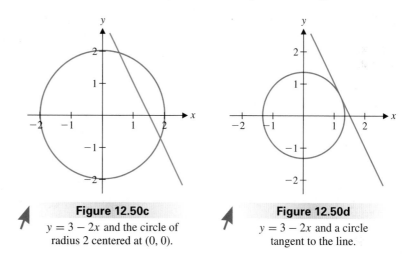

Figure 12.50c

$y = 3 - 2x$ and the circle of radius 2 centered at $(0, 0)$.

Figure 12.50d

$y = 3 - 2x$ and a circle tangent to the line.

We now translate the preceding geometric argument into the language of calculus. We want to minimize the distance from the point (x, y) to the origin. Using the usual distance formula, this distance is given by $\sqrt{x^2 + y^2}$. Before we continue, we must point out that the distance is minimized at exactly the same point at which the square of the distance is minimized. Notice that if we instead minimize the square of the distance, given by $x^2 + y^2$, we avoid the mess created by the presence of the square root in the distance formula. So, instead, we minimize $f(x, y) = x^2 + y^2$, subject to the constraint that the point lie on the line (i.e., that $y = 3 - 2x$) or $g(x, y) = 2x + y - 3 = 0$. We have already argued that at the closest point, the line and circle are tangent. Recall that the gradient vector for a given function is orthogonal to its level curves at any given point. So, for a level curve of f to be tangent to the constraint curve $g(x, y) = 0$, the gradients of f and g must be parallel. That is, at the closest point (x, y) to the origin, we must have $\nabla f(x, y) = \lambda \nabla g(x, y)$ for some constant λ. We solve this equation to find the closest point to the origin in the example that follows.

Example 8.1   Finding a Minimum Distance

Use the relationship $\nabla f(x, y) = \lambda \nabla g(x, y)$ and the constraint $y = 3 - 2x$ to find the point on the line $y = 3 - 2x$ that is closest to the origin.

Solution   For $f(x, y) = x^2 + y^2$, we have $\nabla f(x, y) = \langle 2x, 2y \rangle$ and for $g(x, y) = 2x + y - 3$, we have $\nabla g(x, y) = \langle 2, 1 \rangle$. The vector equation $\nabla f(x, y) = \lambda \nabla g(x, y)$ becomes

$$\langle 2x, 2y \rangle = \lambda \langle 2, 1 \rangle,$$

from which it follows that

$$2x = 2\lambda \quad \text{and} \quad 2y = \lambda.$$

Solving the second equation for λ, we have $\lambda = 2y$. The first equation then gives us $x = \lambda = 2y$. Substituting $x = 2y$ into the constraint equation $y = 3 - 2x$, we have $y = 3 - 2(2y)$, or $5y = 3$. The solution is $y = \frac{3}{5}$, giving us $x = 2y = \frac{6}{5}$. The closest point is then $\left(\frac{6}{5}, \frac{3}{5} \right)$. Look carefully at Figure 12.50d and recognize that this is consistent with our graphical solution.

It turns out that the technique illustrated in example 8.1 can be applied to a wide variety of constrained optimization problems. We will now develop this method, referred to as the **method of Lagrange multipliers.**

Suppose that we want to find maximum or minimum values of the function $f(x, y, z)$, subject to the constraint that $g(x, y, z) = 0$. We assume that both f and g have continuous first partial derivatives. Now, suppose that f has an extremum at (x_0, y_0, z_0) lying on the level surface S defined by $g(x, y, z) = 0$. Let C be any curve lying on the level surface and passing through the point (x_0, y_0, z_0). Assume that C is traced out by the terminal point of the vector-valued function $\mathbf{r}(t) = \langle x(t), y(t), z(t) \rangle$ and that $\mathbf{r}(t_0) = \langle x_0, y_0, z_0 \rangle$. Define a function of the single variable t by

$$h(t) = f(x(t), y(t), z(t)).$$

Notice that if $f(x, y, z)$ has an extremum at (x_0, y_0, z_0), then $h(t)$ must have an extremum at t_0 and so, $h'(t_0) = 0$. From the chain rule, we get that

$$\begin{aligned} 0 = h'(t_0) &= f_x(x_0, y_0, z_0)x'(t_0) + f_y(x_0, y_0, z_0)y'(t_0) + f_z(x_0, y_0, z_0)z'(t_0) \\ &= \langle f_x(x_0, y_0, z_0), f_y(x_0, y_0, z_0), f_z(x_0, y_0, z_0) \rangle \cdot \langle x'(t_0), y'(t_0), z'(t_0) \rangle \\ &= \nabla f(x_0, y_0, z_0) \cdot \mathbf{r}'(t_0). \end{aligned}$$

That is, if $f(x_0, y_0, z_0)$ is an extremum, the gradient of f at (x_0, y_0, z_0) is orthogonal to the tangent vector $\mathbf{r}'(t_0)$. Since C was an arbitrary curve lying on the level surface S, it follows that $\nabla f(x_0, y_0, z_0)$ must be orthogonal to every curve lying on the level surface S and so, too is orthogonal to S. Recall from Theorem 6.2 that ∇g is also orthogonal to the level surface $g(x, y, z) = 0$, so that $\nabla f(x_0, y_0, z_0)$ and $\nabla g(x_0, y_0, z_0)$ must be parallel. This proves the following result.

Theorem 8.1

Suppose that $f(x, y, z)$ and $g(x, y, z)$ are functions with continuous first partial derivatives and $\nabla g(x, y, z) \neq \mathbf{0}$ on the surface $g(x, y, z) = 0$. Suppose that either

(i) the minimum value of $f(x, y, z)$ subject to the constraint $g(x, y, z) = 0$ occurs at (x_0, y_0, z_0); or

(ii) the maximum value of $f(x, y, z)$ subject to the constraint $g(x, y, z) = 0$ occurs at (x_0, y_0, z_0).

Then $\nabla f(x_0, y_0, z_0) = \lambda \nabla g(x_0, y_0, z_0)$ for some constant λ (called a **Lagrange multiplier**).

Note that Theorem 8.1 says that if $f(x, y, z)$ has an extremum at a point (x_0, y_0, z_0) on the surface $g(x, y, z) = 0$, we will have for $(x, y, z) = (x_0, y_0, z_0)$,

$$f_x(x, y, z) = \lambda g_x(x, y, z),$$
$$f_y(x, y, z) = \lambda g_y(x, y, z),$$
$$f_z(x, y, z) = \lambda g_z(x, y, z)$$

and

$$g(x, y, z) = 0.$$

Finding such extrema then boils down to solving these four equations for the four unknowns x, y, z and λ. (Actually, we only need the values of x, y and z.)

It's important to recognize that, just like when we find critical points in unconstrained optimization, this method only produces **candidates** for extrema. Along with finding a solution(s) to the above four equations, you need to verify (graphically as we did in example 8.1 or by some other means) that the solution you found in fact represents the desired optimal point.

Notice that the Lagrange multiplier method we have just developed can also be applied to functions of two variables, by ignoring the third variable in Theorem 8.1. That is, if $f(x, y)$ and $g(x, y)$ have continuous first partial derivatives and $f(x_0, y_0)$ is an extremum of f, where (x_0, y_0) lies on the level curve $g(x, y) = 0$, we must have

$$\nabla f(x_0, y_0) = \lambda \nabla g(x_0, y_0),$$

for some constant λ. Graphically, this says that if $f(x_0, y_0)$ is an extremum, the level curve of f passing through (x_0, y_0) is tangent to the constraint curve $g(x, y) = 0$ at (x_0, y_0). We illustrate this in Figure 12.51. In this case, we end up with the three equations

$$f_x(x, y) = \lambda g_x(x, y), \quad f_y(x, y) = \lambda g_y(x, y) \quad \text{and} \quad g(x, y) = 0$$

for the three unknowns x, y and λ. We illustrate this in the following example.

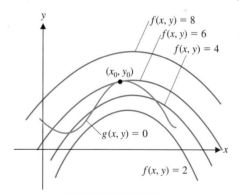

Figure 12.51
Level curve tangent to constraint
curve at an extremum.

Example 8.2 Finding the Optimal Thrust of a Rocket

A rocket is launched with a constant thrust corresponding to an acceleration of u ft/s^2. Ignoring air resistance, the rocket's height after t seconds is given by $f(t, u) = \frac{1}{2}(u - 32)t^2$ feet. Fuel usage for t seconds is proportional to $u^2 t$, so the limited fuel capacity of the rocket can be expressed by an equation of the form $u^2 t = 10{,}000$. Find the value of u that maximizes the height that the rocket reaches.

Solution   From Theorem 8.1, we look for solutions of $\nabla f(t, u) = \lambda \nabla g(t, u)$ where $g(t, u) = u^2 t - 10,000 = 0$ is the constraint equation. We have $\nabla f(t, u) = \langle (u - 32)t, \frac{1}{2}t^2 \rangle$ and $\nabla g(t, u) = \langle u^2, 2ut \rangle$. From Theorem 8.1, we must have

$$\left\langle (u - 32)t, \frac{1}{2}t^2 \right\rangle = \lambda \langle u^2, 2ut \rangle,$$

for some constant λ. It follows that

$$(u - 32)t = \lambda u^2 \quad \text{and} \quad \frac{1}{2}t^2 = \lambda 2ut.$$

Solving both equations for λ, we have

$$\lambda = \frac{(u - 32)t}{u^2} = \frac{\frac{1}{2}t^2}{2ut}.$$

This gives us

$$2u(u - 32)t^2 = \frac{1}{2}t^2 u^2.$$

Solutions include $t = 0$ and $u = 0$. On physical grounds, we can argue that these solutions represent minimum heights, since zero thrust or zero time would correspond to zero height. Canceling the factors of t^2 and u, we have $4(u - 32) = u$. The solution to this is $u = \frac{128}{3}$. With this value of u, the engines can burn for

$$t = \frac{10,000}{u^2} = \frac{10,000}{(128/3)^2} \approx 5.5 \text{ seconds},$$

with the rocket reaching a height of $z = \frac{1}{2}\left(\frac{128}{3} - 32\right)(5.5)^2 \approx 161$ feet.

We left example 8.2 unfinished. (Can you tell what's missing?) It is very difficult to argue that our solution actually represents a **maximum** height. (Could it be a saddle point?) What we do know is that by Theorem 8.1, **if** there is a maximum, we found it. Returning to a discussion of the physical problem, it should be completely reasonable that with a limited amount of fuel, there is a maximum attainable altitude and so, we did indeed find the maximum altitude.

Theorem 8.1 provides another major piece of the optimization puzzle. We now solve a problem where the goal is to optimize a function subject to an inequality constraint of the form $g(x, y) \leq c$. To understand our technique, recall how we solved for absolute extrema of functions of several variables on a closed and bounded region in section 12.7. We found critical points in the interior of the region, and compared the values of the function at the critical points with the maximum and minimum function values on the boundary of the region. To find the extrema of $f(x, y)$ subject to a constraint of the form $g(x, y) \leq c$, we first find the critical points of $f(x, y)$ that satisfy the constraint, then find the extrema of the function on the boundary $g(x, y) = c$ (the constraint curve) and finally, compare the function values. We illustrate this in the following example.

Example 8.3   Optimization with an Inequality Constraint

Suppose that the temperature of a metal plate is given by $T(x, y) = x^2 + 2x + y^2$, for points (x, y) on the elliptical plate defined by $x^2 + 4y^2 \leq 24$. Find the maximum and minimum temperatures on the plate.

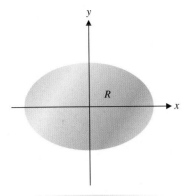

Figure 12.52

A metal plate.

Solution The plate corresponds to the shaded region R shown in Figure 12.52. We first look for critical points of $T(x, y)$ inside the region R. We have $\nabla T(x, y) = \langle 2x + 2, 2y \rangle = \langle 0, 0 \rangle$ if $x = -1$ and $y = 0$. At this point, $T(-1, 0) = -1$. We next look for the extrema of $T(x, y)$ on the ellipse $x^2 + 4y^2 = 24$. We first rewrite the constraint equation as $g(x, y) = x^2 + 4y^2 - 24 = 0$. From Theorem 8.1, any extrema will satisfy the Lagrange multiplier equation: $\nabla T(x, y) = \lambda \nabla g(x, y)$ or

$$\langle 2x + 2, 2y \rangle = \lambda \langle 2x, 8y \rangle = \langle 2\lambda x, 8\lambda y \rangle.$$

This occurs when

$$2x + 2 = 2\lambda x \quad \text{and} \quad 2y = 8\lambda y.$$

Notice that the second equation holds when $y = 0$ or $\lambda = \frac{1}{4}$. If $y = 0$, the constraint $x^2 + 4y^2 = 24$ gives $x = \pm\sqrt{24}$. If $\lambda = \frac{1}{4}$, the first equation becomes $2x + 2 = \frac{1}{2}x$ so that $x = -\frac{4}{3}$. The constraint $x^2 + 4y^2 = 24$ now gives $y = \pm\frac{\sqrt{50}}{3}$. We now need only compare the function values at all of these points (the one interior critical point and the candidates for boundary extrema) as follows:

$$T(-1, 0) = -1,$$
$$T\left(\sqrt{24}, 0\right) = 24 + 2\sqrt{24} \approx 33.8,$$
$$T\left(-\sqrt{24}, 0\right) = 24 - 2\sqrt{24} \approx 14.2,$$
$$T\left(-\frac{4}{3}, \frac{\sqrt{50}}{3}\right) = \frac{14}{3} \approx 4.7$$

and

$$T\left(-\frac{4}{3}, -\frac{\sqrt{50}}{3}\right) = \frac{14}{3} \approx 4.7.$$

From this list, it's easy to identify the minimum value of -1 at the point $(-1, 0)$ and the maximum value of $24 + 2\sqrt{24}$ at the point $(\sqrt{24}, 0)$.

■———

In the following example, we illustrate the use of Lagrange multipliers for functions of three variables. In the course of doing so, we develop an interpretation of the Lagrange multiplier λ.

Example 8.4 Finding an Optimal Level of Production

For a business that produces three products, suppose that when producing x, y and z thousand units of the products, the profit of the company (in thousands of dollars) can be modeled by $P(x, y, z) = 4x + 8y + 6z$. Manufacturing constraints force $x^2 + 4y^2 + 2z^2 \leq 800$. Find the maximum profit for the company. Rework the problem with the constraint $x^2 + 4y^2 + 2z^2 \leq 801$ and use the result to interpret the meaning of λ.

Solution We start with $\nabla P(x, y, z) = \langle 4, 8, 6 \rangle$ and note that there are no critical points. This says that the extrema must lie on the boundary of the constraint region. That is, they must satisfy the constraint equation $g(x, y, z) = x^2 + 4y^2 + 2z^2 - 800 = 0$. From Theorem 8.1, the Lagrange multiplier equation is $\nabla P(x, y, z) = \lambda \nabla g(x, y, z)$ or

$$\langle 4, 8, 6 \rangle = \lambda \langle 2x, 8y, 4z \rangle = \langle 2\lambda x, 8\lambda y, 4\lambda z \rangle.$$

This occurs when

$$4 = 2\lambda x, \quad 8 = 8\lambda y \quad \text{and} \quad 6 = 4\lambda z.$$

From the first equation, we get $x = \frac{2}{\lambda}$. The second equation gives us $y = \frac{1}{\lambda}$ and finally, the third equation gives us $z = \frac{3}{2\lambda}$. From the constraint equation $x^2 + 4y^2 + 2z^2 = 800$, we now have

$$800 = \left(\frac{2}{\lambda}\right)^2 + 4\left(\frac{1}{\lambda}\right)^2 + 2\left(\frac{3}{2\lambda}\right)^2 = \frac{25}{2\lambda^2},$$

so that $\lambda^2 = \frac{25}{1600}$ and

$$\lambda = \frac{1}{8}.$$

(Why did we choose the positive sign for λ? Hint: Think about what x, y and z represent. Since $x > 0$, we must have $\lambda = \frac{2}{x} > 0$.) The only candidate for an extremum is then

$$x = \frac{2}{\lambda} = 16, \quad y = \frac{1}{\lambda} = 8 \quad \text{and} \quad z = \frac{3}{2\lambda} = 12,$$

where the corresponding profit is

$$P(16, 8, 9) = 4(16) + 8(8) + 6(12) = 200.$$

Observe that this is the maximum profit, while $\lambda = -\frac{1}{8}$ gives the minimum value of the profit function. Notice that if the constant on the right-hand side of the constraint equation is changed to 801, the first difference occurs in solving for λ, where we now get

$$801 = \frac{25}{2\lambda^2},$$

so that $\lambda \approx 0.12492$, $x = \frac{2}{\lambda} \approx 16.009997$, $y = \frac{1}{\lambda} \approx 8.004998$ and $z = \frac{3}{2\lambda} \approx 12.007498$. In this case, the maximum profit is

$$P(16.009997, 8.004998, 12.007498) \approx 200.12496.$$

It is interesting to observe that the increase in profit is

$$P(16.009997, 8.004998, 12.007498) - P(16, 8, 9) \approx 200.12496 - 200$$

$$= 0.12496 \approx \lambda.$$

As you might suspect from this observation, the Lagrange multiplier λ actually gives you the instantaneous rate of change of the profit with respect to a change in the production constraint.

We close this section by considering the case of finding the minimum or maximum value of a differentiable function $f(x, y, z)$ subject to two constraints $g(x, y, z) = 0$ and $h(x, y, z) = 0$, where g and h are also differentiable. Notice that for both constraints to be satisfied at a point (x, y, z), the point must lie on both surfaces defined by the constraints. Consequently, in order for there to be a solution, we must assume that the two surfaces intersect. We further assume that ∇g and ∇h are nonzero and are not parallel, so that the two surfaces intersect in a curve C and are not tangent to one another. As we have already seen, if f has an extremum at a point (x_0, y_0, z_0) on a curve C, then $\nabla f(x_0, y_0, z_0)$ must be normal to the curve. Notice that since C lies on both constraint surfaces, $\nabla g(x_0, y_0, z_0)$ and $\nabla h(x_0, y_0, z_0)$ are both orthogonal to C at (x_0, y_0, z_0). This says that $\nabla f(x_0, y_0, z_0)$ must

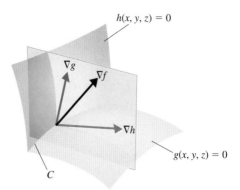

Figure 12.53
Constraint surfaces and the plane
determined by the normal vectors
∇g and ∇h.

lie in the plane determined by $\nabla g(x_0, y_0, z_0)$ and $\nabla h(x_0, y_0, z_0)$ (see Figure 12.53). That is, for $(x, y, z) = (x_0, y_0, z_0)$ and some constants λ and μ (Lagrange multipliers),

$$\nabla f(x, y, z) = \lambda \nabla g(x, y, z) + \mu \nabla h(x, y, z).$$

The method of Lagrange multipliers for the case of two constraints then consists of finding the point (x, y, z) and the Lagrange multipliers λ and μ (for a total of five unknowns) satisfying the five equations defined by:

$$f_x(x, y, z) = \lambda g_x(x, y, z) + \mu h_x(x, y, z)$$
$$f_y(x, y, z) = \lambda g_y(x, y, z) + \mu h_y(x, y, z)$$
$$f_z(x, y, z) = \lambda g_z(x, y, z) + \mu h_z(x, y, z)$$
$$g(x, y, z) = 0$$

and
$$h(x, y, z) = 0.$$

We illustrate the use of Lagrange multipliers for the case of two constraints in the following example.

Example 8.5 Optimization with Two Constraints

The plane $x + y + z = 12$ intersects the paraboloid $z = x^2 + y^2$ in an ellipse. Find the point on the ellipse that is closest to the origin.

Solution We illustrate the intersection of the plane with the paraboloid in Figure 12.54 (on the following page). Observe that minimizing the distance to the origin is equivalent to minimizing $f(x, y, z) = x^2 + y^2 + z^2$ [the **square** of the distance from the point (x, y, z) to the origin]. Further, the constraints may be written as $g(x, y, z) = x + y + z - 12 = 0$ and $h(x, y, z) = x^2 + y^2 - z = 0$. At any extremum, we must have that

$$\nabla f(x, y, z) = \lambda \nabla g(x, y, z) + \mu \nabla h(x, y, z)$$

or

$$\langle 2x, 2y, 2z \rangle = \lambda \langle 1, 1, 1 \rangle + \mu \langle 2x, 2y, -1 \rangle.$$

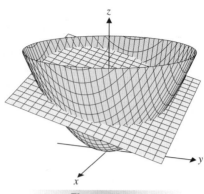

Figure 12.54
Intersection of a paraboloid
and a plane.

Together with the constraint equations, we now have the system of equations

$$2x = \lambda + 2\mu x \tag{8.1}$$

$$2y = \lambda + 2\mu y \tag{8.2}$$

$$2z = \lambda - \mu \tag{8.3}$$

$$x + y + z - 12 = 0 \tag{8.4}$$

and

$$x^2 + y^2 - z = 0. \tag{8.5}$$

From (8.1), we have

$$\lambda = 2x(1 - \mu),$$

while from (8.2), we have

$$\lambda = 2y(1 - \mu).$$

Setting these two expressions for λ equal gives us

$$2x(1 - \mu) = 2y(1 - \mu),$$

from which it follows that either $\mu = 1$ (in which case $\lambda = 0$) or $x = y$. However, if $\mu = 1$ and $\lambda = 0$, we have from (8.3) that $z = -\frac{1}{2}$, which contradicts (8.5). Consequently, the only possibility is to have $x = y$, from which it follows from (8.5) that $z = 2x^2$. Substituting this into (8.4) gives us

$$0 = x + y + z - 12 = x + x + 2x^2 - 12$$
$$= 2x^2 + 2x - 12 = 2(x^2 + x - 6) = 2(x + 3)(x - 2),$$

so that $x = -3$ or $x = 2$. Since $y = x$ and $z = 2x^2$, we have that $(2, 2, 8)$ and $(-3, -3, 18)$ are the only candidates for extrema. Finally, since

$$f(2, 2, 8) = 72 \quad \text{and} \quad f(-3, -3, 18) = 342,$$

the closest point on the intersection of the two surfaces to the origin is $(2, 2, 8)$. By the same reasoning, observe that the furthest point on the intersection of the two surfaces from the origin is $(-3, -3, 18)$. Notice that these are also consistent with what you can see in Figure 12.54.

The method of Lagrange multipliers can be extended in a straightforward fashion to the case of minimizing or maximizing a function subject to any number of constraints.

EXERCISES 12.8

1. Explain why the point of tangency in Figure 12.50d must be the closest point to the origin.

2. Explain why in example 8.1 you know that the critical point found corresponds to the minimum distance and not the maximum distance or a saddle point.

3. In example 8.2, explain in physical terms why there would be a value of u that would maximize the rocket's height. In particular, explain why a larger value of u wouldn't *always* produce a larger height.

4. In example 8.4, we showed that the Lagrange multiplier λ corresponds to the rate of change of profit with respect to a change in production level. Explain how knowledge of this value (positive, negative, small, large) would be useful to a plant manager.

In exercises 5–12, use Lagrange multipliers to find the closest point on the given curve to the indicated point.

5. $y = 3x - 4$, origin

6. $y = 2x + 1$, origin

7. $y = 3 - 2x$, $(4, 0)$

8. $y = x - 2$, $(0, 2)$

9. $y = x^2$, $(3, 0)$

10. $y = x^2$, $(0, 2)$

11. $y = x^2$, $\left(2, \frac{1}{2}\right)$

12. $y = x^2 - 1$, $(1, 2)$

In exercises 13–20, use Lagrange multipliers to find the maximum and minimum of the function $f(x, y)$ subject to the constraint $g(x, y) = c$.

13. $f(x, y) = 4xy$ subject to $x^2 + y^2 = 8$

14. $f(x, y) = 4xy$ subject to $4x^2 + y^2 = 8$

15. $f(x, y) = 4x^2 y$ subject to $x^2 + y^2 = 3$

16. $f(x, y) = 2x^3 y$ subject to $x^2 + y^2 = 4$

17. $f(x, y) = xe^y$ subject to $x^2 + y^2 = 2$

18. $f(x, y) = e^{2x+y}$ subject to $x^2 + y^2 = 5$

19. $f(x, y) = x^2 e^y$ subject to $x^2 + y^2 = 3$

20. $f(x, y) = x^2 y^2$ subject to $x^2 + 4y^2 = 24$

In exercises 21–24, find the maximum and minimum of the function $f(x, y)$ subject to the constraint $g(x, y) \leq c$.

21. $f(x, y) = 4xy$ subject to $x^2 + y^2 \leq 8$

22. $f(x, y) = 4xy$ subject to $4x^2 + y^2 \leq 8$

23. $f(x, y) = 4x^2 y$ subject to $x^2 + y^2 \leq 3$

24. $f(x, y) = 2x^3 y$ subject to $x^2 + y^2 \leq 4$

25. Rework example 8.2 with extra fuel, so that $u^2 t = 11,000$.

26. In exercise 25, compute λ. Comparing solutions to example 8.2 and exercise 25, compute the change in z divided by the change in $u^2 t$.

27. Suppose that the business in example 8.4 has profit function $P(x, y, z) = 3x + 6y + 6z$ and manufacturing constraint $2x^2 + y^2 + 4z^2 \leq 8800$. Maximize the profits.

28. Suppose that the business in example 8.4 has profit function $P(x, y, z) = 3xz + 6y$ and manufacturing constraint $x^2 + 2y^2 + z^2 \leq 6$. Maximize the profits.

29. In exercise 27, show that the Lagrange multiplier gives the rate of change of the profit relative to a change in the production constraint.

30. Use the value of λ (do not solve any equations) to determine the amount of profit if the constraint in exercise 28 is changed to $x^2 + 2y^2 + z^2 \leq 7$.

31. Minimize $2x + 2y$ subject to the constraint $xy = c$ for some constant $c > 0$ and conclude that for a given area, the rectangle with smallest perimeter is the square.

32. As in exercise 31, find the rectangular box of a given volume that has the minimum surface area.

33. Maximize $y - x$ subject to the constraint $x^2 + y^2 = 1$.

34. Maximize e^{x+y} subject to the constraint $x^2 + y^2 = 2$.

35. In the picture, a sailboat is sailing into a crosswind. The wind is blowing out of the north, the sail is at an angle α to the east of due north and at an angle β north of the hull of the boat. The hull, in turn, is at an angle θ to the north of due east. Explain why $\alpha + \beta + \theta = \frac{\pi}{2}$. If the wind is blowing with speed w, then the northward component of the wind's force on the boat is given by $w \sin \alpha \sin \beta \sin \theta$. If this component is positive, the boat can travel "against the wind." Taking $w = 1$ for

convenience, maximize $\sin\alpha\sin\beta\sin\theta$ subject to the constraint $\alpha + \beta + \theta = \frac{\pi}{2}$.

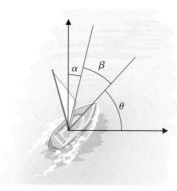

36. Suppose a music company sells two types of speakers. The profit for selling x speakers of style A and y speakers of style B is modeled by $f(x, y) = x^3 + y^3 - 5xy$. The company can't manufacture more than k speakers total in a given month for some constant $k > 5$. Show that the maximum profit is $\frac{k^2(k-5)}{4}$ and show that $\lambda = \frac{df}{dk}$.

37. Consider the problem of finding extreme values of xy^2 subject to $x + y = 0$. Show that the Lagrange multiplier method identifies $(0, 0)$ as a critical point. Show that this point is neither a local minimum nor a local maximum.

38. Make the substitution $y = -x$ in the function $f(x, y) = xy^2$. Show that $x = 0$ is a critical point and determine what type point is at $x = 0$. Explain why the Lagrange multiplier method fails in exercise 37.

39. The production of a company is given by the Cobb-Douglas function $P = 200L^{2/3}K^{1/3}$. Cost constraints on the business

force $2L + 5K \leq 150$. Find the values of the labor L and capital K to maximize production.

40. Maximize the profit $P = 4x + 5y$ of a business given the production possibilities constraint curve $2x^2 + 5y^2 \leq 32{,}500$.

In exercises 41 and 42, you will illustrate the least-cost rule.

41. Minimize the cost function $C = 25L + 100K$, given the production constraint $P = 60L^{2/3}K^{1/3} = 1920$.

42. In exercise 41, show that the minimum cost occurs when the ratio of marginal productivity of labor, $\frac{\partial P}{\partial L}$, to the marginal productivity of capital, $\frac{\partial P}{\partial K}$, equals the ratio of the price of labor, $\frac{\partial C}{\partial L}$, to the price of capital, $\frac{\partial C}{\partial K}$.

43. Minimize $f(x, y, z) = x^2 + y^2 + z^2$, subject to the constraints $x + 2y + 3z = 6$ and $y + z = 0$.

44. Interpret the function $f(x, y, z)$ of exercise 43 in terms of the distance from a point (x, y, z) to the origin. Sketch the two planes given in exercise 43. Interpret exercise 43 as finding the closest point on a line to the origin.

45. Maximize $f(x, y, z) = xyz$, subject to the constraints $x + y + z = 4$ and $x + y - z = 0$.

46. Maximize $f(x, y, z) = 3x + y + 2z$, subject to the constraints $y^2 + z^2 = 1$ and $x + y - z = 1$.

47. Find the points on the intersection of $x^2 + y^2 = 1$ and $x^2 + z^2 = 1$ that are (a) closest to and (b) farthest from the origin.

48. Find the point on the intersection of $x + 2y + z = 2$ and $y = x$ that is closest to the origin.

CHAPTER REVIEW EXERCISES

In exercises 1–10, sketch the graph of $z = f(x, y)$.

1. $f(x, y) = x^2 - y^2$

2. $f(x, y) = \sqrt{x^2 + y^2}$

3. $f(x, y) = 2 - x^2 - y^2$

4. $f(x, y) = \sqrt{2 - x^2 - y^2}$

5. $f(x, y) = \frac{3}{x^2} + \frac{2}{y^2}$

6. $f(x, y) = \frac{x^5}{y}$

7. $f(x, y) = \sin(x^2 y)$

8. $f(x, y) = \sin(y - x^2)$

9. $f(x, y) = 3xe^y - x^3 - e^{3y}$

10. $f(x, y) = 4x^2 e^y - 2x^4 - e^{4y}$

11. In parts a–f, match the functions to the surfaces.
    (a) $f(x, y) = \sin xy$
    (b) $f(x, y) = \sin(x/y)$
    (c) $f(x, y) = \sin\sqrt{x^2 + y^2}$
    (d) $f(x, y) = x \sin y$
    (e) $f(x, y) = \dfrac{4}{2x^2 + 3y^2 - 1}$
    (f) $f(x, y) = \dfrac{4}{2x^2 + 3y^2 + 1}$

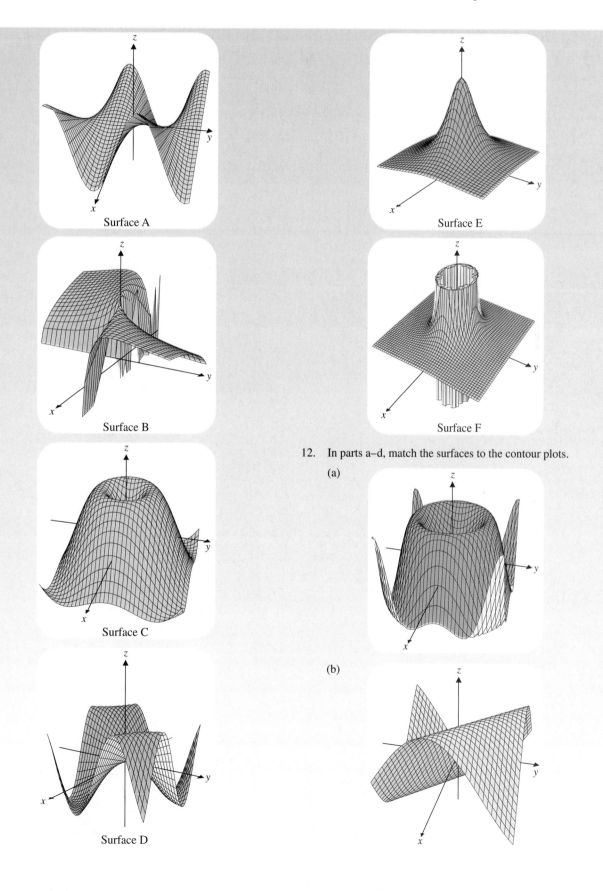

Surface A

Surface B

Surface C

Surface D

Surface E

Surface F

12. In parts a–d, match the surfaces to the contour plots.

(a)

(b)

(c)

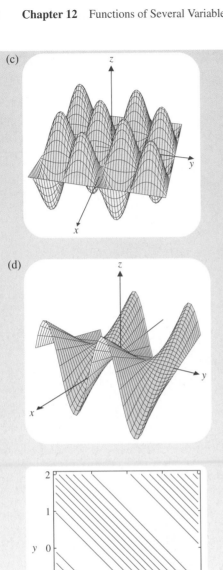

(d)

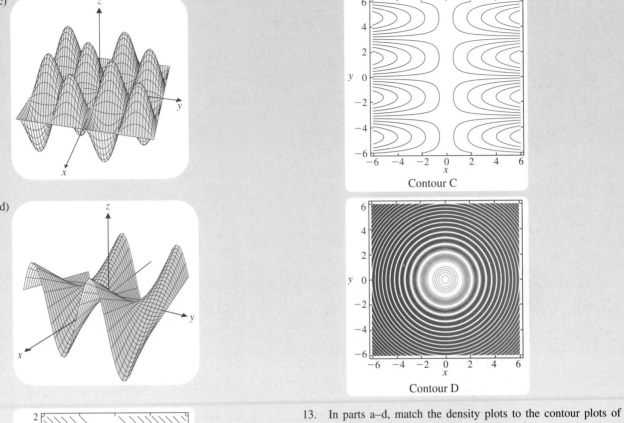

Contour C

Contour D

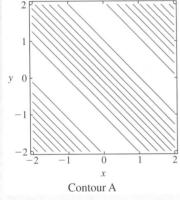

Contour A

13. In parts a–d, match the density plots to the contour plots of exercise 12.

(a)

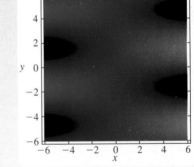

(b)

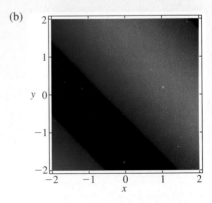

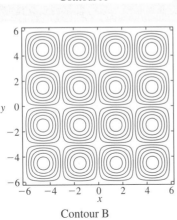

Contour B

(c)

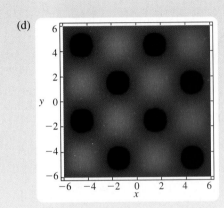

(d)

14. Compute the indicated limit.

(a) $\displaystyle\lim_{(x,y)\to(0,2)} \frac{3x}{y^2+1}$ (b) $\displaystyle\lim_{(x,y)\to(1,\pi)} \frac{xy-1}{\cos xy}$

In exercises 15–18, show that the indicated limit does not exist.

15. $\displaystyle\lim_{(x,y)\to(0,0)} \frac{3x^2 y}{x^4+y^2}$ 16. $\displaystyle\lim_{(x,y)\to(0,0)} \frac{2xy^{3/2}}{x^2+y^3}$

17. $\displaystyle\lim_{(x,y)\to(0,0)} \frac{x^2+y^2}{x^2+xy+y^2}$ 18. $\displaystyle\lim_{(x,y)\to(0,0)} \frac{x^2}{x^2+xy+y^2}$

In exercises 19 and 20, show that the indicated limit exists.

19. $\displaystyle\lim_{(x,y)\to(0,0)} \frac{x^3+xy^2}{x^2+y^2}$ 20. $\displaystyle\lim_{(x,y)\to(0,0)} \frac{3y^2 \ln(x+1)}{x^2+3y^2}$

In exercises 21 and 22, find the region on which the function is continuous.

21. $f(x,y) = 3x^2 e^{4y} - \dfrac{3y}{x}$ 22. $f(x,y) = \sqrt{4-4x^2-y^2}$

In exercises 23–26, find both first-order partial derivatives.

23. $f(x,y) = \dfrac{4x}{y} + xe^{xy}$ 24. $f(x,y) = xe^{xy} + 3y^2$

25. $f(x,y) = 3x^2 y \cos y - \sqrt{x}$

26. $f(x,y) = \sqrt{x^3 y} + 3x - 5$

27. Show that the function $f(x,y) = e^x \sin y$ satisfies **Laplace's equation** $\dfrac{\partial^2 f}{\partial x^2} + \dfrac{\partial^2 f}{\partial y^2} = 0$.

28. Show that the function $f(x,y) = e^x \cos y$ satisfies **Laplace's equation** (see exercise 27).

In exercises 29 and 30, use the chart to estimate the partial derivatives.

29. $\dfrac{\partial f}{\partial x}(0,0)$ and $\dfrac{\partial f}{\partial y}(0,0)$ 30. $\dfrac{\partial f}{\partial x}(10,0)$ and $\dfrac{\partial f}{\partial y}(10,0)$

y \ x	−20	−10	0	10	20
−20	2.4	2.1	0.8	0.5	1.0
−10	2.6	2.2	1.4	1.0	1.2
0	2.7	2.4	2.0	1.6	1.2
10	2.9	2.5	2.6	2.2	1.8
20	3.1	2.7	3.0	2.9	2.7

In exercises 31–34, compute the linear approximation of the function at the given point.

31. $f(x,y) = 3y\sqrt{x^2+5}$ at $(-2,5)$

32. $f(x,y) = \dfrac{x+2}{4y-2}$ at $(2,3)$

33. $f(x,y) = \tan(x+2y)$ at $\left(\pi, \frac{\pi}{2}\right)$

34. $f(x,y) = \ln(x^2+3y)$ at $(4,2)$

In exercises 35 and 36, find the indicated derivatives.

35. $f(x,y) = 2x^4 y + 3x^2 y^2;\ f_{xx}, f_{yy}, f_{xy}$

36. $f(x,y) = x^2 e^{3y} - \sin y;\ f_{xx}, f_{yy}, f_{yyx}$

In exercises 37–40, find an equation of the tangent plane.

37. $z = x^2 y + 2x - y^2$ at $(1,-1,0)$

38. $z = \sqrt{x^2+y^2}$ at $(3,-4,5)$

39. $x^2 + 2xy + y^2 + z^2 = 5$ at $(0,2,1)$

40. $x^2 z - y^2 x + 3y - z = -4$ at $(1,-1,2)$

In exercises 41 and 42, use the chain rule to find the indicated derivative(s).

41. $g'(t)$ where $g(t) = f(x(t), y(t))$, $f(x, y) = x^2 y + y^2$, $x(t) = e^{4t}$, $y(t) = \sin t$

42. $\dfrac{\partial g}{\partial u}$ and $\dfrac{\partial g}{\partial v}$ where $g(u, v) = f(x(u, v), y(u, v))$, $f(x, y) = 4x^2 - y$, $x(u, v) = u^3 v + \sin u$, $y(u, v) = 4v^2$

In exercises 43 and 44, state the chain rule for the general composite function.

43. $g(t) = f(x(t), y(t), z(t), w(t))$

44. $g(u, v) = f(x(u, v), y(u, v))$

In exercises 45 and 46, use implicit differentiation to find $\dfrac{\partial z}{\partial x}$ and $\dfrac{\partial z}{\partial y}$.

45. $x^2 + 2xy + y^2 + z^2 = 1$

46. $x^2 z - y^2 x + 3y - z = -4$

In exercises 47 and 48, find the gradient of the given function at the indicated point.

47. $f(x, y) = 3x \sin 4y - \sqrt{xy}$, (π, π)

48. $f(x, y, z) = 4xz^2 - 3 \cos x + 4y^2$, $(0, 1, -1)$

In exercises 49–52, compute the directional derivative of f at the given point in the direction of the indicated vector.

49. $f(x, y) = x^3 y - 4y^2$, $(-2, 3)$, $\mathbf{u} = \left\langle \frac{3}{5}, \frac{4}{5} \right\rangle$

50. $f(x, y) = x^2 + xy^2$, $(2, 1)$, $\mathbf{u}$ in the direction of $\langle 3, -2 \rangle$

51. $f(x, y) = e^{3xy} - y^2$, $(0, -1)$, $\mathbf{u}$ in the direction from $(2, 3)$ to $(3, 1)$

52. $f(x, y) = \sqrt{x^2 + xy^2}$, $(2, 1)$, $\mathbf{u}$ in the direction of $\langle 1, -2 \rangle$

In exercises 53–56, find the directions of maximum and minimum change of f at the given point, and the values of the maximum and minimum rates of change.

53. $f(x, y) = x^3 y - 4y^2$, $(-2, 3)$

54. $f(x, y) = x^2 + xy^2$, $(2, 1)$

55. $f(x, y) = \sqrt{x^4 + y^4}$, $(2, 0)$

56. $f(x, y) = x^2 + xy^2$, $(1, 2)$

57. Suppose that the elevation on a hill is given by $f(x, y) = 100 - 4x^2 - 2y$. From the site at $(2, 1)$, in which direction will the rain run off?

58. If the temperature at the point (x, y, z) is given by $T(x, y, z) = 70 + 5e^{-z^2}(4x + 3y^{-1})$, find the direction from the point $(1, 2, 1)$ in which the temperature decreases most rapidly.

In exercises 59–62, find all critical points and use Theorem 7.2 (if applicable) to classify them.

59. $f(x, y) = 2x^4 - xy^2 + 2y^2$

60. $f(x, y) = 2x^4 + y^3 - x^2 y$

61. $f(x, y) = 4xy - x^3 - 2y^2$

62. $f(x, y) = 3xy - x^3 y + y^2 - y$

63. The following data show the height and weight of a small number of people. Use the linear model to predict the weight of a $6'2''$ person and a $5'0''$ person. Comment on how accurate you think the model is.

Height (inches)	64	66	70	71
Weight (pounds)	140	156	184	190

64. The following data show the age and income for a small number of people. Use the linear model to predict the income of a 20-year-old and of a 60-year-old. Comment on how accurate you think the model is.

Age (years)	28	32	40	56
Income ($)	36,000	34,000	88,000	104,000

In exercises 65 and 66, find the absolute extrema of the function on the given region.

65. $f(x, y) = 2x^4 - xy^2 + 2y^2$, $0 \le x \le 4$, $0 \le y \le 2$

66. $f(x, y) = 2x^4 + y^3 - x^2 y$, region bounded by $y = 0$, $y = x$ and $x = 2$

In exercises 67–70, use Lagrange multipliers to find the maximum and minimum of the function $f(x, y)$ subject to the constraint $g(x, y) = c$.

67. $f(x, y) = x + 2y$, subject to $x^2 + y^2 = 5$

68. $f(x, y) = 2x^2 y$, subject to $x^2 + y^2 = 4$

69. $f(x, y) = xy$, subject to $x^2 + y^2 = 1$

70. $f(x, y) = x^2 + 2y^2 - 2x$, subject to $x^2 + y^2 = 1$

In exercises 71 and 72, use Lagrange multipliers to find the closest point on the given curve to the indicated point.

71. $y = x^3, (4, 0)$ 72. $y = x^3, (2, 1)$

73. The graph (from the excellent book *Tennis Science for Tennis Players* by Howard Brody) shows the vertical angular acceptance of a tennis serve as a function of height and velocity. With vertical angular acceptance, Brody is measuring a margin of error. For example, if serves with angles ranging from 5° to 8° will land in the service box (for a given height and velocity), the vertical angular acceptance is 3°. For a given height, does the angular acceptance increase or decrease as velocity increases? Explain why this is reasonable. For a given velocity, does the angular acceptance increase or decrease as height increases? Explain why this is reasonable.

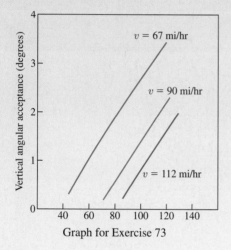

Graph for Exercise 73

74. The graphic in exercise 73 is somewhat like a contour plot. Assuming that angular acceptance is the dependent variable, explain what is different about this plot from the contour plots drawn in this chapter. Which type of plot do you think is easier to read? The accompanying plot shows angular acceptance in terms of a number of variables. Identify the independent variables and compare this plot to the level surfaces drawn in this chapter.

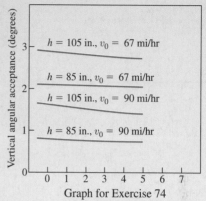

Graph for Exercise 74

75. The horizontal range of a golf ball or baseball depends on the launch angle and the rate of backspin on the ball. The figure below, reprinted from *Keep Your Eye on the Ball* by Watts and Bahill, shows level curves for this relationship for an initial velocity of 110 mph, although the dependent variable (range) is graphed vertically and the level curves represent constant values of one of the independent variables. Estimate the partial derivatives of range at 30° and 1910 rpm and use them to find a linear approximation of range. Predict the range at 25° and 2500 rpm, and also at 40° and 4000 rpm. Discuss the accuracy of each prediction.

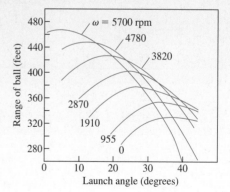

CHAPTER 13

MULTIPLE INTEGRALS

The launch of a spacecraft into orbit is now a familiar event for us. In fact, it's so familiar that it is easy to overlook how remarkable this achievement is. Countless innovations and sophisticated mathematical models form the foundation of modern rocket design. One simple aspect of rocketry that we have already studied in our presentation of the calculus is projectile motion. Although the massively powerful rocket engines may attract most of the attention, the aerodynamic properties of rockets are also absolutely critical.

One property of rocket design that is very apparent, even to the untrained eye, is a rocket's slender profile in the direction of motion. It should make sense intuitively that the wider the rocket is, the more air resistance it will encounter. The total surface area of the rocket plays a part in this, but most significantly, we need to know how much of the surface area is normal to the direction of motion, since this is what produces the bulk of the air resistance. This type of surface area calculation will require new integration skills. In this chapter, we extend the notion of integration to double and triple integrals of functions of several variables and examine a variety of coordinate systems. In section 13.4, we use double integrals to calculate surface areas and in Chapter 14 we use double and triple integrals to compute more complicated surface integrals. These calculations can be used to produce computer simulations of the aerodynamics of rockets, conventional aircraft or even things as mundane as automobiles.

The reduction of air resistance is not the only design issue in rocketry. To discover a different challenge, take your rocket-shaped pen or pencil and launch it into the air. You should quickly recognize that this is not a very stable projectile shape. A real rocket is even more complicated, since its engines generate a very large thrust from the bottom of the rocket and since its center of mass changes while in flight, as the rocket burns its fuel. The calculus in this chapter can help us with a basic design principle. If a rocket starts to tilt while in flight, it will tend to rotate about its center of mass (potentially tumbling end over end). We discuss the center of mass of three-dimensional solids in sections 13.2 and 13.5. Further, in this tilted position, the side of the rocket (where there is far greater surface area than in the nose cone) produces significant air resistance. The force of the air affects the rocket as if it were focused at a single point (see the diagram shown here). This

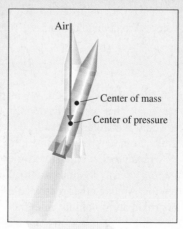

A tilting rocket.

A model rocket.

focus point, called the **center of pressure** is located at the same position as the center of mass of a rocket with the same shape but constant density. If the center of pressure lies below the center of mass, the air resistance will tend to push the rocket back into alignment. If the center of pressure lies above the center of mass, then, the rocket will tend to tumble end over end.

You may have noticed that model rockets usually have large lightweight fins at the bottom. One purpose of the fins is to guarantee that the center of pressure stays below the center of mass of the rocket (see G. Harry Stine's *The Handbook of Model Rocketry*). The calculus we develop in this chapter can be used to gain insight into rocket design and a variety of other situations that we explore throughout the chapter.

13.1   DOUBLE INTEGRALS

Before we introduce the idea of a double integral for a function of two variables, we want to briefly remind you of the definition of definite integral for a function of a single variable and then generalize the definition slightly. Recall that we were led to define the integral while looking for the area A under the graph of a continuous function f defined on an interval $[a, b]$, where $f(x) \geq 0$ on $[a, b]$. Our solution to this problem was to first **partition** the interval $[a, b]$ into n subintervals $[x_{i-1}, x_i]$, for $i = 1, 2, \ldots, n$, of equal width $\Delta x = \dfrac{b - a}{n}$, where

$$a = x_0 < x_1 < \cdots < x_n = b.$$

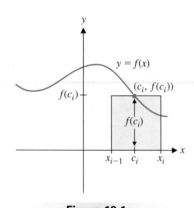

On each subinterval $[x_{i-1}, x_i]$, for $i = 1, 2, \ldots, n$, we approximated the area under the curve by the area of the rectangle of height $f(c_i)$, for some point $c_i \in [x_{i-1}, x_i]$, as indicated in Figure 13.1a. Adding together the areas of all of these rectangles, we found an approximation of the area as indicated in Figure 13.1b:

$$A \approx \sum_{i=1}^{n} f(c_i)\,\Delta x.$$

Figure 13.1a

Approximating the area on the subinterval $[x_{i-1}, x_i]$.

Finally, taking the limit as $n \to \infty$ (which also means that $\Delta x \to 0$), we get the exact area (assuming that the limit exists and is the same for all choices of the evaluation points c_i):

$$A = \lim_{n \to \infty} \sum_{i=1}^{n} f(c_i)\,\Delta x.$$

In Chapter 4, we observed that this limit may exist even when f takes on some negative values. We defined the definite integral as this limit:

$$\int_a^b f(x)\,dx = \lim_{n \to \infty} \sum_{i=1}^{n} f(c_i)\,\Delta x. \tag{1.1}$$

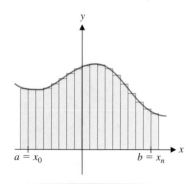

Figure 13.1b

Area under the curve.

Following this definition, we made what might have seemed at the time to be a cryptic comment: that while this definition sufficed for nearly any function you would encounter, it would need to be generalized later. Well, it's now time for us to make this very slight generalization, before we define integrals of functions of several variables.

In general, we must consider partitions that are **irregular** (that is, where not all subintervals have the same width). You may be wondering why we need this kind of generalization. Among other reasons, this is needed for more sophisticated numerical methods for approximating definite integrals. This generalization is also needed for theoretical purposes; this is pursued in a more advanced course. We proceed essentially as above, except that we allow different subintervals to have different widths and define the width of the ith subinterval $[x_{i-1}, x_i]$ to be $\Delta x_i = x_i - x_{i-1}$ (see Figure 13.2 for the case where $n = 7$).

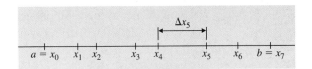

Figure 13.2
Irregular partition of $[a, b]$.

An approximation of the area is then (essentially, as before)

$$A \approx \sum_{i=1}^{n} f(c_i)\, \Delta x_i,$$

for any choice of the evaluation points $c_i \in [x_{i-1}, x_i]$, for $i = 1, 2, \ldots, n$. To get the exact area, we need to let $n \to \infty$, but since the partition is irregular, this alone will not guarantee that all of the Δx_i's will approach zero, unless we take a little extra care, by defining $\|P\|$ (the **norm of the partition**) to be the **largest** of all the Δx_i's. We then arrive at the following more general definition of definite integral.

Definition 1.1

For any function f defined on the interval $[a, b]$, the **definite integral** of f on $[a, b]$ is

$$\int_a^b f(x)\, dx = \lim_{\|P\| \to 0} \sum_{i=1}^{n} f(c_i)\, \Delta x_i,$$

provided the limit exists and is the same for all choices of the evaluation points $c_i \in [x_{i-1}, x_i]$, for $i = 1, 2, \ldots, n$. In this case, we say that f is **integrable** on $[a, b]$.

Here, by saying that the limit in Definition 1.1 equals some value L, we mean that we can make $\sum_{i=1}^{n} f(c_i)\, \Delta x_i$ as close as needed to L, just by making $\|P\|$ sufficiently small. How close must the sum get to L? We must be able to make the sum within any specified distance $\varepsilon > 0$ of L. More precisely, given any $\varepsilon > 0$, there must be a $\delta > 0$ (depending on the choice of ε), such that

$$\left| \sum_{i=1}^{n} f(c_i)\, \Delta x_i - L \right| < \varepsilon,$$

for **every** partition P with $\|P\| < \delta$. Notice that this is only a very slight generalization of our original notion of definite integral. All we have done is to allow the partitions to be irregular and then defined $\|P\|$ to ensure that $\Delta x_i \to 0$, for every i. For virtually all of the functions that you have encountered in this course, this new definition is equivalent to the original one in (1.1).

Realize that you would likely never use Definition 1.1 to compute an area, but your computer or calculator software probably does use irregular partitions to estimate integrals. We will use Definition 1.1 primarily for theoretical reasons (e.g., we want to generalize the notion of integral to functions of several variables).

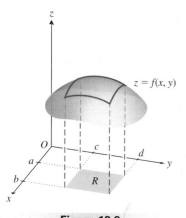

Figure 13.3

Volume under the surface
$z = f(x, y)$.

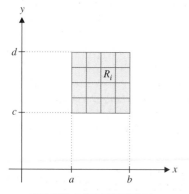

Figure 13.4a

Partition of R.

Double Integrals Over a Rectangle

We developed the definite integral as a natural outgrowth of our method for finding area under a curve in the xy-plane. Likewise, we are guided in our development of the double integral by a corresponding problem. For a function $f(x, y)$, where f is continuous and $f(x, y) \geq 0$ for all $a \leq x \leq b$ and $c \leq y \leq d$, we wish to find the **volume** of the solid lying below the surface $z = f(x, y)$ and above the rectangle $R = \{(x, y) \mid a \leq x \leq b$ and $c \leq y \leq d\}$ in the xy-plane (see Figure 13.3).

We proceed essentially the same way as we did to find the area under a curve. First, we partition the rectangle R by laying down a grid on top of R consisting of n smaller rectangles (see Figure 13.4a). (Note: The rectangles in the grid need not be all of the same size.) Call the smaller rectangles $R_1, R_2, \ldots, R_n$. (The order in which you number them is irrelevant.) For each rectangle R_i ($i = 1, 2, \ldots, n$) in the partition, we want to find an approximation to the volume V_i lying beneath the surface $z = f(x, y)$ and above the rectangle R_i. The sum of these approximate volumes is then an approximation to the total volume. Above each rectangle R_i in the partition, construct a rectangular box whose height is $f(u_i, v_i)$, for some point $(u_i, v_i) \in R_i$ (see Figure 13.4b). Notice that the volume V_i under the surface is approximated by the volume of the box:

$$V_i \approx \text{Height} \times \text{Area of base} = f(u_i, v_i) \, \Delta A_i,$$

where ΔA_i denotes the area of the rectangle R_i.

The total volume is then approximately

$$V \approx \sum_{i=1}^{n} f(u_i, v_i) \, \Delta A_i. \tag{1.2}$$

As in our development of the definite integral in Chapter 4, we call the sum in (1.2) a **Riemann sum.** We illustrate the approximation of the volume under a surface by a Riemann sum in Figures 13.4c and 13.4d. Notice that the larger number of rectangles used in Figure 13.4d appears to give a better approximation of the volume.

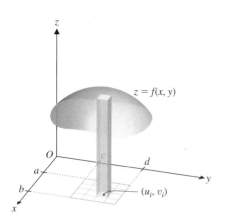

Figure 13.4b

Approximating the volume above
R_i by a rectangular box.

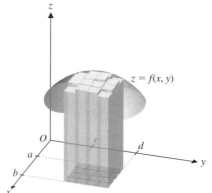

Figure 13.4c

Approximate volume.

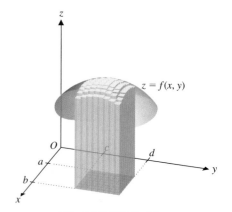

Figure 13.4d

Approximate volume.

Example 1.1   Approximating the Volume Lying Beneath a Surface

Approximate the volume lying beneath the surface $z = x^2 \sin \dfrac{\pi y}{6}$ and above the rectangle $R = \{(x, y) \mid 0 \leq x \leq 6, 0 \leq y \leq 6\}$.

Solution   First, note that f is continuous and $f(x, y) = x^2 \sin \dfrac{\pi y}{6} \geq 0$ on R (see Figure 13.5a). Next, a simple partition of R is a partition into four squares of equal size, as indicated in Figure 13.5b. We choose the evaluation points (u_i, v_i) to be the centers of each of the four squares, that is, $\left(\frac{3}{2}, \frac{3}{2}\right)$, $\left(\frac{9}{2}, \frac{3}{2}\right)$, $\left(\frac{3}{2}, \frac{9}{2}\right)$ and $\left(\frac{9}{2}, \frac{9}{2}\right)$.

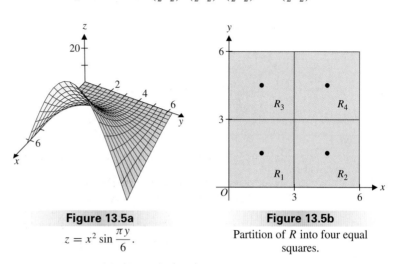

Figure 13.5a

$$z = x^2 \sin \frac{\pi y}{6}.$$

Figure 13.5b

Partition of R into four equal squares.

Since the four squares are the same size, we have $\Delta A_i = 9$, for each i. For $f(x, y) = x^2 \sin \dfrac{\pi y}{6}$, we have from (1.2) that

$$V \approx \sum_{i=1}^{4} f(u_i, v_i)\, \Delta A_i$$

$$= f\left(\frac{3}{2}, \frac{3}{2}\right)(9) + f\left(\frac{9}{2}, \frac{3}{2}\right)(9) + f\left(\frac{3}{2}, \frac{9}{2}\right)(9) + f\left(\frac{9}{2}, \frac{9}{2}\right)(9)$$

$$= 9\left[\left(\frac{3}{2}\right)^2 \sin\left(\frac{\pi}{4}\right) + \left(\frac{9}{2}\right)^2 \sin\left(\frac{\pi}{4}\right) + \left(\frac{3}{2}\right)^2 \sin\left(\frac{3\pi}{4}\right) + \left(\frac{9}{2}\right)^2 \sin\left(\frac{3\pi}{4}\right)\right]$$

$$= \frac{405}{2}\sqrt{2} \approx 286.38.$$

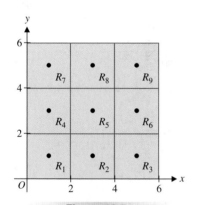

Figure 13.5c

Partition of R into nine equal squares.

We can improve on this approximation by increasing the number of rectangles in the partition. For instance, if we partition R into nine squares of equal size (see Figure 13.5c) and again use the center of each square as the evaluation point, we have $\Delta A_i = 4$ for each i and

$$V \approx \sum_{i=1}^{9} f(u_i, v_i)\, \Delta A_i$$

$$= 4\big[f(1, 1) + f(3, 1) + f(5, 1) + f(1, 3) + f(3, 3) + f(5, 3)$$
$$\qquad + f(1, 5) + f(3, 5) + f(5, 5)\big]$$

$$= 4\left[1^2 \sin\left(\frac{\pi}{6}\right) + 3^2 \sin\left(\frac{\pi}{6}\right) + 5^2 \sin\left(\frac{\pi}{6}\right) + 1^2 \sin\left(\frac{3\pi}{6}\right) + 3^2 \sin\left(\frac{3\pi}{6}\right)\right.$$

$$\left. + 5^2 \sin\left(\frac{3\pi}{6}\right) + 1^2 \sin\left(\frac{5\pi}{6}\right) + 3^2 \sin\left(\frac{5\pi}{6}\right) + 5^2 \sin\left(\frac{5\pi}{6}\right)\right]$$

$$= 280.$$

No. of Squares in Partition	Approximate Volume
4	286.38
9	280.00
36	276.25
144	275.33
400	275.13
900	275.07

NOTES

The choice of the center of each square as the evaluation point, as used in example 1.1 corresponds to the midpoint rule for approximating the value of a definite integral for a function of a single variable (discussed in section 4.7). This choice of evaluation points generally produces a reasonably good approximation.

Remark 1.1

It can be shown that if f is continuous on R, then it is also integrable over R. The proof can be found in more advanced texts.

Continuing in this fashion to divide R into more and more squares of equal size and using the center of each square as the evaluation point, we construct continually better and better approximations of the volume (see the table in the margin). From the table, it appears that a reasonable approximation to the volume is 275.07. As it turns out, the exact volume is $\frac{864}{\pi} \approx 275.02$. (We'll show you how to find this shortly.)

Now, how can we turn (1.2) into an exact formula for volume? Note that it takes more than simply letting $n \to \infty$. We need to have **all** of the rectangles in the partition shrink to zero area. A convenient way of doing this is to define the **norm of the partition** $\|P\|$ to be the largest diagonal of any rectangle in the partition. Note that if $\|P\| \to 0$, then **all** of the rectangles must shrink to zero area. We can now make the volume approximation (1.2) exact:

$$V = \lim_{\|P\| \to 0} \sum_{i=1}^{n} f(u_i, v_i) \Delta A_i,$$

assuming the limit exists and is the same for every choice of the evaluation points. Here, by saying that this limit equals V, we mean that we can make $\sum_{i=1}^{n} f(u_i, v_i) \Delta A_i$ as close as needed to V, just by making $\|P\|$ sufficiently small. More precisely, this says that given any $\varepsilon > 0$, there is a $\delta > 0$ (depending on the choice of ε), such that

$$\left| \sum_{i=1}^{n} f(u_i, v_i) \Delta A_i - V \right| < \varepsilon,$$

for every partition P with $\|P\| < \delta$. In general, we make the following definition, which applies even when the function takes on negative values.

Definition 1.2

For any function $f(x, y)$ defined on a rectangle R in the xy-plane, we define the **double integral** of f over R by

$$\iint\limits_{R} f(x, y)\, dA = \lim_{\|P\| \to 0} \sum_{i=1}^{n} f(u_i, v_i) \Delta A_i,$$

provided the limit exists and is the same for every choice of the evaluation points (u_i, v_i) in R_i, for $i = 1, 2, \ldots, n$. When this happens, we say that f is **integrable** over R.

There's one small problem with this new double integral. Just like when we first defined the definite integral of a function of one variable, we don't yet know how to compute it! For complicated regions R, this is a little bit tricky, but for a rectangle, it's a snap, as we see in the following.

We first consider the special case where $f(x, y) \geq 0$ on the rectangle $R = \{(x, y) \,|\, a \leq x \leq b \text{ and } c \leq y \leq d\}$. Notice that here, $\iint_R f(x, y)\, dA$ represents the volume lying beneath the surface $z = f(x, y)$ and above the region R. Recall that we already know how to compute this volume, from our work in section 5.2. We can do this by slicing the solid with planes parallel to the yz-plane, as indicated in Figure 13.6a. If we denote the area of the cross section of the solid for a given value of x by $A(x)$, then we have from equation (2.1) in section 5.2 that the volume is given by

$$V = \int_a^b A(x)\, dx.$$

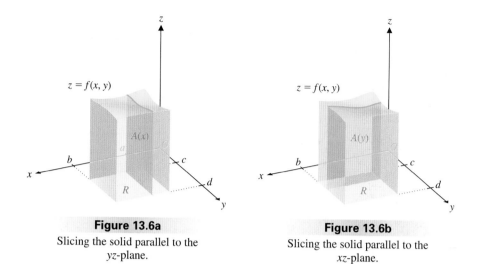

Figure 13.6a
Slicing the solid parallel to the
yz-plane.

Figure 13.6b
Slicing the solid parallel to the
xz-plane.

Now, note that for each **fixed** value of x, the area of the cross section is simply the area under the curve $z = f(x, y)$ for $c \leq y \leq d$, which is given by the integral

$$A(x) = \int_c^d f(x, y)\, dy.$$

This integration is called a **partial integration** with respect to y, since x is held fixed and $f(x, y)$ is integrated with respect to y. This leaves us with

$$V = \int_a^b A(x)\, dx = \int_a^b \left[\int_c^d f(x, y)\, dy \right] dx. \tag{1.3}$$

Likewise, if we instead slice the solid with planes parallel to the xz-plane, as indicated in Figure 13.6b, we get that the volume is given by

$$V = \int_c^d A(y)\, dy = \int_c^d \left[\int_a^b f(x, y)\, dx \right] dy. \tag{1.4}$$

The integrals in (1.3) and (1.4) are called **iterated integrals.** Note that each of these indicates a partial integration with respect to the inner variable (i.e., you first integrate with respect to the inner variable, treating the outer variable as a constant), to be followed by an integration with respect to the outer variable.

For simplicity, we ordinarily write the iterated integrals without the brackets:

$$\int_a^b \left[\int_c^d f(x, y)\, dy \right] dx = \int_a^b \int_c^d f(x, y)\, dy\, dx$$

and

$$\int_c^d \left[\int_a^b f(x, y)\, dx \right] dy = \int_c^d \int_a^b f(x, y)\, dx\, dy.$$

As indicated, these integrals are evaluated inside out, using the methods of integration we've already established for functions of a single variable. This now establishes the following result for the special case where $f(x, y) \geq 0$. The proof of the result for the general case is rather lengthy and we omit it.

HISTORICAL NOTES

Guido Fubini (1879–1943)
Italian mathematician who made wide-ranging contributions to mathematics, physics and engineering. Fubini's early work was in differential geometry, but he quickly diversified his research to include analysis, the calculus of variations, group theory, non-Euclidean geometry and mathematical physics. Mathematics was the family business, as his father was a mathematics teacher and his sons became engineers. Fubini moved to the United States in 1939 to escape the persecution of Jews in Italy. He was working on an engineering textbook inspired by his sons' work when he died.

Theorem 1.1 (Fubini's Theorem)

Suppose that f is integrable over the rectangle $R = \{(x, y) | a \leq x \leq b \text{ and } c \leq y \leq d\}$. Then we can write the double integral of f over R as either of the iterated integrals:

$$\iint\limits_{R} f(x, y)\, dA = \int_a^b \int_c^d f(x, y)\, dy\, dx = \int_c^d \int_a^b f(x, y)\, dx\, dy. \qquad (1.5)$$

Fubini's Theorem simply tells you that you can always rewrite a double integral over a rectangle as either one of a pair of iterated integrals. We illustrate this in the following example.

Example 1.2   Double Integral over a Rectangle

If $R = \{(x, y) | 0 \leq x \leq 2 \text{ and } 1 \leq y \leq 4\}$, evaluate $\iint\limits_{R} (6x^2 + 4xy^3)\, dA$.

Solution   From (1.5), we have

$$\iint\limits_{R} (6x^2 + 4xy^3)\, dA = \int_1^4 \int_0^2 (6x^2 + 4xy^3)\, dx\, dy$$

$$= \int_1^4 \left[\int_0^2 (6x^2 + 4xy^3)\, dx \right] dy$$

$$= \int_1^4 \left(6\frac{x^3}{3} + 4\frac{x^2}{2} y^3 \right) \Bigg|_{x=0}^{x=2} dy$$

$$= \int_1^4 (16 + 8y^3)\, dy$$

$$= \left(16y + 8\frac{y^4}{4} \right) \Bigg|_1^4$$

$$= \left[16(4) + 2(4)^4 \right] - \left[16(1) + 2(1)^4 \right] = 558.$$

Note that we evaluated the first integral above by integrating with respect to x while treating y as a constant. We leave it as an exercise to show that you get the same value by integrating first with respect to y, that is, that

$$\iint\limits_{R} (6x^2 + 4xy^3)\, dA = \int_0^2 \int_1^4 (6x^2 + 4xy^3)\, dy\, dx = 558,$$

also.

Double Integrals Over General Regions

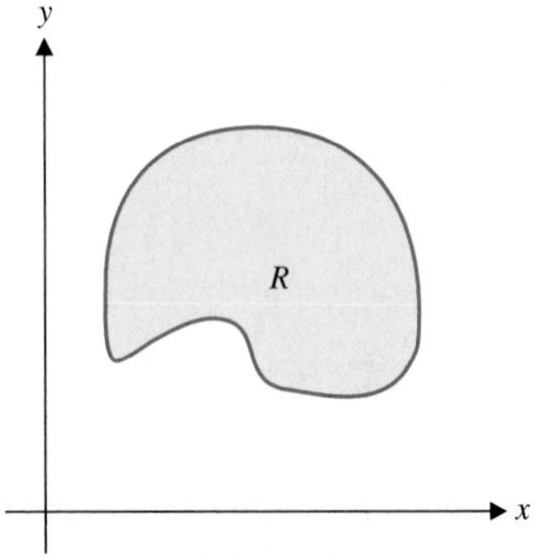

So, what if we wanted to extend the notion of double integral to a bounded, nonrectangular region like the one shown in Figure 13.7a? (Recall that a region is bounded if it fits inside a circle of some finite radius.) We begin, as we did for the case of rectangular regions, by looking for the volume lying beneath the surface $z = f(x, y)$ and lying above the region R, where $f(x, y) \geq 0$ and f is continuous on R. First, notice that the grid we used initially to partition a rectangular region must somehow be modified, since such a rectangular grid won't "fit" a nonrectangular region, as shown in Figure 13.7b.

Figure 13.7a
Nonrectangular region.

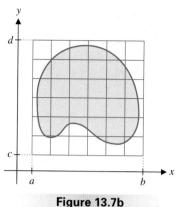

Figure 13.7b
Grid for a general region.

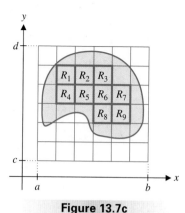

Figure 13.7c
Inner partition.

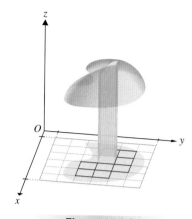

Figure 13.7d
Sample volume box.

We resolve this problem by considering only those rectangular subregions that lie **completely** inside the region R (see Figure 13.7c where we have highlighted these rectangles).

We call the collection of these rectangles an **inner partition** of R. For instance, in the inner partition indicated in Figure 13.7c, there are nine subregions.

From this point on, we proceed essentially as we did for the case of a rectangular region. That is, on each rectangular subregion R_i ($i = 1, 2, \ldots, n$) in an inner partition, we construct a rectangular box of height $f(u_i, v_i)$, for some point $(u_i, v_i) \in R_i$ (see Figure 13.7d for a sample box). The volume V_i under the surface and above R_i is then approximately

$$V_i \approx \text{Height} \times \text{Area of base} = f(u_i, v_i)\, \Delta A_i,$$

where we again denote the area of R_i by ΔA_i. The total volume V lying beneath the surface and above the region R is then approximately

$$V \approx \sum_{i=1}^{n} f(u_i, v_i)\, \Delta A_i. \tag{1.6}$$

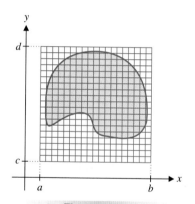

Figure 13.8a
Refined grid.

We define the norm of the inner partition $\|P\|$ to be the length of the largest diagonal of any of the rectangles $R_1, R_2, \ldots, R_n$. Notice that as we make $\|P\|$ smaller and smaller, the inner partition fills in R nicely (see Figure 13.8a) and the approximate volume given by (1.6) should get closer and closer to the actual volume (see Figure 13.8b). We then have

$$V = \lim_{\|P\| \to 0} \sum_{i=1}^{n} f(u_i, v_i)\, \Delta A_i,$$

assuming the limit exists and is the same for every choice of the evaluation points.

More generally, we have the following definition.

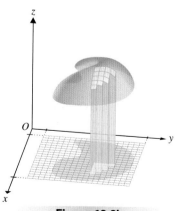

Figure 13.8b
Approximate volume.

Definition 1.3

For any function $f(x, y)$ defined on a bounded region $R \subset \mathbb{R}^2$, we define the **double integral** of f over R by

$$\iint\limits_{R} f(x, y)\, dA = \lim_{\|P\| \to 0} \sum_{i=1}^{n} f(u_i, v_i)\, \Delta A_i, \tag{1.7}$$

provided the limit exists and is the same for every choice of the evaluation points (u_i, v_i) in R_i, for $i = 1, 2, \ldots, n$. In this case, we say that f is **integrable** over R.

The question remains as to how we can calculate a double integral over a nonrectangular region. The answer is a bit more complicated than it was for the case of a rectangular region and depends on the exact form of R. We present two cases separately.

Case 1: R has the form: $R = \{(x, y) | a \leq x \leq b \text{ and } g_1(x) \leq y \leq g_2(x)\}$.
Case 2: R has the form: $R = \{(x, y) | c \leq y \leq d \text{ and } h_1(y) \leq x \leq h_2(y)\}$.

Note that these cases are the same as we used in computing areas between curves in section 5.1. We will start by analyzing regions with a well-defined top and bottom.

Suppose that the region R has the form

$$R = \{(x, y) | a \leq x \leq b \text{ and } g_1(x) \leq y \leq g_2(x)\}.$$

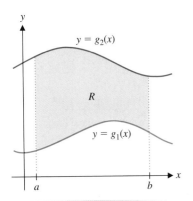

$y = g_2(x)$

R

$y = g_1(x)$

a b

Figure 13.9a

The region R.

See Figure 13.9a for a typical region of this form lying in the first quadrant of the xy-plane. The easiest way to think about this is for the special case where $f(x, y) \geq 0$ on R. Here, the double integral of f over R corresponds to the volume lying beneath the surface $z = f(x, y)$ and above the region R in the xy-plane. Notice that we can find this volume by the method of slicing just as we did for the case of a double integral over a rectangular region, as follows.

First, from Figure 13.9b, observe that for each **fixed** $x \in [a, b]$, the area of the slice lying above the line segment indicated and below the surface $z = f(x, y)$ is given by

$$A(x) = \int_{g_1(x)}^{g_2(x)} f(x, y) \, dy.$$

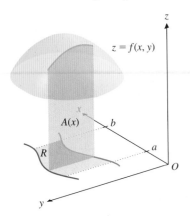

$z = f(x, y)$

$A(x)$ b

R a

O

Figure 13.9b

Volume by slicing.

Recall that if you know the area $A(x)$ of every cross section, then the volume of the solid is given by equation (2.1) in section 5.2 to be

$$V = \int_a^b A(x) \, dx = \int_a^b \int_{g_1(x)}^{g_2(x)} f(x, y) \, dy \, dx.$$

Recognizing the volume as $V = \iint_R f(x, y) \, dA$ proves the following theorem, for the special case where $f(x, y) \geq 0$ on R.

Theorem 1.2

Suppose that $f(x, y)$ is continuous on the region R defined by $R = \{(x, y) | a \leq x \leq b \text{ and } g_1(x) \leq y \leq g_2(x)\}$, for continuous functions g_1 and g_2. Then,

$$\iint_R f(x, y) \, dA = \int_a^b \int_{g_1(x)}^{g_2(x)} f(x, y) \, dy \, dx.$$

Although the general proof of the theorem is beyond the level of this text, the derivation given above for the special case where $f(x, y) \geq 0$ should help to make some sense of why it is true.

Notice that once again, we have managed to write a double integral as an iterated integral. This allows us to use all of our techniques of integration for functions of a single variable to help evaluate double integrals.

We illustrate the process of writing a double integral as an iterated integral in the following example.

Example 1.3 Evaluating a Double Integral

Let R be the region bounded by the graphs of $y = x$, $y = 0$ and $x = 4$. Evaluate

$$\iint\limits_{R} \left(4e^{x^2} - 5\sin y\right) dA.$$

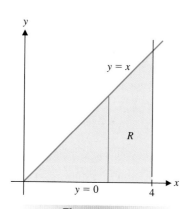

Figure 13.10

The region R.

Solution First, we draw a graph of the region R in Figure 13.10. To help with determining the limits of integration, notice that we have drawn a line segment illustrating the inner limits of integration. The line indicates that for each fixed value of x, the y-values range from 0 up to x. Without specifying the integrand, we have from Theorem 1.2 that

$$\iint\limits_{R} f(x, y) \, dA = \int_0^4 \int_0^x f(x, y) \, dy \, dx. \tag{1.8}$$

Be **very** careful here; there are plenty of traps to fall into. The most common error is to simply look for the minimum and maximum values of x and y and mistakenly write

$$\iint\limits_{R} f(x, y) \, dA = \int_0^4 \int_0^4 f(x, y) \, dy \, dx. \quad \text{This is incorrect!}$$

Compare this last iterated integral to the correct expression in (1.8). Notice that instead of integrating over the region R shown in Figure 13.10, it corresponds to integration over the rectangle: $0 \le x \le 4$, $0 \le y \le 4$. (This is close, but no cigar!) Substituting $f(x, y) = 4e^{x^2} - 5\sin y$, we have from (1.8) that

$$\iint\limits_{R} \left(4e^{x^2} - 5\sin y\right) dA = \int_0^4 \int_0^x \left(4e^{x^2} - 5\sin y\right) dy \, dx$$

$$= \int_0^4 \left(4ye^{x^2} + 5\cos y\right)\Big|_{y=0}^{y=x} dx$$

$$= \int_0^4 \left[\left(4xe^{x^2} + 5\cos x\right) - (0 + 5\cos 0)\right] dx$$

$$= \int_0^4 \left(4xe^{x^2} + 5\cos x - 5\right) dx$$

$$= \left(2e^{x^2} + 5\sin x - 5x\right)\Big|_0^4$$

$$= 2e^{16} + 5\sin 4 - 22 \approx 1.78 \times 10^7.$$

Keep in mind that the inner integration above (with respect to y) is a partial integration with respect to y, so that we hold x fixed.

As with any other integral, iterated integrals often cannot be evaluated symbolically (even with a very good computer algebra system). In such cases, we must rely on approximate methods. The basic rule of thumb here is to evaluate the inner integral symbolically, if at all possible and then use a numerical method (e.g., Simpson's Rule) to approximate the outer integral.

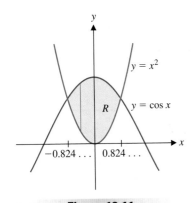

Figure 13.11

The region R.

Example 1.4   Approximate Limits of Integration

Evaluate $\iint\limits_{R} (x^2 + 6y)\, dA$, where R is the region bounded by the graphs of $y = \cos x$ and $y = x^2$.

Solution   As usual, we first draw a graph of the region R (see Figure 13.11). Notice that the inner limits of integration are easy to see from the figure: for each fixed x, y ranges from x^2 up to $\cos x$. But the outer limits of integration are not quite so clear. To find these, we must find the intersections of the two curves by solving the equation $\cos x = x^2$. Of course, we can't solve this exactly, but using a numerical procedure (e.g., Newton's Method or one built into your calculator or computer algebra system), we get approximate intersections of $x \approx \pm 0.82413$. From Theorem 1.2, we now have

$$\iint\limits_{R} (x^2 + 6y)\, dA \approx \int_{-0.82413}^{0.82413} \int_{x^2}^{\cos x} (x^2 + 6y)\, dy\, dx$$

$$= \int_{-0.82413}^{0.82413} \left(x^2 y + 6\frac{y^2}{2} \right) \Bigg|_{y=x^2}^{y=\cos x} dx$$

$$= \int_{-0.82413}^{0.82413} [(x^2 \cos x + 3\cos^2 x) - (x^4 + 3x^4)]\, dx$$

$$\approx 3.659765588,$$

where we have evaluated the last integral approximately, even though it could be done exactly, using integration by parts and a trigonometric identity.

■

Not all double integrals can be computed using the technique of examples 1.3 and 1.4. Often, it is necessary (or at least convenient) to think of the geometry of the region R in a different way.

Suppose that the region R has the form

$$R = \{(x, y)\mid c \le y \le d \text{ and } h_1(y) \le x \le h_2(y)\}.$$

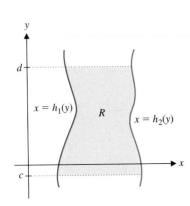

Figure 13.12

Typical region.

See Figure 13.12 for a typical region of this form. Then, much the same as in the first case, we can simply write double integrals as iterated integrals, as in the following theorem.

Theorem 1.3
Suppose that $f(x, y)$ is continuous on the region R defined by $R = \{(x, y)\mid c \le y \le d$ and $h_1(y) \le x \le h_2(y)\}$, for continuous functions h_1 and h_2. Then,

$$\iint\limits_{R} f(x, y)\, dA = \int_{c}^{d} \int_{h_1(y)}^{h_2(y)} f(x, y)\, dx\, dy.$$

Again, the general proof of this theorem is beyond the level of this course, although the reasonableness of this result should be apparent from Theorem 1.2 and the analysis preceding that theorem, for the special case where $f(x, y) \ge 0$ on R.

Example 1.5   Integrating First with Respect to x

Write $\iint\limits_{R} f(x, y)\, dA$ as an iterated integral, where R is the region bounded by the graphs of $x = y^2$ and $x = 2 - y$.

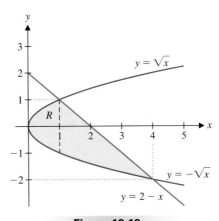

Figure 13.13a
The region R.

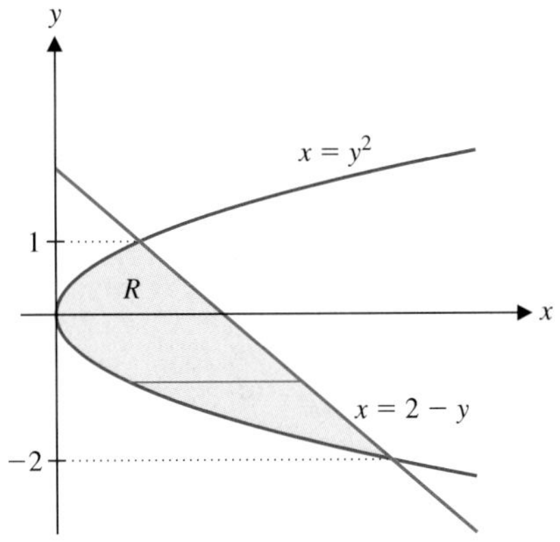

Figure 13.13b
The region R.

Solution   First, we sketch a graph of the region (see Figure 13.13a). Notice that integrating first with respect to y is not a very good choice, since the upper boundary of the region is $y = \sqrt{x}$ for $0 \le x \le 1$ and $y = 2 - x$ for $1 \le x \le 4$. A more reasonable choice is to use Theorem 1.3 and integrate first with respect to x. Notice that in Figure 13.13b, we have included a horizontal line segment indicating the inner limits of integration: for each fixed y, x runs from $x = y^2$ over to $x = 2 - y$. The value of y then runs between the values at the intersections of the two curves. To find these, we solve $y^2 = 2 - y$ or

$$0 = y^2 + y - 2 = (y + 2)(y - 1),$$

so that the intersections are at $y = -2$ and $y = 1$. From Theorem 1.3, we now have

$$\iint\limits_{R} f(x, y)\, dA = \int_{-2}^{1} \int_{y^2}^{2-y} f(x, y)\, dx\, dy.$$

You will often have to choose which variable to integrate with respect to first. Sometimes, you make your choice on the basis of the region. Often, a double integral can be set up either way, but it's much easier to calculate one way than the other. This is the case in the following example.

Example 1.6   Evaluating a Double Integral

Let R be the region bounded by the graphs of $y = \sqrt{x}$, $x = 0$ and $y = 3$. Evaluate $\iint\limits_{R} (2xy^2 + 2y \cos x)\, dA$.

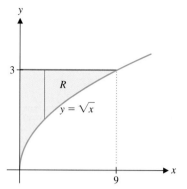

Figure 13.14

The region R.

Solution   First, we draw a graph of the region in Figure 13.14. Notice that from Theorem 1.3, we have

$$\iint\limits_R (2xy^2 + 2y\cos x)\, dA = \int_0^3 \int_0^{y^2} (2xy^2 + 2y\cos x)\, dx\, dy$$

$$= \int_0^3 (x^2 y^2 + 2y\sin x)\Big|_{x=0}^{x=y^2}\, dy$$

$$= \int_0^3 [(y^6 + 2y\sin y^2) - (0 + 2y\sin 0)]\, dy$$

$$= \int_0^3 (y^6 + 2y\sin y^2)\, dy$$

$$= \left(\frac{y^7}{7} - \cos y^2\right)\Big|_0^3$$

$$= \frac{3^7}{7} - \cos 9 + \cos 0 \approx 314.3.$$

On the other hand, integrating with respect to y first, we get

$$\iint\limits_R (2xy^2 + 2y\cos x)\, dA = \int_0^9 \int_{\sqrt{x}}^3 (2xy^2 + 2y\cos x)\, dy\, dx$$

$$= \int_0^9 \left(2x\frac{y^3}{3} + y^2\cos x\right)\Big|_{y=\sqrt{x}}^{y=3}\, dx$$

$$= \int_0^9 \left[\frac{2}{3}x(27 - x^{3/2}) + (3^2 - x)\cos x\right]\, dx,$$

which leaves you with an integration by parts to carry out. We leave the details as an exercise. Which way do you think is easier?

∎

In example 1.6, we saw that changing the order of integration may make a given double integral easier to compute. As we see in the following example, sometimes you will **need** to change the order of integration in order to evaluate a double integral.

Example 1.7   *A Case Where We Must Switch the Order of Integration*

Evaluate the iterated integral $\displaystyle\int_0^1 \int_y^1 e^{x^2}\, dx\, dy$.

Solution   First, note that we cannot evaluate the integral the way it is presently written, as we don't know an antiderivative for e^{x^2}. On the other hand, if we switch the order of integration, the integral becomes quite simple, as follows. First, recognize that for each fixed y on the interval $[0, 1]$, x ranges from y over to 1. We sketch the triangular region of integration in Figure 13.15. If we switch the order of integration, notice that for each fixed x in the interval $[0, 1]$, y ranges from 0 up to x and we get the double iterated integral:

$$\int_0^1 \int_y^1 e^{x^2}\, dx\, dy = \int_0^1 \int_0^x e^{x^2}\, dy\, dx$$

$$= \int_0^1 e^{x^2} y\Big|_{y=0}^{y=x}\, dx$$

$$= \int_0^1 e^{x^2} x\, dx.$$

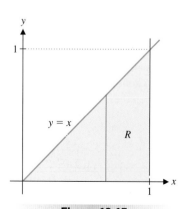

Figure 13.15

The region R.

Notice that we can now evaluate this last integral with the substitution $u = x^2$, since $du = 2x\, dx$ and the first integration has conveniently provided us with the needed factor of x. We have

$$\int_0^1 \int_y^1 e^{x^2}\, dx\, dy = \frac{1}{2} \int_0^1 \underbrace{e^{x^2}}_{e^u}\, \underbrace{(2x)\, dx}_{du}$$

$$= \frac{1}{2} e^{x^2} \Big|_{x=0}^{x=1} = \frac{1}{2}(e^1 - 1).$$

We complete the section by stating several simple properties of double integrals.

Theorem 1.4

Let $f(x, y)$ and $g(x, y)$ be integrable over the region $R \subset \mathbb{R}^2$ and let c be any constant. Then, the following hold:

(i) $\displaystyle\iint_R cf(x, y)\, dA = c \iint_R f(x, y)\, dA,$

(ii) $\displaystyle\iint_R [f(x, y) + g(x, y)]\, dA = \iint_R f(x, y)\, dA + \iint_R g(x, y)\, dA$ and

(iii) if $R = R_1 \cup R_2$, where R_1 and R_2 are nonoverlapping regions (see Figure 13.16), then

$$\iint_R f(x, y)\, dA = \iint_{R_1} f(x, y)\, dA + \iint_{R_2} f(x, y)\, dA.$$

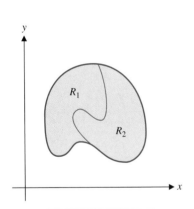

Figure 13.16
$R = R_1 \cup R_2$.

Each of these follows directly from the definition of double integral in (1.7) and their proof is left as an exercise.

EXERCISES 13.1

1. If $f(x, y) \geq 0$ on a region R, then $\iint_R f(x, y)\, dA$ gives the volume of the solid above the region R in the xy-plane and below the surface $z = f(x, y)$. If $f(x, y) \geq 0$ on a region R_1 and $f(x, y) \leq 0$ on a region R_2, discuss the geometric meaning of $\iint_{R_2} f(x, y)\, dA$ and $\iint_R f(x, y)\, dA$, where $R = R_1 \cup R_2$.

2. The definition of $\iint_R f(x, y)\, dA$ requires that the norm of the partition $\|P\|$ approaches 0. Explain why it is not enough to simply require that the number of rectangles n in the partition approaches ∞.

3. When computing areas between curves in section 5.1, we discussed strategies for deciding whether to integrate with respect to x or y. Compare these strategies to those given in this section for deciding which variable to use as the inside variable of a double integral.

4. Suppose you (or your software) are using Riemann sums to approximate a particularly difficult double integral $\iint_R f(x, y)\, dA$. Further, suppose that $R = R_1 \cup R_2$ and the function $f(x, y)$ is nearly constant on R_1 but oscillates wildly on R_2, where R_1 and R_2 are nonoverlapping regions. Explain

why you would need more rectangles in R_2 than R_1 to get equally accurate approximations. Thus, irregular partitions can be used to improve the efficiency of numerical integration routines.

In exercises 5–10, compute the Riemann sum for the given function and region, a partition with n equal-sized rectangles and the given evaluation rule.

5. $f(x, y) = x + 2y^2$, $0 \le x \le 2$, $-1 \le y \le 1$, $n = 4$, evaluate at midpoint

6. $f(x, y) = 4x^2 + y$, $1 \le x \le 5$, $0 \le y \le 2$, $n = 4$, evaluate at midpoint

7. $f(x, y) = x + 2y^2$, $0 \le x \le 2$, $-1 \le y \le 1$, $n = 16$, evaluate at midpoint

8. $f(x, y) = 4x^2 + y$, $1 \le x \le 5$, $0 \le y \le 2$, $n = 16$, evaluate at midpoint

9. $f(x, y) = 3x - y$, $-1 \le x \le 1$, $0 \le y \le 4$, $n = 4$, evaluate at upper right corner

10. $f(x, y) = 3x - y$, $-1 \le x \le 1$, $0 \le y \le 4$, $n = 4$, evaluate at lower left corner

In exercises 11 and 12, compute the Riemann sum for the given function, the irregular partition shown, and midpoint evaluation.

11. $f(x, y) = 3x - y$ 12. $f(x, y) = 2x + y$

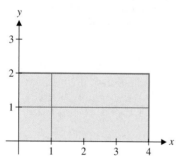

In exercises 13–18, evaluate the double integral.

13. $\iint\limits_{R} (x^2 - 2y) \, dA$, where $R = \{0 \le x \le 2, -1 \le y \le 1\}$

14. $\iint\limits_{R} (2xy + y^3) \, dA$, where $R = \{-1 \le x \le 2, 0 \le y \le 2\}$

15. $\iint\limits_{R} 4xe^{2y} \, dA$, where $R = \{2 \le x \le 4, 0 \le y \le 1\}$

16. $\iint\limits_{R} (3 + 2x - 3y^2) \, dA$, where $R = \{-3 \le x \le 2, 0 \le y \le 4\}$

17. $\iint\limits_{R} (1 - ye^{xy}) \, dA$, where $R = \{0 \le x \le 2, 0 \le y \le 3\}$

18. $\iint\limits_{R} (3x - 4x\sqrt{xy}) \, dA$, where $R = \{0 \le x \le 4, 0 \le y \le 9\}$

In exercises 19–22, sketch the solid whose volume is given by the iterated integral.

19. $\displaystyle\int_{-2}^{2} \int_{0}^{1} (6 - 2x - 3y) \, dy \, dx$

20. $\displaystyle\int_{0}^{2} \int_{-1}^{1} (2 + x + 2y) \, dy \, dx$

21. $\displaystyle\int_{0}^{2} \int_{0}^{3} (x^2 + y^2) \, dy \, dx$

22. $\displaystyle\int_{-1}^{1} \int_{-1}^{1} (4 - x^2 - y^2) \, dy \, dx$

In exercises 23–32, evaluate the iterated integral.

23. $\displaystyle\int_{0}^{1} \int_{0}^{2x} (x + 2y) \, dy \, dx$ 24. $\displaystyle\int_{0}^{2} \int_{0}^{x^2} (x + 3) \, dy \, dx$

25. $\displaystyle\int_{0}^{2} \int_{0}^{4y} (x + 2y) \, dx \, dy$ 26. $\displaystyle\int_{-2}^{2} \int_{0}^{y} (x^3 - 1) \, dx \, dy$

27. $\displaystyle\int_{0}^{1} \int_{0}^{2y} (4x\sqrt{y} + y) \, dx \, dy$ 28. $\displaystyle\int_{0}^{\pi} \int_{0}^{2} (y \sin xy) \, dx \, dy$

29. $\displaystyle\int_{0}^{2} \int_{0}^{2y} e^{y^2} \, dx \, dy$ 30. $\displaystyle\int_{1}^{2} \int_{0}^{2/x} e^{xy} \, dy \, dx$

31. $\displaystyle\int_{1}^{4} \int_{0}^{1/x} \cos xy \, dy \, dx$ 32. $\displaystyle\int_{0}^{1} \int_{0}^{y^2} \frac{3}{4 + y^3} \, dx \, dy$

33. Show that $\displaystyle\int_{0}^{1} \int_{0}^{2x} x^2 \, dy \, dx \ne \int_{0}^{1} \int_{0}^{2y} x^2 \, dx \, dy$.

34. Sketch the solids whose volumes are given in exercise 33 and explain why the volumes are not equal.

In exercises 35–42, find an integral equal to the volume of the solid bounded by the given surfaces and evaluate the integral.

35. $z = x^2 + y^2$, $z = 0$, $y = 1$, $y = 4$, $x = 0$, $x = 3$

36. $z = 3x^2 + 2y$, $z = 0$, $y = 0$, $y = 1$, $x = 1$, $x = 3$

37. $z = x^2 + y^2$, $z = 0$, $y = x^2$, $y = 1$

38. $z = 3x^2 + 2y$, $z = 0$, $y = 1 - x^2$, $y = 0$

39. $z = 6 - x - y$, $z = 0$, $x = 4 - y^2$, $x = 0$

40. $z = 4 - 2y$, $z = 0$, $x = y^4$, $x = 1$

41. $z = y^2$, $z = 0$, $y = 0$, $y = x$, $x = 2$

42. $z = x^2$, $z = 0$, $y = x$, $y = 4$, $x = 0$

In exercises 43–46, approximate the double integral.

43. $\iint\limits_{R} (2x - y)\, dA$, where R is bounded by $y = \sin x$ and $y = 1 - x^2$

44. $\iint\limits_{R} (2x - y)\, dA$, where R is bounded by $y = e^x$ and $y = 2 - x^2$

45. $\iint\limits_{R} e^{x^2}\, dA$, where R is bounded by $y = x^2$ and $y = 1$

46. $\iint\limits_{R} \sqrt{y^2 + 1}\, dA$, where R is bounded by $x = 4 - y^2$ and $x = 0$

In exercises 47–52, change the order of integration.

47. $\displaystyle\int_{0}^{1} \int_{0}^{2x} f(x, y)\, dy\, dx$

48. $\displaystyle\int_{0}^{1} \int_{2x}^{2} f(x, y)\, dy\, dx$

49. $\displaystyle\int_{0}^{2} \int_{2y}^{4} f(x, y)\, dx\, dy$

50. $\displaystyle\int_{0}^{1} \int_{0}^{2y} f(x, y)\, dx\, dy$

51. $\displaystyle\int_{0}^{\ln 4} \int_{e^x}^{4} f(x, y)\, dy\, dx$

52. $\displaystyle\int_{1}^{2} \int_{0}^{\ln y} f(x, y)\, dx\, dy$

In exercises 53–56, evaluate the iterated integral by first changing the order of integration.

53. $\displaystyle\int_{0}^{2} \int_{x}^{2} 2e^{y^2}\, dy\, dx$

54. $\displaystyle\int_{0}^{1} \int_{\sqrt{x}}^{1} \frac{3}{4 + y^3}\, dy\, dx$

55. $\displaystyle\int_{0}^{1} \int_{y}^{1} 3xe^{x^3}\, dx\, dy$

56. $\displaystyle\int_{0}^{1} \int_{\sqrt{y}}^{1} \cos x^3\, dx\, dy$

57. Determine if your CAS can evaluate the integrals $\displaystyle\int_{x}^{2} 2e^{y^2}\, dy$ and $\displaystyle\int_{0}^{2} \int_{x}^{2} 2e^{y^2}\, dy\, dx$.

58. Explain why a CAS can't evaluate the first integral in exercise 57. Based on your result in exercise 57, can your CAS switch orders of integration to evaluate a double integral?

In exercises 59–62, sketch the solid whose volume is described by the given iterated integral.

59. $\displaystyle\int_{0}^{3} \int_{0}^{6-2x} (6 - 2x - y)\, dy\, dx$

60. $\displaystyle\int_{0}^{4} \int_{0}^{4-x} (4 - x - y)\, dy\, dx$

61. $\displaystyle\int_{-2}^{2} \int_{-\sqrt{4-x^2}}^{\sqrt{4-x^2}} (4 - x^2 - y^2)\, dy\, dx$

62. $\displaystyle\int_{0}^{1} \int_{0}^{\sqrt{1-x^2}} (x^2 + y^2)\, dy\, dx$

63. Explain why $\displaystyle\int_{0}^{1} \int_{0}^{2x} f(x, y)\, dy\, dx$ is not generally equal to $\displaystyle\int_{0}^{1} \int_{0}^{2y} f(x, y)\, dx\, dy$.

64. Give an example of a function for which the integrals in exercise 63 are equal. As generally as possible, describe what property such a function must have.

65. Prove Theorem 1.4 part (i).

66. Prove Theorem 1.4 part (ii).

67. Prove Theorem 1.4 part (iii).

68. Compute the iterated integral by sketching a graph and using a basic geometric formula:
$$\int_{-1}^{1} \int_{-\sqrt{1-x^2}}^{\sqrt{1-x^2}} \sqrt{1 - x^2 - y^2}\, dy\, dx.$$

69. Set up a double integral for the volume of the solid bounded by the graphs of $z = 4 - x^2 - y^2$ and $z = x^2 + y^2$. Note that you actually have two tasks. First, the general rule for finding the volume between two surfaces is analogous to the general rule for finding the area between two curves. The greater challenge here is to find the limits of integration.

70. As mentioned in the text, numerical methods for approximating double integrals can be troublesome. The **Monte Carlo method** makes clever use of probability theory to approximate $\iint\limits_{R} f(x, y)\, dA$ for a bounded region R. Suppose, for example, that R is contained within the rectangle $0 \le x \le 1, 0 \le y \le 1$. Generate two random numbers a and b from the uniform distribution on $[0, 1]$; this means that any number between 0 and 1 is in some sense equally likely. Determine whether or not the point (a, b) is in the region R, and then repeat the process a large number of times. If, for example, 64 out of 100 points generated were within R, explain why a reasonable estimate of the area of R is 0.64 times the area of the rectangle $0 \le x \le 1, 0 \le y \le 1$. For each point (a, b) which is within R, compute $f(a, b)$. If the average of all of these function values is 13.6, explain why a reasonable estimate of $\iint\limits_{R} f(x, y)\, dA$ is $(0.64)(13.6) = 8.704$. Use the Monte Carlo method to estimate $\displaystyle\int_{1}^{2} \int_{\ln x}^{\sqrt{x}} \sin(xy)\, dy\, dx$. (Hint: Show that y is between $\ln 1 = 0$ and $\sqrt{2} < 2$.)

13.2   AREA, VOLUME AND CENTER OF MASS

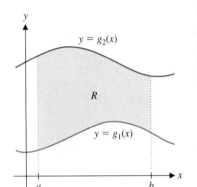

Figure 13.17
The region R.

To use double integrals to solve problems, it's very important that you have a clear picture of what each of the components of the integral represents. For this reason, we want to pause briefly to set up a double iterated integral as a double sum, so that the components of the iterated integral become more clear. Consider the case of a continuous function $f(x, y) \geq 0$ on some region $R \subset \mathbb{R}^2$. If R has the form

$$R = \{(x, y) \mid a \leq x \leq b \text{ and } g_1(x) \leq y \leq g_2(x)\},$$

as indicated in Figure 13.17, then we have from our work in section 13.1 that the volume V lying beneath the surface $z = f(x, y)$ and above the region R is given by

$$V = \int_a^b A(x)\, dx = \int_a^b \int_{g_1(x)}^{g_2(x)} f(x, y)\, dy\, dx. \tag{2.1}$$

Here, for each fixed x, $A(x)$ is the area of the cross section of the solid corresponding to that particular value of x. Our aim is to write the above volume integral in a slightly different way from our derivation in section 13.1. First, notice that by the definition of definite integral, we have that

$$\int_a^b A(x)\, dx = \lim_{\|P_1\| \to 0} \sum_{i=1}^n A(c_i)\, \Delta x_i, \tag{2.2}$$

where P_1 represents a partition of the interval $[a, b]$, c_i is some point in the ith subinterval, $[x_{i-1}, x_i]$ and $\Delta x_i = x_i - x_{i-1}$ (the width of the ith subinterval). For each fixed $x \in [a, b]$, since $A(x)$ is the area of the cross section, we have that

$$A(x) = \int_{g_1(x)}^{g_2(x)} f(x, y)\, dy = \lim_{\|P_2\| \to 0} \sum_{j=1}^m f(x, v_j)\, \Delta y_j, \tag{2.3}$$

where P_2 represents a partition of the interval $[g_1(x), g_2(x)]$, v_j is some point in the jth subinterval $[y_{j-1}, y_j]$ of the partition P_2 and $\Delta y_j = y_j - y_{j-1}$ (the width of the jth subinterval). Putting (2.1), (2.2) and (2.3) together, we get

$$V = \lim_{\|P_1\| \to 0} \sum_{i=1}^n A(c_i)\, \Delta x_i$$

$$= \lim_{\|P_1\| \to 0} \sum_{i=1}^n \left[\lim_{\|P_2\| \to 0} \sum_{j=1}^m f(c_i, v_j)\, \Delta y_j \right] \Delta x_i$$

$$= \lim_{\|P_1\| \to 0} \lim_{\|P_2\| \to 0} \sum_{i=1}^n \sum_{j=1}^m f(c_i, v_j)\, \Delta y_j\, \Delta x_i. \tag{2.4}$$

The double summation in (2.4) is called a **double Riemann sum.** Notice that each term corresponds to the volume of a box of length Δx_i, width Δy_j and height $f(c_i, v_j)$ (see Figure 13.18). Observe that by superimposing the two partitions, we have produced an inner partition of the region R. If we represent this inner partition of R by P and the norm of the partition P by $\|P\|$, the length of the longest diagonal of any rectangle in the partition, we

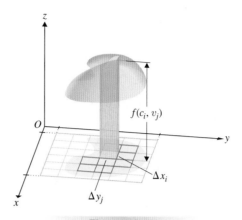

Figure 13.18
Volume of a typical box.

can write (2.4) with only one limit, as

$$V = \lim_{\|P\| \to 0} \sum_{i=1}^{n} \sum_{j=1}^{m} f(c_i, v_j)\, \Delta y_j\, \Delta x_i. \tag{2.5}$$

When you write down an iterated integral representing volume, you can use (2.5) to help identify each of the components as follows:

$$V = \lim_{\|P\| \to 0} \sum_{i=1}^{n} \sum_{j=1}^{m} \underbrace{f(c_i, v_j)}_{\text{height}}\, \underbrace{\Delta y_j}_{\text{width}}\, \underbrace{\Delta x_i}_{\text{length}}$$

$$= \int_{a}^{b} \int_{g_1(x)}^{g_2(x)} \underbrace{f(x, y)}_{\text{height}}\, \underbrace{dy}_{\text{width}}\, \underbrace{dx}_{\text{length}}. \tag{2.6}$$

You should make at least a mental picture of the components of the integral in (2.6), keeping in mind the corresponding components of the Riemann sum. We leave it as an exercise to show that for a region of the form

$$R = \{(x, y) \mid c \le y \le d \text{ and } h_1(y) \le x \le h_2(y)\},$$

we get a corresponding interpretation of the iterated integral:

$$V = \lim_{\|P\| \to 0} \sum_{j=1}^{m} \sum_{i=1}^{n} \underbrace{f(c_i, v_j)}_{\text{height}}\, \underbrace{\Delta x_i}_{\text{length}}\, \underbrace{\Delta y_j}_{\text{width}}$$

$$= \int_{c}^{d} \int_{h_1(y)}^{h_2(y)} \underbrace{f(x, y)}_{\text{height}}\, \underbrace{dx}_{\text{length}}\, \underbrace{dy}_{\text{width}}. \tag{2.7}$$

Observe that for any bounded region $R \subset \mathbb{R}^2$, $\iint_R 1\, dA$, which we sometimes write simply as $\iint_R dA$, gives the volume under the surface $z = 1$ and above the region R in the xy-plane. Since all of the cross sections parallel to the xy-plane are the same, the solid is a cylinder and so, its volume is the product of its height (1) and its cross-sectional area. That is

$$\iint\limits_{R} dA = (1)(\text{Area of } R) = \text{Area of } R. \tag{2.8}$$

So, we now have the option of using a double integral to find the area of a plane region.

Example 2.1   **Using a Double Integral to Find Area**

Find the area of the plane region bounded by the graphs of $x = y^2$, $y - x = 3$, $y = -3$ and $y = 2$ (see Figure 13.19).

Solution   Note that we have indicated in the figure a small rectangle with sides dx and dy, respectively. This helps to indicate the limits for the iterated integral. From (2.8), we have

$$A = \iint\limits_{R} dA = \int_{-3}^{2} \int_{y-3}^{y^2} dx\,dy = \int_{-3}^{2} x \Big|_{x=y-3}^{x=y^2} dy$$

$$= \int_{-3}^{2} [y^2 - (y - 3)]\,dy = \left(\frac{y^3}{3} - \frac{y^2}{2} + 3y \right) \Big|_{-3}^{2}$$

$$= \frac{175}{6}.$$

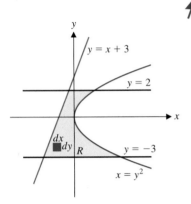

Figure 13.19
The region R.

Think about example 2.1 a little further. Recall that we had worked similar problems in section 5.1 using single integrals. In fact, you might have set up the desired area directly as

$$A = \int_{-3}^{2} [y^2 - (y - 3)]\,dy,$$

exactly as you see in the second line of work above. While we will sometimes use double integrals to more easily solve familiar problems, double integrals will allow us to solve many new problems as well.

We have already developed formulas for calculating the volume of a solid lying below a surface of the form $z = f(x, y)$ and above a region R (of several different forms), lying in the xy-plane. So, what's the problem, then? As you will see in the next few examples, the challenge in setting up the iterated integrals comes in seeing the region R that the solid lies above and then determining the limits of integration for the iterated integrals.

Example 2.2   **Using a Double Integral to Find Volume**

Find the volume of the tetrahedron bounded by the plane $2x + y + z = 2$ and the three coordinate planes.

Solution   First, we need to draw a sketch of the solid. Since the plane $2x + y + z = 2$ intersects the coordinate axes at the points $(1, 0, 0)$, $(0, 2, 0)$ and $(0, 0, 2)$, a sketch is easy to draw. Simply connect the three points of intersection with the coordinate axes and you'll get the graph of the tetrahedron (a four-sided object with all triangular sides) seen in Figure 13.20a. In order to use our volume formula, though, we'll first need to visualize the tetrahedron as a solid lying below a surface of the form $z = f(x, y)$ and

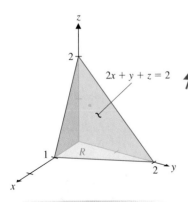

Figure 13.20a
Tetrahedron.

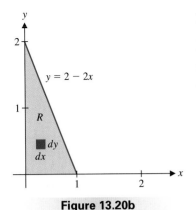

Figure 13.20b

The region R.

lying above some region R in the xy-plane. Notice that the solid lies below the plane $z = 2 - 2x - y$ and above the triangular region R in the xy-plane as indicated in Figure 13.20a. Although we're not simply handed R, you can see that R is the triangular region bounded by the x- and y-axes and the trace of the plane $2x + y + z = 2$ in the xy-plane. The trace is found by simply setting $z = 0 : 2x + y = 2$ (see Figure 13.20b). From (2.6), we get that the volume is

$$V = \int_0^1 \int_0^{2-2x} \underbrace{(2 - 2x - y)}_{\text{height}} \underbrace{dy}_{\text{width}} \underbrace{dx}_{\text{length}}$$

$$= \int_0^1 \left(2y - 2xy - \frac{y^2}{2} \right) \Bigg|_{y=0}^{y=2-2x} dx$$

$$= \int_0^1 \left[2(2 - 2x) - 2x(2 - 2x) - \frac{(2 - 2x)^2}{2} \right] dx$$

$$= \frac{2}{3},$$

where we leave the routine details of the calculation to you.

We cannot emphasize enough the need to draw reasonable sketches of the solid and particularly of the base of the solid in the xy-plane. You may be lucky enough to guess the limits of integration for a few of these problems, but don't be deceived: you need to draw good sketches and look carefully to determine the limits of integration correctly.

Example 2.3 — Finding the Volume of a Solid

Find the volume of the solid lying in the first octant and bounded by the graphs of $z = 4 - x^2$, $x + y = 2$, $x = 0$, $y = 0$ and $z = 0$.

Solution First, draw a sketch of the solid. You should note that $z = 4 - x^2$ is a cylinder (remember: there's no y term), $x + y = 2$ is a plane and $x = 0$, $y = 0$ and $z = 0$ are the coordinate planes (see Figure 13.21a). Notice that the solid lies below the surface $z = 4 - x^2$ and above the triangular region R in the xy-plane formed by the x- and y-axes and the trace of the plane $x + y = 2$ in the xy-plane (i.e., the line $x + y = 2$). This is shown in Figure 13.21b. Although we could integrate with respect to either x or y first, we integrate with respect to x first. From (2.7), we have

$$V = \int_0^2 \int_0^{2-y} \underbrace{(4 - x^2)}_{\text{height}} \underbrace{dx}_{\text{length}} \underbrace{dy}_{\text{width}}$$

$$= \int_0^2 \left(4x - \frac{x^3}{3} \right) \Bigg|_{x=0}^{x=2-y} dy$$

$$= \int_0^2 \left[4(2 - y) - \frac{(2 - y)^3}{3} \right] dy$$

$$= \frac{20}{3}.$$

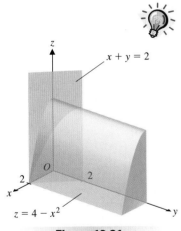

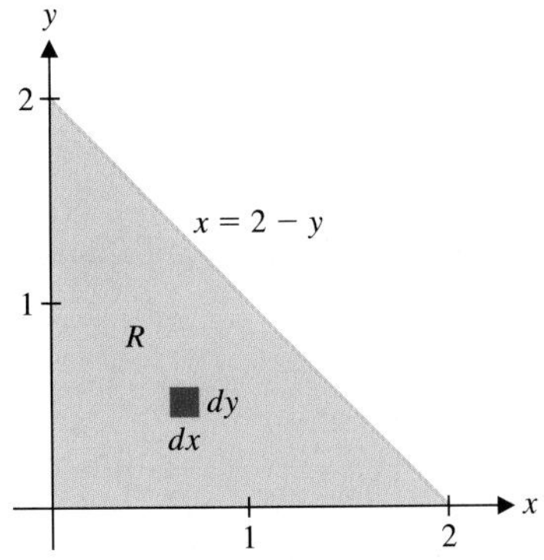

Figure 13.21a

Solid in the first octant.

Figure 13.21b

The region R.

Example 2.4   Finding the Volume of a Solid

Find the volume of the solid lying in the first octant bounded by the graphs of $z = 4 - x^2 - y^2$, $y = 2 - 2x^2$, $x = 0$, $y = 0$ and $z = 0$.

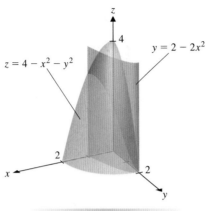

Figure 13.22a	**Figure 13.22b**
Solid in the first octant.	The base of the solid.

Solution   Notice that $z = 4 - x^2 - y^2$ is a paraboloid, opening downward, with vertex at the point $(0, 0, 4)$, $y = 2 - 2x^2$ is a cylinder with axis parallel to the z-axis and $x = 0$, $y = 0$ and $z = 0$ are the yz-, xz- and xy-planes, respectively. We draw a sketch of the solid in Figure 13.22a. From the figure, observe that the solid lies below the paraboloid and above the region R in the first quadrant of the xy-plane bounded by the parabola $y = 2 - 2x^2$ and the x- and y-axes, as shown in Figure 13.22b. From (2.6), we get the volume

$$V = \int_0^1 \int_0^{2-2x^2} \underbrace{(4 - x^2 - y^2)}_{\text{height}} \underbrace{dy}_{\text{width}} \underbrace{dx}_{\text{length}}$$

$$= \int_0^1 \left(4y - x^2 y - \frac{y^3}{3} \right) \Bigg|_{y=0}^{y=2-2x^2} dx$$

$$= \int_0^1 \left[4(2 - 2x^2) - x^2(2 - 2x^2) - \frac{(2 - 2x^2)^3}{3} \right] dx$$

$$= \frac{404}{105} \approx 3.848.$$

Double integrals are used to calculate numerous quantities of interest in applications. One example follows and others can be found in the exercises.

Example 2.5   Estimating Population

Suppose that $f(x, y) = 20{,}000 y e^{-x^2 - y^2}$ models the population density (population per square mile) of a species of small animals, with x and y measured in miles. Estimate the population in the triangular shaped habitat with vertices $(1, 1)$, $(2, 1)$ and $(1, 0)$.

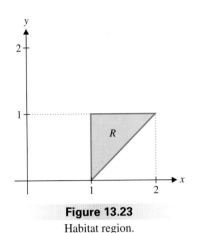

Figure 13.23

Habitat region.

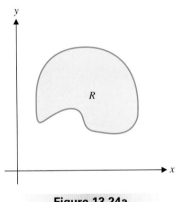

Figure 13.24a

Lamina.

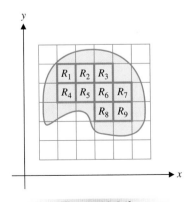

Figure 13.24b

Inner partition of R.

Solution The population in any region R is estimated by

$$\iint_R f(x, y)\, dA = \iint_R 20{,}000 y e^{-x^2-y^2}\, dA.$$

[As a quick check on the reasonableness of this formula, note that $f(x, y)$ is measured in units of population per square mile and the area increment dA carries units of square miles, so that the combination $f(x, y)\, dA$ carries the desired units of population.] Notice that the integrand is $20{,}000 y e^{-x^2-y^2} = 20{,}000 e^{-x^2} y e^{-y^2}$, which suggests that we should integrate with respect to y first. As always, we first sketch a graph of the region R (shown in Figure 13.23). Notice that the line through the points $(1, 0)$ and $(2, 1)$ has the equation $y = x - 1$, so that R extends from $y = x - 1$ up to $y = 1$, as x increases from 1 to 2. We now have

$$\iint_R f(x, y)\, dA = \int_1^2 \int_{x-1}^1 20{,}000 e^{-x^2} y e^{-y^2}\, dy\, dx$$

$$= \int_1^2 10{,}000 e^{-x^2}\left[e^{-(x-1)^2} - e^{-1}\right] dx$$

$$\approx 698,$$

where we approximated the last integral numerically.

Moments and Center of Mass

We close this section by briefly discussing a physical application of double integrals. Consider a thin, flat plate (a **lamina**) in the shape of the region $R \subset \mathbb{R}^2$ whose density (mass per unit area) varies throughout the plate (i.e., some areas of the plate are more dense than others). From an engineering standpoint, it's often important to determine where you could place a support to balance the plate. We call this point the **center of mass** of the lamina. The first task we need to complete here is to find the total mass of the plate. For a real plate, we'd simply place it on a scale, but for our theoretical plate, we'll need to be more clever. Suppose the lamina has the shape of the region R shown in Figure 13.24a and has mass density (mass per unit area) given by the function $\rho(x, y)$. Construct an inner partition of R, as in Figure 13.24b. Notice that if the norm of the partition $\|P\|$ is small, then the density will be nearly constant on each rectangle of the inner partition. So, for each $i = 1, 2, \ldots, n$, pick some point $(u_i, v_i) \in R_i$. Then, the mass m_i of the portion of the lamina corresponding to the rectangle R_i is given approximately by

$$m_i \approx \underbrace{\rho(u_i, v_i)}_{\text{mass/unit area}} \underbrace{\Delta A_i}_{\text{area}},$$

where ΔA_i denotes the area of R_i. The total mass m of the lamina is then given approximately by

$$m \approx \sum_{i=1}^n \rho(u_i, v_i)\, \Delta A_i.$$

Notice that if $\|P\|$ is small, then this should be a reasonable approximation of the total mass.

To get the mass exactly, we take the limit as $\|P\|$ tends to zero, which you should recognize as a double integral:

$$m = \lim_{\|P\|\to 0} \sum_{i=1}^{n} \rho(u_i, v_i)\, \Delta A_i = \iint\limits_{R} \rho(x, y)\, dA. \tag{2.9}$$

Notice that if you want to balance a lamina like the one shown in Figure 13.24a, you'll need to balance it both from left to right and from top to bottom. In the language of our previous discussion of center of mass in section 5.6, we'll need to find the first moments: both left to right (we call this the **moment with respect to the y-axis**) and top to bottom (the **moment with respect to the x-axis**). First, we approximate the moment M_y with respect to the y-axis. Assuming that the mass in the ith rectangle of the partition is concentrated at the point (u_i, v_i), we have

$$M_y \approx \sum_{i=1}^{n} u_i \rho(u_i, v_i)$$

(i.e., the sum of the products of the masses and their directed distances from the y-axis). Taking the limit as $\|P\|$ tends to zero, we get

$$M_y = \lim_{\|P\|\to 0} \sum_{i=1}^{n} u_i \rho(u_i, v_i) = \iint\limits_{R} x\rho(x, y)\, dA. \tag{2.10}$$

Similarly, looking at the sum of the products of the masses and their directed distances from the x-axis, we get the moment M_x with respect to the x-axis,

$$M_x = \lim_{\|P\|\to 0} \sum_{i=1}^{n} v_i \rho(u_i, v_i) = \iint\limits_{R} y\rho(x, y)\, dA. \tag{2.11}$$

The center of mass is the point $(\bar{x}, \bar{y})$ defined by

$$\bar{x} = \frac{M_y}{m} \quad \text{and} \quad \bar{y} = \frac{M_x}{m}. \tag{2.12}$$

Example 2.6   Finding the Center of Mass of a Lamina

Find the center of mass of the lamina in the shape of region bounded by the graphs of $y = x^2$ and $y = 4$, having mass density given by $\rho(x, y) = 1 + 2y + 6x^2$.

Solution   We sketch the region in Figure 13.25. From (2.9), we have that the total mass of the lamina is given by

$$m = \iint\limits_{R} \rho(x, y)\, dA = \int_{-2}^{2} \int_{x^2}^{4} (1 + 2y + 6x^2)\, dy\, dx$$

$$= \int_{-2}^{2} \left(y + 2\frac{y^2}{2} + 6x^2 y \right) \Bigg|_{y=x^2}^{y=4} dx$$

$$= \int_{-2}^{2} [(4 + 16 + 24x^2) - (x^2 + x^4 + 6x^4)]\, dx$$

$$= \frac{1696}{15} \approx 113.1.$$

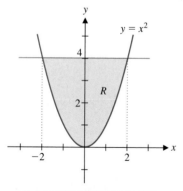

Figure 13.25

Lamina.

We compute the moment M_y from (2.10):

$$M_y = \iint\limits_R x\rho(x, y)\, dA = \int_{-2}^{2}\int_{x^2}^{4} x(1 + 2y + 6x^2)\, dy\, dx$$

$$= \int_{-2}^{2}\int_{x^2}^{4} (x + 2xy + 6x^3)\, dy\, dx$$

$$= \int_{-2}^{2} (xy + xy^2 + 6x^3 y)\Big|_{y=x^2}^{y=4} dx$$

$$= \int_{-2}^{2} [(4x + 16x + 24x^3) - (x^3 + x^5 + 6x^5)]\, dx = 0.$$

Note that from (2.12), this says that the x-coordinate of the center of mass is $\bar{x} = \dfrac{M_y}{m} = \dfrac{0}{113.1} = 0$. This should not surprise you since both the region **and** the mass density are symmetric with respect to the y-axis. [Notice that $\rho(-x, y) = \rho(x, y)$.] Next, from (2.11), we have

$$M_x = \iint\limits_R y\rho(x, y)\, dA = \int_{-2}^{2}\int_{x^2}^{4} y(1 + 2y + 6x^2)\, dy\, dx$$

$$= \int_{-2}^{2}\int_{x^2}^{4} (y + 2y^2 + 6x^2 y)\, dy\, dx$$

$$= \int_{-2}^{2} \left(\frac{y^2}{2} + 2\frac{y^3}{3} + 6x^2\frac{y^2}{2}\right)\Big|_{y=x^2}^{y=4} dx$$

$$= \int_{-2}^{2} \left[\left(8 + \frac{128}{3} + 48x^2\right) - \left(\frac{x^4}{2} + \frac{2}{3}x^6 + 3x^6\right)\right] dx$$

$$= \frac{11,136}{35} \approx 318.2,$$

and so, from (2.12) we have $\bar{y} = \dfrac{M_x}{m} \approx \dfrac{318.2}{113.1} \approx 2.8$. The center of mass is then located at approximately

$$(\bar{x}, \bar{y}) \approx (0, 2.8).$$

■

In example 2.6, we computed the first moments M_y and M_x to find the balance point (center of mass) of the lamina in Figure 13.25. Further physical properties of this lamina can be determined using the **second moments** I_y and I_x. Much as we defined the first moments in equations (2.10) and (2.11), the second moment about the y-axis (often called the **moment of inertia about the y-axis**) of a lamina in the shape of the region R, with density function $\rho(x, y)$ is defined by

$$I_y = \iint\limits_R x^2\rho(x, y)\, dA.$$

Similarly, the second moment about the x-axis (also called the **moment of inertia about the x-axis**) of a lamina in the shape of the region R, with density function $\rho(x, y)$ is defined by

$$I_x = \iint\limits_R y^2\rho(x, y)\, dA.$$

Physics tells us that the larger I_y is, the more difficult it is to rotate the lamina about the y-axis. Similarly, the larger I_x is, the more difficult it is to rotate the lamina about the x-axis. We explore this briefly in the following example.

Example 2.7   Finding the Moments of Inertia of a Lamina

Find the moments of inertia I_y and I_x for the lamina in example 2.6.

Solution   The region R is the same as in example 2.6, so that the limits of integration are the same. We have

$$I_y = \int_{-2}^{2} \int_{x^2}^{4} x^2(1 + 2y + 6x^2)\, dy\, dx$$

$$= \int_{-2}^{2} (20x^2 + 23x^4 - 7x^6)\, dx$$

$$= \frac{2176}{15} \approx 145.07$$

and

$$I_x = \int_{-2}^{2} \int_{x^2}^{4} y^2(1 + 2y + 6x^2)\, dy\, dx$$

$$= \int_{-2}^{2} \left(\frac{448}{3} + 128x^2 - \frac{1}{3}x^6 - \frac{5}{2}x^8 \right) dx$$

$$= \frac{61,952}{63} \approx 983.37.$$

A comparison of the two moments of inertia shows that it is much more difficult to rotate the lamina of Figure 13.25 about the x-axis than about the y-axis. Examine the figure and the density function to be sure that this makes sense to you.

EXERCISES 13.2

1. The double Riemann sum in (2.5) disguises the fact that the order of integration is important. Explain how the order of integration affects the details of the double Riemann sum.

2. Many double integrals can be set up in two steps: first identify the function $f(x, y)$, then identify the two-dimensional region R and set up the limits of integration. Explain how these two steps are separated in examples 2.2, 2.3 and 2.4.

3. The sketches in examples 2.2, 2.3 and 2.4 are essential, but somewhat difficult to draw. Explain each sketch, including which surface should be drawn first, second and so on. Also, when a previously drawn surface is cut in half by a plane, explain how to identify which half of the cut surface to keep.

4. The moment M_y is the moment about the y-axis, but is used to find the x-coordinate of the center of mass. Explain why it is M_y and not M_x that is used to compute the x-coordinate of the center of mass.

In exercises 5–10, use a double integral to compute the area of the region bounded by the curves.

5. $y = x^2, y = 8 - x^2$

6. $y = x^2, y = x + 2$

7. $y = 2x, y = 3 - x, y = 0$

8. $y = 3x, y = 5 - 2x, y = 0$

9. $y = x^2, x = y^2$

10. $y = x^3, y = x^2$

In exercises 11–22, compute the volume of the solid bounded by the given surfaces.

11. $2x + 3y + z = 6$ and the three coordinate planes

12. $x + 2y - 3z = 6$ and the three coordinate planes

13. $z = 4 - x^2 - y^2, z = 0, x = -1, x = 1, y = -1, y = 1$

14. $z = x^2 + y^2, z = 0, x = 0, x = 1, y = 0, y = 1$

15. $z = 1 - y, z = 0, y = 0, x = 1, x = 2$

16. $z = 2 + x, z = 0, x = 0, y = 0, y = 1$

17. $z = 1 - y^2, x + y = 1$ and the three coordinate planes (first octant)

18. $z = 1 - x^2 - y^2, x + y = 1$ and the three coordinate planes

19. $z = x^2 + y^2, z = 0, y = x^2, y = 4$

20. $z = x^2 + y^2 + 1, z = 0, y = x^2, y = 2x + 3$

21. $z = x + 2, z = 0, x = y^2 - 2, x = y$

22. $z = 2x + y + 1, z = 0, x = y^2, x = 1$

In exercises 23–26, set up a double integral for the volume bounded by the given surfaces and estimate it numerically.

23. $z = \sqrt{x^2 + y^2}, y = 4 - x^2$, first octant

24. $z = \sqrt{4 - x^2 - y^2}, x^2 + y^2 = 1$, first octant

25. $z = e^{xy}, x + 2y = 4$ and the three coordinate planes

26. $z = e^{x^2 + y^2}, z = 0$ and $x^2 + y^2 = 4$

In exercises 27–32, find the mass and center of mass of the lamina with the given density.

27. Lamina bounded by $y = x^3$ and $y = x^2, \rho(x, y) = 4$

28. Lamina bounded by $y = x^4$ and $y = x^2, \rho(x, y) = 4$

29. Lamina bounded by $x = y^2$ and $x = 1, \rho(x, y) = y^2 + x + 1$

30. Lamina bounded by $x = y^2$ and $x = 4, \rho(x, y) = y + 3$

31. Lamina bounded by $y = x^2 \ (x > 0), \ y = 4$ and $x = 0$, $\rho(x, y) =$ distance from y-axis

32. Lamina bounded by $y = x^2 - 4$ and $y = 5, \rho(x, y) =$ square of the distance from the y-axis

33. The laminae of exercises 29 and 30 are both symmetric about the x-axis. Explain why it is not true in both exercises that the center of mass is located on the x-axis.

34. Suppose that a lamina is symmetric about the x-axis. State a condition on the density function $\rho(x, y)$ that guarantees that the center of mass is located on the x-axis.

35. Suppose that a lamina is symmetric about the y-axis. State a condition on the density function $\rho(x, y)$ that guarantees that the center of mass is located on the y-axis.

36. Give an example of a lamina that *is* symmetric about the y-axis but that *does not* have its center of mass on the y-axis.

37. Suppose that $f(x, y) = 15{,}000xe^{-x^2-y^2}$ is the population density of a species of small animals. Estimate the population in the triangular region with vertices $(1, 1), (2, 1)$ and $(1, 0)$.

38. Suppose that $f(x, y) = 15{,}000xe^{-x^2-y^2}$ is the population density of a species of small animals. Estimate the population in the region bounded by $y = x^2, y = 0$ and $x = 1$.

39. Suppose that $f(x, t) = 20e^{-t/6}$ is the yearly rate of change of the price per barrel of oil. If x is the number of billions of barrels and t is the number of years since 2000, compute and interpret $\int_0^{10} \int_0^4 f(x, t) \, dt \, dx$.

40. Repeat exercise 39 for $f(x, t) = \begin{cases} 20e^{-t/6}, & \text{if } 0 \le x \le 4 \\ 14e^{-t/6}, & \text{if } x > 4 \end{cases}$.

41. Find the mass and moments of inertia I_y and I_x for a lamina in the shape of the region bounded by $y = x^2$ and $y = 4$ with density $\rho(x, y) = 1$.

42. Find the mass and moments of inertia I_y and I_x for a lamina in the shape of the region bounded by $y = \frac{1}{4}x^2$ and $y = 1$ with density $\rho(x, y) = 4$. Comparing your answers to exercises 41 and 42, you should have found the same mass but different moments of inertia. Use the shapes of the regions to explain why this makes sense.

43. Figure skaters can control their rate of spin ω by varying their body positions, utilizing the principle of **conservation of angular momentum.** This states that in the absence of outside forces, the quantity $I_y \omega$ remains constant. Thus, reducing I_y by

a factor of 2 will increase spin rate by a factor of 2. Compare the spin rates of the following two crude models of a figure skater, the first with arms extended (use $\rho = 1$) and the second with arms raised and legs crossed (use $\rho = 2$).

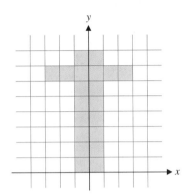

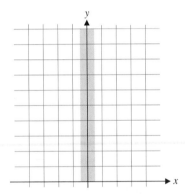

44. Lamina A is in the shape of the rectangle $-1 \leq x \leq 1$ and $-5 \leq y \leq 5$ with density $\rho(x, y) = 1$. It models a diver in the "layout" position. Lamina B is in the shape of the rectangle $-1 \leq x \leq 1$ and $-2 \leq y \leq 2$ with density $\rho(x, y) = 2.5$. It models a diver in the "tuck" position. Find the moment of inertia I_x for each lamina, and explain why divers use the tuck position to do multiple rotation dives.

45. Estimate the moment of inertia about the y-axis of the two ellipses R_1 bounded by $x^2 + 4y^2 = 16$ and R_2 bounded by $x^2 + 4y^2 = 36$. Assume a constant density of $\rho = 1$. R_1 and R_2 can be thought of as models of two tennis racket heads. The rackets have the same shape, but the second racket is much bigger than the first (the difference in size is about the same as the difference between rackets of the 1960s and rackets of the 1990s).

46. For the tennis rackets in exercise 45, a rotation about the y-axis would correspond to the racket twisting in your hand, which is undesirable. Compare the tendency of each racket to twist. As related in Blandig and Monteleone's *What Makes a Boomerang Come Back*, the larger moment of inertia is what motivated a sore-elbowed Howard Head to construct large-headed tennis rackets in the 1970s.

In exercises 47–54, define the average value of $f(x, y)$ on a region R of area a by $\dfrac{1}{a} \displaystyle\iint\limits_R f(x, y) \, dA$.

47. Compute the average value of $f(x, y) = y$ on the region bounded by $y = x^2$ and $y = 4$.

48. Compute the average value of $f(x, y) = y^2$ on the region bounded by $y = x^2$ and $y = 4$.

49. In exercise 47, compare the average value of $f(x, y)$ to the y-coordinate of the center of mass of a lamina with the same shape and constant density.

50. In exercise 48, R extends from $y = 0$ to $y = 4$. Explain why the average value of $f(x, y)$ corresponds to a y-value larger than 2.

51. Compute the average value of $f(x, y) = \sqrt{x^2 + y^2}$ on the region bounded by $y = x^2 - 4$ and $y = 3x$.

52. Interpret the geometric meaning of the average value in exercise 51. (Hint: What does $\sqrt{x^2 + y^2}$ represent geometrically?)

53. Suppose the temperature at the point (x, y) in a region R is given by $T(x, y) = 50 + \cos(2x + y)$, where R is bounded by $y = x^2$ and $y = 8 - x^2$. Estimate the average temperature in R.

54. Suppose the elevation at the point (x, y) in a region R is given by $h(x, y) = 2300 + 50 \sin x \cos y$, where R is bounded by $y = x^2$ and $y = 2x$. Estimate the average elevation in R.

55. Suppose that the function $f(x, y)$ gives the rainfall per unit area at the point (x, y) in a region R. State in words what

(a) $\iint\limits_R f(x, y) \, dA$ and (b) $\dfrac{\iint\limits_R f(x, y) \, dA}{\iint\limits_R 1 \, dA}$ represent.

56. Suppose that the function $p(x, y)$ gives the population density at the point (x, y) in a region R. State in words what

(a) $\iint\limits_R p(x, y) \, dA$ and (b) $\dfrac{\iint\limits_R p(x, y) \, dA}{\iint\limits_R 1 \, dA}$ represent.

57. A triangular lamina has vertices $(0, 0)$, $(0, 1)$ and $(c, 0)$ for some positive constant c. Show that the y-coordinate of the center of mass of the lamina is independent of the constant c.

58. Find the x-coordinate of the center of mass of the lamina of exercise 57 as a function of c.

59. A function $f(x, y)$ is a **joint probability density function** on a region R if $f(x, y) \geq 0$ for all (x, y) in R and $\iint\limits_R f(x, y) \, dA = 1$. Suppose that a person playing darts is aiming at the bull's-eye but is not very accurate. Suppose that the bull's-eye is centered at the origin and the dartboard is the

region R bounded by $x^2 + y^2 = 64$ (units are inches), and the joint density function for the resulting position of the dart is $f(x, y) = ce^{-x^2-y^2}$ for some constant c. Estimate the value of the constant c such that $f(x, y)$ is a joint density function on R. For a region U contained within R, the probability that the dart lands in U is given by $\iint_U f(x, y)\, dA$. Estimate the probability that the dart hits inside the bull's-eye circle $x^2 + y^2 = \frac{1}{4}$. Estimate the probability that the dart accidentally lands in the "triple 20" band bounded by $x^2 + y^2 = 16$, $x^2 + y^2 = 14$, $y = 6.3x$ and $y = -6.3x$. Explain why all of the regions in this exercise would be easily described in polar coordinates. (Then start reading the next section!)

60. 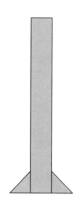 In this exercise, we explore the rocket design issues discussed in the introduction to the chapter. We will work with the crude model shown, where the main tower of the rocket is 1 unit by 8 units and each triangular fin has height 1 and width w. First, find the y-coordinate y_1 of the center of mass, assuming a constant density $\rho(x, y) = 1$. Second, find

the y-coordinate y_2 of the center of mass assuming the following density structure: the top half of the main tower has density $\rho = 1$, the bottom half of the main tower has density $\rho = 2$ and the fins have density $\rho = \frac{1}{4}$. Find the smallest value of w such that the stability criterion of $y_1 < y_2$ is met.

13.3   DOUBLE INTEGRALS IN POLAR COORDINATES

Polar coordinates prove to be particularly useful for dealing with certain double integrals. This happens for several reasons. Most importantly, if the region over which you are integrating is in some way circular, polar coordinates may be exactly what you need for dealing with an otherwise intractable integration problem. For instance, you might need to evaluate

$$\iint_R (x^2 + y^2 + 3)\, dA.$$

This certainly looks simple enough: all you need to do is integrate a polynomial. Nothing could be easier, until we tell you that R is the circle of radius 2, centered at the origin. You think that it still doesn't sound too bad? Take a look at the region in Figure 13.26. We need to write the top half of the circle as the graph of $y = \sqrt{4 - x^2}$ and the bottom half as $y = -\sqrt{4 - x^2}$. The double integral in question now becomes

$$\iint_R (x^2 + y^2 + 3)\, dA = \int_{-2}^{2} \int_{-\sqrt{4-x^2}}^{\sqrt{4-x^2}} (x^2 + y^2 + 3)\, dy\, dx$$

$$= \int_{-2}^{2} \left(x^2 y + \frac{y^3}{3} + 3y \right) \Bigg|_{y=-\sqrt{4-x^2}}^{y=\sqrt{4-x^2}} dx$$

$$= 2 \int_{-2}^{2} \left[(x^2 + 3)\sqrt{4 - x^2} + \frac{1}{3}(4 - x^2)^{3/2} \right] dx. \qquad (3.1)$$

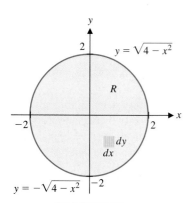

Figure 13.26

A circular region.

We probably don't need to convince you that the integral in (3.1) is most unpleasant. The reason is that we're doing the mathematical equivalent of trying to fit a square peg into a round hole. It's usually not very pretty. On the other hand, as we'll see shortly, this double integral is simple when it's written in polar coordinates. We consider several types of polar regions.

Suppose the region R can be written in the form

$$R = \{(r, \theta) | \alpha \leq \theta \leq \beta \text{ and } g_1(\theta) \leq r \leq g_2(\theta)\},$$

as pictured in Figure 13.27a. Of course, as our first step, we'll need to partition R, but rather than use a rectangular grid, as we have done with rectangular coordinates, we use a partition consisting of a number of concentric circular arcs (these are of the form $r = $ constant) and rays (these have the form $\theta = $ constant). We indicate such a partition of the region R in Figure 13.27b.

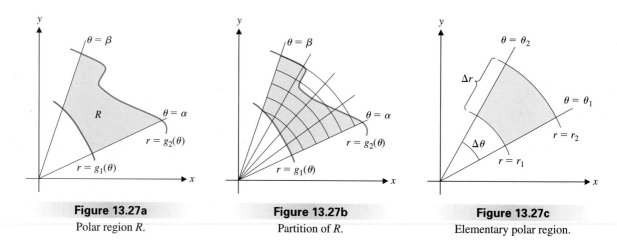

Figure 13.27a	**Figure 13.27b**	**Figure 13.27c**
Polar region R.	Partition of R.	Elementary polar region.

Notice that rather than consisting of rectangles, the "grid" in this case is made up of **elementary polar regions,** each bounded by two circular arcs and two rays (as shown in Figure 13.27c). In an **inner partition,** we include only those elementary polar regions that lie completely inside R.

We pause now briefly to calculate the area ΔA of the elementary polar region indicated in Figure 13.27c. Let $\bar{r} = \frac{1}{2}(r_1 + r_2)$ be the average radius of the two concentric circular arcs $r = r_1$ and $r = r_2$. Recall that the area of a circular sector is given by $A = \frac{1}{2}\theta r^2$, where $r = $ radius and θ is the central angle of the sector. Consequently, we have that

$$\Delta A = \text{Area of outer sector} - \text{Area of inner sector}$$

$$= \frac{1}{2}\Delta\theta r_2^2 - \frac{1}{2}\Delta\theta r_1^2$$

$$= \frac{1}{2}\left(r_2^2 - r_1^2\right)\Delta\theta$$

$$= \frac{1}{2}(r_2 + r_1)(r_2 - r_1)\Delta\theta$$

$$= \bar{r}\,\Delta r\,\Delta\theta. \tag{3.2}$$

As a familiar starting point, we first consider the problem of finding the volume lying beneath a surface $z = f(r, \theta)$, where f is continuous and $f(r, \theta) \geq 0$ on R. Using (3.2), we find that the volume V_i lying beneath the surface $z = f(r, \theta)$ and above the ith elementary polar region in the partition is then approximately the volume of the cylinder:

$$V_i \approx \underbrace{f(r_i, \theta_i)}_{\text{height}}\;\underbrace{\Delta A_i}_{\text{area of base}} = f(r_i, \theta_i)\,r_i\,\Delta r_i\,\Delta\theta_i,$$

where (r_i, θ_i) is a point in R_i with r_i the average radius in R_i. We get an approximation to the total volume V by summing over all the regions in the inner partition:

$$V \approx \sum_{i=1}^{n} f(r_i, \theta_i)\, r_i\, \Delta r_i\, \Delta \theta_i.$$

As we have done a number of times now, we can obtain the exact volume by taking the limit as the norm of the partition $\|P\|$ tends to zero and recognizing the iterated integral:

$$V = \lim_{\|P\| \to 0} \sum_{i=1}^{n} f(r_i, \theta_i)\, r_i\, \Delta r_i\, \Delta \theta_i$$

$$= \int_{\alpha}^{\beta} \int_{g_1(\theta)}^{g_2(\theta)} f(r, \theta)\, r\, dr\, d\theta.$$

In this case, $\|P\|$ is the longest diagonal of any elementary polar region in the inner partition. More generally, we have the following result, which holds regardless of whether or not $f(r, \theta) \geq 0$ on R.

NOTES

Theorem 3.1 says that to write a double integral in polar coordinates, we write $x = r \cos \theta$, $y = r \sin \theta$, find the limits of integration for r and θ and replace dA by $r\, dr\, d\theta$. Be certain not to omit the factor of r in $dA = r\, dr\, d\theta$; this is a very common error.

Theorem 3.1 (Fubini's Theorem)

Suppose that $f(r, \theta)$ is continuous on the region $R = \{(r, \theta) | \alpha \leq \theta \leq \beta \text{ and } g_1(\theta) \leq r \leq g_2(\theta)\}$. Then,

$$\iint_R f(r, \theta)\, dA = \int_{\alpha}^{\beta} \int_{g_1(\theta)}^{g_2(\theta)} f(r, \theta)\, r\, dr\, d\theta. \qquad (3.3)$$

Again, the proof of this result is beyond the level of this text. However, the result should seem reasonable from our development for the case where $f(r, \theta) \geq 0$.

Example 3.1   Computing Area in Polar Coordinates

Find the area inside the curve defined by $r = 2 - 2\sin \theta$.

Solution   First, we sketch a graph of the region in Figure 13.28. From (3.3), we get

$$A = \iint_R \underbrace{dA}_{r\, dr\, d\theta} = \int_0^{2\pi} \int_0^{2 - 2\sin\theta} r\, dr\, d\theta$$

$$= \int_0^{2\pi} \frac{r^2}{2} \bigg|_{r=0}^{r=2-2\sin\theta} d\theta$$

$$= \frac{1}{2} \int_0^{2\pi} [(2 - 2\sin\theta)^2 - 0]\, d\theta = 6\pi,$$

where we have left the details of the final calculation as an exercise.

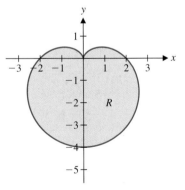

Figure 13.28

$r = 2 - 2\sin \theta$.

We now return to our introductory example and show how the introduction of polar coordinates can dramatically simplify certain double integrals in rectangular coordinates.

Example 3.2   Evaluating a Double Integral in Polar Coordinates

Evaluate $\iint_R (x^2 + y^2 + 3)\, dA$, where R is the circle of radius 2 centered at the origin.

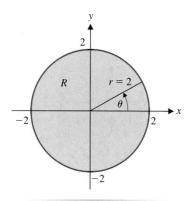

Figure 13.29
The region R.

Solution   First, recall from this section's introduction that in rectangular coordinates as in (3.1), this integral is extremely messy. From the region of integration shown in Figure 13.29, it's easy to see that for each fixed θ, r ranges from 0 (corresponding to the origin) to 2 (corresponding to a point on the circle). Then, in order to go around the circle exactly once, θ ranges from 0 to 2π. Finally, notice that the integrand contains the quantity $x^2 + y^2$, which you should recognize as r^2 in polar coordinates. From (3.3), we now have

$$\iint_R \underbrace{(x^2 + y^2 + 3)}_{r^2 + 3} \underbrace{dA}_{r\,dr\,d\theta} = \int_0^{2\pi} \int_0^2 (r^2 + 3)\, r\, dr\, d\theta$$

$$= \int_0^{2\pi} \int_0^2 (r^3 + 3r)\, dr\, d\theta$$

$$= \int_0^{2\pi} \left(\frac{r^4}{4} + 3\frac{r^2}{2} \right) \Big|_{r=0}^{r=2} d\theta$$

$$= \int_0^{2\pi} \left[\left(\frac{2^4}{4} + 3\frac{2^2}{2} \right) - 0 \right] d\theta$$

$$= 10 \int_0^{2\pi} d\theta = 20\pi.$$

Notice how simple this iterated integral was, as compared to the corresponding integral in rectangular coordinates in (3.1).

When dealing with double integrals, you should always consider whether the region over which you're integrating is in some way circular. If it is a circle or some portion of a circle, consider using polar coordinates.

Example 3.3   Finding Volume Using Polar Coordinates

Find the volume inside the paraboloid $z = 9 - x^2 - y^2$, outside the cylinder $x^2 + y^2 = 4$ and above the xy-plane.

Solution   Notice that the paraboloid has its vertex at the point $(0, 0, 9)$, and the axis of the cylinder is the z-axis (see Figure 13.30a). You should observe that the solid lies below the paraboloid and above the region in the xy-plane lying between the traces of the cylinder and the paraboloid in the xy-plane, that is, between the circles of radius 2 and 3, both centered at the origin. So, for each fixed $\theta \in [0, 2\pi]$, r ranges from 2 to 3. We call such a region a **circular annulus** (see Figure 13.30b). From (3.3), we have

$$V = \iint_R \underbrace{(9 - x^2 - y^2)}_{9 - r^2} \underbrace{dA}_{r\,dr\,d\theta} = \int_0^{2\pi} \int_2^3 (9 - r^2)\, r\, dr\, d\theta$$

$$= \int_0^{2\pi} \int_2^3 (9r - r^3)\, dr\, d\theta$$

$$= \int_0^{2\pi} \left(9\frac{r^2}{2} - \frac{r^4}{4} \right) \Big|_{r=2}^{r=3} d\theta = \frac{25}{2}\pi.$$

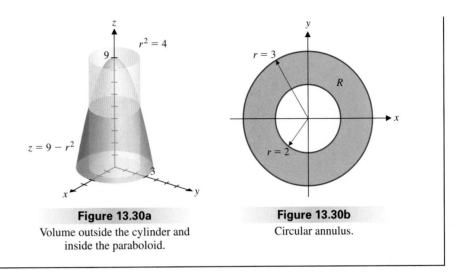

Figure 13.30a
Volume outside the cylinder and
inside the paraboloid.

Figure 13.30b
Circular annulus.

There are actually two things that you should look for when you are considering using polar coordinates for a double integral. The first is most obvious: Is the geometry of the region circular? The other is: Does the integral contain the expression $x^2 + y^2$ (particularly inside of other functions such as square roots, exponentials, etc.)? Since $r^2 = x^2 + y^2$, changing to polar coordinates will often simplify terms of this form.

Example 3.4   Changing a Double Integral to Polar Coordinates

Evaluate the iterated integral $\displaystyle\int_{-1}^{1}\int_{0}^{\sqrt{1-x^2}} x^2(x^2+y^2)^2\,dy\,dx$.

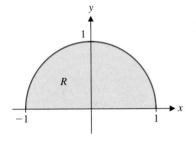

Figure 13.31
The region R.

Solution   First, you should recognize that evaluating this integral in rectangular coordinates is nearly hopeless. (Try it and see why!) On the other hand, it does have a term of the form $x^2 + y^2$, which we discussed above. Even more significantly, the region over which you're integrating turns out to be a semicircle, as follows. Reading the inside limits of integration first, you should see that for each fixed x between -1 and 1, y ranges from $y = 0$ up to $y = \sqrt{1-x^2}$ (the top half of the circle of radius 1 centered at the origin). We sketch the region in Figure 13.31. From (3.3), we have

$$\int_{-1}^{1}\int_{0}^{\sqrt{1-x^2}} x^2(x^2+y^2)^2\,dy\,dx = \iint_{R} \underbrace{x^2}_{r^2\cos^2\theta}\;\underbrace{(x^2+y^2)^2}_{(r^2)^2}\;\underbrace{dA}_{r\,dr\,d\theta}\qquad \text{Since } x = r\cos\theta.$$

$$= \int_{0}^{\pi}\int_{0}^{1} r^7\cos^2\theta\,dr\,d\theta$$

$$= \int_{0}^{\pi} \frac{r^8}{8}\bigg|_{r=0}^{r=1}\cos^2\theta\,d\theta$$

$$= \frac{1}{8}\int_{0}^{\pi} \frac{1}{2}(1+\cos 2\theta)\,d\theta\qquad \begin{array}{l}\text{Since } \cos^2\theta = \\ \tfrac{1}{2}(1+\cos 2\theta).\end{array}$$

$$= \frac{1}{16}\left(\theta + \frac{1}{2}\sin 2\theta\right)\bigg|_{0}^{\pi} = \frac{\pi}{16}.$$

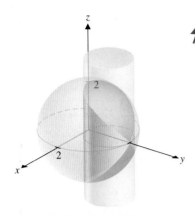

Figure 13.32a

Volume inside the sphere and inside the cylinder.

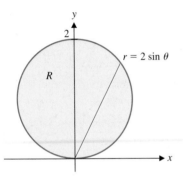

Figure 13.32b

The region R.

Example 3.5 Finding Volume Using Polar Coordinates

Find the volume cut out of the sphere $x^2 + y^2 + z^2 = 4$ by the cylinder $x^2 + y^2 = 2y$.

Solution We show a sketch of the solid in Figure 13.32a. (If you complete the square on the equation of the cylinder, you'll see that it is a circular cylinder of radius 1, whose axis is the line: $x = 0$, $y = 1$, $z = t$.) Notice that equal portions of the volume lie above and below the circle of radius 1 centered at $(0, 1)$, indicated in Figure 13.32b. So, we compute the volume lying below the top hemisphere $z = \sqrt{4 - x^2 - y^2}$ and above the region R indicated in Figure 13.32b and double it. We have

$$V = 2 \iint_R \sqrt{4 - x^2 - y^2}\, dA.$$

Since R is a circle and the integrand includes a term of the form $x^2 + y^2$, we introduce polar coordinates. Notice that since $y = r \sin\theta$, the circle $x^2 + y^2 = 2y$ becomes $r^2 = 2r \sin\theta$ or $r = 2 \sin\theta$. This gives us

$$V = 2 \int_0^\pi \int_0^{2\sin\theta} \sqrt{4 - r^2}\, r\, dr\, d\theta,$$

since the entire circle $r = 2 \sin\theta$ is traced out for $0 \le \theta \le \pi$ and since for each fixed $\theta \in [0, \pi]$, r ranges from $r = 0$ to $r = 2 \sin\theta$. Notice further that by symmetry, we get

$$V = 4 \int_0^{\pi/2} \int_0^{2\sin\theta} \sqrt{4 - r^2}\, r\, dr\, d\theta$$

$$= -2 \int_0^{\pi/2} \left[\frac{2}{3}(4 - r^2)^{3/2} \right]_{r=0}^{r=2\sin\theta} d\theta$$

$$= -\frac{4}{3} \int_0^{\pi/2} \left[(4 - 4\sin^2\theta)^{3/2} - 4^{3/2} \right] d\theta$$

$$= -\frac{32}{3} \int_0^{\pi/2} \left[(\cos^2\theta)^{3/2} - 1 \right] d\theta$$

$$= -\frac{32}{3} \int_0^{\pi/2} (\cos^3\theta - 1)\, d\theta$$

$$= -\frac{64}{9} + \frac{16}{3}\pi \approx 9.644.$$

There are several things to observe here. First, our use of symmetry was crucial. By restricting the integral to the interval $\left[0, \frac{\pi}{2}\right]$, we could write $(\cos^2\theta)^{3/2} = \cos^3\theta$. This is **not** true on the entire interval $[0, \pi]$. (Why not?) Second, if you think that this integral was messy, think about what it looks like in rectangular coordinates. (It's not pretty!)

Finally, we observe that we can also evaluate double integrals in polar coordinates by integrating first with respect to θ. Although such integrals are uncommon (given the way in which we change variables from rectangular to polar coordinates), we provide this for the sake of completeness.

Suppose the region R can be written in the form:

$$R = \{(r, \theta) | a \le r \le b \text{ and } h_1(r) \le \theta \le h_2(r)\},$$

as pictured in Figure 13.33. Then, it can be shown that if $f(r, \theta)$ is continuous on R, we have

$$\iint_R f(r, \theta)\, dA = \int_a^b \int_{h_1(\theta)}^{h_2(\theta)} f(r, \theta)\, r\, d\theta\, dr. \tag{3.4}$$

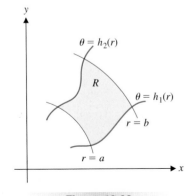

Figure 13.33

The region R.

EXERCISES 13.3

1. Thinking of $dy\,dx$ as representing the area dA of a small rectangle, explain in geometric terms why $dA \neq dr\,d\theta$.

2. In all of the examples in this section, we integrated with respect to r first. It is perfectly legitimate to integrate with respect to θ first. Explain why it is unlikely that it will ever be necessary to do so. [Hint: If θ is on the inside, you need functions of the form $\theta(r)$ for the limits of integration.]

3. Given a double integral in rectangular coordinates as in example 3.2 or 3.4, identify at least two indications that the integral would be easier to evaluate in polar coordinates.

4. In section 9.5, we derived a formula $A = \int_a^b \frac{1}{2}[f(\theta)]^2\,d\theta$ for the area bounded by the polar curve $r = f(\theta)$ and rays $\theta = a$ and $\theta = b$. Discuss how this formula relates to the formula used in example 3.1. Discuss which formula is easier to remember and which formula is more generally useful.

In exercises 5–12, find the area of the region bounded by the given curves.

5. $r = 3 + 2\sin\theta$

6. $r = 2 - 2\cos\theta$

7. $r = 2\sin\theta$

8. $r = 3\cos\theta$

9. one leaf of $r = \sin 3\theta$

10. one leaf of $r = \cos 2\theta$

11. inside $r = 2\sin 3\theta$, outside $r = 1$, first quadrant

12. inside $r = 1$ and outside $r = 2 - 2\cos\theta$

In exercises 13–18, use polar coordinates to evaluate the double integral.

13. $\iint\limits_R \sqrt{x^2 + y^2}\,dA$, where R is the disk $x^2 + y^2 \leq 9$

14. $\iint\limits_R \sqrt{x^2 + y^2 + 1}\,dA$, where R is the disk $x^2 + y^2 \leq 16$

15. $\iint\limits_R e^{-x^2 - y^2}\,dA$, where R is the disk $x^2 + y^2 \leq 4$

16. $\iint\limits_R e^{-\sqrt{x^2 + y^2}}\,dA$, where R is the disk $x^2 + y^2 \leq 1$

17. $\iint\limits_R y\,dA$, where R is bounded by $r = 2 - \cos\theta$

18. $\iint\limits_R x\,dA$, where R is bounded by $r = 1 - \sin\theta$

In exercises 19–22, use the most appropriate coordinate system to evaluate the double integral.

19. $\iint\limits_R (x^2 + y^2)\,dA$, where R is bounded by $x^2 + y^2 = 9$

20. $\iint\limits_R 2xy\,dA$, where R is bounded by $y = 4 - x^2$ and $y = 0$

21. $\iint\limits_R (x^2 + y^2)\,dA$, where R is bounded by $y = x$, $y = 0$ and $x = 2$

22. $\iint\limits_R \cos\sqrt{x^2 + y^2}\,dA$, where R is bounded by $x^2 + y^2 = 9$

In exercises 23–32, use an appropriate coordinate system to compute the volume of the indicated solid.

23. Below $z = x^2 + y^2$, above $z = 0$, inside $x^2 + y^2 = 9$

24. Below $z = x^2 + y^2 - 4$, above $z = 0$, inside $x^2 + y^2 = 9$

25. Below $z = \sqrt{x^2 + y^2}$, above $z = 0$, inside $x^2 + y^2 = 4$

26. Below $z = \sqrt{x^2 + y^2}$, above $z = 0$, inside $x^2 + (y - 1)^2 = 1$

27. Below $z = \sqrt{1 - x^2 - y^2}$, above the xy-plane, inside $x^2 + y^2 = \frac{1}{4}$

28. Below $z = 8 - x^2 - y^2$, above the xy-plane

29. Below $z = 6 - x - y$, in the first octant

30. Below $z = 4 - x^2 - y^2$, between $y = x$, $y = 0$ and $x = 1$

31. Below $z = 4 - x^2 - y^2$, above the xy-plane, between $y = 0$ and $y = x$

32. Below $z = \sqrt{x^2 + y^2}$, below $z = 4$, above the xy-plane, between $y = x$ and $y = 2x$

In exercises 33–38, evaluate the iterated integral by converting to polar coordinates.

33. $\displaystyle\int_{-2}^{2} \int_{-\sqrt{4-x^2}}^{\sqrt{4-x^2}} \sqrt{x^2 + y^2}\,dy\,dx$

34. $\displaystyle\int_{-2}^{2} \int_{0}^{\sqrt{4-x^2}} \sin(x^2 + y^2)\,dy\,dx$

35. $\displaystyle\int_{0}^{2} \int_{-\sqrt{4-x^2}}^{\sqrt{4-x^2}} e^{-x^2 - y^2}\,dy\,dx$

36. $\displaystyle\int_{0}^{2} \int_{-\sqrt{4-x^2}}^{0} y\,dy\,dx$

37. $\displaystyle\int_{0}^{2} \int_{x}^{\sqrt{8-x^2}} (x^2 + y^2)^{3/2}\,dy\,dx$

38. $\displaystyle\int_{0}^{1} \int_{y}^{\sqrt{2-y^2}} \frac{1}{\sqrt{x^2 + y^2}}\,dx\,dy$

In exercises 39–42, compute the probability that a dart lands in the region R, assuming that the probability is given by $\iint\limits_R \frac{1}{\pi} e^{-x^2-y^2} dA.$

39. A double bull's-eye, R is the region inside $r = \frac{1}{4}$ (inch)

40. A single bull's-eye, R bounded by $r = \frac{1}{4}$ and $r = \frac{1}{2}$

41. A triple-20, R bounded by $r = 3\frac{3}{4}$, $r = 4$, $\theta = \frac{9\pi}{20}$ and $\theta = \frac{11\pi}{20}$

42. A double-20, R bounded by $r = 6\frac{1}{4}$, $r = 6\frac{1}{2}$, $\theta = \frac{9\pi}{20}$ and $\theta = \frac{11\pi}{20}$

43. Find the area of the triple-20 region described in exercise 41.

44. Find the area of the double-20 region described in exercise 42.

45. Find the center of mass of a lamina in the shape of $x^2 + (y-1)^2 = 1$ with density $\rho(x, y) = 1/\sqrt{x^2 + y^2}$.

46. Find the center of mass of a lamina in the shape of $r = 2 - 2\cos\theta$ with density $\rho(x, y) = x^2 + y^2$.

47. Suppose that $f(x, y) = 20{,}000 e^{-x^2-y^2}$ is the population density of a species of small animals. Estimate the population in the region bounded by $x^2 + y^2 = 1$.

48. Suppose that $f(x, y) = 15{,}000 e^{-x^2-y^2}$ is the population density of a species of small animals. Estimate the population in the region bounded by $(x-1)^2 + y^2 = 1$.

49. Find the moment of inertia I_y of the circular lamina bounded by $x^2 + y^2 = r^2$, with density $\rho(x, y) = 1$. If the radius doubles, by what factor does the moment of inertia increase?

50. Repeat exercise 49 for the density function $\rho(x, y) = \sqrt{x^2 + y^2}$.

51. Use a double integral to derive the formula for the volume of a sphere of radius a.

52. Use a double integral to derive the formula for the volume of a right circular cone of height h and base radius a. [Hint: Find constants c and d such that $z = c(x^2 + y^2) + d$ is a cone with radius 0 at $z = h$ and radius a at $z = 0$.]

53. Suppose that the following data gives the density of a lamina at different locations. Estimate the mass of the lamina.

$\rho(r, \theta)$ θ	0	$\frac{\pi}{2}$	π	$\frac{3\pi}{2}$	2π
r					
$\frac{1}{2}$	1.0	1.4	1.4	1.2	1.0
1	0.8	1.2	1.0	1.0	0.8
$\frac{3}{2}$	1.0	1.3	1.4	1.3	1.2
2	1.2	1.6	1.6	1.4	1.2

54. One of the most important integrals in probability theory is $\int_{-\infty}^{\infty} e^{-x^2} dx$. Since there is no antiderivative of e^{-x^2} among the elementary functions, we can't evaluate this integral directly. A clever use of polar coordinates is needed. Start by giving the integral a name,

$$\int_{-\infty}^{\infty} e^{-x^2} dx = I.$$

Now, argue that $\int_{-\infty}^{\infty} e^{-y^2} dy = I$ and

$$\int_{-\infty}^{\infty} e^{-x^2} dx \int_{-\infty}^{\infty} e^{-y^2} dy = \int_{-\infty}^{\infty}\int_{-\infty}^{\infty} e^{-x^2-y^2} \, dy\, dx = I^2.$$

Convert the iterated integral to polar coordinates and evaluate it. The desired integral I is simply the square root of the iterated integral. Explain why the same trick can't be used to evaluate $\int_{-1}^{1} e^{-x^2} dx$.

13.4   SURFACE AREA

You may recall that in section 5.4, we devised a method of finding the surface area for a surface of revolution. In this section, we will consider how to find surface area in a more general setting. Suppose that $f(x, y) \geq 0$ and f has continuous first partial derivatives in some region R in the xy-plane. We would like to find a way to calculate the surface area of that portion of the surface $z = f(x, y)$ lying above R. As we have done innumerable times now, we begin by forming an inner partition of R, consisting of the rectangles R_1, $R_2, \ldots, R_n$. Our strategy is to approximate the surface area lying above each R_i, for $i = 1, 2, \ldots, n$ and then sum the individual approximations to obtain an approximation of the total surface area. We proceed as follows.

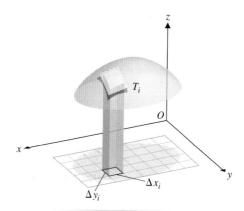

Figure 13.34
Surface area.

For each $i = 1, 2, \ldots, n$, let $(x_i, y_i, 0)$ represent the corner of R_i closest to the origin and construct the tangent plane to the surface $z = f(x, y)$ at the point $(x_i, y_i, f(x_i, y_i))$. Since the tangent plane stays close to the surface near the point of tangency, the area ΔT_i of that portion of the tangent plane that lies above R_i is an approximation to the surface area above R_i (see Figure 13.34). Notice, too that the portion of the tangent plane lying above R_i is a parallelogram, T_i, whose area ΔT_i you should be able to easily compute. Adding together these approximations, we get that the total surface area S is approximately

$$S \approx \sum_{i=1}^{n} \Delta T_i.$$

Also note that as the norm of the partition $\| P \|$ tends to zero, the approximations should approach the exact surface area and so we have

$$S = \lim_{\| P \| \to 0} \sum_{i=1}^{n} \Delta T_i, \qquad (4.1)$$

assuming the limit exists. The only remaining question is how to find the values of ΔT_i, for $i = 1, 2, \ldots, n$. Let the dimensions of R_i be Δx_i and Δy_i and let the vectors $\mathbf{a}_i$ and $\mathbf{b}_i$ form two adjacent sides of the parallelogram T_i, as indicated in Figure 13.35. Recall from our discussion of tangent planes in section 12.4 that the tangent plane is given by

$$z - f(x_i, y_i) = f_x(x_i, y_i)(x - x_i) + f_y(x_i, y_i)(y - y_i). \qquad (4.2)$$

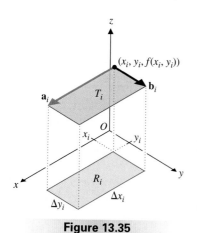

Figure 13.35
Portion of the tangent plane
above R_i.

Look carefully at Figure 13.35; the vector $\mathbf{a}_i$ has its initial point at $(x_i, y_i, f(x_i, y_i))$. Its terminal point is the point on the tangent plane corresponding to $x = x_i + \Delta x_i$ and $y = y_i$. From (4.2), we get that the z-coordinate of the terminal point satisfies

$$z - f(x_i, y_i) = f_x(x_i, y_i)(x_i + \Delta x_i - x_i) + f_y(x_i, y_i)(y_i - y_i)$$
$$= f_x(x_i, y_i) \, \Delta x_i.$$

This says that the vector $\mathbf{a}_i$ is given by

$$\mathbf{a}_i = \langle \Delta x_i, 0, f_x(x_i, y_i) \, \Delta x_i \rangle.$$

Likewise, $\mathbf{b}_i$ has its initial point at $(x_i, y_i, f(x_i, y_i))$, but has its terminal point at the point on the tangent plane corresponding to $x = x_i$ and $y = y_i + \Delta y_i$. Again, using (4.2), we get that the z-coordinate of this point is given by

$$z - f(x_i, y_i) = f_x(x_i, y_i)(x_i - x_i) + f_y(x_i, y_i)(y_i + \Delta y_i - y_i)$$
$$= f_y(x_i, y_i) \, \Delta y_i.$$

Figure 13.36
The parallelogram T_i.

This says that $\mathbf{b}_i$ is given by

$$\mathbf{b}_i = \langle 0, \Delta y_i, f_y(x_i, y_i)\, \Delta y_i \rangle.$$

Notice that ΔT_i is the area of the parallelogram shown in Figure 13.36, which you should recognize as

$$\Delta T_i = \|\mathbf{a}_i\| \|\mathbf{b}_i\| \sin\theta = \|\mathbf{a}_i \times \mathbf{b}_i\|,$$

where θ indicates the angle between $\mathbf{a}_i$ and $\mathbf{b}_i$. We have

$$\mathbf{a}_i \times \mathbf{b}_i = \begin{vmatrix} \mathbf{i} & \mathbf{j} & \mathbf{k} \\ \Delta x_i & 0 & f_x(x_i, y_i)\,\Delta x_i \\ 0 & \Delta y_i & f_y(x_i, y_i)\,\Delta y_i \end{vmatrix}$$

$$= -f_x(x_i, y_i)\, \Delta x_i\, \Delta y_i \mathbf{i} - f_y(x_i, y_i)\, \Delta x_i\, \Delta y_i \mathbf{j} + \Delta x_i\, \Delta y_i \mathbf{k}.$$

This gives us

$$\Delta T_i = \|\mathbf{a}_i \times \mathbf{b}_i\| = \sqrt{[f_x(x_i, y_i)]^2 + [f_y(x_i, y_i)]^2 + 1}\ \underbrace{\Delta x_i\, \Delta y_i}_{\Delta A_i},$$

where $\Delta A_i = \Delta x_i\, \Delta y_i$ is the area of the rectangle R_i. From (4.1), we now have that the total surface area is given by

$$S = \lim_{\|P\| \to 0} \sum_{i=1}^{n} \Delta T_i$$

$$= \lim_{\|P\| \to 0} \sum_{i=1}^{n} \sqrt{[f_x(x_i, y_i)]^2 + [f_y(x_i, y_i)]^2 + 1}\ \Delta A_i.$$

You should recognize this limit as the double integral

Surface area

$$S = \iint_R \sqrt{[f_x(x, y)]^2 + [f_y(x, y)]^2 + 1}\ dA. \tag{4.3}$$

There are several things to note here. First, you can easily show that the surface area formula (4.3) also holds for the case where $f(x, y) \leq 0$ on R. Second, you should note the similarity to the arc length formula derived in section 5.4. Further, recall that $\mathbf{n} = \langle f_x(x, y), f_y(x, y), -1 \rangle$ is a normal vector for the tangent plane to the surface $z = f(x, y)$ at (x, y). With this in mind, recognize that you can think of the integrand in (4.3) as $\|\mathbf{n}\|$, an idea we'll develop more fully in Chapter 14.

Example 4.1   Calculating Surface Area

Find the surface area of that portion of the surface $z = y^2 + 4x$ lying above the triangular region R in the xy-plane with vertices at $(0, 0)$, $(0, 2)$ and $(2, 2)$.

Solution   We show a computer-generated sketch of the surface in Figure 13.37a and the region R in Figure 13.37b. If we take $f(x, y) = y^2 + 4x$, then we have $f_x(x, y) = 4$ and $f_y(x, y) = 2y$. From (4.3), we now have

$$S = \iint_R \sqrt{[f_x(x, y)]^2 + [f_y(x, y)]^2 + 1}\ dA$$

$$= \iint_R \sqrt{4^2 + 4y^2 + 1}\ dA.$$

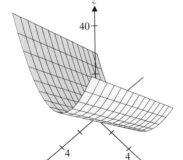

Figure 13.37a
The surface $z = y^2 + 4x$.

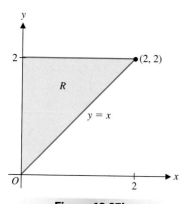

Figure 13.37b

The region R.

Looking carefully at Figure 13.37b, you can read off the limits of integration, to obtain

$$S = \int_0^2 \int_0^y \sqrt{4y^2 + 17} \, dx \, dy$$

$$= \int_0^2 \sqrt{4y^2 + 17} \, x \Big|_{x=0}^{x=y} \, dy$$

$$= \int_0^2 y\sqrt{4y^2 + 17} \, dy$$

$$= \frac{1}{8}(4y^2 + 17)^{3/2} \left(\frac{2}{3}\right) \Big|_0^2$$

$$= \frac{1}{12}\left[[4(2^2) + 17]^{3/2} - [4(0)^2 + 17]^{3/2}\right]$$

$$\approx 9.956.$$

Computing surface area requires more than simply substituting into formula (4.3). You will also need to carefully determine the region over which you're integrating and the best coordinate system to use, as in the following example.

Example 4.2 Finding Surface Area Using Polar Coordinates

Find the surface area of that portion of the paraboloid $z = 1 + x^2 + y^2$ that lies below the plane $z = 5$.

Solution First, note that we have not given you the region of integration; you'll need to determine that from a careful analysis of the graph (see Figure 13.38a). Note that the plane $z = 5$ intersects the paraboloid in a circle of radius 2, parallel to the xy-plane and centered at the point $(0, 0, 5)$ on the z-axis. (Simply plug $z = 5$ into the equation of the paraboloid to see why.) So, the surface area **below** the plane $z = 5$ lies **above** the circle in the xy-plane of radius 2, centered at the origin. We show the region of integration R in Figure 13.38b. Taking $f(x, y) = 1 + x^2 + y^2$, we have $f_x(x, y) = 2x$ and

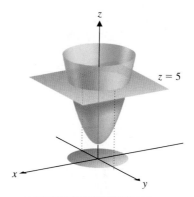

Figure 13.38a

Intersection of the paraboloid with the plane $z = 5$.

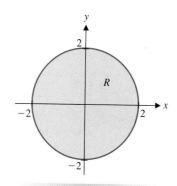

Figure 13.38b

The region R.

$f_y(x, y) = 2y$, so that from (4.3), we have

$$S = \iint\limits_R \sqrt{[f_x(x, y)]^2 + [f_y(x, y)]^2 + 1} \, dA$$

$$= \iint\limits_R \sqrt{4x^2 + 4y^2 + 1} \, dA.$$

Note that since the region of integration is circular and the integrand contains the term $x^2 + y^2$, polar coordinates are indicated. We have

$$S = \iint\limits_R \underbrace{\sqrt{4(x^2 + y^2) + 1}}_{\sqrt{4r^2 + 1}} \underbrace{dA}_{r \, dr \, d\theta}$$

$$= \int_0^{2\pi} \int_0^2 \sqrt{4r^2 + 1} \, r \, dr \, d\theta$$

$$= \frac{1}{8} \int_0^{2\pi} \left(\frac{2}{3}\right)(4r^2 + 1)^{3/2} \Big|_{r=0}^{r=2} d\theta$$

$$= \frac{1}{12} \int_0^{2\pi} (17^{3/2} - 1^{3/2}) \, d\theta$$

$$= \frac{2\pi}{12}(17^{3/2} - 1^{3/2}) \approx 36.18.$$

We must point out that (just as with arc length) most surface area integrals cannot be computed exactly. Most of the time, you must rely on numerical approximations of the integrals. Although your computer algebra system no doubt can approximate even iterated integrals numerically, such approximations can be unreliable. Indeed, numerical methods for iterated integrals are somewhat complicated and we don't have the space to discuss them here. Whenever possible, you should try to evaluate at least one of the iterated integrals and then approximate the second integral numerically (e.g., using Simpson's Rule). This is the situation in the following example.

Example 4.3   *Surface Area That Must be Approximated Numerically*

Find the surface area of that portion of the paraboloid $z = 4 - x^2 - y^2$ that lies above the triangular region R in the xy-plane with vertices at the points $(0, 0)$, $(1, 1)$ and $(1, 0)$.

Solution   We sketch the paraboloid and the region R in Figure 13.39a. Taking $f(x, y) = 4 - x^2 - y^2$, we get $f_x(x, y) = -2x$ and $f_y(x, y) = -2y$. From (4.3), we have

$$S = \iint\limits_R \sqrt{[f_x(x, y)]^2 + [f_y(x, y)]^2 + 1} \, dA$$

$$= \iint\limits_R \sqrt{4x^2 + 4y^2 + 1} \, dA.$$

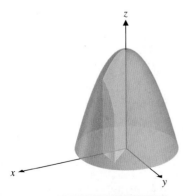

Figure 13.39a
$z = 4 - x^2 - y^2$.

Note that you have little hope of evaluating this double integral in rectangular coordinates. (Think about this!) Even though the region of integration is not circular, we'll try

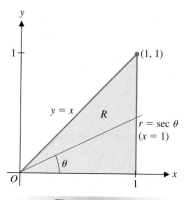

Figure 13.39b

The region R.

polar coordinates, since the integrand contains the term $x^2 + y^2$. We indicate the region R in Figure 13.39b. The difficulty here is in describing the region R in terms of polar coordinates. Look carefully at Figure 13.39b and notice that for each fixed angle θ, the radius r varies from 0 out to a point on the line $x = 1$. Since in polar coordinates $x = r \cos \theta$, the line $x = 1$ corresponds to $r \cos \theta = 1$ or $r = \sec \theta$, in polar coordinates. Further, θ varies from $\theta = 0$ (the x-axis) to $\theta = \frac{\pi}{4}$ (the line $y = x$). The surface area integral now becomes

$$S = \iint\limits_{R} \underbrace{\sqrt{4x^2 + 4y^2 + 1}}_{\sqrt{4r^2 + 1}} \underbrace{dA}_{r \, dr \, d\theta}$$

$$= \int_0^{\pi/4} \int_0^{\sec \theta} \sqrt{4r^2 + 1} \, r \, dr \, d\theta$$

$$= \frac{1}{8} \int_0^{\pi/4} \left(\frac{2}{3}\right)(4r^2 + 1)^{3/2} \Big|_{r=0}^{r=\sec \theta} d\theta$$

$$= \frac{1}{12} \int_0^{\pi/4} \left[(4 \sec^2 \theta + 1)^{3/2} - 1 \right] d\theta$$

$$\approx 0.93078,$$

where we have approximated the value of the final integral, since no exact means of integration was available. You can arrive at this approximation using Simpson's Rule or using your computer algebra system.

EXERCISES 13.4

1. Starting at equation (4.1), there are several ways to estimate ΔT_i. Explain why it is important that we were able to find an approximation of the form $f(x_i, y_i) \, \Delta x_i \, \Delta y_i$.

2. In example 4.3, we evaluated the inner integral before estimating the remaining integral numerically. Discuss the number of calculations that would be necessary to use a rule such as Simpson's Rule to estimate an iterated integral. Explain why we thought it important to evaluate the inner integral first.

In exercises 3–20, evaluate or estimate the surface area.

3. The portion of $z = x^2 + 2y$ between $y = x$, $y = 0$ and $x = 4$

4. The portion of $z = 4y + 3x^2$ between $y = 2x$, $y = 0$ and $x = 2$

5. The portion of $z = 4 - x^2 - y^2$ above the xy-plane

6. The portion of $z = x^2 + y^2$ below $z = 4$

7. The portion of $z = \sqrt{x^2 + y^2}$ below $z = 2$

8. The portion of $z = \sqrt{x^2 + y^2}$ between $y = x^2$ and $y = 4$

9. The portion of $z = e^{x^2 + y^2}$ inside $x^2 + y^2 = 4$

10. The portion of $z = e^{-x^2 - y^2}$ inside $x^2 + y^2 = 1$

11. The portion of $x + 3y + z = 6$ in the first octant

12. The portion of $2x + y + z = 8$ in the first octant

13. The portion of $x - y - 2z = 4$ with $z \leq 0$, $x \geq 0$, $y \leq 0$

14. The portion of $2x + y - 4z = 4$ with $z \leq 0$, $x \geq 0$, $y \geq 0$

15. The portion of $z = x^2 + y^2$ between $y = x$, $y = 1$, $x = 0$

16. The portion of $z = x^2 + y^2$ inside $r = 2 - 2 \cos \theta$

17. The portion of $z = y^2$ below $z = 4$ and between $x = -2, x = 2$

18. The portion of $z = 4 - x^2$ above $z = 0$ and between $y = 0$, $y = 4$

19. The portion of $z = \sqrt{4 - x^2 - y^2}$ above $z = 0$

20. The portion of $z = \sqrt{9 - x^2 - y^2}$ above $z = 0$

21. In exercises 7 and 8, determine the surface area of the cone as a function of the area A of the base R of the solid.

22. Use your solution to exercise 21 to quickly find the surface area of the portion of $z = \sqrt{x^2 + y^2}$ above the rectangle $0 \le x \le 2$, $1 \le y \le 4$.

23. In exercises 13 and 14, determine the surface area of the portion of the plane indicated as a function of the area A of the base R of the solid and the angle θ between the given plane and the xy-plane.

24. Use your solution to exercise 23 to quickly find the surface area of the portion of $z = 1 + y$ above the rectangle $-1 \le x \le 3$, $0 \le y \le 2$.

25. In exercises 17 and 18, determine the surface area of the portion of the cylinder indicated as a function of the arc length L of the base (two-dimensional) curve of the cylinder.

26. Use your solution to exercise 25 to quickly find the surface area of the portion of the cylinder with triangular cross sections parallel to the triangle with vertices $(1, 0, 0)$, $(0, 1, 0)$ and the origin and lying between the planes $z = 0$ and $z = 4$.

27. An old joke tells of the theoretical mathematician hired to improve dairy production who starts his report with the assumption, "Consider a spherical cow." In this exercise, we will approximate an animal's body with ellipsoids. Estimate the volume and surface area of the ellipsoids $16x^2 + y^2 + 4z^2 = 16$ and $16x^2 + y^2 + 4z^2 = 36$. Note that the second ellipsoid retains the proportions of the first ellipsoid, but the length of each dimension is multiplied by $\frac{3}{2}$. Show that the volume increases by a much greater proportion than does the surface area. In general, volume increases as the cube of length (in this case, $\left(\frac{3}{2}\right)^3 = 3.375$) and surface area increases as the square of length (in this case, $\left(\frac{3}{2}\right)^2 = 2.25$). This has implications for the sizes of animals, since volume tends to be proportional to weight and surface area tends to be proportional to strength. Explain why a cow increased in size proportionally by a factor of $\frac{3}{2}$ might collapse under its weight.

28. For a surface $z = f(x, y)$, recall that a normal vector to the tangent plane at $(a, b, f(a, b))$ is $\langle f_x(a, b), f_y(a, b), -1 \rangle$. Show that the surface area formula can be rewritten as

$$\text{Surface area} = \iint\limits_R \frac{\|\mathbf{n}\|}{|\mathbf{n} \cdot \mathbf{k}|} \, dA,$$

where $\mathbf{n}$ is the unit normal vector to the surface. Use this formula to set up a double integral for the surface area of the top half of the sphere $x^2 + y^2 + z^2 = 4$ and compare this to the work required to set up the same integral in exercise 19. (Hint: Use the gradient to compute the normal vector and substitute $z = \sqrt{4 - x^2 - y^2}$ to write the integral in terms of x and y.) For a surface such as $y = 4 - x^2 - z^2$, it is convenient to think of y as the dependent variable and double integrate with respect to x and z. Write out the surface area formula in terms of the normal vector for this orientation and use it to compute the surface area of the portion of $y = 4 - x^2 - z^2$ inside $x^2 + z^2 = 1$ and to the right of the xz-plane.

13.5   TRIPLE INTEGRALS

We initially developed the definite integral of a function of one variable $f(x)$ in order to compute the area under the curve $y = f(x)$. Similarly, we first devised the double integral of a function of two variables $f(x, y)$ as a means of computing the volume lying beneath the surface $z = f(x, y)$. We have no comparable geometric motivation for defining the triple integral of a function of three variables $f(x, y, z)$, since the graph of $u = f(x, y, z)$ is a **hypersurface** in four dimensions. We can't even visualize a graph in four dimensions, since our world is three-dimensional. You might reasonably wonder then, if an integral of a function of three or more variables has any meaning or significance. Briefly, the answer is yes! Integrals of functions of three variables have many very significant applications to studying the three-dimensional world in which we live. We'll consider

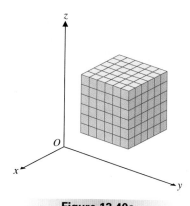

Figure 13.40a

Partition of the box Q.

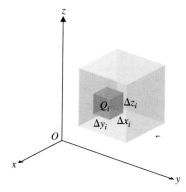

Figure 13.40b

Typical box Q_i.

just one of these applications (finding the center of mass of a solid) at the end of this section.

Since we have no obvious geometric motivation for developing the triple integral of a function of three variables, we pattern our development after our development of the double integral of a function of two variables. We first consider the relatively simple case of a function $f(x, y, z)$ defined on the region Q of space defined by

$$Q = \{(x, y, z) | a \leq x \leq b, c \leq y \leq d \text{ and } r \leq z \leq s\}.$$

That is, Q is a rectangular box in three-dimensional space. As we have done countless times, we begin by partitioning the region Q. We do this by slicing Q by planes parallel to the xy-plane, planes parallel to the xz-plane and planes parallel to the yz-plane. Notice that this divides Q into a number of smaller boxes (see Figure 13.40a). Number the smaller boxes in any order: $Q_1, Q_2, \ldots, Q_n$. For each box Q_i $(i = 1, 2, \ldots, n)$ call the x, y and z dimensions of the box Δx_i, Δy_i and Δz_i, respectively (see Figure 13.40b). Notice that this says that the volume of the box Q_i is $\Delta V_i = \Delta x_i \Delta y_i \Delta z_i$. As we did in both one and two dimensions, we pick any point (u_i, v_i, w_i) in the box Q_i and form the Riemann sum

$$\sum_{i=1}^{n} f(u_i, v_i, w_i) \Delta V_i.$$

In this three dimensional case, we define the norm of the partition $\|P\|$ to be the longest diagonal of any of the boxes Q_i, $i = 1, 2, \ldots, n$. We now define the triple integral of $f(x, y, z)$ over Q as follows.

Definition 5.1

For any function $f(x, y, z)$ defined on the rectangular box Q, we define the **triple integral** of f over Q by

$$\iiint\limits_{Q} f(x, y, z) \, dV = \lim_{\|P\| \to 0} \sum_{i=1}^{n} f(u_i, v_i, w_i) \Delta V_i, \tag{5.1}$$

provided the limit exists and is the same for every choice of evaluation points (u_i, v_i, w_i) in Q_i, for $i = 1, 2, \ldots, n$. When this happens, we say that f is **integrable** over Q.

Remark 5.1

It can be shown that as long as f is continuous over Q, f will be integrable over Q.

Now that we have defined this triple integral, how can we calculate the value of one? The answer should prove to be no surprise. Just as a double integral can be written as two iterated integrals, a triple integral turns out to be equivalent to **three** iterated integrals.

Theorem 5.1 (Fubini's Theorem)

Suppose that $f(x, y, z)$ is continuous on the box Q defined by $Q = \{(x, y, z) | a \leq x \leq b, c \leq y \leq d \text{ and } r \leq z \leq s\}$. Then, we can write the triple integral over Q as a triple iterated integral:

$$\iiint\limits_{Q} f(x, y, z) \, dV = \int_{r}^{s} \int_{c}^{d} \int_{a}^{b} f(x, y, z) \, dx \, dy \, dz. \tag{5.2}$$

As was the case for double integrals, the three iterated integrals in (5.2) are evaluated from the inside out, using partial integrations. That is, in the inner most integral, we hold y and z fixed and integrate with respect to x and in the second integration, we hold z fixed and integrate with respect to y. Notice also that in this simple case (where Q is a rectangular box) the order of the integrations in (5.2) is irrelevant, so that we might just as easily write the triple integral as

$$\iiint_Q f(x, y, z)\, dV = \int_a^b \int_c^d \int_r^s f(x, y, z)\, dz\, dy\, dx,$$

or in any of the four remaining orders.

Example 5.1   Triple Integral over a Rectangular Box

Evaluate the triple integral $\iiint_Q 2xe^y \sin z\, dV$, where Q is the rectangle defined by

$$Q = \{(x, y, z) \mid 1 \le x \le 2,\, 0 \le y \le 1 \text{ and } 0 \le z \le \pi\}.$$

Solution   From (5.2), we have

$$\iiint_Q 2xe^y \sin z\, dV = \int_0^\pi \int_0^1 \int_1^2 2xe^y \sin z\, dx\, dy\, dz$$

$$= \int_0^\pi \int_0^1 e^y \sin z \frac{2x^2}{2} \Big|_{x=1}^{x=2} dy\, dz$$

$$= 3 \int_0^\pi \sin z\, e^y \Big|_{y=0}^{y=1} dz$$

$$= 3(e^1 - 1)(-\cos z) \Big|_{z=0}^{z=\pi}$$

$$= 3(e - 1)(-\cos \pi + \cos 0)$$

$$= 6(e - 1).$$

You should pick one of the other five possible orders of integration and show that you get the same result.

As we did for double integrals, we can define triple integrals for more general regions in three dimensions by using an inner partition of the region. For any bounded solid Q in three dimensions, we partition Q by slicing it with planes parallel to the three coordinate planes. As in the case where Q was a box, these planes form a number of boxes (see Figures 13.41a and 13.41b). In this case, we consider only those boxes $Q_1, Q_2, \ldots, Q_n$ that lie **entirely** in Q, and call this an **inner partition** of the solid Q. For each $i = 1, 2, \ldots, n$, we pick any point $(u_i, v_i, w_i) \in Q_i$ and form the Riemann sum

$$\sum_{i=1}^n f(u_i, v_i, w_i)\, \Delta V_i,$$

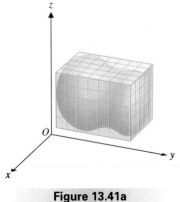

Figure 13.41a
Partition of a solid.

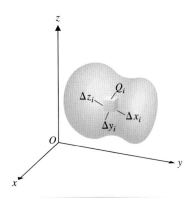

Figure 13.41b

Typical rectangle in inner partition of solid.

where $\Delta V_i = \Delta x_i \, \Delta y_i \, \Delta z_i$ represents the volume of Q_i. We can then define a triple integral over a general region Q as the limit of Riemann sums, as follows.

Definition 5.2

For a function $f(x, y, z)$ defined on the (bounded) solid Q, we define the triple integral of $f(x, y, z)$ over Q by

$$\iiint\limits_Q f(x, y, z) \, dV = \lim_{\|P\| \to 0} \sum_{i=1}^n f(u_i, v_i, w_i) \, \Delta V_i, \qquad (5.3)$$

provided the limit exists and is the same for every choice of the evaluation points (u_i, v_i, w_i) in Q_i, for $i = 1, 2, \ldots, n$. When this happens, we say that f is **integrable** over Q.

You should notice that (5.3) is precisely the same as (5.1), except that in the case of (5.3), we are summing over an inner partition of Q.

The (very) big remaining question is how to evaluate triple integrals over more general regions. As you saw with triple integrals over a box, there are six different orders of integration possible in a triple iterated integral. This makes it difficult to write down a single result that will allow us to evaluate all triple integrals. So, rather than write down an exhaustive list, we'll indicate the general idea by looking at several specific cases. For instance, if the region Q can be written in the form

$$Q = \{(x, y, z) | (x, y) \in R \text{ and } g_1(x, y) \le z \le g_2(x, y)\},$$

where R is some region in the xy-plane (see Figure 13.42), then it can be proved that

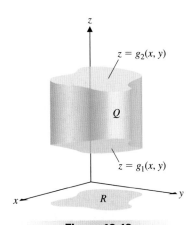

Figure 13.42

Solid with defined top and bottom surfaces.

$$\iiint\limits_Q f(x, y, z) \, dV = \iint\limits_R \int_{g_1(x,y)}^{g_2(x,y)} f(x, y, z) \, dz \, dA. \qquad (5.4)$$

As we have seen before, the innermost integration in (5.4) is a partial integration where we hold x and y fixed and integrate with respect to z and the outer double integral is evaluated using the methods we have already developed in sections 13.1 and 13.3.

Example 5.2 Triple Integral over a Tetrahedron

Evaluate $\iiint\limits_Q 6xy \, dV$, where Q is the tetrahedron bounded by the planes $x = 0$, $y = 0$, $z = 0$ and $2x + y + z = 4$ (see Figure 13.43a).

Solution Notice that each point in the solid lies above the triangular region R in the xy-plane indicated in Figures 13.43a and 13.43b. You can think of R as forming the **base** of the solid. Notice that for each fixed point $(x, y) \in R$, z ranges from $z = 0$ up to $z = 4 - 2x - y$. It helps to draw a vertical line from the base and through the top surface of the solid, as we have indicated in Figure 13.43a. The line first enters the solid on the xy-plane ($z = 0$) and exits the solid on the plane $z = 4 - 2x - y$. This tells you that the innermost limits of integration (given that the first integration is with respect to z)

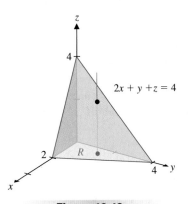

Figure 13.43a

Tetrahedron.

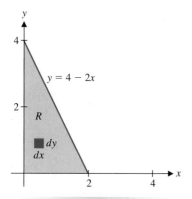

Figure 13.43b

The base of the solid in the
xy-plane.

are $z = 0$ and $z = 4 - 2x - y$. From (5.4), we now have

$$\iiint\limits_{Q} 6xy \, dV = \iint\limits_{R} \int_0^{4-2x-y} 6xy \, dz \, dA$$

$$= \iint\limits_{R} (6xyz) \Big|_{z=0}^{z=4-2x-y} dA$$

$$= \iint\limits_{R} 6xy(4 - 2x - y) \, dA.$$

This leaves us with computing the double integral over the triangular region shown in Figure 13.43b. Notice that for each fixed $x \in [0, 2]$, y ranges from 0 up to $y = 4 - 2x$. We now have

$$\iiint\limits_{Q} 6xy \, dV = \iint\limits_{R} 6xy(4 - 2x - y) \, dA$$

$$= \int_0^2 \int_0^{4-2x} 6xy(4 - 2x - y) \, dy \, dx$$

$$= \int_0^2 6 \left(4x \frac{y^2}{2} - 2x^2 \frac{y^2}{2} - x \frac{y^3}{3} \right) \Big|_{y=0}^{y=4-2x} dx$$

$$= \int_0^2 \left[12x(4 - 2x)^2 - 6x^2(4 - 2x)^2 - 2x(4 - 2x)^3 \right] dx$$

$$= \frac{64}{5},$$

where we leave the details of the last integration to you.

■

The greatest challenge in setting up a triple integral is to get the limits of integration correct. To improve your chances of doing this, you need to take the time to draw a good sketch of the solid and try to identify either the base of the solid in one of the coordinate planes (as we did in example 5.2) or top and bottom boundaries of the solid when both lie above or below the same region R in one of the coordinate planes. This may seem like a lot to keep in mind, but we'll illustrate these ideas generously in the examples that follow and in the exercises. Be sure that you don't rely on making guesses. Guessing may get you through the first several exercises, but will not work in general.

To help determine the limits of integration, once you have identified a base or a top and bottom surface of a solid, draw a line from a representative point in the base (or bottom surface) through the top surface of the solid, as we did in Figure 13.43a. This will indicate the limits for the innermost integral. To illustrate this, we take several different views of example 5.2.

Example 5.3   A Triple Integral Where the First Integration Is with Respect to x

Evaluate $\iiint\limits_{Q} 6xy \, dV$, where Q is the tetrahedron bounded by the planes $x = 0$, $y = 0$, $z = 0$ and $2x + y + z = 4$, as in example 5.2, but this time, integrate first with respect to x.

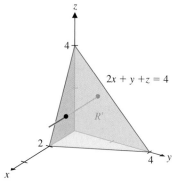

Figure 13.44a

Tetrahedron viewed with base in the yz-plane.

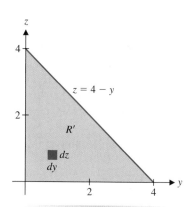

Figure 13.44b

The region R'.

Solution You might object that our only evaluation result for triple integrals (5.4) is for integration with respect to z first. While this is true, you need to realize that x, y and z are simply variables that we represent by letters of the alphabet. Who cares which letter is which? Notice that we can think of the tetrahedron as a solid with its base in the triangular region R' of the yz-plane as indicated in Figure 13.44a. In this case, we draw a line orthogonal to the yz-plane. Notice that it enters the solid in the yz-plane ($x = 0$) and exits in the plane $x = \frac{1}{2}(4 - y - z)$. Adapting (5.4) to this situation (i.e., interchanging the roles of x and z), we have

$$\iiint\limits_Q 6xy\, dV = \iint\limits_{R'} \int_0^{\frac{1}{2}(4-y-z)} 6xy\, dx\, dA$$

$$= \iint\limits_{R'} \left(6\frac{x^2}{2} y \right) \Bigg|_{x=0}^{x=\frac{1}{2}(4-y-z)} dA$$

$$= \iint\limits_{R'} 3\frac{(4-y-z)^2}{4} y\, dA.$$

To evaluate the remaining double integral, we look at the region R' in the yz-plane, as shown in Figure 13.44b. We now have

$$\iiint\limits_Q 6xy\, dV = \frac{3}{4} \int_0^4 \int_0^{4-y} (4-y-z)^2 y\, dz\, dy = \frac{64}{5},$$

where we have left the routine details for you to verify. Finally, we leave it to you to show that we can also write this triple integral as a triple iterated integral where we integrate with respect to y first, as in

$$\iiint\limits_Q 6xy\, dV = \int_0^2 \int_0^{4-2x} \int_0^{4-2x-z} 6xy\, dy\, dz\, dx.$$

We want to emphasize again that the challenge here is to get the correct limits of integration. Calculating the individual integrals is old news. Besides, you can always use a computer algebra system to evaluate the integrals, at least approximately. However, no computer algebra system will set up the limits of integration for you! One thing to keep in mind here is that the innermost limits of integration will always correspond to two three-dimensional surfaces. (You can think of these as the top and the bottom of the solid, if you orient yourself properly.) These limits can involve either or both (or neither) of the two outer variables of integration. The limits of integration for the middle integral will always represent two curves in one of the coordinate planes and can involve only the outermost variable of integration. Realize, too, that this says that once you integrate with respect to a given variable, that variable is eliminated from subsequent integrations (since you've evaluated the result of the integration between two specific values of that variable). Keep these ideas in mind as you work through the examples and exercises and make sure you work lots of problems. Triple integrals can look intimidating at first and **the only way to become proficient at dealing with these is to work plenty of problems!** Multiple integrals form the basis of much of the remainder of the book, so don't skimp on your effort now.

We have not yet given any geometric significance to the triple integral. For a special type of triple integral, we have the following useful interpretation. Recall that for double integrals, we had found that $\iint_R dA$ gives the area of the region R. Similarly, you should observe that if $f(x, y, z) = 1$ for all $(x, y, z) \in Q$, then from (5.3), we have

$$\iiint_Q 1\, dV = \lim_{\|P\| \to 0} \sum_{i=1}^{n} \Delta V_i = V, \tag{5.5}$$

where V is the volume of the solid Q.

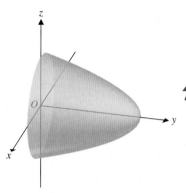

Figure 13.45a

The solid Q.

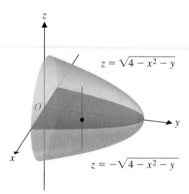

Figure 13.45b

Solid showing projection onto xy-plane.

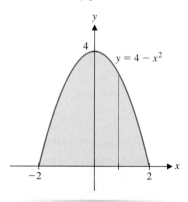

Figure 13.45c

Projection onto the xy-plane.

Example 5.4   Using a Triple Integral to Find Volume

Use a triple integral to find the volume of the solid Q bounded by the graph of $y = 4 - x^2 - z^2$ and the xz-plane.

Solution   Notice that the graph of $y = 4 - x^2 - z^2$ is a paraboloid with vertex at $(0, 4, 0)$, whose axis is the y-axis and that opens toward the negative y-axis. We show the solid in Figure 13.45a. The question is, as usual, how to set up the limits of integration. Without thinking too much about it, we might consider integration with respect to z first. In this case, the projection of the solid onto the xy-plane is the parabola formed by the intersection of the paraboloid with the xy-plane (see Figure 13.45b). Notice from Figure 13.45b that for each fixed x and y, the line through the point $(x, y, 0)$ and perpendicular to the xy-plane enters the solid on the bottom half of the paraboloid ($z = -\sqrt{4 - x^2 - y}$) and exits the solid on the top surface of the paraboloid ($z = \sqrt{4 - x^2 - y}$). This gives you the innermost limits of integration. We get the rest from looking at the projection of the paraboloid onto the xy-plane (see Figure 13.45c).

You should be able to read the outer limits of integration from Figure 13.45c. From (5.5), we get

$$V = \iiint_Q dV = \int_{-2}^{2} \int_{0}^{4-x^2} \int_{-\sqrt{4-x^2-y}}^{\sqrt{4-x^2-y}} dz\, dy\, dx$$

$$= \int_{-2}^{2} \int_{0}^{4-x^2} z \Big|_{z=-\sqrt{4-x^2-y}}^{z=\sqrt{4-x^2-y}} dy\, dx$$

$$= \int_{-2}^{2} \int_{0}^{4-x^2} 2\sqrt{4 - x^2 - y}\, dy\, dx$$

$$= \int_{-2}^{2} (-2)\left(\frac{2}{3}\right)(4 - x^2 - y)^{3/2} \Big|_{y=0}^{y=4-x^2} dx$$

$$= \frac{4}{3} \int_{-2}^{2} (4 - x^2)^{3/2}\, dx = 8\pi.$$

Notice that the last integration here is challenging. (We used a CAS to carry it out.) In such situations, you might wonder if there is an easier way, so let's take another view of this. If you look at Figure 13.45a and turn your head to the side, you should see a paraboloid with a circular base in the xz-plane. This suggests that you might want to integrate first with respect to y. Referring to Figure 13.46a, observe that for each point

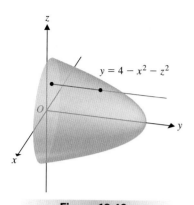

Figure 13.46a
Paraboloid with base in the xz-plane.

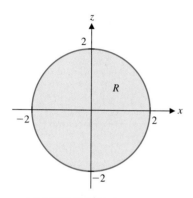

Figure 13.46b
Base of the solid.

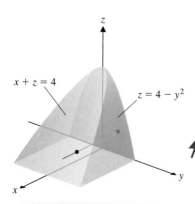

Figure 13.47a
The solid Q.

in the base of the solid in the xz-plane, y ranges from 0 to $4 - x^2 - z^2$. Notice that the base in this case is formed by the intersection of the paraboloid with the xz-plane ($y = 0$): $0 = 4 - x^2 - z^2$ or $x^2 + z^2 = 4$ (i.e., the circle of radius 2 centered at the origin; see Figure 13.46b).

From (5.5), we can now write the volume as

$$V = \iiint_Q dV = \iint_R \int_0^{4-x^2-z^2} dy\, dA$$
$$= \iint_R (4 - x^2 - z^2)\, dA,$$

where R is the disk indicated in Figure 13.46b. Now, you could always set up the remaining double integral in rectangular coordinates (in fact, you'll get the same result as we did above). On the other hand, there is something worth observing here first. Notice that the region R is a circle and the integrand contains the combination of variables $x^2 + z^2$. Does this suggest anything? Hopefully, you'll recognize that both observations suggest polar coordinates. Since the variables here are x and z, we define $x = r\cos\theta$ and $z = r\sin\theta$. From Figure 13.46b, notice that for each fixed angle θ, r runs from 0 to 2. This gives us

$$V = \iint_R \underbrace{(4 - x^2 - z^2)}_{4-r^2} \underbrace{dA}_{r\,dr\,d\theta}$$
$$= \int_0^{2\pi}\int_0^2 (4 - r^2)\, r\, dr\, d\theta$$
$$= -\frac{1}{2}\int_0^{2\pi} \frac{(4-r^2)^2}{2}\Big|_{r=0}^{r=2} d\theta$$
$$= 4\int_0^{2\pi} d\theta = 8\pi.$$

Compare this to our first approach to the problem. Notice that in some sense, viewing the solid as having its base in the xz-plane is more natural. Further, compare the level of difficulty of the integrations each approach required. The first approach required trigonometric substitution or use of a computer algebra system for the final integral. With polar coordinates there are obvious advantages to the second approach.

As you can see from example 5.4, there are clear advantages to considering alternative approaches for calculating triple integrals. There often are several alternative approaches for dealing with a given triple integral. The lesson to learn from example 5.4 is to take a little time to consider your alternatives before jumping into the problem (i.e., look before you leap). Taking an extra moment to look at a sketch of a solid will often produce a better means of solution.

Example 5.5   Using a Triple Integral to Find Volume

Find the volume of the solid bounded by the graphs of $z = 4 - y^2$, $x + z = 4$, $x = 0$ and $z = 0$.

Solution   We show a sketch of the solid in Figure 13.47a. First, observe that we can consider the base of the solid to be the region R formed by the projection of the solid

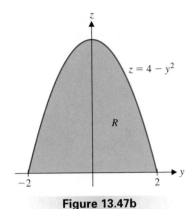

Figure 13.47b

Base R of the solid.

onto the yz-plane ($x = 0$). Notice that this is the region bounded by the parabola $z = 4 - y^2$ and the y-axis (see Figure 13.47b). Then, for each fixed y and z, the corresponding values of x range from 0 to $4 - z$. The volume of the solid is then given by

$$
\begin{aligned}
V = \iiint\limits_{Q} dV &= \iint\limits_{R} \int_{0}^{4-z} dx\, dA \\
&= \int_{-2}^{2} \int_{0}^{4-y^2} \int_{0}^{4-z} dx\, dz\, dy \\
&= \int_{-2}^{2} \int_{0}^{4-y^2} (4 - z)\, dz\, dy \\
&= \int_{-2}^{2} \left(4z - \frac{z^2}{2} \right) \Big|_{z=0}^{|z=4-y^2|} dy \\
&= \int_{-2}^{2} \left[4(4 - y^2) - \frac{1}{2}(4 - y^2)^2 \right] dy \\
&= \frac{128}{5},
\end{aligned}
$$

where we have left the details of the last integration to you.

NOTES

To integrate with respect to z first, you must identify surfaces forming the top and bottom of the solid. To integrate with respect to y first, you must identify surfaces forming (from the standard viewpoint) the right and left sides of the solid. To integrate with respect to x first, you must identify surfaces forming the front and back of the solid. Often, the easiest pair of surfaces to identify will indicate the best choice of variable for the innermost integration.

Mass and Center of Mass

In section 13.2, we discussed finding the mass and center of mass of a lamina (a thin, flat plate). We now pause briefly to extend these results to three dimensions. Suppose that a solid Q has mass density given by $\rho(x, y, z)$ (in units of mass per unit volume). To find the total mass of a solid, we proceed (as we did for laminas) by constructing an inner partition of the solid: $Q_1, Q_2, \ldots, Q_n$. Realize that if each box Q_i is small (see Figure 13.48), then the density should be nearly constant on Q_i and so, it is reasonable to approximate the mass m_i of Q_i by

$$
m_i \approx \underbrace{\rho(u_i, v_i, w_i)}_{\text{mass/unit volume}} \underbrace{\Delta V_i}_{\text{volume}},
$$

for any point $(u_i, v_i, w_i) \in Q_i$, where ΔV_i is the volume of Q_i. The total mass m of Q is then given approximately by

$$
m \approx \sum_{i=1}^{n} \rho(u_i, v_i, w_i)\, \Delta V_i.
$$

Letting the norm of the partition $\|P\|$ approach zero, we get the exact mass, which we recognize as a triple integral:

$$
m = \lim_{\|P\| \to 0} \sum_{i=1}^{n} \rho(u_i, v_i, w_i)\, \Delta V_i = \iiint\limits_{Q} \rho(x, y, z)\, dV. \tag{5.6}
$$

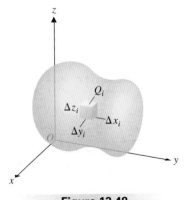

Figure 13.48

One box Q_i of the inner partition of Q.

Now, recall that the center of mass of a lamina was the point at which the lamina will balance. For an object in three dimensions, you can think of this as balancing it left to right (i.e., along the y-axis), front to back (i.e., along the x-axis) and top to bottom (i.e., along the z-axis). To do this, we need to find first moments with respect to each of the three coordinate planes. We define these moments as

$$M_{yz} = \iiint\limits_Q x\rho(x, y, z)\,dV, \quad M_{xz} = \iiint\limits_Q y\rho(x, y, z)\,dV \tag{5.7}$$

and

$$M_{xy} = \iiint\limits_Q z\rho(x, y, z)\,dV, \tag{5.8}$$

the **first moments** with respect to the yz-plane, the xz-plane and the xy-plane, respectively. The **center of mass** is then given by the point $(\bar{x}, \bar{y}, \bar{z})$, where

$$\bar{x} = \frac{M_{yz}}{m}, \quad \bar{y} = \frac{M_{xz}}{m}, \quad \bar{z} = \frac{M_{xy}}{m}. \tag{5.9}$$

Notice that these are a straightforward generalization of the corresponding formulas for the center of mass of a lamina.

Example 5.6   Center of Mass of a Solid

Find the center of mass of the solid of constant mass density ρ bounded by the graphs of the right circular cone $z = \sqrt{x^2 + y^2}$ and the plane $z = 4$ (see Figure 13.49a).

Solution   Notice that the projection R of the solid onto the xy-plane is the disk of radius 4 centered at the origin (see Figure 13.49b). Further, for each $(x, y) \in R$, z ranges from the cone ($z = \sqrt{x^2 + y^2}$) up to the plane $z = 4$. From (5.6), the total mass of the solid is given by

$$m = \iiint\limits_Q \rho(x, y, z)\,dV = \rho \iint\limits_R \int_{\sqrt{x^2+y^2}}^4 dz\,dA$$

$$= \rho \iint\limits_R \left(4 - \sqrt{x^2 + y^2}\right) dA,$$

where R is the disk of radius 4 in the xy-plane, centered at the origin, as indicated in Figure 13.49b. Since the region R is circular and since the integrand contains a term of the form $\sqrt{x^2 + y^2}$, we use polar coordinates for the remaining double integral. We have

$$m = \rho \iint\limits_R \left(4 - \underbrace{\sqrt{x^2 + y^2}}_{r}\right) \underbrace{dA}_{r\,dr\,d\theta}$$

$$= \rho \int_0^{2\pi} \int_0^4 (4 - r)\,r\,dr\,d\theta$$

$$= \rho \int_0^{2\pi} \left(4\frac{r^2}{2} - \frac{r^3}{3}\right)\Bigg|_{r=0}^{r=4} d\theta$$

$$= \rho\left(32 - \frac{4^3}{3}\right)(2\pi) = \frac{64}{3}\pi\rho.$$

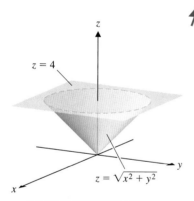

Figure 13.49a
The solid Q.

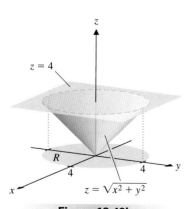

Figure 13.49b
Projection of the solid onto the xy-plane.

From (5.8), we get that the moment with respect to the xy-plane is

$$M_{xy} = \iiint\limits_{Q} z\,\rho(x, y, z)\,dV = \rho \iint\limits_{R} \int_{\sqrt{x^2+y^2}}^{4} z\,dz\,dA$$

$$= \rho \iint\limits_{R} \left.\frac{z^2}{2}\right|_{\sqrt{x^2+y^2}}^{4} dA$$

$$= \frac{\rho}{2} \iint\limits_{R} [16 - (x^2 + y^2)]\,dA.$$

For the same reasons as when we computed the mass, we change to polar coordinates in the double integral to get

$$M_{xy} = \frac{\rho}{2} \iint\limits_{R} \left[16 - \underbrace{(x^2 + y^2)}_{r^2} \right] \underbrace{dA}_{r\,dr\,d\theta}$$

$$= \frac{\rho}{2} \int_{0}^{2\pi} \int_{0}^{4} (16 - r^2)\,r\,dr\,d\theta$$

$$= \frac{\rho}{2} \int_{0}^{2\pi} \left.\left(16\frac{r^2}{2} - \frac{r^4}{4} \right)\right|_{r=0}^{4} d\theta$$

$$= 32\rho(2\pi) = 64\pi\rho.$$

Notice that the solid is symmetric with respect to both the xz-plane and the yz-plane and so, the moments with respect to those planes are zero, since the density is constant. (Why does constant density matter?) That is, $M_{xz} = M_{yz} = 0$. From (5.9), the center of mass is then given by

$$(\bar{x}, \bar{y}, \bar{z}) = \left(\frac{M_{yz}}{m}, \frac{M_{xz}}{m}, \frac{M_{xy}}{m} \right) = \left(0, 0, \frac{64\pi\rho}{64\pi\rho/3} \right) = (0, 0, 3).$$

EXERCISES 13.5

1. Discuss the importance of having a reasonably accurate sketch to help determine the limits (and order) of integration. Identify which features of a sketch are essential and which are not. Discuss whether it's important for your sketch to distinguish between two surfaces like $z = 4 - x^2 - y^2$ and $z = \sqrt{4 - x^2 - y^2}$.

2. In example 5.2, explain why all six orders of integration are equally simple. Given this choice, most people prefer to integrate in the order of example 5.2 ($dz\,dy\,dx$). Discuss the visual advantages of this order.

3. In example 5.4, identify any clues in the problem statement that might indicate that y should be the innermost variable of integration. In example 5.5, identify any clues that might indicate that z should **not** be the innermost variable of integration. (Hint: With how many surfaces is each variable associated?)

4. In example 5.4, we used polar coordinates in x and z. Explain why this is permissible and when it is likely to be convenient to do so. Give an example in which you would want to use polar coordinates in y and z.

In exercises 5–18, evaluate the triple integral $\iiint\limits_{Q} f(x, y, z)\,dV$.

5. $f(x, y, z) = 2x + y - z$,
   $Q = \{(x, y, z)| 0 \le x \le 2, -2 \le y \le 2, 0 \le z \le 2\}$

6. $f(x, y, z) = 2x^2 + y^3$,
   $Q = \{(x, y, z) \,|\, 0 \leq x \leq 3, -2 \leq y \leq 1, 1 \leq z \leq 2\}$

7. $f(x, y, z) = \sqrt{y} - 3z^2$,
   $Q = \{(x, y, z) \,|\, 2 \leq x \leq 3, 0 \leq y \leq 1, -1 \leq z \leq 1\}$

8. $f(x, y, z) = 2xy - 3xz^2$,
   $Q = \{(x, y, z) \,|\, 0 \leq x \leq 2, -1 \leq y \leq 1, 0 \leq z \leq 2\}$

9. $f(x, y, z) = 4yz$, Q is the tetrahedron bounded by $x + 2y + z = 2$ and the coordinate planes

10. $f(x, y, z) = 3x - 2y$, Q is the tetrahedron bounded by $4x + y + 3z = 12$ and the coordinate planes

11. $f(x, y, z) = 3y^2 - 2z$, Q is the tetrahedron bounded by $3x + 2y - z = 6$ and the coordinate planes

12. $f(x, y, z) = 6xz^2$, Q is the tetrahedron bounded by $-2x + y + z = 4$ and the coordinate planes

13. $f(x, y, z) = 2xy$, Q is bounded by $z = 1 - x^2 - y^2$ and $z = 0$

14. $f(x, y, z) = x - y$, Q is bounded by $z = x^2 + y^2$ and $z = 4$

15. $f(x, y, z) = (x^2 + z^2)y^2$, Q is bounded by $x^2 + z^2 = 4$, $y = -2$ and $y = 2$

16. $f(x, y, z) = x^3 e^{yz}$, Q is bounded by $z = 1 - y^2$, $z = 0$, $x = -1$ and $x = 1$

17. $f(x, y, z) = 15$, Q is bounded by $2x + y + z = 4$, $z = 0$, $x = 1 - y^2$ and $x = 0$

18. $f(x, y, z) = x^2 + y^2$, Q is bounded by $z = 6 - x - y$, $x^2 + y^2 = 1$ and $z = 0$

19. Sketch the region Q in exercise 13 and explain why the triple integral equals 0. Would the integral equal 0 if $f(x, y, z) = 2x^2y$?

20. Show that $\iiint\limits_{Q} (z - x)\, dV = 0$, where Q is bounded by $z = 6 - x - y$ and the coordinate planes. Explain geometrically why this is correct.

In exercises 21–32, compute the volume of the solid bounded by the given surfaces.

21. $z = x^2$, $z = 1$, $y = 0$ and $y = 2$

22. $z = 1 - x^2$, $z = 0$, $y = 2$, $y = 4$

23. $z = 1 - y^2$, $z = 0$, $z = 4 - 2x$, $x = 4$

24. $z = y^2$, $z = 1$, $2x + z = 4$, $x = 0$

25. $x = y^2 + z^2$, $x = 4$

26. $x = 9 - y^2 - z^2$, the yz-plane

27. $y = \sqrt{x^2 + z^2}$, $y = 3$

28. $y = 1 - x^2 - z^2$, the xz-plane

29. $y = 4 - x^2$, $z = 0$, $z - y = 6$

30. $x = y^2$, $x = 4$, $x + z = 6$, $x + z = 8$

31. $y = 3 - x$, $y = 0$, $z = x^2$, $z = 1$

32. $x = y^2$, $x = 4$, $z = 2 + x$, $z = 0$

In exercises 33–36, find the mass and center of mass of the solid with density $\rho(x, y, z)$ and the given shape.

33. $\rho(x, y, z) = 4$, solid bounded by $z = x^2 + y^2$, $z = 4$

34. $\rho(x, y, z) = 2 + x$, solid bounded by $z = x^2 + y^2$, $z = 4$

35. $\rho(x, y, z) = 10 + x$, tetrahedron bounded by $x + 3y + z = 6$ and the coordinate planes

36. $\rho(x, y, z) = 1 + x$, tetrahedron bounded by $2x + y + 4z = 4$ and the coordinate planes

37. Explain why the x-coordinate of the center of mass in exercise 33 is zero, but the x-coordinate in exercise 34 is not zero.

38. In exercise 33, if $\rho(x, y, z) = 2 + x^2$, is the x-coordinate of the center of mass zero? Explain.

39. In exercise 9, evaluate the integral in three different ways, using each variable as the innermost variable once.

40. In exercise 10, evaluate the integral in three different ways, using each variable as the innermost variable once.

In exercises 41–46, sketch the solid whose volume is given and rewrite the iterated integral using a different innermost variable.

41. $\displaystyle \int_0^2 \int_0^{4-2y} \int_0^{4-2y-z} dx\, dz\, dy$

42. $\displaystyle \int_0^1 \int_0^{2-2y} \int_0^{2-x-2y} dz\, dx\, dy$

43. $\displaystyle \int_0^1 \int_0^{\sqrt{1-x^2}} \int_0^{\sqrt{1-x^2-y^2}} dz\, dy\, dx$

44. $\int_0^1 \int_0^{1-x^2} \int_0^{2-x} dy\,dz\,dx$

45. $\int_0^2 \int_0^{\sqrt{4-z^2}} \int_{x^2+z^2}^4 dy\,dx\,dz$

46. $\int_0^2 \int_0^{\sqrt{4-z^2}} \int_{\sqrt{y^2+z^2}}^2 dx\,dy\,dz$

47. 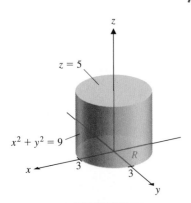 In this exercise, you will examine several models of baseball bats. Sketch the region extending from $y = 0$ to $y = 32$ with distance from the y-axis given by $r = \frac{1}{2} + \frac{3}{128}y$. This should look vaguely like a baseball bat, with 32 representing the 32 inch length of a typical bat. Assume a constant **weight density** of $\rho = 0.39$ ounces per cubic inch. Compute the weight of the bat and the center of mass of the bat. (Hint: Compute the y-coordinate and argue that the x- and z-coordinates are zero.) Sketch each of the following regions, explain what the name means and compute the mass and center of mass. (a) **Long bat:** same as the original except y extends from $y = 0$ to $y = 34$. (b) **Choked up:** y goes from -2 to 30 with $r = \frac{35}{64} + \frac{3}{128}y$. (c) **Corked bat:** same as the original with the cylinder $26 \le y \le 32$ and $0 \le r \le \frac{1}{4}$ removed. (d) **Aluminum bat:** same as the original with the section from $r = 0$ to $r = \frac{3}{8} + \frac{3}{128}y$, $0 \le y \le 32$ removed and density $\rho = 1.56$. Explain why it makes sense that the choked up bat has the center of mass 2 inches to the left of the original bat. Part of the "folklore" of baseball is that batters with aluminum bats can hit "inside" pitches better than batters with traditional wood bats. If "inside" means smaller values of y and the center of mass represents the "sweet spot" of the bat (the best place to hit the ball), discuss whether your calculations support baseball's folk wisdom.

48. In this exercise, we continue with the baseball bats of exercise 47. This time, we want to compute the moment of inertia $\iiint\limits_{Q} y^2 \rho\, dV$ for each of the bats. The smaller the moment of inertia is, the easier it is to swing the bat. Use your calculations to answer the following questions. How much harder is it to swing a slightly longer bat? How much easier is it to swing a bat that has been choked up 2 inches? Does corking really make a noticeable difference in the ease with which a bat can be swung? How much easier is it to swing a hollow aluminum bat, even if it weighs the same as a regular bat?

13.6   CYLINDRICAL COORDINATES

There were several instances in section 13.5 where we found it convenient to introduce polar coordinates in order to evaluate the outer double integral in a triple integral problem. Sometimes, this is more than a mere convenience, as we see in the following example.

Example 6.1   A Triple Integral Requiring Polar Coordinates

Evaluate $\iiint\limits_{Q} e^{x^2+y^2}\,dV$, where Q is the solid bounded by the cylinder $x^2 + y^2 = 9$, the xy-plane and the plane $z = 5$.

Solution   We show a sketch of the solid in Figure 13.50a. This might seem simple enough; certainly the solid is not particularly complicated. Unfortunately, the integral is rather troublesome. Notice that the base of the solid is the circle R of radius 3 centered at the origin and lying in the xy-plane. Further, for each (x, y) in R, z ranges from 0 up to 5. So, we have

$$\iiint\limits_{Q} e^{x^2+y^2}\,dV = \iint\limits_{R} \int_0^5 e^{x^2+y^2}\,dz\,dA$$

$$= 5 \iint\limits_{R} e^{x^2+y^2}\,dA.$$

Figure 13.50a
The solid Q.

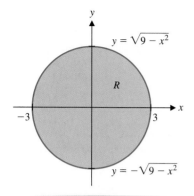

Figure 13.50b

The region R.

The challenge lies in evaluating the remaining double integral. From Figure 13.50b, observe that for each fixed $x \in [-3, 3]$, y ranges from $-\sqrt{9 - x^2}$ (the bottom semicircle) up to $\sqrt{9 - x^2}$ (the top semicircle). We now have

$$\iiint_Q e^{x^2+y^2} \, dV = 5 \iint_R e^{x^2+y^2} \, dA = 5 \int_{-3}^{3} \int_{-\sqrt{9-x^2}}^{\sqrt{9-x^2}} e^{x^2+y^2} \, dy \, dx.$$

Without polar coordinates, we're at a dead end, since we don't know an antiderivative for $e^{x^2+y^2}$. Even the authors' computer algebra system has difficulty with this, giving a nearly indecipherable answer in terms of an integral of the *error* function, which you have likely never seen before. Even so, our computer algebra system could not handle the second integration, except approximately. On the other hand, if we introduce polar coordinates: $x = r \cos \theta$ and $y = r \sin \theta$, we get that for each $\theta \in [0, 2\pi]$, r ranges from 0 up to 3. We now have an integral requiring only a simple substitution:

$$\iiint_Q e^{x^2+y^2} \, dV = 5 \iint_R \underbrace{e^{x^2+y^2}}_{e^{r^2}} \underbrace{dA}_{r \, dr \, d\theta}$$

$$= 5 \int_0^{2\pi} \int_0^3 e^{r^2} r \, dr \, d\theta$$

$$= \frac{5}{2} \int_0^{2\pi} e^{r^2} \Big|_{r=0}^{r=3} \, d\theta$$

$$= 5\pi(e^9 - 1)$$

$$\approx 1.27 \times 10^5,$$

which is a much more acceptable answer.

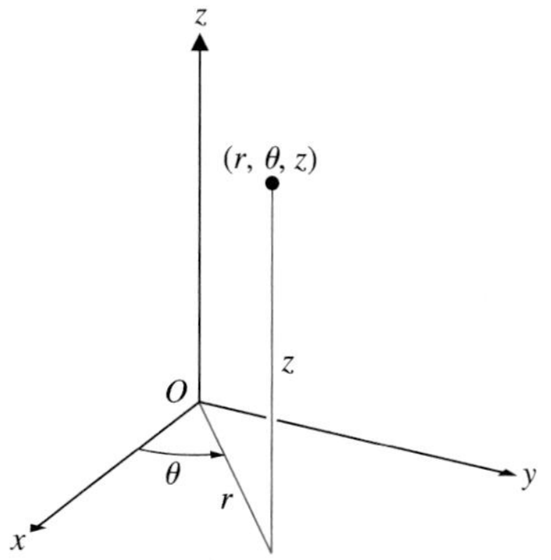

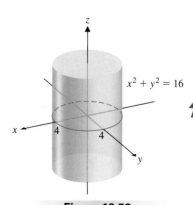

Figure 13.51

Cylindrical coordinates.

The process of replacing two of the variables in a three-dimensional coordinate system by polar coordinates, as we illustrated in example 6.1, is so common that we give this coordinate system a name: cylindrical coordinates.

Specifically, we specify a point $P(x, y, z) \in \mathbb{R}^3$ by specifying polar coordinates for the point $(x, y) \in \mathbb{R}^2 : x = r \cos \theta$ and $y = r \sin \theta$, where $r^2 = x^2 + y^2$ and θ is the angle made by the line segment connecting the origin and the point $(x, y, 0)$ with the positive x-axis, as indicated in Figure 13.51, so that $\tan \theta = \dfrac{y}{x}$. We refer to (r, θ, z) as **cylindrical coordinates** for the point P.

Example 6.2 Equation of a Cylinder in Cylindrical Coordinates

Write the equation for the cylinder $x^2 + y^2 = 16$ (see Figure 13.52) in cylindrical coordinates.

Solution In cylindrical coordinates $r^2 = x^2 + y^2$, so the cylinder becomes $r^2 = 16$ or $r = \pm 4$. But note that since θ is not specified, the equation $r = 4$ describes the same cylinder.

Figure 13.52

The cylinder $r = 4$.

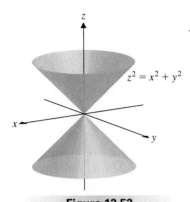

Figure 13.53

The cone $z = r$.

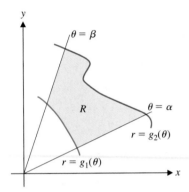

Figure 13.54

The region R.

Example 6.3 **Equation of a Cone in Cylindrical Coordinates**

Write the equation for the cone $z^2 = x^2 + y^2$ (see Figure 13.53) in cylindrical coordinates.

Solution Since $x^2 + y^2 = r^2$, the cone becomes $z^2 = r^2$ or $z = \pm r$. Since r can be both positive and negative, this simplifies to $z = r$.

As we did on a case-by-case basis in several examples in section 13.5 and in example 6.1, we can use cylindrical coordinates to simplify certain triple integrals. For instance, suppose that we can write the solid Q as

$$Q = \{(r, \theta, z) \,|\, (r, \theta) \in R \text{ and } k_1(r, \theta) \leq z \leq k_2(r, \theta)\},$$

where R is the region of the xy-plane defined by

$$R = \{(r, \theta) \,|\, \alpha \leq \theta \leq \beta \text{ and } g_1(\theta) \leq r \leq g_2(\theta)\},$$

as shown in Figure 13.54. Then, notice that from (5.4), we can write

$$\iiint\limits_{Q} f(r, \theta, z) \, dV = \iint\limits_{R} \left[\int_{k_1(r,\theta)}^{k_2(r,\theta)} f(r, \theta, z) \, dz \right] dA.$$

But, the outer double integral is a double integral in polar coordinates, which we already know how to write as an iterated integral. We have

$$\iiint\limits_{Q} f(r, \theta, z) \, dV = \iint\limits_{R} \left[\int_{k_1(r,\theta)}^{k_2(r,\theta)} f(r, \theta, z) \, dz \right] \underbrace{dA}_{r \, dr \, d\theta}$$

$$= \int_{\alpha}^{\beta} \int_{g_1(\theta)}^{g_2(\theta)} \left[\int_{k_1(r,\theta)}^{k_2(r,\theta)} f(r, \theta, z) \, dz \right] r \, dr \, d\theta.$$

This gives us an evaluation formula for triple integrals in cylindrical coordinates:

$$\iiint\limits_{Q} f(r, \theta, z) \, dV = \int_{\alpha}^{\beta} \int_{g_1(\theta)}^{g_2(\theta)} \int_{k_1(r,\theta)}^{k_2(r,\theta)} f(r, \theta, z) \, r \, dz \, dr \, d\theta. \tag{6.1}$$

In setting up triple integrals in cylindrical coordinates, it often helps to visualize the volume element $dV = r \, dz \, dr \, d\theta$ (see Figure 13.55).

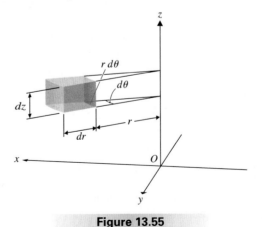

Figure 13.55

Volume element for cylindrical coordinates.

Example 6.4 A Triple Integral in Cylindrical Coordinates

Write $\iiint\limits_{Q} f(r, \theta, z)\, dV$ as a triple iterated integral in cylindrical coordinates if

$$Q = \left\{(x, y, z)\,\middle|\, \sqrt{x^2 + y^2} \le z \le \sqrt{18 - x^2 - y^2}\right\}.$$

Solution Of course, the first task in setting up any iterated multiple integral is to draw a sketch of the region over which you are integrating. Here, notice that $z = \sqrt{x^2 + y^2}$ is the top half of a right circular cone, with vertex at the origin and axis lying along the z-axis. We also have that $z = \sqrt{18 - x^2 - y^2}$ is the top hemisphere of radius $\sqrt{18}$ centered at the origin. So, we are looking for the set of all points lying above the cone and below the hemisphere (see Figure 13.56a). You should recognize that in cylindrical coordinates, the cone is written $z = r$ and the hemisphere becomes $z = \sqrt{18 - r^2}$, since $x^2 + y^2 = r^2$. This says that for each r and θ, z ranges from r up to $\sqrt{18 - r^2}$. Notice that the cone and the hemisphere intersect when

$$\sqrt{18 - r^2} = r$$

or

$$18 - r^2 = r^2,$$

so that

$$18 = 2r^2 \quad \text{or} \quad r = 3.$$

That is, the two surfaces intersect in a circle of radius 3 lying in the plane $z = 3$. The projection of the solid down onto the xy-plane is then the circle of radius 3 centered at the origin (see Figure 13.56b) and we have:

$$\iiint\limits_{Q} f(r, \theta, z)\, dV = \int_0^{2\pi} \int_0^3 \int_r^{\sqrt{18-r^2}} f(r, \theta, z)\, r\, dz\, dr\, d\theta.$$

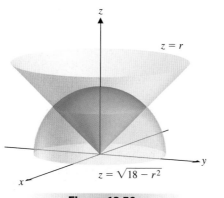

Figure 13.56a

The solid Q.

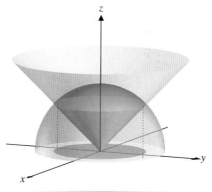

Figure 13.56b

Projection of Q onto the xy-plane.

Very often, you will need to recognize that a triple integral written in rectangular coordinates would be simpler in cylindrical coordinates. You must then recognize how to write the solid in cylindrical coordinates, as well as how to rewrite the integral.

Example 6.5 Changing from Rectangular to Cylindrical Coordinates

Evaluate the triple iterated integral $\displaystyle\int_{-1}^{1} \int_{-\sqrt{1-x^2}}^{\sqrt{1-x^2}} \int_{x^2+y^2}^{2-x^2-y^2} (x^2 + y^2)^{3/2}\, dz\, dy\, dx.$

Solution   Of course, the way this is written, the integral is virtually impossible to evaluate exactly. (Even our computer algebra system had trouble with it.) You should quickly notice that the integrand involves $x^2 + y^2$, which is simply r^2 in cylindrical coordinates. You should also try to visualize the region over which you are integrating. First, from the innermost limits of integration, notice that $z = 2 - x^2 - y^2$ is a paraboloid opening downward, with vertex at the point $(0, 0, 2)$ and $z = x^2 + y^2$ is a paraboloid opening upward with vertex at the origin. So, the solid is some portion of the solid bounded by the two paraboloids. Observe that the two paraboloids intersect when

$$2 - x^2 - y^2 = x^2 + y^2,$$

or

$$1 = x^2 + y^2.$$

That is, the intersection forms a circle of radius 1 lying in the plane $z = 1$ and centered at the point $(0, 0, 1)$. Looking at the outer two integrals, note that for each $x \in [-1, 1]$, y ranges from $-\sqrt{1 - x^2}$ (the bottom semicircle of radius 1 centered at the origin) to $\sqrt{1 - x^2}$ (the top semicircle of radius 1 centered at the origin). Since this corresponds to the projection of the circle of intersection onto the xy-plane, the triple integral is over the entire solid below the one paraboloid and above the other (see Figure 13.57).

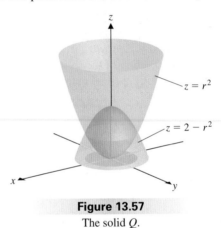

Figure 13.57
The solid Q.

Now, in cylindrical coordinates, the top paraboloid becomes $z = 2 - x^2 - y^2 = 2 - r^2$ and the bottom paraboloid becomes $z = x^2 + y^2 = r^2$. So, for each fixed value of r and θ, z varies from r^2 up to $2 - r^2$. Further, since the projection of the solid onto the xy-plane is the circle of radius 1 centered at the origin, r varies from 0 to 1 and θ varies from 0 to 2π. We can now write the triple integral in cylindrical coordinates as

$$\int_{-1}^{1} \int_{-\sqrt{1-x^2}}^{\sqrt{1-x^2}} \int_{x^2+y^2}^{2-x^2-y^2} (x^2 + y^2)^{3/2} \, dz \, dy \, dx = \int_{0}^{2\pi} \int_{0}^{1} \int_{r^2}^{2-r^2} (r^2)^{3/2} r \, dz \, dr \, d\theta$$

$$= \int_{0}^{2\pi} \int_{0}^{1} \int_{r^2}^{2-r^2} r^4 \, dz \, dr \, d\theta$$

$$= \int_{0}^{2\pi} \int_{0}^{1} r^4 (2 - 2r^2) \, dr \, d\theta$$

$$= 2 \int_{0}^{2\pi} \left(\frac{r^5}{5} - \frac{r^7}{7} \right) \Big|_{r=0}^{r=1} d\theta = \frac{8\pi}{35}.$$

Notice that this was an easy exercise, compared to evaluating the original integral directly.

When converting an iterated integral from rectangular to cylindrical coordinates, you need to carefully visualize the solid over which you are integrating. This entails looking carefully at **all** of the limits of integration (and not just the limits on the innermost integral), as we see in the following example.

Example 6.6 Changing from Rectangular to Cylindrical Coordinates

Evaluate the triple iterated integral $\int_0^1 \int_0^{\sqrt{1-x^2}} \int_{x^2+y^2}^{2-x^2-y^2} (x^2 + y^2)^{3/2}\, dz\, dy\, dx$.

Solution Notice that this is nearly identical to the integral in example 6.5, except that two of the six limits of integration are different. Again, this integral is extremely challenging, as currently written in rectangular coordinates. As in example 6.5, the innermost limits of integration tell you that the solid is bounded below by the paraboloid $z = x^2 + y^2$ and above by the paraboloid $z = 2 - x^2 - y^2$. So, once again, the region of integration is some portion of the solid bounded by the two paraboloids and as we saw in example 6.5, the two paraboloids intersect in a circle of radius 1 lying in the plane $z = 1$ and centered at the point $(0, 0, 1)$ (see Figure 13.58a). In this case, however, notice from the limits of integration on the outer two integrals, that for each fixed x in the interval $[0, 1]$, y ranges from 0 (corresponding to the x-axis) up to $\sqrt{1 - x^2}$ (corresponding to the top half of the circle of radius 1 centered at the origin). That is, the outer two integrals correspond to integration over the quarter circle R indicated in Figure 13.58b.

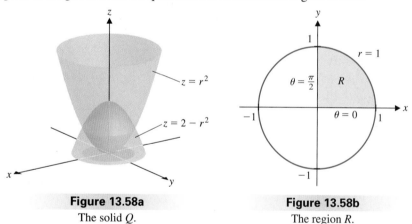

Figure 13.58a	**Figure 13.58b**
The solid Q.	The region R.

Again, in cylindrical coordinates, the top paraboloid becomes $z = 2 - x^2 - y^2 = 2 - r^2$ and the bottom paraboloid becomes $z = x^2 + y^2 = r^2$. So, for each fixed value of r and θ, z varies from r^2 up to $2 - r^2$. Further, since the projection of the solid onto the xy-plane is the portion of the circle of radius 1 centered at the origin lying in the first quadrant, r varies from 0 to 1 and θ varies from 0 to $\frac{\pi}{2}$. We can now write the triple integral in cylindrical coordinates as

$$\int_0^1 \int_0^{\sqrt{1-x^2}} \int_{x^2+y^2}^{2-x^2-y^2} (x^2 + y^2)^{3/2}\, dz\, dy\, dx = \iint_R \int_{x^2+y^2}^{2-x^2-y^2} (x^2 + y^2)^{3/2}\, dz\, dA$$

$$= \int_0^{\pi/2} \int_0^1 \int_{r^2}^{2-r^2} (r^2)^{3/2}\, r\, dz\, dr\, d\theta$$

$$= 2 \int_0^{\pi/2} \left(\frac{r^5}{5} - \frac{r^7}{7} \right)\Bigg|_{r=0}^{r=1} d\theta = \frac{2\pi}{35}.$$

EXERCISES 13.6

1. Using the examples in this section as a guide, make a short list of figures that are easily described in cylindrical coordinates.

2. The three-dimensional solid bounded by $x = a$, $x = b$, $y = c$, $y = d$, $z = e$ and $z = f$ is a rectangular box. Given this, speculate on how **cylindrical coordinates** got its name. Specifically, identify what type of cylinder is the basic figure of cylindrical coordinates.

3. In example 6.4, explain why the outer double integration limits are determined by the intersection of the cone and the hemisphere (and not, for example, by the trace of the hemisphere in the xy-plane).

4. Carefully examine the rectangular and cylindrical limits of integration in example 6.5. Note that in both integrals z is the innermost variable of integration. In this case, explain why the innermost limits of integration for the triple integral in cylindrical coordinates can be obtained by substituting polar coordinates into the innermost limits of integration of the triple integral in rectangular coordinates. Explain why this would not be the case if the order of integration had been changed.

In exercises 5–14, write the given equation in cylindrical coordinates.

5. $x^2 + y^2 = 16$

6. $x^2 + y^2 = 1$

7. $(x-2)^2 + y^2 = 4$

8. $x^2 + (y-3)^2 = 9$

9. $z = x^2 + y^2$

10. $z = \sqrt{x^2 + y^2}$

11. $z = \cos(x^2 + y^2)$

12. $z = e^{-x^2 - y^2}$

13. $y = x$

14. $y = 2x$

In exercises 15–24, set up the triple integral $\iiint_Q f(x, y, z)\, dV$ in cylindrical coordinates.

15. Q is the region above $z = \sqrt{x^2 + y^2}$ and below $z = \sqrt{8 - x^2 - y^2}$.

16. Q is the region above $z = \sqrt{x^2 + y^2}$ and below $z = \sqrt{4 - x^2 - y^2}$.

17. Q is the region above the xy-plane and below $z = 9 - x^2 - y^2$.

18. Q is the region above the xy-plane and below $z = 4 - x^2 - y^2$ in the first octant.

19. Q is the region above $z = x^2 + y^2$ and below $z = 4$.

20. Q is the region above $z = x^2 + y^2$ and below $z = 8 - x^2 - y^2$.

21. Q is the region bounded by $y = 4 - x^2 - z^2$ and $y = 0$.

22. Q is the region bounded by $y = \sqrt{x^2 + z^2}$ and $y = 9$.

23. Q is the region bounded by $x = y^2 + z^2$ and $x = 2 - y^2 - z^2$.

24. Q is the region bounded by $x = \sqrt{y^2 + z^2}$ and $x = 4$.

In exercises 25–36, set up and evaluate the indicated triple integral in the appropriate coordinate system.

25. $\iiint_Q e^{x^2 + y^2}\, dV$, where Q is the region inside $x^2 + y^2 = 4$ and between $z = 1$ and $z = 2$.

26. $\iiint_Q z e^{\sqrt{x^2 + y^2}}\, dV$, where Q is the region inside $x^2 + y^2 = 4$ and between $z = 0$ and $z = 3$.

27. $\iiint_Q (x + z)\, dV$, where Q is the region below $x + 2y + 3z = 6$ in the first octant.

28. $\iiint_Q (y + 2)\, dV$, where Q is the region below $x + z = 4$ in the first octant between $y = 1$ and $y = 2$.

29. $\iiint_Q z\, dV$, where Q is the region between $z = \sqrt{x^2 + y^2}$ and $z = \sqrt{4 - x^2 - y^2}$.

30. $\iiint_Q \sqrt{x^2 + y^2}\, dV$, where Q is the region between $z = \sqrt{x^2 + y^2}$ and $z = 0$ and inside $x^2 + y^2 = 4$.

31. $\iiint_Q (x + y)\, dV$, where Q is the tetrahedron bounded by $x + 2y + z = 4$ and the coordinate planes.

32. $\iiint_Q (2x - y)\, dV$, where Q is the tetrahedron bounded by $3x + y + 2z = 6$ and the coordinate planes.

33. $\iiint_Q e^z\, dV$, where Q is the region inside $x^2 + y^2 = 9$ and between $z = x^2 + y^2$ and $z = 0$.

34. $\iiint_Q \sqrt{x^2 + y^2}\, e^z\, dV$, where Q is the region inside $x^2 + y^2 = 1$ and between $z = (x^2 + y^2)^{3/2}$ and $z = 0$.

35. $\iiint_Q 2x\, dV$, where Q is the region between $z = \sqrt{x^2 + y^2}$ and $z = 0$ and inside $x^2 + (y - 1)^2 = 1$.

36. $\iiint_Q \frac{y}{x}\, dV$, where Q is the region between $z = x^2 + y^2$ and $z = 0$ and inside $(x - 2)^2 + y^2 = 4$.

In exercises 37–42, evaluate the iterated integral after changing coordinate systems.

37. $\displaystyle\int_{-1}^{1}\int_{-\sqrt{1-x^2}}^{\sqrt{1-x^2}}\int_{0}^{\sqrt{x^2+y^2}} 3z^2\, dz\, dy\, dx$

38. $\displaystyle\int_{0}^{1}\int_{-\sqrt{1-x^2}}^{\sqrt{1-x^2}}\int_{0}^{2-x^2-y^2} \sqrt{x^2+y^2}\, dz\, dy\, dx$

39. $\displaystyle\int_{0}^{2}\int_{-\sqrt{4-y^2}}^{\sqrt{4-y^2}}\int_{\sqrt{x^2+y^2}}^{\sqrt{8-x^2-y^2}} 2\, dz\, dx\, dy$

40. $\displaystyle\int_{0}^{1}\int_{0}^{\sqrt{1-x^2}}\int_{1-x^2-y^2}^{4} \sqrt{x^2+y^2}\, dz\, dy\, dx$

41. $\displaystyle\int_{-3}^{3}\int_{-\sqrt{9-x^2}}^{\sqrt{9-x^2}}\int_{0}^{x^2+z^2} (x^2+z^2)\, dy\, dz\, dx$

42. $\displaystyle\int_{-2}^{2}\int_{-\sqrt{4-z^2}}^{\sqrt{4-z^2}}\int_{y^2+z^2}^{4} (y^2+z^2)^{3/2}\, dx\, dy\, dz$

In exercises 43–50, sketch graphs of the cylindrical equations.

43. $z = r$

44. $z = r^2$

45. $z = 4 - r^2$

46. $z = \sqrt{4 - r^2}$

47. $r = 2\sec\theta$

48. $r = 2\sin\theta$

49. $\theta = \pi/4$

50. $r = 4$

In exercises 51–54, find the mass and center of mass of the solid with the given density and bounded by the graphs of the indicated equations.

51. $\rho(x,y,z) = \sqrt{x^2+y^2}$, bounded by $z = \sqrt{x^2+y^2}$ and $z = 4$.

52. $\rho(x,y,z) = e^{-x^2-y^2}$, bounded by $z = \sqrt{4-x^2-y^2}$ and the xy-plane.

53. $\rho(x,y,z) = 4$, between $z = x^2 + y^2$ and $z = 4$ and inside $x^2 + (y-1)^2 = 1$.

54. $\rho(x,y,z) = \sqrt{x^2+z^2}$, bounded by $y = \sqrt{x^2+z^2}$ and $y = \sqrt{8-x^2-z^2}$.

55. Many computer graphing packages will sketch graphs in cylindrical coordinates, with one option being to have r as a function of z and θ. In some cases, the graphs are very familiar. Sketch the following and solve for z to write the equation in the notation of this section: (a) $r = \sqrt{z}$, (b) $r = z^2$, (c) $r = \ln z$, (d) $r = \sqrt{4-z^2}$, (e) $r = z^2\cos\theta$. By leaving z out altogether, some old polar curves get an interesting three-dimensional extension: (f) $r = \sin^2\theta$, $0 \le z \le 4$, (g) $r = 2 - 2\cos\theta$, $0 \le z \le 4$. Many graphs are simply new. Explore the following graphs and others of your own creation: (h) $r = \cos\theta - \ln z$, (i) $r = z^2\ln(\theta+1)$, (j) $r = ze^{\theta/8}$, (k) $r = \theta e^{-z}$.

13.7 SPHERICAL COORDINATES

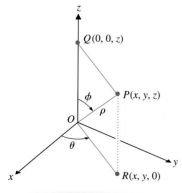

Figure 13.59
Spherical coordinates.

Just as some surfaces are more convenient to describe in cylindrical coordinates than in rectangular coordinates, there is another common coordinate system, called spherical coordinates, that is frequently more convenient than either rectangular or cylindrical coordinates. Further, some triple integrals that cannot be calculated exactly in either rectangular or cylindrical coordinates can be dealt with easily in spherical coordinates.

We can specify a point P with rectangular coordinates (x, y, z) by the corresponding **spherical coordinates** (ρ, ϕ, θ). Here, ρ is defined to be the distance from the origin,

$$\rho = \sqrt{x^2 + y^2 + z^2}. \tag{7.1}$$

Note that specifying the distance a point lies away from the origin specifies a sphere on which the point must lie (i.e., the equation $\rho = \rho_0 > 0$ represents the sphere of radius ρ_0 centered at the origin). To name the specific point on the sphere, we further specify two angles, ϕ and θ, as indicated in Figure 13.59. Notice that ϕ is the angle from the positive z-axis to the vector $\overrightarrow{OP}$ and θ is the angle from the positive x-axis to the vector $\overrightarrow{OR}$, where

R is the point lying in the xy-plane with rectangular coordinates $(x, y, 0)$ (i.e., R is the projection of P onto the xy-plane). You should observe from this description that

$$\rho \geq 0 \quad \text{and} \quad 0 \leq \phi \leq \pi.$$

If you look closely at Figure 13.59, you can see how to relate rectangular and spherical coordinates. Notice that

$$x = \|\overrightarrow{OR}\| \cos \theta = \|\overrightarrow{QP}\| \cos \theta.$$

Looking at the triangle OQP, we find that $\|\overrightarrow{QP}\| = \rho \sin \phi$, so that

$$x = \rho \sin \phi \cos \theta. \tag{7.2}$$

Similarly, we have

$$y = \|\overrightarrow{OR}\| \sin \theta = \rho \sin \phi \sin \theta. \tag{7.3}$$

Finally, focusing again on triangle OQP, we have

$$z = \rho \cos \phi. \tag{7.4}$$

Example 7.1   Converting from Spherical to Rectangular Coordinates

Find rectangular coordinates for the point described by $(8, \pi/4, \pi/3)$ in spherical coordinates.

Solution   We show a sketch of the point in Figure 13.60. From (7.2), (7.3) and (7.4), we have

$$x = 8 \sin \frac{\pi}{4} \cos \frac{\pi}{3} = 8 \left(\frac{\sqrt{2}}{2} \right) \left(\frac{1}{2} \right) = 2\sqrt{2},$$

$$y = 8 \sin \frac{\pi}{4} \sin \frac{\pi}{3} = 8 \left(\frac{\sqrt{2}}{2} \right) \left(\frac{\sqrt{3}}{2} \right) = 2\sqrt{6}$$

and

$$z = 8 \cos \frac{\pi}{4} = 8 \left(\frac{\sqrt{2}}{2} \right) = 4\sqrt{2}.$$

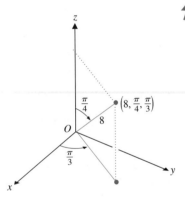

Figure 13.60
The point $\left(8, \frac{\pi}{4}, \frac{\pi}{3}\right)$.

It's often very helpful (especially when dealing with triple integrals) to represent common surfaces in spherical coordinates.

Example 7.2   Equation of a Cone in Spherical Coordinates

Rewrite the equation of the cone $z^2 = x^2 + y^2$ in spherical coordinates.

Solution   Using (7.2), (7.3) and (7.4), the equation of the cone becomes

$$\rho^2 \cos^2 \phi = \rho^2 \sin^2 \phi \cos^2 \theta + \rho^2 \sin^2 \phi \sin^2 \theta$$
$$= \rho^2 \sin^2 \phi (\cos^2 \theta + \sin^2 \theta)$$
$$= \rho^2 \sin^2 \phi.$$

Notice that in order to have $\rho^2 \cos^2 \phi = \rho^2 \sin^2 \phi$, we must either have $\rho = 0$ (this corresponds to the origin) **or** $\cos^2 \phi = \sin^2 \phi$. Notice that for the latter to occur, we must have $\phi = \frac{\pi}{4}$ or $\phi = \frac{3\pi}{4}$. (Recall that $0 \leq \phi \leq \pi$.) Observe that taking $\phi = \frac{\pi}{4}$ (and allowing ρ and θ to be anything) describes the top half of the cone, as shown in Figure 13.61a.

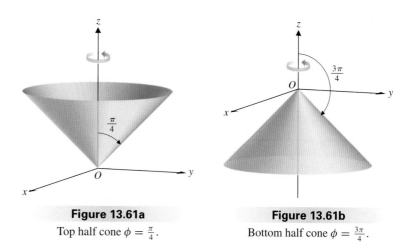

Figure 13.61a

Top half cone $\phi = \frac{\pi}{4}$.

Figure 13.61b

Bottom half cone $\phi = \frac{3\pi}{4}$.

You can think of this as taking a single ray (say in the yz-plane) with $\phi = \frac{\pi}{4}$ and revolving this around the z-axis. (This is the effect of letting θ run from 0 to 2π.) Similarly, $\phi = \frac{3\pi}{4}$ describes the bottom half cone, as seen in Figure 13.61b.

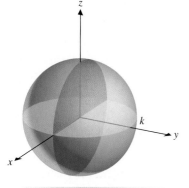

Figure 13.62a

The sphere $\rho = k$.

Notice that, in general, the equation $\rho = k$ (for any constant $k > 0$) represents the sphere of radius k centered at the origin (see Figure 13.62a). The equation $\theta = k$ (for any constant k) represents a vertical half-plane, with its edge along the z-axis (see Figure 13.62b). Further, the equation $\phi = k$ (for any constant k) represents the top half of a cone if $0 < k < \frac{\pi}{2}$ (see Figure 13.63a) and represents the bottom half of a cone if $\frac{\pi}{2} < k < \pi$ (see Figure 13.63b). Finally, note that $\phi = \frac{\pi}{2}$ describes the xy-plane. Can you think of what the equations $\phi = 0$ and $\phi = \pi$ represent?

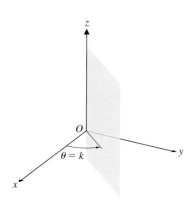

Figure 13.62b

The half plane $\theta = k$.

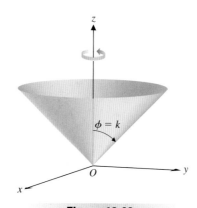

Figure 13.63a

Top half-cone $\phi = k$, where $0 < k < \frac{\pi}{2}$.

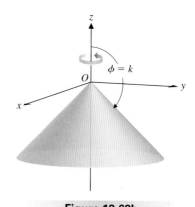

Figure 13.63b

Bottom half-cone $\phi = k$, where $\frac{\pi}{2} < k < \pi$.

Triple Integrals in Spherical Coordinates

Just as polar coordinates are indispensable in calculating double integrals over circular regions, especially when the integrand involves the particular combination of variables $x^2 + y^2$, spherical coordinates are an indispensable aid in dealing with triple integrals over spherical regions, particularly with those where the integrand involves the combination $x^2 + y^2 + z^2$. Integrals of this type are encountered frequently in applications. For instance, consider the triple integral

$$\iiint\limits_{Q} \cos(x^2 + y^2 + z^2)^{3/2} \, dV,$$

where Q is the **unit ball:** $x^2 + y^2 + z^2 \le 1$. No matter which order you choose for the integrations, you will arrive at a triple iterated integral that looks like

$$\int_{-1}^{1} \int_{-\sqrt{1-x^2}}^{\sqrt{1-x^2}} \int_{-\sqrt{1-x^2-y^2}}^{\sqrt{1-x^2-y^2}} \cos(x^2 + y^2 + z^2)^{3/2} \, dz \, dy \, dx.$$

In rectangular coordinates (or cylindrical coordinates, for that matter), you have little hope of calculating this integral exactly. In spherical coordinates, however, this integral is a snap. First, we need to see how to write triple integrals in spherical coordinates.

For the integral $\iiint\limits_{Q} f(\rho, \phi, \theta) \, dV$, we begin, as we have many times before, by constructing an inner partition of the solid Q. But, rather than slicing up Q using planes parallel to the three coordinate planes, we divide Q by slicing it with spheres of the form $\rho = \rho_k$, half-planes of the form $\theta = \theta_k$ and half-cones of the form $\phi = \phi_k$. Notice that instead of subdividing Q into a number of rectangular boxes, this divides Q into a number of spherical wedges of the form:

$$Q_k = \{(\rho, \phi, \theta) \mid \rho_{k-1} \le \rho \le \rho_k, \phi_{k-1} \le \phi \le \phi_k, \theta_{k-1} \le \theta \le \theta_k\},$$

as depicted in Figure 13.64. Here, we have $\Delta\rho_k = \rho_k - \rho_{k-1}$, $\Delta\phi_k = \phi_k - \phi_{k-1}$ and $\Delta\theta_k = \theta_k - \theta_{k-1}$. Notice that Q_k is nearly a rectangular box and so, its volume ΔV_k is approximately the same as that of a rectangular box with the same dimensions:

$$\Delta V_k \approx \Delta\rho_k (\rho_k \, \Delta\phi_k)(\rho_k \sin\phi_k \, \Delta\theta_k)$$
$$= \rho_k^2 \sin\phi_k \, \Delta\rho_k \, \Delta\phi_k \, \Delta\theta_k.$$

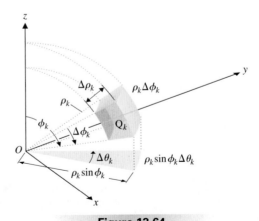

Figure 13.64
The spherical wedge Q_k.

We consider only those wedges that lie completely inside Q, to form an inner partition $Q_1, Q_2, \ldots, Q_n$ of the solid Q. Summing over the inner partition and letting the norm of the partition $\|P\|$ (here, the longest diagonal of any of the wedges in the inner partition) approach zero, we get

$$\iiint_Q f(\rho, \phi, \theta)\, dV = \lim_{\|P\| \to 0} \sum_{k=1}^{n} f(\rho_k, \phi_k, \theta_k)\, \Delta V_k$$

$$= \lim_{\|P\| \to 0} \sum_{k=1}^{n} f(\rho_k, \phi_k, \theta_k) \rho_k^2 \sin \phi_k\, \Delta \rho_k\, \Delta \phi_k\, \Delta \theta_k$$

$$= \iiint_Q f(\rho, \phi, \theta) \rho^2 \sin \phi\, d\rho\, d\phi\, d\theta, \qquad (7.5)$$

where the limits of integration for each of the three iterated integrals are found in much the same way as we have done a number of times before. From (7.5), notice that the volume element in spherical coordinates is given by

$$dV = \rho^2 \sin \phi\, d\rho\, d\phi\, d\theta.$$

We can now return to our introductory example.

Example 7.3 A Triple Integral in Spherical Coordinates

Evaluate the triple integral $\iiint_Q \cos(x^2 + y^2 + z^2)^{3/2}\, dV$, where Q is the unit ball: $x^2 + y^2 + z^2 \le 1$.

Solution Notice that since Q is the unit ball, ρ (the radial distance from the origin) ranges from 0 to 1. Further, the angle ϕ ranges from 0 to π (where $\phi = 0$ starts us at the top of the sphere, $\phi \in [0, \pi/2]$ corresponds to the top hemisphere and $\phi \in [\pi/2, \pi]$ corresponds to the bottom hemisphere). Finally (to get all the way around the sphere), the angle θ ranges from 0 to 2π. From (7.5), we have that since $x^2 + y^2 + z^2 = \rho^2$,

$$\iiint_Q \cos \underbrace{(x^2 + y^2 + z^2)}_{\rho^2}{}^{3/2} \underbrace{dV}_{\rho^2 \sin \phi\, d\rho\, d\phi\, d\theta}$$

$$= \int_0^{2\pi} \int_0^{\pi} \int_0^1 \cos(\rho^2)^{3/2} \rho^2 \sin \phi\, d\rho\, d\phi\, d\theta$$

$$= \frac{1}{3} \int_0^{2\pi} \int_0^{\pi} \int_0^1 \cos(\rho^3)(3\rho^2) \sin \phi\, d\rho\, d\phi\, d\theta$$

$$= \frac{1}{3} \int_0^{2\pi} \int_0^{\pi} \sin(\rho^3) \Big|_{\rho=0}^{\rho=1} \sin \phi\, d\phi\, d\theta$$

$$= \frac{\sin 1}{3} \int_0^{2\pi} \int_0^{\pi} \sin \phi\, d\phi\, d\theta$$

$$= -\frac{\sin 1}{3} \int_0^{2\pi} \cos \phi \Big|_{\phi=0}^{\phi=\pi} d\theta$$

$$= -\frac{\sin 1}{3} \int_0^{2\pi} (\cos \pi - \cos 0)\, d\theta$$

$$= \frac{2}{3}(\sin 1)(2\pi) \approx 3.525.$$

Generally, spherical coordinates are useful in triple integrals when the solid over which you are integrating is in some way spherical and particularly when the integrand contains the term $x^2 + y^2 + z^2$. In the following example, we use spherical coordinates to simplify the calculation of a volume.

Example 7.4 Finding a Volume Using Spherical Coordinates

Find the volume lying inside the sphere $x^2 + y^2 + z^2 = 2z$ and inside the cone $z^2 = x^2 + y^2$.

Solution Notice that by completing the square in the equation of the sphere, we get

$$x^2 + y^2 + (z^2 - 2z + 1) = 1$$

or

$$x^2 + y^2 + (z - 1)^2 = 1.$$

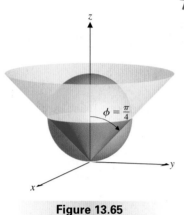

Figure 13.65
The cone $\phi = \frac{\pi}{4}$ and the sphere $\rho = 2\cos\phi$.

You should recognize this as a sphere of radius 1, centered at the point $(0, 0, 1)$. Notice, too that since the sphere sits completely above the xy-plane, only the top half of the cone, $z = \sqrt{x^2 + y^2}$ intersects the sphere. See Figure 13.65 for a sketch of the solid. We suggest that you try to find the volume in question using rectangular coordinates (but, don't waste too much time on it). Because of the spherical geometry, we consider the problem in spherical coordinates (keep in mind that cones have a very simple representation in spherical coordinates). From (7.1) and (7.4) and the original equation of the sphere, we get

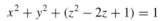

$$\underbrace{x^2 + y^2 + z^2}_{\rho^2} = 2 \underbrace{z}_{\rho\cos\phi}$$

or

$$\rho^2 = 2\rho\cos\phi.$$

This equation is satisfied when $\rho = 0$ (corresponding to the origin) or when $\rho = 2\cos\phi$ (the equation of the sphere in spherical coordinates). For the top half of the cone, we have $z = \sqrt{x^2 + y^2}$, or in spherical coordinates $\phi = \frac{\pi}{4}$, as discussed in example 7.2.

Referring again to Figure 13.65, notice that to stay inside the cone and inside the sphere, we have that for each fixed ϕ and θ, ρ can range from 0 up to $2\cos\phi$. For each fixed θ, to stay inside the cone, ϕ must range from 0 to $\frac{\pi}{4}$. Finally, to get all the way around the solid, θ ranges from 0 to 2π. The volume of the solid is then given by

$$V = \iiint_Q \underbrace{dV}_{\rho^2 \sin\phi \, d\rho \, d\phi \, d\theta}$$

$$= \int_0^{2\pi} \int_0^{\pi/4} \int_0^{2\cos\phi} \rho^2 \sin\phi \, d\rho \, d\phi \, d\theta$$

$$= \int_0^{2\pi} \int_0^{\pi/4} \frac{1}{3}\rho^3 \Big|_{\rho=0}^{\rho=2\cos\phi} \sin\phi \, d\phi \, d\theta$$

$$= \frac{8}{3} \int_0^{2\pi} \int_0^{\pi/4} \cos^3\phi \sin\phi \, d\phi \, d\theta$$

$$= -\frac{8}{3} \int_0^{2\pi} \frac{\cos^4\phi}{4} \Big|_{\phi=0}^{\phi=\pi/4} d\theta = -\frac{8}{12} \int_0^{2\pi} \left(\cos^4\frac{\pi}{4} - 1\right) d\theta$$

$$= -\frac{16\pi}{12}\left(\cos^4\frac{\pi}{4} - 1\right) = -\frac{4\pi}{3}\left(\frac{1}{4} - 1\right) = \pi.$$

Example 7.5

Changing an Integral from Rectangular to Spherical Coordinates

Evaluate the triple iterated integral $\int_{-2}^{2} \int_{0}^{\sqrt{4-x^2}} \int_{0}^{\sqrt{4-x^2-y^2}} (x^2 + y^2 + z^2) \, dz \, dy \, dx$.

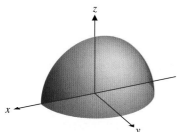

Figure 13.66
The solid Q.

Solution   At first glance, this integral might not seem too difficult, since the integrand is simply a polynomial, but look more closely. The limits of integration make the second and third integrations very messy. There are several things to notice here. First, the integrand contains the combination of variables $x^2 + y^2 + z^2$, which equals ρ^2 in spherical coordinates. Second, the solid over which we are integrating is a portion of a sphere, as follows. Notice that for each x in the interval $[-2, 2]$ indicated by the outermost limits of integration, y varies from 0 (corresponding to the x-axis) to $y = \sqrt{4 - x^2}$ (the top semicircle of radius 2 centered at the origin). Finally, z varies from 0 (corresponding to the xy-plane) up to $z = \sqrt{4 - x^2 - y^2}$ (the top hemisphere of radius 2 centered at the origin). The solid Q over which we are integrating is then the half of the hemisphere that lies above the first and second quadrants of the xy-plane, as illustrated in Figure 13.66. Notice that in spherical coordinates, this portion of the sphere is obtained if we let ρ range from 0 up to 2, ϕ range from 0 up to $\frac{\pi}{2}$ and θ range from 0 to π. The integral then becomes

$$\int_{-2}^{2} \int_{0}^{\sqrt{4-x^2}} \int_{0}^{\sqrt{4-x^2-y^2}} (x^2 + y^2 + z^2) \, dz \, dy \, dx$$

$$= \iiint_{Q} \underbrace{(x^2 + y^2 + z^2)}_{\rho^2} \underbrace{dV}_{\rho^2 \sin\phi \, d\rho \, d\phi \, d\theta}$$

$$= \int_{0}^{\pi} \int_{0}^{\pi/2} \int_{0}^{2} \rho^2 (\rho^2 \sin\phi) \, d\rho \, d\phi \, d\theta$$

$$= \frac{32}{5}\pi,$$

where we leave the details of this relatively simple integration to you.

EXERCISES 13.7

1. Discuss the relationship between the spherical coordinates angles ϕ and θ and the longitude and latitude angles on a map of the earth. Satellites in geosynchronous orbit remain at a constant distance above a fixed point on the earth. Discuss how spherical coordinates could be used to represent the position of the satellite.

2. Explain why any point in $\mathbb{R}^3$ can be represented in spherical coordinates with $\rho \geq 0$, $0 \leq \theta \leq 2\pi$ and $0 \leq \phi \leq \pi$. In particular, explain why it is not necessary to allow $\rho < 0$ or $\pi < \phi \leq 2\pi$. Discuss whether the ranges

$\rho \geq 0$, $0 \leq \theta \leq \pi$ and $0 \leq \phi \leq 2\pi$ would suffice to describe all points.

3. Explain why in spherical coordinates the equation $\theta = k$ represents a half-plane (see Figure 13.62b) and not a whole plane.

4. Using the examples in this section as a guide, make a short list of surfaces that are simple to describe in spherical coordinates.

In exercises 5–12, convert the spherical point (ρ, ϕ, θ) into rectangular coordinates.

5. $(4, 0, \pi)$

6. $\left(4, \frac{\pi}{2}, \pi\right)$

7. $\left(4, \frac{\pi}{2}, 0\right)$

8. $\left(4, \pi, \frac{\pi}{2}\right)$

9. $\left(2, \frac{\pi}{4}, 0\right)$

10. $\left(2, \frac{\pi}{4}, \frac{2\pi}{3}\right)$

11. $\left(\sqrt{2}, \frac{\pi}{4}, \frac{\pi}{4}\right)$

12. $\left(\sqrt{2}, \frac{\pi}{4}, \frac{\pi}{3}\right)$

In exercises 13–20, convert the equation into spherical coordinates.

13. $x^2 + y^2 + z^2 = 9$

14. $x^2 + y^2 + z^2 = 6$

15. $y = x$

16. $z = 0$

17. $z = 2$

18. $x^2 + y^2 + (z - 1)^2 = 1$

19. $z = \sqrt{3(x^2 + y^2)}$

20. $z = -\sqrt{x^2 + y^2}$

In exercises 21–28, sketch the graph of the spherical equation.

21. $\rho = 2$

22. $\rho = 4$

23. $\phi = \frac{\pi}{4}$

24. $\phi = \frac{\pi}{2}$

25. $\theta = 0$

26. $\theta = \frac{\pi}{4}$

27. $\phi = \frac{3\pi}{4}$

28. $\theta = \frac{3\pi}{4}$

In exercises 29–36, sketch the region defined by the given ranges.

29. $0 \le \rho \le 4, 0 \le \phi \le \frac{\pi}{4}, 0 \le \theta \le \pi$

30. $0 \le \rho \le 4, 0 \le \phi \le \frac{\pi}{2}, 0 \le \theta \le 2\pi$

31. $0 \le \rho \le 2, 0 \le \phi \le \frac{\pi}{2}, 0 \le \theta \le \pi$

32. $0 \le \rho \le 2, 0 \le \phi \le \frac{\pi}{4}, \pi \le \theta \le 2\pi$

33. $0 \le \rho \le 3, 0 \le \phi \le \pi, 0 \le \theta \le \pi$

34. $0 \le \rho \le 3, 0 \le \phi \le \frac{3\pi}{4}, 0 \le \theta \le 2\pi$

35. $2 \le \rho \le 3, 0 \le \phi \le \pi, 0 \le \theta \le 2\pi$

36. $2 \le \rho \le 3, 0 \le \phi \le \frac{\pi}{2}, 0 \le \theta \le 2\pi$

In exercises 37–46, set up and evaluate the indicated triple integral in an appropriate coordinate system.

37. $\iiint_Q e^{(x^2+y^2+z^2)^{3/2}} dV$, where Q is bounded by the hemisphere $z = \sqrt{4 - x^2 - y^2}$ and the xy-plane.

38. $\iiint_Q \sqrt{x^2 + y^2 + z^2} \, dV$, where Q is bounded by the hemisphere $z = \sqrt{9 - x^2 - y^2}$ and the xy-plane.

39. $\iiint_Q (x^2 + y^2 + z^2)^{5/2} dV$, where Q is bounded by $x^2 + y^2 + z^2 = 2$, $z \ge 0$ and the xy-plane.

40. $\iiint_Q e^{\sqrt{x^2+y^2+z^2}} dV$, where Q is bounded by $x^2 + y^2 + z^2 = 2$, $z \ge 0$ and the xy-plane.

41. $\iiint_Q (x^2 + y^2 + z^2) \, dV$, where Q is the cube with $0 \le x \le 1$, $1 \le y \le 2$ and $3 \le z \le 4$.

42. $\iiint_Q (x + y + z) \, dV$, where Q is the tetrahedron bounded by $x + 2y + z = 4$ and the coordinate planes.

43. $\iiint_Q (x^2 + y^2) \, dV$, where Q is bounded by $z = 4 - x^2 - y^2$ and the xy-plane.

44. $\iiint_Q e^{x^2+y^2} dV$, where Q is bounded by $x^2 + y^2 = 4$, $z = 0$ and $z = 2$.

45. $\iiint_Q \sqrt{x^2 + y^2 + z^2} \, dV$, where Q is bounded by $z = \sqrt{x^2 + y^2}$ and $z = \sqrt{2 - x^2 - y^2}$.

46. $\iiint_Q (x^2 + y^2 + z^2)^{3/2} dV$, where Q is the region inside both $z = \sqrt{x^2 + y^2}$ and $z = \sqrt{8 - x^2 - y^2}$.

In exercises 47–56, use an appropriate coordinate system to find the volume of the given solid.

47. The region below $x^2 + y^2 + z^2 = 4z$ and above $z = \sqrt{x^2 + y^2}$

48. The region above $z = \sqrt{x^2 + y^2}$ and below $x^2 + y^2 + z^2 = 4$

49. The region inside $z = \sqrt{x^2 + y^2}$ and below $z = 4$

50. The region inside $z = \sqrt{x^2 + y^2}$ and below $z = 6$

51. The region under $z = \sqrt{x^2 + y^2}$ and above the square $-1 \le x \le 1, -1 \le y \le 1$

52. The region bounded by $x + 2y + z = 4$ and the coordinate planes

53. The region below $x^2 + y^2 + z^2 = 4$, above $z = \sqrt{x^2 + y^2}$ in the first octant

54. The region below $x^2 + y^2 + z^2 = 4$, above $z = \sqrt{x^2 + y^2}$, between $y = x$ and $x = 0$ with $y \geq 0$

55. The region below $z = \sqrt{x^2 + y^2}$, above the xy-plane and inside $x^2 + y^2 = 4$

56. The region between $z = 4 - x^2 - y^2$ and the xy-plane

In exercises 57–60, evaluate the iterated integral by changing coordinate systems.

57. $\displaystyle\int_0^1 \int_{-\sqrt{1-x^2}}^{\sqrt{1-x^2}} \int_{-\sqrt{1-x^2-y^2}}^{\sqrt{1-x^2-y^2}} \sqrt{x^2+y^2+z^2}\, dz\, dy\, dx$

58. $\displaystyle\int_{-1}^1 \int_{-\sqrt{1-x^2}}^{\sqrt{1-x^2}} \int_{1-\sqrt{1-x^2-y^2}}^{1+\sqrt{1-x^2-y^2}} (x^2+y^2+z^2)^{3/2}\, dz\, dy\, dx$

59. $\displaystyle\int_{-2}^2 \int_0^{\sqrt{4-x^2}} \int_{\sqrt{x^2+y^2}}^{\sqrt{8-x^2-y^2}} (x^2+y^2+z^2)^{3/2}\, dz\, dy\, dx$

60. $\displaystyle\int_0^4 \int_0^{\sqrt{16-x^2}} \int_{\sqrt{x^2+y^2}}^4 \sqrt{x^2+y^2+z^2}\, dz\, dy\, dx$

61. Find the center of mass of the solid with constant density and bounded by $z = \sqrt{x^2 + y^2}$ and $z = \sqrt{4 - x^2 - y^2}$.

62. Find the center of mass of the solid with constant density in the first quadrant and bounded by $z = \sqrt{x^2 + y^2}$ and $z = \sqrt{4 - x^2 - y^2}$.

63. If you have a graphing utility that will graph surfaces of the form $\rho = f(\phi, \theta)$, graph $\rho = 2\phi$ and $\rho = \left(\phi - \frac{\pi}{2}\right)^2$. Discuss the symmetry that results from the variable θ not appearing in the equation. Discuss the changes in ρ as you move down from $\phi = 0$ to $\phi = \pi$. Using what you have learned, try graphing the following by hand, then compare your sketches to those of your graphing utility: (a) $\rho = e^{-\phi}$, (b) $\rho = e^{\phi}$, (c) $\rho = \sin^2 \phi$ and (d) $\rho = \sin^2\left(\phi - \frac{\pi}{2}\right)$.

64. Use a graphing utility to graph $\rho = 5\cos\theta$ and $\rho = \sqrt{\cos\theta}$. Discuss the symmetry that results from the variable ϕ not appearing in the equation. Discuss the changes in ρ as you move around from $\theta = 0$ to $\theta = 2\pi$. Using what you have learned, try graphing the following by hand, then compare your sketches to those of your graphing utility: (a) $\rho = \sin^2 \theta$, (b) $\rho = \sin^2\left(\theta - \frac{\pi}{2}\right)$, (c) $\rho = e^{\theta}$ and (d) $\rho = e^{-\theta}$.

65. Use a graphing utility to graph each of the following. Adjust the graphing window as needed to get a good idea of what the graph looks like. (a) $\rho = \sin(\phi + \theta)$, (b) $\rho = \phi \sin\theta$, (c) $\rho = \sin^2\theta\cos\phi$, (d) $\rho = 4\cos^2\theta + 2\sin\phi - 3\sin\theta$ and (e) $\rho = 4\cos\theta\sin 5\phi + 3\cos^2\phi$. There are innumerable interesting and unusual graphs in spherical coordinates. Experiment and find your own!

13.8   CHANGE OF VARIABLES IN MULTIPLE INTEGRALS

Recall that one of our most basic tools for evaluating a single definite integral is substitution. For instance, in order to evaluate the integral $\int_0^2 2xe^{x^2+3}\, dx$, you would make the substitution $u = x^2 + 3$. By now, you can probably do this in your head, but let's consider the details one more time. When we make such a substitution, we also need to observe that $du = 2x\, dx$. But, there's one more thing: when you change variables in a definite integral, you must also change the limits of integration to suit the new variable. In this case, when $x = 0$, we have $u = 0^2 + 3 = 3$ and when $x = 2$, $u = 2^2 + 3 = 7$. This leaves us with

$$\int_0^2 2xe^{x^2+3}\, dx = \int_0^2 \underbrace{e^{x^2+3}}_{e^u}\, \underbrace{(2x)\, dx}_{du}$$

$$= \int_3^7 e^u\, du = e^u \Big|_3^7 = e^7 - e^3.$$

The primary reason for making the preceding change of variables was to simplify the integrand, so that it was easier to find an antiderivative. Notice too that we not only transformed the integrand, but we also changed the interval over which we were integrating.

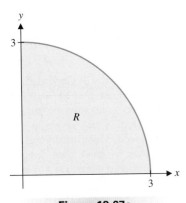

Figure 13.67a
The region of integration in the *xy*-plane.

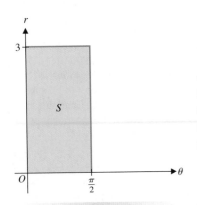

Figure 13.67b
The region of integration in the *rθ*-plane.

You might recognize that we have already implemented changes of variables in multiple integrals in the very special cases of polar coordinates (for double integrals) and cylindrical and spherical coordinates (for triple integrals). There were several reasons for doing this. In the case of double integrals in rectangular coordinates, if the integrand contains the term $x^2 + y^2$ or the region over which you are integrating is in some way circular, then polar coordinates may be indicated. For instance, consider the iterated integral

$$\int_0^3 \int_0^{\sqrt{9-x^2}} \cos(x^2 + y^2)\, dy\, dx.$$

There is really no way to evaluate this integral as it is written in rectangular coordinates. (Try it!) However, you should quickly recognize that the region of integration R is the portion of the circle of radius 3 centered at the origin that lies in the first quadrant (see Figure 13.67a). Since the integrand also includes the term $x^2 + y^2$, it's a good bet that polar coordinates will help. In fact, we have

$$\int_0^3 \int_0^{\sqrt{9-x^2}} \cos(x^2 + y^2)\, dy\, dx = \iint_R \cos(\underbrace{x^2 + y^2}_{r^2})\, \underbrace{dA}_{r\,dr\,d\theta}$$

$$= \int_0^{\pi/2} \int_0^3 \cos(r^2)\, r\, dr\, d\theta,$$

which is now an easy integral to evaluate. Notice that there are two things that happened here. First, we simplified the integrand (into one with a known antiderivative) and second, we transformed the region over which we integrated, as follows. In the *xy*-plane, we integrated over the circular sector indicated in Figure 13.67a. In the *rθ*-plane, we integrated over a (simpler) region: the rectangle S defined by $S = \{(r, \theta)\, |\, 0 \le r \le 3 \text{ and } 0 \le \theta \le \frac{\pi}{2}\}$, as indicated in Figure 13.67b.

Recall that we changed to cylindrical or spherical coordinates in triple integrals for similar reasons. In each case, we ended up simplifying the integrand and transforming the region of integration. The question we face now is: can this be done in general and if so, how can we change variables in a multiple integral? Before we answer this question, we must first explore the concept of transformation in several variables.

A **transformation** T from the *uv*-plane to the *xy*-plane is a function that maps points in the *uv*-plane to points in the *xy*-plane, so that

$$T(u, v) = (x, y),$$

where

$$x = g(u, v) \quad \text{and} \quad y = h(u, v),$$

for some functions g and h. We consider changes of variables in double integrals as defined by a transformation T from a region S in the *uv*-plane onto a region R in the *xy*-plane (see Figure 13.68). We refer to R as the **image** of S under the transformation T. We say that T is **one-to-one** if for every point (x, y) in R there is exactly one point (u, v) in S such that $T(u, v) = (x, y)$. Notice that this says that (at least in principle), we can solve for u and v in terms of x and y. Further, we only consider transformations for which g and h have continuous first partial derivatives in the region S.

The primary reason for introducing a change of variables in a multiple integral is to simplify the calculation of the integral. This is accomplished by simplifying the integrand, the region over which you are integrating or both. Before exploring the effect of a transformation on a multiple integral, we examine several examples of how a transformation can simplify a region in two dimensions.

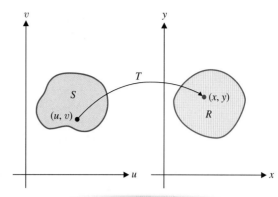

Figure 13.68
The transformation T mapping S
onto R.

Example 8.1 The Transformation of a Simple Region

Let R be the region bounded by the straight lines $y = 2x + 3$, $y = 2x + 1$, $y = 5 - x$ and $y = 2 - x$. Find a transformation T mapping a region S in the uv-plane onto R, where S is a rectangular region, with sides parallel to the u and v-axes.

Solution First, notice that the region R is a parallelogram in the xy-plane (see Figure 13.69a). We can rewrite the equations for the lines forming the boundaries of R as $y - 2x = 3$, $y - 2x = 1$, $y + x = 5$ and $y + x = 2$. This suggests the change of variables

$$u = y - 2x \quad \text{and} \quad v = y + x. \tag{8.1}$$

Observe that the lines forming the boundaries of R then correspond to the lines $u = 3$, $u = 1$, $v = 5$ and $v = 2$, respectively, forming the boundaries of the corresponding region S in the uv-plane (see Figure 13.69b). Solving equations (8.1) for x and y, we have the transformation T defined by

$$x = \frac{1}{3}(v - u) \quad \text{and} \quad y = \frac{1}{3}(2v + u).$$

You should note that the transformation maps the four corners of the rectangle S to the vertices of the parallelogram R, as follows:

$$T(1, 2) = \left(\frac{1}{3}(2 - 1), \frac{1}{3}[2(2) + 1] \right) = \left(\frac{1}{3}, \frac{5}{3} \right),$$

$$T(3, 2) = \left(\frac{1}{3}(2 - 3), \frac{1}{3}[2(2) + 3] \right) = \left(-\frac{1}{3}, \frac{7}{3} \right),$$

$$T(1, 5) = \left(\frac{1}{3}(5 - 1), \frac{1}{3}[2(5) + 1] \right) = \left(\frac{4}{3}, \frac{11}{3} \right)$$

and

$$T(3, 5) = \left(\frac{1}{3}(5 - 3), \frac{1}{3}[2(5) + 3] \right) = \left(\frac{2}{3}, \frac{13}{3} \right).$$

We leave it as an exercise to verify that the above four points are indeed the vertices of the parallelogram R. (To do this, simply solve the system of equations for the points of intersection.)

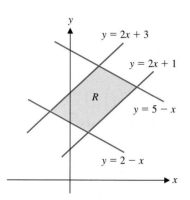

Figure 13.69a
The region R in the xy-plane.

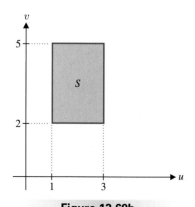

Figure 13.69b
The region S in the uv-plane.

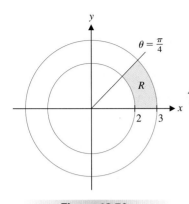

Figure 13.70a

The region R in the xy-plane.

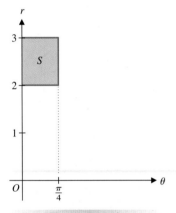

Figure 13.70b

The region S in the $r\theta$-plane.

In the following example, we see how polar coordinates can be used to transform a rectangle in the $r\theta$-plane into a sector of a circular annulus.

Example 8.2   A Transformation Involving Polar Coordinates

Let R be the region inside the circle $x^2 + y^2 = 9$ and outside the circle $x^2 + y^2 = 4$ and lying in the first quadrant between the lines $y = 0$ and $y = x$. Find a transformation T from a rectangular region S in the $r\theta$-plane to the region R.

Solution   First, we picture the region R in Figure 13.70a. Observe that this is a sector of a circular annulus. The obvious transformation is accomplished with polar coordinates. We let $x = r \cos \theta$ and $y = r \sin \theta$, so that $x^2 + y^2 = r^2$. The inner and outer circles forming a portion of the boundary of R then correspond to $r = 2$ and $r = 3$, respectively. Further, the line $y = x$ corresponds to the line $\theta = \frac{\pi}{4}$ and the line $y = 0$ corresponds to the line $\theta = 0$. We show the region S in Figure 13.70b.

Now that we have introduced transformations in two dimensions, we can consider our primary goal for this section: to determine how a change of variables in a multiple integral will affect the integral. We consider the double integral

$$\iint_R f(x, y) \, dA,$$

where f is continuous on R. Further, we assume that R is the image of a region S in the uv-plane under the one-to-one transformation T, so that for every point (x, y) in R, there is exactly one point (u, v) in S for which $T(u, v) = (x, y)$. Recall that we originally constructed the double integral by forming an inner partition of R and taking a limit of the corresponding Riemann sums. We now consider an inner partition of the region S in the uv-plane, consisting of the n rectangles $S_1, S_2, \ldots, S_n$, as depicted in Figure 13.71a. We denote the lower left corner of each rectangle S_i by (u_i, v_i) $(i = 1, 2, \ldots, n)$ and take all of the rectangles to have the same dimensions Δu by Δv, as indicated in Figure 13.71b. Let $R_1, R_2, \ldots, R_n$ be the images of $S_1, S_2, \ldots, S_n$, respectively, under the transformation T and let the points $(x_1, y_1), (x_2, y_2), \ldots, (x_n, y_n)$ be the images of $(u_1, v_1), (u_2, v_2), \ldots, (u_n, v_n)$,

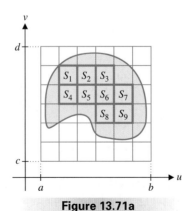

Figure 13.71a

An inner partition of the region S in the uv-plane.

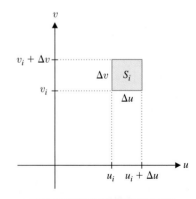

Figure 13.71b

The rectangle S_i.

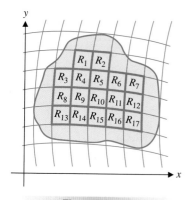

Figure 13.72
Curvilinear inner partition of the region R in the xy-plane.

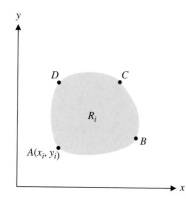

Figure 13.73a
The region R_i.

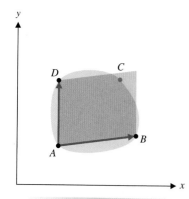

Figure 13.73b
The parallelogram determined by the vectors $\overrightarrow{AB}$ and $\overrightarrow{AD}$.

respectively. Notice that $R_1, R_2, \ldots, R_n$ will then form an inner partition of the region R in the xy-plane (although it will not generally consist of rectangles), as indicated in Figure 13.72. In particular, the image of the rectangle S_i under T is the curvilinear region R_i. From our development of the double integral, we know that

$$\iint\limits_R f(x, y)\, dA \approx \sum_{i=1}^{n} f(x_i, y_i)\, \Delta A_i, \tag{8.2}$$

where ΔA_i is the area of R_i, for $i = 1, 2, \ldots, n$. The only problem with this approximation is that we don't know how to find ΔA_i, since the regions R_i are not generally rectangles. We can, however, find a reasonable approximation, as follows.

Notice that T maps the four corners of S_i: (u_i, v_i), $(u_i + \Delta u, v_i)$, $(u_i + \Delta u, v_i + \Delta v)$ and $(u_i, v_i + \Delta v)$ to four points denoted A, B, C and D, respectively, on the boundary of R_i, as indicated below:

$$(u_i, v_i) \xrightarrow{T} A(g(u_i, v_i), h(u_i, v_i)) = A(x_i, y_i),$$

$$(u_i + \Delta u, v_i) \xrightarrow{T} B(g(u_i + \Delta u, v_i), h(u_i + \Delta u, v_i)),$$

$$(u_i + \Delta u, v_i + \Delta v) \xrightarrow{T} C(g(u_i + \Delta u, v_i + \Delta v), h(u_i + \Delta u, v_i + \Delta v))$$

and

$$(u_i, v_i + \Delta v) \xrightarrow{T} D(g(u_i, v_i + \Delta v), h(u_i, v_i + \Delta v)).$$

We indicate these four points and a typical curvilinear region R_i in Figure 13.73a. Notice that as long as Δu and Δv are small, we can approximate the area of R_i by the area of the parallelogram determined by the vectors $\overrightarrow{AB}$ and $\overrightarrow{AD}$, as indicated in Figure 13.73b. If we consider $\overrightarrow{AB}$ and $\overrightarrow{AD}$ as three-dimensional vectors (with zero $\mathbf{k}$ components), recall from our discussion in section 10.4 that the area of the parallelogram is simply $\|\overrightarrow{AB} \times \overrightarrow{AD}\|$. We will take this as an approximation of the area ΔA_i. First, notice that

$$\overrightarrow{AB} = \langle g(u_i + \Delta u, v_i) - g(u_i, v_i), h(u_i + \Delta u, v_i) - h(u_i, v_i)\rangle \tag{8.3}$$

and

$$\overrightarrow{AD} = \langle g(u_i, v_i + \Delta v) - g(u_i, v_i), h(u_i, v_i + \Delta v) - h(u_i, v_i)\rangle. \tag{8.4}$$

From the definition of partial derivative, we have

$$g_u(u_i, v_i) = \lim_{\Delta u \to 0} \frac{g(u_i + \Delta u, v_i) - g(u_i, v_i)}{\Delta u}.$$

This tells us that for Δu small,

$$g(u_i + \Delta u, v_i) - g(u_i, v_i) \approx g_u(u_i, v_i)\, \Delta u.$$

Likewise, we have

$$h(u_i + \Delta u, v_i) - h(u_i, v_i) \approx h_u(u_i, v_i)\, \Delta u.$$

Similarly, for Δv small, we have

$$g(u_i, v_i + \Delta v) - g(u_i, v_i) \approx g_v(u_i, v_i)\, \Delta v$$

and

$$h(u_i, v_i + \Delta v) - h(u_i, v_i) \approx h_v(u_i, v_i)\, \Delta v.$$

Together with (8.3) and (8.4), these give us

$$\overrightarrow{AB} \approx \langle g_u(u_i, v_i)\, \Delta u, h_u(u_i, v_i)\, \Delta u\rangle = \Delta u \langle g_u(u_i, v_i), h_u(u_i, v_i)\rangle$$

and

$$\overrightarrow{AD} \approx \langle g_v(u_i, v_i)\, \Delta v, h_v(u_i, v_i)\, \Delta v \rangle = \Delta v \langle g_v(u_i, v_i), h_v(u_i, v_i) \rangle.$$

An approximation of the area of R_i is then given by

$$\Delta A_i \approx \|\overrightarrow{AB} \times \overrightarrow{AD}\|, \tag{8.5}$$

where

$$\overrightarrow{AB} \times \overrightarrow{AD} \approx \begin{vmatrix} \mathbf{i} & \mathbf{j} & \mathbf{k} \\ \Delta u g_u(u_i, v_i) & \Delta u h_u(u_i, v_i) & 0 \\ \Delta v g_v(u_i, v_i) & \Delta v h_v(u_i, v_i) & 0 \end{vmatrix}$$

$$= \begin{vmatrix} g_u(u_i, v_i) & h_u(u_i, v_i) \\ g_v(u_i, v_i) & h_v(u_i, v_i) \end{vmatrix} \Delta u\, \Delta v \mathbf{k}. \tag{8.6}$$

For simplicity, we write the determinant as

$$\begin{vmatrix} g_u(u_i, v_i) & h_u(u_i, v_i) \\ g_v(u_i, v_i) & h_v(u_i, v_i) \end{vmatrix} = \begin{vmatrix} g_u(u_i, v_i) & g_v(u_i, v_i) \\ h_u(u_i, v_i) & h_v(u_i, v_i) \end{vmatrix} = \begin{vmatrix} \dfrac{\partial x}{\partial u} & \dfrac{\partial x}{\partial v} \\[2mm] \dfrac{\partial y}{\partial u} & \dfrac{\partial y}{\partial v} \end{vmatrix} (u_i, v_i).$$

We give this determinant a name and introduce some new notation in the following definition.

Definition 8.1

The determinant $\begin{vmatrix} \dfrac{\partial x}{\partial u} & \dfrac{\partial x}{\partial v} \\[2mm] \dfrac{\partial y}{\partial u} & \dfrac{\partial y}{\partial v} \end{vmatrix}$ is referred to as the **Jacobian** of the transformation T and is written using the notation $\dfrac{\partial(x, y)}{\partial(u, v)}$.

From (8.5) and (8.6), we now have (since $\mathbf{k}$ is a unit vector)

$$\Delta A_i \approx \|\overrightarrow{AB} \times \overrightarrow{AD}\| = \left| \frac{\partial(x, y)}{\partial(u, v)} \right| \Delta u\, \Delta v,$$

where the determinant is evaluated at the point (u_i, v_i). From (8.2), we now have

$$\iint\limits_R f(x, y)\, dA \approx \sum_{i=1}^n f(x_i, y_i)\, \Delta A_i \approx \sum_{i=1}^n f(x_i, y_i) \left| \frac{\partial(x, y)}{\partial(u, v)} \right| \Delta u\, \Delta v$$

$$= \sum_{i=1}^n f(g(u_i, v_i), h(u_i, v_i)) \left| \frac{\partial(x, y)}{\partial(u, v)} \right| \Delta u\, \Delta v.$$

You should recognize this last expression as a Riemann sum for the double integral

$$\iint\limits_S f(g(u, v), h(u, v)) \left| \frac{\partial(x, y)}{\partial(u, v)} \right| du\, dv.$$

The preceding analysis is a sketch of the more extensive proof of the following theorem.

Theorem 8.1 (Change of Variables in Double Integrals)

Suppose that the region S in the uv-plane is mapped onto the region R in the xy-plane by the one-to-one transformation T defined by $x = g(u, v)$ and $y = h(u, v)$, where g and h have continuous first partial derivatives on S. If f is continuous on R and the Jacobian $\dfrac{\partial(x, y)}{\partial(u, v)}$ is nonzero on S, then

$$\iint_R f(x, y)\, dA = \iint_S f(g(u, v), h(u, v)) \left| \frac{\partial(x, y)}{\partial(u, v)} \right| du\, dv.$$

We first observe that the change of variables to polar coordinates in a double integral is just a special case of Theorem 8.1.

Example 8.3 Changing Variables to Polar Coordinates

Use Theorem 8.1 to derive the evaluation formula for polar coordinates:

$$\iint_R f(x, y)\, dA = \iint_S f(r \cos \theta, r \sin \theta)\, r\, dr\, d\theta.$$

Solution First, recognize that a change of variables to polar coordinates consists of the transformation from the $r\theta$-plane to the xy-plane defined by $x = r \cos \theta$ and $y = r \sin \theta$. This gives us the Jacobian

$$\frac{\partial(x, y)}{\partial(r, \theta)} = \begin{vmatrix} \dfrac{\partial x}{\partial r} & \dfrac{\partial x}{\partial \theta} \\ \dfrac{\partial y}{\partial r} & \dfrac{\partial y}{\partial \theta} \end{vmatrix} = \begin{vmatrix} \cos \theta & -r \sin \theta \\ \sin \theta & r \cos \theta \end{vmatrix} = r \cos^2 \theta + r \sin^2 \theta = r.$$

By Theorem 8.1, we now have the familiar formula

$$\iint_R f(x, y)\, dA = \iint_S f(r \cos \theta, r \sin \theta) \left| \frac{\partial(x, y)}{\partial(r, \theta)} \right| dr\, d\theta$$

$$= \iint_S f(r \cos \theta, r \sin \theta)\, r\, dr\, d\theta.$$

In the following example, we show how a change of variables can be used to simplify the region of integration (thereby also simplifying the integral).

Example 8.4 Changing Variables to Transform a Region

Evaluate the integral $\iint_R (x^2 + 2xy)\, dA$, where R is the region bounded by the lines $y = 2x + 3$, $y = 2x + 1$, $y = 5 - x$ and $y = 2 - x$.

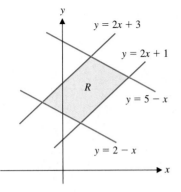

Figure 13.74
The region R.

Solution The difficulty in evaluating this integral is that the region of integration (see Figure 13.74) requires us to break the integral into three pieces. (Think about this some!) An alternative is to find a change of variables corresponding to a transformation from a rectangle in the uv-plane to R in the xy-plane. Recall that we did just this in example 8.1. There, we had found that the change of variables

$$x = \frac{1}{3}(v - u) \quad \text{and} \quad y = \frac{1}{3}(2v + u)$$

mapped the rectangle $S = \{(u, v) \mid 1 \leq u \leq 3 \text{ and } 2 \leq v \leq 5\}$ to R. Notice that the Jacobian of this transformation is

$$\frac{\partial(x, y)}{\partial(u, v)} = \begin{vmatrix} \dfrac{\partial x}{\partial u} & \dfrac{\partial x}{\partial v} \\ \dfrac{\partial y}{\partial u} & \dfrac{\partial y}{\partial v} \end{vmatrix} = \begin{vmatrix} -\dfrac{1}{3} & \dfrac{1}{3} \\ \dfrac{1}{3} & \dfrac{2}{3} \end{vmatrix} = -\frac{1}{3}.$$

By Theorem 8.1, we now have:

$$\iint\limits_{R} (x^2 + 2xy) \, dA = \iint\limits_{S} \left[\frac{1}{9}(v - u)^2 + \frac{2}{9}(v - u)(2v + u) \right] \left| \frac{\partial(x, y)}{\partial(u, v)} \right| \, du \, dv$$

$$= \frac{1}{27} \int_{2}^{5} \int_{1}^{3} [(v - u)^2 + 2(2v^2 - uv - u^2)] \, du \, dv$$

$$= \frac{196}{27},$$

where we leave the calculation of the final (routine) iterated integral to you.

Recall that for single definite integrals, we often must introduce a change of variable in order to find an antiderivative for the integrand. This is also the case in double integrals, as we see with the following example.

Example 8.5
A Change of Variables Required to Find an Antiderivative

Evaluate the double integral $\displaystyle\iint\limits_{R} \frac{e^{(x-y)}}{x + y} \, dA$, where R is the rectangle bounded by the lines $y = x$, $y = x + 5$, $y = 2 - x$ and $y = 4 - x$.

Solution First, notice that although the region over which you are to integrate is simply a rectangle in the xy-plane, its sides are not parallel to the x- and y-axes (see Figure 13.75a). This is the least of your problems right now, though. If you look carefully at the integrand, you'll recognize that you do not know an antiderivative for this integrand (no matter which variable you integrate with respect to first). A straightforward change of variables is to let $u = x - y$ and $v = x + y$. Solving these equations for x and y gives us

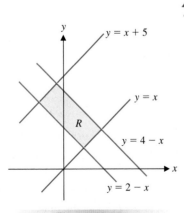

Figure 13.75a
The region R.

$$x = \frac{1}{2}(u + v) \quad \text{and} \quad y = \frac{1}{2}(v - u). \tag{8.7}$$

The Jacobian of this transformation is then

$$\frac{\partial(x, y)}{\partial(u, v)} = \begin{vmatrix} \dfrac{\partial x}{\partial u} & \dfrac{\partial x}{\partial v} \\[2mm] \dfrac{\partial y}{\partial u} & \dfrac{\partial y}{\partial v} \end{vmatrix} = \begin{vmatrix} \dfrac{1}{2} & \dfrac{1}{2} \\[2mm] -\dfrac{1}{2} & \dfrac{1}{2} \end{vmatrix} = \frac{1}{2}.$$

The next issue is to find the region S in the uv-plane which is mapped onto the region R in the xy-plane by this transformation. Remember that the boundary curves of the region S are mapped to the boundary curves of R. From (8.7), we have that $y = x$ corresponds to

$$\frac{1}{2}(v - u) = \frac{1}{2}(u + v) \quad \text{or} \quad u = 0.$$

Likewise, $y = x + 5$ corresponds to

$$\frac{1}{2}(v - u) = \frac{1}{2}(u + v) + 5 \quad \text{or} \quad u = -5,$$

$y = 2 - x$ corresponds to

$$\frac{1}{2}(v - u) = 2 - \frac{1}{2}(u + v) \quad \text{or} \quad v = 2$$

and $y = 4 - x$ corresponds to

$$\frac{1}{2}(v - u) = 4 - \frac{1}{2}(u + v) \quad \text{or} \quad v = 4.$$

This says that the region S in the uv-plane corresponding to the region R in the xy-plane is the rectangle

$$S = \{(u, v) \mid -5 \le u \le 0 \quad \text{and} \quad 2 \le v \le 4\},$$

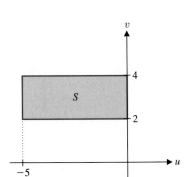

Figure 13.75b

The region S.

as indicated in Figure 13.75b. You can now easily read off the limits of integration in the uv-plane. By Theorem 8.1, we have

$$\iint_R \frac{e^{(x-y)}}{x + y}\, dA = \iint_S \frac{e^u}{v} \left| \frac{\partial(x, y)}{\partial(u, v)} \right| du\, dv = \frac{1}{2} \int_2^4 \int_{-5}^0 \frac{e^u}{v}\, du\, dv$$

$$= \frac{1}{2} \int_2^4 \frac{1}{v} e^u \Big|_{u=-5}^{u=0} dv = \frac{1}{2}(e^0 - e^{-5}) \int_2^4 \frac{1}{v}\, dv$$

$$= \frac{1}{2}(1 - e^{-5}) \ln|v| \Big|_{v=2}^{v=4} = \frac{1}{2}(1 - e^{-5})(\ln 4 - \ln 2)$$

$$\approx 0.34424.$$

Much as we have now done in two dimensions, we can develop a change of variables formula for triple integrals. The proof of the following result can be found in most texts on advanced calculus. We first define the Jacobian of a transformation in three dimensions.

If T is the transformation from a region S of uvw-space onto a region R in xyz-space, defined by $x = g(u, v, w)$, $y = h(u, v, w)$ and $z = l(u, v, w)$, the **Jacobian** of the transformation is the determinant $\dfrac{\partial(x, y, z)}{\partial(u, v, w)}$ defined by

$$\frac{\partial(x, y, z)}{\partial(u, v, w)} = \begin{vmatrix} \dfrac{\partial x}{\partial u} & \dfrac{\partial x}{\partial v} & \dfrac{\partial x}{\partial w} \\[2ex] \dfrac{\partial y}{\partial u} & \dfrac{\partial y}{\partial v} & \dfrac{\partial y}{\partial w} \\[2ex] \dfrac{\partial z}{\partial u} & \dfrac{\partial z}{\partial v} & \dfrac{\partial z}{\partial w} \end{vmatrix}.$$

The following result for triple integrals corresponds to Theorem 8.1.

Theorem 8.2 (Change of Variables in Triple Integrals)

Suppose that the region S in uvw-space is mapped onto the region R in xyz-space by the one-to-one transformation T defined by $x = g(u, v, w)$, $y = h(u, v, w)$ and $z = l(u, v, w)$, where g, h and l have continuous first partial derivatives in S. If f is continuous in R and the Jacobian $\dfrac{\partial(x, y, z)}{\partial(u, v, w)}$ is nonzero in S, then

$$\iiint\limits_{R} f(x, y, z)\, dV = \iiint\limits_{S} f(g(u, v, w), h(u, v, w), l(u, v, w)) \left| \frac{\partial(x, y, z)}{\partial(u, v, w)} \right| du\, dv\, dw.$$

We introduce a change of variables in a triple integral for precisely the same reasons as we do in double integrals: in order to simplify the integrand or the region of integration or both. In the following example, we use Theorem 8.2 to derive the change of variables formula for the conversion from rectangular to spherical coordinates and see that this is simply a special case of the general change of variables process given in Theorem 8.2.

Example 8.6 Deriving the Evaluation Formula for Spherical Coordinates

Use Theorem 8.2 to derive the evaluation formula for triple integrals in spherical coordinates:

$$\iiint\limits_{R} f(x, y, z)\, dV = \iiint\limits_{S} f(\rho \sin\phi \cos\theta, \rho \sin\phi \sin\theta, \rho \cos\phi) \rho^2 \sin\phi\, d\rho\, d\phi\, d\theta.$$

Solution Suppose that the region R in xyz-space is the image of the region S in $\rho\phi\theta$-space under the transformation T defined by the change to spherical coordinates. Recall that we have

$$x = \rho \sin\phi \cos\theta, \quad y = \rho \sin\phi \sin\theta \quad \text{and} \quad z = \rho \cos\phi.$$

The Jacobian of this transformation is then

$$\frac{\partial(x, y, z)}{\partial(\rho, \phi, \theta)} = \begin{vmatrix} \dfrac{\partial x}{\partial \rho} & \dfrac{\partial x}{\partial \phi} & \dfrac{\partial x}{\partial \theta} \\ \dfrac{\partial y}{\partial \rho} & \dfrac{\partial y}{\partial \phi} & \dfrac{\partial y}{\partial \theta} \\ \dfrac{\partial z}{\partial \rho} & \dfrac{\partial z}{\partial \phi} & \dfrac{\partial z}{\partial \theta} \end{vmatrix} = \begin{vmatrix} \sin\phi\cos\theta & \rho\cos\phi\cos\theta & -\rho\sin\phi\sin\theta \\ \sin\phi\sin\theta & \rho\cos\phi\sin\theta & \rho\sin\phi\cos\theta \\ \cos\phi & -\rho\sin\phi & 0 \end{vmatrix}.$$

For the sake of convenience, we expand this determinant according to the third row, rather than the first row. This gives us

$$\frac{\partial(x, y, z)}{\partial(\rho, \phi, \theta)} = \cos\phi \begin{vmatrix} \rho\cos\phi\cos\theta & -\rho\sin\phi\sin\theta \\ \rho\cos\phi\sin\theta & \rho\sin\phi\cos\theta \end{vmatrix}$$

$$+ \rho\sin\phi \begin{vmatrix} \sin\phi\cos\theta & -\rho\sin\phi\sin\theta \\ \sin\phi\sin\theta & \rho\sin\phi\cos\theta \end{vmatrix}$$

$$= \cos\phi(\rho^2 \sin\phi\cos\phi\cos^2\theta + \rho^2 \sin\phi\cos\phi\sin^2\theta)$$

$$+ \rho\sin\phi(\rho\sin^2\phi\cos^2\theta + \rho\sin^2\phi\sin^2\theta)$$

$$= \rho^2 \sin\phi\cos^2\phi + \rho^2 \sin^3\phi = \rho^2 \sin\phi(\cos^2\phi + \sin^2\phi)$$

$$= \rho^2 \sin\phi.$$

From Theorem 8.2, observe that we now have that

$$\iiint\limits_{R} f(x, y, z)\, dV = \iiint\limits_{S} f(\rho\sin\phi\cos\theta, \rho\sin\phi\sin\theta, \rho\cos\phi) \left| \frac{\partial(x, y, z)}{\partial(\rho, \phi, \theta)} \right| d\rho\, d\phi\, d\theta$$

$$= \iiint\limits_{S} f(\rho\sin\phi\cos\theta, \rho\sin\phi\sin\theta, \rho\cos\phi)\rho^2 \sin\phi\, d\rho\, d\phi\, d\theta,$$

where we have used the fact that $0 \le \phi \le \pi$ to write $|\sin\phi| = \sin\phi$. Notice that this is the same evaluation formula as we had developed in section 13.7 and we now see that this is merely a special case of the more general result in Theorem 8.2.

EXERCISES 13.8

1. Explain what is meant by a "rectangular region" in the uv-plane. In particular, explain what is rectangular about the polar region $1 \le r \le 2$ and $0 \le \theta \le \pi$.

2. The order of variables in the Jacobian is not important in the sense that $\left| \dfrac{\partial(x, y)}{\partial(v, u)} \right| = \left| \dfrac{\partial(x, y)}{\partial(u, v)} \right|$ but the order is very important in the sense that $\left| \dfrac{\partial(x, y)}{\partial(u, v)} \right| \neq \left| \dfrac{\partial(u, v)}{\partial(x, y)} \right|$. Give a geometric explanation of why $\left| \dfrac{\partial(x, y)}{\partial(u, v)} \right| \left| \dfrac{\partial(u, v)}{\partial(x, y)} \right| = 1$.

In exercises 3–14, find a transformation from a rectangular region S in the uv-plane to the region R.

3. R is bounded by $y = 4x + 2$, $y = 4x + 5$, $y = 3 - 2x$ and $y = 1 - 2x$.

4. R is bounded by $y = 2x - 1$, $y = 2x + 5$, $y = 1 - 3x$ and $y = -1 - 3x$.

5. R is bounded by $y = 1 - 3x$, $y = 3 - 3x$, $y = x - 1$ and $y = x - 3$.

6. R is bounded by $y = 2x - 1$, $y = 2x + 1$, $y = 3$ and $y = 1$.

7. R is inside $x^2 + y^2 = 4$, outside $x^2 + y^2 = 1$ and in the first quadrant.

8. R is inside $x^2 + y^2 = 4$, outside $x^2 + y^2 = 1$ and in the first quadrant between $y = x$ and $x = 0$.

9. R is inside $x^2 + y^2 = 9$, outside $x^2 + y^2 = 4$ and between $y = x$ and $y = -x$ with $y \geq 0$.

10. R is inside $x^2 + y^2 = 9$ with $x \geq 0$.

11. R is bounded by $y = x^2$, $y = x^2 + 2$, $y = 4 - x^2$ and $y = 2 - x^2$ with $x \geq 0$.

12. R is bounded by $y = x^2$, $y = x^2 + 2$, $y = 3 - x^2$ and $y = 1 - x^2$ with $x \leq 0$.

13. R is bounded by $y = e^x$, $y = e^x + 1$, $y = 3 - e^x$ and $y = 5 - e^x$.

14. R is bounded by $y = 2x^2 + 1$, $y = 2x^2 + 3$, $y = 2 - x^2$ and $y = 4 - x^2$ with $x \geq 0$.

In exercises 15–24, evaluate the double integral.

15. $\iint\limits_R (y - 4x)\, dA$, where R is given in exercise 3.

16. $\iint\limits_R (y + 3x)\, dA$, where R is given in exercise 4.

17. $\iint\limits_R (y + 3x)^2\, dA$, where R is given in exercise 5.

18. $\iint\limits_R e^{y-x}\, dA$, where R is given in exercise 6.

19. $\iint\limits_R x\, dA$, where R is given in exercise 7.

20. $\iint\limits_R e^{y-e^x}\, dA$, where R is given in exercise 13.

21. $\iint\limits_R \dfrac{e^{y-4x}}{y + 2x}\, dA$, where R is given in exercise 3.

22. $\iint\limits_R \dfrac{e^{y+3x}}{y - 2x}\, dA$, where R is given in exercise 4.

23. $\iint\limits_R (x + y)\, dA$, where R is given in exercise 3.

24. $\iint\limits_R (x + 2y)\, dA$, where R is given in exercise 4.

In exercises 25–28, find the Jacobian of the given transformation.

25. $x = ue^v$, $y = ue^{-v}$ 26. $x = 2uv$, $y = 3u - v$

27. $x = u/v$, $y = v^2$ 28. $x = 4u + v^2$, $y = 2uv$

In exercises 29 and 30, find a transformation from a (three-dimensional) rectangular region S in uvw-space to the solid Q.

29. Q is bounded by $x + y + z = 1$, $x + y + z = 2$, $x + 2y = 0$, $x + 2y = 1$, $y + z = 2$ and $y + z = 4$.

30. Q is bounded by $x + z = 1$, $x + z = 2$, $2y + 3z = 0$, $2y + 3z = 1$, $y + 2z = 2$ and $y + 2z = 4$.

In exercises 31 and 32, find the volume of the given solid.

31. Q in exercise 29   32. Q in exercise 30

33. In Theorem 8.1, we required that the Jacobian be nonzero. To see why this is necessary, consider a transformation where $x = u - v$ and $y = 2v - 2u$. Show that the Jacobian is zero. Then try solving for u and v.

34. Compute the Jacobian for the spherical-like transformation $x = \rho \sin\phi$, $y = \rho \cos\phi \cos\theta$ and $z = \rho \cos\phi \sin\theta$.

35. Transformations are involved in many important applications of mathematics. The **direct linear transformation** discussed in this exercise is used by Titleist golf researchers Gobush, Pelletier and Days to study the motion of golf balls (see *Science and Golf II*, 1996). Bright dots are drawn onto golf balls. The dots are tracked by a pair of cameras as the ball is hit. The challenge is to use this information to reconstruct the exact position of the ball at various times, allowing the researchers to estimate the speed, spin rate and launch angle of the ball. In the direct linear transformation model developed by Abdel-Aziz and Karara, a dot at actual position (x, y, z) will appear at pixel (u_1, v_1) of camera number one's digitized image where

$$u_1 = \frac{c_{11}x + c_{21}y + c_{31}z + c_{41}}{d_{11}x + d_{21}y + d_{31}z + 1} \quad \text{and}$$

$$v_1 = \frac{c_{51}x + c_{61}y + c_{71}z + c_{81}}{d_{11}x + d_{21}y + d_{31}z + 1},$$

for constants $c_{11}, c_{21}, \ldots, c_{81}$ and d_{11}, d_{21} and d_{31}. Similarly, camera 2 "sees" this dot at pixel (u_2, v_2) where

$$u_2 = \frac{c_{12}x + c_{22}y + c_{32}z + c_{42}}{d_{12}x + d_{22}y + d_{32}z + 1} \quad \text{and}$$

$$v_2 = \frac{c_{52}x + c_{62}y + c_{72}z + c_{82}}{d_{12}x + d_{22}y + d_{32}z + 1},$$

for a different set of constants $c_{12}, c_{22}, \ldots, c_{82}$ and d_{12}, d_{22} and d_{32}. The constants are determined by taking a series of measurements of motionless balls to calibrate the model. Given that the model for each camera consists of eleven constants, explain why in theory, six different measurements would more than suffice to determine the constants. In reality, more measurements are taken and a least squares criterion is used to find the best fit of the model to the data. Suppose that this procedure gives us the model

$$u_1 = \frac{2x + y + z + 1}{x + y + 2z + 1}, \quad v_1 = \frac{3x + z}{x + y + 2z + 1}$$

$$u_2 = \frac{x + z + 6}{2x + 3z + 1}, \quad v_2 = \frac{4x + y + 3}{2x + 3z + 1}$$

If the screen coordinates of a dot are $(u_1, v_1) = (0, -3)$ and $(u_2, v_2) = (5, 0)$, solve for the actual position (x, y, z) of the

dot. Actually, a dot would not show up as a single pixel, but as a somewhat blurred image over several pixels. The dot is officially located at the nearest pixel to the center of mass of the pixels involved. Suppose that a dot's image activates the following pixels: (34, 42), (35, 42), (32, 41), (33, 41), (34, 41), (35, 41), (36, 41), (34, 40), (35, 40), (36, 40) and (36, 39). Find the center of mass of these pixels and round off to determine the "location" of the dot.

CHAPTER REVIEW EXERCISES

In exercises 1 and 2, compute the Riemann sum for the given function and region, a partition with n equal-sized rectangles and the given evaluation rule.

1. $f(x, y) = 5x - 2y$, $1 \leq x \leq 3$, $0 \leq y \leq 1$, $n = 4$, evaluate at midpoint

2. $f(x, y) = 4x^2 + y$, $0 \leq x \leq 1$, $1 \leq y \leq 3$, $n = 4$, evaluate at midpoint

In exercises 3–10, evaluate the double integral.

3. $\iint_R (4x + 9x^2 y^2) \, dA$, where $R = \{0 \leq x \leq 3, 1 \leq y \leq 2\}$

4. $\iint_R 2e^{4x+2y} \, dA$, where $R = \{0 \leq x \leq 1, 0 \leq y \leq 1\}$

5. $\iint_R e^{-x^2-y^2} \, dA$, where $R = \{1 \leq x^2 + y^2 \leq 4\}$

6. $\iint_R 2xy \, dA$, where R is bounded by $y = x$, $y = 2 - x$ and $y = 0$

7. $\int_{-1}^{1} \int_{x^2}^{2x} (2xy - 1) \, dy \, dx$ 8. $\int_{0}^{1} \int_{2x}^{2} (3y^2 x + 4) \, dy \, dx$

9. $\iint_R xy \, dA$, where R is bounded by $r = 2 \cos \theta$

10. $\iint_R \sin(x^2 + y^2) \, dA$, where R is bounded by $x^2 + y^2 = 4$

In exercises 11 and 12, approximate the double integral.

11. $\iint_R 4xy \, dA$, where R is bounded by $y = x^2 - 4$ and $y = \ln x$

12. $\iint_R 6x^2 y \, dA$, where R is bounded by $y = \cos x$ and $y = x^2 - 1$

In exercises 13–22, compute the volume of the solid.

13. Bounded by $z = 1 - x^2$, $z = 0$, $y = 0$, $y = 1$

14. Bounded by $z = 4 - x^2 - y^2$, $z = 0$, $x = 0$, $x + y = 1$

15. Between $z = x^2 + y^2$ and $z = 8 - x^2 - y^2$

16. Under $z = e^{\sqrt{x^2+y^2}}$ and inside $x^2 + y^2 = 4$

17. Bounded by $x + 2y + z = 8$ and the coordinate planes

18. Bounded by $x + 5y + 7z = 1$ and the coordinate planes

19. Bounded by $z = \sqrt{x^2 + y^2}$ and $z = 4$

20. Bounded by $x = \sqrt{y^2 + z^2}$ and $x = 2$

21. Between $z = \sqrt{x^2 + y^2}$ and $x^2 + y^2 + z^2 = 4$

22. Inside $x^2 + y^2 + z^2 = 4z$ and below $z = 1$

23. Under $z = 6 - x^2 - y^2$, inside $x^2 + y^2 = 1$

24. Under $z = x$, inside $r = \cos \theta$

In exercises 25 and 26, change the order of integration.

25. $\int_{0}^{2} \int_{0}^{x^2} f(x, y) \, dy \, dx$

26. $\int_{0}^{2} \int_{x^2}^{4} f(x, y) \, dy \, dx$

In exercises 27 and 28, convert to polar coordinates and evaluate the integral

27. $\displaystyle\int_0^2 \int_{-\sqrt{4-x^2}}^{\sqrt{4-x^2}} 2x \, dy \, dx$

28. $\displaystyle\int_0^2 \int_0^{\sqrt{4-x^2}} 2\sqrt{x^2+y^2} \, dy \, dx$

In exercises 29–32, find the mass and center of mass.

29. The lamina bounded by $y = 2x$, $y = x$ and $x = 2$, $\rho(x, y) = 2x$

30. The lamina bounded by $y = x$, $y = 4 - x$ and $y = 0$, $\rho(x, y) = 2y$

31. The solid bounded by $z = 1 - x^2$, $z = 0$, $y = 0$, $y + z = 2$, $\rho(x, y, z) = 2$

32. The solid bounded by $x = \sqrt{y^2 + z^2}$, $x = 2$, $\rho(x, y, z) = 3x$

In exercises 33 and 34, use a double integral to find the area.

33. Bounded by $y = x^2$, $y = 2 - x$ and $y = 0$

34. One leaf of $r = \sin 4\theta$

In exercises 35 and 36, find the average value of the function on the indicated region.

35. $f(x, y) = x^2$, region bounded by $y = 2x$, $y = x$ and $x = 1$

36. $f(x, y) = \sqrt{x^2 + y^2}$, region bounded by $x^2 + y^2 = 1$, $x = 0$, $y = 0$

In exercises 37–42, evaluate or estimate the surface area.

37. The portion of $z = 2x + 4y$ between $y = x$, $y = 2$ and $x = 0$

38. The portion of $z = x^2 + 6y$ between $y = x^2$ and $y = 4$

39. The portion of $z = xy$ inside $x^2 + y^2 = 8$, first octant

40. The portion of $z = \sin(x^2 + y^2)$ inside $x^2 + y^2 = \pi$

41. The portion of $z = \sqrt{x^2 + y^2}$ below $z = 4$

42. The portion of $x + 2y + 3z = 6$ in the first octant

In exercises 43–50, set up the triple integral $\iiint_Q f(x, y, z) dV$ in an appropriate coordinate system. If $f(x, y, z)$ is given, evaluate the integral.

43. $f(x, y, z) = z(x + y)$,
    $Q = \{(x, y, z) \,|\, 0 \le x \le 2, -1 \le y \le 1, -1 \le z \le 1\}$

44. $f(x, y, z) = 2xye^{yz}$,
    $Q = \{(x, y, z) \,|\, 0 \le x \le 2, 0 \le y \le 1, 0 \le z \le 1\}$

45. $f(x, y, z) = \sqrt{x^2 + y^2 + z^2}$, Q is above $z = \sqrt{x^2 + y^2}$ and below $x^2 + y^2 + z^2 = 4$.

46. $f(x, y, z) = 3x$, Q is the region below $z = \sqrt{x^2 + y^2}$, above $z = 0$ and inside $x^2 + y^2 = 4$.

47. Q is bounded by $x + y + z = 6$, $z = 0$, $y = x$, $y = 2$ and $x = 0$.

48. Q is the region below $z = \sqrt{4 - x^2 - y^2}$, above $z = 0$ and inside $x^2 + y^2 = 1$.

49. Q is the region below $z = \sqrt{4 - x^2 - y^2}$ and above $z = 0$.

50. Q is the region below $z = 6 - x - y$, above $z = 0$ and inside $x^2 + y^2 = 8$.

In exercises 51–54, evaluate the integral after changing coordinate systems.

51. $\displaystyle\int_0^1 \int_x^{\sqrt{2-x^2}} \int_0^{\sqrt{x^2+y^2}} e^z \, dz \, dy \, dx$

52. $\displaystyle\int_0^{\sqrt{2}} \int_y^{\sqrt{4-y^2}} \int_0^2 4z \, dz \, dx \, dy$

53. $\displaystyle\int_{-1}^1 \int_0^{\sqrt{1-x^2}} \int_{\sqrt{x^2+y^2}}^{\sqrt{2-x^2-y^2}} \sqrt{x^2+y^2+z^2} \, dz \, dy \, dx$

54. $\displaystyle\int_{-2}^2 \int_0^{\sqrt{4-y^2}} \int_0^{\sqrt{4-x^2-y^2}} dz \, dy \, dx$

In exercises 55–60, write the given equation in (a) cylindrical and (b) spherical coordinates.

55. $y = 3$ 56. $x^2 + y^2 = 9$

57. $x^2 + y^2 + z^2 = 4$ 58. $y = x$

59. $z = \sqrt{x^2 + y^2}$ 60. $z = 4$

In exercises 61–66, sketch the graph.

61. $r = 4$

62. $\rho = 4$

63. $\theta = \frac{\pi}{4}$

64. $\phi = \frac{\pi}{4}$

65. $r = 2 \cos \theta$

66. $\rho = 2 \sec \phi$

In exercises 67 and 68, find a transformation from a rectangular region S in the uv-plane to the region R.

67. R bounded by $y = 2x - 1$, $y = 2x + 1$, $y = 2 - 2x$ and $y = 4 - 2x$

68. R inside $x^2 + y^2 = 9$, outside $x^2 + y^2 = 4$ and in the second quadrant

In exercises 69 and 70, evaluate the double integral.

69. $\iint\limits_{R} e^{y-2x} \, dA$, where R is given in exercise 67.

70. $\iint\limits_{R} (y + 2x)^3 \, dA$, where R is given in exercise 67.

In exercises 71 and 72, find the Jacobian of the given transformation.

71. $x = u^2 v$, $y = 4u + v^2$

72. $x = 4u - 5v$, $y = 2u + 3v$

CHAPTER 14

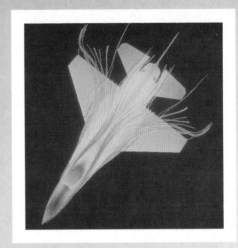

VECTOR CALCULUS

The Volkswagen Beetle was one of the most beloved and recognizable cars of the 1950s, 1960s and 1970s. Volkswagen's announcement that they would release a redesigned Beetle in 1998 created quite a stir in the automotive world. The new Beetle has a similar look to the classic Beetle, but has been modernized to improve gas mileage, safety, handling and overall performance. The calculus that we introduce in this chapter will provide you with some of the basic tools necessary for designing and analyzing automobiles, aircraft and other types of complex machinery.

Think about how you might redesign an automobile to improve its aerodynamic performance. Engineers have identified many important principles of aerodynamics, but the design of a complicated structure like a car still has an element of trial and error. Before high-speed computers were available, an engineer's options were to build a small-scale or full-scale model of a new design and then test it in a wind tunnel. Unfortunately, small-scale models aren't always faithful enough to the original design to provide useful information. Full-scale models give more reliable information, but are prohibitively expensive to build, particularly if you have 20 or 30 new ideas you'd like to try.

The old Beetle.

With modern computers, wind tunnel tests can be accurately simulated by sophisticated programs. Designers can inexpensively analyze numerous ideas, including unusual innovations that have never been tried before. Mathematical models give engineers the opportunity to thoroughly test anything from minor modifications to radical changes.

The new Beetle.

The calculus that goes into a computer simulation of a wind tunnel is beyond what you've seen so far. Such simulations must keep track of the air velocity at each point on and around a car. A function assigning a vector (e.g., a velocity vector) to each point in space is called a vector field, which we introduce in section 14.1. To determine where vortices and turbulence occur in a fluid flow, you must compute line integrals, which are discussed in sections 14.2 and 14.3. The curl and divergence, introduced in section 14.5, allow you to analyze the rotational and linear properties of a fluid flow. Other properties of three-dimensional objects, such as mass and moments of inertia for a thin shell (such as a dome of a building), require the evaluation of surface integrals, which we develop in section 14.6. The relationships among line integrals, surface integrals, double integrals and triple integrals are explored in the remaining sections of the chapter.

In the case of the redesigned Volkswagen Beetle, computer simulations resulted in numerous improvements over the original. One measure of a vehicle's aerodynamic efficiency is its drag coefficient. Without getting into the technicalities, the lower its drag coefficient is, the less the velocity of the car is reduced by air resistance. The original Beetle has a drag coefficient of 0.46 (as reported by Robertson and Crowe in *Engineering Fluid Mechanics*). By comparison, a low-slung (and quite aerodynamic) 1985 Chevrolet Corvette has a drag coefficient of 0.34. Volkswagen's specification sheet for the new Beetle lists a drag coefficient of 0.38, representing a considerable reduction in air drag from the original Beetle. Thanks to careful mathematical analysis, Volkswagen successfully improved the performance of the Beetle while retaining the distinctive shape of the original car.

14.1 VECTOR FIELDS

To analyze the flight characteristics of an airplane, engineers use wind tunnel tests to provide vital information about the flow of air over the wings and around the fuselage of the aircraft. As you can imagine, to model such a test mathematically, we need to be able to describe the velocity of the air at various points throughout the tunnel. The only problem with this is that velocity is a vector. So, we need to define a function that assigns a vector to each point in space. Such a function would have both a multidimensional domain (like the functions of Chapters 12 and 13) and a multidimensional range (like the vector-valued functions introduced in Chapter 11). We call such a function a **vector field.** Although vector fields in higher dimensions can be very useful, we will focus here on vector fields in two and three dimensions.

Definition 1.1

A **vector field** in the plane is a function $\mathbf{F}(x, y)$ mapping points in $\mathbb{R}^2$ into the set of two-dimensional vectors V_2. We write

$$\mathbf{F}(x, y) = \langle f_1(x, y), f_2(x, y) \rangle = f_1(x, y)\mathbf{i} + f_2(x, y)\mathbf{j},$$

for scalar functions $f_1(x, y)$ and $f_2(x, y)$. In space, a **vector field** is a function $\mathbf{F}(x, y, z)$ mapping points in $\mathbb{R}^3$ into the set of three-dimensional vectors V_3. In this case, we write

$$\mathbf{F}(x, y, z) = \langle f_1(x, y, z), f_2(x, y, z), f_3(x, y, z) \rangle$$
$$= f_1(x, y, z)\mathbf{i} + f_2(x, y, z)\mathbf{j} + f_3(x, y, z)\mathbf{k},$$

for scalar functions $f_1(x, y, z)$, $f_2(x, y, z)$ and $f_3(x, y, z)$.

To describe a two-dimensional vector field graphically, we draw a collection of the vectors $\mathbf{F}(x, y)$ for various points (x, y) in the domain, in each case drawing the vector so that its initial point is located at (x, y). We illustrate this in the following example.

Example 1.1   Plotting a Vector Field

For the vector field $\mathbf{F}(x, y) = \langle x + y, 3y - x \rangle$, evaluate (a) $\mathbf{F}(1, 0)$, (b) $\mathbf{F}(0, 1)$ and (c) $\mathbf{F}(-2, 1)$. Plot each vector $\mathbf{F}(x, y)$ using the point (x, y) as the initial point.

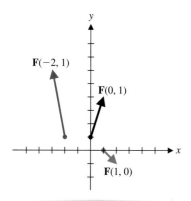

Figure 14.1
Values of **F**(*x*, *y*).

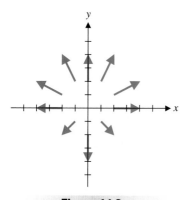

Figure 14.2a
F(*x*, *y*) = ⟨*x*, *y*⟩.

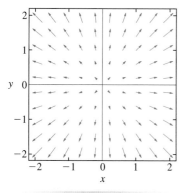

Figure 14.2b
F(*x*, *y*) = ⟨*x*, *y*⟩.

Solution (a) Taking $x = 1$ and $y = 0$, we have $\mathbf{F}(1, 0) = \langle 1 + 0, 0 - 1 \rangle = \langle 1, -1 \rangle$. In Figure 14.1, we have plotted the vector $\langle 1, -1 \rangle$ with its initial point located at the point $(1, 0)$, so that its terminal point is located at $(2, -1)$.

(b) Taking $x = 0$ and $y = 1$, we have $\mathbf{F}(0, 1) = \langle 0 + 1, 3 - 0 \rangle = \langle 1, 3 \rangle$. In Figure 14.1, we have also indicated the vector $\langle 1, 3 \rangle$ taking the point $(0, 1)$ as its initial point, so that its terminal point is located at $(1, 4)$.

(c) With $x = -2$ and $y = 1$, we have $\mathbf{F}(-2, 1) = \langle -2 + 1, 3 + 2 \rangle = \langle -1, 5 \rangle$. In Figure 14.1, the vector $\langle -1, 5 \rangle$ is plotted by placing its initial point at $(-2, 1)$ and its terminal point at $(-3, 6)$.

■

Graphing vector fields poses something of a problem. Notice that the graph of a two-dimensional vector field would be **four**-dimensional (i.e., two independent variables plus two dimensions for the vectors). Likewise, the graph of a three-dimensional vector field would be **six**-dimensional. Despite this, we can visualize many of the important properties of a vector field by plotting a number of values of the vector field as we had started to do in Figure 14.1. In general, by the **graph of a vector field** $\mathbf{F}(x, y)$ we will mean a two-dimensional graph with vectors $\mathbf{F}(x, y)$ plotted with their initial point located at (x, y), for a variety of points (x, y). If you are plotting vectors by hand, a few points per quadrant is usually sufficient. Many graphing calculators and computer algebra systems have commands to graph vector fields. Notice in the following example that the vectors in the computer-generated graphs are not drawn to the correct length. Instead, some software packages automatically shrink or enlarge all of the vectors proportionally to a size that avoids cluttering up the overall graph.

Example 1.2 Graphing Vector Fields

Graph the vector fields $\mathbf{F}(x, y) = \langle x, y \rangle$, $\mathbf{G}(x, y) = \dfrac{\langle x, y \rangle}{\sqrt{x^2 + y^2}}$ and $\mathbf{H}(x, y) = \langle y, -x \rangle$ and identify any patterns.

Solution Your task is to choose a variety of points (x, y), evaluate the vector field at these points and plot the vectors using (x, y) as the initial point. Notice that in the following table we have chosen points on the axes and in each of the four quadrants.

(x, y)	$(2, 0)$	$(1, 2)$	$(2, 1)$	$(0, 2)$	$(-1, 2)$
$\langle x, y \rangle$	$\langle 2, 0 \rangle$	$\langle 1, 2 \rangle$	$\langle 2, 1 \rangle$	$\langle 0, 2 \rangle$	$\langle -1, 2 \rangle$
(x, y)	$(-2, 1)$	$(-2, 0)$	$(-1, -1)$	$(0, -2)$	$(1, -1)$
$\langle x, y \rangle$	$\langle -2, 1 \rangle$	$\langle -2, 0 \rangle$	$\langle -1, -1 \rangle$	$\langle 0, -2 \rangle$	$\langle 1, -1 \rangle$

The vectors indicated in the table are plotted in Figure 14.2a. A computer-generated plot of the vector field is shown in Figure 14.2b. Notice that the vectors drawn here have not been drawn to scale in order to improve the readability of the graph.

In both plots, notice that the vectors all point away from the origin and increase in length as the points get farther from the origin. In fact, the initial point (x, y) lies a distance $\sqrt{x^2 + y^2}$ from the origin and the vector $\langle x, y \rangle$ has length $\sqrt{x^2 + y^2}$. So, the length of each vector corresponds to the distance from its initial point to the origin. This gives us an important clue about the graph of $\mathbf{G}(x, y)$. Although the formula may look messy, notice that $\mathbf{G}(x, y)$ is the same as $\mathbf{F}(x, y)$ except for the division by $\sqrt{x^2 + y^2}$, which is the magnitude of the vector $\langle x, y \rangle$. Recall that dividing a vector by its

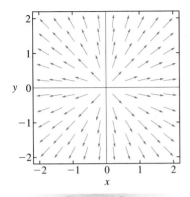

Figure 14.2c

$$\mathbf{G}(x, y) = \frac{\langle x, y \rangle}{\sqrt{x^2 + y^2}}.$$

magnitude yields a unit vector in the same direction. Thus, $\mathbf{G}(x, y)$ always has the same direction as $\mathbf{F}(x, y)$, but produces unit vectors. A computer-generated plot of $\mathbf{G}(x, y)$ is shown in Figure 14.2c.

Note that although $\mathbf{H}(x, y)$ looks somewhat like $\mathbf{F}(x, y)$, we need to recompute the vectors to plot. We have

(x, y)	$(2, 0)$	$(1, 2)$	$(2, 1)$	$(0, 2)$	$(-1, 2)$
$\langle y, -x \rangle$	$\langle 0, -2 \rangle$	$\langle 2, -1 \rangle$	$\langle 1, -2 \rangle$	$\langle 2, 0 \rangle$	$\langle 2, 1 \rangle$
(x, y)	$(-2, 1)$	$(-2, 0)$	$(-1, -1)$	$(0, -2)$	$(1, -1)$
$\langle y, -x \rangle$	$\langle 1, 2 \rangle$	$\langle 0, 2 \rangle$	$\langle -1, 1 \rangle$	$\langle -2, 0 \rangle$	$\langle -1, -1 \rangle$

The vectors indicated in the table are plotted in Figure 14.3a and a computer-generated plot of $\mathbf{H}(x, y)$ is shown in Figure 14.3b. Notice that if you think of $\mathbf{H}(x, y)$ as representing the velocity field for a fluid in motion, the vectors suggest a circular rotation of the fluid. Recall that tangent lines to a circle are perpendicular to radius lines. The radius vector from the origin to the point (x, y) is $\langle x, y \rangle$, which is perpendicular to the vector $\langle y, -x \rangle$ since $\langle x, y \rangle \cdot \langle y, -x \rangle = 0$. Also, notice that the vectors are not of constant size. As for $\mathbf{F}(x, y)$, we compute the length of the vector $\langle y, -x \rangle$ as $\sqrt{x^2 + y^2}$, equal to the distance from the origin to the initial point (x, y).

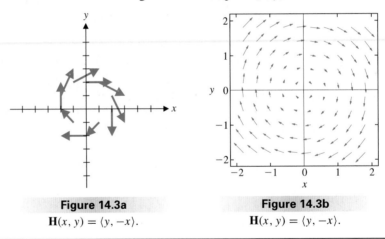

Figure 14.3a

$\mathbf{H}(x, y) = \langle y, -x \rangle.$

Figure 14.3b

$\mathbf{H}(x, y) = \langle y, -x \rangle.$

The ideas in example 1.2 are very important, but we must admit that most vector fields are too complicated to effectively draw by hand. The following example will give you more practice at relating the component functions of a vector field to its graph.

Example 1.3   Matching Vector Fields to Graphs

Match the vector fields $\mathbf{F}(x, y) = \langle y^2, x - 1 \rangle$, $\mathbf{G}(x, y) = \langle y + 1, e^{x/6} \rangle$ and $\mathbf{H}(x, y) = \langle y^3, x^2 - 1 \rangle$ to the graphs shown.

Solution   There is no general procedure for matching vector fields to their graphs. Instead, you should look for special features of the components of the vector fields and try to locate these in the graphs. For instance, the first component of $\mathbf{F}(x, y)$ is $y^2 \geq 0$, so the vectors $\mathbf{F}(x, y)$ will never point to the left. The fourth quadrants of both Graph A and Graph C have vectors with negative first components, so Graph B must be the graph

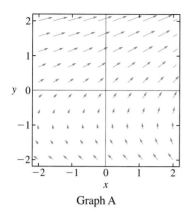

Graph A

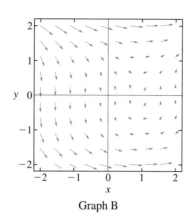

Graph B

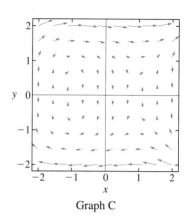

Graph C

of $\mathbf{F}(x, y)$. Notice also that the vectors in Graph B have small vertical components near $x = 1$, where the second component of $\mathbf{F}(x, y)$ equals zero. Similarly, the second component of $\mathbf{G}(x, y)$ is $e^{x/6} > 0$, so the vectors $\mathbf{G}(x, y)$ will always point upward. Graph A is the only one of these graphs with this property. Notice also that the vectors in Graph A are almost vertical near $y = -1$, where the first component of $\mathbf{G}(x, y)$ equals zero. That leaves Graph C for $\mathbf{H}(x, y)$, but let's check to be sure this is reasonable. Observe that the first component of $\mathbf{H}(x, y)$ is y^3, which is negative for $y < 0$ and positive for $y > 0$. The vectors will point to the left for $y < 0$ and to the right for $y > 0$, as seen in Graph C. Further, the vectors in Graph C have small vertical components near $x = 1$ and $x = -1$, where the second component of $\mathbf{H}(x, y)$ equals zero.

As you might imagine, vector fields in space are typically more difficult to sketch than vector fields in the plane, but the idea is the same. That is, pick a variety of representative points and plot the vector $\mathbf{F}(x, y, z)$ with its initial point located at (x, y, z). Unfortunately, the difficulties associated with representing three-dimensional vectors on two-dimensional paper reduce the usefulness of these graphs.

Example 1.4   Graphing a Vector Field in Space

Graph the vector field $\mathbf{F}(x, y, z) = \dfrac{\langle -x, -y, -z \rangle}{(x^2 + y^2 + z^2)^{3/2}}$.

Solution   In Figure 14.4, we show a computer-generated plot of the vector field $\mathbf{F}(x, y, z)$.

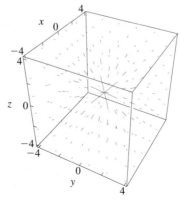

Figure 14.4

Gravitational force field.

Notice that the vectors all point toward the origin, getting larger as you near the origin (where the field is actually undefined). You should get the sense of an attraction to the origin that gets stronger the closer you get. In fact, you might have recognized that $\mathbf{F}(x, y, z)$ describes the gravitational force field for an object located at the origin.

The force field shown in Figure 14.4 indicates that an object acted on by this force field will be drawn toward the origin. However, this can be misleading, in that a given object need not move in a straight path toward the origin. For instance, an object with initial position $(2, 0, 0)$ and initial velocity $\langle 0, 2, 0 \rangle$ will spiral in toward the origin. To more accurately sketch the *path* followed by an object, we need additional information. Notice that in many cases, we can think of velocity as not explicitly depending on time, but instead depending on location. For instance, imagine watching a mountain stream with waterfalls and whirlpools that don't change (significantly) over time. In this case, the motion of a leaf dropped into the stream would depend on **where** you drop the leaf, rather than **when** you drop the leaf. This says that we should consider the velocity of the stream as a function of location. That is, the velocity of any particle located at the point (x, y) in the stream can be described by a vector field $\mathbf{F}(x, y) = \langle f_1(x, y), f_2(x, y) \rangle$, called the **velocity field.** The path of any given particle in the flow starting at the point (x_0, y_0) is then the curve traced out by $\langle x(t), y(t) \rangle$, where $x(t)$ and $y(t)$ are the solutions of the differential equations $x'(t) = f_1(x(t), y(t))$, $y'(t) = f_2(x(t), y(t))$ with initial conditions $x(t_0) = x_0$ and $y(t_0) = y_0$. In these cases, we can use the velocity field to construct **flow lines,** which indicate the path followed by a particle starting at a given point in the flow. (We'll see how to do this shortly.)

In practice, one way to visualize the velocity field of a given process is to plot a number of velocity vectors at a single instant in time. Figure 14.5 shows the velocity field of Pacific Ocean currents in March 1998. The picture is color-coded for temperature, with a band of water swinging up from South America to the Pacific northwest representing "El Niño." The velocity field gives you information about how the warmer and cooler areas of ocean water are likely to change. Since El Niño is associated with significant changes in temperature and rainfall, the ability to predict its movement can be crucial.

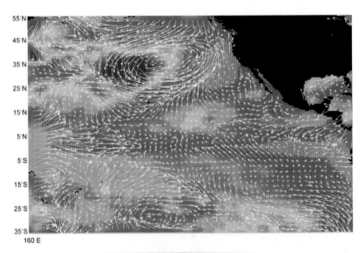

Figure 14.5
Velocity field for Pacific Ocean currents (March 1998).

Example 1.5   Graphing Vector Fields and Flow Lines

Graph the vector fields $\langle y, -x \rangle$ and $\langle 2, 1 + 2xy \rangle$ and for each, sketch in flow lines through the points $(0, 1)$, $(0, -1)$ and $(1, 1)$.

Solution   We have already graphed the vector field $\langle y, -x \rangle$ in example 1.2 and show a computer-generated graph of the vector field Figure 14.6a.

Notice that in Figure 14.6a, the plotted vectors nearly join together as concentric circles. In Figure 14.6b, we have superimposed circular paths that stay tangent to the velocity field and pass through the points $(0, 1)$, $(0, -1)$ and $(1, 1)$. (Notice that the first two of these paths are the same.) It isn't difficult to verify that the flow lines are indeed circles, as follows. Observe that a circle of radius a centered at the origin with a clockwise orientation (as indicated) can be described by the endpoint of the vector-valued function $\mathbf{r}(t) = \langle a \sin t, a \cos t \rangle$. The velocity vector $\mathbf{r}'(t) = \langle a \cos t, -a \sin t \rangle$ gives a tangent vector to the curve for each t. If we eliminate the parameter, the velocity field for the position vector $\mathbf{r} = \langle x, y \rangle$ is given by $\mathbf{T} = \langle y, -x \rangle$, which is the vector field we are presently plotting.

The vector field $\langle 2, 1 + 2xy \rangle$ is a little more complicated. We show a computer-generated graph of the field in Figure 14.7a, where the vectors indicate some parabolic-like paths, and in Figure 14.7b, we sketch two of these paths through the points $(0, 1)$ and $(0, -1)$. However, the vectors in Figure 14.7a also indicate some paths that look more like cubics, such as the path through $(0, 0)$ sketched in Figure 14.7b. In this case, though, it's more difficult to determine equations for the flow lines. As it turns out, they are not parabolic or cubic. We'll explore this further in the exercises.

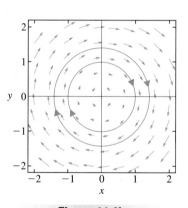

Figure 14.6a

$\langle y, -x \rangle$.

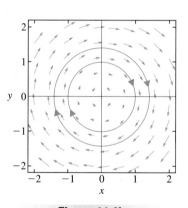

Figure 14.6b

Flow lines: $\langle y, -x \rangle$.

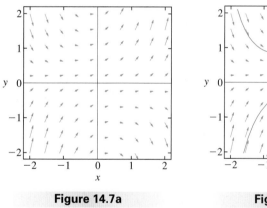

Figure 14.7a

$\langle 2, 1 + 2xy \rangle$.

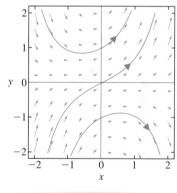

Figure 14.7b

Flow lines: $\langle 2, 1 + 2xy \rangle$.

A good sketch of a vector field can make it easy to visualize at least some of the flow lines. However, even a great sketch can't replace the information available in an exact equation for the flow lines. To see how we might solve for an equation of a flow line, note that if $\mathbf{F}(x, y) = \langle f_1(x, y), f_2(x, y) \rangle$ is a velocity field and $\langle x(t), y(t) \rangle$ is the position function, then $x'(t) = f_1(x, y)$ and $y'(t) = f_2(x, y)$. By the chain rule, we have

$$\frac{dy}{dx} = \frac{dy/dt}{dx/dt} = \frac{y'(t)}{x'(t)} = \frac{f_2(x, y)}{f_1(x, y)}. \tag{1.1}$$

Equation (1.1) is a first order differential equation for the unknown function $y(x)$. We refer you to section 6.5, where we developed a technique for solving one group of differential equations, called **separable equations.** In section 6.6, we presented a method (Euler's method) for approximating the solution of any first order differential equation passing through a given point.

Example 1.6 Using a Differential Equation to Construct Flow Lines

Construct the flow lines for the vector field $\langle y, -x \rangle$.

Solution From (1.1), we know that the flow lines are solutions of the differential equation

$$\frac{dy}{dx} = -\frac{x}{y}.$$

You may recall from our discussion in section 6.5 that this differential equation is separable and can be solved as follows. We first rewrite the equation as

$$y \frac{dy}{dx} = -x.$$

Integrating both sides with respect to x gives us

$$\int y \frac{dy}{dx} \, dx = -\int x \, dx,$$

so that

$$\frac{y^2}{2} = -\frac{x^2}{2} + k.$$

Multiplying both sides by 2 and replacing the constant $2k$ by c, we have

$$y^2 = -x^2 + c$$

or

$$x^2 + y^2 = c.$$

That is, for any choice of the constant $c > 0$, the solution corresponds to a circle centered at the origin. The vector field and the flow lines are then exactly as plotted in Figures 14.6a and 14.6b.

In the following example, we illustrate the use of Euler's method for constructing an approximate flow line.

Example 1.7 Using Euler's Method to Approximate Flow Lines

Use Euler's method with $h = 0.05$ to approximate the flow line for the vector field $\langle 2, 1 + 2xy \rangle$ passing through the point $(0, 1)$, for $0 \le x \le 1$.

Solution Recall that for the differential equation $y' = f(x, y)$ and for a given value of h, Euler's method produces a sequence of approximate values of the solution function $y = y(x)$ corresponding to the points $x_i = x_0 + ih$, for $i = 1, 2, \ldots$. Specifically, starting from an initial point (x_0, y_0), where $y_0 = y(x_0)$, we construct the approximate values $y_i \approx y(x_i)$, where the y_i's are determined iteratively from the equation

$$y_{i+1} = y_i + hf(x_i, y_i), \quad i = 0, 1, 2, \ldots.$$

Since the flow line must pass through the point $(0, 1)$, we start with $x_0 = 0$ and $y_0 = 1$. Further, here we have the differential equation

$$\frac{dy}{dx} = \frac{1 + 2xy}{2} = \frac{1}{2} + xy = f(x, y).$$

Notice that in this case (unlike example 1.6), this differential equation is not separable and you do not know how to solve it exactly. For Euler's method, we then have

$$y_{i+1} = y_i + hf(x_i, y_i) = y_i + 0.05 \left(\frac{1}{2} + x_i y_i \right),$$

with $x_0 = 0$, $y_0 = 1$. For the first two steps, we have

$$y_1 = y_0 + 0.05 \left(\frac{1}{2} + x_0 y_0 \right) = 1 + 0.05(0.5) = 1.025,$$

$x_1 = 0.05$,

$$y_2 = y_1 + 0.05 \left(\frac{1}{2} + x_1 y_1 \right) = 1.025 + 0.05(0.5 + 0.05125) = 1.0525625$$

and $x_2 = 0.1$. Continuing on in this fashion, we get the sequence of approximate values indicated in the following table.

x_i	0	0.05	0.10	0.15	0.20	0.25	0.30
y_i	1	1.025	1.0526	1.0828	1.1159	1.1521	1.1915
x_i	0.35	0.40	0.45	0.50	0.55	0.60	0.65
y_i	1.2344	1.2810	1.3316	1.3866	1.4462	1.5110	1.5813
x_i	0.70	0.75	0.80	0.85	0.90	0.95	1.00
y_i	1.6577	1.7404	1.8310	1.9293	2.0363	2.1529	2.2801

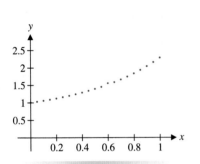

Figure 14.8
Approximate flow line
through (0, 1).

A plot of these points is shown in Figure 14.8. Compare this path to the top curve (also through the point (0, 1)) shown in Figure 14.7b.

An important type of vector field with which we already have some experience is the **gradient field,** where the vector field is the gradient of some scalar function. Because of the importance of gradient fields, there are a number of terms associated with them. Notice that in the following definition, we do not specify the number of independent variables, since the terms can be applied to functions of two, three or more variables.

Definition 1.2
For any scalar function f, the vector field $\mathbf{F} = \nabla f$ is called the **gradient field** for the function f. We call f a **potential function** for $\mathbf{F}$. Whenever $\mathbf{F} = \nabla f$, for some scalar function f, we refer to $\mathbf{F}$ as a **conservative vector field.**

It's worth noting that if you read about conservative vector fields and potentials in some applied areas (such as physics and engineering), you will sometimes see the function $-f$ referred to as the potential function. This is a minor difference in terminology, only. In this text (as is traditional in mathematics), we will consistently refer to f as the potential function. Rest assured that everything we say here about conservative vector fields is also true in these applications areas. The only slight difference may be that we call f the potential function, while others refer to $-f$ as the potential function.

As you will see, it's a simple matter to find the gradient field corresponding to a given scalar function.

Example 1.8   Finding Gradient Fields

Find the gradient fields corresponding to the functions (a) $f(x, y) = x^2 y - e^y$ and (b) $g(x, y, z) = \dfrac{1}{x^2 + y^2 + z^2}$ and sketch the fields.

Solution   (a) We first compute the partial derivatives $\dfrac{\partial f}{\partial x} = 2xy$ and $\dfrac{\partial f}{\partial y} = x^2 - e^y$ so that

$$\nabla f(x, y) = \left\langle \frac{\partial f}{\partial x}, \frac{\partial f}{\partial y} \right\rangle = \langle 2xy, x^2 - e^y \rangle.$$

A computer-generated graph of $\nabla f(x, y)$ is shown in Figure 14.9a.

(b) For $g(x, y, z) = (x^2 + y^2 + z^2)^{-1}$, we have the partial derivative

$$\frac{\partial g}{\partial x} = -(x^2 + y^2 + z^2)^{-2}(2x) = -\frac{2x}{x^2 + y^2 + z^2}.$$

We can use symmetry (think about this!) to conclude that $\dfrac{\partial g}{\partial y} = -\dfrac{2y}{x^2 + y^2 + z^2}$ and $\dfrac{\partial g}{\partial z} = -\dfrac{2z}{x^2 + y^2 + z^2}$. This gives us

$$\nabla g(x, y, z) = \left\langle \frac{\partial g}{\partial x}, \frac{\partial g}{\partial y}, \frac{\partial g}{\partial z} \right\rangle = -\frac{2\langle x, y, z \rangle}{x^2 + y^2 + z^2}.$$

A computer-generated graph of $\nabla g(x, y, z)$ is shown in Figure 14.9b.

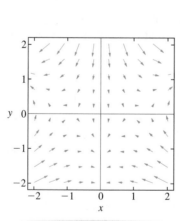

Figure 14.9a
$\nabla(x^2 y - e^y)$.

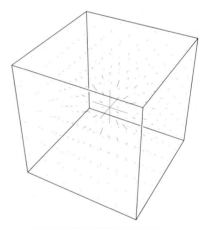

Figure 14.9b
$\nabla\left(\dfrac{1}{x^2 + y^2 + z^2}\right)$.

You will discover that many calculations involving vector fields simplify dramatically if the vector field is a gradient field (i.e., if the vector field is conservative). To take full advantage of these simplifications, you will need to be able to construct a potential function that generates a given conservative field. The technique introduced in the following example will work for most of the examples in this chapter.

Example 1.9 Finding Potential Functions

Determine whether each of the following vector fields is conservative. If it is, find a corresponding potential function $f(x, y)$: (a) $\mathbf{F}(x, y) = \langle 2xy - 3, x^2 + \cos y \rangle$ and (b) $\mathbf{G}(x, y) = \langle 3x^2y^2 - 2y, x^2y - 2x \rangle$.

Solution   The idea here is to try to construct a potential function. In the process of trying to construct one, we may instead recognize that there is no potential function for the given vector field. For (a), if $f(x, y)$ is a potential function for $\mathbf{F}(x, y)$, we must have that

$$\nabla f(x, y) = \mathbf{F}(x, y) = \langle 2xy - 3, x^2 + \cos y \rangle.$$

In this case, we have

$$\frac{\partial f}{\partial x} = 2xy - 3 \quad \text{and} \quad \frac{\partial f}{\partial y} = x^2 + \cos y. \tag{1.2}$$

Integrating the first of these two equations with respect to x and treating y as a constant, we get

$$f(x, y) = \int (2xy - 3)\,dx = x^2y - 3x + g(y). \tag{1.3}$$

Here, observe that we have added an arbitrary function of y alone, $g(y)$, rather than a *constant* of integration, because any function of y is treated as a constant when integrating with respect to x. Differentiating the expression for $f(x, y)$ with respect to y gives us

$$\frac{\partial f}{\partial y}(x, y) = x^2 + g'(y) = x^2 + \cos y,$$

from (1.2). This gives us $g'(y) = \cos y$, so that

$$g(y) = \int \cos y\,dy = \sin y + c.$$

From (1.3), we now have

$$f(x, y) = x^2y - 3x + \sin y + c,$$

where c is an arbitrary constant. Since we have been able to construct a potential function, the vector field $\mathbf{F}(x, y)$ is conservative.

(b) Again, we assume that there is a potential function f for $\mathbf{G}(x, y)$ and try to construct it. In this case, we have

$$\nabla f(x, y) = \mathbf{G}(x, y) = \langle 3x^2y^2 - 2y, x^2y - 2x \rangle,$$

so that

$$\frac{\partial f}{\partial x} = 3x^2y^2 - 2y \quad \text{and} \quad \frac{\partial f}{\partial y} = x^2y - 2x. \tag{1.4}$$

Integrating the first equation in (1.4) with respect to x, we have

$$f(x, y) = \int (3x^2y^2 - 2y)\,dx = x^3y^2 - 2yx + g(y),$$

where g is an arbitrary function of y. Differentiating this with respect to y, we have

$$\frac{\partial f}{\partial y}(x, y) = 2x^3y - 2x + g'(y) = x^2y - 2x,$$

from (1.4). Solving for $g'(y)$, we get

$$g'(y) = x^2 y - 2x - 2x^3 y + 2x = x^2 y - 2x^3 y,$$

which is impossible, since $g(y)$ is a function of y alone. We then conclude that there is no potential function for $\mathbf{G}(x, y)$ and so, the vector field is not conservative.

Remark 1.1

To find a potential function, you can either integrate $\dfrac{\partial f}{\partial x}$ with respect to x or integrate $\dfrac{\partial f}{\partial y}$ with respect to y. Before choosing which one to integrate, think about which integral will be easier to compute. In section 14.3, we introduce a simple method for determining whether or not a vector field is conservative.

EXERCISES 14.1

1. Compare hand-drawn sketches of the vector fields $\langle y, -x \rangle$ and $\langle 10y, -10x \rangle$. In particular, describe which graph is easier to interpret. Computer-generated graphs of these vector fields are identical when the software "scales" the vector field by dividing out the 10. It may seem odd that computers don't draw accurate graphs, but explain why the software programmers chose to scale the vector fields.

2. The gravitational force field is an example of an "inverse square law." That is, the magnitude of the gravitational force is inversely proportional to the square of the distance from the origin. Explain why the $\frac{3}{2}$ exponent in the denominator of example 1.4 is correct for an inverse square law.

3. Explain why each vector in a vector field graph is tangent to a flow line. Explain why this means that a flow line is essentially a large number of small (scaled) vector field vectors joined together.

4. In example 1.9(b), explain why the presence of the x's in the expression for $g'(y)$ proves that there is no potential function.

In exercises 5–14, sketch several vectors in the vector field by hand and verify your sketch with a CAS.

5. $\mathbf{F}(x, y) = \langle -y, x \rangle$

6. $\mathbf{F}(x, y) = \dfrac{\langle -y, x \rangle}{\sqrt{x^2 + y^2}}$

7. $\mathbf{F}(x, y) = \langle 0, x^2 \rangle$

8. $\mathbf{F}(x, y) = \langle 2x, 0 \rangle$

9. $\mathbf{F}(x, y) = 2y\mathbf{i} + \mathbf{j}$

10. $\mathbf{F}(x, y) = -\mathbf{i} + y^2\mathbf{j}$

11. $\mathbf{F}(x, y, z) = \langle 0, z, 1 \rangle$

12. $\mathbf{F}(x, y, z) = \langle 2, 0, 0 \rangle$

13. $\mathbf{F}(x, y, z) = \dfrac{\langle x, y, z \rangle}{\sqrt{x^2 + y^2 + z^2}}$

14. $\mathbf{F}(x, y, z) = \dfrac{\langle x, y, z \rangle}{x^2 + y^2 + z^2}$

15. Match the vector fields with their graphs.

$$\mathbf{F}_1(x, y) = \dfrac{\langle x, y \rangle}{\sqrt{x^2 + y^2}}, \quad \mathbf{F}_2(x, y) = \langle x, y \rangle,$$

$$\mathbf{F}_3(x, y) = \langle e^y, x \rangle, \quad \mathbf{F}_4(x, y) = \langle e^y, y \rangle$$

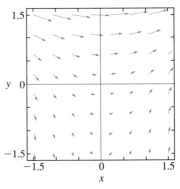

Graph A

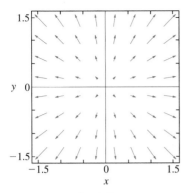

Graph B

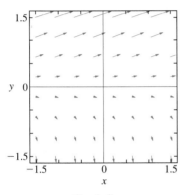

Graph C

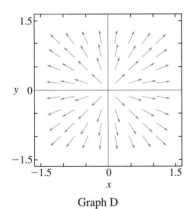

Graph D

16. Show that $\dfrac{\mathbf{r}}{r^n} = \dfrac{\langle x, y\rangle}{(x^2+y^2)^{n/2}}$ is conservative for any integer n.

In exercises 17–26, find the gradient field corresponding to f. Use a CAS to graph it.

17. $f(x, y) = x^2 + y^2$ 18. $f(x, y) = x^2 - y^2$

19. $f(x, y) = \sqrt{x^2 + y^2}$ 20. $f(x, y) = \sin(x^2 + y^2)$

21. $f(x, y) = xe^{-y}$ 22. $f(x, y) = y \sin x$

23. $f(x, y, z) = \sqrt{x^2 + y^2 + z^2}$ 24. $f(x, y, z) = xyz$

25. $f(x, y, z) = x^2 y + yz$ 26. $f(x, y, z) = (x - y)^2 + z$

In exercises 27–38, determine whether or not the vector field is conservative. If it is, find a potential function.

27. $\langle y, x\rangle$ 28. $\langle 2, y\rangle$

29. $\langle y, -x\rangle$ 30. $\langle y, 1\rangle$

31. $\langle x - 2xy\rangle\mathbf{i} + \langle y^2 - x^2\rangle\mathbf{j}$ 32. $\langle x^2 - y\rangle\mathbf{i} + \langle x - y\rangle\mathbf{j}$

33. $\langle y \sin xy, x \sin xy\rangle$ 34. $\langle y \cos x, \sin x - y\rangle$

35. $\langle 4x - z, 3y + z, y - x\rangle$ 36. $\langle z^2 + 2xy, x^2 - z, 2xz - 1\rangle$

37. $\langle y^2 z^2 - 1, 2xyz^2, 4z^3\rangle$ 38. $\langle z^2 + 2xy, x^2 + 1, 2xz - 3\rangle$

In exercises 39–46, find equations for the flow lines.

39. $\langle 2, \cos x\rangle$ 40. $\langle x^2, 2\rangle$

41. $\langle 2y, 3x^2\rangle$ 42. $\left\langle \frac{1}{y}, 2x\right\rangle$

43. $y\mathbf{i} + xe^y\mathbf{j}$ 44. $e^{-x}\mathbf{i} + 2x\mathbf{j}$

45. $\langle y, y^2 + 1\rangle$ 46. $\langle 2, y^2 + 1\rangle$

47. Suppose that $f(x)$, $g(y)$ and $h(z)$ are continuous functions. Show that $\langle f(x), g(y), h(z)\rangle$ is conservative by finding a potential function.

48. Show that $\langle k_1, k_2\rangle$ is conservative for constants k_1 and k_2.

In exercises 49–54, use the notation $\mathbf{r} = \langle x, y\rangle$ and $r = \|\mathbf{r}\| = \sqrt{x^2 + y^2}$.

49. Show that $\nabla(r) = \dfrac{\mathbf{r}}{r}$. 50. Show that $\nabla(r^2) = 2\mathbf{r}$.

51. Find $\nabla(r^3)$.

52. Use exercises 49–51 to conjecture the value of $\nabla(r^n)$ for any positive integer n. Prove that your answer is correct.

53. Show that $\dfrac{\langle 1, 1\rangle}{r}$ is *not* conservative.

54. Show that $\dfrac{\langle -y, x\rangle}{r^2}$ is conservative by finding a potential function. Show that the potential function can be thought of as the polar angle θ.

55. The current in a wire produces a magnetic field $\mathbf{B} = \dfrac{k\langle -y, x\rangle}{r^2}$. The constant k is the **strength** of the current. Draw a sketch showing a wire and its magnetic field.

56. Show that the vector field $\mathbf{F}(x, y) = \langle y, x\rangle$ has potential function $f(x, y) = xy$. The curves $f(x, y) = c$ for constants c are called **equipotential curves.** Sketch equipotential curves for several constants (positive and negative). Find the flow lines for this vector field and show that the flow lines and equipotential curves intersect at right angles. This situation is

common. To further develop these relationships, show that the potential function and the flow function $g(x, y) = \frac{1}{2}(y^2 - x^2)$ are both solutions of **Laplace's equation** $\nabla^2 u = 0$ where $\nabla^2 u = u_{xx} + u_{yy}$.

57.   In example 1.5, we graphed the flow lines for the vector field $\langle 2, 1 + 2xy \rangle$ and mentioned that finding equations for the flow lines was beyond what's been presented in the text. We develop a method for finding the flow lines here by solving **linear ordinary differential equations.** We will illustrate for an easier vector field, $\langle x, 2x - y \rangle$. First, note that if $x'(t) = x$ and $y'(t) = 2x - y$, then

$$\frac{dy}{dx} = \frac{2x - y}{x} = 2 - \frac{y}{x}.$$

The flow lines will be the graphs of functions $y(x)$ such that $y'(x) = 2 - \frac{y}{x}$, or $y' + \frac{1}{x}y = 2$. The left-hand side of the equation should look a little like a product rule. Our main goal is to multiply by a term called the **integrating factor** to make the left-hand side exactly a product rule derivative. It turns out that for the equation $y' + f(x)y = g(x)$, an integrating factor is $e^{\int f(x)\,dx}$. In the present case, we have $e^{\int 1/x\,dx} = e^{\ln x} = x$. (We have chosen the integration constant to be 0 to keep the integrating factor simple.) Multiply the entire equation by x and show that $xy' + y = 2x$. Show that $xy' + y = (xy)'$. From $(xy)' = 2x$, integrate to get $xy = x^2 + c$, or $y = x + \frac{c}{x}$. To find a flow line passing through the point $(1, 2)$, show that $c = 1$ and thus $y = x + \frac{1}{x}$. To find a flow line passing through the point $(1, 1)$, show that $c = 0$ and thus $y = x$. Sketch the vector field and highlight the curves $y = x + \frac{1}{x}$ and $y = x$.

14.2   LINE INTEGRALS

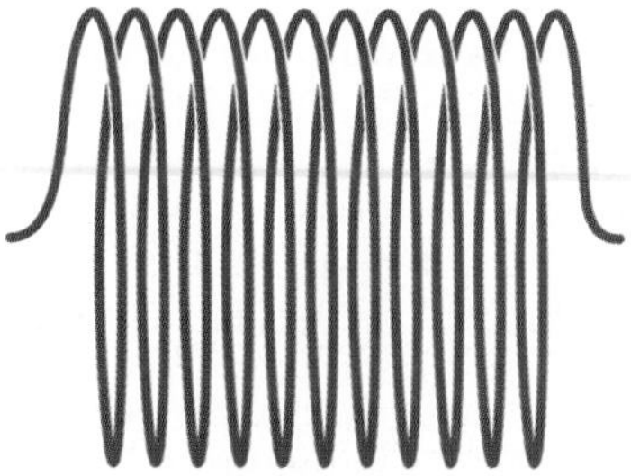

Recall that in section 5.6, we used integration to find the mass of a thin rod with variable mass density. There, we had observed that if the rod extends from $x = a$ to $x = b$ and has mass density function $\rho(x)$, then the mass of the rod is given by $\int_a^b \rho(x)\,dx$. This definition is fine for objects that are essentially one-dimensional. But what if we wanted to find the mass of a helical spring (see Figure 14.10)? Calculus is remarkable in that the same technique can solve a wide variety of problems. As you should expect by now, we will derive a solution by first approximating the curve with line segments and then passing to a limit.

In this three-dimensional setting, the density function has the form $\rho(x, y, z)$. We assume that the object is in the shape of a curve C in three dimensions with endpoints (a, b, c) and (d, e, f). Further, we assume that the curve is **oriented,** which means that there is a direction to the curve. For example, the curve C might start at (a, b, c) and end at (d, e, f). We first partition the curve into n pieces with endpoints $(a, b, c) = (x_0, y_0, z_0)$, $(x_1, y_1, z_1), (x_2, y_2, z_2), \ldots, (x_n, y_n, z_n) = (d, e, f)$, as indicated in Figure 14.11. We will use the shorthand P_i to denote the point (x_i, y_i, z_i), and C_i for the section of the curve C extending from P_{i-1} to P_i, for each $i = 1, 2, \ldots, n$. Our initial objective is to approximate the mass of the portion of the object along C_i. Note that if the segment C_i is small enough, we can consider the density to be constant on C_i. In this case, the mass of this segment would simply be the product of the density and the length of C_i. For some point (x_i^*, y_i^*, z_i^*) on C_i, we approximate the density on C_i by $\rho(x_i^*, y_i^*, z_i^*)$. The mass of the section C_i is then approximately

$$\rho(x_i^*, y_i^*, z_i^*)\,\Delta s_i,$$

where Δs_i represents the arc length of C_i. The mass m of the entire object is then approximately the sum of the masses of the sections,

$$m \approx \sum_{i=1}^{n} \rho(x_i^*, y_i^*, z_i^*)\,\Delta s_i.$$

You should expect that this approximation will improve as we divide the curve into more and more segments that are shorter and shorter in length. In particular, if we take the norm of the

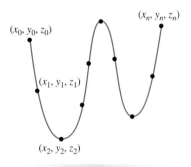

Figure 14.10

A helical spring.

(x_n, y_n, z_n)

(x_0, y_0, z_0)

(x_1, y_1, z_1)

(x_2, y_2, z_2)

Figure 14.11

Partitioned curve.

partition $\|P\|$ to be the maximum of the arc lengths Δs_i ($i = 1, 2, \ldots, n$), then we have

$$m = \lim_{\|P\| \to 0} \sum_{i=1}^{n} \rho(x_i^*, y_i^*, z_i^*) \, \Delta s_i, \tag{2.1}$$

provided the limit exists and is the same for every choice of the evaluation points (x_i^*, y_i^*, z_i^*) ($i = 1, 2, \ldots, n$).

You might recognize that (2.1) looks like the limit of a Riemann sum (an integral). As it turns out, this limit arises naturally in numerous applications. We pause now to give this limit a name and identify some useful properties.

Definition 2.1

The **line integral of $f(x, y, z)$ with respect to arc length** along the oriented curve C in three-dimensional space, written $\int_C f(x, y, z) \, ds$, is defined by

$$\int_C f(x, y, z) \, ds = \lim_{\|P\| \to 0} \sum_{i=1}^{n} f(x_i^*, y_i^*, z_i^*) \, \Delta s_i,$$

provided the limit exists and is the same for all choices of evaluation points.

We define line integrals of functions $f(x, y)$ of two variables along an oriented curve C in the xy-plane in a similar way. Often, the curve C is specified by parametric equations or you can use your skills developed in Chapters 9 and 11 to construct parametric equations for the curve. In these cases, the following result allows us to evaluate the line integral as a definite integral of a function of one variable.

Theorem 2.1 (Evaluation Theorem)

Suppose that $f(x, y, z)$ is continuous in a region D containing the curve C and that C is described parametrically by $(x(t), y(t), z(t))$, for $a \leq t \leq b$, where $x(t), y(t)$ and $z(t)$ have continuous first derivatives. Then,

$$\int_C f(x, y, z) \, ds = \int_a^b f(x(t), y(t), z(t)) \sqrt{[x'(t)]^2 + [y'(t)]^2 + [z'(t)]^2} \, dt.$$

Suppose that $f(x, y)$ is continuous in a region D containing the curve C and that C is described parametrically by $(x(t), y(t))$, for $a \leq t \leq b$, where $x(t)$ and $y(t)$ have continuous first derivatives. Then

$$\int_C f(x, y) \, ds = \int_a^b f(x(t), y(t)) \sqrt{[x'(t)]^2 + [y'(t)]^2} \, dt.$$

Proof

We prove the result for the case of a curve in two dimensions and leave the three-dimensional case as an exercise. From Definition 2.1 (adjusted for the two-dimensional case), we have

$$\int_C f(x, y) \, ds = \lim_{\|P\| \to 0} \sum_{i=1}^{n} f(x_i^*, y_i^*) \, \Delta s_i, \tag{2.2}$$

where Δs_i represents the arc length of the section of the curve C between (x_{i-1}, y_{i-1}) and (x_i, y_i). Choose $t_0, t_1, \ldots, t_n$ so that $x(t_i) = x_i$ and $y(t_i) = y_i$, for $i = 0, 1, \ldots, n$.

Notice that we can approximate the arc length of such a small section of the curve by the straight-line distance:

$$\Delta s_i \approx \sqrt{(x_i - x_{i-1})^2 + (y_i - y_{i-1})^2}.$$

Further, since $x(t)$ and $y(t)$ have continuous first derivatives, we have by the Mean Value Theorem (as in the derivation of the arc length formula in section 9.3), that

$$\Delta s_i \approx \sqrt{(x_i - x_{i-1})^2 + (y_i - y_{i-1})^2} \approx \sqrt{[x'(t_i^*)]^2 + [y'(t_i^*)]^2}\, \Delta t_i,$$

for some $t_i^* \in (t_{i-1}, t_i)$. Together with (2.2), this gives us

$$\int_C f(x, y)\, ds = \lim_{\|P\| \to 0} \sum_{i=1}^{n} f(x(t_i^*), y(t_i^*)) \sqrt{[x'(t_i^*)]^2 + [y'(t_i^*)]^2}\, \Delta t_i$$

$$= \int_a^b f(x(t), y(t)) \sqrt{[x'(t)]^2 + [y'(t)]^2}\, dt.$$

We refer to any curve C satisfying the hypotheses of Theorem 2.1 as **smooth.** That is, C is smooth if it can be described parametrically by $x = x(t)$, $y = y(t)$ and $z = z(t)$, for $a \le t \le b$, where $x(t)$, $y(t)$ and $z(t)$ all have continuous first derivatives on the interval $[a, b]$. Similarly, a plane curve is smooth if it can be parameterized by $x = x(t)$ and $y = y(t)$, for $a \le t \le b$, where $x(t)$ and $y(t)$ have continuous first derivatives on the interval $[a, b]$.

Notice that for curves in space, Theorem 2.1 says essentially that the arc length element ds can be replaced by

$$ds = \sqrt{[x'(t)]^2 + [y'(t)]^2 + [z'(t)]^2}\, dt. \tag{2.3}$$

The term $\sqrt{[x'(t)]^2 + [y'(t)]^2 + [z'(t)]^2}$ in the integral should be very familiar, having been present in our integral representations of both arc length and surface area. Likewise, for curves in the plane, the arc length element is

$$ds = \sqrt{[x'(t)]^2 + [y'(t)]^2}\, dt. \tag{2.4}$$

Example 2.1   Finding the Mass of a Helical Spring

Find the mass of a spring in the shape of the helix defined parametrically by $x = 2\cos t$, $y = t$, $z = 2\sin t$, for $0 \le t \le 6\pi$, with density $\rho(x, y, z) = 2y$

Solution   A graph of the helix is shown in Figure 14.12. The density is

$$\rho(x, y, z) = 2y = 2t,$$

and from (2.3), the arc length element ds is given by

$$ds = \sqrt{[x'(t)]^2 + [y'(t)]^2 + [z'(t)]^2}\, dt = \sqrt{(-2\sin t)^2 + (1)^2 + (2\cos t)^2}\, dt = \sqrt{5}\, dt,$$

where we have used the identity $4\sin^2 t + 4\cos^2 t = 4$. By Theorem 2.1, we have

$$\text{mass} = \int_C \rho(x, y, z)\, ds = \int_0^{6\pi} \underbrace{2t}_{\rho(x, y, z)}\, \underbrace{\sqrt{5}\, dt}_{ds} = 2\sqrt{5} \int_0^{6\pi} t\, dt$$

$$= 2\sqrt{5}\, \frac{(6\pi)^2}{2} = 36\pi^2 \sqrt{5}.$$

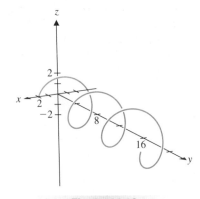

Figure 14.12
The helix $x = 2\cos t$, $y = t$, $z = 2\sin t$, $0 \le t \le 6\pi$.

We should point out that example 2.1 is very unusual in at least one respect: we were able to compute the integral exactly. Most line integrals of the type defined in Definition 2.1 are too complicated to evaluate exactly and will need to be approximated with some numerical method, as in the following example.

Example 2.2　Evaluating a Line Integral with Respect to Arc Length

Evaluate the line integral $\int_C (2x^2 - 3yz)\, ds$, where C is the curve defined parametrically by $x = \cos t$, $y = \sin t$, $z = \cos t$ with $0 \le t \le 2\pi$.

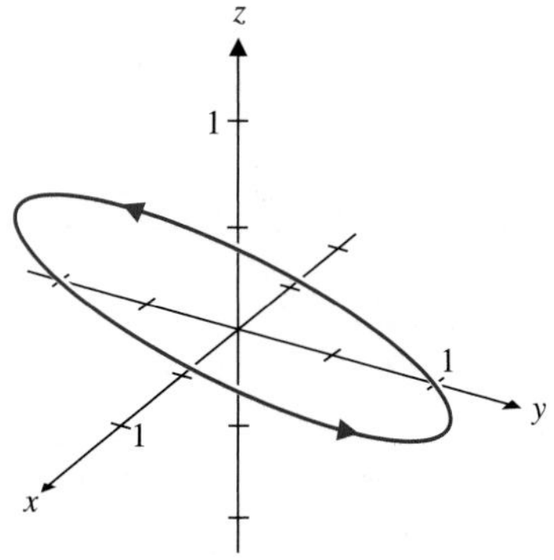

Solution　A graph of C is shown in Figure 14.13. The integrand is

$$f(x, y, z) = 2x^2 - 3yz = 2\cos^2 t - 3\sin t \cos t.$$

From (2.3), the arc length element is given by

$$ds = \sqrt{[x'(t)]^2 + [y'(t)]^2 + [z'(t)]^2}\, dt$$

$$= \sqrt{(-\sin t)^2 + (\cos t)^2 + (-\sin t)^2}\, dt = \sqrt{1 + \sin^2 t}\, dt,$$

where we have used the identity $\sin^2 t + \cos^2 t = 1$. By Theorem 2.1, we now have

$$\int_C (2x^2 - 3yz)\, ds = \int_0^{2\pi} \underbrace{(2\cos^2 t - 3\sin t \cos t)}_{2x^2 - 3yz}\,\underbrace{\sqrt{1 + \sin^2 t}\, dt}_{ds} \approx 6.9922,$$

where we approximated the last integral numerically.

◼

In the following (two-dimensional) example, you must find parametric equations for the curve before evaluating the line integral. Also, you will discover an important fact about the orientation of the curve C.

Example 2.3　Evaluating a Line Integral with Respect to Arc Length

Evaluate the line integral $\int_C 2x^2 y\, ds$, where C is (a) the portion of the parabola $y = x^2$ from $(-1, 1)$ to $(2, 4)$ and (b) the portion of the parabola $y = x^2$ from $(2, 4)$ to $(-1, 1)$.

Solution　(a) A sketch of the curve is shown in Figure 14.14a. Notice that you can use $x = t$ as the parameter in this case, since the curve is already written explicitly in terms of x. We can then describe the curve parametrically by $x = t$ and $y = t^2$ with $-1 \le t \le 2$. Using this, the integrand becomes $2x^2 y = 2t^2 t^2 = 2t^4$ and from (2.4), the arc length element is

$$ds = \sqrt{[x'(t)]^2 + [y'(t)]^2}\, dt = \sqrt{1 + 4t^2}\, dt.$$

The integral is now written as

$$\int_C 2x^2 y\, ds = \int_{-1}^{2} \underbrace{2t^4}_{2x^2 y}\,\underbrace{\sqrt{1 + 4t^2}\, dt}_{ds} \approx 45.391,$$

where we have again evaluated the integral numerically (although in this case a good CAS can give you an exact answer).

(b) A sketch of the curve is shown in Figure 14.14b. Notice that the curve is the same as in part (a), except that the orientation is backward. In this case, we represent the curve with the parametric equations $x = -t$ and $y = t^2$, where t runs from $t = -2$ to

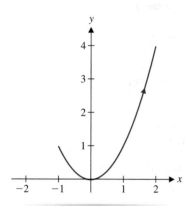

Figure 14.13

The curve $x = \cos t$, $y = \sin t$, $z = \cos t$ with $0 \le t \le 2\pi$.

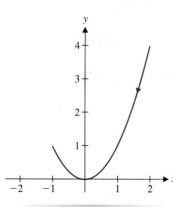

Figure 14.14a

$y = x^2$ from $(-1, 1)$ to $(2, 4)$.

Figure 14.14b

$y = x^2$ from $(2, 4)$ to $(-1, 1)$.

$t = 1$. Observe that everything else in the integral remains the same and we have

$$\int_C 2x^2 y \, ds = \int_{-2}^{1} 2t^4 \sqrt{1 + 4t^2} \, dt \approx 45.391,$$

as before.

Notice that in example 2.3, the line integral was the same no matter which orientation we took for the curve. It turns out that this is true in general for all line integrals defined by Definition 2.1 (i.e., line integrals with respect to arc length).

Look carefully at the hypotheses of Theorem 2.1. Notice that you can use this evaluation theorem to rewrite a line integral only when the curve C is smooth. Unfortunately, this is often not the case in practice; many curves of interest are not smooth. Fortunately, we can extend the result of Theorem 2.1 to the case where C is a union of a finite number of smooth curves:

$$C = C_1 \cup C_2 \cup \cdots \cup C_n,$$

where each of $C_1, C_2, \ldots, C_n$ are smooth. We call such a curve C **piecewise-smooth.** Notice that if C_1 and C_2 are oriented curves and the endpoint of C_1 is the same as the initial point of C_2, then the curve $C_1 \cup C_2$ is an oriented curve with the same initial point as C_1 and the same endpoint as C_2 (see Figure 14.15). The following results should not seem surprising. Here, for an oriented curve C in two or three dimensions, the curve $-C$ denotes the same curve as C, but with the opposite orientation.

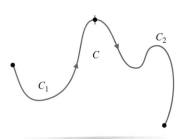

Figure 14.15
$C = C_1 \cup C_2.$

Theorem 2.2

Suppose that $f(x, y, z)$ is a continuous function in some region D containing the oriented curve C. Then, if C is piecewise-smooth, with $C = C_1 \cup C_2 \cup \cdots \cup C_n$, where $C_1, C_2, \ldots, C_n$ are all smooth, we have
(i)

$$\int_{-C} f(x, y, z) \, ds = \int_C f(x, y, z) \, ds$$

and (ii)

$$\int_C f(x, y, z) \, ds = \int_{C_1} f(x, y, z) \, ds + \int_{C_2} f(x, y, z) \, ds + \cdots + \int_{C_n} f(x, y, z) \, ds.$$

We leave the proof of the theorem as an exercise. Notice that the corresponding result will also be true in two dimensions. We use part (ii) of Theorem 2.2 in the following example.

Example 2.4   Evaluating a Line Integral over a Piecewise-Smooth Curve

Evaluate the line integral $\int_C (3x - y) \, ds$, where C is the line segment from $(1, 2)$ to $(3, 3)$, followed by the portion of the circle $x^2 + y^2 = 18$ traversed from $(3, 3)$ clockwise around to $(3, -3)$.

Solution   A graph of the curve is shown in Figure 14.16. Notice that we'll need to evaluate the line integral separately over the line segment C_1 and the quarter-circle C_2.

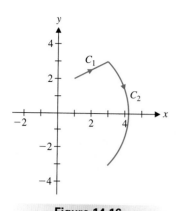

Figure 14.16

Piecewise-smooth curve.

Further, although C is not smooth, it is piecewise smooth, since C_1 and C_2 are both smooth. Now, observe that we can write C_1 parametrically as $x = 1 + (3 - 1)t = 1 + 2t$ and $y = 2 + (3 - 2)t = 2 + t$, for $0 \le t \le 1$. Also, on C_1, the integrand is given by

$$3x - y = 3(1 + 2t) - (2 + t) = 1 + 5t$$

and from (2.4), the arc length element is

$$ds = \sqrt{(2)^2 + (1)^2}\, dt = \sqrt{5}\, dt.$$

Putting this together, we have

$$\int_{C_1} f(x, y)\, ds = \int_0^1 \underbrace{(1 + 5t)}_{f(x,\, y)}\, \underbrace{\sqrt{5}\, dt}_{ds} = \frac{7}{2}\sqrt{5}. \tag{2.5}$$

Next, for C_2, the usual parametric equations for a circle of radius r oriented counterclockwise are $x(t) = r \cos t$ and $y(t) = r \sin t$. In the present case, the radius is $\sqrt{18}$ and the curve is oriented clockwise, which means that $y(t)$ has the opposite sign from the usual orientation. So, parametric equations for C_2 are $x(t) = \sqrt{18} \cos t$ and $y(t) = -\sqrt{18} \sin t$. Notice, too that the initial point $(3, 3)$ corresponds to the angle $-\frac{\pi}{4}$ and the endpoint $(3, -3)$ corresponds to the angle $\frac{\pi}{4}$. Finally, on C_2, the integrand is given by

$$3x - y = 3\sqrt{18} \cos t + \sqrt{18} \sin t$$

and the arc length element is

$$ds = \sqrt{(-\sqrt{18} \sin t)^2 + (-\sqrt{18} \cos t)^2}\, dt = \sqrt{18}\, dt,$$

where we have again used the fact that $\sin^2 t + \cos^2 t = 1$. This gives us

$$\int_{C_2} f(x, y)\, ds = \int_{-\pi/4}^{\pi/4} \underbrace{(3\sqrt{18} \cos t + \sqrt{18} \sin t)}_{f(x,\, y)}\, \underbrace{\sqrt{18}\, dt}_{ds} = 54\sqrt{2}. \tag{2.6}$$

Combining the integrals over the two curves, we have from (2.5) and (2.6) that

$$\int_C f(x, y)\, ds = \int_{C_1} f(x, y)\, ds + \int_{C_2} f(x, y)\, ds = \frac{7}{2}\sqrt{5} + 54\sqrt{2}.$$

So far, we have discussed how to calculate line integrals for curves described parametrically and we have given one application of line integrals (calculation of mass). In the exercises, we discuss further applications. We now develop a geometric interpretation of the line integral. The key observation here is that the line integral $\int_C f(x, y)\, ds$ has a great deal in common with the definite integral $\int_a^b f(x)\, dx$. Whereas $\int_a^b f(x)\, dx$ corresponds to a limit of sums of the heights of the function $f(x)$ above or below the x-axis for an interval $[a, b]$ of the x-axis, the line integral $\int_C f(x, y)\, ds$ corresponds to a limit of the sums of the heights of the function $f(x, y)$ above or below the xy-plane for a curve C lying in the xy-plane. Consider the depiction of this interpretation shown in Figures 14.17a and 14.17b (on the following page). You should recall that for $f(x) \ge 0$, $\int_a^b f(x)\, dx$ measures the area under the curve $y = f(x)$ on the interval $[a, b]$, shaded in Figure 14.17a. Likewise, if $f(x, y) \ge 0$, $\int_C f(x, y)\, ds$ measures the surface area of the shaded surface indicated in Figure 14.17b. In general, $\int_a^b f(x)\, dx$ measures signed area [positive if $f(x) > 0$ and negative if $f(x) < 0$] and the line integral $\int_C f(x, y)\, ds$ measures the (signed) surface area of the surface formed by vertical segments from the xy-plane to the graph of $z = f(x, y)$.

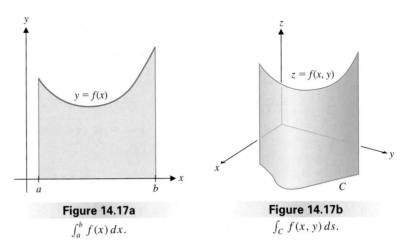

Figure 14.17a
$\int_a^b f(x)\,dx.$

Figure 14.17b
$\int_C f(x, y)\,ds.$

The following result, whose proof we leave as a straightforward exercise, gives some geometric significance to line integrals in both two and three dimensions.

Theorem 2.3
For any piecewise smooth curve C (in two or three dimensions), $\int_C 1\,ds$ gives the arc length of the curve C.

We have now extended the geometry and certain applications of definite integrals to line integrals of the form $\int_C f(x, y)\,ds$. However, not all of the properties of definite integrals can be extended to line integrals as we have defined them thus far. As we will see, a careful consideration of the calculation of work will force us to define several alternative versions of the line integral.

Recall that if a constant force f is exerted to move an object a distance d in a straight line, the work done is given by $W = f \cdot d$. In section 5.6, we extended this to a variable force $f(x)$ applied to an object as it moves in a straight line from $x = a$ to $x = b$. In this case, we found that the work done by the force is given by

$$W = \int_a^b f(x)\,dx.$$

We now want to extend this idea to find the work done as an object moves along a curve in three dimensions. Remember that in three dimensions, force vectors are given by the values of vector fields (force fields). Here, we want to compute the work done on an object by a force field $\mathbf{F}(x, y, z)$, as the object moves along a curve C. Unfortunately, as it turns out, our present notion of line integral (the line integral with respect to arc length) does not help in this case. As we did for finding mass, we need to start from scratch and so, partition the curve C into n segments $C_1, C_2, \ldots, C_n$. Notice that on each segment C_i ($i = 1, 2, \ldots, n$), if the segment is small and $\mathbf{F}$ is continuous, then $\mathbf{F}$ will be nearly constant on C_i and so, we can approximate $\mathbf{F}$ by its value at some point (x_i^*, y_i^*, z_i^*) on C_i. The work done along C_i (call it W_i) is then approximately the same as the product of the component of the force $\mathbf{F}(x_i^*, y_i^*, z_i^*)$ in the direction of the unit tangent vector $\mathbf{T}(x, y, z)$ to C at (x_i^*, y_i^*, z_i^*) and the distance traveled. That is,

$$W_i \approx \mathbf{F}(x_i^*, y_i^*, z_i^*) \cdot \mathbf{T}(x_i^*, y_i^*, z_i^*)\,\Delta s_i,$$

where Δs_i is the arc length of the segment C_i. Now, if C_i can be represented parametrically by $x = x(t)$, $y = y(t)$ and $z = z(t)$, for $a \leq t \leq b$ and assuming that C_i is smooth, we have

$$W_i \approx \mathbf{F}(x_i^*, y_i^*, z_i^*) \cdot \mathbf{T}(x_i^*, y_i^*, z_i^*) \, \Delta s_i$$

$$= \frac{\mathbf{F}(x_i^*, y_i^*, z_i^*) \cdot \langle x'(t_i^*), y'(t_i^*), z'(t_i^*) \rangle}{\sqrt{[x'(t_i^*)]^2 + [y'(t_i^*)]^2 + [z'(t_i^*)]^2}} \sqrt{[x'(t_i^*)]^2 + [y'(t_i^*)]^2 + [z'(t_i^*)]^2} \, \Delta t$$

$$= \mathbf{F}(x_i^*, y_i^*, z_i^*) \cdot \langle x'(t_i^*), y'(t_i^*), z'(t_i^*) \rangle \, \Delta t,$$

where $(x_i^*, y_i^*, z_i^*) = (x(t_i^*), y(t_i^*), z(t_i^*))$. Next, if

$$\mathbf{F}(x, y, z) = \langle F_1(x, y, z), F_2(x, y, z), F_3(x, y, z) \rangle,$$

we have

$$W_i \approx \langle F_1(x_i^*, y_i^*, z_i^*), F_2(x_i^*, y_i^*, z_i^*), F_3(x_i^*, y_i^*, z_i^*) \rangle \cdot \langle x'(t_i^*), y'(t_i^*), z'(t_i^*) \rangle \, \Delta t.$$

Adding together the approximations of the work done over the various segments of C, we have that the total work done is approximately

$$W \approx \sum_{i=1}^{n} \langle F_1(x_i^*, y_i^*, z_i^*), F_2(x_i^*, y_i^*, z_i^*), F_3(x_i^*, y_i^*, z_i^*) \rangle \cdot \langle x'(t_i^*), y'(t_i^*), z'(t_i^*) \rangle \, \Delta t.$$

Finally, taking the limit as the norm of the partition of C approaches zero, we arrive at

$$W = \lim_{\|P\| \to 0} \sum_{i=1}^{n} \mathbf{F}(x_i^*, y_i^*, z_i^*) \cdot \langle x'(t_i^*), y'(t_i^*), z'(t_i^*) \rangle \, \Delta t$$

$$= \lim_{\|P\| \to 0} \sum_{i=1}^{n} [F_1(x_i^*, y_i^*, z_i^*)x'(t_i^*)\Delta t + F_2(x_i^*, y_i^*, z_i^*)y'(t_i^*) \, \Delta t$$

$$+ F_3(x_i^*, y_i^*, z_i^*)z'(t_i^*) \, \Delta t]$$

$$= \int_a^b F_1(x(t), y(t), z(t))x'(t) \, dt + \int_a^b F_2(x(t), y(t), z(t))y'(t) \, dt$$

$$+ \int_a^b F_3(x(t), y(t), z(t))z'(t) \, dt. \tag{2.7}$$

We now define line integrals corresponding to each of the three integrals in (2.7).

In the following definition, the notation is the same as in Definition 2.1, with the added terms $\Delta x_i = x_i - x_{i-1}$, $\Delta y_i = y_i - y_{i-1}$ and $\Delta z_i = z_i - z_{i-1}$.

Definition 2.2

The **line integral of $f(x, y, z)$ with respect to x** along the oriented curve C in three-dimensional space is written as $\int_C f(x, y, z) \, dx$ and is defined by

$$\int_C f(x, y, z) \, dx = \lim_{\|P\| \to 0} \sum_{i=1}^{n} f(x_i^*, y_i^*, z_i^*) \, \Delta x_i,$$

provided the limit exists and is the same for all choices of evaluation points.

Likewise, we define the **line integral of $f(x, y, z)$ with respect to y** along C by

$$\int_C f(x, y, z) \, dy = \lim_{\|P\| \to 0} \sum_{i=1}^{n} f(x_i^*, y_i^*, z_i^*) \, \Delta y_i$$

and the **line integral of $f(x, y, z)$ with respect to z** along C by

$$\int_C f(x, y, z) \, dz = \lim_{\|P\| \to 0} \sum_{i=1}^{n} f(x_i^*, y_i^*, z_i^*) \, \Delta z_i.$$

In each case, the line integral is defined whenever the corresponding limit exists and is the same for all choices of evaluation points.

If we have a parametric representation of the curve C, then we can rewrite each line integral as a definite integral. The proof of the following result is very similar to that of Theorem 2.1 and we leave it as an exercise.

Theorem 2.4 (Evaluation Theorem)

Suppose that $f(x, y, z)$ is continuous in a region D containing the curve C and that C is described parametrically by $x = x(t)$, $y = y(t)$ and $z = z(t)$, where t ranges from $t = a$ to $t = b$ and $x(t)$, $y(t)$ and $z(t)$ have continuous first derivatives. Then,

$$\int_C f(x, y, z)\, dx = \int_a^b f(x(t), y(t), z(t))\, x'(t)\, dt$$

$$\int_C f(x, y, z)\, dy = \int_a^b f(x(t), y(t), z(t))\, y'(t)\, dt \quad \text{and}$$

$$\int_C f(x, y, z)\, dz = \int_a^b f(x(t), y(t), z(t))\, z'(t)\, dt.$$

Before returning to the calculation of work, we will examine some simpler examples. Recall that the line integral along a given curve with respect to arc length will not change if we traverse the curve in the opposite direction. On the other hand, as we'll see, line integrals with respect to x, y or z change sign when the orientation of the curve changes. You will observe this property in the following example.

Example 2.5   Calculating a Line Integral in Space

Compute the line integral $\int_C (4xz + 2y)\, dx$, where C is the line segment (a) from $(2, 1, 0)$ to $(4, 0, 2)$ and (b) from $(4, 0, 2)$ to $(2, 1, 0)$.

Solution   First, we find parametric equations for C. For part (a), notice that you can use

$$x = 2 + (4 - 2)t = 2 + 2t,$$
$$y = 1 + (0 - 1)t = 1 - t \quad \text{and}$$
$$z = 0 + (2 - 0)t = 2t,$$

for $0 \le t \le 1$. The integrand is then

$$4xz + 2y = 4(2 + 2t)(2t) + 2(1 - t) = 16t^2 + 14t + 2$$

and the element dx is given by

$$dx = x'(t)\, dt = 2\, dt.$$

From the evaluation theorem, the line integral is now given by

$$\int_C (4xz + 2y)\, dx = \int_0^1 \underbrace{(16t^2 + 14t + 2)}_{4xz + 2y}\, \underbrace{(2)\, dt}_{dx} = \frac{86}{3}.$$

For part (b), you can use the fact that the line segment connects the same two points as in part (a), but in the opposite direction. The same parametric equations will then work, with the single change that t will run from $t = 1$ to $t = 0$. This gives us

$$\int_C (4xz + 2y)\, dx = \int_1^0 (16t^2 + 14t + 2)(2)\, dt = -\frac{86}{3},$$

where you should recall that reversing the order of integration changes the sign of the integral.

The following result corresponds to Theorem 2.2 for line integrals with respect to arc length, but pay special attention to the minus sign in part (i). We state the theorem for line integrals with respect to x, with corresponding results holding true for line integrals with respect to y or z, as well. We leave the proof as an exercise.

Theorem 2.5

Suppose that $f(x, y, z)$ is a continuous function in some region D containing the oriented curve C. Then, the following hold.

(i) If C is piecewise-smooth, then

$$\int_{-C} f(x, y, z)\, dx = -\int_C f(x, y, z)\, dx.$$

(ii) If $C = C_1 \cup C_2 \cup \cdots \cup C_n$, where $C_1, C_2, \ldots, C_n$ are all smooth, then

$$\int_C f(x, y, z)\, dx = \int_{C_1} f(x, y, z)\, dx + \int_{C_2} f(x, y, z)\, dx + \cdots + \int_{C_n} f(x, y, z)\, dx.$$

Line integrals with respect to x, y and z can be very simple when the curve C consists of line segments parallel to the coordinate axes. The following example explores this case.

Example 2.6 Calculating a Line Integral in Space

Compute $\int_C 4x\, dy + 2y\, dz$, where C consists of the line segment from $(0, 1, 0)$ to $(0, 1, 1)$, followed by the line segment from $(0, 1, 1)$ to $(2, 1, 1)$ and followed by the line segment from $(2, 1, 1)$ to $(2, 4, 1)$.

Solution The curve consists of three line segments. (We show a sketch of the curves in Figure 14.18.) Notice that parametric equations for the first segment C_1 are $x = 0$, $y = 1$ and $z = t$ with $0 \le t \le 1$. On this segment, we have $dy = 0\, dt$ and $dz = 1\, dt$. On the second segment C_2, parametric equations are $x = 2t$, $y = 1$ and $z = 1$ with $0 \le t \le 1$. On this segment, we have $dy = dz = 0\, dt$. On the third segment, parametric equations are $x = 2$, $y = 3t + 1$ and $z = 1$ with $0 \le t \le 1$. On this segment, we have $dy = 3\, dt$ and $dz = 0\, dt$. Putting this all together, we have

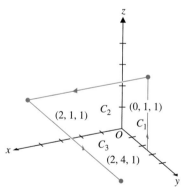

z

$(2, 1, 1)$ C_2 $(0, 1, 1)$

C_1

O

x C_3

$(2, 4, 1)$

y

Figure 14.18
The path C.

$$\int_C 4x\, dy + 2y\, dz = \int_{C_1} 4x\, dy + 2y\, dz + \int_{C_2} 4x\, dy + 2y\, dz + \int_{C_3} 4x\, dy + 2y\, dz$$

$$= \int_0^1 [\underbrace{4(0)}_{4x}\ \underbrace{(0)}_{y'(t)} + \underbrace{2(1)}_{2y}\ \underbrace{(1)}_{z'(t)}]\, dt + \int_0^1 [\underbrace{4(2t)}_{4x}\ \underbrace{(0)}_{y'(t)} + \underbrace{2(1)}_{2y}\ \underbrace{(0)}_{z'(t)}]\, dt$$

$$+ \int_0^1 [\underbrace{4(2)}_{4x}\ \underbrace{(3)}_{y'(t)} + \underbrace{2(3t + 1)}_{2y}\ \underbrace{(0)}_{z'(t)}]\, dt$$

$$= \int_0^1 26\, dt = 26.$$

Notice that a line integral will turn out to be zero if the integrand simplifies to 0 or if the variable of integration is constant along the curve. For instance, if z is constant on some curve, then the change in z (given by dz) will be 0 on that curve.

Recall that our motivation for introducing line integrals with respect to the three coordinate variables was to compute the work done by a force field while moving an object along a curve. From (2.7), we have that the work performed by the force field $\mathbf{F}(x, y, z) = \langle F_1(x, y, z), F_2(x, y, z), F_3(x, y, z) \rangle$ along the curve defined parametrically by $x = x(t)$, $y = y(t)$, $z = z(t)$, for $a \leq t \leq b$ is given by

$$W = \int_a^b F_1(x(t), y(t), z(t)) x'(t)\, dt + \int_a^b F_2(x(t), y(t), z(t)) y'(t)\, dt$$

$$+ \int_a^b F_3(x(t), y(t), z(t)) z'(t)\, dt.$$

You should now recognize that we can rewrite each of the three terms in this expression for work using Theorem 2.4, to obtain

$$W = \int_C F_1(x, y, z)\, dx + \int_C F_2(x, y, z)\, dy + \int_C F_3(x, y, z)\, dz.$$

We now introduce some notation to write such a combination of line integrals in a simpler form.

Suppose that a vector field $\mathbf{F}(x, y, z) = \langle F_1(x, y, z), F_2(x, y, z), F_3(x, y, z) \rangle$. We define

$$d\mathbf{r} = dx\, \mathbf{i} + dy\, \mathbf{j} + dz\, \mathbf{k} \quad \text{or} \quad d\mathbf{r} = \langle dx, dy, dz \rangle.$$

We now define the line integral

$$\int_C \mathbf{F}(x, y, z) \cdot d\mathbf{r} = \int_C F_1(x, y, z)\, dx + \int_C F_2(x, y, z)\, dy + \int_C F_3(x, y, z)\, dz.$$

In the case where $\mathbf{F}(x, y, z)$ is a force field, the work done by $\mathbf{F}$ in moving a particle along the curve C can be written simply as

$$\boxed{W = \int_C \mathbf{F}(x, y, z) \cdot d\mathbf{r}.} \tag{2.8}$$

Notice how the different parts of $\int_C \mathbf{F}(x, y, z) \cdot d\mathbf{r}$ correspond to our knowledge of work. The only way in which the x-component of force affects the work done is when the object moves in the x-direction (i.e., when $dx \neq 0$). Similarly, the y-component of force only contributes to the work when $dy \neq 0$ and the z-component of force only contributes to the work when $dz \neq 0$.

Example 2.7   Computing Work

Compute the work done by the force field $\mathbf{F}(x, y, z) = \langle 4y, 2xz, 3y \rangle$ acting on an object as it moves along the helix defined parametrically by $x = 2\cos t$, $y = 2\sin t$ and $z = 3t$, from the point $(2, 0, 0)$ to the point $(-2, 0, 3\pi)$.

Solution   From (2.8), the work is given by

$$W = \int_C \mathbf{F}(x, y, z) \cdot d\mathbf{r} = \int_C 4y\, dx + 2xz\, dy + 3y\, dz.$$

We have already provided parametric equations for the curve, but not the range of t-values. Notice that from $z = 3t$, you can determine that $(2, 0, 0)$ corresponds to $t = 0$

and $(-2, 0, 3\pi)$ corresponds to $t = \pi$. Substituting in for x, y, z and $dx = -2\sin t\, dt$, $dy = 2\cos t\, dt$ and $dz = 3\, dt$, we have

$$W = \int_C 4y\, dx + 2xz\, dy + 3y\, dz$$

$$= \int_0^\pi [\underbrace{4(2\sin t)}_{4y}\underbrace{(-2\sin t)}_{x'(t)} + \underbrace{2(2\cos t)(3t)}_{2xz}\underbrace{(2\cos t)}_{y'(t)} + \underbrace{3(2\sin t)}_{3y}\underbrace{(3)}_{z'(t)}]\, dt$$

$$= \int_0^\pi (-16\sin^2 t + 24t\cos^2 t + 18\sin t)\, dt = 36 - 8\pi + 6\pi^2,$$

where we used a computer algebra system to evaluate the final integral.

We compute the work performed by a two-dimensional vector field in the same way as we did in three dimensions, as we illustrate in the following example.

Example 2.8   **Computing Work**

Compute the work done by the force field $\mathbf{F}(x, y) = \langle y, -x\rangle$ acting on an object as it moves along the parabola $y = x^2 - 1$ from $(1, 0)$ to $(-2, 3)$.

Solution   From (2.8), the work is given by

$$\int_C \mathbf{F}(x, y)\cdot d\mathbf{r} = \int_C y\, dx - x\, dy.$$

Notice that we can use $x = t$ and $y = t^2 - 1$ as parametric equations for the curve, with t ranging from $t = 1$ to $t = -2$. In this case, $dx = 1\, dt$ and $dy = 2t\, dt$ and the work is given by

$$W = \int_C y\, dx - x\, dy = \int_1^{-2} [(t^2 - 1)(1) - (t)(2t)]\, dt = \int_1^{-2} (-t^2 - 1)\, dt = 6.$$

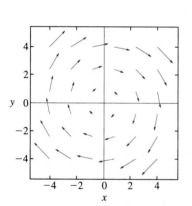

Figure 14.19
$\mathbf{F}(x, y) = \langle y, -x\rangle$.

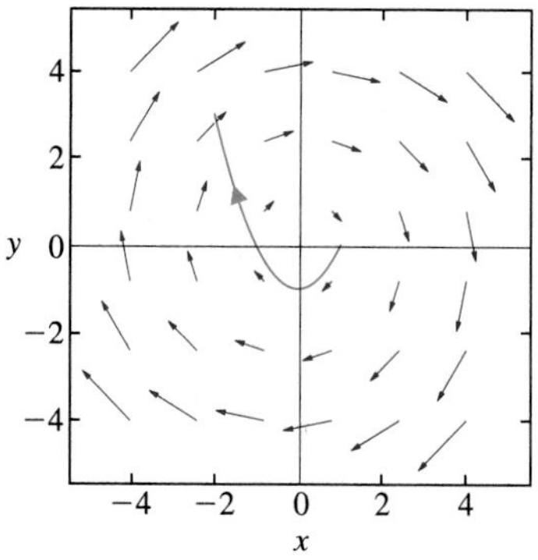

Figure 14.20
$\mathbf{F}(x, y) = \langle y, -x\rangle$ and
$x = t, y = t^2 - 1, -2 \le t \le 1.$

A careful look at example 2.8 graphically will show us an important geometric interpretation of the work line integral. First, graph the vector field $\mathbf{F}(x, y) = \langle y, -x\rangle$. A computer-generated graph is shown in Figure 14.19. If we think of $\mathbf{F}(x, y)$ as describing the velocity field for a fluid in motion, notice that this vector field describes a clockwise rotation of the fluid. In Figure 14.20, we superimpose the curve $x = t$, $y = t^2 - 1$, $-2 \le t \le 1$, onto the vector field $\mathbf{F}(x, y)$. Notice that an object moving along the curve from $(1, 0)$ to $(-2, 3)$ is generally moving in the same direction as that indicated by the vectors in the vector field. If $\mathbf{F}(x, y)$ represents a force field, then the force pushes an object moving along C, adding energy to it and therefore doing positive work. If the curve were oriented in the opposite direction, the force would oppose the motion of the object, thereby doing negative work. Although this analysis can't help us to compute the amount of work done, we can often use it to determine whether the work is positive or negative.

Example 2.9   **Determining the Sign of a Line Integral Graphically**

In each of the following graphs, an oriented curve is superimposed onto a vector field $\mathbf{F}(x, y)$. Determine if $\int_C \mathbf{F}(x, y)\cdot d\mathbf{r}$ is positive or negative.

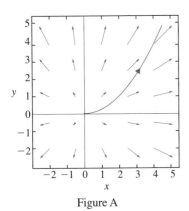

Figure A

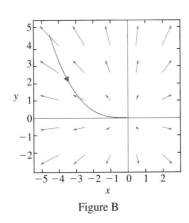

Figure B

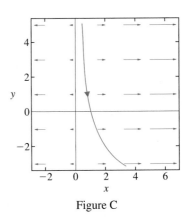

Figure C

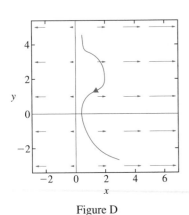

Figure D

Solution   In Figure A, the curve is oriented in the same direction as the vectors, so the force is making a positive contribution to the object's motion. The work done by the force is then positive. In Figure B, the curve is oriented in the opposite direction as the vectors, so the force is making a negative contribution to the object's motion. The work done by the force is then negative. In Figure C, the force field vectors are purely horizontal. Since both the curve and the force vectors point to the right, the work is positive. Finally, in Figure D, the force field is the same as in Figure C, but the curve is more complicated. It is important here to notice that the force vectors are horizontal. Then the work done as the object moves to the right is exactly cancelled when the object doubles back to the left. Comparing the initial and end points, the object has made a net movement to the right (the same direction as the vector field), so the work done is again positive.

EXERCISES 14.2

1. It is important to understand why $\int_C f\,ds = \int_{-C} f\,ds$. Think of f as being a density function and the line integral as giving the mass of an object. Explain why the integrals must be equal.

2. For example 2.3, part (a), a different set of parametric equations is $x = -t$ and $y = t^2$ with t running from $t = 1$ to $t = -2$. In light of the Evaluation Theorem, explain why we couldn't use these parametric equations.

3. Explain in words why Theorem 2.5(i) is true. In particular, explain in terms of approximating sums why the integrals in Theorem 2.5(i) have opposite signs but the integrals in Theorem 2.2(i) are the same.

4. In example 2.9(d), we noted that the force vectors are horizontal. Explain why this allows us to ignore the vertical component of the curve. Also, explain why the work would be the same for *any* curve with the same initial and terminal points.

In exercises 5–28, evaluate the line integral.

5. $\int_C 2x\,ds$, where C is the line segment from $(1, 2)$ to $(3, 5)$

6. $\int_C (x - y)\,ds$, where C is the line segment from $(1, 0)$ to $(3, 1)$

7. $\int_C (3x + y)\,ds$, where C is the line segment from $(5, 2)$ to $(1, 1)$

8. $\int_C 2xy\,ds$, where C is the line segment from $(1, 2)$ to $(-1, 0)$

9. $\int_C 2x\,dx$, where C is the line segment from $(0, 2)$ to $(2, 6)$

10. $\int_C 3y^2\,dy$, where C is the line segment from $(2, 0)$ to $(1, 3)$

11. $\int_C 3x\,ds$, where C is the quarter-circle $x^2 + y^2 = 4$ from $(2, 0)$ to $(0, 2)$

12. $\int_C (3x - y)\,ds$, where C is the quarter-circle $x^2 + y^2 = 9$ from $(0, 3)$ to $(3, 0)$

13. $\int_C 2x\,dx$, where C is the quarter-circle $x^2 + y^2 = 4$ from $(2, 0)$ to $(0, 2)$

14. $\int_C 3y^2\,dy$, where C is the quarter-circle $x^2 + y^2 = 4$ from $(0, 2)$ to $(-2, 0)$

15. $\int_C 3y\,dx$, where C is the half-ellipse $x^2 + 4y^2 = 4$ from $(0, 1)$ to $(0, -1)$ with $x \geq 0$

16. $\int_C x^2\,dy$, where C is the ellipse $4x^2 + y^2 = 4$ oriented counterclockwise

17. $\int_C 3y\,ds$, where C is the portion of $y = x^2$ from $(0, 0)$ to $(2, 4)$

18. $\int_C 2x\,ds$, where C is the portion of $y = x^2$ from $(-2, 4)$ to $(2, 4)$

19. $\int_C 2x\,dx$, where C is the portion of $y = x^2$ from $(2, 4)$ to $(0, 0)$

20. $\int_C 3y^2\,dy$, where C is the portion of $y = x^2$ from $(2, 4)$ to $(0, 0)$

21. $\int_C 3y\,dx$, where C is the portion of $x = y^2$ from $(1, 1)$ to $(4, 2)$

22. $\int_C (x + y)\,dy$, where C is the portion of $x = y^2$ from $(1, 1)$ to $(1, -1)$

23. $\int_C 3x\,ds$, where C is the line segment from $(0, 0)$ to $(1, 0)$ followed by the quarter-circle to $(0, 1)$

24. $\int_C 2y\,ds$, where C is the portion of $y = x^2$ from $(0, 0)$ to $(2, 4)$ followed by the line segment to $(3, 0)$

25. $\int_C 4z\,ds$, where C is the line segment from $(1, 0, 1)$ to $(2, -2, 2)$

26. $\int_C xz\,ds$, where C is the line segment from $(2, 1, 0)$ to $(2, 0, 2)$

27. $\int_C 4(x - z)z\,dx$, where C is the portion of $y = x^2$ in the plane $z = 2$ from $(1, 1, 2)$ to $(2, 4, 2)$

28. $\int_C z\,ds$, where C is the intersection of $x^2 + y^2 = 4$ and $z = 0$ (oriented clockwise as viewed from above)

In exercises 29–40, compute the work done by the force F along the curve C.

29. $\mathbf{F}(x, y) = \langle 2x, 2y \rangle$, C is the line segment from $(3, 1)$ to $(5, 4)$

30. $\mathbf{F}(x, y) = \langle 2y, -2x \rangle$, C is the line segment from $(4, 2)$ to $(0, 4)$

31. $\mathbf{F}(x, y) = \langle 2x, 2y \rangle$, C is the quarter-circle from $(4, 0)$ to $(0, 4)$

32. $\mathbf{F}(x, y) = \langle 2y, -2x \rangle$, C is the upper half-circle from $(-3, 0)$ to $(3, 0)$

33. $\mathbf{F}(x, y) = \langle 2, x \rangle$, C is the portion of $y = x^2$ from $(0, 0)$ to $(1, 1)$

34. $\mathbf{F}(x, y) = \langle 0, xy \rangle$, C is the portion of $y = x^3$ from $(0, 0)$ to $(1, 1)$

35. $\mathbf{F}(x, y) = \langle 3x, 2 \rangle$, C is the line segment from $(0, 0)$ to $(0, 1)$ followed by the line segment to $(4, 1)$

36. $\mathbf{F}(x, y) = \langle y, x \rangle$, C is the square from $(0, 0)$ to $(1, 0)$ to $(1, 1)$ to $(0, 1)$ to $(0, 0)$

37. $\mathbf{F}(x, y, z) = \langle y, 0, z \rangle$, C is the triangle from $(0, 0, 0)$ to $(2, 1, 2)$ to $(2, 1, 0)$ to $(0, 0, 0)$

38. $\mathbf{F}(x, y, z) = \langle z, y, 0 \rangle$, C is the line segment from $(1, 0, 2)$ to $(2, 4, 2)$

39. $\mathbf{F}(x, y, z) = \langle xy, 3z, 1 \rangle$, C is the helix $x = \cos t$, $y = \sin t$, $z = 2t$ from $(1, 0, 0)$ to $(0, 1, \pi)$

40. $\mathbf{F}(x, y, z) = \langle z, 0, 3x^2 \rangle$, C is the quarter-ellipse $x = 2\cos t$, $y = 3\sin t$, $z = 1$ from $(2, 0, 1)$ to $(0, 3, 1)$

In exercises 41–46, use the graph to determine if the work done is positive, negative or zero.

41.

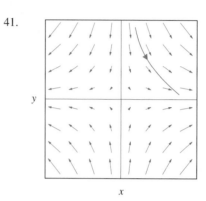

42.

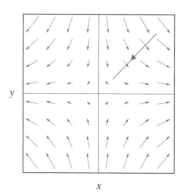

43.

44.

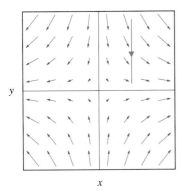

45.

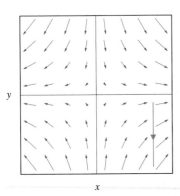

46.

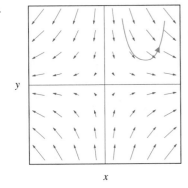

In exercises 47–56, use the formulas $m = \int_C \rho\,ds$, $\bar{x} = \frac{1}{m}\int_C x\rho\,ds$, $\bar{y} = \frac{1}{m}\int_C y\rho\,ds$, $I = \int_C w^2\rho\,ds$.

47. Compute the mass m of a rod with density $\rho(x, y) = x$ in the shape of $y = x^2, 0 \le x \le 3$.

48. Compute the mass m of a rod with density $\rho(x, y) = y$ in the shape of $y = 4 - x^2, 0 \le x \le 2$.

49. Compute the center of mass $(\bar{x}, \bar{y})$ of the rod of exercise 47.

50. Compute the center of mass $(\bar{x}, \bar{y})$ of the rod of exercise 48.

51. Compute the moment of inertia I for rotating the rod of exercise 47 about the y-axis. Here, w is the distance from the point (x, y) to the y-axis.

52. Compute the moment of inertia I for rotating the rod of exercise 48 about the x-axis. Here, w is the distance from the point (x, y) to the x-axis.

53. Compute the moment of inertia I for rotating the rod of exercise 47 about the line $y = 9$. Here, w is the distance from the point (x, y) to $y = 9$.

54. Compute the moment of inertia I for rotating the rod of exercise 48 about the line $x = 2$. Here, w is the distance from the point (x, y) to $x = 2$.

55. Compute the mass m of the helical spring $x = \cos 2t$, $y = \sin 2t, z = t, 0 \le t \le \pi$, with density $\rho = z^2$.

56. Compute the mass m of the ellipse $x = 4\cos t$, $y = 4\sin t$, $z = 4\cos t, 0 \le t \le 2\pi$, with density $\rho = 4$.

57. Show that the center of mass in exercises 49 and 50 is not located at a point on the rod. Explain why this means that our previous interpretation of center of mass as a balance point is no longer valid. Instead, the center of mass is the point about which the object rotates when a torque is applied.

58. Suppose a torque is applied to the rod in exercise 47 such that the rod rotates but has no other motion. Find parametric equations for the position of the part of the rod that starts at the point $(1, 1)$.

In exercises 59–64, find the surface area extending from the given curve to the given surface.

59. Above the quarter-circle from $(2, 0, 0)$ to $(0, 2, 0)$ up to the surface $z = x^2 + y^2$

60. Above the portion of $y = x^2$ from $(0, 0, 0)$ to $(2, 4, 0)$ up to the surface $z = x^2 + y^2$

61. Above the line segment from $(2, 0, 0)$ to $(-2, 0, 0)$ up to the surface $z = 4 - x^2 - y^2$

62. Above the line segment from $(1, 1, 0)$ to $(-1, 1, 0)$ up to the surface $z = \sqrt{x^2 + y^2}$

63. Above the unit square $x \in [0, 1]$, $y \in [0, 1]$ up to the plane $z = 4 - x - y$

64. Above the ellipse $x^2 + 4y^2 = 4$ up to the plane $z = 4 - x$

In exercises 65 and 66, estimate the line integrals (a) $\int_C f\,ds$, **(b)** $\int_C f\,dx$ **and (c)** $\int_C f\,dy$.

65.

(x, y)	$(0, 0)$	$(1, 0)$	$(1, 1)$	$(1.5, 1.5)$
$f(x, y)$	2	3	3.6	4.4

(x, y)	$(2, 2)$	$(3, 2)$	$(4, 1)$
$f(x, y)$	5	4	4

66.

(x, y)	$(0, 0)$	$(1, -1)$	$(2, 0)$	$(3, 1)$
$f(x, y)$	1	0	-1.2	0.4

(x, y)	$(4, 0)$	$(3, -1)$	$(2, -2)$
$f(x, y)$	1.5	2.4	2

67. Prove Theorem 2.1 in the case of a curve in three dimensions.

68. Prove Theorem 2.2.

69. Prove Theorem 2.3.

70. Prove Theorem 2.4.

71. Prove Theorem 2.5.

72. Look carefully at the solutions to exercises 9–10, 13–14 and 19–20. Compare the solutions to integrals of the form $\int_a^b 2x\,dx$ and $\int_c^d 3y^2\,dy$. Formulate a rule for evaluating line integrals of the form $\int_C f(x)\,dx$ and $\int_C g(y)\,dy$. If the curve C is a closed curve (e.g., a square or a circle), evaluate the line integrals $\int_C f(x)\,dx$ and $\int_C g(y)\,dy$.

14.3 INDEPENDENCE OF PATH AND CONSERVATIVE VECTOR FIELDS

Now that you have computed a number of line integrals, you're probably thinking that these are among the most complicated integrals you've run into. They're not necessarily technically challenging, but there are a lot of steps to carry out to evaluate one of these. You first need to parameterize the curve and then rewrite the line integral as a definite integral. After all of this, you must still evaluate the resulting definite integral using any necessary means. Unfortunately, this process is unavoidable for many line integrals. There is some good news, though. In this section, we will look at a group of line integrals that are the same along every curve connecting two given endpoints. We'll determine the circumstances under which this occurs and see that when this does happen, there is a particularly simple way to evaluate the integral.

We begin with a simple observation. Consider the line integral $\int_{C_1} \mathbf{F} \cdot d\mathbf{r}$, where $\mathbf{F}(x, y) = \langle 2x, 3y^2 \rangle$ and C_1 is the straight line segment joining the two points $(0, 0)$ and $(1, 2)$ (see Figure 14.21a). To parameterize the curve, we take $x = t$ and $y = 2t$, for $0 \le t \le 1$. We then have

$$\int_{C_1} \mathbf{F} \cdot d\mathbf{r} = \int_{C_1} \langle 2x, 3y^2 \rangle \cdot \langle dx, dy \rangle$$

$$= \int_{C_1} 2x\,dx + 3y^2\,dy$$

$$= \int_0^1 [2t + 12t^2(2)]\,dt = 9,$$

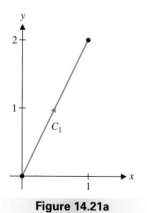

Figure 14.21a
The path C_1.

where we have left the details of the final (routine) calculation to you. For the same vector field $\mathbf{F}(x, y)$, consider now $\int_{C_2} \mathbf{F} \cdot d\mathbf{r}$, where C_2 is made up of the horizontal line segment from $(0, 0)$ to $(1, 0)$ followed by the vertical line segment from $(1, 0)$ to $(1, 2)$

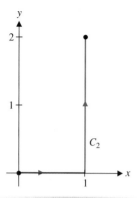

Figure 14.21b
The path C_2.

(see Figure 14.21b). In this case, we have

$$\int_{C_2} \mathbf{F} \cdot d\mathbf{r} = \int_{C_2} \langle 2x, 3y^2 \rangle \cdot \langle dx, dy \rangle$$

$$= \int_0^1 2x \, dx + \int_0^2 3y^2 \, dy = 9,$$

where we have again left the final details to you. Look carefully at these two line integrals. Although the integrands are the same and the endpoints of the two curves are the same, the curves followed are quite different. You should try computing this line integral over several additional curves from $(0, 0)$ to $(1, 2)$. You will find that each line integral has the same value: 9. It turns out that this integral is an example of one that is the same along any curve from $(0, 0)$ to $(1, 2)$.

Let C be any piecewise-smooth curve, traced out by the endpoint of the vector-valued function $\mathbf{r}(t)$, for $a \leq t \leq b$. In this context, we usually refer to a curve connecting two given points as a **path**. We say that the line integral $\int_C \mathbf{F} \cdot d\mathbf{r}$ is **independent of path** if the integral is the same for every path connecting the two endpoints. Before we see when this happens, we need the following definition.

Definition 3.1

A region $D \subset \mathbb{R}^n$ (for $n \geq 2$) is called **connected** if every pair of points in D can be connected by a piecewise-smooth curve lying entirely in D.

Figure 14.22a
Connected region.

In Figure 14.22a, we show a region in $\mathbb{R}^2$ that is connected and in Figure 14.22b, we indicate a region that is not connected. We are now in a position to prove the following result concerning integrals that are independent of path. While we state and prove the result for line integrals in the plane, the result is valid in any number of dimensions.

Theorem 3.1

Suppose that the vector field $\mathbf{F}(x, y) = \langle M(x, y), N(x, y) \rangle$ is continuous on the open, connected region $D \subset \mathbb{R}^2$. Then, the line integral $\int_C \mathbf{F}(x, y) \cdot d\mathbf{r}$ is independent of path in D if and only if the vector field $\mathbf{F}$ is conservative.

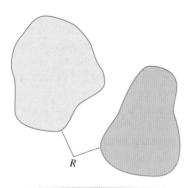

Figure 14.22b
Not connected.

Proof

Recall that a vector field $\mathbf{F}$ is conservative whenever $\mathbf{F} = \nabla f$, for some scalar function f (called a potential function for $\mathbf{F}$). You should recognize that there are several things to prove here.

First, suppose that $\mathbf{F}$ is conservative, with $\mathbf{F}(x, y) = \nabla f(x, y)$. Then

$$\mathbf{F}(x, y) = \langle M(x, y), N(x, y) \rangle = \nabla f(x, y) = \langle f_x(x, y), f_y(x, y) \rangle$$

and so, we must have

$$M(x, y) = f_x(x, y) \quad \text{and} \quad N(x, y) = f_y(x, y).$$

Let $A(x_1, y_1)$ and $B(x_2, y_2)$ be any two points in D and let C be any smooth path from A to B, lying in D and defined parametrically by $C: x = g(t), y = h(t)$, where $t_1 \leq t \leq t_2$. (You can extend this proof to any piecewise-smooth path in the obvious way.)

Then, we have

$$\int_C \mathbf{F}(x, y) \cdot d\mathbf{r} = \int_C M(x, y)\, dx + N(x, y)\, dy$$

$$= \int_C f_x(x, y)\, dx + f_y(x, y)\, dy$$

$$= \int_{t_1}^{t_2} [f_x(g(t), h(t))g'(t) + f_y(g(t), h(t))h'(t)]\, dt. \qquad (3.1)$$

Notice that since f_x and f_y were assumed to be continuous, we have by the chain rule that

$$\frac{d}{dt}[f(g(t), h(t))] = f_x(g(t), h(t))g'(t) + f_y(g(t), h(t))h'(t),$$

which is the integrand in (3.1). By the Fundamental Theorem of Calculus, we now have

$$\int_C \mathbf{F}(x, y) \cdot d\mathbf{r} = \int_{t_1}^{t_2} [f_x(g(t), h(t))g'(t) + f_y(g(t), h(t))h'(t)]\, dt$$

$$= \int_{t_1}^{t_2} \frac{d}{dt}[f(g(t), h(t))]\, dt$$

$$= f(g(t_2), h(t_2)) - f(g(t_1), h(t_1))$$

$$= f(x_2, y_2) - f(x_1, y_1).$$

In particular, this says that the value of the integral depends only on the value of the potential function at the two endpoints of the curve and not on the particular path followed. That is, the line integral is independent of path, as desired.

Next, we need to prove that if the integral is independent of path, then the vector field must be conservative. So, suppose that $\int_C \mathbf{F}(x, y) \cdot d\mathbf{r}$ is independent of path in D. For any points (u, v) and $(x_0, y_0) \in D$, define the function

$$f(u, v) = \int_{(x_0, y_0)}^{(u,v)} \mathbf{F}(x, y) \cdot d\mathbf{r}.$$

(We are using the variables u and v, since the variables x and y inside the integral are dummy variables and cannot be used both inside and outside the line integral.) Notice that since the line integral is independent of path in D, we need not specify a path over which to integrate; it's the same over every path in D. Further, since D is open, there is a disk centered at (u, v) and lying completely inside D. Pick any point (x_1, v) in the disk with $x_1 < u$ and let C_1 be any path from (x_0, y_0) to (x_1, v) lying in D. So, in particular, if we integrate over the path consisting of C_1 followed by the horizontal path C_2 indicated in Figure 14.23, we must have

$$f(u, v) = \int_{(x_0, y_0)}^{(x_1, v)} \mathbf{F}(x, y) \cdot d\mathbf{r} + \int_{(x_1, v)}^{(u,v)} \mathbf{F}(x, y) \cdot d\mathbf{r}. \qquad (3.2)$$

Observe that the first integral in (3.2) is independent of u. So, taking the partial derivative of both sides of (3.2) with respect to u, we get

$$f_u(u, v) = \frac{\partial}{\partial u} \int_{(x_0, y_0)}^{(x_1, v)} \mathbf{F}(x, y) \cdot d\mathbf{r} + \frac{\partial}{\partial u} \int_{(x_1, v)}^{(u,v)} \mathbf{F}(x, y) \cdot d\mathbf{r}$$

$$= 0 + \frac{\partial}{\partial u} \int_{(x_1, v)}^{(u,v)} \mathbf{F}(x, y) \cdot d\mathbf{r}$$

$$= \frac{\partial}{\partial u} \int_{(x_1, v)}^{(u,v)} M(x, y)\, dx + N(x, y)\, dy.$$

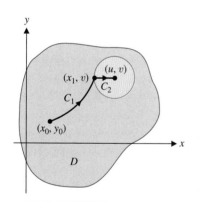

Figure 14.23
First path.

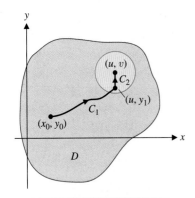

Figure 14.24

Second path.

Notice that on the second portion of the indicated path, y is a constant and so, $dy = 0$. This gives us

$$f_u(u, v) = \frac{\partial}{\partial u} \int_{(x_1,v)}^{(u,v)} M(x, y)\,dx + N(x, y)\,dy = \frac{\partial}{\partial u} \int_{(x_1,v)}^{(u,v)} M(x, y)\,dx.$$

Finally, from the second form of the Fundamental Theorem of Calculus, we have

$$f_u(u, v) = \frac{\partial}{\partial u} \int_{(x_1,v)}^{(u,v)} M(x, y)\,dx = M(u, v). \tag{3.3}$$

Similarly, pick any point (u, y_1) in the disk centered at (u, v) with $y_1 < v$ and let C_1 be any path from (x_0, y_0) to (u, y_1) lying in D. Then, integrating over the path consisting of C_1 followed by the vertical path C_2 indicated in Figure 14.24, we find that

$$f(u, v) = \int_{(x_0,y_0)}^{(u,y_1)} \mathbf{F}(x, y) \cdot d\mathbf{r} + \int_{(u,y_1)}^{(u,v)} \mathbf{F}(x, y) \cdot d\mathbf{r}. \tag{3.4}$$

In this case, the first integral is independent of v. So, differentiating both sides of (3.4) with respect to v, we have

$$f_v(u, v) = \frac{\partial}{\partial v} \int_{(x_0,y_0)}^{(u,y_1)} \mathbf{F}(x, y) \cdot d\mathbf{r} + \frac{\partial}{\partial v} \int_{(u,y_1)}^{(u,v)} \mathbf{F}(x, y) \cdot d\mathbf{r}$$

$$= 0 + \frac{\partial}{\partial v} \int_{(u,y_1)}^{(u,v)} \mathbf{F}(x, y) \cdot d\mathbf{r}$$

$$= \frac{\partial}{\partial v} \int_{(u,y_1)}^{(u,v)} M(x, y)\,dx + N(x, y)\,dy$$

$$= \frac{\partial}{\partial v} \int_{(u,y_1)}^{(u,v)} N(x, y)\,dy = N(u, v), \tag{3.5}$$

by the second form of the Fundamental Theorem of Calculus, where we have used the fact that on the second part of the indicated path, x is a constant, so that $dx = 0$. Replacing u and v by x and y, respectively, in (3.3) and (3.5) establishes that

$$\mathbf{F}(x, y) = \langle M(x, y), N(x, y) \rangle = \langle f_x(x, y), f_y(x, y) \rangle = \nabla f(x, y),$$

so that $\mathbf{F}$ is conservative in D. ∎

Notice that in the course of the first part of the proof of Theorem 3.1, we also proved the following result, which corresponds to the Fundamental Theorem of Calculus for definite integrals.

Theorem 3.2 (Fundamental Theorem for Line Integrals)
Suppose that $\mathbf{F}(x, y) = \langle M(x, y), N(x, y) \rangle$ is continuous in the open, connected region $D \subset \mathbb{R}^2$ and that C is any piecewise-smooth curve lying in D, with initial point (x_1, y_1) and terminal point (x_2, y_2). Then, if $\mathbf{F}$ is conservative, with $\mathbf{F}(x, y) = \nabla f(x, y)$, we have

$$\int_C \mathbf{F}(x, y) \cdot d\mathbf{r} = f(x, y) \Big|_{(x_1,y_1)}^{(x_2,y_2)} = f(x_2, y_2) - f(x_1, y_1).$$

You should quickly recognize the advantages presented by Theorem 3.2. For a conservative vector field, you don't need to parameterize the path to compute a line integral; you need only find a potential function and then simply evaluate the potential function between the endpoints of the curve. We illustrate this in the following example.

Example 3.1   A Line Integral That is Independent of Path

Show that for $\mathbf{F}(x, y) = \langle 2xy - 3, x^2 + 4y^3 + 5 \rangle$, the line integral $\int_C \mathbf{F}(x, y) \cdot d\mathbf{r}$ is independent of path. Then, evaluate the line integral for any curve C with initial point at $(-1, 2)$ and terminal point at $(2, 3)$.

Solution   From Theorem 3.1, the line integral is independent of path if and only if the vector field $\mathbf{F}(x, y)$ is conservative. So, we look for a potential function for $\mathbf{F}$, that is, a function $f(x, y)$ for which

$$\mathbf{F}(x, y) = \langle 2xy - 3, x^2 + 4y^3 + 5 \rangle = \nabla f(x, y) = \langle f_x(x, y), f_y(x, y) \rangle.$$

Of course, this occurs when

$$f_x = 2xy - 3 \quad \text{and} \quad f_y = x^2 + 4y^3 + 5. \tag{3.6}$$

Integrating the first of these two equations with respect to x (note that we might just as easily integrate the second one with respect to y), we get

$$f(x, y) = \int (2xy - 3)\, dx = x^2 y - 3x + g(y), \tag{3.7}$$

where $g(y)$ is some arbitrary function of y alone. (Recall that we get an arbitrary *function* of y instead of a *constant* of integration, since we are integrating a function of x and y with respect to x.) Differentiating this with respect to y, we get

$$f_y(x, y) = x^2 + g'(y).$$

Notice that from (3.6), we already have an expression for f_y. Setting these two expressions equal, we get

$$x^2 + g'(y) = x^2 + 4y^3 + 5,$$

where if we subtract x^2 from both sides, we get

$$g'(y) = 4y^3 + 5.$$

Finally, integrating this last expression with respect to y gives us

$$g(y) = y^4 + 5y + c.$$

We now have from (3.7) that

$$f(x, y) = x^2 y - 3x + y^4 + 5y + c$$

is a potential function for $\mathbf{F}(x, y)$, for any constant c. Now that we have found a potential function, we have by Theorem 3.2 that for any path from $(-1, 2)$ to $(2, 3)$,

$$\int_C \mathbf{F}(x, y) \cdot d\mathbf{r} = f(x, y) \Big|_{(-1,2)}^{(2,3)}$$

$$= [2^2(3) - 3(2) + 3^4 + 5(3) + c] - [2 + 3 + 2^4 + 5(2) + c]$$

$$= 71.$$

Notice that when we evaluated the line integral in example 3.1, the constant c in the expression for the potential function dropped out. For this reason, we usually leave the constant off when we write down a potential function.

We consider a curve C to be **closed** if its two endpoints are the same. That is, for a plane curve C defined parametrically by

$$C = \{(x, y) \mid x = g(t), y = h(t), a \le t \le b\},$$

C is closed if $(g(a), h(a)) = (g(b), h(b))$. The following result provides us with an important connection between conservative vector fields and line integrals along closed curves.

Theorem 3.3

Suppose that $\mathbf{F}(x, y)$ is continuous in the open, connected region $D \subset \mathbb{R}^2$. Then $\mathbf{F}$ is conservative if and only if $\int_C \mathbf{F}(x, y) \cdot d\mathbf{r} = 0$ for every piecewise-smooth closed curve C lying in D.

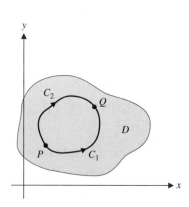

Figure 14.25a

Curves C_1 and C_2.

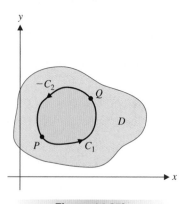

Figure 14.25b

The closed curve formed by $C_1 \cup (-C_2)$.

Proof

Suppose that $\int_C \mathbf{F}(x, y) \cdot d\mathbf{r} = 0$ for every piecewise-smooth closed curve C lying in D. Take any two points P and Q lying in D and let C_1 and C_2 be any two piecewise-smooth closed curves from P to Q that lie in D, as indicated in Figure 14.25a. (Note that since D is connected, there always exist such curves.) Then, the curve C consisting of C_1 followed by $-C_2$ is a piecewise-smooth closed curve lying in D, as indicated in Figure 14.25b. It now follows that

$$0 = \int_C \mathbf{F}(x, y) \cdot d\mathbf{r} = \int_{C_1} \mathbf{F}(x, y) \cdot d\mathbf{r} + \int_{-C_2} \mathbf{F}(x, y) \cdot d\mathbf{r}$$

$$= \int_{C_1} \mathbf{F}(x, y) \cdot d\mathbf{r} - \int_{C_2} \mathbf{F}(x, y) \cdot d\mathbf{r}, \quad \text{From Theorem 2.5}$$

so that

$$\int_{C_1} \mathbf{F}(x, y) \cdot d\mathbf{r} = \int_{C_2} \mathbf{F}(x, y) \cdot d\mathbf{r}.$$

Since C_1 and C_2 were any two curves from P to Q, we have that $\int_C \mathbf{F}(x, y) \cdot d\mathbf{r}$ is independent of path and so, $\mathbf{F}$ is conservative by Theorem 3.1. The second half of the theorem (that $\mathbf{F}$ conservative implies $\int_C \mathbf{F}(x, y) \cdot d\mathbf{r} = 0$ for every piecewise-smooth closed curve C lying in D) is a simple consequence of Theorem 3.2 and is left as an exercise.

■

You have already seen that line integrals are not always independent of path. Said differently, not all vector fields are conservative. Given this fact, it would be helpful to have a simple way of deciding whether or not a line integral is independent of path before going through the process of trying to construct a potential function. It's a little like drilling for oil; it would be nice to have some reason to believe that there is actually some oil there, before going to all the trouble of drilling an oil well. Fortunately, there is a simple way to determine when a line integral is independent of path.

Note that by Theorem 3.1, if $\mathbf{F}(x, y) = \langle M(x, y), N(x, y) \rangle$ is continuous on the open, connected region D, and the line integral $\int_C \mathbf{F}(x, y) \cdot d\mathbf{r}$ is independent of path, then $\mathbf{F}$ must be conservative. That is, there is a function $f(x, y)$ for which $\mathbf{F}(x, y) = \nabla f(x, y)$ or

$$M(x, y) = f_x(x, y) \quad \text{and} \quad N(x, y) = f_y(x, y).$$

Differentiating the first equation with respect to y and the second equation with respect to x, we have

$$M_y(x, y) = f_{xy}(x, y) \quad \text{and} \quad N_x(x, y) = f_{yx}(x, y).$$

Notice now that if M_y and N_x are continuous in D, then the mixed second partial derivatives $f_{xy}(x, y)$ and $f_{yx}(x, y)$ must be the same in D, by Theorem 3.1 in Chapter 12. We must then have that

$$M_y(x, y) = N_x(x, y),$$

for all (x, y) in D. As it turns out, if we further assume that D is **simply-connected** (this means that every closed curve in D encloses only points in D), then the converse of this result is also true [i.e., $\int_C \mathbf{F}(x, y) \cdot d\mathbf{r}$ is independent of path whenever $M_y = N_x$ in D]. Before stating the result, we illustrate a simply-connected region in Figure 14.26a and a region that is not simply-connected in Figure 14.26b. You can think about simply-connected regions as regions that have no holes. We can now state the following result.

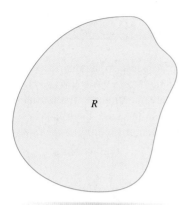

Figure 14.26a
Simply-connected.

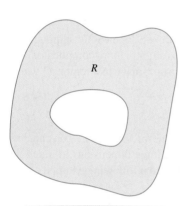

Figure 14.26b
Not simply-connected.

Theorem 3.4

Suppose that $M(x, y)$ and $N(x, y)$ have continuous first partial derivatives on a simply-connected region D. Then, $\int_C M(x, y)\,dx + N(x, y)\,dy$ is independent of path if and only if $M_y(x, y) = N_x(x, y)$ for all (x, y) in D.

We have already proved that independence of path implies that $M_y(x, y) = N_x(x, y)$ for all (x, y) in D. We must postpone the proof of the second half of the theorem until our presentation of Green's Theorem in section 14.4.

Example 3.2   Testing a Line Integral for Independence of Path

Determine whether or not the line integral $\int_C (e^{2x} + x \sin y)\,dx + (x^2 \cos y)\,dy$ is independent of path.

Solution   In this case, we have

$$M_y = \frac{\partial}{\partial y}(e^{2x} + x \sin y) = x \cos y$$

and

$$N_x = \frac{\partial}{\partial x}(x^2 \cos y) = 2x \cos y.$$

Although M_y and N_x are close, they're not the same. So, by Theorem 3.4, the line integral is not independent of path.

CONSERVATIVE VECTOR FIELDS

Before moving on to three-dimensional vector fields, we pause to summarize the results we have developed for two-dimensional vector fields $\mathbf{F}(x, y) = \langle M(x, y), N(x, y) \rangle$, where we assume that $M(x, y)$ and $N(x, y)$ have continuous first partial derivatives on an open, simply-connected region $D \subset \mathbb{R}^2$. In this case, the following five statements are equivalent, meaning that for a given vector field either all five statements are true or all five statements are false.

1. $\mathbf{F}(x, y)$ is conservative in D.
2. $\int_C \mathbf{F} \cdot d\mathbf{r}$ is independent of path in D.
3. $\int_C \mathbf{F} \cdot d\mathbf{r} = 0$ for every piecewise-smooth closed curve C lying in D.
4. $\mathbf{F}(x, y)$ is a gradient field ($\mathbf{F} = \nabla f$ for some potential function f).
5. $M_y(x, y) = N_x(x, y)$, for all $(x, y) \in D$.

Much of the preceding analysis can be extended to higher dimensions, although the test for when a line integral is independent of path becomes slightly more complicated. For a three-dimensional vector field $\mathbf{F}(x, y, z)$, we say $\mathbf{F}$ is **conservative** in a region D whenever there is a scalar function $f(x, y, z)$ for which

$$\mathbf{F}(x, y, z) = \nabla f(x, y, z).$$

As in two dimensions, f is called a **potential function** for the vector field $\mathbf{F}$. You can construct a potential function for a conservative vector field in three dimensions in much the same way as you did in two dimensions. We illustrate this in the following example.

Example 3.3   Showing That a Three-Dimensional Vector Field is Conservative

Show that the vector field $\mathbf{F}(x, y, z) = \langle 4xe^z, \cos y, 2x^2e^z \rangle$ is conservative.

Solution   We need to find a potential function $f(x, y, z)$ for which

$$\mathbf{F}(x, y, z) = \langle 4xe^z, \cos y, 2x^2e^z \rangle = \nabla f(x, y, z)$$
$$= \langle f_x(x, y, z), f_y(x, y, z), f_z(x, y, z) \rangle.$$

Of course, this will occur if and only if

$$f_x = 4xe^z, \quad f_y = \cos y \quad \text{and} \quad f_z = 2x^2e^z. \tag{3.8}$$

Integrating the first of these equations with respect to x, we have

$$f(x, y, z) = \int 4xe^z \, dx = 2x^2e^z + g(y, z),$$

where $g(y, z)$ is an arbitrary function of y and z alone. Note that since y and z are treated as constants when integrating or differentiating with respect to x, we add an arbitrary function of y and z (instead of an arbitrary constant) after a partial integration with respect to x. Differentiating this expression with respect to y, we have

$$f_y(x, y, z) = g_y(y, z) = \cos y,$$

from the second equation in (3.8). Integrating $g_y(y, z)$ with respect to y now gives us

$$g(y, z) = \int \cos y \, dy = \sin y + h(z),$$

where $h(z)$ is an arbitrary function of z alone. Notice that here, we got an arbitrary function of z alone, since we were integrating $g(y, z)$ (a function of y and z alone) with respect to y. This now gives us

$$f(x, y, z) = 2x^2 e^z + g(y, z) = 2x^2 e^z + \sin y + h(z).$$

Differentiating this last expression with respect to z yields

$$f_z(x, y, z) = 2x^2 e^z + h'(z) = 2x^2 e^z,$$

from the third equation in (3.8). This gives us that $h'(z) = 0$, so that $h(z)$ is a constant (we'll choose it to be 0). We now have that a potential function for $\mathbf{F}(x, y, z)$ is

$$f(x, y, z) = 2x^2 e^z + \sin y$$

and so, $\mathbf{F}$ is conservative.

■─────

Most of what we have presented here in two dimensions generalizes directly to three dimensions. We summarize the main results in the following theorem.

Theorem 3.5

Suppose that the vector field $\mathbf{F}(x, y, z)$ is continuous on the open, connected region $D \subset \mathbb{R}^3$. Then, the line integral $\int_C \mathbf{F}(x, y, z) \cdot d\mathbf{r}$ is independent of path if and only if the vector field $\mathbf{F}$ is conservative, that is, $\mathbf{F}(x, y, z) = \nabla f(x, y, z)$ for some scalar function f (a potential function for $\mathbf{F}$). Further, for any piecewise-smooth curve C lying in D, with initial point (x_1, y_1, z_1) and terminal point (x_2, y_2, z_2), we have

$$\int_C \mathbf{F}(x, y, z) \cdot d\mathbf{r} = f(x, y, z) \Big|_{(x_1, y_1, z_1)}^{(x_2, y_2, z_2)} = f(x_2, y_2, z_2) - f(x_1, y_1, z_1).$$

EXERCISES 14.3

1. You have seen two different methods of determining if a line integral is independent of path, one in example 3.1 and the other in example 3.2. If you have reason to believe that a line integral will be independent of path, explain which method you would prefer to use.

2. In the situation of exercise 1, if you doubt that a line integral is independent of path, explain which method you would prefer to use. If you have no evidence as to whether the line integral is or isn't independent of path, explain which method you would prefer to use.

3. In section 14.1, we introduced conservative vector fields and stated that some calculations simplified when the vector field is conservative. Discuss one important example of this.

4. Our definition of independence of path only applies to line integrals of the form $\int_C \mathbf{F} \cdot d\mathbf{r}$. Explain why an arc

length line integral $\int_C f \, ds$ would not be independent of path (unless $f = 0$).

In exercises 5–16, determine if F is conservative. If it is, find a potential function f.

5. $\mathbf{F}(x, y) = \langle 2xy - 1, x^2 \rangle$

6. $\mathbf{F}(x, y) = \langle 3x^2 y^2, 2x^3 y - y \rangle$

7. $\mathbf{F}(x, y) = \left\langle \frac{1}{y} - 2x, y - \frac{x}{y^2} \right\rangle$

8. $\mathbf{F}(x, y) = \langle \sin y - x, x \cos y \rangle$

9. $\mathbf{F}(x, y) = \langle e^{xy} - 1, xe^{xy} \rangle$

10. $\mathbf{F}(x, y) = \langle e^y - 2x, xe^y - x^2 y \rangle$

11. $\mathbf{F}(x, y) = \langle ye^{xy}, xe^{xy} + \cos y \rangle$

12. $\mathbf{F}(x, y) = \langle y \cos xy - 2xy, x \cos xy - x^2 \rangle$

13. $\mathbf{F}(x, y, z) = \langle z^2 + 2xy, x^2 + 1, 2xz - 3 \rangle$

14. $\mathbf{F}(x, y, z) = \langle y^2 - x, 2xy + \sin z, y \cos z \rangle$

15. $\mathbf{F}(x, y, z) = \langle y^2 z^2 + x, y + 2xyz^2, 2xy^2 z \rangle$

16. $\mathbf{F}(x, y, z) = \langle 2xe^{yz} - 1, x^2 + e^{yz}, x^2 ye^{yz} \rangle$

In exercises 17–22, show that the line integral is independent of path and evaluate the integral.

17. $\int_C 2xy \, dx + (x^2 - 1) \, dy$, where C runs from $(1, 0)$ to $(3, 1)$

18. $\int_C 3x^2 y^2 \, dx + (2x^3 y - 4) \, dy$, where C runs from $(1, 2)$ to $(-1, 1)$

19. $\int_C ye^{xy} \, dx + (xe^{xy} - 2y) \, dy$, where C runs from $(1, 0)$ to $(0, 4)$

20. $\int_C \left(2xe^{x^2} - 2y \right) dx + (2y - 2x) \, dy$, where C runs from $(1, 2)$ to $(-1, 1)$

21. $\int_C \left(z^2 + 2xy \right) dx + x^2 \, dy + 2xz \, dz$, where C runs from $(2, 1, 3)$ to $(4, -1, 0)$

22. $\int_C (2x \cos z - x^2) \, dx + (z - 2y) \, dy + (y - x^2 \sin z) \, dz$, where C runs from $(3, -2, 0)$ to $(1, 0, \pi)$

In exercises 23–34, evaluate $\int_C \mathbf{F} \cdot d\mathbf{r}$.

23. $\mathbf{F}(x, y) = \langle x^2 + 1, y^3 - 3y + 2 \rangle$, C is the top half-circle from $(-4, 0)$ to $(4, 0)$

24. $\mathbf{F}(x, y) = \left\langle xe^{x^2} - 2, \sin y \right\rangle$, C is the portion of the parabola $y = x^2$ from $(-2, 4)$ to $(2, 4)$

25. $\mathbf{F}(x, y, z) = \langle x^2, y^2, z^2 \rangle$, C is the top half-circle from $(1, 4, -3)$ to $(1, 4, 3)$

26. $\mathbf{F}(x, y, z) = \langle \cos x, \sqrt{y} + 1, 4z^3 \rangle$, C is the quarter-circle from $(2, 0, 3)$ to $(2, 3, 0)$

27. $\mathbf{F}(x, y, z) = \dfrac{\langle x, y, z \rangle}{\sqrt{x^2 + y^2 + z^2}}$, C runs from $(1, 3, 2)$ to $(2, 1, 5)$

28. $\mathbf{F}(x, y, z) = \dfrac{\langle x, y, z \rangle}{x^2 + y^2 + z^2}$, C runs from $(2, 0, 0)$ to $(0, 1, -1)$

29. $\mathbf{F}(x, y) = \langle 3x^2 y + 1, 3xy^2 \rangle$, C is the bottom half-circle from $(1, 0)$ to $(-1, 0)$

30. $\mathbf{F}(x, y) = \langle 4xy - 2x, 2x^2 - x \rangle$, C is the portion of the parabola $y = x^2$ from $(-2, 4)$ to $(2, 4)$

31. $\mathbf{F}(x, y) = \left\langle y^2 e^{xy^2} - y, 2xye^{xy^2} - x - 1 \right\rangle$, C is the line segment from $(2, 3)$ to $(3, 0)$

32. $\mathbf{F}(x, y) = \langle 2ye^{2x} + y^3, e^{2x} + 3xy^2 \rangle$, C is the line segment from $(4, 3)$ to $(1, -3)$

33. $\mathbf{F}(x, y) = \left\langle \frac{1}{y} - e^{2x}, 2y - \frac{x}{y^2} \right\rangle$, C is the circle $x^2 + y^2 = 16$

34. $\mathbf{F}(x, y) = \langle 3y - \sqrt{y/x}, 3x - \sqrt{x/y} \rangle$, C is the ellipse $4(x - 4)^2 + 9(y - 4)^2 = 36$

In exercises 35–40, use the graph to determine whether or not the vector field is conservative.

35.

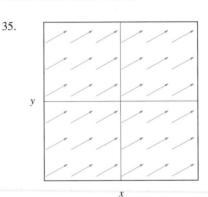

36.

37.

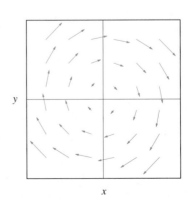

38.

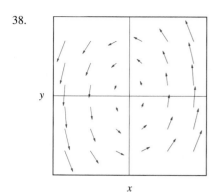

39.

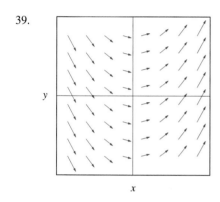

40.

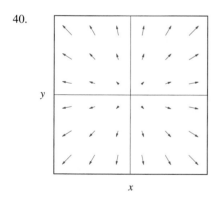

In exercises 41–44, show that the line integral is not independent of path by finding two paths that give different values of the integral.

41. $\int_C y\,dx - x\,dy$, where C goes from $(-2, 0)$ to $(2, 0)$

42. $\int_C 2\,dx + x\,dy$, where C goes from $(1, 4)$ to $(2, -2)$

43. $\int_C y\,dx - 3\,dy$, where C goes from $(-2, 2)$ to $(0, 0)$

44. $\int_C y^2\,dx + x^2\,dy$, where C goes from $(0, 0)$ to $(1, 1)$

In exercises 45–48, label each statement as True or False and briefly explain.

45. If $\mathbf{F}$ is conservative, then $\int_C \mathbf{F} \cdot d\mathbf{r} = 0$.

46. If $\int_C \mathbf{F} \cdot d\mathbf{r}$ is independent of path, then $\mathbf{F}$ is conservative.

47. If $\mathbf{F}$ is conservative, then $\int_C \mathbf{F} \cdot d\mathbf{r} = 0$ for any closed curve C.

48. If $\mathbf{F}$ is conservative, then $\int_C \mathbf{F} \cdot d\mathbf{r}$ is independent of path.

49. Let $\mathbf{F}(x, y) = \dfrac{1}{x^2 + y^2}\langle -y, x\rangle$. Find a potential function f for $\mathbf{F}$ and carefully note any restrictions on the domain of f. Let C be the unit circle and show that $\int_C \mathbf{F} \cdot d\mathbf{r} = 2\pi$. Explain why the Fundamental Theorem for Line Integrals does not apply to this calculation. Quickly explain how to compute $\int_C \mathbf{F} \cdot d\mathbf{r}$ over the circle $(x - 2)^2 + (y - 3)^2 = 1$.

50. Finish the proof of Theorem 3.3 by showing that if $\mathbf{F}$ is conservative in an open connected region $D \subset \mathbb{R}^2$, then $\int_C \mathbf{F} \cdot d\mathbf{r} = 0$ for all piecewise-smooth closed curves C lying in D.

51. Determine whether or not each region is simply connected. (a) $\{(x, y) : x^2 + y^2 < 2\}$ (b) $\{(x, y) : 1 < x^2 + y^2 < 2\}$.

52. Determine whether or not each region is simply connected. (a) $\{(x, y) : 1 < x < 2\}$ (b) $\{(x, y) : 1 < x^2 < 2\}$.

53. For closed curves, we can take advantage of portions of a line integral that will equal zero. For example, if C is a closed curve, explain why you can simplify $\int_C (x + y^2)\,dx + (y^2 + x)\,dy$ to $\int_C y^2\,dx + x\,dy$. In general, explain why the $f(x)$ and $g(y)$ terms can be dropped in the line integral $\int_C (f(x) + y^2)\,dx + (x + g(y))\,dy$. Describe which other terms can be dropped in the line integral over a closed curve. Use the example

$$\int_C (x^3 + y^2 + x^2 y^2 + \cos y)\,dx$$
$$+ (y^2 + 2xy - x\sin y + x^3 y)\,dy$$

to help organize your thinking.

54. In this exercise, we explore a basic principle of physics called **conservation of energy.** Start with the work integral $\int_C \mathbf{F} \cdot d\mathbf{r}$, where the position function $\mathbf{r}(t)$ is a continuously differentiable function of time. Substitute Newton's second law: $\mathbf{F} = m\dfrac{d\mathbf{v}}{dt}$ and $d\mathbf{r} = \mathbf{r}'(t)\,dt = \mathbf{v}\,dt$ and show that $\int_C \mathbf{F} \cdot d\mathbf{r} = \Delta K$. Here, K is **kinetic energy** defined by $K = \frac{1}{2} m\|\mathbf{v}\|^2$ and ΔK is the change of kinetic energy from the initial point of C to the terminal point of C. Next, assume that $\mathbf{F}$ is conservative with $\mathbf{F} = -\nabla f$, where the function f represents **potential energy.** Show that $\int_C \mathbf{F} \cdot d\mathbf{r} = -\Delta f$ where Δf equals the change in potential energy from the initial point of C to the terminal point of C. Conclude that under these hypotheses (conservative force, continuous acceleration) the net change in energy $\Delta K + \Delta f$ equals 0.

14.4 GREEN'S THEOREM

HISTORICAL NOTES

George Green (1793–1841)
English mathematician who
discovered Green's Theorem.
Green was self-taught, receiving
only two years of schooling before
going to work in his father's bakery
at age 9. He continued to work in
and eventually take over the family
mill while teaching himself
mathematics. In 1828, he published
an essay in which he gave potential
functions their name and applied
them to the study of electricity and
magnetism. This little-read essay
introduced Green's Theorem and
the so-called Green's functions
used in the study of partial
differential equations. Green was
admitted to Cambridge University
at age 40 and published several
papers before his early death from
illness. The significance of his
original essay remained unknown
until shortly after his death.

In this section, we develop a connection between certain line integrals around a closed
curve and double integrals over the region enclosed by the curve. At first glance, you might
think this a strange and abstract connection, one that only a mathematician could care
about. Actually, the reverse is true; Green's Theorem is a significant result with far-
reaching implications. It is of fundamental importance in the analysis of fluid flows and in
the theories of electricity and magnetism.

Before stating the main result, we need to briefly define some terminology. Recall that
for a plane curve C defined parametrically by

$$C = \{(x, y) | x = g(t), y = h(t), a \leq t \leq b\},$$

C is closed if its two endpoints are the same, i.e., $(g(a), h(a)) = (g(b), h(b))$. A curve C
is **simple** if it does not intersect itself, except at the endpoints. We illustrate a simple closed
curve in Figure 14.27a and a closed curve that is not simple in Figure 14.27b.

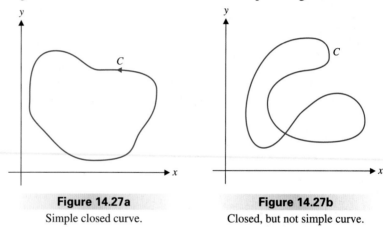

Figure 14.27a	**Figure 14.27b**
Simple closed curve.	Closed, but not simple curve.

We say that a simple closed curve C has **positive orientation** if the region R enclosed by
C stays to the left of C, as the curve is traversed; a curve has **negative orientation** if the
region R stays to the right of C. In Figures 14.28a and 14.28b, we illustrate a simple closed
curve with positive orientation and one with negative orientation, respectively.

We use the notation

$$\oint_C \mathbf{F}(x, y) \cdot d\mathbf{r}$$

to denote a line integral along a simple closed curve C oriented in the positive direction.

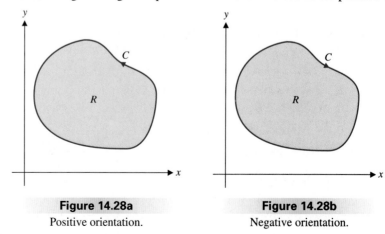

Figure 14.28a	**Figure 14.28b**
Positive orientation.	Negative orientation.

We can now state the main result of the section.

Theorem 4.1 (Green's Theorem)

Let C be a piecewise-smooth, simple closed curve in the plane with positive orientation and let R be the region enclosed by C, together with C. Suppose that $M(x, y)$ and $N(x, y)$ are continuous and have continuous first partial derivatives in some open region D, with $R \subset D$. Then,

$$\oint_C M(x, y)\, dx + N(x, y)\, dy = \iint_R \left(\frac{\partial N}{\partial x} - \frac{\partial M}{\partial y} \right) dA.$$

You can find a general proof of Green's Theorem in a more advanced text. We prove it here only for a special case.

Proof

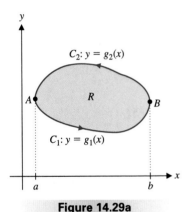

Figure 14.29a

The region R.

Here, we assume that the region R can be written in the form

$$R = \{(x, y) \mid a \le x \le b \text{ and } g_1(x) \le y \le g_2(x)\},$$

as illustrated in Figure 14.29a. Notice that we can divide C into the two pieces indicated in Figure 14.29a:

$$C = C_1 \cup C_2,$$

where C_1 is the bottom portion of the curve, defined by

$$C_1 = \{(x, y) \mid a \le x \le b, y = g_1(x)\}$$

and C_2 is the top portion of the curve, defined by

$$C_2 = \{(x, y) \mid a \le x \le b, y = g_2(x)\},$$

where the orientation is as indicated in the figure. From the evaluation theorem for line integrals (Theorem 2.4), we then have

$$\oint_C M(x, y)\, dx = \int_{C_1} M(x, y)\, dx + \int_{C_2} M(x, y)\, dx$$

$$= \int_a^b M(x, g_1(x))\, dx - \int_a^b M(x, g_2(x))\, dx$$

$$= \int_a^b [M(x, g_1(x)) - M(x, g_2(x))]\, dx, \qquad (4.1)$$

where the minus sign in front of the second integral comes from our traversing C_2 "backward" (i.e., from right to left). On the other hand, notice that we can write

$$\iint_R \frac{\partial M}{\partial y}\, dA = \int_a^b \int_{g_1(x)}^{g_2(x)} \frac{\partial M}{\partial y}\, dy\, dx$$

$$= \int_a^b M(x, y) \Big|_{y=g_1(x)}^{y=g_2(x)}\, dx \qquad \text{By the Fundamental Theorem of Calculus}$$

$$= \int_a^b [M(x, g_2(x)) - M(x, g_1(x))]\, dx.$$

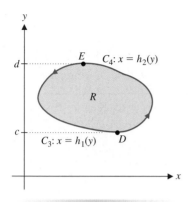

Figure 14.29b

The region R.

Together with (4.1), this gives us

$$\oint_C M(x, y)\, dx = -\iint_R \frac{\partial M}{\partial y}\, dA. \tag{4.2}$$

We now assume that we can also write the region R in the form

$$R = \{(x, y) | c \le y \le d \text{ and } h_1(y) \le x \le h_2(y)\}.$$

Here, we write $C = C_3 \cup C_4$, as illustrated in Figure 14.29b. In this case, notice that we can write

$$\oint_C N(x, y)\, dy = \int_{C_3} N(x, y)\, dy + \int_{C_4} N(x, y)\, dy$$

$$= -\int_c^d N(h_1(y), y)\, dy + \int_c^d N(h_2(y), y)\, dy$$

$$= \int_c^d [N(h_2(y), y) - N(h_1(y), y)]\, dy, \tag{4.3}$$

where the minus sign in front of the first integral accounts for us traversing C_3 "backward" (in this case, from top to bottom). Further, notice that

$$\iint_R \frac{\partial N}{\partial x}\, dA = \int_c^d \int_{h_1(y)}^{h_2(y)} \frac{\partial N}{\partial x}\, dx\, dy$$

$$= \int_c^d [N(h_2(y), y) - N(h_1(y), y)]\, dy.$$

Together with (4.3), this gives us

$$\oint_C N(x, y)\, dy = \iint_R \frac{\partial N}{\partial x}\, dA. \tag{4.4}$$

Adding together (4.2) and (4.4), we have

$$\oint_C M(x, y)\, dx + N(x, y)\, dy = \iint_R \left(\frac{\partial N}{\partial x} - \frac{\partial M}{\partial y} \right) dA,$$

as desired.

■

Although the significance of Green's Theorem lies in the connection it provides between line integrals and double integrals in more theoretical settings, we illustrate the result here by using it to simplify the calculation of a line integral.

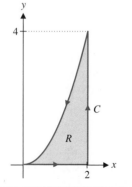

Figure 14.30

The region R.

Example 4.1 **Using Green's Theorem**

Use Green's Theorem to rewrite and evaluate $\oint_C (x^2 + y^3)\, dx + 3xy^2\, dy$, where C consists of the portion of $y = x^2$ from $(2, 4)$ to $(0, 0)$, followed by the line segments from $(0, 0)$ to $(2, 0)$ and from $(2, 0)$ to $(2, 4)$.

Solution We indicate the curve C and the enclosed region R in Figure 14.30. Notice that C is a piecewise-smooth, simple closed curve with positive orientation. Further, for

$M(x, y) = x^2 + y^3$ and $N(x, y) = 3xy^2$, M and N are continuous and have continuous first partial derivatives everywhere. Green's Theorem then says that

$$\oint_C (x^2 + y^3)\, dx + 3xy^2\, dy = \iint_R \left(\frac{\partial N}{\partial x} - \frac{\partial M}{\partial y} \right) dA$$

$$= \iint_R (3y^2 - 3y^2)\, dA = 0.$$

Notice that in example 4.1, since the integrand of the double integral was zero, evaluating the double integral was far easier than evaluating the line integral directly. There is another simple way of thinking of the line integral in example 4.1. Notice that you can write this as $\oint_C \mathbf{F}(x, y) \cdot d\mathbf{r}$, where $\mathbf{F}(x, y) = \langle x^2 + y^3, 3xy^2 \rangle$. Notice further that $\mathbf{F}$ is conservative [with potential function $f(x, y) = \frac{1}{3}x^3 + xy^3$] and so, by Theorem 3.3 in section 14.3, the line integral of $\mathbf{F}$ over any piecewise-smooth, closed curve must be zero.

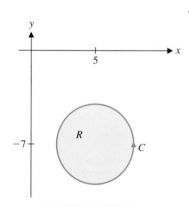

Figure 14.31
The region R.

Example 4.2 Evaluating a Challenging Line Integral with Green's Theorem

Evaluate the line integral $\oint_C (7y - e^{\sin x})\, dx + [15x - \sin(y^3 + 8y)]\, dy$, where C is the circle of radius 3 centered at the point $(5, -7)$, as shown in Figure 14.31.

Solution First, notice that it will be virtually impossible to evaluate the line integral directly. (Think about this some, but don't spend too much time on it!) However, taking $M(x, y) = 7y - e^{\sin x}$ and $N(x, y) = 15x - \sin(y^3 + 8y)$, notice that M and N are continuous and have continuous first partial derivatives everywhere. So, we may apply Green's Theorem, which gives us

$$\oint_C (7y - e^{\sin x})\, dx + [15x - \sin(y^3 + 8y)]\, dy = \iint_R \left(\frac{\partial N}{\partial x} - \frac{\partial M}{\partial y} \right) dA$$

$$= \iint_R (15 - 7)\, dA$$

$$= 8 \iint_R dA = 72\pi,$$

where $\iint_R dA$ is simply the area inside the region R, $\iint_R dA = \pi(3)^2 = 9\pi$.

If you look at example 4.2 critically, you might suspect that the integrand was chosen carefully so that the line integral was impossible to evaluate directly, but so that the integrand of the double integral was trivial. That's true; we did cook up the problem simply to illustrate the power of Green's Theorem. More significantly, Green's Theorem provides us with a wealth of interesting observations. One of these is as follows. Suppose that C is a piecewise-smooth, simple closed curve enclosing the region R. Then, taking $M(x, y) = 0$

and $N(x, y) = x$, we have

$$\oint_C x \, dy = \iint_R \left(\frac{\partial N}{\partial x} - \frac{\partial M}{\partial y} \right) dA = \iint_R dA,$$

which is simply the area of the region R. Alternatively, notice that if we take $M(x, y) = -y$ and $N(x, y) = 0$, we have

$$\oint_C -y \, dx = \iint_R \left(\frac{\partial N}{\partial x} - \frac{\partial M}{\partial y} \right) dA = \iint_R dA,$$

which is again the area of R. Putting these last two results together, we also have

$$\iint_R dA = \frac{1}{2} \oint_C x \, dy - y \, dx. \tag{4.5}$$

> **Example 4.3**   Using Green's Theorem to Find Area
>
> Find the area enclosed by the ellipse $\dfrac{x^2}{a^2} + \dfrac{y^2}{b^2} = 1$.
>
> **Solution**   First, observe that the ellipse corresponds to the simple closed curve C defined parametrically by
>
> $$C = \{(x, y) \mid x = a \cos t, \, y = b \sin t, \, 0 \le t \le 2\pi\},$$
>
> where $a, b > 0$. You should also observe that C is smooth and positively oriented (see Figure 14.32). From (4.5), we have that the area A of the ellipse is given by
>
> $$A = \frac{1}{2} \oint_C x \, dy - y \, dx = \frac{1}{2} \int_0^{2\pi} [(a \cos t)(b \cos t) - (b \sin t)(-a \sin t)] \, dt$$
>
> $$= \frac{1}{2} \int_0^{2\pi} (ab \cos^2 t + ab \sin^2 t) \, dt = \pi ab.$$

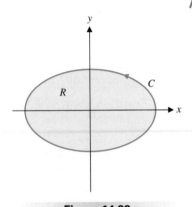

Figure 14.32
Elliptical region R.

> **Example 4.4**   Using Green's Theorem to Evaluate a Line Integral
>
> Evaluate the line integral $\oint_C (e^x + 6xy) \, dx + (8x^2 + \sin y^2) \, dy$, where C is the positively-oriented boundary of the region bounded by the circles of radii 1 and 3, centered at the origin and lying in the first quadrant, as indicated in Figure 14.33.
>
> **Solution**   Notice that since C consists of four distinct pieces, evaluating the line integral directly by parameterizing the curve is probably not a good choice. On the other hand, since C is a piecewise-smooth, simple closed curve, we have by Green's Theorem that
>
> $$\oint_C (e^x + 6xy) \, dx + (8x^2 + \sin y^2) \, dy = \iint_R \left[\frac{\partial}{\partial x}(8x^2 + \sin y^2) - \frac{\partial}{\partial y}(e^x + 6xy) \right] dA$$
>
> $$= \iint_R (16x - 6x) \, dA = \iint_R 10x \, dA,$$
>
> where R is the region between the two circles and lying in the first quadrant. Notice that

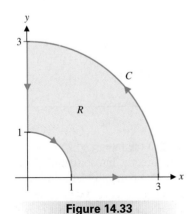

Figure 14.33
The region R.

this is easy to compute using polar coordinates, as follows:

$$\oint_C (e^x + 6xy)\,dx + (8x^2 + \sin y^2)\,dy = \iint_R 10 \underbrace{x}_{r\cos\theta} \underbrace{dA}_{r\,dr\,d\theta}$$

$$= \int_0^{\pi/2}\int_1^3 (10r\cos\theta)\,r\,dr\,d\theta$$

$$= \int_0^{\pi/2} \cos\theta \left.\frac{10r^3}{3}\right|_{r=1}^{r=3} d\theta$$

$$= \frac{10}{3}(3^3 - 1^3)\sin\theta \Big|_0^{\pi/2}$$

$$= \frac{260}{3}.$$

You should notice that in example 4.4, Green's Theorem is not a mere convenience; rather, it is a virtual necessity. Evaluating the line integral directly would prove a very significant challenge. (Go ahead and try it to see what we mean.)

For simplicity, we often will use the notation ∂R to refer to the boundary of the region R, oriented in the positive direction. Using this notation, the conclusion of Green's Theorem is written as

$$\oint_{\partial R} M(x, y)\,dx + N(x, y)\,dy = \iint_R \left(\frac{\partial N}{\partial x} - \frac{\partial M}{\partial y}\right) dA.$$

We can extend Green's Theorem to the case where a region is not simply-connected (i.e., where the region has one or more holes). We must emphasize that when dealing with such regions, the integration is taken over the **entire** boundary of the region (not just the outermost portion of the boundary!) and that the boundary curve is traversed in the positive direction, always keeping the region to the left. For instance, for the region R illustrated in Figure 14.34a with a single hole, notice that the boundary of R, ∂R consists of two separate curves, C_1 and C_2, where C_2 is traversed clockwise, in order to keep the orientation positive on all of the boundary. Since the region is not simply-connected, we may not apply Green's Theorem directly. Rather, we first make two horizontal slits in the region, as indicated in Figure 14.34b, dividing R into the two simply-connected regions R_1 and R_2.

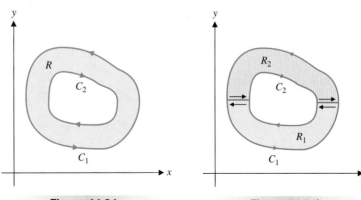

Figure 14.34a
Region with a hole.

Figure 14.34b
$R = R_1 \cup R_2$.

Notice that we can then apply Green's Theorem in each of R_1 and R_2 separately. Adding the double integrals over R_1 and R_2 gives us the double integral over all of R. We have

$$\iint\limits_R \left(\frac{\partial N}{\partial x} - \frac{\partial M}{\partial y}\right) dA = \iint\limits_{R_1} \left(\frac{\partial N}{\partial x} - \frac{\partial M}{\partial y}\right) dA + \iint\limits_{R_2} \left(\frac{\partial N}{\partial x} - \frac{\partial M}{\partial y}\right) dA$$

$$= \oint_{\partial R_1} M(x, y)\,dx + N(x, y)\,dy$$

$$+ \oint_{\partial R_2} M(x, y)\,dx + N(x, y)\,dy.$$

Further, since the line integrals over the common portions of ∂R_1 and ∂R_2 (i.e., the slits) are traversed in the opposite direction (one way on ∂R_1 and the other on ∂R_2), the line integrals over these portions will cancel out, leaving only the line integrals over C_1 and C_2. This gives us

$$\iint\limits_R \left(\frac{\partial N}{\partial x} - \frac{\partial M}{\partial y}\right) dA = \oint_{\partial R_1} M(x, y)\,dx + N(x, y)\,dy + \oint_{\partial R_2} M(x, y)\,dx + N(x, y)\,dy$$

$$= \oint_{C_1} M(x, y)\,dx + N(x, y)\,dy + \oint_{C_2} M(x, y)\,dx + N(x, y)\,dy$$

$$= \oint_C M(x, y)\,dx + N(x, y)\,dy.$$

This says that Green's Theorem also holds for regions with a single hole. Of course, we can repeat the preceding argument to extend Green's Theorem to regions with any **finite** number of holes.

> **Example 4.5**   An Application of Green's Theorem
>
> For $\mathbf{F}(x, y) = \dfrac{1}{x^2 + y^2}\langle -y, x\rangle$, show that $\oint_C \mathbf{F}(x, y) \cdot d\mathbf{r} = 2\pi$, for every simple closed curve C enclosing the origin.
>
> **Solution**   Let C be any simple closed curve enclosing the origin and let C_1 be the circle of radius $a > 0$, centered at the origin (and positively oriented), where a is taken to be sufficiently small so that C_1 is completely enclosed by C, as illustrated in Figure 14.35. Further, let R be the region bounded between the curves C and C_1 (and including the curves themselves). Applying our extended version of Green's Theorem in R, we have
>
>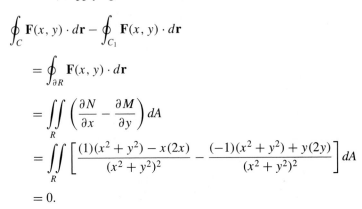

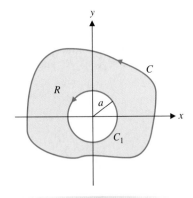

Figure 14.35

The region R.

This gives us

$$\oint_C \mathbf{F}(x, y) \cdot d\mathbf{r} = \oint_{C_1} \mathbf{F}(x, y) \cdot d\mathbf{r}.$$

Now, we chose C_1 to be a circle because we can easily parameterize a circle and then evaluate the line integral around C_1 explicitly. Notice that C_1 can be expressed parametrically by $x = a \cos t$, $y = a \sin t$, for $0 \leq t \leq 2\pi$. This leaves us with an integral that we can easily evaluate, as follows:

$$\oint_C \mathbf{F}(x, y) \cdot d\mathbf{r} = \oint_{C_1} \mathbf{F}(x, y) \cdot d\mathbf{r}$$

$$= \oint_{C_1} \frac{-y}{x^2 + y^2} \, dx + \frac{x}{x^2 + y^2} \, dy$$

$$= \int_0^{2\pi} \frac{(-a \sin t)(-a \sin t) + (a \cos t)(a \cos t)}{a^2 \cos^2 t + a^2 \sin^2 t} \, dt$$

$$= \int_0^{2\pi} dt = 2\pi.$$

Notice that without Green's Theorem, proving a result such as that developed in example 4.5 would be elusive.

Now that we have Green's Theorem, we are in a position to prove the second half of Theorem 3.4. For convenience, we restate the theorem here.

Theorem 4.2

Suppose that $M(x, y)$ and $N(x, y)$ have continuous first partial derivatives on a simply-connected region D. Then, $\int_C M(x, y) \, dx + N(x, y) \, dy$ is independent of path if and only if $M_y(x, y) = N_x(x, y)$ for all (x, y) in D.

Proof

Recall that in section 14.3, we proved the first part of the theorem: that if $\int_C M(x, y) \, dx + N(x, y) \, dy$ is independent of path, then it follows that $M_y(x, y) = N_x(x, y)$ for all (x, y) in D. We now prove that if $M_y(x, y) = N_x(x, y)$ for all (x, y) in D, then it follows that the line integral is independent of path. Let S be any piecewise-smooth, simple closed curve with positive orientation lying in D. Since D is simply-connected, notice that the region R enclosed by S is completely contained in D, so that $M_y(x, y) = N_x(x, y)$ for all (x, y) in R. From Green's Theorem, we now have that

$$\oint_S M(x, y) \, dx + N(x, y) \, dy = \iint_R \left(\frac{\partial N}{\partial x} - \frac{\partial M}{\partial y} \right) dA = 0.$$

That is, for every piecewise-smooth, simple closed curve S lying in D, we have

$$\oint_S M(x, y) \, dx + N(x, y) \, dy = 0.$$

It now follows from Theorem 3.3 that $\mathbf{F}(x, y) = \langle M(x, y), N(x, y) \rangle$ must be conservative in D. Finally, it follows from Theorem 3.1 that $\int_C M(x, y) \, dx + N(x, y) \, dy$ is independent of path.

EXERCISES 14.4

1. Given a line integral to evaluate, briefly describe the circumstances under which you should think about using Green's Theorem to replace the line integral with a double integral. Comment on the properties of the curve C and the functions involved.

2. In example 4.1, Green's Theorem allowed us to quickly show that the line integral equals 0. Following the example, we noted that this was the line integral for a conservative force field. Discuss which method (Green's Theorem, conservative field) you would recommend trying first to determine if a line integral equals 0.

3. Equation (4.5) shows how to compute area as a line integral. Using example 4.3 as a guide, explain why we wrote the area as $\frac{1}{2}\oint_C x\,dy - y\,dx$ instead of $\oint_C x\,dy$ or $\oint_C -y\,dx$.

4. Suppose that you drive a car to a variety of places for shopping and then return home. If your path formed a simple closed curve, explain how you could use (4.5) to estimate the area enclosed by your path. (Hint: If $\langle x, y \rangle$ represents position, what does $\langle x', y' \rangle$ represent?)

In exercises 5–8, evaluate the indicated line integral (a) directly and (b) using Green's Theorem.

5. $\oint_C (x^2 - y)\,dx + y^2\,dy$, where C is the circle $x^2 + y^2 = 1$ oriented counterclockwise

6. $\oint_C (y^2 + x)\,dx + (3x + 2xy)\,dy$, where C is the circle $x^2 + y^2 = 4$ oriented clockwise

7. $\oint_C x^2\,dx - x^3\,dy$, where C is the square from $(0, 0)$ to $(0, 2)$ to $(2, 2)$ to $(2, 0)$ to $(0, 0)$

8. $\oint_C (y^2 - 2x)\,dx + x^2\,dy$, where C is the square from $(0, 0)$ to $(1, 0)$ to $(1, 1)$ to $(0, 1)$ to $(0, 0)$

In exercises 9–24, use Green's Theorem to evaluate the indicated line integral.

9. $\oint_C xe^{2x}\,dx - 3x^2y\,dy$, where C is the rectangle from $(0, 0)$ to $(3, 0)$ to $(3, 2)$ to $(0, 2)$ to $(0, 0)$

10. $\oint_C ye^{2x}\,dx + x^2y^2\,dy$, where C is the rectangle from $(-2, 0)$ to $(3, 0)$ to $(3, 2)$ to $(-2, 2)$ to $(-2, 0)$

11. $\oint_C \left(\frac{x}{x^2 + 1} - y \right)dx + (3x - 4\tan y)\,dy$, where C is the portion of $y = x^2$ from $(-1, 1)$ to $(1, 1)$ followed by the portion of $y = 2 - x^2$ from $(1, 1)$ to $(-1, 1)$

12. $\oint_C (xy - e^{2x})\,dx + (2x^2 - 4y^2)\,dy$, where C is formed by $y = x^2$ and $y = 8 - x^2$ oriented clockwise

13. $\oint_C (\tan x - y^3)\,dx + (x^3 - \sin y)\,dy$, where C is the circle $x^2 + y^2 = 2$ oriented counterclockwise

14. $\oint_C \left(\sqrt{x^2 + 1} - x^2y \right)dx + (xy^2 - y^{5/3})\,dy$, where C is the circle $x^2 + y^2 = 4$ oriented clockwise

15. $\oint_C \mathbf{F} \cdot d\mathbf{r}$, where $\mathbf{F} = \langle x^3 - y, x + y^3 \rangle$ and C is formed by $y = x^2$ and $y = x$ oriented positively

16. $\oint_C \mathbf{F} \cdot d\mathbf{r}$, where $\mathbf{F} = \langle y^2 + 3x^2y, xy + x^3 \rangle$ and C is formed by $y = x^2$ and $y = 2x$ oriented positively

17. $\oint_C \mathbf{F} \cdot d\mathbf{r}$, where $\mathbf{F} = \left\langle e^{x^2} - y, e^{2x} + y \right\rangle$ and C is formed by $y = 1 - x^2$ and $y = 0$ oriented positively

18. $\oint_C \mathbf{F} \cdot d\mathbf{r}$, where $\mathbf{F} = \langle xe^{xy} + y, ye^{xy} + 2x \rangle$ and C is formed by $y = x^2$ and $y = 4$ oriented positively

19. $\oint_C (y^3 - \ln x)\,dx + \left(\sqrt{y^2 + 1} + 3x \right)dy$, where C is formed by $x = y^2$ and $x = 4$ oriented positively

20. $\oint_C (y\sec^2 x - 2)\,dx + (\tan x - 4y^2)\,dy$, where C is formed by $x = 1 - y^2$ and $x = 0$ oriented positively

21. $\oint_C x^2\,dx + 2x\,dy + (z - 2)\,dz$, where C is the triangle from $(0, 0, 2)$ to $(2, 0, 2)$ to $(2, 2, 2)$ to $(0, 0, 2)$

22. $\oint_C 4y\,dx + y^3\,dy + z^4\,dz$, where C is $x^2 + y^2 = 4$ oriented clockwise in the plane $z = 0$

23. $\oint_C \mathbf{F} \cdot d\mathbf{r}$, where $\mathbf{F} = \langle x^3 - y^4, e^{x^2 + z^2}, x^2 - 16y^2z^2 \rangle$ and C is $x^2 + z^2 = 1$ oriented positively in the plane $y = 0$

24. $\oint_C \mathbf{F} \cdot d\mathbf{r}$, where $\mathbf{F} = \langle x^3 - y^2z, \sqrt{x^2 + z^2}, 4xy - z^4 \rangle$ and C is formed by $z = 1 - x^2$ and $z = 0$ oriented positively in the plane $y = 2$

In exercises 25–30, use a line integral to compute the area of the given region.

25. The ellipse $4x^2 + y^2 = 16$

26. The ellipse $4x^2 + y^2 = 4$

27. The region bounded by $x^{2/3} + y^{2/3} = 1$. (Hint: Let $x = \cos^3 t$ and $y = \sin^3 t$)

28. The region bounded by $x^{2/5} + y^{2/5} = 1$

29. The region bounded by $y = x^2$ and $y = 4$

30. The region bounded by $y = x^2$ and $y = 2x$

31. Use Green's Theorem to show that the center of mass of the region bounded by the curve C with constant density is given by $\bar{x} = \frac{1}{2A} \oint_C x^2 \, dy$ and $\bar{y} = -\frac{1}{2A} \oint_C y^2 \, dx$, where A is the area of the region.

32. Use the result of exercise 31 to find the center of mass of the region in exercise 30.

33. Use the result of exercise 31 to find the center of mass of the region bounded by the curve traced out by $\langle t^3 - t, 1 - t^2 \rangle$, for $-1 \le t \le 1$.

34. Use the result of exercise 31 to find the center of mass of the region bounded by the curve traced out by $\langle t^2 - t, t^3 - t \rangle$, for $0 \le t \le 1$.

35. Use Green's Theorem to prove the change of variables formula

$$\iint\limits_R dA = \iint\limits_S \left| \frac{\partial(x, y)}{\partial(u, v)} \right| du \, dv,$$

where $x = x(u, v)$ and $y = y(u, v)$ are functions with continuous partial derivatives.

36. For $\mathbf{F} = \frac{1}{x^2 + y^2} \langle -y, x \rangle$ and C any circle of radius $r > 0$ not containing the origin, show that $\oint_C \mathbf{F} \cdot d\mathbf{r} = 0$.

In exercises 37–40, use the technique of example 4.5 to evaluate the line integral.

37. $\oint_C \mathbf{F} \cdot d\mathbf{r}$, where $\mathbf{F} = \left\langle \frac{x}{x^2 + y^2}, \frac{y}{x^2 + y^2} \right\rangle$ and C is any positively oriented simple closed curve containing the origin

38. $\oint_C \mathbf{F} \cdot d\mathbf{r}$, where $\mathbf{F} = \left\langle \frac{y^2 - x^2}{(x^2 + y^2)^2}, \frac{-2xy}{(x^2 + y^2)^2} \right\rangle$ and C is any positively oriented simple closed curve containing the origin

39. $\oint_C \mathbf{F} \cdot d\mathbf{r}$, where $\mathbf{F} = \left\langle \frac{x^3}{x^4 + y^4}, \frac{y^3}{x^4 + y^4} \right\rangle$ and C is any positively oriented simple closed curve containing the origin

40. $\oint_C \mathbf{F} \cdot d\mathbf{r}$, where $\mathbf{F} = \left\langle \frac{y^2 x}{x^4 + y^4}, \frac{-x^2 y}{x^4 + y^4} \right\rangle$ and C is any positively oriented simple closed curve containing the origin

41. Where is $\mathbf{F}(x, y) = \left\langle \frac{2x}{x^2 + y^2}, \frac{2y}{x^2 + y^2} \right\rangle$ defined? Show that $M_y = N_x$ everywhere the partial derivatives are defined. If C is a simple closed curve enclosing the origin, does Green's Theorem guarantee that $\oint_C \mathbf{F} \cdot d\mathbf{r} = 0$? Explain.

42. If $\mathbf{F}(x, y) = \left\langle \frac{2x}{x^2 + y^2}, \frac{2y}{x^2 + y^2} \right\rangle$ and C is a simple closed curve in the fourth quadrant, does Green's Theorem guarantee that $\oint_C \mathbf{F} \cdot d\mathbf{r} = 0$? Explain.

43. Evaluate $\oint_C \mathbf{F} \cdot d\mathbf{r}$, where

$$\mathbf{F} = \left\langle \frac{-y}{(x^2 + y^2)^2}, \frac{x}{(x^2 + y^2)^2} \right\rangle$$

and C is the circle $x^2 + y^2 = a^2$. Use this result and Green's Theorem to show that $\iint\limits_R \frac{-2}{(x^2 + y^2)^2} \, dA$ diverges, where R is the disk $x^2 + y^2 \le 1$.

14.5 CURL AND DIVERGENCE

We have seen that Green's Theorem relates the line integral of a function over the boundary of a plane region R to the double integral of a related function over the region R. As you have seen, in some cases the line integral is easier to evaluate, while in other cases the double integral is easier. The flexibility provided by Green's Theorem is invaluable when trying to evaluate the difficult integrals that arise in science and engineering. More significantly, Green's Theorem provides us with a connection between physical quantities measured on the boundary of a plane region with related quantities in the interior of the region. The goal of the rest of the chapter is to extend Green's Theorem to results that relate triple integrals, double integrals and line integrals. The first step in this is to understand the vector operations of curl and divergence introduced in this section.

Both the curl and divergence are generalizations of the notion of derivative that are applied to vector fields. Both directly measure important physical quantities related to a vector field $\mathbf{F}(x, y, z)$.

Definition 5.1

The **curl** of the vector field $\mathbf{F}(x, y, z) = \langle F_1(x, y, z), F_2(x, y, z), F_3(x, y, z) \rangle$ is the vector field

$$\operatorname{curl} \mathbf{F} = \left(\frac{\partial F_3}{\partial y} - \frac{\partial F_2}{\partial z} \right) \mathbf{i} + \left(\frac{\partial F_1}{\partial z} - \frac{\partial F_3}{\partial x} \right) \mathbf{j} + \left(\frac{\partial F_2}{\partial x} - \frac{\partial F_1}{\partial y} \right) \mathbf{k},$$

defined at all points at which all the indicated partial derivatives exist.

An easy way to remember curl $\mathbf{F}$ is to use cross product notation, as follows. Notice that using a determinant, we can write

$$\nabla \times \mathbf{F} = \begin{vmatrix} \mathbf{i} & \mathbf{j} & \mathbf{k} \\ \dfrac{\partial}{\partial x} & \dfrac{\partial}{\partial y} & \dfrac{\partial}{\partial z} \\ F_1 & F_2 & F_3 \end{vmatrix}$$

$$= \left(\frac{\partial F_3}{\partial y} - \frac{\partial F_2}{\partial z} \right) \mathbf{i} - \left(\frac{\partial F_3}{\partial x} - \frac{\partial F_1}{\partial z} \right) \mathbf{j} + \left(\frac{\partial F_2}{\partial x} - \frac{\partial F_1}{\partial y} \right) \mathbf{k}$$

$$= \left\langle \frac{\partial F_3}{\partial y} - \frac{\partial F_2}{\partial z}, \frac{\partial F_1}{\partial z} - \frac{\partial F_3}{\partial x}, \frac{\partial F_2}{\partial x} - \frac{\partial F_1}{\partial y} \right\rangle = \operatorname{curl} \mathbf{F}, \tag{5.1}$$

whenever all of the indicated partial derivatives are defined.

Example 5.1   Computing the Curl of a Vector Field

Compute curl $\mathbf{F}$ for (a) $\mathbf{F}(x, y, z) = \langle x^2 y, 3x - yz, z^3 \rangle$ and (b) $\mathbf{F}(x, y, z) = \langle x^3 - y, y^5, e^z \rangle$.

Solution   Using the cross product notation in (5.1), we have that for (a):

$$\operatorname{curl} \mathbf{F} = \nabla \times \mathbf{F} = \begin{vmatrix} \mathbf{i} & \mathbf{j} & \mathbf{k} \\ \dfrac{\partial}{\partial x} & \dfrac{\partial}{\partial y} & \dfrac{\partial}{\partial z} \\ x^2 y & 3x - yz & z^3 \end{vmatrix}$$

$$= \left(\frac{\partial(z^3)}{\partial y} - \frac{\partial(3x - yz)}{\partial z} \right) \mathbf{i} - \left(\frac{\partial(z^3)}{\partial x} - \frac{\partial(x^2 y)}{\partial z} \right) \mathbf{j}$$

$$+ \left(\frac{\partial(3x - yz)}{\partial x} - \frac{\partial(x^2 y)}{\partial y} \right) \mathbf{k}$$

$$= (0 + y)\mathbf{i} - (0 - 0)\mathbf{j} + (3 - x^2)\mathbf{k} = \langle y, 0, 3 - x^2 \rangle.$$

Similarly, for part (b), we have

$$\operatorname{curl} \mathbf{F} = \nabla \times \mathbf{F} = \begin{vmatrix} \mathbf{i} & \mathbf{j} & \mathbf{k} \\ \dfrac{\partial}{\partial x} & \dfrac{\partial}{\partial y} & \dfrac{\partial}{\partial z} \\ x^3 - y & y^5 & e^z \end{vmatrix}$$

$$= \left(\frac{\partial(e^z)}{\partial y} - \frac{\partial(y^5)}{\partial z} \right) \mathbf{i} - \left(\frac{\partial(e^z)}{\partial x} - \frac{\partial(x^3 - y)}{\partial z} \right) \mathbf{j} + \left(\frac{\partial(y^5)}{\partial x} - \frac{\partial(x^3 - y)}{\partial y} \right) \mathbf{k}$$

$$= (0 - 0)\mathbf{i} - (0 - 0)\mathbf{j} + (0 + 1)\mathbf{k} = \langle 0, 0, 1 \rangle.$$

Notice that in part (b) of example 5.1 the only term that contributes to the curl is the term $-y$ in the **i**-component of $\mathbf{F}(x, y, z)$. This illustrates an important property of the curl. Terms in the **i**-component of the vector field involving x only will not contribute to the curl, nor will terms in the **j**-component involving y only nor terms in the **k**-component involving z only. You can use these observations to simplify some calculations of the curl. For instance, notice that

$$\operatorname{curl}\langle x^3, \sin^2 y, \sqrt{z^2 + 1} + x^2 \rangle = \operatorname{curl}\langle 0, 0, x^2 \rangle$$

$$= \nabla \times \langle 0, 0, x^2 \rangle = \langle 0, -2x, 0 \rangle.$$

The simplification discussed above gives an important hint about what the curl measures, since the variables must get "mixed up" to produce a nonzero curl. In the following example, we further investigate the meaning of the curl of a vector field.

Example 5.2   Interpreting the Curl of a Vector Field

Compute the curl of (a) $\mathbf{F}(x, y, z) = x\mathbf{i} + y\mathbf{j}$ and (b) $\mathbf{F}(x, y, z) = y\mathbf{i} - x\mathbf{j}$ and interpret each graphically.

Solution   For (a), we have

$$\nabla \times \mathbf{F} = \begin{vmatrix} \mathbf{i} & \mathbf{j} & \mathbf{k} \\ \dfrac{\partial}{\partial x} & \dfrac{\partial}{\partial y} & \dfrac{\partial}{\partial z} \\ x & y & 0 \end{vmatrix} = \langle 0 - 0, -(0 - 0), 0 - 0 \rangle = \langle 0, 0, 0 \rangle.$$

For (b), we have

$$\nabla \times \mathbf{F} = \begin{vmatrix} \mathbf{i} & \mathbf{j} & \mathbf{k} \\ \dfrac{\partial}{\partial x} & \dfrac{\partial}{\partial y} & \dfrac{\partial}{\partial z} \\ y & -x & 0 \end{vmatrix} = \langle 0 - 0, -(0 - 0), -1 - 1 \rangle = \langle 0, 0, -2 \rangle.$$

Graphs of these vector fields in two dimensions are shown in Figures 14.36a and 14.36b, respectively. It is helpful to think of each of these vector fields as the velocity field for a fluid in motion. In this case, the vectors indicated in the graph of the velocity field are called flow lines and indicate the direction of flow of the fluid. For the vector field $\langle x, y \rangle$, observe that the flow lines are straight, with no rotation and in (a) we found that curl $\mathbf{F} = \mathbf{0}$. By contrast, the flow lines for the vector field $\langle y, -x \rangle$ are circular with a clockwise rotation and in (b) we computed a nonzero curl. In particular, notice that if you curl the fingers of your right hand along the flow lines with your fingertips pointing in the direction of the flow, your thumb will point into the page in the direction of $-\mathbf{k}$, which has the same direction as

$$\operatorname{curl}\langle y, -x \rangle = \nabla \times \langle y, -x \rangle = -2\mathbf{k}.$$

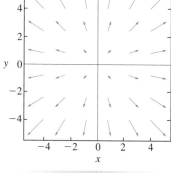

Figure 14.36a

Graph of $\langle x, y \rangle$.

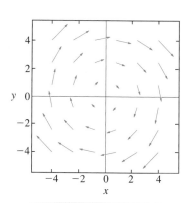

Figure 14.36b

Graph of $\langle y, -x \rangle$.

As we will see through our discussion of Stokes' Theorem in section 14.8, the curl of a vector field at a point always gives a vector parallel to the axis of rotation of the flow lines near the point, with direction determined by the right-hand rule. If $\nabla \times \mathbf{F} = \mathbf{0}$, we say that the vector field is **irrotational** at that point. (That is, the fluid does not tend to rotate near the point.)

We noted earlier that there is no contribution to the curl of a vector field $\mathbf{F}(x, y, z)$ from terms in the **i**-component of $\mathbf{F}$ that involve x only, as well as from terms in the

j-component of **F** involving y only and terms in the **k**-component of **F** involving z only. By contrast, these terms make important contributions to the **divergence** of a vector field, the other major vector operation introduced in this section.

Definition 5.2

The **divergence** of the vector field $\mathbf{F}(x, y, z) = \langle F_1(x, y, z), F_2(x, y, z), F_3(x, y, z) \rangle$ is the scalar function

$$\operatorname{div} \mathbf{F}(x, y, z) = \frac{\partial F_1}{\partial x} + \frac{\partial F_2}{\partial y} + \frac{\partial F_3}{\partial z},$$

defined at all points at which all the indicated partial derivatives exist.

NOTES

Take care to note that, while the curl of a vector field is another vector field, the divergence of a vector field is a scalar function.

Much as we wrote the curl using cross product notation, note that we can write the divergence of a vector field using dot product notation, as follows:

$$\nabla \cdot \mathbf{F} = \left\langle \frac{\partial}{\partial x}, \frac{\partial}{\partial y}, \frac{\partial}{\partial z} \right\rangle \cdot \langle F_1, F_2, F_3 \rangle = \frac{\partial F_1}{\partial x} + \frac{\partial F_2}{\partial y} + \frac{\partial F_3}{\partial z} = \operatorname{div} \mathbf{F}(x, y, z). \quad (5.2)$$

Example 5.3 Computing the Divergence of a Vector Field

Compute $\operatorname{div} \mathbf{F}$ for (a) $\mathbf{F}(x, y, z) = \langle x^2 y, 3x - yz, z^3 \rangle$ and (b) $\mathbf{F}(x, y, z) = \langle x^3 - y, z^5, e^y \rangle$.

Solution For (a), we have from (5.2) that

$$\operatorname{div} \mathbf{F} = \nabla \cdot \mathbf{F} = \frac{\partial (x^2 y)}{\partial x} + \frac{\partial (3x - yz)}{\partial y} + \frac{\partial (z^3)}{\partial z} = 2xy - z + 3z^2.$$

For (b), we have from (5.2) that

$$\operatorname{div} \mathbf{F} = \nabla \cdot \mathbf{F} = \frac{\partial (x^3 - y)}{\partial x} + \frac{\partial (z^5)}{\partial y} + \frac{\partial (e^y)}{\partial z} = 3x^2 + 0 + 0 = 3x^2.$$

Notice that in part (b) of example 5.3 the only term contributing to the divergence is the x^3 term in the **i**-component of **F**. Further, observe that in general, the divergence of $\mathbf{F}(x, y, z)$ is not affected by terms in the **i**-component of **F** that do not involve x, terms in the **j**-component of **F** that do not involve y and terms in the **k**-component of **F** that do not involve z. Returning to the two-dimensional vector fields of example 5.2, we can develop a graphical interpretation of the divergence.

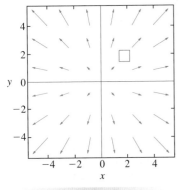

Figure 14.37a
Graph of $\langle x, y \rangle$.

Example 5.4 Interpreting the Divergence of a Vector Field

Compute the divergence of (a) $\mathbf{F}(x, y) = x\mathbf{i} + y\mathbf{j}$ and (b) $\mathbf{F}(x, y) = y\mathbf{i} - x\mathbf{j}$ and interpret each graphically.

Solution For (a), we have $\nabla \cdot \mathbf{F} = \dfrac{\partial (x)}{\partial x} + \dfrac{\partial (y)}{\partial y} = 2$. For (b), we have $\nabla \cdot \mathbf{F} = \dfrac{\partial (y)}{\partial x} + \dfrac{\partial (-x)}{\partial y} = 0$. Graphs of the vector fields in (a) and (b) are shown in Figures 14.37a

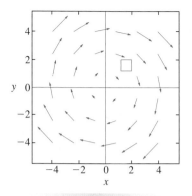

Figure 14.37b
Graph of $\langle y, -x \rangle$.

and 14.37b, respectively. Notice the boxes that we have superimposed on the graph of each vector field. If $\mathbf{F}(x, y)$ represents the velocity field of a fluid in motion, try to use the graphs to estimate the net flow of fluid into or out of each box. For $\langle y, -x \rangle$, the fluid is rotating in circular paths, so that the velocity of any particle on a given circle centered at the origin is a constant. This suggests that the flow into the box should equal the flow out of the box and the net flow is 0, which you'll notice is also the value of the divergence of this velocity field. By contrast, for the vector field $\langle x, y \rangle$, notice that the arrows coming into the box are shorter than the arrows exiting the box. This says that the net flow *out of* the box is positive (i.e., there is more fluid exiting the box than entering the box). Notice that in this case, the divergence is positive.

We'll show in section 14.7 (using the Divergence Theorem) that the divergence of a vector field at a point (x, y, z) corresponds to the net flow of fluid out of a small box centered at (x, y, z). If $\nabla \cdot \mathbf{F}(x, y, z) > 0$, more fluid exits the box than enters (as illustrated in Figure 14.37a) and we call the point (x, y, z) a **source.** If $\nabla \cdot \mathbf{F}(x, y, z) < 0$, more fluid enters the box than exits and we call the point (x, y, z) a **sink.** If $\nabla \cdot \mathbf{F}(x, y, z) = 0$, throughout some region D, then we say that the vector field $\mathbf{F}$ is **source-free** or **incompressible.**

We have now used the "del" operator ∇ for three different derivative-like operations. The gradient of a scalar function f is the vector field ∇f, the curl of a vector field $\mathbf{F}$ is the vector field $\nabla \times \mathbf{F}$ and the divergence of a vector field $\mathbf{F}$ is the scalar function $\nabla \cdot \mathbf{F}$. Pay special attention to the different roles of scalar and vector functions in these operations. An analysis of the possible combinations of these operations will give us further insight into the properties of vector fields.

Example 5.5 Vector Fields and Scalar Functions Involving the Gradient

If $f(x, y, z)$ is a scalar function and $\mathbf{F}(x, y, z)$ is a vector field, determine whether each operation is a scalar function, a vector field or undefined: (a) $\nabla \times (\nabla f)$, (b) $\nabla \times (\nabla \cdot \mathbf{F})$, (c) $\nabla \cdot (\nabla f)$.

Solution Examine each of these expressions one step at a time, working from the inside out. In (a), ∇f is a vector field, so the curl of ∇f is defined and gives a vector field. In (b), $\nabla \cdot \mathbf{F}$ is a scalar function, so the curl of $\nabla \cdot \mathbf{F}$ is undefined. In (c), ∇f is a vector field, so the divergence of ∇f is defined and gives a scalar function.

We can say more about the two operations defined in example 5.5 parts (a) and (c). If f has continuous second-order partial derivatives, then $\nabla f = \langle f_x, f_y, f_z \rangle$ and the divergence of the gradient is the scalar function

$$\nabla \cdot (\nabla f) = \left\langle \frac{\partial}{\partial x}, \frac{\partial}{\partial y}, \frac{\partial}{\partial z} \right\rangle \cdot \langle f_x, f_y, f_z \rangle = f_{xx} + f_{yy} + f_{zz}.$$

This combination of second partial derivatives arises in many important applications in physics and engineering. We call $\nabla \cdot (\nabla f)$ the **Laplacian** of f and typically use the shorthand notation

$$\nabla \cdot (\nabla f) = \nabla^2 f = f_{xx} + f_{yy} + f_{zz}$$

or $\Delta f = \nabla^2 f$.

Using the same notation, the curl of the gradient of a scalar function f is

$$\nabla \times (\nabla f) = \begin{vmatrix} \mathbf{i} & \mathbf{j} & \mathbf{k} \\ \dfrac{\partial}{\partial x} & \dfrac{\partial}{\partial y} & \dfrac{\partial}{\partial z} \\ f_x & f_y & f_z \end{vmatrix} = \langle f_{zy} - f_{yz}, f_{xz} - f_{zx}, f_{yx} - f_{xy} \rangle = \langle 0, 0, 0 \rangle,$$

assuming the mixed partial derivatives are equal. (We've seen that this occurs whenever all of the second order partial derivatives are continuous in some open region.) Recall that if $\mathbf{F} = \nabla f$, then we call $\mathbf{F}$ a conservative field. The result $\nabla \times (\nabla f) = \mathbf{0}$ proves the following theorem, which gives us a simple way for determining when a given three-dimensional vector field is not conservative.

Theorem 5.1

Suppose that $\mathbf{F}(x, y, z) = \langle F_1(x, y, z), F_2(x, y, z), F_3(x, y, z) \rangle$ is a vector field whose components F_1, F_2 and F_3 have continuous first-order partial derivatives throughout an open region $D \subset \mathbb{R}^3$. If $\mathbf{F}$ is conservative, then $\nabla \times \mathbf{F} = \mathbf{0}$.

We can often use Theorem 5.1 to determine that a given vector field is not conservative, as we illustrate in the following example.

Example 5.6   Determining When a Vector Field is Conservative

Use Theorem 5.1 to determine if the following vector fields are conservative: (a) $\mathbf{F} = \langle \cos x - z, y^2, xz \rangle$ and (b) $\mathbf{F} = \langle 2xz, 3z^2, x^2 + 6yz \rangle$.

Solution   For (a), we have

$$\nabla \times \mathbf{F} = \begin{vmatrix} \mathbf{i} & \mathbf{j} & \mathbf{k} \\ \dfrac{\partial}{\partial x} & \dfrac{\partial}{\partial y} & \dfrac{\partial}{\partial z} \\ \cos x - z & y^2 & xz \end{vmatrix} = \langle 0 - 0, -1 - z, 0 - 0 \rangle \neq \mathbf{0},$$

and so, by Theorem 5.1, $\mathbf{F}$ is not conservative.

For (b), we have

$$\nabla \times \mathbf{F} = \begin{vmatrix} \mathbf{i} & \mathbf{j} & \mathbf{k} \\ \dfrac{\partial}{\partial x} & \dfrac{\partial}{\partial y} & \dfrac{\partial}{\partial z} \\ 2xz & 3z^2 & x^2 + 6yz \end{vmatrix} = \langle 6z - 6z, 2x - 2x, 0 - 0 \rangle = \mathbf{0}.$$

Notice that in this case, Theorem 5.1 does not tell us whether or not $\mathbf{F}$ is conservative. However, you might notice that

$$\mathbf{F}(x, y, z) = \langle 2xz, 3z^2, x^2 + 6yz \rangle = \nabla(x^2 z + 3yz^2).$$

Since we have found a potential function for $\mathbf{F}$, we now see that it is indeed a conservative field.

Following example 5.6, you might be wondering whether or not the converse of Theorem 5.1 is true. That is, if $\nabla \times \mathbf{F} = \mathbf{0}$, must it follow that $\mathbf{F}$ is conservative? The answer to this is, "NO." We had an important clue to this in example 4.5. There, we saw that for the two dimensional vector field $\mathbf{F}(x, y) = \dfrac{1}{x^2 + y^2}\langle -y, x\rangle$, $\oint_C \mathbf{F}(x, y) \cdot d\mathbf{r} = 2\pi$, for every simple closed curve C enclosing the origin. We follow up on this idea in the following example.

Example 5.7   An Irrotational Vector Field That Is not Conservative

For $\mathbf{F}(x, y, z) = \dfrac{1}{x^2 + y^2}\langle -y, x, 0\rangle$, show that $\nabla \times \mathbf{F} = \mathbf{0}$, but that $\mathbf{F}$ is not conservative.

Solution   First, notice that

$$\nabla \times \mathbf{F} = \begin{vmatrix} \mathbf{i} & \mathbf{j} & \mathbf{k} \\ \dfrac{\partial}{\partial x} & \dfrac{\partial}{\partial y} & \dfrac{\partial}{\partial z} \\ \dfrac{-y}{x^2 + y^2} & \dfrac{x}{x^2 + y^2} & 0 \end{vmatrix}$$

$$= \mathbf{i}\begin{vmatrix} \dfrac{\partial}{\partial y} & \dfrac{\partial}{\partial z} \\ \dfrac{x}{x^2 + y^2} & 0 \end{vmatrix} - \mathbf{j}\begin{vmatrix} \dfrac{\partial}{\partial x} & \dfrac{\partial}{\partial z} \\ \dfrac{-y}{x^2 + y^2} & 0 \end{vmatrix} + \mathbf{k}\begin{vmatrix} \dfrac{\partial}{\partial x} & \dfrac{\partial}{\partial y} \\ \dfrac{-y}{x^2 + y^2} & \dfrac{x}{x^2 + y^2} \end{vmatrix}$$

$$= \mathbf{k}\left[\dfrac{\partial}{\partial x}\left(\dfrac{x}{x^2 + y^2}\right) + \dfrac{\partial}{\partial y}\left(\dfrac{y}{x^2 + y^2}\right)\right]$$

$$= \mathbf{k}\left[\dfrac{(x^2 + y^2) - 2x^2}{(x^2 + y^2)^2} + \dfrac{(x^2 + y^2) - 2y^2}{(x^2 + y^2)^2}\right] = \mathbf{0},$$

so that $\mathbf{F}$ is irrotational at every point at which it's defined (i.e., everywhere but at the origin). However, in example 4.5, we already showed that $\oint_C \mathbf{F}(x, y, z) \cdot d\mathbf{r} = 2\pi$, for every simple, closed curve C lying in the xy-plane and enclosing the origin. Given this, it follows from Theorem 3.3 that $\mathbf{F}$ cannot be conservative, since if it were, we would need to have $\oint_C \mathbf{F}(x, y, z) \cdot d\mathbf{r} = 0$.

The problem with example 5.7 was that the vector field in question had a singularity (i.e., a point where one or more of the components of the vector field blows up to ∞). Even though the curves we considered did not pass through the singularity (the origin), they in some sense "enclosed" the origin. This is enough to make the converse of Theorem 5.1 false. As it turns out, the converse is true only with some additional hypotheses. Specifically, we can say the following.

Theorem 5.2
Suppose that $\mathbf{F}(x, y, z) = \langle F_1(x, y, z), F_2(x, y, z), F_3(x, y, z)\rangle$ is a vector field whose components F_1, F_2 and F_3 have continuous first partial derivatives throughout all of $\mathbb{R}^3$. Then, $\mathbf{F}$ is conservative if and only if $\nabla \times \mathbf{F} = \mathbf{0}$.

Notice that half of this theorem is already known from Theorem 5.1. Also, notice that we required that the components of $\mathbf{F}$ have continuous first partial derivatives throughout **all** of $\mathbb{R}^3$. (Observe that this requirement was not satisfied by the vector field in example 5.7.)

The other half of the theorem requires the additional sophistication of Stokes' Theorem and we will prove a more general version of this in section 14.8.

CONSERVATIVE VECTOR FIELDS

We can now summarize a number of equivalent properties for three-dimensional vector fields. Suppose that $\mathbf{F}(x, y, z) = \langle F_1(x, y, z), F_2(x, y, z), F_3(x, y, z) \rangle$ is a vector field whose components F_1, F_2 and F_3 have continuous first partial derivatives throughout all of $\mathbb{R}^3$. Then the following are equivalent:

1. $\mathbf{F}(x, y, z)$ is conservative.
2. $\int_C \mathbf{F} \cdot d\mathbf{r}$ is independent of path.
3. $\int_C \mathbf{F} \cdot d\mathbf{r} = 0$ for every piecewise-smooth closed curve C.
4. $\nabla \times \mathbf{F} = \mathbf{0}$.
5. $\mathbf{F}(x, y, z)$ is a gradient field ($\mathbf{F} = \nabla f$ for some potential function f).

We close this section by rewriting Green's Theorem in terms of the curl and divergence. In sections 14.7 and 14.8, we extend these formulas to three-dimensional regions.

First, suppose that $\mathbf{F}(x, y) = \langle M(x, y), N(x, y), 0 \rangle$ is a vector field, for some functions $M(x, y)$ and $N(x, y)$. Suppose that R is a region in the xy-plane whose boundary curve C is piecewise-smooth, positively oriented, simple and closed and that M and N are continuous and have continuous first partial derivatives in some open region D, where $R \subset D$. Then, from Green's Theorem, we have

$$\iint_R \left(\frac{\partial N}{\partial x} - \frac{\partial M}{\partial y} \right) dA = \oint_C M\, dx + N\, dy.$$

Notice that the integrand of the double integral, $\dfrac{\partial N}{\partial x} - \dfrac{\partial M}{\partial y}$, is the $\mathbf{k}$ component of $\nabla \times \mathbf{F}$. Further, since $dz = 0$ on any curve lying in the xy-plane, we have

$$\oint_C M\, dx + N\, dy = \oint_C \mathbf{F} \cdot d\mathbf{r}.$$

Thus, we can write Green's Theorem in the form

$$\oint_C \mathbf{F} \cdot d\mathbf{r} = \iint_R (\nabla \times \mathbf{F}) \cdot \mathbf{k}\, dA.$$

This is generalized to Stokes' Theorem in section 14.8.

To take Green's Theorem in yet another direction, suppose that $\mathbf{F}$ and R are as defined above and suppose that C is traced out by the endpoint of the vector-valued function $\mathbf{r}(t) = \langle x(t), y(t) \rangle$, for $a \leq t \leq b$, where $x(t)$ and $y(t)$ have continuous first derivatives for $a \leq t \leq b$. Recall that the unit tangent vector to the curve is given by

$$\mathbf{T}(t) = \left\langle \frac{x'(t)}{\|\mathbf{r}'(t)\|}, \frac{y'(t)}{\|\mathbf{r}'(t)\|} \right\rangle.$$

It's then easy to verify that the exterior unit normal vector to C at any point (i.e., the unit normal vector that points out of R) is given by

$$\mathbf{n}(t) = \left\langle \frac{y'(t)}{\|\mathbf{r}'(t)\|}, \frac{-x'(t)}{\|\mathbf{r}'(t)\|} \right\rangle$$

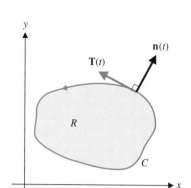

Figure 14.38
Unit tangent and exterior unit
normal vectors to R.

(see Figure 14.38). Now, from Theorem 2.1, we have

$$\oint_C \mathbf{F} \cdot \mathbf{n} \, ds = \int_a^b (\mathbf{F} \cdot \mathbf{n})(t) \|\mathbf{r}'(t)\| \, dt$$

$$= \int_a^b \left[\frac{M(x(t), y(t)) y'(t)}{\|\mathbf{r}'(t)\|} - \frac{N(x(t), y(t)) x'(t)}{\|\mathbf{r}'(t)\|} \right] \|\mathbf{r}'(t)\| \, dt$$

$$= \int_a^b [M(x(t), y(t)) y'(t) \, dt - N(x(t), y(t)) x'(t) \, dt]$$

$$= \oint_C M(x, y) \, dy - N(x, y) \, dx$$

$$= \iint_R \left(\frac{\partial M}{\partial x} + \frac{\partial N}{\partial y} \right) dA,$$

from Green's Theorem. Finally, recognize that the integrand of the double integral is the divergence of $\mathbf{F}$ and this gives us another vector form of Green's Theorem:

$$\oint_C \mathbf{F} \cdot \mathbf{n} \, ds = \iint_R \nabla \cdot \mathbf{F}(x, y) \, dA. \tag{5.3}$$

This form of Green's Theorem is generalized to the Divergence Theorem in section 14.7.

EXERCISES 14.5

1. Suppose that $\nabla \times \mathbf{F} = \langle 2, 0, 0 \rangle$. Describe what the graph of the vector field $\mathbf{F}$ looks like. Explain how the graph of the vector field $\mathbf{G}$ with $\nabla \times \mathbf{G} = \langle 20, 0, 0 \rangle$ compares.

2. If $\nabla \cdot \mathbf{F} > 0$ at a point P and $\mathbf{F}$ is the velocity field of a fluid, explain why the word **source** is a good choice for what's happening at P. Explain why **sink** is a good word if $\nabla \cdot \mathbf{F} < 0$.

3. You now have two ways of determining whether or not a vector field is conservative: try to find the potential or see if the curl equals $\mathbf{0}$. If you have reason to believe that the vector field is conservative, explain which test you prefer.

4. In the text, we discussed geometrical interpretations of the divergence and curl. Discuss the extent to which the divergence and curl are analogous to tangential and normal components of acceleration.

In exercises 5–16, find the curl and divergence of the given vector field.

5. $x^2 \mathbf{i} - 3xy \mathbf{j}$

6. $y^2 \mathbf{i} + 4x^2 y \mathbf{j}$

7. $2xz \mathbf{i} - 3y \mathbf{k}$

8. $x^2 \mathbf{i} - 3yx \mathbf{j} + x \mathbf{k}$

9. $\langle xy, yz, x^2 \rangle$

10. $\langle xe^z, yz^2, x + y \rangle$

11. $\langle x^2, y - z, xe^y \rangle$

12. $\langle y, x^2 y, 3z + y \rangle$

13. $\langle 3yz, x^2, x \cos y \rangle$

14. $\langle y^2, x^2 e^z, \cos xy \rangle$

15. $\langle 2xz, y + z^2, zy^2 \rangle$

16. $\langle xy^2, 3y^2 z^2, 2x - zy^3 \rangle$

In exercises 17–30, determine if the given vector field is conservative and/or incompressible.

17. $\langle 2x, 2yz^2, 2y^2 z \rangle$

18. $\langle 2xy, x^2 - 3y^2 z^2, 1 - 2zy^3 \rangle$

19. $\langle 3yz, x^2, x \cos y \rangle$

20. $\langle y^2, x^2 e^z, \cos xy \rangle$

21. $\langle \sin z, z^2 e^{yz^2}, x \cos z + 2yze^{yz^2} \rangle$

22. $\langle 2xy \cos z, x^2 \cos z - 3y^2 z, -x^2 y \sin z - y^3 \rangle$

23. $\langle z^2 - 3ye^{3x}, z^2 - e^{3x}, 2z\sqrt{xy} \rangle$

24. $\langle 2xz, 3y, x^2 - y \rangle$

25. $\langle xy^2, 3xz, 4 - zy^2 \rangle$

26. $\langle x, y, 1 - 3z \rangle$

27. $\langle 4x, 3y^3, e^z \rangle$

28. $\langle \sin x, 2y^2, \sqrt{z} \rangle$

29. $\langle -2xy, z^2 \cos yz^2 - x^2, 2yz \cos yz^2 \rangle$

30. $\langle e^y, xe^y + z^2, 2yz - 1 \rangle$

31. Label each expression as a scalar quantity, a vector quantity or undefined, if f is a scalar function and $\mathbf{F}$ is a vector field.
    (a) $\nabla \cdot (\nabla f)$ (b) $\nabla \times (\nabla \cdot \mathbf{F})$ (c) $\nabla(\nabla \times \mathbf{F})$
    (d) $\nabla(\nabla \cdot \mathbf{F})$ (e) $\nabla \times (\nabla f)$

32. Label each expression as a scalar quantity, a vector quantity or undefined, if f is a scalar function and $\mathbf{F}$ is a vector field.
    (a) $\nabla(\nabla f)$ (b) $\nabla \cdot (\nabla \cdot \mathbf{F})$ (c) $\nabla \cdot (\nabla \times \mathbf{F})$
    (d) $\nabla \times (\nabla \mathbf{F})$ (e) $\nabla \times (\nabla \times (\nabla \times \mathbf{F}))$

33. If $\mathbf{r} = \langle x, y, z \rangle$, prove that $\nabla \times \mathbf{r} = \mathbf{0}$ and $\nabla \cdot \mathbf{r} = 3$.

34. If $\mathbf{r} = \langle x, y, z \rangle$ and $r = \|\mathbf{r}\|$, prove that $\nabla \cdot (r\mathbf{r}) = 4r$.

In exercises 35–40, conjecture whether the divergence at point P is positive, negative or zero.

35.
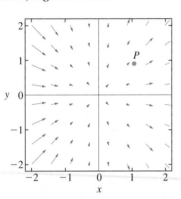

36.
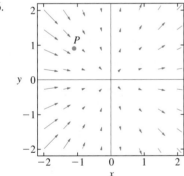

37.
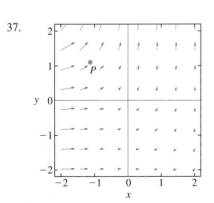

38.
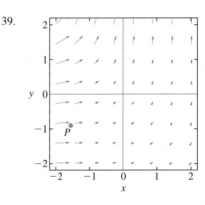

39.
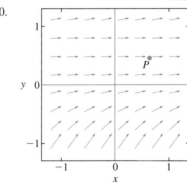

40.

41. If the $\mathbf{j}$-component, $\dfrac{\partial F_1}{\partial z} - \dfrac{\partial F_3}{\partial x}$, of the curl of $\mathbf{F}$ is positive, show that there is a closed curve C such that $\int_C \mathbf{F} \cdot d\mathbf{r} \neq 0$.

42. If the $\mathbf{k}$-component, $\dfrac{\partial F_2}{\partial x} - \dfrac{\partial F_1}{\partial y}$, of the curl of $\mathbf{F}$ is positive, show that there is a closed curve C such that $\int_C \mathbf{F} \cdot d\mathbf{r} \neq 0$.

43. Prove Green's first identity: For $C = \partial R$,

$$\iint_R f \, \nabla^2 g \, dA = \int_C f(\nabla g) \cdot \mathbf{n} \, ds - \iint_R (\nabla f \, \cdot \nabla g) \, dA.$$

[Hint: Use the vector form of Green's Theorem in (5.3) applied to $\mathbf{F} = f \, \nabla g$.]

44. Prove Green's second identity: For $C = \partial R$,

$$\iint_R (f \nabla^2 g - g \nabla^2 f)\, dA = \int_C (f \nabla g - g \nabla f) \cdot \mathbf{n}\, ds.$$

(Hint: Use Green's first identity from exercise 43.)

45. If f is a scalar function and $\mathbf{F}$ a vector field, show that

$$\nabla \cdot (f\mathbf{F}) = \nabla f \cdot \mathbf{F} + f(\nabla \cdot \mathbf{F}).$$

46. If f is a scalar function and $\mathbf{F}$ a vector field, show that

$$\nabla \times (f\mathbf{F}) = \nabla f \times \mathbf{F} + f(\nabla \times \mathbf{F}).$$

47. If $\nabla \cdot \mathbf{F} = 0$, we say that $\mathbf{F}$ is **solenoidal.** If $\nabla^2 f = 0$, show that ∇f is both solenoidal and irrotational.

48. If $\mathbf{F}$ and $\mathbf{G}$ are irrotational, prove that $\mathbf{F} \times \mathbf{G}$ is solenoidal. (Refer to exercise 47.)

49. If f is a scalar function, $\mathbf{r} = \langle x, y \rangle$ and $r = \|\mathbf{r}\|$, show that

$$\nabla f(r) = f'(r)\, \frac{\mathbf{r}}{r}.$$

50. If f is a scalar function, $\mathbf{r} = \langle x, y \rangle$ and $r = \|\mathbf{r}\|$, show that

$$\nabla^2 f(r) = f''(r) + \frac{1}{r}\, f'(r).$$

51. Compute the Laplacian Δf for $f(x, y, z) = \sqrt{x^2 + y^2 + z^2}$.

52. Compute the Laplacian Δf for $f(x, y, z) = \dfrac{1}{x^2 + y^2 + z^2}$.

53. Suppose that $\mathbf{F}(x, y) = \langle x^2, y^2 - 4x \rangle$ represents the velocity field of a fluid in motion. For a small box centered at (x, y), determine whether the flow into the box is greater than, less than or equal to the flow out of the box. (a) $(x, y) = (0, 0)$ and (b) $(x, y) = (1, 0)$.

54. Repeat exercise 53 for (a) $(x, y) = (1, 1)$ and (b) $(x, y) = (0, 1)$.

55. Give an example of a vector function $\mathbf{F}$ such that $\nabla \cdot \mathbf{F}$ is a positive function of y only.

56. Give an example of a vector function $\mathbf{F}$ such that $\nabla \times \mathbf{F}$ is a function of x only.

57. In some calculus and engineering books, you will find the vector identity

$$\nabla \times (\mathbf{F} \times \mathbf{G}) = (\mathbf{G} \cdot \nabla)\mathbf{F} - \mathbf{G}(\nabla \cdot \mathbf{F}) - (\mathbf{F} \cdot \nabla)\mathbf{G} + \mathbf{F}(\nabla \cdot \mathbf{G})$$

Which two of the four terms on the right-hand side look like they should be undefined? Write out the left-hand side as completely as possible, group it into four terms, identify the two familiar terms on the right-hand side, and then define the unusual terms on the right-hand side. (Hint: The notation makes sense as a generalization of the definitions in this section.)

58. Prove the vector formula

$$\nabla \times (\nabla \times \mathbf{F}) = \nabla(\nabla \cdot \mathbf{F}) - \nabla^2 \mathbf{F}.$$

As in exercise 57, a major part of the problem is to decipher an unfamiliar notation.

59. Maxwell's laws relate an electric field $\mathbf{E}(t)$ to a magnetic field $\mathbf{H}(t)$. In a region with no charges and no current, the laws state that $\nabla \cdot \mathbf{E} = 0$, $\nabla \cdot \mathbf{H} = 0$, $\nabla \times \mathbf{E} = -\mu \mathbf{H}_t$ and $\nabla \times \mathbf{H} = \mu \mathbf{E}_t$. From these laws, prove that

$$\nabla \times (\nabla \times \mathbf{E}) = -\mu^2 \mathbf{E}_{tt}$$

and

$$\nabla \times (\nabla \times \mathbf{H}) = -\mu^2 \mathbf{H}_{tt}.$$

14.6   SURFACE INTEGRALS

Few architectural structures are as impressive as domes. Whether it is the ceiling of the Sistine Chapel, the dome of a college library or the massive roof of the Toronto SkyDome, looking up at a domed ceiling can be awe-inspiring. One impressive aspect of a dome is its lack of visible support. As you might imagine, this is also a feature of domes that worries architects, who must be certain that the weight of the dome is properly supported. A critical part of an architect's calculation is the mass of the dome.

How would you compute the mass of a dome? You have already seen how to use double integrals to compute the mass of a two-dimensional lamina and triple integrals to find the mass of a three-dimensional solid. However, a dome is a three-dimensional structure more like a thin shell (a surface) than a solid. We hope you're one step ahead of us on this one: if you don't know how to find the mass of the dome exactly, you can try to

Figure 14.39
Partition of a surface.

approximate its mass by slicing it into a number of small sections and estimating the mass of each section. In Figure 14.39, we show a curved surface that has been divided into a number of pieces. If the pieces are small enough, notice that the density of each piece will be approximately constant.

So, first subdivide (partition) the surface into n smaller pieces, $S_1, S_2, \ldots, S_n$. Next, let $\rho(x, y, z)$ be the density function (measured in units of mass per unit area), and for each $i = 1, 2, \ldots, n$, let (x_i, y_i, z_i) be a point on the section S_i and let ΔS_i be the surface area of S_i. The mass of the section S_i is then given approximately by $\rho(x_i, y_i, z_i)\Delta S_i$. The total mass m of the surface is given approximately by the sum of these approximate masses,

$$m \approx \sum_{i=1}^{n} \rho(x_i, y_i, z_i)\,\Delta S_i.$$

You should expect that the exact mass is given by the limit of these sums as the size of the pieces gets smaller and smaller. We define the **diameter** of a section S_i as the maximum distance between any two points on S_i and the norm of the partition $\|P\|$ as the maximum of the diameters of the S_i's. Then we have that

$$m = \lim_{\|P\| \to 0} \sum_{i=1}^{n} \rho(x_i, y_i, z_i)\,\Delta S_i.$$

This limit is an example of a new type of integral, the **surface integral,** which is the focus of this section.

Definition 6.1

The **surface integral** of a function $g(x, y, z)$ over a surface $S \subset \mathbb{R}^3$, written $\iint_S g(x, y, z)\, dS$ is given by

$$\iint_S g(x, y, z)\, dS = \lim_{\|P\| \to 0} \sum_{i=1}^{n} g(x_i, y_i, z_i)\,\Delta S_i,$$

provided the limit exists and is the same for all choices of the evaluation points (x_i, y_i, z_i).

Notice how our development of the surface integral parallels our development of the line integral. Whereas the line integral extended a single integral over an interval to an integral over a curve in three dimensions, the surface integral extends a double integral over a two-dimensional region to an integral over a two-dimensional surface in three dimensions. In both cases, we are "curving" our domain into three dimensions.

Now that we have defined the surface integral, how can we calculate one? The basic idea is to rewrite a surface integral as a double integral and then evaluate the double integral using existing techniques. To convert a given surface integral into a double integral, you will have two main tasks:

1. Write the integrand $g(x, y, z)$ as a function of two variables.
2. Write the surface area element dS in terms of the area element dA.

We will develop a general rule for step (2) before considering specific examples.

Consider a surface such as the one pictured in Figure 14.39. For the sake of simplicity, we assume that the surface is the graph of the equation $z = f(x, y)$, where f has continuous first partial derivatives in some region R in the xy-plane. Notice that for an inner partition $R_1, R_2, \ldots, R_n$ of R, if we take the point $(x_i, y_i, 0)$ as the point in R_i closest to the origin, then the portion of the surface S_i lying above R_i will differ very little from the portion T_i of the tangent plane to the surface at $(x_i, y_i, f(x_i, y_i))$ lying above R_i. More to the point, the surface area of S_i will be approximately the same as the area of the parallelogram T_i. In Figure 14.40, we have indicated the portion T_i of the tangent plane lying above R_i.

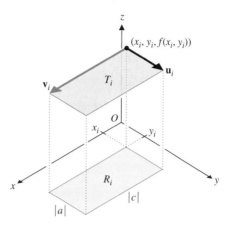

Figure 14.40
Portion of the tangent plane lying above R_i.

Let the vectors $\mathbf{u}_i = \langle 0, a, b \rangle$ and $\mathbf{v}_i = \langle c, 0, d \rangle$ form two adjacent sides of the parallelogram T_i, as indicated in Figure 14.40. Notice that since $\mathbf{u}_i$ and $\mathbf{v}_i$ lie in the tangent plane, $\mathbf{n}_i = \mathbf{u}_i \times \mathbf{v}_i = \langle ad, bc, -ac \rangle$ is a normal vector to the tangent plane. We saw in Chapter 10 that the area of the parallelogram can be written as

$$\Delta S_i = \|\mathbf{u}_i \times \mathbf{v}_i\| = \|\mathbf{n}_i\|.$$

We further observe that the area of R_i is given by $\Delta A_i = |ac|$ and $\mathbf{n}_i \cdot \mathbf{k} = -ac$, so that $|\mathbf{n}_i \cdot \mathbf{k}| = |ac|$. We can now write

$$\Delta S_i = \frac{|ac| \, \|\mathbf{n}_i\|}{|ac|} = \frac{\|\mathbf{n}_i\|}{|\mathbf{n}_i \cdot \mathbf{k}|} \, \Delta A_i,$$

since $ac \neq 0$. The corresponding expression relating the surface area element dS and the area element dA is then

$$dS = \frac{\|\mathbf{n}\|}{|\mathbf{n} \cdot \mathbf{k}|} \, dA.$$

In the exercises, we will ask you to derive similar formulas for the cases where the surface S is written as a function of x and z or as a function of y and z.

We will consider two main cases of surface integrals. In the first, the surface is defined by a function $z = f(x, y)$. In the second, the surface is defined by parametric equations $x = x(u, v)$, $y = y(u, v)$ and $z = z(u, v)$. In each case, your primary task will be to determine a normal vector to use in the general conversion formula for dS.

If S is the surface $z = f(x, y)$, recall from our discussion in section 12.4 that a normal vector to S is given by $\mathbf{n} = \langle f_x, f_y, -1 \rangle$. This is a convenient normal vector for our purposes, since $|\mathbf{n} \cdot \mathbf{k}| = 1$. With $\|\mathbf{n}\| = \sqrt{(f_x)^2 + (f_y)^2 + 1}$, we have the following result.

Theorem 6.1 (Evaluation Theorem)
If the surface S is given by $z = f(x, y)$ for (x, y) in the region $R \subset \mathbb{R}^2$, where f has continuous first partial derivatives, then

$$\iint_S g(x, y, z)\, dS = \iint_R g(x, y, f(x, y)) \sqrt{(f_x)^2 + (f_y)^2 + 1}\, dA.$$

Proof

From the definition of surface integral in Definition 6.1, we have

$$\iint_S g(x, y, z)\, dS = \lim_{\|P\| \to 0} \sum_{i=1}^{n} g(x_i, y_i, z_i)\, \Delta S_i$$

$$= \lim_{\|P\| \to 0} \sum_{i=1}^{n} g(x_i, y_i, z_i) \frac{\|\mathbf{n}_i\|}{|\mathbf{n}_i \cdot \mathbf{k}|}\, \Delta A_i$$

$$= \lim_{\|P\| \to 0} \sum_{i=1}^{n} g(x_i, y_i, f(x_i, y_i)) \sqrt{(f_x)^2 + (f_y)^2 + 1}\, \Big|_{(x_i, y_i)} \Delta A_i$$

$$= \iint_R g(x, y, f(x, y)) \sqrt{(f_x)^2 + (f_y)^2 + 1}\, dA,$$

as desired.

Think about this theorem carefully! It tells us that we can evaluate a surface integral by evaluating a related double integral. To convert the surface integral into a double integral, you must first substitute for z in the function $g(x, y, z)$. This is done by substituting in the equation of the surface $z = f(x, y)$. Also, you must replace the surface area element dS with $\|\mathbf{n}\|\, dA$, which for the surface $z = f(x, y)$ is given by

$$dS = \|\mathbf{n}\|\, dA = \sqrt{(f_x)^2 + (f_y)^2 + 1}\, dA. \tag{6.1}$$

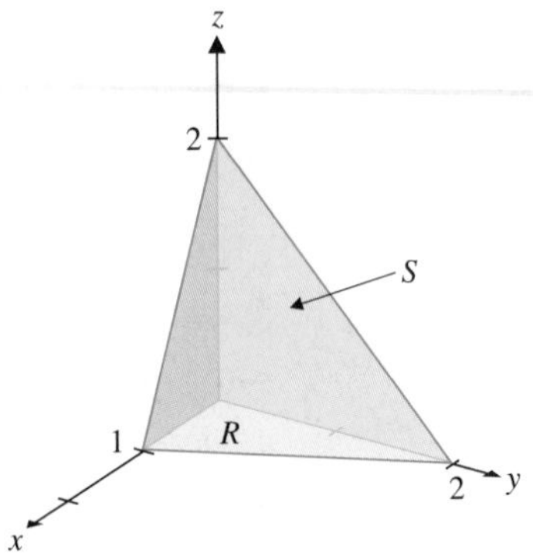

Figure 14.41a

$z = 2 - 2x - y$.

Example 6.1 ~~Evaluating a Surface Integral~~

Evaluate $\iint_S 3z\, dS$, where the surface S is the portion of the plane $2x + y + z = 2$ lying in the first octant.

Solution   On S, we have $z = 2 - 2x - y$, so we must evaluate $\iint_S 3(2 - 2x - y)\, dS$. Note that a normal vector to the plane $2x + y + z = 2$ is $\mathbf{n} = \langle 2, 1, 1 \rangle$, so that in this case, the element of surface area is given by (6.1) to be

$$dS = \|\mathbf{n}\|\, dA = \sqrt{6}\, dA.$$

From the Evaluation Theorem, we then have

$$\iint_S 3(2 - 2x - y)\, dS = \iint_R 3(2 - 2x - y)\sqrt{6}\, dA,$$

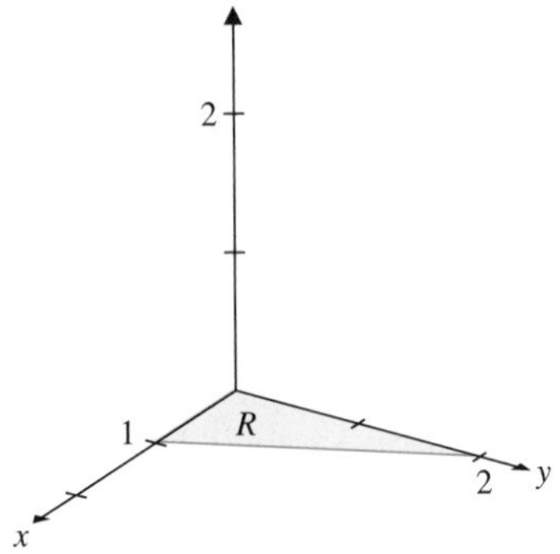

Figure 14.41b

The projection R of the surface S onto the xy-plane.

where R is the projection of the surface onto the xy-plane. A graph of the surface S is shown in Figure 14.41a. In this case, notice that R is the triangle indicated in Figure 14.41b. The triangle is bounded by $x = 0$, $y = 0$ and the line $2x + y = 2$ (the intersection of the plane $2x + y + z = 2$ with the plane $z = 0$). If we integrate with respect to y first, we have inside integration limits of $y = 0$ and $y = 2 - 2x$, with x ranging from 0 to 1. This gives us

$$\iint\limits_S 3(2 - 2x - y)\, dS = \iint\limits_R 3(2 - 2x - y)\sqrt{6}\, dA$$

$$= \int_0^1 \int_0^{2-2x} 3\sqrt{6}(2 - 2x - y)\, dy\, dx$$

$$= 2\sqrt{6},$$

where we leave the details of the integration as an exercise.

In the following example, we will need to rewrite the double integral using polar coordinates.

Example 6.2   Evaluating a Surface Integral Using Polar Coordinates

Evaluate $\iint\limits_S z\, dS$, where the surface S is the portion of the paraboloid $z = 4 - x^2 - y^2$ lying above the xy-plane.

Solution   Substituting $z = 4 - x^2 - y^2$, we have

$$\iint\limits_S z\, dS = \iint\limits_S (4 - x^2 - y^2)\, dS.$$

In this case, a normal vector to the surface $z = 4 - x^2 - y^2$ is $\mathbf{n} = \langle -2x, -2y, -1 \rangle$, so that

$$dS = \|\mathbf{n}\|\, dA = \sqrt{4x^2 + 4y^2 + 1}\, dA.$$

This gives us

$$\iint\limits_S (4 - x^2 - y^2)\, dS = \iint\limits_R (4 - x^2 - y^2)\sqrt{4x^2 + 4y^2 + 1}\, dA.$$

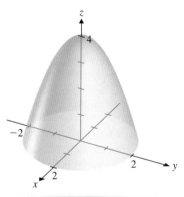

Figure 14.42

$z = 4 - x^2 - y^2$.

Here, the region R is the intersection of the paraboloid with the xy-plane, which is the circle $x^2 + y^2 = 4$ (see Figure 14.42). With a circular region of integration and the term $x^2 + y^2$ appearing (twice!) in the integrand, you had better be thinking about polar coordinates. We have $4 - x^2 - y^2 = 4 - r^2$, $\sqrt{4x^2 + 4y^2 + 1} = \sqrt{4r^2 + 1}$ and $dA = r\, dr\, d\theta$. For the circle $x^2 + y^2 = 4$, r ranges from 0 to 2 and θ ranges from 0 to 2π. Then, we have

$$\iint\limits_S (4 - x^2 - y^2)\, dS = \iint\limits_R (4 - x^2 - y^2)\sqrt{4x^2 + 4y^2 + 1}\, dA$$

$$= \int_0^{2\pi} \int_0^2 (4 - r^2)\sqrt{4r^2 + 1}\, r\, dr\, d\theta$$

$$= \frac{289}{60}\pi\sqrt{17} - \frac{41}{60}\pi,$$

where we leave the details of the final integration to you.

Parametric Representation of Surfaces

In the remainder of this section, we study parametric representations of surface integrals. Before applying parametric equations to the computation of surface integrals, we need a better understanding of surfaces that have been defined parametrically. You have already seen many parametric surfaces, even if you didn't think of them in parametric terms. For instance, you can think of the equation $z = \sqrt{x^2 + y^2}$ as a parametric representation of a cone with parameters x and y. Recall that we can write this cone in cylindrical coordinates as $z = r$, $0 \leq \theta \leq 2\pi$, where the parameters are r and θ. Similarly, the equation $\rho = 4$, $0 \leq \theta \leq 2\pi$ and $0 \leq \phi \leq \pi$, is a parametric representation of the sphere $x^2 + y^2 + z^2 = 16$, with parameters θ and ϕ. It will be helpful to review these graphs as well as to look at some new ones.

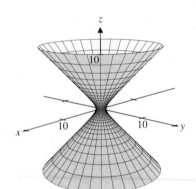

Figure 14.43
$x^2 + y^2 - z^2 = 4$.

Example 6.3   Sketching a Surface Defined Parametrically

Sketch the surface defined parametrically by $x = 2\cos u \cosh v$, $y = 2\sin u \cosh v$ and $z = 2\sinh v$, $0 \leq u \leq 2\pi$ and $-\infty < v < \infty$.

Solution   A sketch such as the one we show in Figure 14.43 can be obtained from a computer algebra system. Notice that this looks like a hyperboloid of one sheet wrapped around the z-axis. To verify that this is correct, observe that

$$x^2 + y^2 - z^2 = 4\cos^2 u \cosh^2 v + 4\sin^2 u \cosh^2 v - 4\sinh^2 v$$
$$= 4(\cos^2 u + \sin^2 u)\cosh^2 v - 4\sinh^2 v$$
$$= 4\cosh^2 v - 4\sinh^2 v = 4,$$

where we have used the identities $\cos^2 u + \sin^2 u = 1$ and $\cosh^2 v - \sinh^2 v = 1$. Recall that the graph of $x^2 + y^2 - z^2 = 4$ is indeed a hyperboloid of one sheet.

Most of the time when we use parametric representations of surfaces, the task is the opposite of that in example 6.3. Given a particular surface, we may need to find a convenient parametric representation of the surface. The general form for parametric equations representing a surface in three dimensions is $x = x(u, v)$, $y = y(u, v)$ and $z = z(u, v)$ for $u_1 \leq u \leq u_2$ and $v_1 \leq v \leq v_2$. The parameters u and v can correspond to familiar coordinates (x and y, or r and θ, for instance), or less familiar expressions. Keep in mind that to fully describe a surface, you will need to define two parameters.

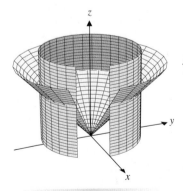

Figure 14.44a

The cone $z = \sqrt{x^2 + y^2}$ and the cylinder $x^2 + y^2 = 4$.

Example 6.4   Finding Parametric Representations of a Surface

Find simple parametric representations for (a) the portion of the cone $z = \sqrt{x^2 + y^2}$ inside the cylinder $x^2 + y^2 = 4$ and (b) the portion of the sphere $x^2 + y^2 + z^2 = 16$ inside of the cone $z = \sqrt{x^2 + y^2}$.

Solution   It is important to realize that both parts (a) and (b) have numerous solutions. (In fact, every surface can be represented parametrically in an infinite number of ways.) The solutions we show here are among the simplest and most useful, but they are **not** the only reasonable solutions. In (a), the repeated appearance of the term $x^2 + y^2$ suggests that cylindrical coordinates (r, θ, z) might be convenient. A sketch of the surface is shown in Figure 14.44a. Notice that the cone $z = \sqrt{x^2 + y^2}$ becomes $z = r$ in cylindrical coordinates. Recall that in cylindrical coordinates, $x = r \cos \theta$ and $y = r \sin \theta$. Notice that the parameters r and θ have ranges determined by the cylinder $x^2 + y^2 = 4$, so that $0 \leq r \leq 2$ and $0 \leq \theta \leq 2\pi$. Parametric equations for the cone are then $x = r \cos \theta$, $y = r \sin \theta$, $z = r$ with $0 \leq r \leq 2$ and $0 \leq \theta \leq 2\pi$.

The surface in part (b) is a portion of a sphere, which suggests (what else?) spherical coordinates. Recall that we convert from spherical coordinates to rectangular coordinates by using $x = \rho \sin \phi \cos \theta$, $y = \rho \sin \phi \sin \theta$ and $z = \rho \cos \phi$. We only want two parameters, so one of the spherical coordinates must be eliminated. Notice that on the sphere $x^2 + y^2 + z^2 = 16$, the distance ρ from the origin is constant and so, an equation of the sphere is $\rho = 4$. Using this, we have the parametric representation of the sphere: $x = 4 \sin \phi \cos \theta$, $y = 4 \sin \phi \sin \theta$ and $z = 4 \cos \phi$, where $0 \leq \theta \leq 2\pi$ and $0 \leq \phi \leq \pi$. To find the portion of the sphere inside the cone, we must observe that the cone can be described in spherical coordinates as $\phi = \frac{\pi}{4}$. Referring to Figure 14.44b, note that the portion of the sphere inside the cone is then described by $x = 4 \sin \phi \cos \theta$, $y = 4 \sin \phi \sin \theta$ and $z = 4 \cos \phi$ where $0 \leq \theta \leq 2\pi$ and $0 \leq \phi \leq \frac{\pi}{4}$.

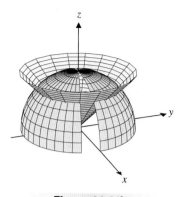

Figure 14.44b

The portion of the sphere inside the cone.

Suppose that we have a parametric representation for the surface S: $x = x(u, v)$, $y = y(u, v)$ and $z = z(u, v)$, defined on the rectangle $R = \{(u, v) | a \leq u \leq b$ and $c \leq v \leq d\}$ in the uv-plane. It is often convenient to use parametric equations to evaluate the surface integral $\iint\limits_{S} f(x, y, z) \, dS$. Of course, to do this, we must substitute for x, y and z to rewrite the integrand in terms of the parameters u and v, as

$$g(u, v) = f(x(u, v), y(u, v), z(u, v)).$$

However, this is only a small piece of the puzzle, as we must also write the surface area element dS in terms of the area element dA for the uv-plane. Unfortunately, we can't use (6.1) here, since this only holds for the case of a surface written in the form $z = f(x, y)$. Instead, we'll need to back up just a bit.

First, notice that the position vector for points on the surface S is $\mathbf{r}(u, v) = \langle x(u, v), y(u, v), z(u, v) \rangle$. We define the vectors $\mathbf{r}_u$ and $\mathbf{r}_v$ (the subscripts denote partial derivatives) by:

$$\mathbf{r}_u(u, v) = \langle x_u(u, v), y_u(u, v), z_u(u, v) \rangle$$

and

$$\mathbf{r}_v(u, v) = \langle x_v(u, v), y_v(u, v), z_v(u, v) \rangle.$$

Notice that for any fixed (u, v), both of the vectors $\mathbf{r}_u(u, v)$ and $\mathbf{r}_v(u, v)$ lie in the tangent plane to S at the point $(x(u, v), y(u, v), z(u, v))$. So, unless these two vectors are parallel, we can find a normal vector to the tangent plane by computing their cross product. That is, $\mathbf{n} = \mathbf{r}_u \times \mathbf{r}_v$ is a normal vector to the surface at the point $(x(u, v), y(u, v), z(u, v))$. We say that the surface S is **smooth** if $\mathbf{r}_u$ and $\mathbf{r}_v$ are continuous and $\mathbf{r}_u \times \mathbf{r}_v \neq \mathbf{0}$, for all $(u, v) \in R$. (This says that the surface will not have any corners.) We say that S is **piecewise-smooth** if we can write $S = S_1 \cup S_2 \cup \cdots \cup S_n$, for some smooth surfaces $S_1, S_2, \ldots, S_n$.

As we have done many times now, we partition the rectangle R in the uv-plane. For each rectangle R_i in the partition, let (u_i, v_i) be the closest point in R_i to the origin, as indicated in Figure 14.45a. Notice that each of the sides of R_i gets mapped to a curve in xyz-space, so that R_i gets mapped to a curvilinear region S_i in xyz-space, as indicated in Figure 14.45b (on the following page). Observe that if we locate their initial points at the point $P_i(x(u_i, v_i), y(u_i, v_i), z(u_i, v_i))$, the vectors $\mathbf{r}_u(u_i, v_i)$ and $\mathbf{r}_v(u_i, v_i)$ lie tangent to two adjacent curved sides of S_i. So, we can approximate the area ΔS_i of S_i by the area of the parallelogram T_i

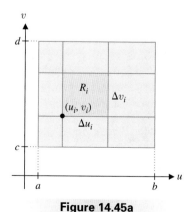

Figure 14.45a

Partition of parameter domain (uv-plane).

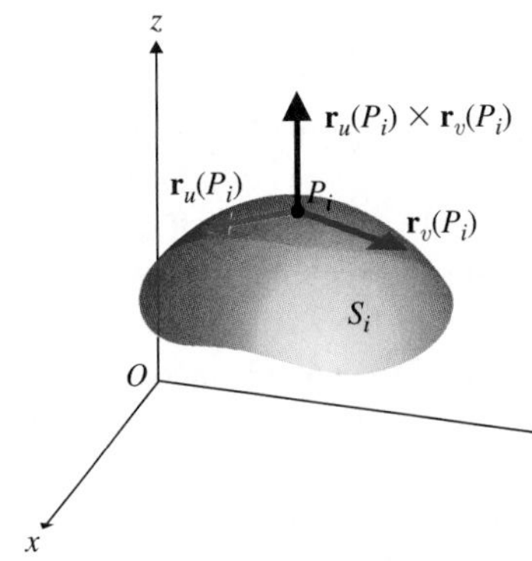

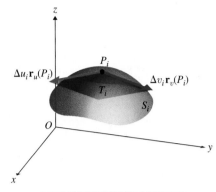

Figure 14.45b

Curvilinear region S_i.

Figure 14.45c

The parallelogram T_i.

whose sides are formed by the vectors $\Delta u_i \mathbf{r}_u(u_i, v_i)$ and $\Delta v_i \mathbf{r}_v(u_i, v_i)$ (see Figure 14.45c). As we know, the area of the parallelogram is given by the magnitude of the cross product

$$\|\Delta u_i \mathbf{r}_u(u_i, v_i) \times \Delta v_i \mathbf{r}_v(u_i, v_i)\| = \|\mathbf{r}_u(u_i, v_i) \times \mathbf{r}_v(u_i, v_i)\| \Delta u_i \Delta v_i$$
$$= \|\mathbf{r}_u(u_i, v_i) \times \mathbf{r}_v(u_i, v_i)\| \Delta A_i,$$

where ΔA_i is the area of the rectangle R_i. We then have that

$$\Delta S_i \approx \|\mathbf{r}_u(u_i, v_i) \times \mathbf{r}_v(u_i, v_i)\| \Delta A_i$$

and it follows that the element of surface area can be written as

$$dS = \|\mathbf{r}_u \times \mathbf{r}_v\| \, dA. \tag{6.2}$$

Notice that this corresponds closely to (6.1), as $\mathbf{r}_u \times \mathbf{r}_v$ is a normal vector to S. Finally, we developed (6.2) in the comparatively simple case where the parameter domain R (that is, the domain in the uv-plane) was a rectangle. If the parameter domain is not a rectangle, you should recognize that we can do the same thing by constructing an inner partition of the region. We can now evaluate surface integrals using parametric equations, as in the following example.

Example 6.5   Evaluating a Surface Integral Using Spherical Coordinates

Evaluate $\iint\limits_S (3x^2 + 3y^2 + 3z^2) \, dS$, where S is the sphere $x^2 + y^2 + z^2 = 4$.

Solution   Since the surface is a sphere and the integrand contains the term $x^2 + y^2 + z^2$, spherical coordinates are indicated. Notice that the sphere is described by $\rho = 2$ and so, on the surface of the sphere, the integrand becomes $3(x^2 + y^2 + z^2) = 12$. Further, we can describe the sphere $\rho = 2$ with the parametric equations $x = 2 \sin \phi \cos \theta$, $y = 2 \sin \phi \sin \theta$ and $z = 2 \cos \phi$, for $0 \le \theta \le 2\pi$ and $0 \le \phi \le \pi$. This says that the sphere is traced out by the endpoint of the vector-valued function

$$\mathbf{r}(\phi, \theta) = \langle 2 \sin \phi \cos \theta, 2 \sin \phi \sin \theta, 2 \cos \phi \rangle.$$

We then have the partial derivatives

$$\mathbf{r}_\theta = \langle -2 \sin \phi \sin \theta, 2 \sin \phi \cos \theta, 0 \rangle$$

and

$$\mathbf{r}_\phi = \langle 2 \cos \phi \cos \theta, 2 \cos \phi \sin \theta, -2 \sin \phi \rangle.$$

We leave it as an exercise to show that a normal vector to the surface is given by

$$\mathbf{n} = \mathbf{r}_\theta \times \mathbf{r}_\phi = \langle -4 \sin^2 \phi \cos \theta, -4 \sin^2 \phi \sin \theta, -4 \sin \phi \cos \phi \rangle,$$

so that $\|\mathbf{n}\| = 4\,|\sin\phi|$. Equation (6.2) now gives us $dS = 4\,|\sin\phi|\,dA$, so that

$$\iint_S (3x^2 + 3y^2 + 3z^2)\,dS = \iint_R (12)(4)\,|\sin\phi|\,dA$$

$$= \int_0^{2\pi}\int_0^{\pi} 48\sin\phi\,d\phi\,d\theta$$

$$= 192\pi,$$

where we used the fact that for $0 \le \phi \le \pi$, $\sin\phi \ge 0$ to replace $|\sin\phi|$ by $\sin\phi$.

If you did the calculation of dS in example 6.5, you may not think that parametric equations lead to simple solutions. (That's why we didn't show all of the details!) However, recall that for changing a triple integral from rectangular to spherical coordinates, you replace $dx\,dy\,dz$ by $\rho^2\sin\phi\,d\rho\,d\phi\,d\theta$. In example 6.5, we have $\rho^2 = 4$ and $dS = 4\sin\phi\,dA$. Looks familiar now, doesn't it? This shortcut is valuable, since surface integrals over spheres are reasonably common.

So, when you evaluate a surface integral, what have you computed? We close the section with two examples. The first is familiar: observe that the surface integral of the function $f(x, y, z) = 1$ over the surface S is simply the surface area of S. That is,

$$\iint_S 1\,dS = \text{Surface Area of } S.$$

The proof of this follows directly from the definition of the surface integral and is left as an exercise.

Example 6.6   Using a Surface Integral to Compute Surface Area

Compute the surface area of the portion of the hyperboloid $x^2 + y^2 - z^2 = 4$ between $z = 0$ and $z = 2$.

Solution   We need to evaluate $\iint_S 1\,dS$. Recall from example 6.3 that we can write the hyperboloid parametrically by $x = 2\cos u\cosh v$, $y = 2\sin u\cosh v$ and $z = 2\sinh v$. (If you didn't remember this, you could always derive parametric equations in the following way. To get a circular cross section of radius 2 in the xy-plane, start with $x = 2\cos u$ and $y = 2\sin u$. To get a hyperbola in the xz- or yz-plane, multiply x and y by $\cosh v$ and set $z = \sinh v$.) We have $0 \le u \le 2\pi$ to get the circular cross sections and $0 \le v \le \sinh^{-1}1(\approx 0.88)$. The hyperboloid is traced out by the endpoint of the vector-valued function

$$\mathbf{r}(u, v) = \langle 2\cos u\cosh v, 2\sin u\cosh v, 2\sinh v\rangle,$$

so that

$$\mathbf{r}_u = \langle -2\sin u\cosh v, 2\cos u\cosh v, 0\rangle$$

and

$$\mathbf{r}_v = \langle 2\cos u\sinh v, 2\sin u\sinh v, 2\cosh v\rangle.$$

This gives us the normal vector

$$\mathbf{n} = \mathbf{r}_u \times \mathbf{r}_v = \langle 4\cos u\cosh^2 v, 4\sin u\cosh^2 v, -4\cosh v\sinh v\rangle,$$

where $\|\mathbf{n}\| = 4\cosh v\sqrt{\cosh^2 v + \sinh^2 v}$. We now have

$$\iint\limits_{S} 1\, dS = \iint\limits_{R} 4\cosh v\sqrt{\cosh^2 v + \sinh^2 v}\, dA$$

$$= \int_{0}^{\sinh^{-1} 1}\int_{0}^{2\pi} 4\cosh v\sqrt{\cosh^2 v + \sinh^2 v}\, du\, dv$$

$$\approx 31.95,$$

where we evaluated the final integral numerically.

Our final example of a surface integral requires some preliminary discussion. We must first define an **oriented** (or **two-sided**) **surface.** We say that a surface S is **oriented** (or **orientable**) if it has a unit normal vector $\mathbf{n}$ at each point (x, y, z) not on the boundary of the surface and if $\mathbf{n}$ is a continuous function of (x, y, z). Further, we assume that S has two identifiable sides (a top and a bottom or an inside and an outside). To orient such a surface, we choose a consistent direction for all normal vectors to point. For instance, a sphere is a two-sided surface; the two sides of the surface are the inside and the outside. Notice that you can't get from the inside to the outside without passing through the sphere. The **positive orientation** for the sphere (or for any other **closed** surface) is to choose outward normal vectors (normal vectors pointing away from the interior).

All of the surfaces we have seen so far in this course are two-sided, but it's not difficult to construct a one-sided surface. Perhaps the most famous example of a one-sided surface is the **Möbius strip,** named after the German mathematician A. F. Möbius. You can easily construct a Möbius strip by taking a long rectangular strip of paper, giving it a half-twist and then taping the short edges together, as illustrated in Figures 14.46a through 14.46c. Notice that if you started painting the strip, you would eventually return to your starting point, having painted both "sides" of the strip, but without having crossed any edges. This says that the Möbius strip has no inside and no outside and is therefore not orientable.

One reason we need to be able to orient a surface is to compute the **flux** of a vector field. It's easiest to visualize the flux for a vector field representing the velocity field for a fluid in motion. In this context, the flux measures the net flow rate of the fluid across the surface, that is, from the inside to the outside. (Notice that for this to make sense, the surface

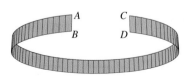

Figure 14.46a
A long, thin strip.

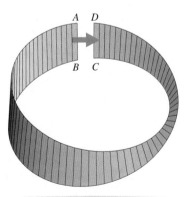

Figure 14.46b
Make one half-twist.

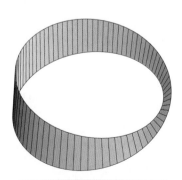

Figure 14.46c
A Möbius strip.

must have two identifiable sides. That is, the surface must be orientable.) The orientation of the surface lets us distinguish one direction from the other. In general, we have the following definition.

Definition 6.2

Let $\mathbf{F}(x, y, z)$ be a continuous vector field defined on an oriented surface S with unit normal vector $\mathbf{n}$. The **surface integral of F over** S (or the **flux of F over** S) is given by $\iint_S \mathbf{F} \cdot \mathbf{n} \, dS$.

Think carefully about the role of the unit normal vector in this definition. Notice that since $\mathbf{n}$ is a unit vector, the integrand $\mathbf{F} \cdot \mathbf{n}$ gives (at any given point on S) the component of $\mathbf{F}$ in the direction of $\mathbf{n}$. So, if $\mathbf{F}$ represents the velocity field for a fluid in motion, $\mathbf{F} \cdot \mathbf{n}$ corresponds to the component of the velocity that moves the fluid across the surface (from one side to the other). Also, note that $\mathbf{F} \cdot \mathbf{n}$ can be positive or negative, depending on which normal vector we have chosen. (Keep in mind that at each point on a surface, there are two unit normal vectors; one pointing toward each side of the surface.) You should recognize that this is why we need to have an oriented surface.

Example 6.7   Computing the Flux of a Vector Field

Compute the flux of the vector field $\mathbf{F}(x, y, z) = \langle x, y, 0 \rangle$ over the portion of the paraboloid $z = x^2 + y^2$ below $z = 4$ (oriented with upward-pointing normal vectors).

Solution   First, observe that at any given point, the normal vectors for the paraboloid $z = x^2 + y^2$ are $\pm \langle 2x, 2y, -1 \rangle$. For the normal vector to point upward, we need a positive z-component. In this case,

$$\mathbf{m} = -\langle 2x, 2y, -1 \rangle = \langle -2x, -2y, 1 \rangle$$

is such a normal vector. However, the definition of flux requires a unit normal vector. A unit vector pointing in the same direction as $\mathbf{m}$ is

$$\mathbf{n} = \frac{\langle -2x, -2y, 1 \rangle}{\sqrt{4x^2 + 4y^2 + 1}}.$$

Before computing $\mathbf{F} \cdot \mathbf{n}$, we use the normal vector $\mathbf{m}$ to write the surface area increment dS in terms of dA. From (6.1), we have

$$dS = \|\mathbf{m}\| \, dA = \sqrt{4x^2 + 4y^2 + 1} \, dA.$$

Putting this all together gives us

$$\iint_S \mathbf{F} \cdot \mathbf{n} \, dS = \iint_R \langle x, y, 0 \rangle \cdot \frac{\langle -2x, -2y, 1 \rangle}{\sqrt{4x^2 + 4y^2 + 1}} \sqrt{4x^2 + 4y^2 + 1} \, dA$$

$$= \iint_R \langle x, y, 0 \rangle \cdot \langle -2x, -2y, 1 \rangle \, dA$$

$$= \iint_R (-4x^2 - 4y^2) \, dA,$$

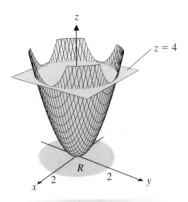

Figure 14.47
$z = x^2 + y^2$.

where the region R is the projection of the portion of the paraboloid under consideration onto the xy-plane. Note how the square roots arising from the calculation of $\|\mathbf{n}\|$ and dS cancelled out one another. Look at the graph in Figure 14.47 and recognize that this projection is bounded by the circle $x^2 + y^2 = 4$. You should quickly realize that the double integral should be set up in polar coordinates. We have

$$\iint_S \mathbf{F} \cdot \mathbf{n}\, dS = \iint_R (-4x^2 - 4y^2)\, dA$$

$$= \int_0^{2\pi} \int_0^2 (-4r^2) r\, dr\, d\theta = -32\pi.$$

EXERCISES 14.6

1. For definition 6.1, we defined the partition of a surface and took the limit as the norm of the partition tends to to 0. Explain why it would not be sufficient to have the number of segments in the partition tend to ∞. (Hint: The pieces of the partition don't have to be the same size.)

2. In example 6.2, you could alternatively start with cylindrical coordinates and use a parametric representation like we did in example 6.5. Discuss which method you think would be simpler.

3. Explain in words why $\iint_S 1\, dS$ equals the surface area of S. (Hint: Although you are supposed to explain in words, you will need to refer to Riemann sums.)

4. For example 6.7, sketch a graph showing the surface S and several normal vectors to the surface. Also, show several vectors in the graph of the vector field $\mathbf{F}$. Explain why the flux is negative.

In exercises 5–12, find a parametric representation of the surface.

5. $z = 3x + 4y$
6. $x^2 + y^2 + z^2 = 4$
7. $x^2 + y^2 - z^2 = 1$
8. $z^2 = x^2 + y^2$
9. The portion of $x^2 + y^2 = 4$ from $z = 0$ to $z = 2$
10. The portion of $y^2 + z^2 = 9$ from $x = -1$ to $x = 1$
11. The portion of $z = 4 - x^2 - y^2$ above the xy-plane
12. The portion of $z = x^2 + y^2$ below $z = 4$

In exercises 13–20, sketch a graph of the parametric surface.

13. $x = u, y = v, z = u^2 + 2v^2$
14. $x = u, y = v, z = 4 - u^2 - v^2$
15. $x = u\cos v, y = u\sin v, z = u^2$
16. $x = u\cos v, y = u\sin v, z = u$
17. $x = u, y = \sin u\cos v, z = \sin u\sin v$
18. $x = \cos u\cos v, y = u, z = \cos u\sin v$
19. $x = 2\sin u\cos v, y = 2\sin u\sin v, z = 2\cos u$
20. $x = u\cos v, y = u\sin v, z = v$

21. Match the parametric equations with the surface.
    (a) $x = u\cos v, y = u\sin v, z = v^2$
    (b) $x = v, y = u\cos v, z = u\sin v$
    (c) $x = u, y = u\cos v, z = u\sin v$

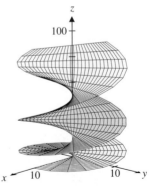

Surface A

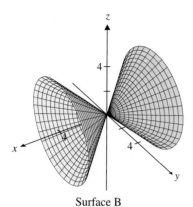

Surface B

Surface C

22. In example 6.5, show that

$$\mathbf{r}_\theta \times \mathbf{r}_\phi = \langle -4\sin^2\phi\cos\theta, -4\sin^2\phi\sin\theta, -4\sin\phi\cos\phi \rangle$$

and then show that $\|\mathbf{n}\| = 4\,|\sin\phi|$.

In exercises 23–30, find the surface area of the given surface.

23. The portion of the cone $z = \sqrt{x^2 + y^2}$ below the plane $z = 4$

24. The portion of the paraboloid $z = x^2 + y^2$ below the plane $z = 4$

25. The portion of the plane $3x + 2y + z = 6$ inside the cylinder $x^2 + y^2 = 4$

26. The portion of the plane $x + 2y + z = 4$ above the region bounded by $y = x^2$ and $y = 1$

27. The portion of the cone $z = \sqrt{x^2 + y^2}$ above the triangle with vertices $(0, 0)$, $(1, 0)$ and $(1, 1)$

28. The portion of the paraboloid $z = x^2 + y^2$ inside the cylinder $x^2 + y^2 = 4$

29. The portion of the hemisphere $z = \sqrt{4 - x^2 - y^2}$ above the plane $z = 1$

30. The portion of $y = 4 - x^2$ with $y \geq 0$ and between $z = 0$ and $z = 2$

In exercises 31–40, set up a double integral and evaluate the surface integral $\iint_S g(x, y, z)\,dS$.

31. $\iint_S x\,dS$, S is the portion of the plane $z = 2x + 3y$ above the rectangle $1 \leq x \leq 2$, $1 \leq y \leq 3$

32. $\iint_S (x + y)\,dS$, S is the portion of the plane $z = 4x + y$ above the region bounded by $y = x^2$ and $y = 1$

33. $\iint_S y\,dS$, S is the portion of the paraboloid $z = x^2 + y^2$ below $z = 4$

34. $\iint_S (x + y)\,dS$, S is the portion of the paraboloid $z = 4 - x^2 - y^2$ above the xy-plane

35. $\iint_S (x^2 + y^2)\,dS$, S is the portion of the paraboloid $z = 4 - x^2 - y^2$ above the xy-plane

36. $\iint_S \sqrt{x^2 + y^2}\,dS$, S is the hemisphere $z = \sqrt{9 - x^2 - y^2}$

37. $\iint_S z\,dS$, S is the portion of the cone $z = \sqrt{x^2 + y^2}$ below the plane $z = 4$

38. $\iint_S z^2\,dS$, S is the portion of the cone $z = \sqrt{x^2 + y^2}$ above the rectangle $0 \leq x \leq 2$, $-1 \leq y \leq 2$

39. $\iint_S (x^2 + y^2 + z^2)\,dS$, S is the hemisphere $z = \sqrt{4 - x^2 - y^2}$

40. $\iint_S \sqrt{x^2 + y^2 + z^2}\,dS$, S is the sphere $x^2 + y^2 + z^2 = 9$

In exercises 41–52, evaluate the flux integral $\iint_S \mathbf{F} \cdot \mathbf{n}\,dS$.

41. $\mathbf{F} = \langle x, y, z \rangle$, S is the portion of $z = 4 - x^2 - y^2$ above the xy-plane (**n** upward)

42. $\mathbf{F} = \langle y, -x, 1 \rangle$, S is the portion of $z = x^2 + y^2$ below $z = 4$ (**n** downward)

43. $\mathbf{F} = \langle y, -x, z \rangle$, S is the portion of $z = \sqrt{x^2 + y^2}$ below $z = 3$ (**n** downward)

44. $\mathbf{F} = \langle 0, 1, y \rangle$, S is the portion of $z = \sqrt{x^2 + y^2}$ inside $x^2 + y^2 = 4$ (**n** downward)

45. $\mathbf{F} = \langle xy, y^2, z \rangle$, S is the boundary of the unit cube with $0 \leq x \leq 1, 0 \leq y \leq 1, 0 \leq z \leq 1$, (**n** outward)

46. $\mathbf{F} = \langle y, z, 0 \rangle$, S is the boundary of the box with $0 \leq x \leq 2$, $0 \leq y \leq 3, 0 \leq z \leq 1$, (**n** outward)

47. $\mathbf{F} = \langle 1, 0, z \rangle$, S is the boundary of the region bounded above by $z = 1 - x^2 - y^2$ and below by $z = 0$ (**n** outward)

48. $\mathbf{F} = \langle 0, 1, y \rangle$, S is the boundary of the region bounded above by $z = 1 - x^2 - y^2$ and below by $z = 0$ (**n** outward)

49. $\mathbf{F} = \langle yx, 1, x \rangle$, S is the portion of $z = 2 - x - y$ above the square $0 \le x \le 1, 0 \le y \le 1$ (**n** upward)

50. $\mathbf{F} = \langle y, 3, z \rangle$, S is the portion of $z = x^2 + y^2$ above the triangle with vertices $(0, 0)$, $(0, 1)$, $(1, 1)$ (**n** downward)

51. $\mathbf{F} = \langle y, 0, 2 \rangle$, S is the boundary of the region bounded above by $z = \sqrt{8 - x^2 - y^2}$ and below by $z = \sqrt{x^2 + y^2}$ (**n** outward)

52. $\mathbf{F} = \langle 3, z, y \rangle$, S is the boundary of the region between $z = 8 - 2x - y$ and $z = \sqrt{x^2 + y^2}$ and inside $x^2 + y^2 = 1$ (**n** outward)

In exercises 53–56, find the mass and center of mass of the region.

53. The portion of the plane $3x + 2y + z = 6$ inside the cylinder $x^2 + y^2 = 4$, $\rho(x, y, z) = x^2 + 1$

54. The portion of the plane $x + 2y + z = 4$ above the region bounded by $y = x^2$ and $y = 1$, $\rho(x, y, z) = y$

55. The hemisphere $z = \sqrt{1 - x^2 - y^2}$, $\rho(x, y, z) = 1 + x$

56. The portion of the paraboloid $z = x^2 + y^2$ inside the cylinder $x^2 + y^2 = 4$, $\rho(x, y, z) = z$

57. State the formula converting a surface integral into a double integral for a projection into the yz-plane.

58. State the formula converting a surface integral into a double integral for a projection into the xz-plane.

In exercises 59–66, use the formulas of exercises 57 and 58 to evaluate the surface integral.

59. $\iint_S z \, dS$, where S is the portion of $x^2 + y^2 = 1$ with $x \ge 0$ and z between $z = 1$ and $z = 2$

60. $\iint_S yz \, dS$, where S is the portion of $x^2 + y^2 = 1$ with $x \ge 0$ and z between $z = 1$ and $z = 4 - y$

61. $\iint_S (y^2 + z^2) \, dS$, where S is the portion of the paraboloid $x = 9 - y^2 - z^2$ in front of the yz-plane

62. $\iint_S (y^2 + z^2) \, dS$, where S is the hemisphere $x = \sqrt{4 - y^2 - z^2}$

63. $\iint_S x^2 \, dS$, where S is the portion of the paraboloid $y = x^2 + z^2$ to the left of the plane $y = 1$

64. $\iint_S (x^2 + z^2) \, dS$, where S is the hemisphere $y = \sqrt{4 - x^2 - z^2}$

65. $\iint_S 4 \, dS$, where S is the portion of $y = 1 - x^2$ with $y \ge 0$ and between $z = 0$ and $z = 2$

66. $\iint_S (x^2 + z^2) \, dS$, where S is the portion of $y = \sqrt{4 - x^2}$ between $z = 1$ and $z = 4$

67. Explain the following result geometrically. The flux integral of $\mathbf{F}(x, y, z) = \langle x, y, z \rangle$ across the cone $z = \sqrt{x^2 + y^2}$ is 0.

68. In geometric terms, determine if the flux integral of $\mathbf{F}(x, y, z) = \langle x, y, z \rangle$ across the hemisphere $z = \sqrt{1 - x^2 - y^2}$ is 0.

69. If $x = 3 \sin u \cos v$, $y = 3 \cos u$ and $z = 3 \sin u \sin v$, show that $x^2 + y^2 + z^2 = 9$. Explain why this equation doesn't guarantee that the parametric surface defined is the entire sphere, but it does guarantee that all points on the surface are also on the sphere. In this case, the parametric surface is the entire sphere. To verify this in graphical terms, sketch a picture showing geometric interpretations of the "spherical coordinates" u and v. To see what problems can occur, sketch the surface defined by $x = 3 \sin \dfrac{u^2}{u^2 + 1} \cos v$, $y = 3 \cos \dfrac{u^2}{u^2 + 1}$ and $z = 3 \sin \dfrac{u^2}{u^2 + 1} \sin v$. Explain why you do not get the entire sphere. To see a more subtle example of the same problem, sketch the surface $x = \cos u \cosh v$, $y = \sinh v$, $z = \sin u \cosh v$. Use identities to show that $x^2 - y^2 + z^2 = 1$ and identify the surface. Then sketch the surface $x = \cos u \cosh v$, $y = \cos u \sinh v$, $z = \sin u$ and use identities to show that $x^2 - y^2 + z^2 = 1$. Explain why the second surface is not the entire hyperboloid. Explain in words and pictures exactly what the second surface is.

14.7 THE DIVERGENCE THEOREM

Recall that at the end of section 14.5, we had rewritten Green's Theorem in terms of the divergence of a two-dimensional vector field. We had found there (see equation 5.3) that

$$\oint_C \mathbf{F} \cdot \mathbf{n}\, ds = \iint_R \nabla \cdot \mathbf{F}(x, y)\, dA.$$

Here, R is a region in the xy-plane enclosed by a piecewise-smooth, positively oriented, simple closed curve C. Further, $\mathbf{F}(x, y) = \langle M(x, y), N(x, y), 0 \rangle$, where $M(x, y)$ and $N(x, y)$ are continuous and have continuous first partial derivatives in some open region D in the xy-plane, with $R \subset D$.

Now that we have studied surface integrals over the boundary of a surface in $\mathbb{R}^3$, you might wonder whether we can extend this two-dimensional result to three dimensions. In fact, it extends in exactly the way you might expect. That is, for a solid region $Q \subset \mathbb{R}^3$ bounded by the surface ∂Q, it turns out that we have

$$\iint_{\partial Q} \mathbf{F} \cdot \mathbf{n}\, dS = \iiint_Q \nabla \cdot \mathbf{F}(x, y, z)\, dV.$$

This result is referred to as the **Divergence Theorem** (or **Gauss' Theorem**). This theorem has great significance in a variety of settings. One convenient context in which to think of the Divergence Theorem is in the case where $\mathbf{F}$ represents the velocity field of a fluid in motion. In this case, it says that the total flux of the velocity field across the boundary of the solid is equal to the triple integral of the divergence of the velocity field over the solid. We now state the result.

Theorem 7.1 (Divergence Theorem)

Suppose that $Q \subset \mathbb{R}^3$ is bounded by the closed surface ∂Q and that $\mathbf{n}(x, y, z)$ denotes the exterior unit normal vector to ∂Q. Then, if the components of $\mathbf{F}(x, y, z)$ have continuous first partial derivatives in Q, we have

$$\iint_{\partial Q} \mathbf{F} \cdot \mathbf{n}\, dS = \iiint_Q \nabla \cdot \mathbf{F}(x, y, z)\, dV.$$

Although we have stated the theorem in the general case, we prove the result only for the case where the solid Q is fairly simple. A proof for the general case can be found in a more advanced text.

Proof

For $\mathbf{F}(x, y, z) = \langle M(x, y, z), N(x, y, z), P(x, y, z) \rangle$, the divergence of $\mathbf{F}$ is

$$\nabla \cdot \mathbf{F}(x, y, z) = \frac{\partial M}{\partial x} + \frac{\partial N}{\partial y} + \frac{\partial P}{\partial z}.$$

We then have that

$$\iiint_Q \nabla \cdot \mathbf{F}(x, y, z)\, dV = \iiint_Q \frac{\partial M}{\partial x}\, dV + \iiint_Q \frac{\partial N}{\partial y}\, dV + \iiint_Q \frac{\partial P}{\partial z}\, dV. \quad (7.1)$$

Further, we can write the flux integral as

$$\iint\limits_{\partial Q} \mathbf{F} \cdot \mathbf{n} \, dS = \iint\limits_{\partial Q} M(x, y, z)\mathbf{i} \cdot \mathbf{n} \, dS + \iint\limits_{\partial Q} N(x, y, z)\mathbf{j} \cdot \mathbf{n} \, dS$$
$$+ \iint\limits_{\partial Q} P(x, y, z)\mathbf{k} \cdot \mathbf{n} \, dS. \qquad (7.2)$$

Looking carefully at (7.1) and (7.2), observe that the theorem will follow if we can show that

$$\iiint\limits_{Q} \frac{\partial M}{\partial x} \, dV = \iint\limits_{\partial Q} M(x, y, z)\mathbf{i} \cdot \mathbf{n} \, dS, \qquad (7.3)$$

$$\iiint\limits_{Q} \frac{\partial N}{\partial y} \, dV = \iint\limits_{\partial Q} N(x, y, z)\mathbf{j} \cdot \mathbf{n} \, dS \qquad (7.4)$$

and

$$\iiint\limits_{Q} \frac{\partial P}{\partial z} \, dV = \iint\limits_{\partial Q} P(x, y, z)\mathbf{k} \cdot \mathbf{n} \, dS. \qquad (7.5)$$

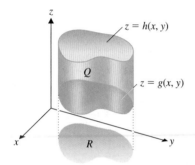

Figure 14.48a

The solid Q.

As you might imagine, the proofs of (7.3), (7.4) and (7.5) are all virtually identical (and all fairly long). Consequently, we prove only one of these three equations here. In order to prove (7.5), we assume that we can describe the solid Q as follows:

$$Q = \{(x, y, z) | g(x, y) \leq z \leq h(x, y), \text{ for } (x, y) \in R\},$$

where R is some region in the xy-plane, as illustrated in Figure 14.48a. (We can prove (7.3) and (7.4) by making corresponding assumptions regarding Q.) Now, notice from Figure 14.48a that there are three distinct surfaces that make up the boundary of Q. In Figure 14.48b, we have labeled these surfaces S_1 (the bottom surface), S_2 (the top surface) and S_3 (the lateral surface), where we have also indicated exterior normal vectors to each of the surfaces.

Notice that on the lateral surface S_3, the $\mathbf{k}$ component of the exterior unit normal $\mathbf{n}$ is zero and so, the flux integral of $P(x, y, z)\mathbf{k}$ over S_3 is zero. This gives us

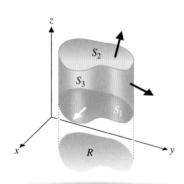

Figure 14.48b

The surfaces S_1, S_2 and S_3 and several exterior normal vectors.

$$\iint\limits_{\partial Q} P(x, y, z)\mathbf{k} \cdot \mathbf{n} \, dS = \iint\limits_{S_1} P(x, y, z)\mathbf{k} \cdot \mathbf{n} \, dS + \iint\limits_{S_2} P(x, y, z)\mathbf{k} \cdot \mathbf{n} \, dS. \quad (7.6)$$

In order to prove the result, we need to rewrite the two integrals on the right side of (7.6) as double integrals over the region R in the xy-plane. First, you must notice that on the surface S_1 (the bottom surface), the exterior unit normal $\mathbf{n}$ points downward (i.e., it has a negative $\mathbf{k}$ component). Now, S_1 is defined by

$$S_1 = \{(x, y, z) | z = g(x, y), \text{ for } (x, y) \in R\}.$$

If we define $k_1(x, y, z) = z - g(x, y)$, then the exterior unit normal on S_1 is given by

$$\mathbf{n} = \frac{-\nabla k_1}{\|\nabla k_1\|} = \frac{g_x(x, y)\mathbf{i} + g_y(x, y)\mathbf{j} - \mathbf{k}}{\sqrt{[g_x(x, y)]^2 + [g_y(x, y)]^2 + 1}}$$

and

$$\mathbf{k} \cdot \mathbf{n} = \frac{-1}{\sqrt{[g_x(x, y)]^2 + [g_y(x, y)]^2 + 1}},$$

since the unit vectors $\mathbf{i}, \mathbf{j}$ and $\mathbf{k}$ are all mutually orthogonal. We now have

$$\iint_{S_1} P(x,y,z)\,\mathbf{k}\cdot\mathbf{n}\,dS = -\iint_{S_1} \frac{P(x,y,z)}{\sqrt{[g_x(x,y)]^2 + [g_y(x,y)]^2 + 1}}\,dS$$

$$= -\iint_{R} \frac{P(x,y,g(x,y))}{\sqrt{[g_x(x,y)]^2 + [g_y(x,y)]^2 + 1}}$$

$$\cdot\sqrt{[g_x(x,y)]^2 + [g_y(x,y)]^2 + 1}\,dA$$

$$= -\iint_{R} P(x,y,g(x,y))\,dA, \tag{7.7}$$

thanks to the two square roots canceling out one another. In a similar way, notice that on S_2 (the top surface), the exterior unit normal $\mathbf{n}$ points upward (i.e., it has a positive $\mathbf{k}$ component). Since S_2 corresponds to the portion of the surface $z = h(x,y)$ for $(x,y) \in R$, if we take $k_2(x,y) = z - h(x,y)$, we have that on S_2,

$$\mathbf{n} = \frac{\nabla k_2}{\|\nabla k_2\|} = \frac{-h_x(x,y)\mathbf{i} - h_y(x,y)\mathbf{j} + \mathbf{k}}{\sqrt{[h_x(x,y)]^2 + [h_y(x,y)]^2 + 1}}$$

and so,

$$\mathbf{k}\cdot\mathbf{n} = \frac{1}{\sqrt{[h_x(x,y)]^2 + [h_y(x,y)]^2 + 1}}.$$

We now have

$$\iint_{S_2} P(x,y,z)\mathbf{k}\cdot\mathbf{n}\,dS = \iint_{S_2} \frac{P(x,y,z)}{\sqrt{[h_x(x,y)]^2 + [h_y(x,y)]^2 + 1}}\,dS$$

$$= \iint_{R} \frac{P(x,y,h(x,y))}{\sqrt{[h_x(x,y)]^2 + [h_y(x,y)]^2 + 1}}$$

$$\cdot\sqrt{[h_x(x,y)]^2 + [h_y(x,y)]^2 + 1}\,dA$$

$$= \iint_{R} P(x,y,h(x,y))\,dA. \tag{7.8}$$

Putting together (7.6), (7.7) and (7.8) gives us

$$\iint_{\partial Q} P(x,y,z)\mathbf{k}\cdot\mathbf{n}\,dS = \iint_{S_1} P(x,y,z)\mathbf{k}\cdot\mathbf{n}\,dS + \iint_{S_2} P(x,y,z)\mathbf{k}\cdot\mathbf{n}\,dS$$

$$= \iint_{R} P(x,y,h(x,y))\,dA - \iint_{R} P(x,y,g(x,y))\,dA$$

$$= \iint_{R} [P(x,y,h(x,y)) - P(x,y,g(x,y))]\,dA$$

$$= \iint_{R} \int_{g(x,y)}^{h(x,y)} \frac{\partial P}{\partial z}\,dz\,dA \quad \text{by the Fundamental Theorem of Calculus}$$

$$= \iiint_{Q} \frac{\partial P}{\partial z}\,dV,$$

which proves (7.5). With appropriate assumptions on Q, we can similarly prove (7.3) and (7.4). This proves the theorem for the special case where the solid Q can be described as indicated.

Example 7.1 Applying the Divergence Theorem

Let Q be the solid bounded by the paraboloid $z = 4 - x^2 - y^2$ and the xy-plane. Find the flux of the vector field $\mathbf{F}(x, y, z) = \langle x^3, y^3, z^3 \rangle$ over the surface ∂Q.

Solution   We show a sketch of the solid in Figure 14.49. Notice that to compute the flux directly, we must consider the two different portions of ∂Q (the surface of the paraboloid and its base in the xy-plane) separately. Alternatively, observe that the divergence of $\mathbf{F}$ is given by

$$\nabla \cdot \mathbf{F}(x, y, z) = \nabla \cdot \langle x^3, y^3, z^3 \rangle = 3x^2 + 3y^2 + 3z^2.$$

From the Divergence Theorem, we now have that the flux of $\mathbf{F}$ over ∂Q is given by

$$\iint_{\partial Q} \mathbf{F} \cdot \mathbf{n} \, dS = \iiint_Q \nabla \cdot \mathbf{F}(x, y, z) \, dV$$

$$= \iiint_Q (3x^2 + 3y^2 + 3z^2) \, dV.$$

If we rewrite the triple integral in cylindrical coordinates, we get

$$\iint_{\partial Q} \mathbf{F} \cdot \mathbf{n} \, dS = \iiint_Q (3x^2 + 3y^2 + 3z^2) \, dV$$

$$= 3 \int_0^{2\pi} \int_0^2 \int_0^{4-r^2} (r^2 + z^2) \, r \, dz \, dr \, d\theta$$

$$= 3 \int_0^{2\pi} \int_0^2 \left(r^2 z + \frac{z^3}{3} \right) \Big|_{z=0}^{z=4-r^2} r \, dr \, d\theta$$

$$= 3 \int_0^{2\pi} \int_0^2 \left[r^3 (4 - r^2) + \frac{1}{3}(4 - r^2)^3 r \right] dr \, d\theta$$

$$= 96\pi,$$

where we have left the details of the final integrations as a straightforward exercise.

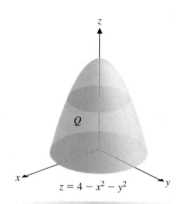

Figure 14.49
The solid Q.

Notice that in example 7.1, we were able to use the Divergence Theorem to replace a very messy surface integral calculation by a comparatively simple double integral. In the following example, we are able to use the Divergence Theorem to prove a general result regarding the flux of a certain vector field over any surface. Notice that you would not be able to prove such a result by trying to directly calculate the surface integral over an unspecified surface.

Example 7.2 Applying the Divergence Theorem

Prove that the flux of the vector field $\mathbf{F}(x, y, z) = \langle 3y \cos z, x^2 e^z, x \sin y \rangle$ is zero over any closed surface ∂Q enclosing a solid region Q.

Solution Notice that in this case, the divergence of $\mathbf{F}$ is

$$\nabla \cdot \mathbf{F}(x, y, z) = \nabla \cdot \langle 3y \cos z, x^2 e^z, x \sin y \rangle$$

$$= \frac{\partial}{\partial x}(3y \cos z) + \frac{\partial}{\partial y}(x^2 e^z) + \frac{\partial}{\partial z}(x \sin y) = 0.$$

From the Divergence Theorem, we then have that the flux of $\mathbf{F}$ over ∂Q is given by

$$\iint\limits_{\partial Q} \mathbf{F} \cdot \mathbf{n} \, dS = \iiint\limits_{Q} \nabla \cdot \mathbf{F}(x, y, z) \, dV$$

$$= \iiint\limits_{Q} 0 \, dV = 0,$$

for any solid region $Q \subset \mathbb{R}^3$.

■

Recall from our discussion in section 4.4 that for a function $f(x)$ of a single variable, if f is continuous on the interval $[a, b]$, then the average value of f on $[a, b]$ is given by

$$f_{\text{ave}} = \frac{1}{b - a} \int_a^b f(x) \, dx.$$

It's not hard to extend this result to the case where $f(x, y, z)$ is a continuous function on the region $Q \subset \mathbb{R}^3$. In this case, the **average value** of f on Q is given by

$$f_{\text{ave}} = \frac{1}{V} \iiint\limits_{Q} f(x, y, z) \, dV,$$

where V is the volume of Q. Further, notice that if f_{ave} represents the average value of f on Q, then by continuity, there must be a point $P(a, b, c) \in Q$ at which f equals its average value, that is, where

$$f(P) = \frac{1}{V} \iiint\limits_{Q} f(x, y, z) \, dV.$$

This says that if $\mathbf{F}(x, y, z)$ has continuous first partial derivatives on Q, then div $\mathbf{F}$ is continuous on Q and so, there is a point $P(a, b, c) \in Q$ for which

$$(\nabla \cdot \mathbf{F})\big|_P = \frac{1}{V} \iiint\limits_{Q} \nabla \cdot \mathbf{F}(x, y, z) \, dV$$

$$= \frac{1}{V} \iint\limits_{\partial Q} \mathbf{F}(x, y, z) \cdot \mathbf{n} \, dS,$$

by the Divergence Theorem. Finally, observe that since the surface integral represents the flux of $\mathbf{F}$ over the surface ∂Q, then $(\nabla \cdot \mathbf{F})\big|_P$ represents the flux per unit volume over ∂Q.

In particular, for any point $P_0(x_0, y_0, z_0)$ in the interior of Q (i.e., in Q, but not on ∂Q), let S_a be the surface of the sphere of radius a, centered at P_0, where a is sufficiently small so that S_a lies completely inside Q. From what we have above, there must be some point P_a in the interior of S_a for which

$$(\nabla \cdot \mathbf{F})\big|_{P_a} = \frac{1}{V_a} \iint\limits_{S_a} \mathbf{F}(x, y, z) \cdot \mathbf{n} \, dS,$$

where V_a is the volume of the sphere ($V_a = \frac{4}{3}\pi a^3$). Finally, taking the limit as $a \to 0$, we have by the continuity of $\nabla \cdot \mathbf{F}$ that

$$(\nabla \cdot \mathbf{F})\big|_{P_0} = \lim_{a \to 0} \frac{1}{V_a} \iint\limits_{S_a} \mathbf{F}(x, y, z) \cdot \mathbf{n} \, dS$$

or

$$\boxed{\operatorname{div} \mathbf{F}(P_0) = \lim_{a \to 0} \frac{1}{V_a} \iint\limits_{S_a} \mathbf{F}(x, y, z) \cdot \mathbf{n} \, dS.} \qquad (7.9)$$

In other words, the divergence of a vector field at a point P_0 is the limiting value of the flux per unit volume over a sphere centered at P_0, as the radius of the sphere tends to zero.

In the case where $\mathbf{F}$ represents the velocity field for a fluid in motion, (7.9) provides us with an interesting (and important) interpretation of the divergence of a vector field. In this case, notice that if $\operatorname{div} \mathbf{F}(P_0) > 0$, this says that the flux per unit volume at P_0 is positive. From (7.9), this means that for a sphere S_a of sufficiently small radius centered at P_0, the net (outward) flux through the surface of S_a is positive. For an incompressible fluid (think of a liquid), this says that more fluid is passing out through the surface of S_a than is passing in through the surface. For an incompressible fluid, this can only happen if there is a source somewhere in S_a, where additional fluid is coming into the flow. Likewise, if $\operatorname{div} \mathbf{F}(P_0) < 0$, there must be a sphere S_a for which the net (outward) flux through the surface of S_a is negative. This says that more fluid is passing in through the surface than is flowing out. Once again, for an incompressible fluid, this can only occur if there is a sink somewhere in S_a, where fluid is leaving the flow. For this reason, in incompressible fluid flow, a point where $\operatorname{div} \mathbf{F}(P) > 0$ is called a **source** and a point where $\operatorname{div} \mathbf{F}(P) < 0$ is called a **sink**. Notice that for an incompressible fluid flow with no sources or sinks, we must have that $\operatorname{div} \mathbf{F}(P) = 0$ throughout the flow.

Example 7.3   Finding the Flux of an Inverse Square Field

Show that the flux of an inverse square field over every closed surface enclosing the origin is a constant.

Solution   Suppose that S is a closed surface forming the boundary of the solid region Q, where the origin lies in the interior of Q and suppose that $\mathbf{F}$ is an inverse square field. That is,

$$\mathbf{F}(x, y, z) = \frac{c}{\|\mathbf{r}\|^3}\mathbf{r},$$

where $\mathbf{r} = \langle x, y, z \rangle$, $\|\mathbf{r}\| = \sqrt{x^2 + y^2 + z^2}$ and c is a constant. Before you rush to apply the Divergence Theorem, notice that $\mathbf{F}$ is **not** continuous in Q, since $\mathbf{F}$ is undefined at

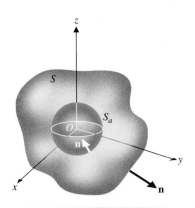

Figure 14.50

The region Q_a.

the origin and so, we cannot apply the theorem in Q. Notice though, that if we could somehow exclude the origin from the region, then we could apply the theorem. A very common method of doing this is to "punch out" a sphere S_a of radius a centered at the origin, where a is sufficiently small that S_a is completely contained in the interior of Q (see Figure 14.50). That is, if we define Q_a to be the set of all points inside Q, but outside of S_a (so that Q_a corresponds to Q, where the sphere S_a has been "punched out"), we can now apply the Divergence Theorem on Q_a. Before we do that, notice that the boundary of Q_a consists of the two surfaces S and S_a. We now have

$$\iiint\limits_{Q_a} \nabla \cdot \mathbf{F} \, dV = \iint\limits_{S} \mathbf{F} \cdot \mathbf{n} \, dS + \iint\limits_{S_a} \mathbf{F} \cdot \mathbf{n} \, dS.$$

We leave it as an exercise to show that for any inverse square field, $\nabla \cdot \mathbf{F} = 0$. This now gives us

$$\iint\limits_{S} \mathbf{F} \cdot \mathbf{n} \, dS = -\iint\limits_{S_a} \mathbf{F} \cdot \mathbf{n} \, dS. \tag{7.10}$$

Since the integral on the right side of (7.10) is taken over the surface of a sphere centered at the origin, we should be able to easily calculate it. You need to be careful, though, to note that the exterior normals here point **out** of Q_a and so, the normal on the right side of (7.10) must point **toward** the origin. That is,

$$\mathbf{n} = \frac{-1}{\|\mathbf{r}\|}\mathbf{r} = \frac{-1}{a}\mathbf{r},$$

since $\|\mathbf{r}\| = a$ on S_a. We now have from (7.10) that

$$\iint\limits_{S} \mathbf{F} \cdot \mathbf{n} \, dS = -\iint\limits_{S_a} \frac{c}{a^3} \mathbf{r} \cdot \left(\frac{-1}{a}\mathbf{r}\right) dS$$

$$= \frac{c}{a^4} \iint\limits_{S_a} \mathbf{r} \cdot \mathbf{r} \, dS$$

$$= \frac{c}{a^4} \iint\limits_{S_a} \|\mathbf{r}\|^2 \, dS$$

$$= \frac{c}{a^2} \iint\limits_{S_a} dS = \frac{c}{a^2}(4\pi a^2) = 4\pi c, \qquad \text{Since } \|\mathbf{r}\| = a$$

since $\iint\limits_{S_a} dS$ simply gives the surface area of the sphere, $4\pi a^2$. Notice that this says that over any closed surface enclosing the origin, the flux of an inverse square field is a constant: $4\pi c$.

◼

The principle derived in example 7.3 is called **Gauss' Law** for inverse square fields and has many important applications, notably in the theory of electricity and magnetism. The method we used to derive Gauss' Law, whereby we punched out a disk surrounding the discontinuity of the integrand, is a common technique used in applying the Divergence Theorem to a variety of important cases where the integrand is discontinuous. In particular, such applications to discontinuous vector fields are quite important in the field of differential equations.

EXERCISES 14.7

1. If **F** is the velocity field of a fluid, explain what $\mathbf{F} \cdot \mathbf{n}$ represents and then what $\iint_{\partial Q} \mathbf{F} \cdot \mathbf{n} \, dS$ represents.

2. If **F** is the velocity field of a fluid, explain what $\nabla \cdot \mathbf{F}$ represents and then what $\iiint_Q \nabla \cdot \mathbf{F} \, dV$ represents.

3. Use your answers to exercises 1 and 2 to explain in physical terms why the Divergence Theorem makes sense.

4. For fluid flowing through a pipe, give one example each of a source and a sink.

In exercises 5–8, verify the Divergence Theorem by computing both integrals.

5. $\mathbf{F} = \langle 2xz, y^2, -xz \rangle$, Q is the cube $0 \le x \le 1$, $0 \le y \le 1$, $0 \le z \le 1$

6. $\mathbf{F} = \langle x, y, z \rangle$, Q is the ball $x^2 + y^2 + z^2 \le 1$

7. $\mathbf{F} = \langle xz, zy, 2z^2 \rangle$, Q is bounded by $z = 1 - x^2 - y^2$ and $z = 0$

8. $\mathbf{F} = \langle x^2, 2y, -x^2 \rangle$, Q is the tetrahedron bounded by $x + 2y + z = 4$ and the coordinate planes

In exercises 9–20, use the Divergence Theorem to compute $\iint_{\partial Q} \mathbf{F} \cdot \mathbf{n} \, dS$.

9. Q is bounded by $x + y + 2z = 2$ (first octant) and the coordinate planes, $\mathbf{F} = \langle 2x - y^2, 4xz - 2y, xy^3 \rangle$.

10. Q is bounded by $4x + 2y + z = 4$ (first octant) and the coordinate planes, $\mathbf{F} = \langle x^2 - y^2z, x \sin z, 4y^2 \rangle$.

11. Q is the cube $-1 \le x \le 1$, $-1 \le y \le 1$, $-1 \le z \le 1$, $\mathbf{F} = \langle 4y^2, 3z - \cos x, z^3 - x \rangle$.

12. Q is the rectangular box $0 \le x \le 2$, $1 \le y \le 2$, $-1 \le z \le 2$, $\mathbf{F} = \langle y^3 - 2x, e^{xz}, 4z \rangle$.

13. Q is bounded by $z = x^2 + y^2$ and $z = 4$, $\mathbf{F} = \langle x^3, y^3 - z, xy^2 \rangle$.

14. Q is bounded by $z = 4 - x^2 - y^2$ and the xy-plane, $\mathbf{F} = \langle z^3, x^2y, y^2z \rangle$.

15. Q is bounded by $z = \sqrt{x^2 + y^2}$ and $z = 4$, $\mathbf{F} = \langle y^3, x + z^2, z + y^2 \rangle$.

16. Q is bounded by $z = \sqrt{x^2 + y^2}$, $z = 1$ and $z = 2$, $\mathbf{F} = \langle x^3, x^2z^2, 3y^2z \rangle$.

17. Q is bounded by $x^2 + y^2 = 1$, $z = 0$ and $z = 1$, $\mathbf{F} = \langle x - y^3, x^2 \sin z, 3z \rangle$.

18. Q is bounded by $x^2 + y^2 = 4$, $z = 1$ and $z = 8 - y$, $\mathbf{F} = \langle y^2z, 2y - e^z, \sin x \rangle$.

19. Q is bounded by $z = \sqrt{1 - x^2 - y^2}$ and $z = 0$, $\mathbf{F} = \langle x^3, y^3, z^3 \rangle$.

20. Q is bounded by $z = \sqrt{4 - x^2 - y^2}$ and $z = 0$, $\mathbf{F} = \langle x^3, y^3, z^3 \rangle$.

In exercises 21–32, find the flux of F over ∂Q.

21. Q is bounded by $z = \sqrt{x^2 + y^2}$ and $z = \sqrt{2 - x^2 - y^2}$, $\mathbf{F} = \langle x^2, z^2 - x, y^3 \rangle$.

22. Q is bounded by $z = \sqrt{x^2 + y^2}$ and $z = \sqrt{8 - x^2 - y^2}$, $\mathbf{F} = \langle 3xz^2, y^3, 3zx^2 \rangle$.

23. Q is bounded by $z = \sqrt{x^2 + y^2}$, $x^2 + y^2 = 1$ and $z = 0$, $\mathbf{F} = \langle y^2, x^2z, z^2 \rangle$.

24. Q is bounded by $z = x^2 + y^2$ and $z = 8 - x^2 - y^2$, $\mathbf{F} = \langle 3y^2, 4x^3, 2z - x^2 \rangle$.

25. Q is bounded by $x^2 + z^2 = 1$, $y = 0$ and $y = 1$, $\mathbf{F} = \langle z - y^3, 2y - \sin z, x^2 - z \rangle$.

26. Q is bounded by $y^2 + z^2 = 4$, $x = 1$ and $x = 8 - y$, $\mathbf{F} = \langle x^2z, 2y - e^z, \sin x \rangle$.

27. Q is bounded by $x = y^2 + z^2$ and $x = 4$, $\mathbf{F} = \langle x^3, y^3 - z, z^3 - y^2 \rangle$.

28. Q is bounded by $y = 4 - x^2 - z^2$ and the xz-plane, $\mathbf{F} = \langle z^2x, x^2y, y^2x \rangle$.

29. Q is bounded by $3x + 2y + z = 6$ and the coordinate planes, $\mathbf{F} = \langle y^2x, 4x^2 \sin z, 3 \rangle$.

30. Q is bounded by $x + 2y + 3z = 12$ and the coordinate planes, $\mathbf{F} = \langle x^2y, 3x, 4y - x^2 \rangle$.

31. Q is bounded by $z = 1 - x^2$, $z = -3$, $y = -2$ and $y = 2$, $\mathbf{F} = \langle x^2, y^3, x^3y^2 \rangle$.

32. Q is bounded by $z = 1 - x^2$, $z = 0$, $y = 0$ and $x + y = 4$, $\mathbf{F} = \langle y^3, x^2 - z, z^2 \rangle$.

33. Prove equation (7.4).

34. Sketch a picture analogous to Figure 14.48a showing the surfaces used in the proof of equation (7.4).

35. Coulomb's Law for an electrostatic field applied to a point charge q at the origin gives us $\mathbf{E}(\mathbf{r}) = q \frac{\mathbf{r}}{r^3}$, where $r = \|\mathbf{r}\|$. Let Q be bounded by the sphere $x^2 + y^2 + z^2 = a^2$ for some constant $a > 0$. Show that the flux of **E** over ∂Q equals $4\pi q$. Discuss the fact that the flux does not depend on the value of a.

36. Show that for any inverse square field (see example 7.3), the divergence is 0.

37. Prove Green's first identity in three dimensions (see exercise 43 in section 14.5 for Green's first identity in two dimensions):

$$\iiint_Q f\,\nabla^2 g\,dV = \iint_{\partial Q} f(\nabla g)\cdot \mathbf{n}\,dS - \iiint_Q (\nabla f \cdot \nabla g)\,dV.$$

(Hint: Use the Divergence Theorem applied to $\mathbf{F} = f\,\nabla g$.)

38. Prove Green's second identity in three dimensions (see exercise 44 in section 14.5 for Green's second identity in two dimensions):

$$\iiint_Q (f\,\nabla^2 g - g\nabla^2 f)\,dV = \iint_{\partial Q} (f\,\nabla g - g\nabla f)\cdot \mathbf{n}\,dS.$$

(Hint: Use Green's first identity from exercise 37.)

39.  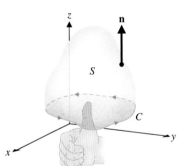 In this exercise, we develop the **continuity equation,** one of the most important results in vector calculus.

Suppose that a fluid has density ρ (a scalar function of space and time) and velocity $\mathbf{v}$ with no sources or sinks. Argue that the rate of change of the mass m of the fluid contained in a region Q can be written as $\dfrac{dm}{dt} = \iiint_Q \dfrac{\partial \rho}{\partial t}\,dV$. Next, explain why the only way that the mass can change is for fluid to cross the boundary of $Q\,(\partial Q)$. Argue that $\dfrac{dm}{dt} = -\iint_{\partial Q} (\rho \mathbf{v})\cdot \mathbf{n}\,dS$.

In particular, explain why the minus sign in front of the surface integral is needed. Use the Divergence Theorem to rewrite this expression as a triple integral over Q. Explain why the two triple integrals must be equal. Since the integration is taken over arbitrary solids Q, the integrands must be equal to each other. Conclude that the continuity equation holds:

$$\nabla \cdot (\rho \mathbf{v}) + \frac{\partial \rho}{\partial t} = 0.$$

14.8　STOKES' THEOREM

Recall that after introducing the curl in section 14.5, we observed that for a piecewise smooth, positively oriented, simple closed curve C in the xy-plane enclosing the region R, we could rewrite Green's Theorem in the vector form

$$\oint_C \mathbf{F}\cdot d\mathbf{r} = \iint_R (\nabla \times \mathbf{F})\cdot \mathbf{k}\,dA, \tag{8.1}$$

where $\mathbf{F}(x, y)$ is a vector field of the form $\mathbf{F}(x, y) = \langle M(x, y), N(x, y), 0\rangle$. In this section, we generalize this result to the case of a vector field defined on a surface in three dimensions. First, we need to introduce the notion of the orientation of a closed curve in three dimensions. Suppose that S is an oriented surface (that is, S can be viewed as having two sides, say, a top side as defined by the exterior normal vectors to S and a bottom side). If S is bounded by the simple closed curve C, we determine the orientation of C using a right-hand rule like the one used to determine the direction of a cross-product of two vectors. Align the thumb of your right hand so that it points in the direction of the exterior unit normals to S. Then if you curl your fingers, they will indicate the **positive orientation** on C, as indicated in Figure 14.51a. If the orientation of C is opposite that indicated by the curling of the fingers on your right hand, as shown in Figure 14.51b, we say that C has **negative orientation.** The vector form of Green's Theorem in (8.1) generalizes as follows.

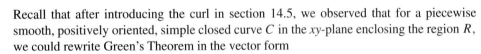

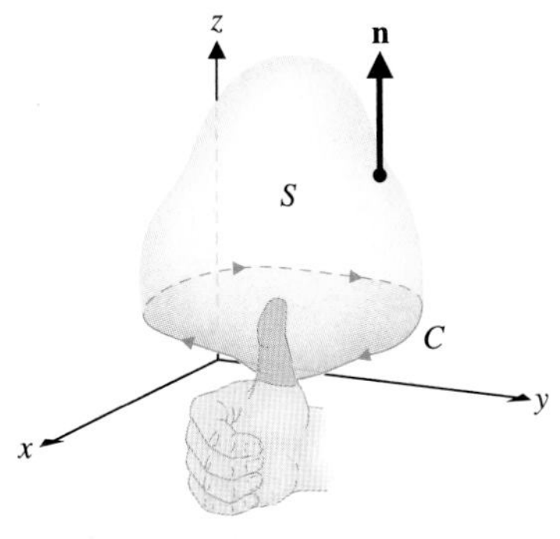

Figure 14.51a
Positive orientation.

Figure 14.51b
Negative orientation.

Theorem 8.1 (Stokes' Theorem)

Suppose that S is an oriented, piecewise-smooth surface, bounded by the simple closed, piecewise-smooth boundary curve ∂S having positive orientation. Let $\mathbf{F}(x, y, z)$ be a vector field whose components have continuous first partial derivatives in some open region containing S. Then,

$$\int_{\partial S} \mathbf{F}(x, y, z)\cdot d\mathbf{r} = \iint_S (\nabla \times \mathbf{F})\cdot \mathbf{n}\,dS. \tag{8.2}$$

Notice right away that the vector form of Green's Theorem (8.1) is a special case of (8.2), as follows. If S is simply a region in the xy-plane, then a unit normal to the surface at every point on S is the vector $\mathbf{k}$. Further, if ∂S has positive orientation, then the *exterior* unit normal vector to S is $\mathbf{n} = \mathbf{k}$, at every point in the region. Further, $dS = dA$ (i.e., the surface area of the plane region is simply the area) and (8.2) simplifies to (8.1). The proof of Stokes' Theorem for the special case considered below hinges on Green's Theorem and the chain rule.

One important interpretation of Stokes' Theorem arises in the case where $\mathbf{F}$ represents a force field. Note that in this case, the integral on the left side of (8.2) corresponds to the work done by the force field $\mathbf{F}$ as the point of application moves along the boundary of S. Likewise, the right side of (8.2) represents the net flux of the curl of $\mathbf{F}$ over the surface S. A general proof of Stokes' Theorem can be found in more advanced texts. We prove it here only for a special case of the surface S.

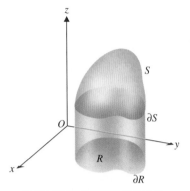

Figure 14.52
The surface S and its projection R onto the xy-plane.

Proof (Special Case)

We consider here the special case where S is a surface of the form

$$S = \{(x, y, z) | z = f(x, y), \text{ for } (x, y) \in R\},$$

where R is a region in the xy-plane with piecewise-smooth boundary ∂R, where $f(x, y)$ has continuous first partial derivatives and for which ∂R is the projection of the boundary of the surface ∂S onto the xy-plane (see Figure 14.52).

Let $\mathbf{F}(x, y, z) = \langle M(x, y, z), N(x, y, z), P(x, y, z) \rangle$. We then have

$$\nabla \times \mathbf{F} = \begin{vmatrix} \mathbf{i} & \mathbf{j} & \mathbf{k} \\ \dfrac{\partial}{\partial x} & \dfrac{\partial}{\partial y} & \dfrac{\partial}{\partial z} \\ M & N & P \end{vmatrix}$$

$$= \left(\frac{\partial P}{\partial y} - \frac{\partial N}{\partial z} \right) \mathbf{i} + \left(\frac{\partial M}{\partial z} - \frac{\partial P}{\partial x} \right) \mathbf{j} + \left(\frac{\partial N}{\partial x} - \frac{\partial M}{\partial y} \right) \mathbf{k}.$$

Note that an exterior normal vector at any point on S is given by

$$\mathbf{m} = \langle -f_x(x, y), -f_y(x, y), 1 \rangle$$

and so, the exterior unit normal is

$$\mathbf{n} = \frac{\langle -f_x(x, y), -f_y(x, y), 1 \rangle}{\sqrt{[f_x(x, y)]^2 + [f_y(x, y)]^2 + 1}}.$$

Since $dS = \sqrt{[f_x(x, y)]^2 + [f_y(x, y)]^2 + 1}\, dA$, we now have

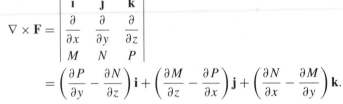

$$\iint_S (\nabla \times \mathbf{F}) \cdot \mathbf{n}\, dS$$

$$= \iint_R \left[-\left(\frac{\partial P}{\partial y} - \frac{\partial N}{\partial z} \right) f_x - \left(\frac{\partial M}{\partial z} - \frac{\partial P}{\partial x} \right) f_y + \left(\frac{\partial N}{\partial x} - \frac{\partial M}{\partial y} \right) \right]_{z=f(x,y)} dA.$$

Equation (8.2) is now equivalent to

$$\int_{\partial S} M\,dx + N\,dy + P\,dz$$

$$= \iint_R \left[-\left(\frac{\partial P}{\partial y} - \frac{\partial N}{\partial z} \right) f_x - \left(\frac{\partial M}{\partial z} - \frac{\partial P}{\partial x} \right) f_y + \left(\frac{\partial N}{\partial x} - \frac{\partial M}{\partial y} \right) \right]_{z=f(x,y)} dA. \quad (8.3)$$

We will now show that

$$\int_{\partial S} M(x, y, z)\, dx = -\iint_R \left(\frac{\partial M}{\partial y} + \frac{\partial M}{\partial z} f_y \right)_{z=f(x,y)} dA. \tag{8.4}$$

Suppose that the boundary of R is described parametrically by

$$\partial R = \{(x, y) \,|\, x = x(t), y = y(t), a \le t \le b\}.$$

Then, the boundary of S is described parametrically by

$$\partial S = \{(x, y, z) \,|\, x = x(t), y = y(t), z = f(x(t), y(t)), a \le t \le b\}$$

and we have

$$\int_{\partial S} M(x, y, z)\, dx = \int_a^b M(x(t), y(t), f(x(t), y(t))) x'(t)\, dt.$$

Now, notice that for $m(x, y) = M(x, y, f(x, y))$, this gives us

$$\int_{\partial S} M(x, y, z)\, dx = \int_a^b m(x(t), y(t)) x'(t)\, dt = \int_{\partial R} m(x, y)\, dx. \tag{8.5}$$

From Green's Theorem, we know that

$$\int_{\partial R} m(x, y)\, dx = -\iint_R \frac{\partial m}{\partial y}\, dA. \tag{8.6}$$

But from the chain rule,

$$\frac{\partial m}{\partial y} = \frac{\partial}{\partial y} M(x, y, f(x, y)) = \left(\frac{\partial M}{\partial y} + \frac{\partial M}{\partial z} f_y \right)_{z=f(x,y)}.$$

Putting this together with (8.5) and (8.6) gives us

$$\int_{\partial S} M(x, y, z)\, dx = -\iint_R \frac{\partial m}{\partial y}\, dA = -\iint_R \left(\frac{\partial M}{\partial y} + \frac{\partial M}{\partial z} f_y \right)_{z=f(x,y)} dA,$$

which is (8.4). Similarly, you can show that

$$\int_{\partial S} N(x, y, z)\, dy = \iint_R \left(\frac{\partial N}{\partial x} + \frac{\partial N}{\partial z} f_x \right)_{z=f(x,y)} dA \tag{8.7}$$

and

$$\int_{\partial S} P(x, y, z)\, dz = \iint_R \left(\frac{\partial P}{\partial x} f_y - \frac{\partial P}{\partial y} f_x \right)_{z=f(x,y)} dA. \tag{8.8}$$

Putting together (8.4), (8.7) and (8.8) now gives us (8.3) which proves Stokes' Theorem for this special case of the surface.

Example 8.1　　Using Stokes' Theorem to Evaluate a Line Integral

Evaluate $\int_C \mathbf{F} \cdot d\mathbf{r}$, for $\mathbf{F}(x, y, z) = \langle -y, x^2, z^3 \rangle$, where C is the intersection of the circular cylinder $x^2 + y^2 = 4$ and the plane $x + z = 3$, oriented so that it is traversed counterclockwise when viewed from the positive z-axis.

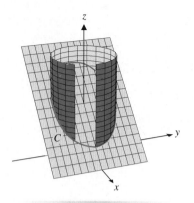

Figure 14.53

Intersection of the plane and the cylinder producing the curve C.

Solution   First, notice that C is an ellipse, as indicated in Figure 14.53. Unfortunately, C is rather difficult to parameterize, which makes the direct evaluation of the line integral somewhat difficult. Instead, we can use Stokes' Theorem to evaluate the integral. First, we calculate the curl of **F**:

$$\nabla \times \mathbf{F} = \begin{vmatrix} \mathbf{i} & \mathbf{j} & \mathbf{k} \\ \dfrac{\partial}{\partial x} & \dfrac{\partial}{\partial y} & \dfrac{\partial}{\partial z} \\ -y & x^2 & z^3 \end{vmatrix} = (2x+1)\mathbf{k}.$$

Notice that on the surface S, consisting of the portion of the plane $x + z = 3$ enclosed by C, we have the exterior unit normal

$$\mathbf{n} = \frac{1}{\sqrt{2}}\langle 1, 0, 1 \rangle.$$

From Stokes' Theorem, we now have

$$\int_C \mathbf{F} \cdot d\mathbf{r} = \iint_S (\nabla \times \mathbf{F}) \cdot \mathbf{n}\, dS = \frac{1}{\sqrt{2}} \iint_R \underbrace{(2x+1)}_{(\nabla \times \mathbf{F})\cdot\mathbf{n}} \underbrace{\sqrt{2}\, dA}_{dS},$$

where R is the circle of radius 2, centered at the origin (i.e., the projection of S onto the xy-plane). Introducing polar coordinates, we have

$$\int_C \mathbf{F} \cdot d\mathbf{r} = \iint_R (2x+1)\, dA = \int_0^{2\pi} \int_0^2 (2r\cos\theta + 1) r\, dr\, d\theta$$

$$= \int_0^{2\pi} \int_0^2 (2r^2 \cos\theta + r)\, dr\, d\theta$$

$$= \int_0^{2\pi} \left(2\frac{r^3}{3}\cos\theta + \frac{r^2}{2} \right) \Bigg|_{r=0}^{r=2} d\theta$$

$$= \int_0^{2\pi} \left(\frac{16}{3}\cos\theta + 2 \right) d\theta = 4\pi,$$

where we have left the final details of the calculation to you.

■

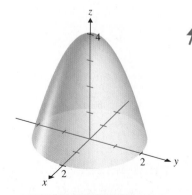

Figure 14.54

$z = 4 - x^2 - y^2$.

Example 8.2   Using Stokes' Theorem to Evaluate a Surface Integral

Evaluate $\iint_S (\nabla \times \mathbf{F}) \cdot \mathbf{n}\, dS$, where $\mathbf{F}(x, y, z) = \langle e^{z^2}, 4z - y, 8x\sin y \rangle$ and where S is the portion of the paraboloid $z = 4 - x^2 - y^2$ above the xy-plane.

Solution   In Figure 14.54, we show the paraboloid S. Notice that the boundary curve is simply the circle $x^2 + y^2 = 4$ lying in the xy-plane. By Stokes' Theorem, we have

$$\iint_S (\nabla \times \mathbf{F}) \cdot \mathbf{n}\, dS = \int_{\partial S} \mathbf{F}(x, y, z) \cdot d\mathbf{r}$$

$$= \int_{\partial S} e^{z^2}\, dx + (4z - y)\, dy + 8x\sin y\, dz.$$

Now, we can parameterize ∂S by $x = 2\cos t$, $y = 2\sin t$, $z = 0$, $0 \leq t \leq 2\pi$. This says that on ∂S, we have $dx = -2\sin t$, $dy = 2\cos t$ and $dz = 0$. In view of this, we have

$$\iint\limits_S (\nabla \times \mathbf{F}) \cdot \mathbf{n}\, dS = \int_{\partial S} e^{z^2}\, dx + (4z - y)\, dy + 8x \sin y\, dz$$

$$= \int_0^{2\pi} \{e^0(-2\sin t) + [4(0) - 2\sin t](2\cos t)\}\, dt = 0,$$

where we leave the (straightforward) details of the calculation to you.

■

In the following example, we consider the same surface integral as in example 8.2, but over a different surface. Although the surfaces are different, they have the same boundary curve, so that they must have the same value.

Example 8.3   Using Stokes' Theorem to Evaluate a Surface Integral

Evaluate $\iint\limits_S (\nabla \times \mathbf{F}) \cdot \mathbf{n}\, dS$, where $\mathbf{F}(x, y, z) = \langle e^{z^2}, 4z - y, 8x \sin y \rangle$ and where S is the hemisphere $z = \sqrt{4 - x^2 - y^2}$.

Solution   We show a sketch of the surface S in Figure 14.55. Notice that although this is not the same surface as in example 8.2, the two surfaces have the same boundary curve: the circle $x^2 + y^2 = 4$ lying in the xy-plane. Just as in example 8.2, we then have

$$\iint\limits_S (\nabla \times \mathbf{F}) \cdot \mathbf{n}\, dS = 0.$$

■

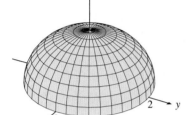

Figure 14.55

$z = \sqrt{4 - x^2 - y^2}$.

Much as we used the Divergence Theorem in section 14.7 to give an interpretation of the meaning of the divergence of a vector field, we can use Stokes' Theorem to give some meaning to the curl of a vector field. (You may have been wondering why we would call $\nabla \times \mathbf{F}$ the curl of $\mathbf{F}$; we can give you an answer to this now.) Suppose once again that $\mathbf{F}(x, y, z)$ represents the velocity field for a fluid in motion. Let S be an oriented surface located in the fluid flow, with positively oriented boundary curve C. Suppose further that C is traced out by the endpoint of the vector-valued function $\mathbf{r}(t)$ for $a \leq t \leq b$. Notice that the closer the direction of $\mathbf{F}$ is to the direction of $\frac{d\mathbf{r}}{dt}$, the larger its component is in the direction of $\frac{d\mathbf{r}}{dt}$ (see Figure 14.56). In other words, the closer the direction of $\mathbf{F}$ is to the direction of $\frac{d\mathbf{r}}{dt}$, the larger $\mathbf{F} \cdot \frac{d\mathbf{r}}{dt}$ will be. Now, recall that $\frac{d\mathbf{r}}{dt}$ points in the direction of the unit tangent vector along C. Then, since

$$\int_C \mathbf{F} \cdot d\mathbf{r} = \int_a^b \mathbf{F} \cdot \frac{d\mathbf{r}}{dt}\, dt,$$

it follows that the closer the direction of $\mathbf{F}$ is to the direction of $\frac{d\mathbf{r}}{dt}$ along C, the larger $\int_C \mathbf{F} \cdot d\mathbf{r}$ will be. This says that $\int_C \mathbf{F} \cdot d\mathbf{r}$ measures the tendency of the fluid to flow around or **circulate** around C. For this reason, we refer to $\int_C \mathbf{F} \cdot d\mathbf{r}$ as the **circulation** of $\mathbf{F}$ around C.

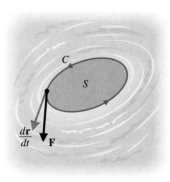

Figure 14.56

The surface S in a fluid flow.

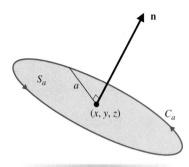

Figure 14.57
The disk S_a.

For any point (x, y, z) in the fluid flow, let S_a be a disk of radius a centered at (x, y, z), with exterior unit normal vector **n**, as indicated in Figure 14.57 and let C_a be the boundary of S_a. Then, by Stokes' Theorem, we have

$$\int_{C_a} \mathbf{F} \cdot d\mathbf{r} = \iint_{S_a} (\nabla \times \mathbf{F}) \cdot \mathbf{n} \, dS. \tag{8.9}$$

Notice that the average value of a function f on the surface S_a is given by

$$f_{\text{ave}} = \frac{1}{\pi a^2} \iint_{S_a} f(x, y, z) \, dS.$$

Further, if f is continuous on S_a, there must be some point P_a on S_a at which f equals its average value, that is, where

$$f(P_a) = \frac{1}{\pi a^2} \iint_{S_a} f(x, y, z) \, dS.$$

In particular, if the velocity field **F** has continuous first partial derivatives throughout S_a, then it follows from equation (8.9) that for some point P_a on S_a,

$$(\nabla \times \mathbf{F})(P_a) \cdot \mathbf{n} = \frac{1}{\pi a^2} \iint_{S_a} (\nabla \times \mathbf{F}) \cdot \mathbf{n} \, dS = \frac{1}{\pi a^2} \int_{C_a} \mathbf{F} \cdot d\mathbf{r}. \tag{8.10}$$

Notice that the expression on the far right of (8.10) corresponds to the circulation of **F** around C_a per unit area. Taking the limit as $a \to 0$, we have by the continuity of curl **F** that

$$(\nabla \times \mathbf{F})(x, y, z) \cdot \mathbf{n} = \lim_{a \to 0} \frac{1}{\pi a^2} \int_{C_a} \mathbf{F} \cdot d\mathbf{r}. \tag{8.11}$$

Read equation (8.11) very carefully. Notice that it says that at any given point, the component of curl **F** in the direction of **n** is the limiting value of the circulation per unit area around circles of radius a centered at that point (and normal to **n**), as the radius a tends to zero. In this sense, $(\nabla \times \mathbf{F}) \cdot \mathbf{n}$ measures the tendency of the fluid to rotate about an axis aligned with the vector **n**. You can visualize this by thinking of a small paddle wheel with axis parallel to **n**, which is immersed in the fluid flow (see Figure 14.58). Notice that the circulation per unit area is greatest (so that the paddle wheel moves fastest) when **n** points in the direction of $\nabla \times \mathbf{F}$.

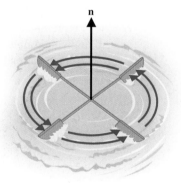

Figure 14.58
Paddle wheel.

If $\nabla \times \mathbf{F} = \mathbf{0}$ at every point in a fluid flow, we say that the flow is **irrotational,** since the circulation about every point is zero. In particular, notice that if the velocity field **F** is a constant vector throughout the fluid flow, then

$$\text{curl } \mathbf{F} = \nabla \times \mathbf{F} = \mathbf{0},$$

everywhere in the fluid flow and so, the flow is irrotational. Physically, this says that there are no eddies in such a flow.

Notice, too that by Stokes' Theorem, if curl $\mathbf{F} = \mathbf{0}$ at every point in some region D, then we must have that

$$\oint_C \mathbf{F} \cdot d\mathbf{r} = 0,$$

for every simple closed curve contained in the region D. In other words, the circulation is zero around every such curve C lying in the region D. It turns out that by suitably restricting the type of regions $D \subset \mathbb{R}^3$ we consider, the converse is also true. That is, if $\oint_C \mathbf{F} \cdot d\mathbf{r} = 0$, for every simple closed curve C contained in the region D, then we must have that curl $\mathbf{F} = \mathbf{0}$ at every point in D. This result is true for regions in space that are simply-connected. Recall that in the plane a region is said to be simply-connected whenever every closed curve contained in the region encloses only points in the region (that is, the region contains no holes). In three dimensions, the situation is slightly more complicated. A region D in $\mathbb{R}^3$ is called **simply-connected** whenever every simple closed curve C lying in D can be continuously shrunk to a point without crossing the boundary of D. Notice that the interior of a sphere or a rectangular box is simply-connected, but any region with one or more cavities is not simply-connected. Be careful not to confuse connected with simply-connected. Recall that a connected region is one where every two points contained in the region can be connected with a path that is completely contained in the region. We illustrate connected and simply-connected two-dimensional regions in Figures 14.59a to 14.59c. We can now state the complete theorem.

Figure 14.59a
Connected and simply-connected.

Figure 14.59b
Connected but not simply-connected.

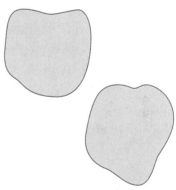

Figure 14.59c
Simply-connected but not connected.

Theorem 8.2

Suppose that $\mathbf{F}(x, y, z)$ is a vector field whose components have continuous first partial derivatives throughout the simply-connected region $D \subset \mathbb{R}^3$. Then, curl $\mathbf{F} = \mathbf{0}$ in D if and only if $\oint_C \mathbf{F} \cdot d\mathbf{r} = 0$, for every simple closed curve C contained in the region D.

Proof

We have already shown that when curl $\mathbf{F} = \mathbf{0}$ in D, we also have $\oint_C \mathbf{F} \cdot d\mathbf{r} = 0$, for every simple closed curve C contained in the region D. Conversely, suppose now that $\oint_C \mathbf{F} \cdot d\mathbf{r} = 0$ for every simple closed curve C contained in the region D. Suppose further that curl $\mathbf{F} \neq \mathbf{0}$ at some point $(x, y, z) \in D$. Since the components of $\mathbf{F}$ have continuous first partial derivatives, curl $\mathbf{F}$ must be continuous in D and so, there must be a sphere of radius $a_0 > 0$, centered at (x, y, z), throughout whose interior S, curl $\mathbf{F} \neq \mathbf{0}$. Let S_a be the disk of radius $a < a_0$ centered at (x, y, z) and oriented so that the exterior normal to the surface has the same direction as curl $\mathbf{F}(x, y, z)$. Notice that since $a < a_0$, S_a will be completely contained in S. If C_a is the boundary of S_a, then we have by Stokes' Theorem that

$$\int_{C_a} \mathbf{F} \cdot d\mathbf{r} = \iint_{S_a} (\nabla \times \mathbf{F}) \cdot \mathbf{n} \, dS > 0,$$

since $\mathbf{n}$ was chosen to be parallel to $\nabla \times \mathbf{F}$. This contradicts the assumption that $\oint_C \mathbf{F} \cdot d\mathbf{r} = 0$, for every simple closed curve C contained in the region D. It now follows that curl $\mathbf{F} = \mathbf{0}$ throughout D.

Recall that we had observed earlier that a vector field is conservative in a given region if and only if $\oint_C \mathbf{F} \cdot d\mathbf{r} = 0$, for every simple closed curve C contained in the region. Theorem 8.2 has then established the following result.

Theorem 8.3

Suppose that $\mathbf{F}(x, y, z)$ has continuous first partial derivatives in a simply-connected region D. Then, the following statements are equivalent.

  (i) $\mathbf{F}$ is conservative in D. That is, for some scalar function $f(x, y, z)$, $\mathbf{F} = \nabla f$;
 (ii) $\int_C \mathbf{F} \cdot d\mathbf{r}$ is independent of path in D;
(iii) $\mathbf{F}$ is irrotational (i.e., curl $\mathbf{F} = \mathbf{0}$) in D; and
(iv) $\oint_C \mathbf{F} \cdot d\mathbf{r} = 0$, for every simple closed curve C contained in D.

EXERCISES 14.8

1. Describe circumstances (e.g., example 8.1) in which the surface integral of Stokes' Theorem will be simpler than the line integral.

2. Describe circumstances (e.g., example 8.2) in which the line integral of Stokes' Theorem will be simpler than the surface integral.

3. The surfaces in examples 8.2 and 8.3 have the same boundary curve C. Explain why all surfaces above the xy-plane with the boundary C will share the same value of

$$\iint_S (\nabla \times \mathbf{F}) \cdot \mathbf{n} \, dS.$$

What would change if the surface was below the xy-plane?

4. Explain why part (iv) of Theorem 8.3 follows immediately from part (ii). Explain why parts (ii) and (iii) follow immediately from part (i).

In exercises 5–8, verify Stokes' Theorem by computing both integrals.

5. S is the portion of $z = 4 - x^2 - y^2$ above the xy-plane, $\mathbf{F} = \langle zx, 2y, z^3 \rangle$.

6. S is the portion of $z = 1 - x^2 - y^2$ above the xy-plane, $\mathbf{F} = \langle x^2 z, xy, xz^2 \rangle$.

7. S is the portion of $z = \sqrt{4 - x^2 - y^2}$ above the xy-plane, $\mathbf{F} = \langle 2x - y, yz^2, y^2 z \rangle$.

8. S is the portion of $z = \sqrt{1 - x^2 - y^2}$ above the xy-plane, $\mathbf{F} = \langle 2x, z^2 - x, xz^2 \rangle$.

In exercises 9–20, use Stokes' Theorem to compute $\iint_S (\nabla \times \mathbf{F}) \cdot \mathbf{n} \, dS$.

9. S is the portion of the tetrahedron bounded by $x + y + 2z = 2$ and the coordinate planes with $z > 0$, $\mathbf{n}$ upward, $\mathbf{F} = \langle zy^4 - y^2, y - x^3, z^2 \rangle$.

10. S is the portion of the tetrahedron bounded by $x + y + 4z = 8$ and the coordinate planes with $z > 0$, $\mathbf{n}$ upward, $\mathbf{F} = \langle y^2, y + 2x, z^2 \rangle$.

11. S is the portion of $z = 1 - x^2 - y^2$ above the xy-plane with $\mathbf{n}$ upward, $\mathbf{F} = \langle zx^2, ze^{xy^2} - x, x \ln y^2 \rangle$.

12. S is the portion of $z = \sqrt{4 - x^2 - y^2}$ above the xy-plane with $\mathbf{n}$ upward, $\mathbf{F} = \langle zx^2, ze^{xy^2} - x, x \ln y^2 \rangle$.

13. S is the portion of the tetrahedron in exercise 9 with $y > 0$, $\mathbf{n}$ to the right, $\mathbf{F} = \langle zy^4 - y^2, y - x^3, z^2 \rangle$.

14. S is the portion of $y = x^2 + z^2$ with $y \leq 2$, $\mathbf{n}$ to the left, $\mathbf{F} = \langle xy, 4xe^{z^2}, yz + 1 \rangle$.

15. S is the portion of the unit cube $0 \leq x \leq 1, 0 \leq y \leq 1, 0 \leq z \leq 1$ with $z < 1$, $\mathbf{n}$ upward, $\mathbf{F} = \langle xyz, 4x^2y^3 - z, 8 \cos xz^2 \rangle$.

16. S is the portion of the unit cube $0 \leq x \leq 1, 0 \leq y \leq 1, 0 \leq z \leq 1$ with $z < 1$, $\mathbf{n}$ downward, $\mathbf{F} = \langle xyz, 4x^2y^3 - z, 8 \cos xz^2 \rangle$.

17. S is the portion of the cone $z = \sqrt{x^2 + y^2}$ below the sphere $x^2 + y^2 + z^2 = 2$, $\mathbf{n}$ downward, $\mathbf{F} = \langle x^2 + y^2, ze^{x^2+y^2}, e^{x^2+z^2} \rangle$.

18. S is the portion of the cone $z = \sqrt{x^2 + y^2}$ inside the cylinder $x^2 + y^2 = 2$, $\mathbf{n}$ downward, $\mathbf{F} = \langle zx, x^2 + y^2, z^2 - y^2 \rangle$.

19. S is the portion of the paraboloid $y = 4 - x^2 - z^2$ with $y > 0$, $\mathbf{n}$ to the right, $\mathbf{F} = \langle yx^2z, x^2 \cos y, x \rangle$.

20. S is the portion of the paraboloid $x = y^2 + z^2$ with $x \leq 4$, $\mathbf{n}$ to the back, $\mathbf{F} = \langle y^2z, x - 4, y \rangle$.

In exercises 21–26, use Stokes' Theorem to evaluate $\int_C \mathbf{F} \cdot d\mathbf{r}$.

21. C is the boundary of the portion of $z = 4 - x^2 - y^2$ above the xy-plane, $\mathbf{F} = \langle x^2 e^x - y, \sqrt{y^2 + 1}, z^3 \rangle$.

22. C is the boundary of the portion of $z = x^2 + y^2$ below $z = 4$, $\mathbf{F} = \langle x^2, y^4 - x, z^2 \sin z \rangle$.

23. C is the intersection of $z = x^2 + y^2$ and $z = 8 - y$ oriented clockwise from above, $\mathbf{F} = \langle 2x^2, 4y^2, e^{8z^2} \rangle$.

24. C is the intersection of $x^2 + y^2 = 1$ and $z = x - y$ oriented clockwise from above, $\mathbf{F} = \langle \cos x^2, \sin y^2, \tan z^2 \rangle$.

25. C is the triangle from $(0, 1, 0)$ to $(0, 0, 4)$ to $(2, 0, 0)$, $\mathbf{F} = \langle x^2 + 2xy^3z, 3x^2y^2z - y, x^2y^3 \rangle$.

26. C is the square from $(0, 2, 2)$ to $(2, 2, 2)$ to $(2, 2, 0)$ to $(0, 2, 0)$, $\mathbf{F} = \langle x^2, y^3 + x, 3y^2 \cos z \rangle$.

27. Show that $\oint_C (f \nabla f) \cdot d\mathbf{r} = 0$ for any closed curve C and twice differentiable function f.

28. Show that $\oint_C (f \nabla g + g \nabla f) \cdot d\mathbf{r} = 0$ for any closed curve C and twice differentiable functions f and g.

29. Let $\mathbf{F}(x, y) = \langle M(x, y), N(x, y) \rangle$ be a vector field whose components M and N have continuous first partial derivatives in all of $\mathbb{R}^2$. Show that $\nabla \cdot \mathbf{F} = 0$ if and only if $\int_C \mathbf{F} \cdot \mathbf{n} \, ds = 0$ for all closed curves C. (Hint: Use a vector form of Green's Theorem.)

30. Under the assumptions of exercise 29, show that $\int_C \mathbf{F} \cdot \mathbf{n} \, ds = 0$ for all closed curves C if and only if $\int_C \mathbf{F} \cdot \mathbf{n} \, ds$ is path-independent.

31. Under the assumptions of exercise 29, show that $\nabla \cdot \mathbf{F} = 0$ if and only if $\mathbf{F}$ has a **stream function** $g(x, y)$ such that $M(x, y) = g_y(x, y)$ and $N(x, y) = -g_x(x, y)$.

32. Combine the results of exercises 29–31 to state a theorem analogous to Theorem 8.3.

33. If S_1 and S_2 are surfaces satisfying the hypotheses of Stokes' Theorem and $\partial S_1 = \partial S_2$, under what circumstances can you conclude that

$$\iint_{S_1} (\nabla \times \mathbf{F}) \cdot \mathbf{n} \, dS = \iint_{S_2} (\nabla \times \mathbf{F}) \cdot \mathbf{n} \, dS?$$

34. Give an example where the two surface integrals of exercise 33 are not equal.

35. The **circulation** of a vector field $\mathbf{F}$ around the curve C is defined by $\int_C \mathbf{F} \cdot d\mathbf{r}$. Show that the curl $\nabla \times \mathbf{F}(0, 0, 0)$ is in the same direction as the normal to the plane in which the circulation around the origin is a maximum. Relate this to the interpretation of the curl given in section 14.5.

36. The Fundamental Theorem of Calculus can be thought of as relating the values of the function on the boundary of a region (interval) to the sum of the derivative values of the function within the region. Explain what this statement means, and then explain why the same statement can be applied to Theorem 3.2, Green's Theorem, the Divergence Theorem and Stokes' Theorem. In each case, carefully state what the "region" is, what its boundary is and what type derivative is involved.

CHAPTER REVIEW EXERCISES

In exercises 1 and 2, sketch several vectors in the velocity field by hand, and verify your sketch with a CAS.

1. $\langle x, -y \rangle$

2. $\langle 0, 2y \rangle$

3. Match the vector fields with their graphs. $\mathbf{F}_1(x, y) = \langle \sin x, y \rangle$, $\mathbf{F}_2(x, y) = \langle \sin y, x \rangle$, $\mathbf{F}_3(x, y) = \langle y^2, 2x \rangle$, $\mathbf{F}_4(x, y) = \langle 3, x^2 \rangle$

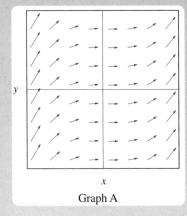

Graph A

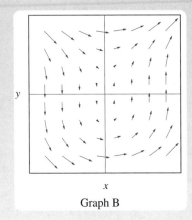

Graph B

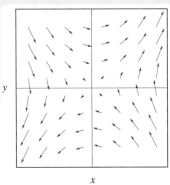

Graph C

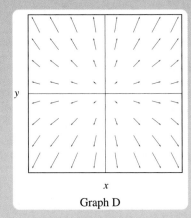

Graph D

4. Find the gradient field corresponding to f. Use a CAS to graph it.
    a. $f(x, y) = \ln \sqrt{x^2 + y^2}$
    b. $f(x, y) = e^{-x^2 - y^2}$

In exercises 5–8, determine whether or not the vector field is conservative. If it is, find a potential function.

5. $\langle y - 2xy^2, x - 2yx^2 + 1 \rangle$ 6. $\langle y^2 + 2e^{2y}, 2xy + 4xe^{2y} \rangle$

7. $\langle 2xy - 1, x^2 + 2xy \rangle$ 8. $\langle y \cos xy - y, x \cos xy - x \rangle$

In exercises 9 and 10, find equations for the flow lines.

9. $\left\langle y, \dfrac{2x}{y} \right\rangle$ 10. $\left\langle \dfrac{3}{x}, y \right\rangle$

In exercises 11 and 12, use the notation $\mathbf{r} = \langle x, y \rangle$ and $r = \|\mathbf{r}\| = \sqrt{x^2 + y^2}$.

11. Show that $\nabla(\ln r) = \dfrac{\mathbf{r}}{r^2}$.   12. Show that $\nabla\left(\dfrac{1}{r}\right) = -\dfrac{\mathbf{r}}{r^3}$.

In exercises 13–18, evaluate the line integral.

13. $\int_C 3y\, dx$, where C is the line segment from $(2, 3)$ to $(4, 3)$

14. $\int_C (x^2 + y^2)\, ds$, where C is the half-circle $x^2 + y^2 = 16$ from $(4, 0)$ to $(-4, 0)$ with $y \geq 0$

15. $\int_C \sqrt{x^2 + y^2}\, ds$, where C is the circle $x^2 + y^2 = 9$ oriented clockwise

16. $\int_C (x - y)\, ds$, where C is the portion of $y = x^3$ from $(1, 1)$ to $(-1, -1)$

17. $\int_C 2x\, dx$, where C is the upper half-circle from $(2, 0)$ to $(-2, 0)$ followed by the line segment to $(2, 0)$

18. $\int_C 3y^2\, dy$, where C is the portion of $y = x^2$ from $(-1, 1)$ to $(1, 1)$ followed by the line segment to $(-1, 1)$

In exercises 19–22, compute the work done by the force F along the curve C.

19. $\mathbf{F}(x, y) = \langle x, -y \rangle$, C is the circle $x^2 + y^2 = 4$ oriented counterclockwise

20. $\mathbf{F}(x, y) = \langle y, -x \rangle$, C is the circle $x^2 + y^2 = 4$ oriented counterclockwise

21. $\mathbf{F}(x, y) = \langle 2, 3x \rangle$, C is the quarter-circle from $(2, 0)$ to $(0, 2)$ followed by the line segment to $(0, 0)$

22. $\mathbf{F}(x, y) = \langle y, -x \rangle$, C is the square from $(-2, 0)$ to $(2, 0)$ to $(2, 4)$ to $(-2, 4)$ to $(-2, 0)$

In exercises 23 and 24, use the graph to determine if the work done is positive, negative or zero.

23.

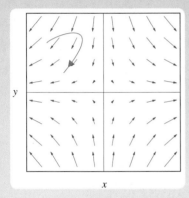

24.

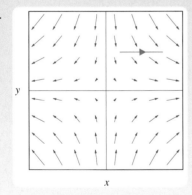

In exercises 25 and 26, find the mass of the indicated object.

25. A spring in the shape of $\langle \cos 3t, \sin 3t, 4t \rangle$, $0 \le t \le 2\pi$, $\rho(x, y, z) = 4$.

26. The portion of $z = x^2 + y^2$ under $z = 4$ with $\rho(x, y) = 12$

In exercises 27 and 28, show that the integral is independent of path and evaluate the integral.

27. $\int_C (3x^2 y - x) \, dx + x^3 \, dy$, where C runs from $(2, -1)$ to $(4, 1)$.

28. $\int_C (y^2 - x^2) \, dx + (2xy + 1) \, dy$, where C runs from $(3, 2)$ to $(1, 3)$.

In exercises 29–32, evaluate $\int_C \mathbf{F} \cdot d\mathbf{r}$.

29. $\mathbf{F}(x, y) = \langle 2xy + y \sin x + e^{x+y}, e^{x+y} - \cos x + x^2 \rangle$, C is the quarter-circle from $(0, 3)$ to $(3, 0)$.

30. $\mathbf{F}(x, y) = \langle 2y + y^3 + \frac{1}{2}\sqrt{y/x}, 3xy^2 + \frac{1}{2}\sqrt{x/y} \rangle$, C is the top half-circle from $(1, 3)$ to $(3, 3)$.

31. $\mathbf{F}(x, y, z) = \langle 2xy, x^2 - y, 2z \rangle$, C runs from $(1, 3, 2)$ to $(2, 1, -3)$.

32. $\mathbf{F}(x, y, z) = \langle yz - x, xz - y, xy - z \rangle$, C runs from $(2, 0, 0)$ to $(0, 1, -1)$.

In exercises 33 and 34, use the graph to determine whether or not the vector field is conservative.

33.

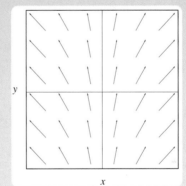

34.

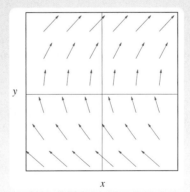

In exercises 35–40, use Green's Theorem to evaluate the indicated line integral.

35. $\oint_C \mathbf{F} \cdot d\mathbf{r}$, where $\mathbf{F} = \langle x^3 - y, x + y^3 \rangle$ and C is formed by $y = x^2$ and $y = x$ oriented positively.

36. $\oint_C \mathbf{F} \cdot d\mathbf{r}$, where $\mathbf{F} = \langle y^2 + 3x^2 y, xy + x^3 \rangle$ and C is formed by $y = x^2$ and $y = 2x$ oriented positively.

37. $\oint_C \tan x^2\, dx + x^2\, dy$, where C is the triangle from $(0, 0)$ to $(1, 1)$ to $(2, 0)$ to $(0, 0)$.

38. $\oint_C x^2 y\, dx + \ln \sqrt{1 + y^2}\, dy$, where C is the triangle from $(0, 0)$ to $(2, 2)$ to $(0, 2)$ to $(0, 0)$.

39. $\oint_C \mathbf{F} \cdot d\mathbf{r}$, where $\mathbf{F} = \langle 3x^2, 4y^3 - z, z^2 \rangle$ and C is formed by $z = y^2$ and $z = 4$ oriented positively in the yz-plane.

40. $\oint_C \mathbf{F} \cdot d\mathbf{r}$, where $\mathbf{F} = \langle 4y^2, 3x^2, 8z \rangle$ and C is $x^2 + y^2 = 4$ oriented positively in the plane $z = 3$.

In exercises 41 and 42, use a line integral to compute the area of the given region.

41. The ellipse $4x^2 + 9y^2 = 36$

42. The region bounded by $y = \sin x$ and the x-axis for $0 \le x \le \pi$

In exercises 43–46, find the curl and divergence of the given vector field.

43. $x^3 \mathbf{i} - y^3 \mathbf{j}$ 44. $y^3 \mathbf{i} - x^3 \mathbf{j}$

45. $\langle 2x, 2yz^2, 2y^2 z \rangle$

46. $\langle 2xy, x^2 - 3y^2 z^2, 1 - 2zy^3 \rangle$

In exercises 47–50, determine if the given vector field is conservative and/or incompressible.

47. $\langle 2x - y^2, z^2 - 2xy, xy^2 \rangle$

48. $\langle y^2 z, x^2 - 3z^2 y, z^3 - y \rangle$

49. $\langle 4x - y, 3 - x, 2 - 4z \rangle$

50. $\langle 4, 2xy^3, z^4 - x \rangle$

In exercises 51 and 52, conjecture whether the divergence at point P is positive, negative or zero.

51.

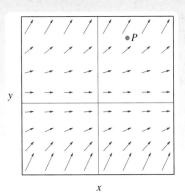

52.

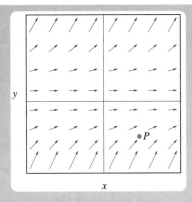

In exercises 53 and 54, sketch a graph of the parametric surface.

53. $x = u^2,\ y = v^2,\ z = u + 2v$

54. $x = (3 + 2\cos u)\cos v,\ y = (3 + 2\cos u)\sin v,\ z = 2\cos u$

55. Match the parametric equations with the surfaces.
   (a) $x = u^2,\ y = u + v,\ z = v^2$
   (b) $x = u^2,\ y = u + v,\ z = v$
   (c) $x = u,\ y = u + v,\ z = v^2$

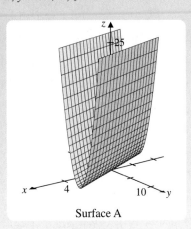

Surface A

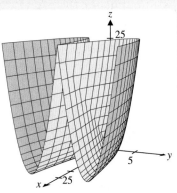

Surface B

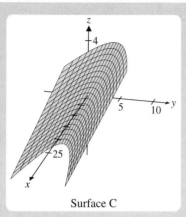

Surface C

56. Find a parametric representation of $x^2 + y^2 + z^2 = 9$.

In exercises 57 and 58, find the surface area.

57. The portion of the paraboloid $z = x^2 + y^2$ between the cylinders $x^2 + y^2 = 1$ and $x^2 + y^2 = 4$

58. The portion of the paraboloid $z = 9 - x^2 - y^2$ between the cylinders $x^2 + y^2 = 1$ and $x^2 + y^2 = 4$

In exercises 59–64, evaluate the surface integral $\iint_S f(x, y, z)\, dS$.

59. $\iint_S (x - y)\, dS$, where S is the portion of the plane $3x + 2y + z = 12$ in the first octant

60. $\iint_S (x^2 + y^2)\, dS$, where S is the portion of $y = 4 - x^2$ above the xy-plane, $y \geq 0$, and below $z = 2$

61. $\iint_S (4x + y + 3z)\, dS$, where S is the portion of the plane $4x + y + 3z = 12$ inside $x^2 + y^2 = 1$

62. $\iint_S (x - z)\, dS$, where S is the portion of the cylinder $x^2 + z^2 = 1$ above the xy-plane between $y = 1$ and $y = 2$

63. $\iint_S yz\, dS$, where S is the portion of the cone $y = \sqrt{x^2 + z^2}$ to the left of $y = 3$

64. $\iint_S (x^2 + z^2)\, dS$, where S is the portion of the paraboloid $x = y^2 + z^2$ behind the plane $x = 4$

In exercises 65 and 66, find the mass and center of mass of the solid.

65. The portion of the paraboloid $z = x^2 + y^2$ below the plane $z = 4$, $\rho(x, y, z) = 2$

66. The portion of the cone $z = \sqrt{x^2 + y^2}$ below the plane $z = 4$, $\rho(x, y, z) = z$

In exercises 67–70, use the Divergence Theorem to compute $\iint_Q \mathbf{F} \cdot \mathbf{n}\, dS$.

67. Q is bounded by $x + 2y + z = 4$ (first octant) and the coordinate planes, $\mathbf{F} = \langle y^2 z, y^2 - \sin z, 4y^2 \rangle$.

68. Q is the cube $-1 \leq x \leq 1$, $-1 \leq y \leq 1$, $-1 \leq z \leq 1$, $\mathbf{F} = \langle 4x, 3z, 4y^2 - x \rangle$.

69. Q is bounded by $z = 1 - y^2$, $z = 0$, $x = 0$ and $x + z = 4$, $\mathbf{F} = \langle 2xy, z^3 + 7yx, 4xy^2 \rangle$.

70. Q is bounded by $z = \sqrt{4 - x^2}$, $z = 0$, $y = 0$ and $y + z = 6$, $\mathbf{F} = \langle y^2, 4yz, 2xy \rangle$.

In exercises 71 and 72, find the flux of F over ∂Q.

71. Q is bounded by $z = \sqrt{x^2 + y^2}$, $x^2 + y^2 = 4$ and $z = 0$, $\mathbf{F} = \langle xz, yz, x^2 - z \rangle$.

72. Q is bounded by $z = x^2 + y^2$ and $z = 2 - x^2 - y^2$, $\mathbf{F} = \langle 4x, x^2 - 2y, 3z + x^2 \rangle$.

In exercises 73–76, use Stokes' Theorem, if appropriate, to compute $\iint_S (\nabla \times \mathbf{F}) \cdot \mathbf{n}\, dS$.

73. S is the portion of the tetrahedron bounded by $x + y + 2z = 2$ and the coordinate planes in front of the yz-plane, $\mathbf{F} = \langle zy^4 - y^2, y - x^3, z^2 \rangle$.

74. S is the portion of $z = x^2 + y^2$ below $z = 4$, $\mathbf{F} = \langle z^2 - x, 2y, z^3 xy \rangle$.

75. S is the portion of the cone $z = \sqrt{x^2 + y^2}$ below $x + 2y + 3z = 24$, $\mathbf{F} = \langle 4x^2, 2ye^{2y}, \sqrt{z^2 + 1} \rangle$.

76. S is the portion of the paraboloid $y = x^2 + 4z^2$ to the left of $y = 8 - z$, $\mathbf{F} = \langle xe^{3x}, 4y^{2/3}, z^2 + 2 \rangle$.

In exercises 77 and 78, use Stokes' Theorem to evaluate $\int_C \mathbf{F} \cdot d\mathbf{r}$.

77. C is the triangle from $(0, 1, 0)$ to $(1, 0, 0)$ to $(0, 0, 20)$, $\mathbf{F} = \langle 2xy \cos z, y^2 + x^2 \cos z, z - x^2 y \sin z \rangle$.

78. C is the square from $(0, 0, 2)$ to $(1, 0, 2)$ to $(1, 1, 2)$ to $(0, 1, 2)$, $\mathbf{F} = \langle x^3 + yz, y^2, z^2 \rangle$.

79. In exercise 57 of section 14.1, we developed a technique for finding equations for flow lines of certain vector fields. The field $\langle 2, 1 + 2xy \rangle$ from example 1.5 is such a vector field, but the calculus is more difficult. First, show that the differential equation is $y' - xy = \frac{1}{2}$ and show that an integrating factor is $e^{-x^2/2}$. The flow lines come from equations of the form $y = e^{x^2/2} \int \frac{1}{2} e^{-x^2/2}\, dx + c e^{x^2/2}$. Unfortunately, there is no elementary function equal to $\int \frac{1}{2} e^{-x^2/2}\, dx$. It can help to write this in the form $y = e^{x^2/2} \int_0^x \frac{1}{2} e^{-u^2/2}\, du + c e^{x^2/2}$. In this form, show that $c = y(0)$. In example 1.5, the curve passing through $(0, 1)$ is $y = e^{x^2/2} \int_0^x \frac{1}{2} e^{-u^2/2}\, du + e^{x^2/2}$. Graph this function and compare it to the path shown in Figure 14.7b. Find an equation for and plot the curve through $(0, -1)$. To find the curve through $(1, 1)$, change the limits of integration and rewrite the solution. Plot this curve and compare to Figure 14.7b.

CHAPTER 1O

Exercises Section 10.1

5.

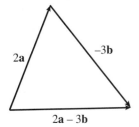

7.

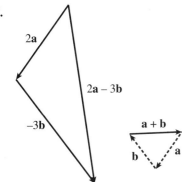

9. $\langle 5, 3 \rangle$, $\langle -4, 6 \rangle$, $\langle 6, 12 \rangle$, $\sqrt{290}$

11. $4\mathbf{i} + \mathbf{j}$, $-5\mathbf{i} + 4\mathbf{j}$, $3\mathbf{i} + 6\mathbf{j}$, $5\sqrt{10}$

13. $\langle -3, 3 \rangle$, $\langle -4, 0 \rangle$, $\langle -5, 9 \rangle$, $4\sqrt{13}$

15. $-2\mathbf{i} + 3\mathbf{j}$, $-12\mathbf{i} + 4\mathbf{j}$, $6\mathbf{i} + 5\mathbf{j}$, $4\sqrt{5}$

17.

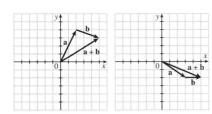

19. parallel **21.** not **23.** parallel **25.** not

27. $\langle 3, 1 \rangle$ **29.** $\langle -3, -3 \rangle$ **31.** $\langle 2, -3 \rangle$

33. $\langle -2, -3 \rangle$ **35.** (a) $\left\langle \frac{4}{5}, -\frac{3}{5} \right\rangle$ (b) $5\left\langle \frac{4}{5}, -\frac{3}{5} \right\rangle$

37. (a) $\dfrac{1}{\sqrt{5}}\mathbf{i} - \dfrac{2}{\sqrt{5}}\mathbf{j}$ (b) $2\sqrt{5}\left\langle \dfrac{1}{\sqrt{5}}, -\dfrac{2}{\sqrt{5}} \right\rangle$

39. (a) $\mathbf{i}$ (b) $4\langle 1, 0 \rangle$ **41.** (a) $\left\langle \dfrac{3}{\sqrt{10}}, \dfrac{1}{\sqrt{10}} \right\rangle$

  (b) $\sqrt{10}\left\langle \dfrac{3}{\sqrt{10}}, \dfrac{1}{\sqrt{10}} \right\rangle$ **43.** (a) $\left\langle -\frac{3}{5}, \frac{4}{5} \right\rangle$ (b) $5\left\langle -\frac{3}{5}, \frac{4}{5} \right\rangle$

45. $\frac{9}{5}\mathbf{i} + \frac{12}{5}\mathbf{j}$ **47.** $\langle 2\sqrt{29}, 5\sqrt{29} \rangle$ **49.** $\langle 4, 0 \rangle$

51. 10 pounds down, 20 pounds to the right

53. 190 pounds up, 30 pounds to the right

55. $\langle 13, 17 \rangle$; right and up

57. $\langle -80\sqrt{14}, 20 \rangle$ or 3.8° north of west

59. $\langle 20, 20\sqrt{399} \rangle$ or 2.9° east of north **61.** 10 feet

63. $20\sqrt{101}$ pounds at 5.7° **65.** 7, 1, 5

69. $\|\mathbf{a} + \mathbf{b}\| = \sqrt{58} < \sqrt{13} + \sqrt{17} = \|\mathbf{a}\| + \|\mathbf{b}\|$

71. $\mathbf{a} = c\mathbf{b}$ $(c > 0)$; $\mathbf{a} \perp \mathbf{b}$; $\|\mathbf{a} + \mathbf{b}\|^2 > \|\mathbf{a}\|^2 + \|\mathbf{b}\|^2$ when $\mathbf{a} = c\mathbf{b}$ for $c > 0$ or when the angle between $\mathbf{a}$ and $\mathbf{b}$ in the triangle formed by $\mathbf{a}$, $\mathbf{b}$, and $\mathbf{a} + \mathbf{b}$ is obtuse, $\|\mathbf{a} + \mathbf{b}\|^2 < \|\mathbf{a}\|^2 + \|\mathbf{b}\|^2$ when $\mathbf{a} = c\mathbf{b}$ for $c < 0$ or when the angle between $\mathbf{a}$ and $\mathbf{b}$ in the triangle formed by $\mathbf{a}$, $\mathbf{b}$, and $\mathbf{a} + \mathbf{b}$ is acute, $\|\mathbf{a} + \mathbf{b}\|^2 = \|\mathbf{a}\|^2 + \|\mathbf{b}\|^2$ when $\mathbf{a} \perp \mathbf{b}$.

Exercises Section 10.2

5. (a)

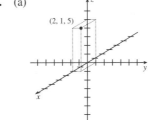

(b)

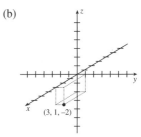

(c)

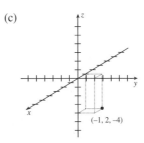

7.

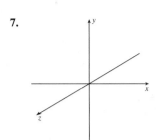

9. 5   **11.** 3   **13.** $\sqrt{38}$

15. $\langle 3, 4, -2 \rangle, \langle -1, -8, -2 \rangle, 2\sqrt{66}$

17. $\langle 3, 3, 4 \rangle, \langle -13, -9, -4 \rangle, 14$

19. $8\mathbf{i} + 4\mathbf{k}, -12\mathbf{i} - 4\mathbf{j} + 4\mathbf{k}, 2\sqrt{186}$

21. (a) $\pm\dfrac{1}{\sqrt{14}}\langle 3, 1, 2 \rangle$ (b) $\sqrt{14}\left\langle \dfrac{3}{\sqrt{14}}, \dfrac{1}{\sqrt{14}}, \dfrac{1}{\sqrt{14}} \right\rangle$

23. (a) $\pm\dfrac{1}{\sqrt{14}}\langle 1, -2, 3 \rangle$ (b) $2\sqrt{14}\left\langle \dfrac{1}{\sqrt{14}}, -\dfrac{2}{\sqrt{14}}, \dfrac{3}{\sqrt{14}} \right\rangle$

25. (a) $\pm\dfrac{1}{3}(2\mathbf{i} - \mathbf{j} + 2\mathbf{k})$ (b) $3\left(\dfrac{2}{3}\mathbf{i} - \dfrac{1}{3}\mathbf{j} + \dfrac{2}{3}\mathbf{k}\right)$

27. (a) $\pm\dfrac{1}{\sqrt{2}}\langle 1, 0, -1 \rangle$ (b) $2\sqrt{2}\left\langle \dfrac{1}{\sqrt{2}}, 0, -\dfrac{1}{\sqrt{2}} \right\rangle$

29. $\langle 4, 4, -2 \rangle$ **31.** $\dfrac{2}{\sqrt{5}}\langle 2, 0, -1 \rangle$

33. $\dfrac{4}{\sqrt{14}}(2\mathbf{i} - \mathbf{j} + 3\mathbf{k})$

35. $(x - 3)^2 + (y - 1)^2 + (z - 4)^2 = 4$

37. $(x - 2)^2 + y^2 + (z + 3)^2 = 9$

39. $(x - \pi)^2 + (y - 1)^2 + (z + 3)^2 = 5$

41. sphere, center $(1, 0, -2)$, radius 2

43. sphere, center $(0, 1, -2)$, radius 3

45. sphere, center $(1, 0, 2)$, radius $\sqrt{5}$

47. plane parallel to xz-plane   **49.** plane parallel to xy-plane

51. plane parallel to yz-plane   **53.** $y = 0$ **55.** $x = 0$

61. $\langle 2, -1, 1 \rangle, \langle 4, -2, 2 \rangle$, yes

63. 1000 pounds up, $\dfrac{150}{\sqrt{2}}$ pounds west, $\dfrac{150}{\sqrt{2}}$ pounds south

Exercises Section 10.3

5. 10   **7.** -8 **9.** -3 **11.** 10   **13.** 1

15. $\cos^{-1}\dfrac{1}{\sqrt{26}} \approx 1.37$ **17.** $\cos^{-1}\dfrac{-2}{\sqrt{10}} \approx 2.26$

19. $\cos^{-1}\dfrac{-8}{\sqrt{234}} \approx 2.12$ **21.** yes   **23.** no   **25.** yes

27. possible answer: $\langle 1, 2 \rangle$ **29.** possible answer: $\langle 1, 4, 0 \rangle$

31. possible answer: $\mathbf{j} + 2\mathbf{k}$ **33.** $2, \left(\frac{6}{5}, \frac{8}{5}\right)$

35. $\frac{9}{5}, \frac{9}{25}(4\mathbf{i} - 3\mathbf{j})$ **37.** $2, \frac{2}{3}\langle 1, 2, 2 \rangle$

39. $-\frac{8}{5}, -\frac{8}{25}\langle 0, -3, 4 \rangle$

41. 105,600 foot-pounds   **45.** 920 foot-pounds

47. (a) false (b) true (c) true (d) false (e) false

49. $\mathbf{a} = c\mathbf{b}$ **53.** $\cos^{-1}\left(-\dfrac{1}{3}\right) \approx 109.5°$

57. $-\dfrac{200}{3\sqrt{14}} \simeq -17.8$ Newtons

59. -347 pounds   **61.** $\mathbf{v} \cdot \mathbf{n} = 0, w\cos\theta$ **63.** 45°

65. \$190,000; monthly revenue

Exercises Section 10.4

5. 1   **7.** 4   **9.** $\langle 4, -3, -2 \rangle$ **11.** $\langle -9, -4, 1 \rangle$

13. $\langle 0, 4, 1 \rangle$ **15.** $\langle 4, -2, 8 \rangle$ **17.** $\pm\dfrac{1}{\sqrt{69}}\langle 8, 1, -2 \rangle$

19. $\pm\dfrac{1}{\sqrt{46}}\langle -3, -6, 1 \rangle$ **21.** $\pm\dfrac{1}{\sqrt{154}}\langle -1, -3, 12 \rangle$

23. $\sin^{-1}\dfrac{7}{\sqrt{85}} \approx 0.86$ **25.** $\sin^{-1}\dfrac{13}{\sqrt{170}} \approx 1.49$

27. $\sqrt{\frac{7}{2}} \approx 1.87$ **29.** $\sqrt{\frac{61}{5}} \approx 3.49$

31. 9.4 foot-pounds   **33.** 10 foot-pounds   **35.** up

37. left   **39.** down, left   **41.** down   **43.** ball rises

45. ball drops, moves left   **47.** ball drops   **49.** no effect

51. ball rises   **53.** ball rises, lands softly

55. ball rises, curves right   **57.** false   **59.** false

61. true   **63.** 5   **65.** $\dfrac{11\sqrt{3}}{2}$ **67.** 10   **69.** 0

71. $-\mathbf{i}$ **73.** $-3\mathbf{j}$ **75.** yes   **77.** no

Exercises Section 10.5

5. (a) $x = 1 + 2t, y = 2 - t, z = -3 + 4t$

(b) $\dfrac{x - 1}{2} = \dfrac{y - 2}{-1} = \dfrac{z + 3}{4}$

7. (a) $x = 2 + 2t, y = 1 - t, z = 3 + t$

(b) $\dfrac{x - 2}{2} = \dfrac{y - 1}{-1} = \dfrac{z - 3}{1}$

9. (a) $x = 1 - 3t, y = 4, z = 1 + t$

(b) $\dfrac{x - 1}{-3} = \dfrac{z - 1}{1}, y = 4$

11. (a) $x = 3 + 3t, y = 1 - 4t, z = -1 + 2t$

(b) $\dfrac{x - 3}{3} = \dfrac{y - 1}{-4} = \dfrac{z + 1}{2}$

13. (a) $x = 2 - 4t, y = -t, z = 1 + 2t$

(b) $\dfrac{x - 2}{-4} = \dfrac{y}{-1} = \dfrac{z - 1}{2}$

15. (a) $x = 1 + 2t, y = 2 - t, z = -1 + 3t$

(b) $\dfrac{x - 1}{2} = \dfrac{y - 2}{-1} = \dfrac{z + 1}{3}$

17. $\cos^{-1}\dfrac{-13}{\sqrt{234}} \approx 2.59$ **19.** perpendicular   **21.** parallel

23. intersect   **25.** parallel

27. $2(x - 1) - (y - 3) + 5(z - 2) = 0$

29. $-3(x + 2) + 2z = 0$

31. $2(x - 2) - 7y - 3(z - 3) = 0$

33. $2(x + 2) + 6(y - 2) - 3z = 0$

35. $3(x - 2) - (y - 1) + 2(z + 1) = 0$

37. $-2x + 4(y + 2) = 0$

39. $(x - 1) - (y - 2) + (z - 1) + 0$

41.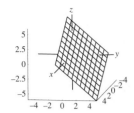

43.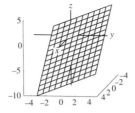

45.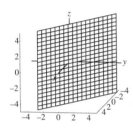

47.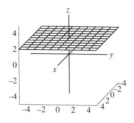

49.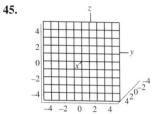

51.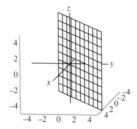

53. $x = t$, $y = \frac{5}{3}t - \frac{4}{3}$, $z = \frac{1}{3}t - \frac{8}{3}$

55. $x = 4t + 11$, $y = -3t - 8$, $z = t$ **57.** $\frac{2}{3}$ **59.** $\frac{2}{\sqrt{3}}$

61. $\frac{3}{\sqrt{6}}$ **65.** $-4(x - 4) + 2(z - 3) = 0$

67. intersect at $(3, 4, 4)$, collide

Exercises Section 10.6

5.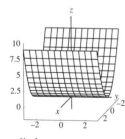

cylinder

7.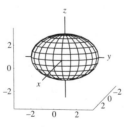

ellipsoid

9.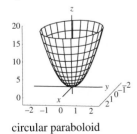

circular paraboloid

11.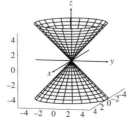

elliptic cone

13.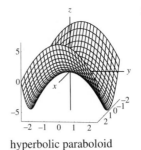

hyperbolic paraboloid

15.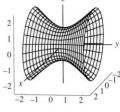

hyperboloid of 1 sheet

17.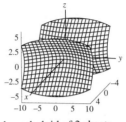

hyperboloid of 2 sheets

19.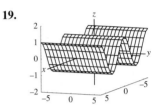

cylinder

21.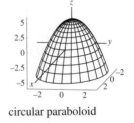

circular paraboloid

23.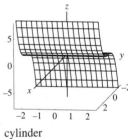

cylinder

25.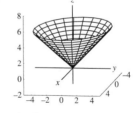

circular cone

27.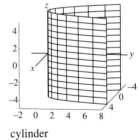

cylinder

29.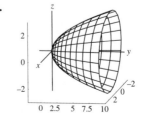

circular paraboloid

31.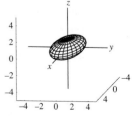

ellipsoid

33.

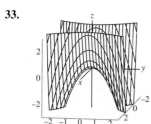
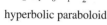

hyperbolic paraboloid

35.

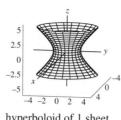

hyperboloid of 1 sheet

37.

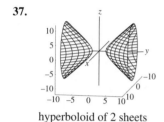

hyperboloid of 2 sheets

39.

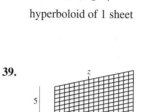

cylinder

41.

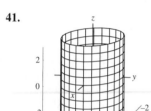

cylinder

43.

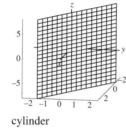

circular paraboloid

45.

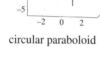

47.

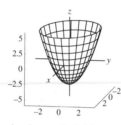

49. $(0, 1, -1)$ and $(0, -1, -1)$

57. exercise 7: $x = \sin s \cos t, y = 3 \sin s \sin t, z = 2 \cos s$;
exercise 9: $x = \frac{1}{2} s \cos t, y = \frac{1}{2} s \sin t, z = s^2$;
exercise 11: $x = \frac{1}{2}\sqrt{s^2} \cos t, y = \sqrt{s^2} \sin t, z = s^2$

59. possible answer: $x = s \cos t, y = s \sin t, z = 4 - s^2$

Chapter 10 Review Exercises

1. $\langle -1, 3 \rangle, \langle 4, 0 \rangle, 5$ **3.** $6\mathbf{i} + 5\mathbf{j}, -16\mathbf{i} + 12\mathbf{j} + 8\mathbf{k}, 2\sqrt{94}$
5. neither   **7.** parallel   **9.** $\langle -1, -2, 3 \rangle$

11. $\left\langle \dfrac{1}{\sqrt{5}}, \dfrac{2}{\sqrt{5}} \right\rangle$ **13.** $\dfrac{1}{3\sqrt{3}}(5\mathbf{i} + \mathbf{j} - \mathbf{k})$

15. $\left\langle -\frac{3}{5}, 0, \frac{4}{5} \right\rangle$ **17.** $\sqrt{46}$ **19.** $\dfrac{2}{\sqrt{3}}(\mathbf{i} - \mathbf{j} + \mathbf{k})$

21. $\langle 20\sqrt{609}, 80 \rangle$ or 9.2° north of east

23. $x^2 + (y + 2)^2 + z^2 = 36$ **25.** 0   **27.** -8

29. $\cos^{-1} \dfrac{1}{\sqrt{84}} \approx 1.46$ **31.** $\dfrac{1}{\sqrt{6}}, \frac{1}{6}(\mathbf{i} + 2\mathbf{j} + \mathbf{k})$

33. $\langle -2, 1, 4 \rangle$ **35.** $-4\mathbf{i} + 4\mathbf{j} - 8\mathbf{k}$

37. $\pm \dfrac{1}{\sqrt{21}}\langle -2, 1, 4 \rangle$ **39.** 1700 foot-pounds   **41.** 3

43. $\sqrt{41}$ **45.** $\frac{25}{2}$ foot-pounds

47. (a) $x = 2 - 2t, y = -1 + 3t, z = -3$

(b) $\dfrac{x - 2}{-2} = \dfrac{y + 1}{3}, z = -3$

49. (a) $x = 2 + 2t, y = -1 + \frac{1}{2}t, z = 1 - 3t$

(b) $\dfrac{x - 2}{2} = 2(y + 1) = \dfrac{z - 1}{-3}$

51. $\cos^{-1} \dfrac{5}{\sqrt{30}} \approx 0.42$ **53.** skew

55. $4(x + 5) + y - 2(z - 1) = 0$

57. $4(x - 2) - (y - 1) + 2(z - 3) = 0$

59.

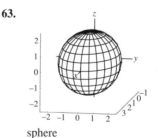

elliptic paraboloid

61.

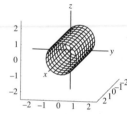

cylinder

63.

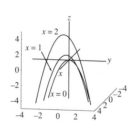

sphere

65.

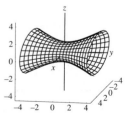

plane

67.

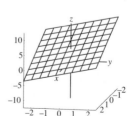

plane

69.

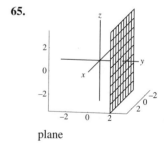

hyperboloid of 1 sheet

71.

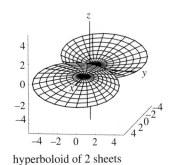

hyperboloid of 2 sheets

CHAPTER 11

Exercises Section 11.1

5.

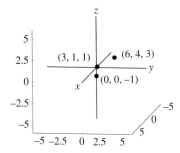

7.

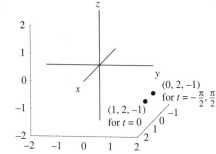

9.

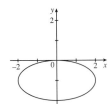

11.

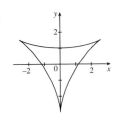

13.

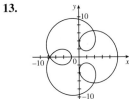

15.

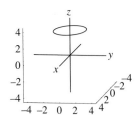

17.

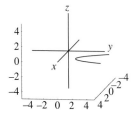

19.

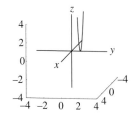

21.

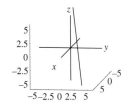

23.

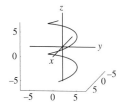

25.

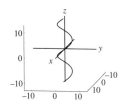

27.

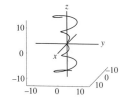

29.

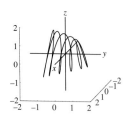

31.

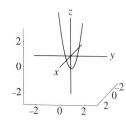

33.

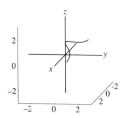

35. (a) F (b) C (c) E (d) A (e) B (f) D
37. 10.54 **39.** 21.56 **41.** 9.57
43. $\cos 2t = \cos t^2 - \sin t^2$
47. same except $-\infty < x < \infty$, $-1 \le x \le 1$, $0 \le x$
49. periodic, not enough points

Exercises Section 11.2

5. $\langle -1, 1, 0 \rangle$ **7.** $\langle 1, 1, -1 \rangle$ **9.** does not exist
11. $t \ne 1$ **13.** $t \ne \dfrac{n\pi}{2}$ (n odd) **15.** $t \ge 0$
17. $\left\langle 4t^3, \dfrac{1}{2\sqrt{t+1}}, -\dfrac{6}{t^3} \right\rangle$ **19.** $\langle \cos t, 2t \cos t^2, -\sin t \rangle$

21. $\langle 2te^{t^2}, 2t, 2\sec 2t \tan 2t \rangle$

23.

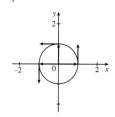

25.

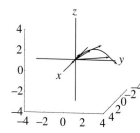

27. $\langle \frac{3}{2}t^2 - t, \frac{2}{3}t^{3/2} \rangle + \mathbf{c}$ **29.** $\langle \frac{1}{3}\sin 3t, -\cos t, \frac{1}{4}e^{4t} \rangle + \mathbf{c}$

31. $\langle \frac{1}{2}e^{t^2}, 3\sin t - 3t\cos t, \frac{3}{2}\ln(t^2 + 1) \rangle + \mathbf{c}$

33. $\langle -\frac{2}{3}, \frac{3}{2} \rangle$ **35.** $\langle 4\ln 3, 1 - e^{-2}, e^2 + 1 \rangle$ **37.** all t

39. $t = 0$ **43.** $t = 0$ **45.** $t = \frac{n\pi}{4}$ (n odd)   **51.** false

Exercises Section 11.3

5. $\langle -10\sin 2t, 10\cos 2t \rangle, \langle -20\cos 2t, -20\sin 2t \rangle$

7. $\langle 25, -32t + 15 \rangle, \langle 0, -32 \rangle$

9. $\langle (4 - 8t)e^{-2t}, -4e^{-2t}, -32t \rangle, \langle (-16 + 16t)e^{-2t}, 8e^{-2t}, -32 \rangle$

11. $\langle 10t + 3, -16t^2 + 4t + 8 \rangle$ **13.** $\langle 5t, -16t^2 + 16 \rangle$

15. $\langle 10t, -3e^{-t} - 3, -16t^2 + 4t + 20 \rangle$

17. $\langle \frac{1}{6}t^3 + 12t + 5, -4t, -8t^2 + 2 \rangle$

19. $-160\langle \cos 2t, \sin 2t \rangle$ **21.** $-960\langle \cos 4t, \sin 4t \rangle$

23. $\langle -120\cos 2t, -200\sin 2t \rangle$ **25.** $\langle 60, 0 \rangle$

27. $\frac{1875}{16} \approx 117$ feet, $\frac{625\sqrt{3}}{4} \approx 271$ feet, 100 ft/s

29. 210 feet, $400 + 40\sqrt{105} \approx 810$ feet, $8\sqrt{410} \approx 162$ ft/s

31. 810 feet, $1600 + 720\sqrt{5} \approx 3210$ feet, $8\sqrt{1610} \approx 321$ ft/s

33. quadruples   **37.** $\langle 60\sqrt{3}t, 3 + 60t - 16t^2 \rangle$, no

39. $\langle 130t, 6 - 16t^2 \rangle$, 2.59 feet   **41.** $\langle 120t, 8 - 16t^2 \rangle$, no

43. 3.86 sec   **45.** $\langle 271, 117, 0 \rangle$

47. $a = 100, b = -1, c = 10$ **49.** 56.57 ft/s

51. 1275.5 m

Exercises Section 11.4

5. $x = 2\cos\left(\frac{s}{2}\right), y = 2\sin\left(\frac{s}{2}\right), 0 \le s \le 4\pi$

7. $x = \frac{3}{5}s, y = \frac{4}{5}s, 0 \le s \le 5$

9. $\langle 1, 0 \rangle, \frac{1}{\sqrt{13}}\langle 3, -2 \rangle, \frac{1}{\sqrt{13}}\langle 3, 2 \rangle$

11. $\langle 0, 1 \rangle, \langle 1, 0 \rangle, \langle -1, 0 \rangle$

13. $\frac{1}{\sqrt{13}}\langle 3, 0, 2 \rangle, \frac{1}{\sqrt{13}}\langle 3, 0, 2 \rangle, \frac{1}{\sqrt{13}}\langle 3, 0, 2 \rangle$

15.

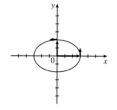

17.

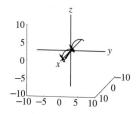

19. $2^{-3/2} \approx 0.3536$ **21.** 0   **23.** $6(37)^{-3/2} \approx 0.0267$

25. 1   **27.** smaller   **29.** $\frac{8}{25}, \frac{8}{25}$ **31.** 1, $\frac{1}{27}$

33. max at $\langle 0, \pm 3 \rangle$, min at $\langle \pm 2, 0 \rangle$

35. max at $(0, -3)$, no min   **37.** 0   **39.** 0

41. curve straightens   **43.** false   **45.** true   **49.** $\frac{2}{3}$, 10

51. $\frac{1}{\sqrt{45}}, \frac{1}{\sqrt{45}}e^{-2}$ **53.** $\frac{25}{52}$, both have constant curvature

Exercises Section 11.5

5. $\langle 1, 0 \rangle$ and $\langle 0, 1 \rangle$, $\frac{1}{\sqrt{5}}\langle 1, 2 \rangle$ and $\frac{1}{\sqrt{5}}\langle -2, 1 \rangle$

7. $\langle 0, 1 \rangle$ and $\langle -1, 0 \rangle$, $\langle -1, 0 \rangle$ and $\langle 0, -1 \rangle$

9. $\frac{1}{\sqrt{5}}\langle 0, 1, 2 \rangle$ and $\langle -1, 0, 0 \rangle$, $\frac{1}{\sqrt{5}}\langle 0, 1, -2 \rangle$ and $\langle 1, 0, 0 \rangle$

11. $\frac{1}{\sqrt{2}}\langle 1, 0, 1 \rangle$ and $\langle 0, 1, 0 \rangle$,

$\frac{1}{\sqrt{6}}\langle 1, 2, 1 \rangle$ and $\frac{1}{\sqrt{3}}\langle -1, 1, -1 \rangle$

13. $x^2 + \left(y - \frac{1}{2}\right)^2 = \frac{1}{4}$ **15.** $x^2 + y^2 = 1$

17. $a_T = -\frac{64}{\sqrt{5}}$ and $a_N = \frac{32}{\sqrt{5}}$, $a_T = \frac{64}{\sqrt{5}}$ and $a_N = \frac{32}{\sqrt{5}}$

19. $a_T = 0$ and $a_N = \sqrt{20}$,

$a_T = \frac{2\pi}{\sqrt{16 + \pi^2}}$ and $a_N = 4\sqrt{\frac{20 + \pi^2}{16 + \pi^2}}$

21. neither; increasing   **23.** $a_T = 0$ and $a_N = a$

25. $\frac{1}{\sqrt{5}}\langle 2, -1, 0 \rangle, \frac{1}{\sqrt{5}}\langle 2, -1, 0 \rangle$

27. $\frac{1}{\sqrt{1 + 16\pi^2}}\langle 0, -1, 4\pi \rangle, \frac{1}{\sqrt{1 + 16\pi^2}}\langle 0, 1, 4\pi \rangle$

29. true   **31.** true   **33.** $10,000\pi^2\langle -\cos\pi t, -\sin\pi t \rangle$

35. $40,000\pi^2\langle -\cos 2\pi t, -\sin 2\pi t \rangle$ **37.** doubles

39. $1 < \frac{3}{2}, |-1| > \left| -\frac{1}{\sqrt{2}} \right|$

Chapter 11 Review Exercises

1.

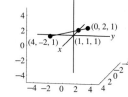

3.

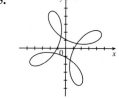

5.

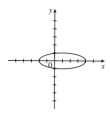

7.

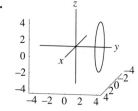

9.
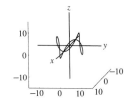

11.
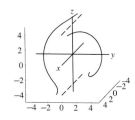

13. (a) B (b) C (c) A (d) F (e) D (f) E
15. $2\pi\sqrt{37}$ **17.** $\langle 0, e^2, -1 \rangle$ **19.** $t \neq 0$

21. $\left\langle \dfrac{t}{\sqrt{t^2+1}}, 4\cos 4t, \dfrac{1}{t} \right\rangle$

23. $\left\langle -\frac{1}{4}e^{-4t}, -t^{-2}, 2t^2 - t \right\rangle + \mathbf{c}$ **25.** $\langle 0, 2, 2 \rangle$
27. $\langle -8\sin 2t, 8\cos 2t, 4 \rangle, \langle -16\cos 2t, -16\sin 2t, 0 \rangle$
29. $\langle t^2 + 4t + 2, -16t^2 + 1 \rangle$
31. $\langle 4t + 2, -16t^2 + 3t + 6 \rangle$
33. $\langle 0, -128 \rangle$
35. $25(2 - \sqrt{3}) \approx 6.70$ feet, 100 feet, 80 ft/s

37. $\dfrac{1}{\sqrt{2}}\langle -1, 1, 0 \rangle, \dfrac{1}{\sqrt{e^{-4}+1}}\langle -e^{-2}, 1, 0 \rangle$

39. $\dfrac{1}{2}, \dfrac{4}{3\sqrt{3}}$ **41.** 0, 0

43. $\dfrac{1}{\sqrt{2}}\langle 0, 1, 1 \rangle, \langle -1, 0, 0 \rangle$

45. $a_T = 0, a_N = 2; a_T = \sqrt{2}, a_N = \sqrt{2}$
47. $345,600\langle -\cos 6t, -\sin 6t \rangle$

CHAPTER 12

Exercises Section 12.1

5. $y \neq -x$ **7.** $x + y + 2 > 0$ **9.** $f \geq 0$
11. $-1 \leq f \leq 1$ **13.** $f \geq -1$ **15.** 3, 3
17. (a) 312 (b) 331 (c) 350 (d) 19 feet

19.
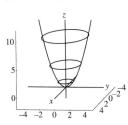

21.
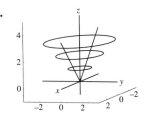

For **23–37**, one view is shown.

23.
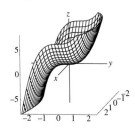

25.
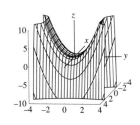

27.
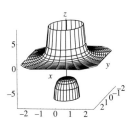

29.
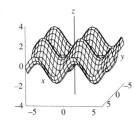

31.
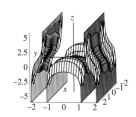

33.
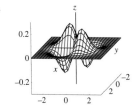

35.
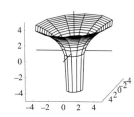

37.
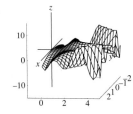

39.
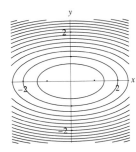

41.
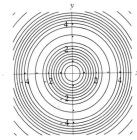

43.
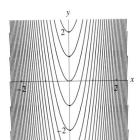

45.
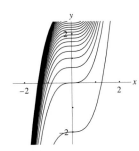

47.
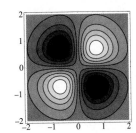

49.
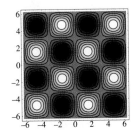

51. (a) B (b) D (c) A (d) F (e) C (f) E
53. (a) A (b) D (c) C (d) B
55.

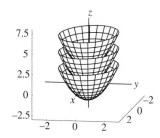

57. (a) B (b) A
59. no height visible   **63.** any point on the line $(x, \sqrt{3}x, 0)$
65. upper left, restaurants, roads
67. left of center, power is increasing away from frame
69. max $= 3.942$, min $= -0.57$, HS
71. 60 mph, impossible

Exercises Section 12.2

5. 3   **7.** $-\frac{1}{2}$ **9.** 2
11. Along $x = 0$, $L_1 = 0$; along $y = 0$, $L_2 = 3$, therefore L does not exist.
13. Along $x = 0$, $L_1 = 0$; along $y = x$, $L_2 = 2$, therefore L does not exist.
15. Along $x = 0$, $L_1 = 0$; along $x = y^2$, $L_2 = 1$, therefore L does not exist.
17. Along $x = 0$, $L_1 = 0$; along $y^3 = x$, $L_2 = \frac{1}{2}$, therefore L does not exist.
19. Along $x = 0$, $L_1 = 0$; along $y = x$, $L_2 = \frac{1}{2}$, therefore L does not exist.
21. Along $x = 1$, $L_1 = 0$; along $y = x + 1$, $L_2 = \frac{1}{2}$, therefore L does not exist.
23. Along $x = 0$, $L_1 = 0$; along $x^2 = y^2 + z^2$, $L_2 = \frac{3}{2}$, therefore L does not exist.
25. Along $y = -x$, L_1 does not exist, therefore L does not exist.
27. 0   **29.** 0   **31.** 2   **33.** 0
39. all x, y **41.** $x^2 + y^2 \le 9$ **43.** $x^2 - y < 3$
45. $y \ne 0$ **47.** $x^2 + y^2 + z^2 \ge 4$ **49.** $\frac{1}{2}$ **51.** true
53. false   **57.** 1   **59.** 0

Exercises Section 12.3

5. $f_x = 3x^2 - 4y^2$, $f_y = -8xy + 4y^3$
7. $f_x = 2xe^y$, $f_y = x^2e^y - 4$
9. $f_x = 2x \sin xy + x^2y \cos xy$, $f_y = x^3 \cos xy - 9y^2$
11. $f_x = \frac{4}{y}e^{x/y} + \frac{y}{x^2}$, $f_y = -\frac{4x}{y^2}e^{x/y} - \frac{1}{x}$
13. $f_x = 3 \sin y + 12x^2y^2z$, $f_y = 3x \cos y + 8x^3yz$, $f_z = 4x^3y^2$
15. $f_x = \frac{-2x}{(x^2 + y^2 + z^2)^{3/2}}$, $f_y = \frac{-2y}{(x^2 + y^2 + z^2)^{3/2}}$,
$f_z = \frac{-2z}{(x^2 + y^2 + z^2)^{3/2}}$

17. $\dfrac{\partial^2 f}{\partial x^2} = 6x$, $\dfrac{\partial^2 f}{\partial y^2} = -8x$, $\dfrac{\partial^2 f}{\partial y \partial x} = -8y$
19. $f_{xx} = 12x^2 - 6y^3$, $f_{xy} = -18xy^2$, $f_{xyy} = -36xy$
21. $f_{xx} = 6xy^2$, $f_{yz} = -\cos yz + yz \sin yz$, $f_{xyz} = 0$
23. $f_{xx} = 4y^2e^{2xy}$, $f_{yy} = 4x^2e^{2xy} - \dfrac{2z^2}{y^3} - xz \sin y$,
$f_{yyzz} = -\dfrac{4}{y^3}$
25. $f_{ww} = 2xy - z^2e^{wz}$, $f_{wxy} = 2w$, $f_{wwxyz} = 0$
27.

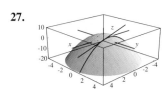

29.

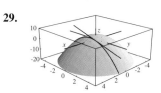

31.

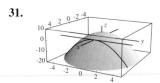

33. $\dfrac{nRV^3}{PV^3 - n^2aV + 2n^3ab}$ Hint: Hold pressure constant.
37. h **39.** $(0, 0, 0) = $ min
41. $\left(\dfrac{m\pi}{2}, \dfrac{n\pi}{2}, 1\right) = $ max for m, n odd;
$\left(\dfrac{m\pi}{2}, \dfrac{n\pi}{2}, -1\right) = $ min for m, n odd;
$(m\pi, n\pi, 0)$ neither max nor min
43. 4, 2   **45.** 1, $-\frac{2}{3}$ **47.** 1.4, -2.4
49. 2.2, 0.0195
53. $\dfrac{\partial^2 f}{\partial x^2} = -n^2\pi^2 \sin n\pi x \cos n\pi ct$,
$\dfrac{\partial^2 f}{\partial t^2} = -c^2n^2\pi^2 \sin n\pi x \cos n\pi ct$
55. $-\dfrac{5}{1 + I}V$, $-\dfrac{0.5}{1 + 0.1(1 - T)}V$, inflation
57. $\cos x \cos t$, $-\sin x \sin t$ **61.** $\left(\dfrac{R}{R_1}\right)^2$
63. 400, $\frac{1}{4}$, decrease by 27
63. concavity of intersection of $z = f(x, y)$ with $y = y_0$ at $x = x_0$

Exercises Section 12.4

5. $4(x - 2) + 2(y - 1) - (z - 4) = 0$ **7.** $z = 1$
9. $-x - z = 0$ **11.** $6(x + 2) + 4(y - 3) - (z - 4) = 0$
13. $-\frac{3}{5}(x + 3) + \frac{4}{5}(y - 4) - (z - 5) = 0$

15. $2(x - 1) - (y - 2) - (z - 2) = 0$ **17.** x

19. $-x$ **21.** $-3x$ **23.** $-8 + 12(x + 2) + y - 8(z - 1)$

25. $11 - 12(w + 2) + 4(x - 3) + 12(y - 1) + 2z$

27. 3 vs. 3.0017, 3.1 vs. 3.1, 3.1 vs. 3.1016

29. 0 vs. 0, -0.1 vs. -0.0998, -0.1 vs. -0.0988

31. 1.5552 ± 0.6307 **33.** 3.85   **35.** 4.03

37. $2y\Delta x + (2x + 2y)\Delta y + (2\Delta y)\Delta x + (\Delta y)\Delta y$

39. $2x\Delta x + 2y\Delta y + (\Delta x)\Delta x + (\Delta y)\Delta y$

41. yes   **43.** $(ye^x + \cos x)\, dx + e^x\, dy$

45. $f_x(0, 0) = f_y(0, 0) = 0$ **49.** $6 + 4x + 2y$

51. $3 + x - \frac{2}{3}y$

53. $-9 + 1.4(t - 10) - 2.4(s - 10)$; -13.4

Exercises Section 12.5

5. $(2t + t^2 + 1 - \cos e^t)e^t$

7. $\dfrac{\partial g}{\partial u} = 512u^6(3u^2 - v\cos u)(u^3 - v\sin u)$
   $+ 1536u^5(u^3 - v\sin u)^2$;

   $\dfrac{\partial g}{\partial v} = 512u^6 \sin u(v\sin u - u^3)$

9. $g'(t) = \dfrac{\partial f}{\partial x}x'(t) + \dfrac{\partial f}{\partial y}y'(t) + \dfrac{\partial f}{\partial z}z'(t)$

11. $\dfrac{\partial g}{\partial u} = \dfrac{\partial f}{\partial x}\dfrac{\partial x}{\partial u} + \dfrac{\partial f}{\partial y}\dfrac{\partial y}{\partial u}$, $\dfrac{\partial g}{\partial v} = \dfrac{\partial f}{\partial x}\dfrac{\partial x}{\partial v} + \dfrac{\partial f}{\partial y}\dfrac{\partial y}{\partial v}$,

   $\dfrac{\partial g}{\partial w} = \dfrac{\partial f}{\partial x}\dfrac{\partial x}{\partial w} + \dfrac{\partial f}{\partial y}\dfrac{\partial y}{\partial w}$

13. -0.6271 **15.** 0.0587

19. $\dfrac{\partial^2 f}{\partial x^2}[x'(t)]^2 + 2\dfrac{\partial^2 f}{\partial y\partial x}x'(t)y'(t) + \dfrac{\partial^2 f}{\partial y^2}[y'(t)]^2$
   $+ \dfrac{\partial f}{\partial x}x''(t) + \dfrac{\partial f}{\partial y}y''(t)$

21. $\dfrac{\partial^2 f}{\partial x^2}\left(\dfrac{\partial x}{\partial u}\right)^2 + \dfrac{\partial^2 f}{\partial y\partial x}\dfrac{\partial y}{\partial u}\dfrac{\partial x}{\partial u} + \dfrac{\partial f}{\partial x}\dfrac{\partial^2 x}{\partial u^2}$
   $+ \dfrac{\partial^2 f}{\partial x\partial y}\dfrac{\partial x}{\partial u}\dfrac{\partial y}{\partial u} + \dfrac{\partial^2 f}{\partial y^2}\left(\dfrac{\partial y}{\partial u}\right)^2 + \dfrac{\partial f}{\partial y}\dfrac{\partial^2 y}{\partial u^2}$

23. $\dfrac{\partial z}{\partial x} = \dfrac{-6xz}{3x^2 + 6z^2 - 3y}$, $\dfrac{\partial z}{\partial y} = \dfrac{3z}{3x^2 + 6z^2 - 3y}$

25. $\dfrac{\partial z}{\partial x} = \dfrac{3yze^{xyz} - 4z^2 + \cos y}{8xz - 3xye^{xyz}}$, $\dfrac{\partial z}{\partial y} = \dfrac{3xze^{xyz} - x\sin y}{8xz - 3xye^{xyz}}$

31. 2 points; halved

Exercises Section 12.6

5. $\langle 2x + 4y^2, 8xy - 5y^4\rangle$

7. $\langle e^{xy^2} + xy^2e^{xy^2}, 2x^2ye^{xy^2} - 2y\sin y^2\rangle$

9. $\langle -8e^{-8} - 2, -16e^{-8}\rangle$ **11.** $\left(\frac{4}{5}, -\frac{3}{5}\right)$ **13.** $\langle 0, 0, -1\rangle$

15. $2 + 6\sqrt{3}$ **17.** $2 - 6\sqrt{3}$ **19.** $\dfrac{17}{5\sqrt{13}}$ **21.** $-3\sqrt{5}$

23. 0   **25.** $-\dfrac{6}{\sqrt{5}}$ **27.** $-\dfrac{12}{\sqrt{5}}$ **29.** $-\dfrac{3}{\sqrt{29}}$

31. $\langle 4, -3\rangle, \langle -4, 3\rangle, 5, -5$

33. $\langle 16, -4\rangle, \langle -16, 4\rangle, \sqrt{272}, -\sqrt{272}$

35. $\langle 1, 0\rangle, \langle -1, 0\rangle, 1, -1$

37. $\left\langle\frac{3}{2}, -\frac{1}{8}\right\rangle, \left\langle-\frac{3}{2}, \frac{1}{8}\right\rangle, \dfrac{\sqrt{145}}{8}, -\dfrac{\sqrt{145}}{8}$

39. $\langle 16, 4, 24\rangle, \langle -16, -4, -24\rangle, \sqrt{848}, -\sqrt{848}$

41. parallel   **45.** $2(x - 1) + 3(y + 1) - z = 0$

47. $-2(x + 1) + 4(y - 2) + 2(z - 1) = 0$

49. $(0, 0, 0), (1, 1, -1), (-1, -1, -1)$

51.   **53.**

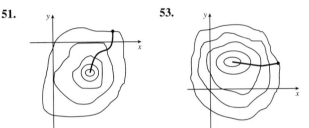

55. possible answer: $\left\langle-\frac{4}{3}, -2\right\rangle$ **57.** $\langle 2, -2\rangle$

59. $\langle -\tan 10°, \tan 6°\rangle \approx \langle -0.176, 0.105\rangle$, 11.6°   **61.** $\langle 8, 4\rangle$

63. $\left\langle 10e^{-8}, \frac{5}{16}e^{-8}, \frac{25}{4}e^{-8}\right\rangle$ **65.** $\langle 0.8, 0.3, -0.004\rangle$

67. $6x + 2$

Exercises Section 12.7

5. $(0, 0)$ saddle

7. $(0, 0)$ saddle, $(1, 1)$ relative min

9. $(0, 1)$ relative min, $(\pm 2, -1)$ saddle

11. $(0, 0)$ max

13. $(0, 0)$ saddle, $(1, 1)$ and $(-1, -1)$ relative min

15. $(\pm\sqrt{2}, 0)$ saddle

17. $(0, 0)$ saddle, $\pm\left(\sqrt{\frac{1}{2}}, \sqrt{\frac{1}{2}}\right)$ relative max,

   $\pm\left(\sqrt{\frac{1}{2}}, -\sqrt{\frac{1}{2}}\right)$ relative min

19. $(2.82, 0.17)$ relative min, $(-2.84, -0.18)$ saddle, $(0.51, 0.99)$ saddle

21. $(\pm 1, 0)$ relative max, $\left(0, -\sqrt{\frac{3}{2}}\right)$ relative max, $\left(0, \sqrt{\frac{3}{2}}\right)$ relative min, $(0, 0)$ saddle, $\left(\pm\dfrac{\sqrt{19}}{3\sqrt{3}}, -\dfrac{2}{3}\right)$ saddle

23. $1.37x - 2.80$ **25.** 9176   **27.** 247, 104

29. (a) 1.29 (b) 2.75

31. $(-0.3210, -0.5185), (-0.1835, -0.4269)$

33. $(0.9044, 0.8087), (3.2924, -0.3853)$

35. $(0, 0)$ is a saddle point

37. $f(2, 0) = 4$, $f(2, 2) = -2$

39. $f(3, 0) = 9$, $f(0, 0) = 0$

41. $f(0, y) = f(x, 0) = 0$ min

43. $f(-1, 2) = -4$ min   **51.** false   **53.** false

55. extrema at $\left(\pm\dfrac{\pi}{2}, \pm\dfrac{\pi}{2}\right)$, saddles at $(\pm n\pi, \pm n\pi)$

57. extrema at $\pm(1, 1)$, saddle $(0, 0)$

59. extrema $(\pm 0.1, 0.1)$, saddle $(0, 0)$

61. $d(x, y) = \sqrt{(x-3)^2 + (y+2)^2 + (3 - x^2 - y^2)^2}$, $(1.55, -1.03)$

63. $(1.6, 0.8, -2.4)$ **65.** $(1, 0), f(1, 0) < f(-10, 0)$

Exercises Section 12.8

5. $x = \frac{6}{5}, y = -\frac{2}{5}$ **7.** $x = 2, y = -1$

9. $x = 1, y = 1$ **11.** $x = 1, y = 1$

13. $\max = f(2, 2) = f(-2, -2) = 16$, $\min = f(2, -2) = f(-2, 2) = -16$

15. $\max = f(\pm\sqrt{2}, 1) = 8, \min = f(\pm\sqrt{2}, -1) = -8$

17. $\max = f(1, 1) = e, \min = f(-1, 1) = -e$

19. $\max = f(\pm\sqrt{2}, 1) = 2e, \min = f(0, \pm\sqrt{3}) = 0$

21. $\max = f(2, 2) = f(-2, -2) = 16$ $\min = f(2, -2) = f(-2, 2) = -16$

23. $\max = f(\pm\sqrt{2}, 1) = 8, \min = f(\pm\sqrt{2}, -1) = -8$

25. $u = \frac{128}{3}, z = 195$ feet   **27.** $P(20, 80, 20) = 660$

29. $P\left(\sqrt{\frac{8801}{22}}, 4\sqrt{\frac{8801}{22}}, \sqrt{\frac{8801}{22}}\right) \approx 660.0374989$ $660 + \lambda = 660.0375$

31. $x = y$ **33.** $f\left(-\dfrac{\sqrt{2}}{2}, \dfrac{\sqrt{2}}{2}\right) = \sqrt{2}$

35. $\alpha = \beta = \theta = \dfrac{\pi}{6}; f\left(\dfrac{\pi}{6}, \dfrac{\pi}{6}, \dfrac{\pi}{6}\right) = \dfrac{1}{8}$

39. $L = 50, K = 10$ **41.** $C(L, K) = C(64, 8) = 2400$

43. $f(4, -2, 2) = 24$ **45.** $f(1, 1, 2) = 2$

47. (a) $(\pm 1, 0, 0)$ (b) $(0, \pm 1, \pm 1)$

Chapter 12 Review Exercises

1.

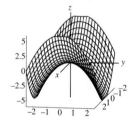

3.

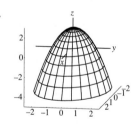

5.

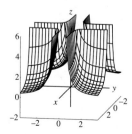

7.

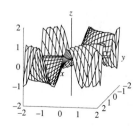

9.

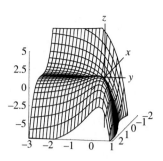

11. (a) D   (b) B   (c) C   (d) A   (e) F   (f) E

13. (a) C   (b) A   (c) D   (d) B   **21.** $x \neq 0$

23. $f_x = \dfrac{4}{y} + (1 + xy)e^{xy}, f_y = -\dfrac{4x}{y^2} + x^2 e^{xy}$

25. $f_x = 6xy\cos y - \dfrac{1}{2\sqrt{x}}, f_y = 3x^2\cos y - 3x^2 y\sin y$

29. $-0.04, 0.06$ **31.** $45 - 10(x + 2) + 9(y - 5)$

33. $(x - \pi) + 2\left(y - \dfrac{\pi}{2}\right)$

35. $f_{xx} = 24x^2 y + 6y^2, f_{yy} = 6x^2, f_{xy} = 8x^3 + 12xy$

37. $3(y + 1) - z = 0$

39. $4x + 4(y - 2) + 2(z - 1) = 0$

41. $8e^{8t}\sin t + (e^{8t} + 2\sin t)\cos t$

43. $g'(t) = \dfrac{\partial f}{\partial x}x'(t) + \dfrac{\partial f}{\partial y}y'(t) + \dfrac{\partial f}{\partial z}z'(t) + \dfrac{\partial f}{\partial w}w'(t)$

45. $\dfrac{\partial z}{\partial x} = -\dfrac{x + y}{z}, \dfrac{\partial z}{\partial y} = -\dfrac{x + y}{z}$ **47.** $\left\langle -\frac{1}{2}, 12\pi - \frac{1}{2}\right\rangle$

49. -4 **51.** $-\dfrac{7}{\sqrt{5}}$ **53.** $\pm\dfrac{1}{\sqrt{145}}\langle 9, -8\rangle, \pm 4\sqrt{145}$

55. $\pm\langle 1, 0\rangle, \pm 4$ **57.** $\langle 16, 2\rangle$

59. $(0, 0)$ relative minimum, $(2, \pm 8)$ saddles

61. $\left(\frac{4}{3}, \frac{4}{3}\right)$ relative max, $(0, 0)$ saddle   **63.** $212, 112$

65. $f(4, 0) = 512, f(0, 0) = 0$

67. $f(1, 2) = 5, f(-1, -2) = -5$

69. $f\left(\sqrt{\frac{1}{2}}, \sqrt{\frac{1}{2}}\right) = f\left(-\sqrt{\frac{1}{2}}, -\sqrt{\frac{1}{2}}\right) = \frac{1}{2}$, $f\left(\sqrt{\frac{1}{2}}, -\sqrt{\frac{1}{2}}\right) = f\left(-\sqrt{\frac{1}{2}}, \sqrt{\frac{1}{2}}\right) = -\frac{1}{2}$

71. $(1, 1)$ **73.** decreases, increases

CHAPTER 13

Exercises Section 13.1

5. 6 **7.** $\frac{13}{2}$ **9.** -12 **11.** 40 **13.** $\frac{16}{3}$

15. $12(e^2 - 1)$ **17.** $\frac{19}{2} - \frac{1}{2}e^6$

19.

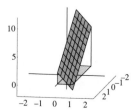

21.

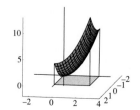

23. 2   **25.** $\frac{128}{3}$ **27.** $\frac{62}{21}$ **29.** $e^4 - 1$

31. $2 \ln 2 \sin 1$ **33.** $\frac{1}{2} \neq \frac{2}{3}$

35. $\int_0^3 \int_1^4 (x^2 + y^2)\, dy\, dx = 90$

37. $\int_{-1}^1 \int_{x^2}^1 (x^2 + y^2)\, dy\, dx = \frac{88}{105}$

39. $\int_{-2}^2 \int_0^{4-y^2} (6 - x - y)\, dx\, dy = \frac{704}{15}$

41. $\int_0^2 \int_0^x y^2\, dy\, dx = \frac{4}{3}$ **43.** -1.5945 **45.** 1.6697

47. $\int_0^2 \int_{y/2}^1 f(x, y)\, dx\, dy$ **49.** $\int_0^4 \int_0^{x/2} f(x, y)\, dy\, dx$

51. $\int_1^4 \int_0^{\ln y} f(x, y)\, dx\, dy$

53. $\int_0^2 \int_0^y 2e^{y^2}\, dx\, dy = e^4 - 1$

55. $\int_0^1 \int_0^x 3xe^{x^3}\, dy\, dx = e - 1$

59. 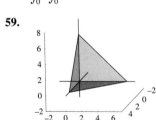   **61.**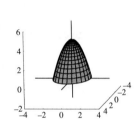

63. different domains

Exercises Section 13.2

5. $\int_{-2}^2 \int_{x^2}^{8-x^2} 1\, dy\, dx = \frac{64}{3}$ **7.** $\int_0^2 \int_{y/2}^{3-y} 1\, dx\, dy = 3$

9. $\int_0^1 \int_{x^2}^{\sqrt{x}} 1\, dy\, dx = \frac{1}{3}$ **11.** 6   **13.** $\frac{40}{3}$

15. $\frac{1}{2}$ **17.** $\frac{5}{12}$ **19.** $\frac{8576}{105}$ **21.** $\frac{36}{5}$

23. $\int_0^2 \int_0^{4-x^2} \sqrt{x^2 + y^2}\, dy\, dx = 10.275$

25. $\int_0^4 \int_0^{2-x/2} e^{xy}\, dy\, dx = 9.003$

27. $m = \frac{1}{3}, \bar{x} = \frac{3}{5}, \bar{y} = \frac{12}{35}$
29. $m = \frac{12}{5}, \bar{x} = \frac{41}{63}, \bar{y} = 0$ **31.** $m = 4, \bar{x} = \frac{16}{15}, \bar{y} = \frac{8}{3}$
33. In exercise 30, $\rho(x, y)$ is not x-axis symmetric.
35. $\rho(-x, y) = \rho(x, y)$ **37.** 1164
39. $1200(1 - e^{-2/3}) \approx 583.899$
41. $m = \frac{32}{3}, I_y = \frac{128}{15}, I_x = \frac{512}{7}$
43. $I_y = \frac{68}{3}, \frac{5}{3}$; second spin rate 13.6 times faster
45. $100.531, 508.938$

47. $\frac{12}{5}$ **49.** same
51. 3.792   **53.** 50.113
55. (a) total rainfall in region
    (b) average rainfall per unit area in region
57. $\bar{y} = \frac{1}{3}$

Exercises Section 13.3

5. 11π **7.** π **9.** $\frac{\pi}{12}$ **11.** $\frac{\pi}{9} + \frac{\sqrt{3}}{6}$ **13.** 18π

15. $\pi - \pi e^{-4}$ **17.** 0   **19.** $\frac{81\pi}{2}$ **21.** $\frac{16}{3}$

23. $\frac{81\pi}{2}$ **25.** $\frac{16\pi}{3}$ **27.** $\frac{8 - 3\sqrt{3}}{12}\pi$ **29.** 36

31. π **33.** $\int_0^{2\pi} \int_0^2 r^2\, dr\, d\theta = \frac{16\pi}{3}$

35. $\int_{-\pi/2}^{\pi/2} \int_0^2 re^{-r^2}\, dr\, d\theta = \frac{\pi}{2}(1 - e^{-4})$

37. $\int_{\pi/4}^{\pi/2} \int_0^{2\sqrt{2}} r^4\, dr\, d\theta = \frac{32\pi}{5}$ **39.** $1 - e^{-1/16} \approx 0.06$

41. $\frac{1}{20}(e^{-225/16} - e^{-16}) \approx 0.000000033$ **43.** $\frac{31\pi}{320}$

45. $\bar{x} = 0, \bar{y} = \frac{2}{3}$ **47.** $20{,}000\pi(1 - e^{-1}) \approx 39{,}717$

49. $\frac{\pi r^4}{4}$ **51.** $V = 2 \int_0^{2\pi} \int_0^a \sqrt{a^2 - r^2}\, dr\, d\theta = \frac{4\pi a^3}{3}$

Exercises Section 13.4

3. 46.831   **5.** 36.177   **7.** $4\sqrt{2}\pi$ **9.** 583.769
11. $6\sqrt{11}$ **13.** $4\sqrt{6}$ **15.** 0.931   **17.** 37.174
19. 25.133   **21.** $\sqrt{2}A$ **23.** $\frac{A}{|\cos\theta|}$ **25.** $4L$

Exercises Section 13.5

5. 16   **7.** $-\frac{2}{3}$ **9.** $\frac{4}{15}$ **11.** $\frac{171}{5}$ **13.** 0

15. $\frac{128\pi}{3}$ **17.** 00   **19.** symmetry, yes

21. $\int_0^2 \int_{-1}^1 \int_{x^2}^1 dz\, dx\, dy = \frac{8}{3}$

23. $\int_{-1}^1 \int_0^{1-y^2} \int_{2-z/2}^4 dx\, dz\, dy = \frac{44}{15}$

25. $\int_{-2}^2 \int_{-\sqrt{4-y^2}}^{\sqrt{4-y^2}} \int_{y^2+z^2}^4 dx\, dz\, dy = 8\pi$

27. $\int_{-3}^3 \int_{-\sqrt{9-x^2}}^{\sqrt{9-x^2}} \int_{\sqrt{x^2+z^2}}^3 dy\, dz\, dx = 9\pi$

29. $\int_{-\sqrt{10}}^{\sqrt{10}} \int_{-6}^{4-x^2} \int_0^{y+6} dz\, dy\, dx = \frac{160\sqrt{3}}{10}$

31. $\int_{-1}^1 \int_{x^2}^1 \int_0^{3-x} dy\, dz\, dx = 4$

33. $m = 32\pi, \bar{x} = \bar{y} = 0, \bar{z} = \frac{8}{3}$

35. $m = 138, \overline{x} = \frac{186}{115}, \overline{y} = \frac{56}{115}, \overline{z} = \frac{168}{115}$

37. right side is heaver in #34

39. $\int_0^1 \int_0^{2-2y} \int_0^{2-x-2y} 4yz \, dz \, dx \, dy$

$= \int_0^1 \int_0^{2-2y} \int_0^{2-2y-z} 4yz \, dx \, dz \, dy$

$= \int_0^2 \int_0^{2-x} \int_0^{1-x/2-z/2} 4yz \, dy \, dz \, dx$

41. $\int_0^2 \int_0^{4-2y} \int_0^{4-2y-x} dz \, dx \, dy$

43. $\int_0^1 \int_0^{\sqrt{1-x^2}} \int_0^{\sqrt{1-x^2-z^2}} dy \, dz \, dx$

45. $\int_0^2 \int_{x^2}^4 \int_0^{\sqrt{y-x^2}} dz \, dy \, dx$

Exercises Section 13.6

5. $r = 4$ **7.** $r = 4\cos\theta$ **9.** $z = r^2$

11. $z = \cos(r^2)$ **13.** $\theta = \frac{\pi}{4}$

15. $\int_0^{2\pi} \int_0^2 \int_r^{\sqrt{8-r^2}} rf(r\cos\theta, r\sin\theta, z) \, dz \, dr \, d\theta$

17. $\int_0^{2\pi} \int_0^3 \int_0^{9-r^2} rf(r\cos\theta, r\sin\theta, z) \, dz \, dr \, d\theta$

19. $\int_0^{2\pi} \int_0^2 \int_{r^2}^4 rf(r\cos\theta, r\sin\theta, z) \, dz \, dr \, d\theta$

21. $\int_0^{2\pi} \int_0^2 \int_0^{4-r^2} rf(r\cos\theta, y, r\sin\theta) \, dy \, dr \, d\theta$

23. $\int_0^{2\pi} \int_0^1 \int_{r^2}^{2-r^2} rf(x, r\cos\theta, r\sin\theta) \, dx \, dr \, d\theta$

25. $\int_0^{2\pi} \int_0^2 \int_1^2 re^{r^2} \, dz \, dr \, d\theta = \pi(e^4 - 1)$

27. $\int_0^2 \int_0^{3-3z/2} \int_0^{6-2y-3z} (x+z) \, dx \, dy \, dz = 12$

29. $\int_0^{2\pi} \int_0^{\sqrt{2}} \int_r^{\sqrt{4-r^2}} zr \, dz \, dr \, d\theta = 2\pi$

31. $\int_0^2 \int_0^{4-2y} \int_0^{4-x-2y} (x+y) \, dz \, dx \, dy = 8$

33. $\int_0^{2\pi} \int_0^3 \int_0^{r^2} re^z \, dz \, dr \, d\theta = \pi(e^9 - 10)$

35. $\int_0^{\pi} \int_0^{2\sin\theta} \int_0^r 2r^2\cos\theta \, dz \, dr \, d\theta = 0$

37. $\int_0^{2\pi} \int_0^1 \int_0^r 3z^2 r \, dz \, dr \, d\theta = \frac{2\pi}{5}$

39. $\int_0^{\pi} \int_0^2 \int_r^{\sqrt{8-r^2}} 2r \, dz \, dr \, d\theta = \frac{32\pi}{3}(\sqrt{2} - 1)$

41. $\int_0^{2\pi} \int_0^3 \int_0^{r^2} r^3 \, dy \, dr \, d\theta = 243\pi$

43.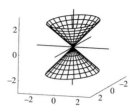

45.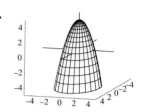

47.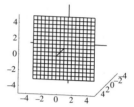

49.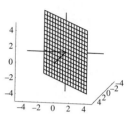

51. $m = \frac{128\pi}{3}, \overline{x} = \overline{y} = 0, \overline{z} = \frac{16}{5}$

53. $m = 10\pi, \overline{x} = 0, \overline{y} = \frac{4}{5}, \overline{z} = \frac{38}{15}$

Exercises Section 13.7

5. $(0, 0, 4)$ **7.** $(4, 0, 0)$ **9.** $(\sqrt{2}, 0, \sqrt{2})$

11. $\left(\frac{\sqrt{2}}{2}, \frac{\sqrt{2}}{2}, 1\right)$ **13.** $\rho = 3$ **15.** $\theta = \frac{\pi}{4}$ or $\frac{5\pi}{4}$

17. $\rho\cos\phi = 2$ **19.** $\phi = \frac{\pi}{6}$

21. 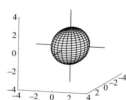   **23.**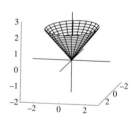

25. 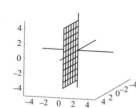   **27.**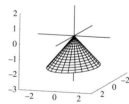

29.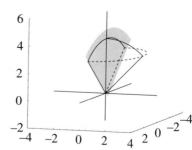

31.

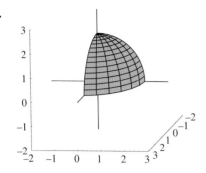

33.

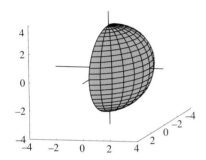

35.

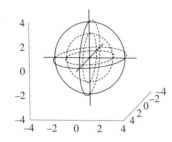

37. $\int_0^{2\pi} \int_0^{\pi/2} \int_0^2 e^{\rho^3} \rho^2 \sin\phi \, d\rho \, d\phi \, d\theta = \frac{2}{3}\pi(e^8 - 1)$

39. $\int_0^{2\pi} \int_0^{\pi/2} \int_0^{\sqrt{2}} \rho^7 \sin\phi \, d\rho \, d\phi \, d\theta = 4\pi$

41. $\int_0^1 \int_1^2 \int_3^4 (x^2 + y^2 + z^2) \, dz \, dy \, dx = 15$

43. $\int_0^{2\pi} \int_0^2 \int_0^{4-r^2} r^3 \, dz \, dr \, d\theta = \frac{32\pi}{3}$

45. $\int_0^{2\pi} \int_0^{\pi/4} \int_0^{\sqrt{2}} \rho^3 \sin\phi \, d\rho \, d\phi \, d\theta = (2 - \sqrt{2})\pi$

47. $\int_0^{2\pi} \int_0^{\pi/4} \int_0^{4\cos\phi} \rho^2 \sin\phi \, d\rho \, d\phi \, d\theta = 8\pi$

49. $\int_0^{2\pi} \int_0^4 \int_r^4 r \, dz \, dr \, d\theta = \frac{64\pi}{3}$

51. $\int_{-1}^1 \int_{-1}^1 \int_0^{\sqrt{x^2+y^2}} dz \, dy \, dx = 3.061$

53. $\int_0^{\pi/2} \int_0^{\pi/4} \int_0^2 \rho^2 \sin\phi \, d\rho \, d\phi \, d\theta = \frac{4 - 2\sqrt{2}}{3}\pi$

55. $\int_0^{2\pi} \int_0^2 \int_0^r r \, dz \, dr \, d\theta = \frac{16\pi}{3}$

57. $\int_{-\pi/2}^{\pi/2} \int_0^{\pi} \int_0^1 \rho^3 \sin\phi \, d\rho \, d\phi \, d\theta = \frac{\pi}{2}$

59. $\int_0^{\pi} \int_0^{\pi/4} \int_0^{\sqrt{2}} \rho^5 \sin\phi \, d\rho \, d\phi \, d\theta = \frac{256 - 128\sqrt{2}}{3}\pi$

61. $\bar{x} = \bar{y} = 0, \bar{z} = \frac{3}{4} + \frac{3\sqrt{2}}{8}$

Exercises Section 13.8

3. $x = \frac{1}{6}(v - u), y = \frac{1}{3}(u + 2v), 2 \le u \le 5, 1 \le v \le 3$

5. $x = \frac{1}{4}(u - v), y = \frac{1}{4}(u + 3v), 1 \le u \le 3, -3 \le v \le -1$

7. $x = r\cos\theta, y = r\sin\theta, 1 \le r \le 2, 0 \le \theta \le \frac{\pi}{2}$

9. $x = r\cos\theta, y = r\sin\theta, 2 \le r \le 3, \frac{\pi}{4} \le \theta \le \frac{3\pi}{4}$

11. $x = \sqrt{\frac{1}{2}(v - u)}, y = \frac{1}{2}(u + v), 0 \le u \le 2, 2 \le v \le 4$

13. $x = \ln\left(\frac{1}{2}(v - u)\right), y = \frac{1}{2}(u + v), 0 \le u \le 1, 3 \le v \le 5$

15. $\frac{7}{2}$ **17.** $\frac{13}{3}$ **19.** $\frac{7}{3}$ **21.** $\frac{1}{6}(e^5 - e^2)\ln 3$

23. $\frac{9}{4}$ **25.** $-2u$ **27.** 2

29. $x = u - w, y = \frac{1}{2}(-u + v + w), z = \frac{1}{2}(u - v + w),$
$1 \le u \le 2, 0 \le v \le 1, 2 \le w \le 4$ **31.** 1

Chapter 13 Review Exercises

1. 18 **3.** 207 **5.** $(e^{-1} - e^{-4})\pi$ **7.** $\frac{2}{3}$ **9.** 0

11. -19.92 **13.** $\frac{4}{3}$ **15.** 16π **17.** $\frac{128}{3}$

19. $\frac{64\pi}{3}$ **21.** $\frac{1}{3}(16 - 8\sqrt{2})\pi$ **23.** $\frac{11\pi}{2}$

25. $\int_0^4 \int_{\sqrt{y}}^2 f(x, y) \, dx \, dy$

27. $\int_{-\pi/2}^{\pi/2} \int_0^2 2r^2 \cos\theta \, dr \, d\theta = \frac{32}{3}$

29. $m = \frac{16}{3}, \bar{x} = \frac{3}{2}, \bar{y} = \frac{9}{4}$

31. $m = \frac{64}{15}, \bar{x} = 0, \bar{y} = \frac{23}{28}, \bar{z} = \frac{5}{14}$

33. $\int_0^1 \int_{\sqrt{y}}^{2-y} dx \, dy = \frac{5}{6}$

35. $\frac{1}{2}$ **37.** $2\sqrt{21}$ **39.** $\frac{13\pi}{3}$ **41.** $16\pi\sqrt{2}$

43. $\int_0^2 \int_{-1}^1 \int_{-1}^1 z(x + y) \, dz \, dy \, dx = 0$

45. $\int_0^{2\pi} \int_0^{\pi/4} \int_0^2 \rho^3 \sin\phi \, d\rho \, d\phi \, d\theta = \left(8 - 4\sqrt{2}\right)\pi$

47. $\int_0^2 \int_x^2 \int_0^{6-x-y} f(x, y, z)\, dz\, dy\, dx$

49. $\int_0^{2\pi} \int_0^{\pi/2} \int_0^2 f(\rho \sin\phi \cos\theta, \rho \sin\phi \sin\theta, \rho \cos\phi)$
$\qquad \cdot \rho^2 \sin\phi\, d\rho\, d\phi\, d\theta$

51. $\int_{\pi/4}^{\pi/2} \int_0^{\sqrt{2}} \int_0^r e^z r\, dz\, dr\, d\theta = \dfrac{e^{\sqrt{2}}(\sqrt{2}-1)\pi}{4}$

53. $\int_0^\pi \int_0^{\pi/4} \int_0^{\sqrt{2}} \rho^3 \sin\phi\, d\rho\, d\phi\, d\theta = \left(1 - \dfrac{\sqrt{2}}{2}\right)\pi$

55. (a) $r \sin\theta = 3$ (b) $\rho \sin\phi \sin\theta = 3$

57. (a) $r^2 + z^2 = 4$ (b) $\rho = 2$

59. (a) $z = r$ (b) $\phi = \dfrac{\pi}{4}$

61.

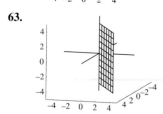

63.

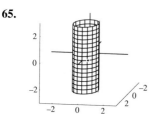

65.

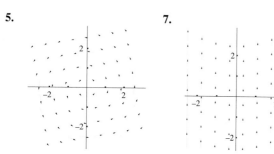

67. $x = \frac{1}{4}(v - u), y = \frac{1}{2}(u + v), -1 \le u \le 1, 2 \le v \le 4$

69. $\frac{1}{2}(e - e^{-1})$ **71.** $4uv^2 - 4u^2$

CHAPTER 14

Exercises Section 14.1

5. 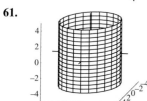   **7.**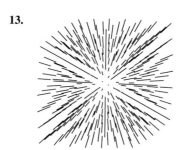

9. 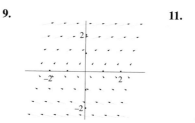   **11.**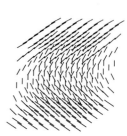

13.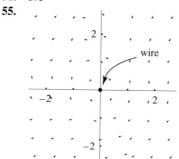

15. $\mathbf{F}_1 = D, \mathbf{F}_2 = B, \mathbf{F}_3 = A, \mathbf{F}_4 = C$ **17.** $\langle 2x, 2y \rangle$

19. $\dfrac{\langle x, y \rangle}{\sqrt{x^2 + y^2}}$ **21.** $\langle e^{-y}, -xe^{-y} \rangle$ **23.** $\dfrac{\langle x, y, z \rangle}{\sqrt{x^2 + y^2 + z^2}}$

25. $\langle 2xy, x^2 + z, y \rangle$ **27.** $f(x, y) = xy + c$ **29.** not

31. $f(x, y) = \frac{1}{2}x^2 - x^2 y + \frac{1}{3}y^3 + c$

33. $f(x, y) = -\cos xy + c$

35. $f(x, y, z) = 2x^2 - xz + \frac{3}{2}y^2 + yz + c$ **37.** not

39. $y = \frac{1}{2}\sin x + c$ **41.** $y^2 = x^3 + c$

43. $(y + 1)e^{-y} = -\frac{1}{2}x^2 + c$ **45.** $y^2 + 1 = ce^{2x}$

47. $f(x, y, z) = \int_0^x f(u)\, du + \int_0^y g(u)\, du + \int_0^z h(u)\, du + c$

51. $3r\mathbf{r}$

55.

wire

Exercises Section 14.2

5. $4\sqrt{13}$ **7.** $\frac{21}{2}\sqrt{17}$ **9.** 4   **11.** 12   **13.** -4

15. 3π **17.** 25.41 **19.** -4 **21.** 14   **23.** $\frac{9}{2}$

25. $6\sqrt{6}$ **27.** -4 **29.** 31   **31.** 0   **33.** $\frac{8}{3}$

35. 26   **37.** 0   **39.** $4\pi - \frac{19}{3}$ **41.** positive

43. zero   **45.** negative   **47.** 18.67

49. $\overline{x} = 2.227, \overline{y} = 5.324$

51. 99.41   **53.** 359.9   **55.** $\dfrac{\pi^3}{3}\sqrt{5}$ **59.** 4π

61. $\frac{32}{3}$ **63.** 12   **65.** (a) 22.1   (b) 15.35   (c) 3.65

Exercises Section 14.3

5. $f(x, y) = x^2 y - x + c$ **7.** $f(x, y) = \dfrac{x}{y} - x^2 + \dfrac{1}{2}y^2 + c$

9. not   **11.** $f(x, y) = e^{xy} + \sin y + c$

13. $f(x, y, z) = xz^2 + x^2 y + y - 3z + c$

15. $f(x, y, z) = xy^2 z^2 + \frac{1}{2}x^2 + \frac{1}{2}y^2 + c$

17. $f(x, y) = x^2 y - y; 8$ **19.** $f(x, y) = e^{xy} - y^2; -16$

21. $f(x, y, z) = xz^2 + x^2 y; -38$ **23.** $\frac{152}{3}$ **25.** 18

27. $\sqrt{30} - \sqrt{14}$ **29.** -2 **31.** $10 - e^{18}$ **33.** 0

35. yes   **37.** no   **39.** no   **45.** false   **47.** true

49. $\tan^{-1}\left(\dfrac{y}{x}\right) + c, x \neq 0; 0$

51. (a) Simply connected   (b) Not simply connected

Exercises Section 14.4

5. π **7.** 16   **9.** -54 **11.** $\frac{32}{3}$ **13.** 6π

15. $\frac{1}{3}$ **17.** $\frac{4}{3} + \frac{1}{2}e^2 + \frac{3}{2}e^{-2}$ **19.** $\frac{32}{5}$ **21.** 4

23. 0   **25.** 8π **27.** $\frac{3}{8}\pi$ **29.** $\frac{32}{3}$

33. $\bar{x} = 0, \bar{y} = \frac{4}{7}$ **37.** 0   **39.** 0

41. $\{(x, y) \in \mathbb{R}^2 | (x, y) \neq (0, 0)\}$; No

Exercises Section 14.5

5. $\langle 0, 0, -3y \rangle, -x$ **7.** $\langle -3, 2x, 0 \rangle, 2z$

9. $\langle -y, -2x, -x \rangle, y + z$ **11.** $\langle xe^y + 1, -e^y, 0 \rangle, 2x + 1$

13. $\langle -x \sin y, 3y - \cos y, 2x - 3z \rangle, 0$

15. $\langle 2yz - 2z, 2x, 0 \rangle, 2z + 1 + y^2$ **17.** conservative

19. incompressible   **21.** conservative

23. neither   **25.** incompressible

27. conservative   **29.** conservative

31. (a) scalar   (b) undefined   (c) undefined
(d) vector   (e) vector

35. positive   **37.** zero   **39.** negative

51. $-\dfrac{x^2}{(x^2 + y^2 + z^2)^{3/2}} - \dfrac{y^2}{(x^2 + y^2 + z^2)^{3/2}}$
$-\dfrac{z^2}{(x^2 + y^2 + z^2)^{3/2}} + \dfrac{3}{\sqrt{x^2 + y^2 + z^2}}$

53. (a) Equal   (b) Greater

55. $\mathbf{F} = z\mathbf{i} + y^2\mathbf{j} + x\mathbf{k}, \nabla \cdot \mathbf{F} = 2y$

Exercises Section 14.6

5. $x = x, y = y, z = 3x + 4y$

7. $x = \cos u \cosh v, y = \sin u \cosh v, z = \sinh v,$
$0 \leq u \leq 2\pi, -\infty < v < c$

9. $x = 2\cos\theta, y = 2\sin\theta, z = z, 0 \leq \theta \leq 2\pi, 0 \leq z \leq 2$

11. $x = r\cos\theta, y = r\sin\theta, z = 4 - r^2,$
$0 \leq \theta \leq 2\pi, 0 \leq r \leq 2$

13.

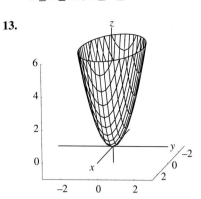

15.

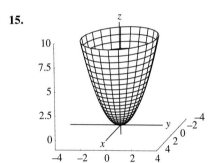

17.

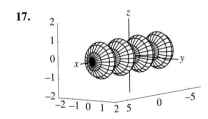

19.

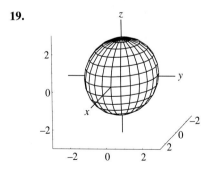

21. (a) A   (b) C   (c) B   **23.** $16\pi\sqrt{2}$

25. $4\pi\sqrt{14}$ **27.** $\frac{1}{2}\sqrt{2}$ **29.** 4π

31. $\displaystyle\int_1^3 \int_1^2 x\sqrt{14}\, dx\, dy = 3\sqrt{14}$

33. $\displaystyle\int_0^{2\pi}\int_0^2 r^2\sin\theta\sqrt{4r^2+1}\,dr\,d\theta=0$

35. $\displaystyle\int_0^{2\pi}\int_0^2 r^3\sqrt{4r^2+1}\,dr\,d\theta=\dfrac{391\sqrt{17}+1}{60}\pi$

37. $\displaystyle\int_0^{2\pi}\int_0^4 \sqrt{2}r^2\,dr\,d\theta=\dfrac{128\sqrt{2}}{3}\pi$

39. $\displaystyle\int_0^{2\pi}\int_0^{\pi/2} 4(4\sin\phi)\,d\phi\,d\theta\ 32\pi$

41. 24π **43.** -18π **45.** $\frac{5}{2}$ **47.** $\frac{\pi}{2}$ **49.** $\frac{7}{4}$

51. 0 **53.** $m=8\sqrt{14}\pi,\ \overline{x}=\overline{y}=0,\ \overline{z}=6$
55. $m=2\pi,\ \overline{x}=\frac{1}{3},\ \overline{y}=0,\ \overline{z}=\frac{1}{2}$

57. $\displaystyle\iint\limits_S g(x,y,z)\,dS$

$\displaystyle=\iint\limits_R g(f(y,z),y,z)\sqrt{(f_y)^2+(f_z)^2+1}\,dA$

where S is given by $x=f(y,z)$ for (y,z) in region $R:\mathbb{R}^2$.

59. $\dfrac{3\pi}{2}$ **61.** 198.8π

63. 0.474π **65.** 23.66

67. Flow lines don't cross boundary.

Exercises Section 14.7

5. $\frac{3}{2}$ **7.** π **9.** 0 **11.** 8 **13.** 32π

15. $\dfrac{64\pi}{3}$ **17.** 4π **19.** $\dfrac{6\pi}{5}$ **21.** 0 **23.** $\dfrac{\pi}{2}$

25. π **27.** 224π **29.** $\frac{27}{5}$ **31.** $\frac{512}{3}$

Exercises Section 14.8

5. 0 **7.** 4π **9.** $-\frac{4}{3}$ **11.** $-\pi$ **13.** 0
15. 1 **17.** 0 **19.** -4π **21.** 4π **23.** 0
25. 0 **33.** Both surfaces S_1 and S_2 should be positively oriented, or both negatively oriented.

Chapter 14 Review Exercises

1.

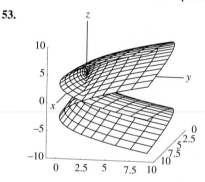

3. $\mathbf{F}_1=\text{D},\ \mathbf{F}_2=\text{C},\ \mathbf{F}_3=\text{B},\ \mathbf{F}_4=\text{A}$
5. $f(x,y)=xy-x^2y^2+y+c$ **7.** not
9. $y^3=3x^2+c$ **13.** 18 **15.** 18π **17.** 0
19. 0 **21.** $3\pi-4$ **23.** zero   **25.** 40π
27. 66 **29.** 3 **31.** 10 **33.** conservative   **35.** $\frac{1}{3}$
37. -2 **39.** $\frac{32}{3}$ **41.** 6π
43. $\langle 0,0,0\rangle,\ 3x^2-3y^2$ **45.** $\langle 0,0,0\rangle,\ 2+2z^2+2y^2$
47. neither   **49.** both   **51.** positive

53.

55. (a) B (b) C (c) A   **57.** $\frac{1}{6}(17^{3/2}-5^{3/2})\pi$
59. $-8\sqrt{14}$ **61.** $4\pi\sqrt{26}$ **63.** 0
65. $m=\frac{1}{3}(17\sqrt{17}-1)\pi,\ \overline{x}=\overline{y}=0,\ \overline{z}=\dfrac{1+391\sqrt{17}}{10(17\sqrt{17}-1)}$
67. $\frac{16}{3}$ **69.** $\frac{304}{5}$ **71.** $\dfrac{8\pi}{3}$ **73.** 0 **75.** 0
77. 0

BIBLIOGRAPHY

This section is part bibliography and part acknowledgments, focusing on applications more than on theoretical mathematics. We don't intend to give a comprehensive list of related books that are worth reading. Instead, we wish to give credit to many of the sources from which we have derived inspiration. Even this list is nowhere near complete, as we could not possibly list all of the books or thank all of the professors and students from whom we've learned mathematics. We have organized the list by chapter and topic.

Chapter 10

Walking friction
[1] Alexander, R. McNeill, *Exploring Biomechanics,* Scientific American Library, 1992.

Helicopter hovering: see the interesting scenario described in
[2] Krauss, Lawrence, *Beyond Star Trek,* Basic Books, 1997.

Torque: see or an introductory physics text or
[3] Brancazio, Peter, *Sport Science,* Simon and Schuster, 1984. (Brancazio enthusiastically uses sports to illustrate physics concepts and physics to explain sports techniques. Thoroughly enjoyable.)

Magnus force: see [3], also
[4] Watts, Robert and A. Terry Bahill, *Keep Your Eye on the Ball,* W. H. Freeman, 1990.
[5] deMestre, Neville, *The Mathematics of Projectiles in Sports,* Cambridge University Press, 1990.

Golf ellipsoids: from the paper "Experimental Determination of Inertia Ellipsoids" by S.H. Johnson in
[6] Cochran, A.J. and M.R. Farrally, eds., *Science and Golf II,* E & FN Spon, 1994. (A collection of research papers on a variety of golf-related topics and at varying degrees of complexity. This is the proceedings of the 1994 World Scientific Congress of Golf.)

Chapter 11

Vector projectile motion: see [5].

Kepler's laws
[7] Peterson, Ivars, *Newton's Clock,* W.H. Freeman, 1993.

Chapter 12

Effect of humidity on projectile: see [3].

Density plots: created using Mathematica.

Tennis graphs
[8] Schrier, Eric and William Allman, eds., *Newton at the Bat,* Charles Scribner's Sons, 1987. (An interesting collection of essays on a variety of sports.)

Grade average predictions: thanks to Dr. Dan Larsen, Roanoke College.

Digital pictures, Lambert shading
[9] Friedhoff, Richard and William Benzon, *Visualization,* Abrams Publishers, 1989.

Interpretation of mixed partial derivative: adapted from this article.
[10] McCartin, Brian, "What is f_{xy}?" *PRIMUS,* March 1998.

Baseball graphs: see [4].

Golf dimples: see [3] and
[11] Griffing, David F., *The Dynamics of Sports,* The Dalog Company, 1982.

Gauge of manufactured metal: thanks to Tom Burns of General Electric.

Mental calculations: one of numerous interesting examples in this exploration of how the human brain does mathematics.
[12] Dehaene, Stanislas, *The Number Sense,* Oxford University Press, 1997.

Football strategy: football is almost as carefully analyzed as baseball.

[13]   Carroll, Bob et al., *The Hidden Game of Pro Football,* Warner Books, 1998.

Optimal thrust of rocket: adapted from this textbook for the calculus of variations, an upper-level course closely related to control theory.

[14]   Smith, Donald, *Variational Methods in Optimization,* Prentice-Hall, 1974.

Sailboat steering: the design of racing yachts has become one of the most competitive engineering problems in sports.

[15]   Stein, Sherman and Anthony Barcellos, *Calculus,* 5th edition, McGraw-Hill, 1992.

Chapter 13

Model rockets
[16]   Stine, G. Harry, *Handbook of Model Rocketry,* 6th edition, John Wiley and Sons, 1994.

Monte Carlo simulation: the interaction between random and nonrandom processes can be surprising.

[17]   Ripley, Brian D., *Stochastic Simulation,* John Wiley and Sons, 1987.

Moment of inertia: the principle behind many sports phenomena. See [3], [4], [11] and the following general interest science book.

[18]   Blanding, Sharon and John Monteleone, *What Makes a Boomerang Come Back,* Longmeadow Press, 1992.

Baseball bats: see [3] and [4].

Direct linear transformations: see the paper "Video Monitoring System to Measure Initial Launch Characteristics of Golf Balls" by W. Gobush, D. Pelletier and C. Days in [6]. The technique itself is derived from

[19]   Abdel-Aziz, Y.I. and H.M. Karara, "Direct Linear Transformation into Object Space Coordinates in Close-Range Photogrammetry," Symposium on Close-Range Photogrammetry, University of Illinois at Urbana-Champaign, 1971.

Chapter 14

Volkswagen Beetle: the information on the new Beetle came from the Volkswagen website. The older drag coefficients are from this physics book.

[20]   Roberson, John and Clayton Crowe, *Engineering Fluid Mechanics,* 3rd edition, Houghton Mifflin, 1985.

Vector calculus: a different organization of the material in this chapter can be found in vector calculus books such as

[21]   Schey, H.M., *Div, Grad, Curl and All That,* 3rd edition, W. W. Norton, 1997.

CREDITS

CHAPTER 10

Opener: Philip Gould/Corbis, PG006587; 798: The Granger Collection; 817: Culver Pictures; 846: Jeremy Hoare/PhotoDisc; 846: Neil Beer/PhotoDisc

CHAPTER 11

Opener: George Hall/Corbis, HL001043; 899: The Granger Collection

CHAPTER 12

Opener: PhotoDisc, SP000855; 12.12a, 12.12b: WW2010 Project, Department of Atmospheric Sciences, University of Illinois at Urbana-Champaign; 12.12c, 12.12d: Jet Propulsion Laboratory, California Institute of Technology; 923 (top & bottom): Roland Minton; 996: The Granger Collection

CHAPTER 13

1011 (top): PhotoDisc, vol. 29; 1012: PhotoDisc, vol. 43; 1018: University of St. Andrews; 1084: The Granger Collection

CHAPTER 14

Opener: Dale E. Boyer/Courtesy NASA/Photo Researchers, Inc.; 1095 (center): Corbis, DD001338; 1095 (bottom): Volkswagen of America Inc.; 1134: George Green Memorial Fund; 1162: University of St. Andrews; 1176: The Granger Collection

DERIVATIVE FORMULAS

GENERAL RULES

$$\frac{d}{dx}[f(x) + g(x)] = f'(x) + g'(x)$$

$$\frac{d}{dx}[f(x) - g(x)] = f'(x) - g'(x)$$

$$\frac{d}{dx}[cf(x)] = cf'(x)$$

$$\frac{d}{dx}[f(g(x))] = f'(g(x))g'(x)$$

$$\frac{d}{dx}[f(x)g(x)] = f'(x)g(x) + f(x)g'(x)$$

$$\frac{d}{dx}\left[\frac{f(x)}{g(x)}\right] = \frac{f'(x)g(x) - f(x)g'(x)}{[g(x)]^2}$$

POWER RULES

$$\frac{d}{dx}(x^n) = nx^{n-1}$$

$$\frac{d}{dx}(c) = 0$$

$$\frac{d}{dx}(cx) = c$$

$$\frac{d}{dx}(\sqrt{x}) = \frac{1}{2\sqrt{x}}$$

EXPONENTIAL

$$\frac{d}{dx}[e^x] = e^x$$

$$\frac{d}{dx}[a^x] = a^x \ln a$$

$$\frac{d}{dx}\left[e^{u(x)}\right] = e^{u(x)}u'(x)$$

$$\frac{d}{dx}[e^{rx}] = r\,e^{rx}$$

TRIGONOMETRIC

$$\frac{d}{dx}(\sin x) = \cos x$$

$$\frac{d}{dx}(\cos x) = -\sin x$$

$$\frac{d}{dx}(\tan x) = \sec^2 x$$

$$\frac{d}{dx}(\cot x) = -\csc^2 x$$

$$\frac{d}{dx}(\sec x) = \sec x \tan x$$

$$\frac{d}{dx}(\csc x) = -\csc x \cot x$$

INVERSE TRIGONOMETRIC

$$\frac{d}{dx}(\sin^{-1} x) = \frac{1}{\sqrt{1 - x^2}}$$

$$\frac{d}{dx}(\cos^{-1} x) = -\frac{1}{\sqrt{1 - x^2}}$$

$$\frac{d}{dx}(\tan^{-1} x) = \frac{1}{1 + x^2}$$

$$\frac{d}{dx}(\cot^{-1} x) = -\frac{1}{1 + x^2}$$

$$\frac{d}{dx}(\sec^{-1} x) = \frac{1}{|x|\sqrt{x^2 - 1}}$$

$$\frac{d}{dx}(\csc^{-1} x) = -\frac{1}{|x|\sqrt{x^2 - 1}}$$

HYPERBOLIC

$$\frac{d}{dx}(\sinh x) = \cosh x$$

$$\frac{d}{dx}(\cosh x) = \sinh x$$

$$\frac{d}{dx}(\tanh x) = \text{sech}^2 x$$

$$\frac{d}{dx}(\coth x) = -\text{csch}^2 x$$

$$\frac{d}{dx}(\text{sech}\, x) = -\text{sech}\, x \tanh x$$

$$\frac{d}{dx}(\text{csch}\, x) = -\text{csch}\, x \coth x$$

INVERSE HYPERBOLIC

$$\frac{d}{dx}(\sinh^{-1} x) = \frac{1}{\sqrt{1 + x^2}}$$

$$\frac{d}{dx}(\cosh^{-1} x) = \frac{1}{\sqrt{x^2 - 1}}$$

$$\frac{d}{dx}(\tanh^{-1} x) = \frac{1}{1 - x^2}$$

$$\frac{d}{dx}(\coth^{-1} x) = \frac{1}{1 - x^2}$$

$$\frac{d}{dx}(\text{sech}^{-1} x) = -\frac{1}{x\sqrt{1 - x^2}}$$

$$\frac{d}{dx}(\text{csch}^{-1} x) = -\frac{1}{|x|\sqrt{x^2 + 1}}$$

TABLE OF INTEGRALS

FORMS INVOLVING $a + bu$

1. $\displaystyle\int \frac{1}{a+bu}\,du = \frac{1}{b}\ln|a+bu| + c$

2. $\displaystyle\int \frac{u}{a+bu}\,du = \frac{1}{b^2}(a+bu - a\ln|a+bu|) + c$

3. $\displaystyle\int \frac{u^2}{a+bu}\,du = \frac{1}{2b^3}[(a+bu)^2 - 4a(a+bu) + 2a^2\ln|a+bu|] + c$

4. $\displaystyle\int \frac{1}{u(a+bu)}\,du = \frac{1}{a}\ln\left|\frac{u}{a+bu}\right| + c$

5. $\displaystyle\int \frac{1}{u^2(a+bu)}\,du = \frac{b}{a^2}\ln\left|\frac{a+bu}{u}\right| - \frac{1}{au} + c$

FORMS INVOLVING $(a + bu)^2$

6. $\displaystyle\int \frac{1}{(a+bu)^2}\,du = \frac{-1}{b(a+bu)} + c$

7. $\displaystyle\int \frac{u}{(a+bu)^2}\,du = \frac{1}{b^2}\left(\frac{a}{a+bu} + \ln|a+bu|\right) + c$

8. $\displaystyle\int \frac{u^2}{(a+bu)^2}\,du = \frac{1}{b^3}\left(a+bu - \frac{a^2}{a+bu} - 2a\ln|a+bu|\right) + c$

9. $\displaystyle\int \frac{1}{u(a+bu)^2}\,du = \frac{1}{a(a+bu)} + \frac{1}{a^2}\ln\left|\frac{u}{a+bu}\right| + c$

10. $\displaystyle\int \frac{1}{u^2(a+bu)^2}\,du = \frac{2b}{a^3}\ln\left|\frac{a+bu}{u}\right| - \frac{a+2bu}{a^2u(a+bu)} + c$

FORMS INVOLVING $\sqrt{a + bu}$

11. $\displaystyle\int u\sqrt{a+bu}\,du = \frac{2}{15b^2}(3bu - 2a)(a+bu)^{3/2} + c$

12. $\displaystyle\int u^2\sqrt{a+bu}\,du = \frac{2}{105b^3}(15b^2u^2 - 12abu + 8a^2)(a+bu)^{3/2} + c$

13. $\displaystyle\int u^n\sqrt{a+bu}\,du = \frac{2}{b(2n+3)}u^n(a+bu)^{3/2} - \frac{2na}{b(2n+3)}\int u^{n-1}\sqrt{a+bu}\,du$

14. $\displaystyle\int \frac{\sqrt{a+bu}}{u}\,du = 2\sqrt{a+bu} + a\int \frac{1}{u\sqrt{a+bu}}\,du$

15. $\displaystyle\int \frac{\sqrt{a+bu}}{u^n}\,du = \frac{-1}{a(n-1)}\frac{(a+bu)^{3/2}}{u^{n-1}} - \frac{(2n-5)b}{2a(n-1)}\int \frac{\sqrt{a+bu}}{u^{n-1}}\,du,\ n\neq 1$

16a. $\displaystyle\int \frac{1}{u\sqrt{a+bu}}\,du = \frac{1}{\sqrt{a}}\ln\left|\frac{\sqrt{a+bu}-\sqrt{a}}{\sqrt{a+bu}+\sqrt{a}}\right| + c,\ a > 0$

16b. $\displaystyle\int \frac{1}{u\sqrt{a+bu}}\,du = \frac{2}{\sqrt{-a}}\tan^{-1}\sqrt{\frac{a+bu}{-a}} + c,\ a < 0$

17. $\displaystyle\int \frac{1}{u^n\sqrt{a+bu}}\,du = \frac{-1}{a(n-1)}\frac{\sqrt{a+bu}}{u^{n-1}} - \frac{(2n-3)b}{2a(n-1)}\int \frac{1}{u^{n-1}\sqrt{a+bu}}\,du,\ n\neq 1$

18. $\displaystyle\int \frac{u}{\sqrt{a+bu}}\,du = \frac{2}{3b^2}(bu-2a)\sqrt{a+bu} + c$

19. $\displaystyle\int \frac{u^2}{\sqrt{a+bu}}\,du = \frac{2}{15b^3}(3b^2u^2 - 4abu + 8a^2)\sqrt{a+bu} + c$

20. $\displaystyle\int \frac{u^n}{\sqrt{a+bu}}\,du = \frac{2}{(2n+1)b}u^n\sqrt{a+bu} - \frac{2na}{(2n+1)b}\int \frac{u^{n-1}}{\sqrt{a+bu}}\,du$

FORMS INVOLVING $\sqrt{a^2 + u^2}$, $a > 0$

21. $\displaystyle\int \sqrt{a^2+u^2}\,du = \frac{1}{2}u\sqrt{a^2+u^2} + \frac{1}{2}a^2\ln|u + \sqrt{a^2+u^2}| + c$

22. $\displaystyle\int u^2\sqrt{a^2+u^2}\,du = \frac{1}{8}u(a^2+2u^2)\sqrt{a^2+u^2} - \frac{1}{8}a^4\ln|u + \sqrt{a^2+u^2}| + c$

23. $\displaystyle\int \frac{\sqrt{a^2+u^2}}{u}\,du = \sqrt{a^2+u^2} - a\ln\left|\frac{a+\sqrt{a^2+u^2}}{u}\right| + c$

24. $\displaystyle\int \frac{\sqrt{a^2+u^2}}{u^2}\,du = \ln|u+\sqrt{a^2+u^2}| - \frac{\sqrt{a^2+u^2}}{u} + c$

25. $\displaystyle\int \frac{1}{\sqrt{a^2+u^2}}\,du = \ln|u+\sqrt{a^2+u^2}| + c$

26. $\displaystyle\int \frac{u^2}{\sqrt{a^2+u^2}}\,du = \frac{1}{2}u\sqrt{a^2+u^2} - \frac{1}{2}a^2\ln|u+\sqrt{a^2+u^2}| + c$

27. $\displaystyle\int \frac{1}{u\sqrt{a^2+u^2}}\,du = \frac{1}{a}\ln\left|\frac{u}{a+\sqrt{a^2+u^2}}\right| + c$

28. $\displaystyle\int \frac{1}{u^2\sqrt{a^2+u^2}}\,du = -\frac{\sqrt{a^2+u^2}}{a^2u} + c$

FORMS INVOLVING $\sqrt{a^2 - u^2}$, $a > 0$

29. $\displaystyle\int \sqrt{a^2-u^2}\,du = \frac{1}{2}u\sqrt{a^2-u^2} + \frac{1}{2}a^2\sin^{-1}\frac{u}{a} + c$

30. $\displaystyle\int u^2\sqrt{a^2-u^2}\,du = \frac{1}{8}u(2u^2-a^2)\sqrt{a^2-u^2} + \frac{1}{8}a^4\sin^{-1}\frac{u}{a} + c$

31. $\displaystyle\int \frac{\sqrt{a^2-u^2}}{u}\,du = \sqrt{a^2-u^2} - a\ln\left|\frac{a+\sqrt{a^2-u^2}}{u}\right| + c$

32. $\displaystyle\int \frac{\sqrt{a^2-u^2}}{u^2}\,du = -\frac{\sqrt{a^2-u^2}}{u} - \sin^{-1}\frac{u}{a} + c$

33. $\displaystyle\int \frac{1}{\sqrt{a^2-u^2}}\,du = \sin^{-1}\frac{u}{a} + c$

34. $\displaystyle\int \frac{1}{u\sqrt{a^2-u^2}}\,du = -\frac{1}{a}\ln\left|\frac{a+\sqrt{a^2-u^2}}{u}\right| + c$

35. $\displaystyle\int \frac{u^2}{\sqrt{a^2-u^2}}\,du = -\frac{1}{2}u\sqrt{a^2-u^2} + \frac{1}{2}a^2\sin^{-1}\frac{u}{a} + c$

36. $\displaystyle\int \frac{1}{u^2\sqrt{a^2-u^2}}\,du = -\frac{\sqrt{a^2-u^2}}{a^2u} + c$

FORMS INVOLVING $\sqrt{u^2 - a^2}$, $a > 0$

37. $\int \sqrt{u^2 - a^2}\, du = \frac{1}{2}u\sqrt{u^2 - a^2} - \frac{1}{2}a^2 \ln\left|u + \sqrt{u^2 - a^2}\right| + c$

38. $\int u^2\sqrt{u^2 - a^2}\, du = \frac{1}{8}u(2u^2 - a^2)\sqrt{u^2 - a^2} - \frac{1}{8}a^4 \ln\left|u + \sqrt{u^2 - a^2}\right| + c$

39. $\int \frac{\sqrt{u^2 - a^2}}{u}\, du = \sqrt{u^2 - a^2} - a\sec^{-1}\frac{|u|}{a} + c$

40. $\int \frac{\sqrt{u^2 - a^2}}{u^2}\, du = \ln\left|u + \sqrt{u^2 - a^2}\right| - \frac{\sqrt{u^2 - a^2}}{u} + c$

41. $\int \frac{1}{\sqrt{u^2 - a^2}}\, du = \ln\left|u + \sqrt{u^2 - a^2}\right| + c$

42. $\int \frac{u^2}{\sqrt{u^2 - a^2}}\, du = \frac{1}{2}u\sqrt{u^2 - a^2} + \frac{1}{2}a^2 \ln\left|u + \sqrt{u^2 - a^2}\right| + c$

43. $\int \frac{1}{u\sqrt{u^2 - a^2}}\, du = \frac{1}{a}\sec^{-1}\frac{|u|}{a} + c$

44. $\int \frac{1}{u^2\sqrt{u^2 - a^2}}\, du = \frac{\sqrt{u^2 - a^2}}{a^2 u} + c$

FORMS INVOLVING $\sqrt{2au - u^2}$

45. $\int \sqrt{2au - u^2}\, du = \frac{1}{2}(u - a)\sqrt{2au - u^2} + \frac{1}{2}a^2 \cos^{-1}\left(\frac{a - u}{a}\right) + c$

46. $\int u\sqrt{2au - u^2}\, du = \frac{1}{6}(2u^2 - au - 3a^2)\sqrt{2au - u^2} + \frac{1}{2}a^3 \cos^{-1}\left(\frac{a - u}{a}\right) + c$

47. $\int \frac{\sqrt{2au - u^2}}{u}\, du = \sqrt{2au - u^2} + a\cos^{-1}\left(\frac{a - u}{a}\right) + c$

48. $\int \frac{\sqrt{2au - u^2}}{u^2}\, du = -\frac{2\sqrt{2au - u^2}}{u} - \cos^{-1}\left(\frac{a - u}{a}\right) + c$

49. $\int \frac{1}{\sqrt{2au - u^2}}\, du = \cos^{-1}\left(\frac{a - u}{a}\right) + c$

50. $\int \frac{u}{\sqrt{2au - u^2}}\, du = -\sqrt{2au - u^2} + a\cos^{-1}\left(\frac{a - u}{a}\right) + c$

51. $\int \frac{u^2}{\sqrt{2au - u^2}}\, du = -\frac{1}{2}(u + 3a)\sqrt{2au - u^2} + \frac{3}{2}a^2 \cos^{-1}\left(\frac{a - u}{a}\right) + c$

52. $\int \frac{1}{u\sqrt{2au - u^2}}\, du = -\frac{\sqrt{2au - u^2}}{au} + c$

FORMS INVOLVING $\sin u$ OR $\cos u$

53. $\int \sin u\, du = -\cos u + c$

54. $\int \cos u\, du = \sin u + c$

55. $\int \sin^2 u\, du = \frac{1}{2}u - \frac{1}{2}\sin u \cos u + c$

56. $\int \cos^2 u\, du = \frac{1}{2}u + \frac{1}{2}\sin u \cos u + c$

57. $\int \sin^3 u\, du = -\frac{2}{3}\cos u - \frac{1}{3}\sin^2 u \cos u + c$

58. $\int \cos^3 u\, du = \frac{2}{3}\sin u + \frac{1}{3}\sin u \cos^2 u + c$

59. $\int \sin^n u\, du = -\frac{1}{n}\sin^{n-1} u \cos u + \frac{n-1}{n}\int \sin^{n-2} u\, du$

60. $\int \cos^n u\, du = \frac{1}{n}\cos^{n-1} u \sin u + \frac{n-1}{n}\int \cos^{n-2} u\, du$

61. $\int u \sin u\, du = \sin u - u\cos u + c$

62. $\int u \cos u\, du = \cos u + u\sin u + c$

63. $\int u^n \sin u\, du = -u^n \cos u + n\int u^{n-1}\cos u\, du + c$

64. $\int u^n \cos u\, du = u^n \sin u - n\int u^{n-1}\sin u\, du + c$

65. $\int \frac{1}{1 + \sin u}\, du = \tan u - \sec u + c$

66. $\int \frac{1}{1 - \sin u}\, du = \tan u + \sec u + c$

67. $\int \frac{1}{1 + \cos u}\, du = -\cot u + \csc u + c$

68. $\int \frac{1}{1 - \cos u}\, du = -\cot u - \csc u + c$

69. $\int \sin(mu)\sin(nu)\, du = \frac{\sin(m - n)u}{2(m - n)} - \frac{\sin(m + n)u}{2(m + n)} + c$

70. $\int \cos(mu)\cos(nu)\, du = \frac{\sin(m - n)u}{2(m - n)} + \frac{\sin(m + n)u}{2(m + n)} + c$

71. $\int \sin(mu)\cos(nu)\, du = \frac{\cos(n - m)u}{2(n - m)} - \frac{\cos(m + n)u}{2(m + n)} + c$

72. $\int \sin^m u \cos^n u\, du = -\frac{\sin^{m-1} u \cos^{n+1} u}{m + n} + \frac{m - 1}{m + n}\int \sin^{m-2} u \cos^n u\, du$

FORMS INVOLVING OTHER TRIGONOMETRIC FUNCTIONS

73. $\int \tan u\, du = -\ln|\cos u| + c = \ln|\sec u| + c$

74. $\int \cot u\, du = \ln|\sin u| + c$

75. $\int \sec u\, du = \ln|\sec u + \tan u| + c$

76. $\int \csc u\, du = \ln|\csc u - \cot u| + c$

77. $\int \tan^2 u\, du = \tan u - u + c$

78. $\int \cot^2 u\, du = -\cot u - u + c$

79. $\int \sec^2 u\, du = \tan u + c$

80. $\int \csc^2 u\, du = -\cot u + c$